ATOMIC INNER-SHELL PHYSICS

PHYSICS OF ATOMS AND MOLECULES

ATOM—MOLECULE COLLISION THEORY: A Guide for the Experimentalist
Edited by Richard B. Bernstein

ATOMIC INNER-SHELL PHYSICS
Edited by Bernd Crasemann

ATOMS IN ASTROPHYSICS
Edited by P. G. Burke, W. B. Eissner, D. G. Hummer, and I. C. Percival

AUTOIONIZATION: Recent Developments and Applications
Edited by Aaron Temkin

COHERENCE AND CORRELATION IN ATOMIC COLLISIONS
Edited by H. Kleinpoppen and J. F. Williams

COLLISIONS OF ELECTRONS WITH ATOMS AND MOLECULES
G. F. Drukarev

DENSITY MATRIX THEORY AND APPLICATIONS
Karl Blum

ELECTRON AND PHOTON INTERACTIONS WITH ATOMS
Edited by H. Kleinpoppen and M. R. C. McDowell

ELECTRON—ATOM AND ELECTRON—MOLECULE COLLISIONS
Edited by Juergen Hinze

ELECTRON-MOLECULE COLLISIONS
Edited by Isao Shimamura and Kazuo Takayanagi

INNER-SHELL AND X-RAY PHYSICS OF ATOMS AND SOLIDS
Edited by Derek J. Fabian, Hans Kleinpoppen, and Lewis M. Watson

INTRODUCTION TO THE THEORY OF LASER—ATOM INTERACTIONS
Marvin H. Mittleman

ISOTOPE SHIFTS IN ATOMIC SPECTRA
W. H. King

PROGRESS IN ATOMIC SPECTROSCOPY, Parts A, B, and C
Edited by W. Hanle, H. Kleinpoppen, and H. J. Beyer

VARIATIONAL METHODS IN ELECTRON—ATOM SCATTERING THEORY
R. K. Nesbet

ATOMIC INNER-SHELL PHYSICS

EDITED BY

BERND CRASEMANN
University of Oregon
Eugene, Oregon

PLENUM PRESS • NEW YORK AND LONDON

Library of Congress Cataloging in Publication Data

Main entry under title:

Atomic inner-shell physics.

(Physics of atoms and molecules)
Includes bibliographies and index.
1. Atomic transition probabilities. 2. Inner-shell ionization. 3. Quantum electrodynamics.
4. X-rays. I. Crasemann, Bernd. II. Series.
QC454.N8A76 1985 539.7 85-12357

ISBN-13: 978-1-4612-9472-6 e-ISBN-13: 978-1-4613-2417-1
DOI: 10.1007/978-1-4613-2417-1

Sotcover reprint of the hardcover 1st edition 1985

A Division of Plenum Publishing Corporation
233 Spring Street, New York, N.Y. 10013

CONTRIBUTORS

T. ÅBERG
Laboratory of Physics, Helsinki University of Technology, Finland
R. ANHOLT
Department of Physics, Stanford University, Stanford, California
H. BACKE
Institut für Physik, Johannes Gutenberg-Universität, Mainz, Federal Republic of Germany
G. L. BORCHERT
Institut für Kernphysik, Kernforschungsanlage, Jülich, Federal Republic of Germany
T. N. CHANG
Physics Department, University of Southern California, Los Angeles, California
MAU HSIUNG CHEN
Lawrence Livermore Laboratory, Livermore, California
K. T. CHENG
Physics Division, Argonne National Laboratory, Argonne, Illinois
BERND CRASEMANN
Department of Physics and Chemical Physics Institute, University of Oregon, Eugene, Oregon
RICHARD D. DESLATTES
Quantum Metrology Group, National Bureau of Standards, Washington, D.C.
I. J. FENG
Varian Associates, Palo Alto, California
P. G. HANSEN
Institute of Physics, University of Aarhus, Aarhus, Denmark
W. R. JOHNSON
Department of Physics, University of Notre Dame, Notre Dame, Indiana
B. JONSON
CERN-ISOLDE, CERN, Geneva, Switzerland
ERNEST G. KESSLER, JR.
Quantum Metrology Group, National Bureau of Standards, Washington, D.C.
GEORGE C. KING
Department of Physics, Schuster Laboratory, University of Manchester, Manchester, England
LYNN KISSEL
Test Planning and Diagnostics Division, Sandia National Laboratories, Albuquerque, New Mexico
W. MEHLHORN
Fakultät für Physik, Universität Freiburg, Freiburg, Federal Republic of Germany
B. MÜLLER
Institut für Theoretische Physik, Johann Wolfgang Goethe-Universität, Frankfurt am Main, Federal Republic of Germany

A. NIEHAUS
Fysisch Laboratorium, Rijksuniversiteit Utrecht, Utrecht, The Netherlands

R. H. PRATT
Department of Physics and Astronomy, University of Pittsburgh, Pittsburgh, Pennsylvania

FRANK H. READ
Department of Physics, Schuster Laboratory, University of Manchester, Manchester, England

O. W. B. SCHULT
Institut für Kernphysik, Kernforschungsanlage, Jülich, Federal Republic of Germany

J. TULKKI
Laboratory of Physics, Helsinki University of Technology, Finland

P. VINCENT
Department of Physics and Astronomy, Rutgers University, New Brunswick, New Jersey.
Present address: Brookhaven National Laboratory, Upton, New York

FRANÇOIS WUILLEUMIER
LURE et Laboratoire de Spectroscopie Atomique et Ionique, Orsay, France

Preface

The physics of atomic inner shells has undergone significant advances in recent years. Fast computers and new experimental tools, notably synchrotron-radiation sources and heavy-ion accelerators, have greatly enhanced the scope of problems that are accessible. The level of research activity is growing substantially; added incentives are provided by the importance of inner-shell processes in such diverse areas as plasma studies, astrophysics, laser technology, biology, medicine, and materials science.

The main reason for all this exciting activity in atomic inner-shell physics, to be sure, lies in the significance of the fundamental problems that are coming within grasp. The large energies of many inner-shell processes cause relativistic and quantum-electrodynamic effects to become strong. Unique opportunities exist for delicate tests of such phenomena as the screening of the electron self-energy and the limits of validity of the present form of the frequency-dependent Breit interaction, to name but two. The many-body problem, which pervades virtually all of physics, presents somewhat less intractable aspects in the atomic inner-shell regime: correlations are relatively weak so that they can be treated perturbatively, and the basic potential is simple and known! The dynamics of inner-shell processes are characterized by exceedingly short lifetimes and high transition rates that strain perturbation theory to its limits and obliterate the traditional separation of excitation and deexcitation. These factors are only now being explored, as are interference phenomena between the various channels.

This book was planned in response to a perceived need for a collection of monographs that cover important aspects of inner-shell physics in substantial depth. The volume is thus intended to complement, rather than compete with, the series of proceedings of recent international conferences on the subject,* and to supplement an earlier collection of reviews.†

The selection of topics in the present volume is necessarily somewhat arbitrary and reflects the editor's personal view of where some of the most important and exciting developments in the field are taking place. Part A, which deals with structure and transitions, begins with three chapters on fundamental theoretical aspects of quantum-electrodynamic, relativistic, and many-body effects in inner-shell processes. Two general reviews follow, one on Auger spectrometry and the other on precise determinations of hole-state energies, that describe these broad areas of experimental effort and compare recent results with relevant theory. Chapter 6 contains an extensive review of a more specialized and less widely-known subject that overlaps both atomic and nuclear physics, viz., the causes of energy shifts in K X rays. Synchrotron radiation has already exerted a major impact on the study of atomic and molecular inner shells, and can be expected to become even more important in the near future; this topic is surveyed in Chapter 7. A thorough review of inner-shell studies by electron energy-loss measurements follows in Chapter 8, and Part A concludes with a monograph on coherence effects in electron emission from atoms, including interference phenomena that were observed only recently and that exemplify intriguing aspects of the quantum theory.

Part B of the book deals with energetic phenomena that arise in collisions of photons and charged particles with atoms. Chapters 10 and 11 describe inelastic and Rayleigh scattering of X rays, respectively. Electron–atom bremsstrahlung is reviewed in Chapter 12, and in the last three chapters the production of photons and positrons in nuclear reactions and heavy-ion collisions is discussed. We hope that the surveys contained in this book will convey to the reader a sense of the intense current research activity in the physics of the inner atom, and stimulate ideas for new investigations of some of the important as yet unsolved problems.

The editor is grateful to the contributors for their painstaking efforts to present thorough and carefully reasoned reviews, and for their patience with frustrating delays that often seem to be unavoidable in the production of a work of this kind. He thanks Stacey Sorensen for advice on computer generation of indices, and the Air Force Office of Scientific Research for grant support.

BERND CRASEMANN
Eugene, Oregon

* *Proceedings of the International Conference on Inner-Shell Ionization Phenomena and Future Applications, April 17–22, 1972,* edited by R. W. Fink, S. T. Manson, J. M. Palms, and P. Venugopala Rao. United States Atomic Energy Commission Report No. CONF-720404, 1973 (unpublished).

Proceedings of the International Conference on X-Ray Processes in Matter, edited by T. Åberg, E. Källne, and R. Manne. *Phys. Fenn.* **9**, Supplement S1 (1974).

International Conference on the Physics of X-Ray Spectra, Program and Extended Abstracts, edited by R. D. Deslattes. National Bureau of Standards, Gaithersburg, 1976 (unpublished).

Proceedings of the Second International Conference on Inner-Shell Ionization Phenomena, Freiburg 1976, edited by W. Mehlhorn and R. Brenn. Fakultät für Physik, Universität Freiburg, 1976 (unpublished).

Proceedings of the International Conference on X-Ray and XUV Spectroscopy, edited by T. Sagawa, *Jpn. J. Appl. Phys.* **17**, Suppl. 17-2 (1978).

Inner-Shell and X-Ray Physics of Atoms and Solids, edited by D. J. Fabian, H. Kleinpoppen, and L. M. Watson (Plenum, New York, 1981).

X-Ray and Atomic Inner-Shell Physics—1982, edited by B. Crasemann, AIP Conference Proceedings No. 94 (American Institute of Physics, New York, 1982).

† *Atomic Inner-Shell Processes*, edited by B. Crasemann (Academic, New York, 1975).

Contents

PART A: ATOMIC STRUCTURE AND TRANSITIONS

CHAPTER 6

MECHANISMS FOR ENERGY SHIFTS OF ATOMIC *K* X RAYS

P. G. Hansen, B. Jonson, G. L. Borchert, and O. W. B. Schult

CHAPTER 7

ATOMIC PHYSICS RESEARCH WITH SYNCHROTRON RADIATION

Bernd Crasemann and François Wuilleumier

CHAPTER 8

INVESTIGATIONS OF INNER-SHELL STATES BY THE ELECTRON ENERGY-LOSS TECHNIQUE AT HIGH RESOLUTION

George C. King and Frank H. Read

CHAPTER 9

COHERENCE EFFECTS IN ELECTRON EMISSION BY ATOMS

A. Niehaus

PART B: SCATTERING AND COLLISION-INDUCED PROCESSES

CHAPTER 10

INELASTIC X-RAY SCATTERING INCLUDING RESONANCE PHENOMENA

T. Åberg and J. Tulkki

CHAPTER 11
RAYLEIGH SCATTERING: ELASTIC PHOTON SCATTERING BY BOUND ELECTRONS
Lynn Kissel and R. H. Pratt

CHAPTER 12
ELECTRON–ATOM BREMSSTRAHLUNG
R. H. Pratt and I. J. Feng

CHAPTER 13
X-RAY AND BREMSSTRAHLUNG PRODUCTION IN NUCLEAR REACTIONS
R. Anholt

CHAPTER 14
POSITRON PRODUCTION IN HEAVY-ION COLLISIONS
H. Backe and B. Müller

PART A

Atomic Structure and Transitions

CHAPTER 1

Relativistic and Quantum Electrodynamic Effects on Atomic Inner Shells

W. R. Johnson and K. T. Cheng

1. INTRODUCTION

In this chapter our aim is to describe efforts to develop a consistent and accurate theory of atomic inner shells which includes both relativistic and quantum-electrodynamic (QED) effects. We begin with a brief historical description of these efforts in Sec. 1 and continue with more detailed discussions of various theoretical considerations.

The first modern attempt to understand inner-electron binding energies in systems with high nuclear charge was that of Brenner and Brown (1953), who calculated K absorption edges in several heavy atoms. In this seminal work the K-shell electrons were described by Dirac–Coulomb wave functions. The Dirac–Coulomb energy of each K-shell electron was corrected for its electromagnetic interaction with the other K-shell electron and for its electrostatic interaction with the cloud of outer electrons. The potential of the outer electrons was approximated by $V(r) = a + b(r)$, where the parameters a and $b(r)$ were determined from nonrelativistic Hartee–Fock (HF) calculations. The contributions of a and $b(r)$ to the K-shell binding energy were corrected for relativistic effects. The calculations of Brenner and Brown agreed well (to 0.2%) with the existing experimental data even though several of the effects now known to be important in determing K-shell binding energies were ignored. Among these effects are the QED corrections (self-energy and vacuum polarization), the corrections for the

W. R. Johnson • Department of Physics, University of Notre Dame, Notre Dame, Indiana 46556. K. T. Cheng • Physics Division, Argonne National Laboratory, Argonne, Illinois 60439.

rearrangement of the electron cloud, and the finite-nuclear-size corrections. Part of the success of Brenner and Brown was due to the fact that the terms ignored in their calculation tend to cancel. The terms included in their calculation, although determined by arguments somewhat crude by today's standards, agree rather well with more recent theoretical values.

In present-day calculations of inner-shell binding energies, the initial estimates are commonly taken from the Dirac–Fock (DF) theory in which the electrostatic interactions between the atomic electrons as well as the one-electron relativistic effects are accounted for in the self-consistent-field (SCF) approximation. Although the relativistic SCF equations were already known at the time of the Brenner and Brown paper from the pioneering work of Swirles (1935), Brenner and Brown had available just one numerical study of those equations by Williams (1940), who solved the relativistic SCF equations without exchange for Cu^+. The difficulties faced by early workers in treating the relativistic equations are obvious. Firstly, there are pairs of radial functions $P(r)$ and $Q(r)$ describing large and small components of the Dirac orbitals rather than a single Schrödinger radial function $P(r)$. Secondly, there are pairs of Dirac orbitals describing subshells with $j = l \pm 1/2$ for each subshell (n, l). Lastly, a large number of subshells must be considered for those heavy atomic systems for which relativistic effects are significant.

It was not until large-scale computing machines became available that significant progress was made on solutions to the relativistic SCF equations for really large atomic systems. Mayers (1957) published the first such study, which was a numerical solutions of the relativistic equations without exchange for Hg and Hg^{++}. This calculation was carried out on the EDSAC computer at Cambridge. Several years later, Cohen (1960) published the results of similar calculations for a number of heavy atoms, W, Pt, Hg, Hg^{++}, and U, done on an IBM-704 computer. There was substantial disagreement between the calculations of Cohen and those of Mayers, illustrating the delicacy required in treating such large-scale numerical problems. Furthermore, although the relativistic SCF equations with exchange were known from the work of Swirles (1935), no attempt was made in these early studies to include exchange because of the complexity of the resulting equations.

In 1959, the problem of K-shell binding energies was again addressed by Mayers, Brown, and Sanderson (1959). In their paper the initial approximation was taken from a relativistic SCF calculation including exchange for Hg. The DF eigenvalue for the $1s$ electron, which in the frozen-orbital approximation is the K-shell binding energy, was corrected for the magnetic interaction between the $1s$ electrons. Included in the calculation was the $1s$ Lamb shift in Hg, which had been previously determined by Brown and Mayers (1959) based on the theoretical expressions developed by Brown, Langer, and Schaefer (1959). In addition, the effects of finite nuclear size and of rearrangement were estimated and included. The results were within 0.15% of the experimental K absorption edge. Again, although the ingredients for an accurate calculation were present, the remarkable agreement between theory an experiment was fortuitous since the actual numerical values for some of the contributions were incorrect. Specifically the $1s$ Lamb shift in Hg is now known to be only one-third as large as the value given by Brown and Mayers, while the rearrangement energy is about twice their estimated value. Furthermore, the magnetic interaction between the K electrons and the outer electrons, ignored by Mayers, Brown, and Sanderson, proves to be important in precise calculations.

The relativistic SCF theory was reexamined in 1961 by Grant (Grant, 1961, 1965, 1970; Cooper, 1965), who employed the techniques of Racah algebra to bring the DF equations into a compact, easily used form. In Grant's work expressions were derived for the corrections to the DF wave functions caused by the unretarded magnetic interaction between electrons. Grant also calculated the K-shell binding energy in Hg using Mayers' numerical result (Mayers, 1957) for the $1s$ eigenvalue in Hg, and obtained a theoretical value even closer to experiment than that of Mayers, Brown, and Sanderson (1959). Again, the close agreement was partially fortuitous since it includes an incorrect value for the Lamb shift and ignores both finite nuclear size and rearrangement effects. Even though the numerical value of the K-shell binding energy in Grant's paper was in error, the elegant theoretical techniques introduced in that work to treat the angular momentum questions arising in the relativistic theory have had an important impact on later work.

No further relativistic SCF calculations appeared until 1965, when the extensive work of Liberman, Waber, and Cromer (1965) was published. In these calculations the exchange terms in the DF equations were approximated through Slater's averaging procedure (Slater, 1951), which is based on a free-electron model. Liberman, Waber, and Cromer made extensive comparisons with previous calculations and with experimental energy levels for elements ranging from Fe to U. The comparisons with experiment for the first time went beyond the K shell, and illustrated the potential of the relativistic SCF equations to describe quantitatively the structure of L and M shells as well. More extensive calculations of this type were published in 1971 by Lu *et al.* (1971), who determined energy eigenvalues and potentials for all atoms with nuclear charge in the range $Z=1$ to $Z=108$. The Dirac–Fock–Slater (DFS) method pioneered by Liberman, Waber, and Cromer has been used extensively to account for relativistic effects in atomic, molecular, and solid state processes; for example, in inner-shell photoionization studies (Scofield, 1973) and in studies of coherent scattering of energetic photons by inner electrons (Johnson and Cheng, 1976), where details of the treatment of exchange effects are not particularly important.

The problem of the *K*-shell binding energy in Hg was reconsidered in 1967 by Coulthard (1967), who performed DF calculations with exchange for Hg and Hg^+ (for Hg^+, off-diagonal Lagrange multipliers were included to ensure the orthogonality of the DF orbitals). The results of Coulthard's calculations, in which finite-nuclear-size effects as well as magnetic interactions between the *K* electrons and the outer electrons were ignored, was further from experimental energies than Grant's energies (Grant, 1961) determined six years earlier. Nevertheless, Coulthard's calculations were, as far as they went, correct. Furthermore, Coulthard's calculation for Hg agreed well with the previous calculation of Mayers (1957), indicating that the purely numerical aspects of the DF equations were under control.

In the same year two other DF calculations appeared. First, there was the work of Kim (1967), who solved the DF equations using the finite-basis scheme developed by Roothaan (1951) to treat nonrelativistic atomic systems. Kim solved the DF equations for several light atoms: He, Be, and Ne. This work is particularly noteworthy since for the first time the effects of retardation were considered along with those of magnetic interactions. In the same year, Smith and Johnson (1967) gave solutions to the DF equations, including the retarded electromagnetic interaction for He, Be, Ne, Ar, and Cu^+. Because of the limitations on computer storage and speed, it was impractical to attack large atoms using the finite-basis methods developed by Kim or directly solving the equations written by Smith and Johnson. These limitations were overcome later by Desclaux (1973), who performed complete DF studies for atoms with $Z=1$ to $Z=120$, and by Mann and Waber (1973), who published complete DF results for the lanthanides, including magnetic, retardation, and rearrangement effects.

Desiderio and Johnson (1971) reexamined the numerical calculations of the electron self-energy for heavy atoms. The work of Brown, Langer, and Schaefer (1959) was used as a guide and the earlier calculations of Brown and Mayers (1959) were redone making use of the availability of large high-speed computers. For Hg, Desiderio and Johnson obtained a value of 15 Ry for the 1*s* self-energy instead of 41 Ry as given earlier by Brown and Mayers. The *K*-shell binding energy was recalculated using this self-energy. In addition, the DF energy including the effects of rearrangement was calculated following a method similar to that of Coulthard (1967), except that finite-nuclear-size effects were included. Magnetic and retardation effects (including rearrangement corrections) were considered. The resulting value of 6108.08 Ry agreed closely with the experimental result of 6108.39 ± 0.06 Ry listed by Bearden and Burr (1967). In the same year, some improvement was achieved by Mann and Johnson (1971), who recalculated the magnetic and retardation contributions to *K*-shell binding energies using the frequency-dependent Breit interaction, which follows from QED. It is interesting to note that the magnetic interaction of the *K* electron with the outer electrons, which had been ignored previously, was found to account for nearly two-thirds of the *K*–*K* magnetic interaction. Similar studies were made of the *K*-shell binding energy in Fm ($Z = 100$) by Freedman, Porter, and Mann (1972), and also by Fricke, Desclaux, and Waber (1972). Since correlation effects and perhaps higher-order QED effects become important at the level of 1 eV, further improvements in the understanding of inner-shell binding energies will be very difficult indeed.

Despite obvious difficulties, there has been continuing interest in the study of inner-shell binding energies during the past decade, and there has been significant progress in such theoretical studies. One of the most complete studies of inner-shell binding energies along the lines described above is that of Huang *et al.* (1976), who used analysis similar to that employed by Rosén and Lindgren (1968) to calculate binding energies for all subshells of all elements from $Z = 2$ to $Z = 106$ on the basis of DFS wave functions. These calculations include the contributions of magnetic and retardation effects as well as those of electron self-energy and vacuum polarization. Even though the calculations are based on DFS wave functions, they give energies in excellent agreement with corresponding DF calculations, since corrections are made for the approximate treatment of exchange. These rather complete tabulations serve as a theoretical standard with which Deslattes and co-workers (Deslattes *et al.*, 1979; Deslattes and Kessler, 1980; Kessler *et al.*, 1982) have compared their recent high-precision X-ray measurements.

At the present time, various computer codes are available for DF calculations. Especially noteworthy are the multiconfiguration Dirac–Fock (MCDF) codes of Desclaux (1975) and of Grant and co-workers (Grant *et al.*, 1980; McKenzie *et al.*, 1980). Both of these codes provide options for including the finite-nuclear-size effect, the Breit interaction, the vacuum polarization, and the self-energy correction.

In the paragraphs below we describe in greater detail the various considerations that arise in connection with determining the structure of inner shells. Our point of departure will be the DF method, which we discuss in Sec. 2. The roles of the Breit interaction and of QED effects are described in Sec. 3 and 4, respectively. We do not touch on several important theories, such as the relativistic $1/Z$ expansion method (Doyle, 1969). Furthermore, we omit any discussion of electron–electron correlation, since this topic is covered in detail in Chaps. 3 and 4.

2. REVIEW OF THE DIRAC–FOCK METHOD

The DF equations for an $\mathcal{N}$-electron system are based on an approximate configuration-space Hamiltonian

$$H=\sum_{i=1}^{\mathcal{N}} h_i+\sum_{i>j}\frac{e^2}{r_{ij}}, \tag{1}$$

consisting of a sum of one-electron terms

$$h=c\boldsymbol{\alpha}\cdot\mathbf{p}+(\beta-1)\,mc^2-\frac{Ze^2}{r}, \tag{2}$$

which account for each electron's kinetic energy and for its interaction with the nucleus, and a sum of two-electron terms which describe the electron–electron Coulomb repulsion. In Eq. (2), $\mathbf{p}=-i\hbar\boldsymbol{\nabla}$ is the momentum operator in configuration space while $\boldsymbol{\alpha}$ and β are 4×4 Dirac matrices.

The fact that the one-electron Dirac Hamiltonian h of Eq. (2) has negative- as well as positive-energy solutions leads into difficulties in constructing a many-body theory based on Eq. (1). One such difficulty, pointed out by Brown and Ravenhall (1951) and reemphasized by Sucher (1980), is that H may have no normalizable eigenfunctions. Another difficulty, discussed by Mittleman (1980), is that the Hamiltonian H is not quite the configuration-space Hamiltonian which one obtains by an analysis based on QED. Despite such fundamental problems, H has been used as the basis for relativistic SCF calculations, and the resulting SCF energies and wave functions have been applied with a remarkable degree of success to explain the spectra and structure of atoms with high nuclear charge. The difficulties associated with Eq. (1) and the reason for the success of the applications have been the subject of several reviews (Mittleman, 1981; Sucher, 1983). In the following, we proceed in a traditional way and ignore the subtleties associated with this Hamiltonian.

We approximate the $\mathcal{N}$-electron wave function by a single Slater determinant constructed from $\mathcal{N}$ single-particle central-field Dirac orbitals. Multiconfiguration wave functions consisting of superpositions of Slater determinants are now commonly used in SCF calculations to determine atomic structure (Cheng *et al.*, 1979). Nevertheless, since the important features of inner-shell physics can be understood using wave functions constructed from a single Slater determinant, we restrict our present discussion to this simpler case. A typical Dirac orbital in the Slater determinant is written

$$u_{n\kappa m}(\mathbf{r})=\frac{1}{r}\begin{pmatrix} P_{n\kappa}(r)\,\Omega_{\kappa m}(\hat{r}) \\ iQ_{n\kappa}(r)\,\Omega_{-\kappa m}(\hat{r})\end{pmatrix}, \tag{3}$$

where $P_{n\kappa}(r)$ and $Q_{n\kappa}(r)$ are radial functions for the large and small components of the orbital, and where $\Omega_{\kappa m}(\hat{r})$ is a two-component spherical spinor (Akhiezer and Berestetskii, 1965). The subscripts n, κ, and m are the principal, angular, and magnetic quantum numbers of the orbital, respectively, with $\kappa = \mp(j+1/2)$ for $j = l \pm 1/2$, where j and l are the total and orbital angular momenta. Electrons with the same values of n and κ are energy degenerate and constitute an atomic subshell. There are $2j+1$ electrons in a closed subshell (n, κ). The large-component radial function $P_{n\kappa}(r)$ reduces to the corresponding Schrödinger radial function $P_{nl}(r)$ in the nonrelativistic limit, while $Q_{n\kappa}(r)$ vanishes nonrelativistically.

We assume that the various orbitals in the Slater determinant are orthonormal. This assumption can be formulated in terms of constraints on the radial functions, viz.,

$$\int_0^\infty (P_{n\kappa} P_{n'\kappa} + Q_{n\kappa} Q_{n'\kappa})\, dr = \delta_{nn'}. \tag{4}$$

With these constraints in mind, the expectation value of H can be readily worked out in terms of radial Dirac orbitals as

$$E_{\mathrm{DF}} = \langle H \rangle = \sum_{a,m_a} I_a + \frac{1}{2} \sum_{a,m_a,b,m_b,L} [a_L(\kappa_a m_a, \kappa_b m_b)\, F_L(a, b) - b_L(\kappa_a m_a, \kappa_b m_b)\, G_L(a, b)]. \tag{5}$$

In Eq. (5) we use a single index a to designate the subshell (n_a, κ_a). The contribution to E_{DF} from the one-electron part of H, designated by I_a, is

$$I_a = \int_0^\infty dr \left\{ Q_a \frac{dP_a}{dr} - P_a \frac{dQ_a}{dr} + \frac{2\kappa_a}{r} P_a Q_a \right] - 2mc^2 Q_a^2 - \frac{Ze^2}{r}(P_a^2 + Q_a^2) \bigg\}. \tag{6}$$

The symbols $F_L(a, b)$ and $G_L(a, b)$ in Eq. (5) designate two-electron radial integrals

$$F_L(a, b) = R_L(ab; ab), \tag{7a}$$

$$G_L(a, b) = R_L(ab; ba), \tag{7b}$$

where

$$R_L(ab; cd) = e^2 \int_0^\infty dr \int_0^\infty dr' \frac{r_<^L}{r_>^{L+1}} (P_a P_c + Q_a Q_c)(P_b P_d + Q_b Q_d)', \tag{7c}$$

with $r_> = \sup(r, r')$ and $r_< = \inf(r, r')$. In Eq. (5), the coefficients a_L and b_L which arise from angular integration are

$$a_L(\kappa_a m_a, \kappa_b m_b) = (-1)^{j_a - m_a} \begin{pmatrix} j_a & L & j_a \\ -m_a & 0 & m_a \end{pmatrix} \langle \kappa_a \| C^L \| \kappa_a \rangle$$
$$\times (-1)^{j_b - m_b} \begin{pmatrix} j_b & L & j_b \\ -m_b & 0 & m_b \end{pmatrix} \langle \kappa_b \| C^L \| \kappa_b \rangle, \tag{8a}$$

$$b_L(\kappa_a m_a, \kappa_b m_b) = \begin{pmatrix} j_a & L & j_b \\ -m_a & m_a - m_b & m_b \end{pmatrix}^2 \langle \kappa_a \| C^L \| \kappa_b \rangle^2, \tag{8b}$$

where the one-electron reduced matrix element $\langle a \| C^L \| b \rangle$ of the spherical tensor C^L_q is

$$\langle a \| C^L \| b \rangle = (-1)^{j_a + 1/2} [(2j_a + 1)(2j_b + 1)]^{1/2} \begin{pmatrix} j_a & L & j_b \\ 1/2 & 0 & -1/2 \end{pmatrix} \pi(l_a L l_b), \tag{9}$$

with

$$\pi(l_a L l_b) = \begin{cases} 1 & \text{for } l_a + L + l_b \text{ even}, \\ 0 & \text{for } l_a + L + l_b \text{ odd}. \end{cases} \tag{10}$$

In Eqs. (8) and (9), the large parentheses designate 3-j symbols. The range of L in Eq. (5) is limited by the angular momentum selection rules implicit in Eqs. (9) and (10).

For an atom consisting of closed subshells one can easily carry out the summations over m_b, to find

$$\sum_{m_b} a_L(\kappa_a m_a, \kappa_b m_b) = (2j_b + 1)\, \delta_{L0}, \tag{11a}$$

$$\sum_{m_b} b_L(\kappa_a m_a, \kappa_b m_b) = (2j_b + 1)\, \Lambda(\kappa_a L \kappa_b), \tag{11b}$$

where

$$\Lambda(\kappa_a L \kappa_b) = \frac{\langle \kappa_a \| C^L \| \kappa_b \rangle^2}{(2j_a + 1)(2j_b + 1)}$$
$$= \begin{pmatrix} j_a & L & j_b \\ 1/2 & 0 & -1/2 \end{pmatrix}^2 \pi(l_a L l_b). \tag{12}$$

For systems with closed subshells, and for systems with closed subshells plus or minus one electron, the DF energy is independent of the magnetic quantum numbers. For other open-shell systems, the expression (5) must be averaged over m_a and m_b to obtain a value of $E_{\rm DF}$ which depends only on the radial functions. The averaging scheme adopted is to replace the sum over m_b in Eq. (5) for terms with $b \neq a$ by

$$\sum_{m_b} (\cdots) \to \frac{q_b}{2j_b+1} \sum_{m_b} (\cdots) \qquad \text{for} \quad b \neq a, \tag{13}$$

where q_b is the number of electrons in subshell b. For those terms for which $b = a$, the sum is replaced by

$$\sum_{m_b} (\cdots) \to \frac{q_b - 1}{2j_b} \sum_{m_b} (\cdots) \qquad \text{for} \quad b = a, \tag{14}$$

since there are only $q_b - 1$ nonzero contributions out of $2j_b$ possibilities in this case due to the Pauli exclusion principle. Using a single-Slater-determinant wave function and adopting the above averaging procedure, we can write the total energy from Eq. (5) as

$$\begin{aligned} \bar{E}_{\rm DF} = {} & \sum_a q_a I_a + \tfrac{1}{2} \sum_{b \neq a} q_a q_b \left[F_0(a, b) - \sum_L \Lambda(\kappa_a L \kappa_b)\, G_L(a, b) \right] \\ & + \tfrac{1}{2} \sum_a q_a (q_a - 1) \left[F_0(a, a) - \frac{2j_a + 1}{2j_a} \sum_{L \geq 2} \Lambda(\kappa_a L \kappa_a)\, F_L(a, a) \right]. \end{aligned} \tag{15}$$

Equation (15) gives the correct m-independent value of $E_{\rm DF}$ for closed-subshell systems as well as for systems with closed subshells plus or minus one electron, where averaging is unnecessary.

To obtain the radial DF equations from Eq. (15) we seek extrema of $\bar{E}_{\rm DF}$ with respect to the radial functions $P_a(r)$ and $Q_a(r)$. Before writing out the DF equations, it is convenient to introduce the following notation. We let $F_a(r)$ be the two-component radial function

$$F_a(r) = \begin{pmatrix} P_a(r) \\ Q_a(r) \end{pmatrix}, \tag{16}$$

and we introduce the Hartree Y function

$$Y_L(a, b, r) = r \int_0^\infty dr' \, \frac{r_<^L}{r_>^{L+1}} F_a^\dagger(r') \, F_b(r'). \tag{17}$$

The equations resulting from the variation of $\bar{E}_{\rm DF}$ with respect to $P_a(r)$ and $Q_a(r)$ can be written compactly as

$$(H_a + V_{\rm DF}) F_a = \varepsilon_a F_a + \sum_{b \neq a} q_b \varepsilon_{ab} F_b \delta_{\kappa_a \kappa_b}. \tag{18}$$

The terms ε_a and ε_{ab} on the right-hand side of Eq. (18) are Lagrange multipliers introduced to ensure that the orthonormality constraints (4) are satisfied. The 2×2 radial Hamiltonian H_a is

$$H_a = \begin{pmatrix} -\dfrac{Ze^2}{r} & -\hbar c\left(\dfrac{d}{dr} - \dfrac{\kappa_a}{r}\right) \\ \hbar c\left(\dfrac{d}{dr} + \dfrac{\kappa_a}{r}\right) & -2mc^2 - \dfrac{Ze^2}{r} \end{pmatrix}, \tag{19}$$

and the DF potential is

$$\begin{aligned} V_{\mathrm{DF}} F_a = \sum_{b \neq a} q_b \left[\frac{e^2}{r} Y_0(b, b, r) F_a - \sum_L \Lambda(\kappa_a L \kappa_b) \frac{e^2}{r} Y_L(b, a, r) F_b \right] \\ + (q_a - 1)\left[\frac{e^2}{r} Y_0(a, a, r) - \frac{2j_a + 1}{2j_a} \sum_{L \geqslant 2} \Lambda(\kappa_a L \kappa_a) \frac{e^2}{r} Y_L(a, a, r) \right] F_a. \end{aligned} \tag{20}$$

The radial DF equations (18) can be solved numerically in much the same way as the corresponding nonrelativitic HF equations (Froese Fischer, 1977).

The diagonal Lagrange multiplier ε_a in the DF equations is the DF eigenvalue, and is an approximation to the binding energy of an electron in subshell a. The off-diagonal Lagrange multiplier ε_{ab} in Eq. (18) connects subshells a and b with the same angular quantum numbers ($\kappa_a = \kappa_b$). If the two subshells do not have the same occupation numbers ($q_a \neq q_b$), then we have

$$\varepsilon_{ab} = \varepsilon_{ba} = (X_{ba} - X_{ab})/(q_b - q_a), \tag{21}$$

where

$$X_{\alpha\beta} = \left(1 - \frac{q_\alpha}{2j_\alpha + 1}\right)\left[R_0(\alpha\alpha, \alpha\beta) - \frac{2j_\alpha + 1}{2j_\alpha} \sum_{L \geqslant 2} \Lambda(\kappa_\alpha L \kappa_\alpha)\, R_L(\alpha\alpha, \alpha\beta) \right]. \tag{22}$$

If the two subshells have the same occupation numbers ($q_a = q_b$), Eq. (21) is not applicable, and special investigations are in general required (Froese Fischer, 1977). There is, however, an important special case, namely, when both subshells a and b are closed. In that case, one can simply set $\varepsilon_{ab} = \varepsilon_{ba} = 0$ and still maintain orthogonality. Thus, no off-diagonal Lagrange multiplier is required between closed shells.

We can obtain some notion of the importance of relativistic effects in heavy atoms by comparing the DF and HF eigenvalues for K-shell electrons with experimentally determined K binding energies. This is done in Fig. 1, where the DF eigenvalues are taken from the calculation by Desclaux (1973), the HF eigenvalues from the work of Froese Fischer (1977), and the experimental values from the tabulation by Bearden and Burr (1967). One should first notice that the DF and HF eigenvalues agree very well with each other but not with measured energies at low Z. This is to be expected, since relativistic effects are small for light atoms, while rearrangement and correlation effects, which influence the difference between theory and measurement, are significant. At high Z, where rearrangement and correlation are expected to be relatively less important, but where relativistic corrections become significant, the DF approximation comes into close agreement with measurement, while the error in the HF approximation grows dramatically. One of our principal aims in the following sections is to trace out the degree to which the residual discrepancy between the DF eigenvalues and experimental measurements shown in Fig. 1 can be accounted for theoretically.

For an N-electron atom the binding energy of an electron in subshell a is the difference between the energy $E(N)$ of the atom, and the energy $E_a(N-1)$ of the ion formed when an electron is removed from subshell a. In the DF approximation, the binding energy is

$$\Delta_a = [E_a(N-1) - E(N)]_{\mathrm{DF}}. \tag{23}$$

This approximation to the binding energy, which is referred to as the Δ(SCF) approximation, provides values in good agreement with

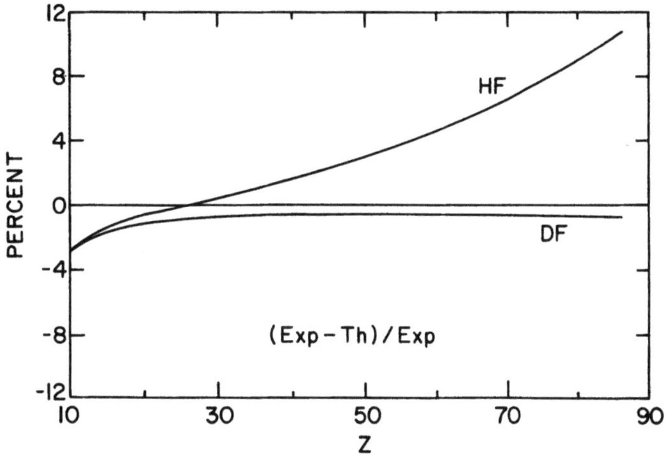

FIGURE 1. Relative difference between the "frozen-orbital" values of the K-shell binding energies and measurements listed by Bearden and Burr (1967). HF results are from Froese-Fischer (1977), and DF results are from Desclaux (1973).

measurement for inner shells, as will be shown below. If we make the additional approximation that the ionic orbitals used to evaluate $E_a(\mathcal{N}-1)$ are identical with the DF orbitals used to evaluate $E(\mathcal{N})$ (i.e., if we disregard the effects of orbital rearrangements in the ion), then we find from Eqs. (15) and (18)

$$\begin{aligned}\Delta_a \approx & -I_a - \sum_{b \neq a} q_b \left[F_0(a, b) - \sum_L \Lambda(\kappa_a L \kappa_b)\, G_L(a, b) \right] \\ & - (q_a - 1) \left[F_0(a, a) - \frac{2j_a + 1}{2j_a} \sum_{L \geq 2} \Lambda(\kappa_a L \kappa_a)\, F_L(a, a) \right] \\ = & -\varepsilon_a. \end{aligned} \tag{24}$$

This results, called Koopmans' theorem (Koopmans, 1933), provides the justification for using the DF and HF eigenvalues in the comparisons shown in Fig. 1.

One can go beyond Koopmans' theorem in making the comparison and account for rearrangement by making separate DF calculations for the atom and for the ion and then subtracting the resulting total energies to find the SCF approximation Δ_a to the binding energy given in Eq. (23). One can write Δ_a in the form

$$\Delta_a = -\varepsilon_a + R_a, \tag{25}$$

where R_a is the relaxation correction to the "frozen-orbital" energy $-\varepsilon_a$ given by Koopmans' theorem. As has been shown by Rosén and Lindgren (1968), R_a is necessarily a negative quantity; inclusion of R_a reduces the binding energy by ~ 7 Ry for the K shell of Hg. To provide some further examples of the influence of rearrangement on inner shells we compare "frozen-orbital" energies $-\varepsilon_a$ and "relaxed-orbital" energies Δ_a with experimental K- and L-shell binding energies for several rare-gas atoms in Table I. For the K shell of Ar, the 1% discrepancy in the "frozen-orbital" approximation is reduced to $\sim 0.1\%$, while in Kr and in Xe the discrepancies are reduced from 0.6% to 0.2% and from 0.6% to 0.4%, respectively. For the L shells, the situation is similar: rearrangement corrections substantially improve agreement with experiment.

In addition to the rearrangement effects, one must consider the effects of electron correlation which arise when one goes beyond the DF approximation. For nonrelativistic systems, Pickup and Goscinski (1973) have examined both rearrangement and correlation energies to second order in perturbation theory, while Born *et al.* (1978) have extended such calculations to third order. The experience gained from these perturbation studies indicates that rearrangement corrections are considerably larger than

TABLE I

Binding energies (in eV) of inner-shell electrons of rare-gas atoms.

Atom	Z	Shell	Frozen[a]	Relaxed[b]	Experiment
Ar	18	K	3241.6	3209.2	3206.0[c]
		L_{I}	337.7	327.2	326.37(5)[d]
		L_{II}	262.1	250.5	250.70(5)[d]
		L_{III}	259.8	248.3	248.60(5)[d]
Kr	36	K	14413.6	14359.3	14325.6(8)[e]
		L_{I}	1961.4	1933.6	1921.2(6)[e]
		L_{II}	1765.3	1735.0	1727.2(5)[e]
		L_{III}	1711.0	1681.3	1674.9(5)[e]
Xe	54	K	34756.3	34690.0	34561.4(11)[e]
		L_{I}	5509.4	5472.4	5453.8(4)[e]
		L_{II}	5161.5	5120.4	5103.7(4)[e]
		L_{III}	4835.6	4796.0	4782.2(4)[e]

[a] Desclaux (1973).
[b] Beatham *et al.* (1980).
[c] Deslattes (1969).
[d] Thomas and Shaw (1974).
[e] Bearden and Burr (1967).

correlation corrections for deep inner shells. Correlation corrections, which have the same sign as rearrangement corrections for inner shells, gradually increase in importance and change sign as one moves outward from the nucleus, so that for the outermost shells there is often a significant cancellation between the two effects. Because of this cancellation, the binding energy $-\varepsilon_a$ determined from Koopmans' theorem often agrees better with measurement for outer shells than the value Δ_a obtained from the Δ(SCF) prescription.

In the graph shown in Fig. 1, as well as in the comparisons in Table I, the effects of finite nuclear size have been incorporated in the one-electron Hamiltonian by replacing the nuclear Coulomb potential with that of a distributed charge $\rho(\mathbf{r})$, such that

$$-\frac{Ze^2}{r} \rightarrow e \int \frac{\rho(\mathbf{r}')}{|\mathbf{r}-\mathbf{r}'|}\, d^3r'. \tag{26}$$

The nuclear charge density is often taken as a uniform charge distribution with radius R. For light atoms for which the finite-nuclear-size effects are essentially negligible, the approximation

$$R = 1.2A^{1/3}\ \text{fm} \tag{27}$$

is often adequate, where A is the nucleon number of the nucleus. For heavier atoms, the expression (Johnson and Soff, unpublished)

$$R = 1.08A^{1/3} + 0.74 \pm 0.06 \text{ fm}, \tag{28}$$

which is a more recent version of the Hofstadter–Collard formula (Elton, 1967), gives more accurate nuclear radii. This latter expression is derived by fitting the available experimental data from electron–nucleus scattering measurements (de Jager *et al.*, 1974) and from muonic x-ray data (Engfer *et al.*, 1974). A careful treatment of the nuclear-size correction requires that the nuclear radius be taken directly from experimental data, rather than from either of the above formulas.

A more realistic model of the nuclear charge distribution is the two-parameter Fermi-type distribution. Practically all available DF codes (Desclaux, 1975; Grant *et al.*, 1980; etc.) have built-in options to choose from either a uniform or Fermi-type nuclear distribution. Usually, the difference between the finite-nuclear-size corrections as obtained from the one-parameter uniform distribution model and those from the two-parameter Fermi distribution model are negligible. In fact, Chen *et al.* (1981) have shown that these differences amount to just a few eV for the K shells of heavy atoms with $Z = 80$–100. However, if one chooses the parameters so that the two models give the same root-mean-square nuclear radius, it has been found that the two finite-nuclear-size corrections thus obtained are essentially identical, even for very heavy atoms (Johnson and Soff, unpublished). As a result, it appears that there is really no need to use the more complicated Fermi distributions in atomic-binding-energy studies.

Having accounted for relativistic, rearrangement, and finite-nuclear-size effects, there still remain important differences between theoretical and experimental inner-shell binding energies, principally due to neglect of the Breit interaction between electrons and of QED corrections. We turn our attention to these topics in the following two sections.

3. BREIT INTERACTION

The effect of the exchange of a virtual photon between two orbital electrons leads to a modification of the instantaneous Coulomb interaction e^2/r_{ij} in the Hamiltonian H of Eq. (1). This correction was considered heuristically by Gaunt (1929), and later from the point of view of QED by Breit (1929). The interaction given by Breit can be written

$$b_{ij} = -\frac{e^2}{r_{ij}} \boldsymbol{\alpha}_i \cdot \boldsymbol{\alpha}_j + \frac{e^2}{2r_{ij}} [\boldsymbol{\alpha}_i \cdot \boldsymbol{\alpha}_j - (\boldsymbol{\alpha}_i \cdot \hat{r}_{ij})(\boldsymbol{\alpha}_j \cdot \hat{r}_{ij})], \tag{29}$$

where $\boldsymbol{\alpha}_i$ and $\boldsymbol{\alpha}_j$ are the usual Dirac matrices. The first term in Eq. (29) is the instantaneous magnetic interaction between two Dirac currents, which was also given by Gaunt, and the second term is a retardation correction to the electric interaction. Since $\boldsymbol{\alpha}$ is the electron velocity operator, the Breit interaction leads to corrections of $O(v^2/c^2) \simeq O(\alpha^2 Z^2)$ compared with the Coulomb interaction, and is therefore most important for the inner shells of heavy atoms.

Starting from Eq. (29), the Breit correction to the many-electron Hamiltonian H of Eq. (1) is

$$H_{\mathrm{Br}} = \tfrac{1}{2} \sum_{i>j} b_{ij}. \tag{30}$$

The correction to the energy of an atomic state due to H_{Br} is easily calculated in first-order perturbation theory, assuming that the state is described as in Sec. 2 by a single-Slater-determinant wave function [the multiconfiguration case has been discussed by Grant and Pyper (1976)]. The resulting energy shift is (Mann and Johnson, 1971)

$$E_{\mathrm{Br}} = \tfrac{1}{2} \sum_{a,b} (L_{ab} + M_{ab}) + \tfrac{1}{2} \sum_{a,b} R_{ab}, \tag{31}$$

where

$$L_{ab} = - \sum_{m_a, m_b, L} a_L(-\kappa_a m_a, \kappa_b m_b) \frac{4\kappa_a \kappa_b}{L(L+1)} S_{ab}^L, \tag{32}$$

$$M_{ab} = \sum_{m_a, m_b, L} \left\{ b_L(\kappa_a m_a, \kappa_b m_b) \left(\frac{L}{2L-1} T_{ab}^{L,L-1} + \frac{L+1}{2L+3} T_{ab}^{L,L+1} \right) + b_L(-\kappa_a m_a, \kappa_b m_b) \frac{(\kappa_a + \kappa_b)^2}{L(L+1)} T_{ab}^{LL} \right\}, \tag{33}$$

$$R_{ab} = - \sum_{m_a, m_b, L} b_L(\kappa_a m_a, \kappa_b m_b) \left[\frac{L^2}{(2L+1)(2L-1)} T_{ab}^{L,L-1} + \frac{(L+1)^2}{(2L+1)(2L+3)} T_{ab}^{L,L+1} + \frac{L(L+1)}{2L+1} X_{ab}^L \right]. \tag{34}$$

The angular coefficients $a_L(-\kappa_a m_a, \kappa_b m_b)$ and $b_L(-\kappa_a m_a, \kappa_b m_b)$ that appear in Eqs. (32) and (33) are identical to the corresponding coefficients $a_L(\kappa_a m_a, \kappa_b m_b)$ and $b_L(\kappa_a m_a, \kappa_b m_b)$ in Eq. (8), except that the parity selection rule shown in Eq. (10) must be modified by replacing L by $L+1$. The quantities S_{ab}^L, $T_{ab}^{L,L'}$, and X_{ab}^L are radial integrals, given by

$$S_{ab}^L = e^2 \int_0^\infty dr \int_0^\infty dr' \frac{r_<^L}{r_>^{L+1}} V_{aa}(r) V_{bb}(r'), \tag{35}$$

$$T_{ab}^{L,L'} = e^2 \int_0^\infty dr \int_0^\infty dr' \frac{r_<^{L'}}{r_>^{L'+1}} W_{ab}^{L'}(r)\, W_{ab}^{L'}(r') \qquad L' = L, L \pm 1, \tag{36}$$

$$X_{ab}^{L} = e^2 \int_0^\infty dr \int_0^r dr' \left(\frac{r_<^{L-1}}{r_>^{L}} - \frac{r_<^{L+1}}{r_>^{L+2}} \right) W_{ab}^{L+1}(r)\, W_{ab}^{L-1}(r'), \tag{37}$$

with

$$V_{ab}(r) = P_a(r)\, Q_b(r) + Q_a(r)\, P_b(r), \tag{38}$$

$$U_{ab}(r) = P_a(r)\, Q_b(r) - Q_a(r)\, P_b(r), \tag{39}$$

$$W_{ab}^{L-1}(r) = U_{ab}(r) + \frac{\kappa_b - \kappa_a}{L} V_{ab}(r), \tag{40}$$

$$W_{ab}^{L}(r) = V_{ab}(r), \tag{41}$$

$$W_{ab}^{L+1}(r) = -U_{ab}(r) + \frac{\kappa_b - \kappa_a}{L+1} V_{ab}(r). \tag{42}$$

The quantities L_{ab} and M_{ab} in Eq. (31) are the direct and exchange contributions to the first-order energy arising from the first term in Eq. (29), i.e., the magnetic interaction. There is no direct contribution from the retardation correction; the quantity R_{ab} designates the exchange retardation contribution.

Averaging over magnetic substates as explained in Sec. 2, we obtain an expression for the Breit correction (Mann and Johnson, 1971):

$$\bar{E}_{\rm Br} = \frac{1}{2} \sum_{a \neq b} q_a q_b (\bar{M}_{ab} + \bar{R}_{ab}) + \frac{1}{2} \sum_a q_a (q_a - 1) \frac{2j_a + 1}{2j_a} \bar{M}_{aa}, \tag{43}$$

where

$$\bar{M}_{ab} = \sum_L \left\{ \Lambda(\kappa_a L \kappa_b) \left(\frac{L}{2L-1} T_{ab}^{L,L-1} + \frac{L+1}{2L+3} T_{ab}^{L,L+1} \right) + \Lambda(-\kappa_a L \kappa_b) \frac{(\kappa_a + \kappa_b)^2}{L(L+1)} T_{ab}^{L,L} \right\}, \tag{44}$$

and

$$\bar{R}_{ab} = -\sum_L \Lambda(\kappa_a L \kappa_b) \left[\frac{L^2}{(2L+1)(2L-1)} T_{ab}^{L,L-1} + \frac{(L+1)^2}{(2L+1)(2L+3)} T_{ab}^{L,L+1} + \frac{L(L+1)}{2L+1} X_{ab}^{L} \right]. \tag{45}$$

The terms $\bar{M}_{ab}$ and $\bar{R}_{ab}$ in the sum over subshells in Eq. (43) are averages of the magnetic and retardation interactions given by Eqs. (33) and (34). The average contribution from the direct magnetic interaction term $\bar{L}_{ab}$ vanishes. Just as in the expression (15) for the DF average energy, the expression in Eq. (43) is an exact result, within the DF framework, for the Breit correction for atoms with closed shells and for atoms with closed shells plus or minus one electron.

In columns 3–5 of Table II, we list the DF energy together with the magnetic and retardation parts of the Breit correction for rare-gas atoms. The magnitudes of the DF, magnetic, and retardation energies for the ground states of neutral atoms with $Z=2$–100 are shown for comparison purposes in Fig. 2.

As mentioned in the introduction to this section, the Breit interaction is due to the exchange of a virtual photon between two orbital electrons. For the direct interaction between two electrons a and b, no energy is exchanged, and the wave number of the virtual photon is zero ($k=0$). For the exchange interaction, the photon energy $\hbar\omega$ is the difference between the orbital energies ε_a and ε_b of the two electrons; the corresponding wave number is $k=|\varepsilon_b-\varepsilon_a|/\hbar c$. The generalization of the Breit interaction (29) which includes the dependence on photon frequency (Mann and Johnson, 1971) can be written

$$b_{ij}(k) = -\frac{e^2}{r_{ij}}[\boldsymbol{\alpha}_i\cdot\boldsymbol{\alpha}_j\cos kr_{ij} - (\boldsymbol{\alpha}_i\cdot\boldsymbol{\nabla}_i)(\boldsymbol{\alpha}_j\cdot\boldsymbol{\nabla}_j)(1-\cos kr_{ij})/k^2]. \tag{46}$$

For the direct two-electron matrix elements in which $k=0$, Eq. (46) reduces to Eq. (29), while for exchange matrix elements, k-dependent corrections to Eq. (29) appear. These corrections are largest for the inner

TABLE II

Comparison of the DF energies with the magnetic and retardation parts of the Breit correction for the ground states of rare-gas atoms (Mann and Johnson, 1971). Energies in Ry units.

Atom	Z	E_{DF}	E_{Br}^{mag}	E_{Br}^{ret}	$E_{Br}^{mag}(k)$	$E_{Br}^{ret}(k)$
He	2	−5.7236	0.00013	0.	0.00013	0.
Ne	10	−257.3838	0.0.3508	−0.00180	0.03507	−0.00179
Ar	18	−1057.3673	0.28686	−0.02214	0.28636	−0.02194
Kr	36	−5577.7193	3.15621	−0.30272	3.13165	−0.29278
Xe	54	−14893.7970	12.88098	−1.33080	12.66281	−1.24193
Rn	86	−47204.9084	65.75845	−6.96351	63.29303	−5.93143

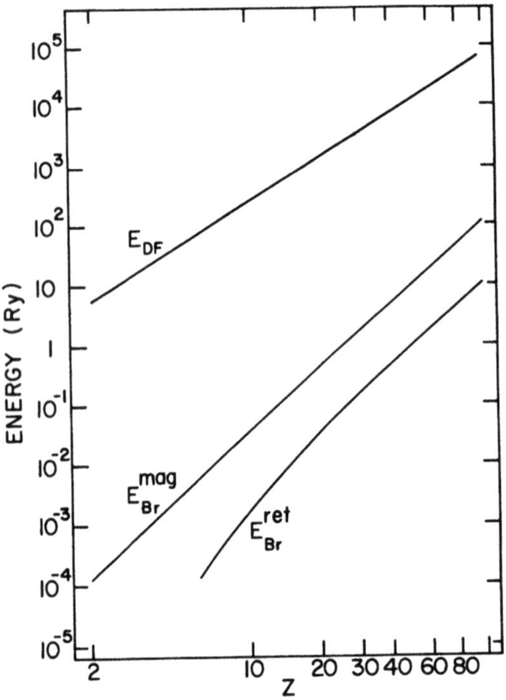

FIGURE 2. The magnitudes of the DF, magnetic, and retardation energies for the ground states of neutral atoms as functions of nuclear charge Z.

shells of heavy atoms and lead to shifts in energy levels of $O(k^2r^2) \simeq O(\alpha^2 Z^2)$ compared to the frequency-independent part of the Breit interaction.

The reduction of the matrix elements of $b_{ij}(k)$ to radial form follows essentially the steps described in the previous paragraphs. One major difference is that the quantity $r_<^L/r_>^{L+1}$ in the integrals $T_{ab}^{L,L'}$ in Eq. (36) must be replaced by products of spherical Bessel functions (Mann and Johnson, 1971):

$$\frac{r_<^L}{r_>^{L+1}} \to -k(2L+1)\, j_L(kr_<)\, y_L(kr_>). \tag{47}$$

It is easily shown that the limiting value of the right-hand side for $k=0$ is $r_<^L/r_>^{L+1}$. Furthermore, it is necessary to replace the integral X_{ab}^L in Eq. (37) by the rather complicated expression

$$X_{ab}^L \to \int_0^\infty dr \int_0^r dr' \left\{ -2\left[kj_{L-1}(kr_<)\, y_{L+1}(kr_>) + \frac{2L+1}{k^2}\frac{r_<^L}{r_>^{L+1}} \right] \right.$$

$$\left. \times W_{ab}^{L+1}(r)\, W_{ab}^{L-1}(r') - 2kj_{L+1}(kr_<)\, y_{L-1}(kr_>)\, W_{ab}^{L-1}(t)\, W_{ab}^{L+1}(r') \right\}. \tag{48}$$

The formulas for the average energy shift presented previously apply just as well to the frequency-dependent case, with the changes noted in Eqs. (47) and (48). A similar analysis for the multiconfiguration case has been made by Grant and McKenzie (1980). Numerical values of the frequency-dependent Breit corrections to the energies of closed-shell atoms are presented in columns 6 and 7 of Table II.

As discussed in Sec. 2, the magnitude of the DF eigenvalue $|\varepsilon_a|$ can be interpreted as a "frozen-orbital" approximation to the binding energy of an electron is subshell a. The Breit correction to the binding energy can be worked out easily in the "frozen-orbital" approximation, from Eq. (43), as

$$\Delta\varepsilon_a = \sum_{b \neq a} q_b(\bar{M}_{ab} + \bar{R}_{ab}) + (q_a - 1)\frac{2j_a + 1}{2j_a}\bar{M}_{aa}. \tag{49}$$

To provide an indication of the relative size of the Breit correction to binding energies, we list in Table III the DF eigenvalues of Hg ($Z = 80$)

TABLE III

Frozen-orbital contributions to electron binding energies for Hg ($Z = 80$). The magnetic and retardation energies are evaluated using the frequency-dependent version of the Breit interaction (Mann and Johnson, 1971). Energies in Ry units.

Subshell[a]	$-\varepsilon_{\mathrm{DF}}$	$-\Delta\varepsilon_{\mathrm{Br}}^{\mathrm{mag}}(k)$	$-\Delta\varepsilon_{\mathrm{r}}^{\mathrm{ret}}(k)$
1s	6148.6546	−24.3548	1.6064
2s	1100.5281	−3.4662	0.3697
2p*	1053.7019	−5.3248	0.5107
2p	910.3115	−3.5144	0.4902
3s	266.2319	−0.8127	0.0976
3p*	245.2758	−1.1892	0.1257
3p	213.0895	−0.7887	0.1158
3d*	178.8730	−0.6990	0.0875
3	172.0400	−0.5628	0.0850
4s	61.2979	−0.2137	0.0285
4p*	52.2476	−0.2935	0.0342
4p	44.3769	−0.1907	0.0302
4d*	29.5933	−0.1492	0.0206
4d	28.1050	−0.1185	0.0199
4f*	8.9458	−0.0650	0.0090
4f	8.6234	−0.0527	0.0088
5s	10.1064	−0.0424	0.0058
5p*	7.0757	−0.0501	0.0060
5p	5.6839	−0.0308	0.0050
5d*	1.3001	−0.0142	0.0020
5d	1.1493	−0.0106	0.0019
6s	0.6561	−0.0036	0.0005

[a] Subshell nl^* means $j = l - 1/2$ and subshell nl means $j = l + 1/2$.

TABLE IV

Comparison between the relaxed (Δ_{1s}) and the frozen ($-\varepsilon_{1s}$) values of the $1s$ binding energy for Hg (Mann and Johnson, 1971). Energies in Ry units.

	E_{DF}	$E_{Br}^{mag}(k)$	$E_{Br}^{t}(k)$	Total
$Hg^+(1s^{-1})$	−33156.521	25.678	−3.187	−33134.030
Hg	−39298.167	49.006	−4.670	−39253.831
$\Delta_{1s}(Hg^- – Hg)$	6141.646	−23.328	1.483	6119.801
$-\varepsilon_{1s}(Hg)$	6148.655	−24.355	1.606	6125.906
Experiment[a]				6108.39(6)

[a] Bearden and Burr (1967).

together with the corresponding corrections due to the Breit interaction. The frequency-dependent version of the Breit interaction has been used to evaluate the contributions in Table III. One sees that these corrections are most substantial for inner shells and that throughout the shells, the retardation corrections amount to ~10% of the magnetic corrections.

The effects of rearrangement on the binding energy were discussed in Sec. 2. There are, of course, similar rearrangement effects on the Breit interaction. To include these effects, it is necessary to evaluate separately the Breit corrections to the energies of the atom and the ion, and to subtract the two results. In Table IV, we illustrate the effects of rearrangement on the Breit interaction by comparing the relaxed and unrelaxed values of the K binding energy for Hg. Although the Breit corrections are several orders of magnitude smaller than the DF energy, the relaxation effect is seen to be relatively more important for the Breit correction than for the DF energy. From Table IV it is also apparent that there is a residual discrepancy between theory and experiment of about 11 Ry. This discrepancy is due mainly to QED corrections, which are discussed next.

4. QUANTUM ELECTRODYNAMIC CORRECTIONS

Here we describe the influence of the two dominant QED corrections, electron self-energy and vacuum polarization, on the binding energies of inner electrons in heavy atoms. The approach taken is heuristic; radiative corrections for atomic systems are taken to be those of a collection of electrons moving in a self-consistent potential. What is remarkable is that this simple approach to the complex problem of radiative corrections to a many-body system leads to predictions in relatively precise agreement with measurement.

Of the two dominant radiative corrections, the electron self-energy is the largest. The self-energy for an electron with quantum numbers (n, l, j) moving in the Coulomb field of the nucleus is

$$\Delta E_{nlj} = \frac{\alpha}{\pi} \frac{(\alpha Z)^4}{n^3} F(\alpha Z)\, mc^2, \tag{50}$$

where $F(\alpha Z)$ is a dimensionless function of the nuclear charge. For light atoms, an expansion of $F(\alpha Z)$ in powers of αZ has been developed using perturbation theory

$$\begin{aligned} F(\alpha Z) = \frac{4}{3} \Big\{ & \left[\ln(\alpha Z)^{-2} + \frac{11}{24} \right] \delta_{l0} + \frac{3}{8} \frac{C_{lj}}{2l+1} + L_{nl} \\ & + \alpha Z A_{50} + (\alpha Z)^2 [A_{62} \ln^2(\alpha Z)^{-2} + A_{61} \ln(\alpha Z)^{-2}] \\ & + (\alpha Z)^2 G(\alpha Z) \Big\}. \end{aligned} \tag{51}$$

The quantity L_{nl} in Eq. (51) is called the Bethe logarithm. For low values of n and l, L_{nl} is given by

$$L_{nl} = \begin{cases} -2.9841285 & 1s \\ -2.8117699 & 2s \\ 0.0300167 & 2p \end{cases}$$

More complete tables of L_{nl} are provided, for example, by Klarsfeld and Maquet (1973). The quantity C_{lj} in Eq. (51) is the coefficient of the anomalous-moment contribution to the self-energy; it is

$$C_{lj} = \begin{cases} -1/l & j = l - 1/2, \\ 1/(l+1) & j = l + 1/2. \end{cases}$$

The terms on the second and third lines of Eq. (51) are called structure corrections (Erickson, 1977). The leading coefficients of the structure corrections are

$$A_{50} = 3\pi \left(1 + \frac{11}{128} - \frac{1}{2} \ln 2 \right) \delta_{l0},$$

$$A_{61} = \begin{cases} 7 \ln 2 - 63/80 & 1s \\ 4 \ln 2 + 67/40 & 2s \\ 103/240 & 2p_{1/2} \\ 29/120 & 2p_{3/2}, \end{cases}$$

$$A_{62} = -\tfrac{3}{4} \delta_{l0}.$$

The remaining higher-order structure corrections are represented by the quantity $G(\alpha Z)$, which has been investigated analytically by Erickson and Yennie (1965), by Erickson (1971), and more recently by Sapirstein (1981).

One difficulty encountered in applying the analytical expressions above to high-Z systems is that the structure corrections represented by the term $G(\alpha Z)$ on the third line in Eq. (51) are comparable in size to the leading terms for $Z > 20$; the perturbation series does not appear to converge. An alternative, which is applicable to high-Z systems, is the direct numerical evaluation of the self-energy. One approach is that laid out by Brown, Langer, and Schaefer (1959). It was first used by Brown and Mayers (1959) to calculate the self-energy of the $1s$ electron in Hg. Later a similar approach was adopted by Desiderio and Johnson (1971), who investigated the $1s$ self-energy for elements in the range $Z = 70$–90. For the case of Hg the value predicted by Desiderio and Johnson, 15 Ry, is substantially smaller than that obtained by Brown and Mayers, 41 Ry, and agrees well with more recent results (Mohr, 1974). An alternative approach was developed by Mohr (1974, 1975, 1982), who calculated the self-energy for $1s$, $2s$, $2p_{1/2}$ and $2p_{3/2}$ electrons in hydrogenic ions with nuclear charges ranging from $Z = 10$ to $Z = 110$. Values of $F(\alpha Z)$ from Mohr's work are listed in column 3 of Table V.

For atomic electrons, the values of $F(\alpha Z)$ are modified by screening of the nuclear charge and by finite-nuclear-size effects. The influence of screening on the $1s$ self-energy was addressed by Desiderio and Johnson (1971), who use a DFS potential rather than a Coulomb potential in their calculations. The resulting values of $F(\alpha Z)$ are $\sim 2\%$ lower than the corresponding Coulomb values, in the interval $Z = 70$–90.

Finite-nuclear-size effects on the $1s$ self-energy were examined by Cheng and Johnson (1976). These calculations were carried out using a uniform nuclear charge distribution of radius $R = 1.2A^{1/3}$ fm. The deviations of $F(\alpha Z)$ from the Coulomb field values are shown in Fig. 3, where comparisons are made with point-nucleus values from the work of Mohr (1974). As one sees from Fig. 3, the effects of finite nuclear size are relatively small for low-Z systems, but become very important in the superheavy region above $Z = 100$. In fact, in the point-nucleus case, $F(\alpha Z)$ actually develops a logarithmic singularity at $Z = 137$ (Labzovskii, 1971). These finite-nuclear-size corrections are also listed in column 4 of Table V. Further calculations of $F(\alpha Z)$ including the effects of finite nuclear size have been carried out by Soff *et al.* (1983), who determined the self-energy of $1s$ electrons for Z up to 170 where the $1s$ state starts to merge into the negative-energy continuum (Pieper and Greiner, 1969).

The vacuum-polarization correction to the energy of an electron moving in the nuclear Coulomb field has been discussed in great detail by Wichmann and Kroll (1956), who showed that to an excellent

TABLE V

Hydrogenic values of the functions $F(\alpha Z)$ and $F_u(\alpha Z)$ for the electron self-energy and the vacuum polarization, respectively. ΔF^{fs} and ΔF_u^{fs} are the finite-nuclear-size corrections, and ΔF_u^{WK} are the Wichmann–Kroll corrections. Except for the values of $F(\alpha Z)$, which are taken from Mohr (1974, 1975, 1982), all data are from Johnson and Soff (unpublished).

Z	Electron[a]	$F(\alpha Z)$	ΔF^{fs}	$F_u(\alpha Z)$	ΔF_u^{WK}	ΔF_u^{s}
20	1s	3.2460	−0.0001	−0.2409	0.0009	0.0000
	2s	3.5063	−0.0001	−0.2446	0.0009	0.000
	2p*	−0.0922	0.0000	−0.0014	0.0000	0.0000
	2p	0.1436	0.0000	−0.0003	0.0000	0.0000
40	1s	2.1351	−0.0006	−0.2421	0.0033	0.0003
	2s	2.4450	−0.0007	−0.2572	0.0034	0.0003
	2p*	−0.0308	0.0000	−0.0063	0.0001	0.0000
	2p	0.1794	0.0000	−0.0009	0.0000	0.0000
60	1s	1.6838	−0.0030	−0.2675	0.0071	0.0014
	1s	2.0948	−0.0037	−0.3071	0.0079	0.0017
	2p*	0.0549	−0.0001	−0.0188	0.0007	0.0001
	2p	0.2215	0.0000	−0.0019	0.0002	0.0000
80	1s	1.5032	−0.0118	−0.3310	0.0134	0.0066
	2s	2.0650	−0.0162	−0.4267	0.0164	0.0091
	2p*	0.1884	−0.0015	−0.0518	0.0028	0.0006
	2p	0.2671	0.0000	−0.0032	0.0004	0.0000
100	1s	1.5317	−0.0460	−0.4890	0.0252	0.0367
	1s	2.3870	−0.0717	−0.7435	0.0357	0.0616
	2p*	0.4530	−0.0136	−0.1608	0.0108	0.0085
	2p	0.3141	0.0000	−0.0050	0.0099	0.0000

[a] $2p^* = 2p_{1/2}$; $2p = 2p_{3/2}$.

approximation the effects of vacuum polarization are accounted for by the Uehling potential (1935),

$$V_u(r) = -\frac{\alpha}{3\pi}\frac{Ze^2}{r}\int_1^\infty dt\,(t^2-1)^{1/2}\left(\frac{1}{t^4}+\frac{2}{t^2}\right)e^{-2mc\,tr/\hbar}. \tag{52}$$

The shift in energy of an electron caused by the Uehling potential can be expressed in a form similar to Eq. (50):

$$\Delta E_u = \frac{\alpha}{\pi}\frac{(\alpha Z)^4}{n^3}F_u(\alpha Z)\,mc^2. \tag{53}$$

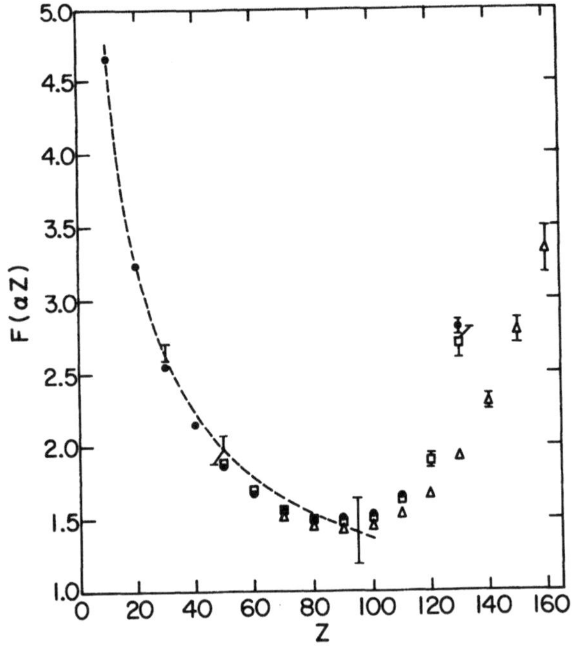

FIGURE 3. The function $F(\alpha Z)$ for the self-energies of the $1s$ electrons. Dashed line: Erickson (1971); closed circles: Mohr (1974); open squares: hydrogenic values of Cheng and Johnson (1976); open triangles: finite nuclear size values of Cheng and Johnson (1976).

For low values of αZ, $F_u(\alpha Z)$ is given by the approximation

$$F_u(\alpha Z) = \begin{cases} -\left(\dfrac{4}{15} - \dfrac{5}{48}\pi\alpha Z\right) & 1s \quad \text{and} \quad 2s \\ -\dfrac{9}{140}(\alpha Z)^2 & 2p_{1/2} \\ -\dfrac{2}{140}(\alpha Z)^2 & 2p_{3/2}. \end{cases}$$

Numerical values of $F_u(\alpha Z)$ determined by calculating the expectation value of the Uehling potential with Dirac–Coulomb wave functions are listed in column 5 Table V.

Wichmann and Kroll also discussed structure corrections to the Uehling potential, analogous to the structure corrections to the self-energy. These corrections are expressed as power series in $(\alpha Z)^2$ for the Laplace transform of the vacuum-polarization charge density. From this charge density one can obtain a potential (Blomqvist, 1972; Huang, 1976) and predict corrections to the energy shift resulting from the structue corrections to the Uehling potential. The contribution to $F_u(\alpha Z)$ resulting from the lowest-order Wichmann–Kroll corrections are shown in column 6 of Table V. In contrast to the self-energy, higher-order terms in the series for the vacuum-polarization form factor $F_u(\alpha Z)$ are found to the well behaved, so that the corresponding analytical results are applicable even for high Z.

The influence of finite-nuclear-size effects on the vacuum-polarization energy shift is accounted for by generalizing the Uehling potential from a point charge to a distributed charge (Gyulassy, 1974). The finite-nuclear-size corrections to the Uehling potential, which arise from a uniform charge distribution of radius $1.2A^{1/3}$ fm, are listed in column 7 of Table V.

All quantities in Table V refer to one-electron ions. Adopting the heuristic approach mentioned at the beginning of this section, one can modify the values in Table V to account for the effect of electron screening, by replacing the Coulomb potential by a self-consistent potential. With the exception of the calculation of Desiderio and Johnson (1971), this procedure has not been generally followed. Rather, the practice has been to use values for the self-energy interpolated from the one-electron hydrogenic results, with Z replaced by an effective nuclear charge Z^* chosen to account for screening. The corresponding values for the vacuum polarization are determined, in practice, by evaluating the energy shift from the Uehling potential using self-consistent wave functions. While these approximations leave something to be desired, they do give first-order estimates of the size of radiative corrections to the levels of many-electron atoms.

The sizes of these QED corrections for the 1*s* electrons in several heavy atoms are illustrated in Table VI. In column 3 of Table VI we list

TABLE VI

K-shell binding energies for several heavy atoms. Here, Δ_{DF} are the DF energies including the frequency dependent Breit interactions and the relaxation effects, and $\Delta_{LS} = -E_{LS}$ are the QED corrections from the self-energy and vacuum polarization. Energies in Ry units.

Atom	Z	Δ_{DF} [a]	Δ_{LS} [b]	Δ_{tot}	Δ_{exp} [c]
W	74	6119.15	−8.73	5110.42	5110.46(2)
Hg	80	6119.80	−11.36	6108.44	6108.39(6)
Pb	82	6481.06	−12.35	6468.71	6468.67(5)
Rn	86	7247.44	−14.51	7232.93	7233.08(90)

[a] Mann and Johnson (1971).
[b] Desiderio and Johnson (1971).
[c] Bearden and Burr (1967). Photoelectric work functions have been added to the published values.

theoretical values of the 1*s* binding energies that include the Breit interaction and rearrangement effects. Theoretical values of the 1*s* Lamb shift taken from Desiderio and Johnson (1971) are listed in column 4, and the resulting theoretical binding energies including the 1*s* Lamb shift are given in column 5. Comparing the theoretical and experimental values of the 1*s* binding energies, one sees that the residual discrepancy is at the 1-eV level, which is the size expected for inner-shell correlation effects (Cowan, 1967).

It might be said that, while no really satisfactory approach to the problem of evaluating the radiative corrections to many-electron atoms has yet been developed, the existing calculations based on a simple heuristic model are capable of explaining the inner-shell binding energies of atoms at the level of a few eV, at which level the effects of electron–electron correlation become important. Accurate values for the *L*-shell QED effects are more difficult to obtain because of the uncertainties associated with screening corrections. Nevertheless, a comparison (Mohr, 1983) of the existing theoretical calculation of *L*-shell binding energies (Huang *et al.*, 1976) with measurements again reveals that the approach described above accounts for the QED effects to the level of a few eV.

5. CONCLUSION

The problem of determining inner-shell binding energies in many-electron atoms can be viewed as a problem in QED. Presently, no coherent and accurate solution to this problem exists. Rather, what we have given in this chapter is a calculational prescription, fragments of which are taken from QED. Although this prescription leads to results which agree very precisely with experiment, it appears unlikely that further extensions along the lines described here will lead to significant improvement in the understanding of inner-shell binding energies. The reason is that this prescription is based on the configuration-space Hamiltonian in Eq. (1), which is not a result of QED (Mittleman, 1980; Sucher, 1980). Furthermore, rigorous solutions of the many-electron problem based on Eq. (1) may not exist. Modifications of this Hamiltonian have been proposed which avoid the pitfalls mentioned above (Mittleman, 1981; Sucher, 1983), but to date, practical calculations based on these modifications have not been pursued. In summary, although the techniques described in this chapter lead to accurate values of inner-shell binding energies, a systematic approach to the problem of calculating these energies based on QED provides a challenge for the future.

ACKNOWLEDGMENTS

The work of WRJ was supported in part by the National Science Foundation under grant No. PHY83-08136. The work of KTC was supported by the U.S. Department of Energy, Office of Basic Energy Sciences, under contract No. W-31-109-Eng-38.

REFERENCES

AKHIEZER, A. I., AND V. B. BERESTETSKII, 1965, *Quantum Electrodynamics* (Interscience, New York), Sec. 11.1.
BEARDEN, J. A., AND A. F. BURR, 1967, *Rev. Mod. Phys.* **39**, 125.
BEATHAM, N., I. P. GRANT, B. J. MCKENZIE, AND S. J. ROSE, 1980, *Phys. Scr.* **21**, 423.
BLOMQVIST, J., 1972, *Nucl. Phys.* **B48**, 95.
BORN, G., H. A. KURTZ, AND Y. ÖHRN, 1978, *J. Chem. Phys.* **68**, 74.
BREIT, G., 1929, *Phys. Rev.* **34**, 553.
BRENNER, S., AND G. E. BROWN, 1953, *Proc. R. Soc. (London)* **A218**, 422.
BROWN, G. E., AND D. F. MAYERS, 1959, *Proc. R. Soc. (London)* **A251**, 105.
BROWN, G. E., AND D. G. RAVENHALL, 1951, *Proc. R. Soc. (London)* **A208**, 552.
BROWN, G. E., J. S. LANGER, AND G. W. SCHAEFER, 1959, *Proc. R. Soc. (London)* **A251**, 92.
CHEN, M. H., B. CRASEMANN, M. AOYAGI, K.-N. HUANG, AND H. MARK, 1981, *At. Data Nucl. Data Tables* **26**, 561.
CHENG, K. T., AND W. R. JOHNSON, 1976, *Phys. Rev. A* **14**, 1943.
CHENG, K. T., Y.-K. KIM, AND J. P. DESCLAUX, 1979, *At. Data Nucl. Data Tables* **24**, 111.
COHEN, S., 1960, *Phys. Rev.* **118**, 489.
COOPER, J. R. A., 1965, *Proc. Phys. Soc. London* **86**, 529.
COULTHARD, M. A., 1967, *Proc. Phys. Soc. London* **91**, 44.
COWAN, R. D., 1967, *Phys. Rev.* **163**, 54.
DE JAGER, C. W., H. DE VRIES, AND C. DE VRIES, 1974, *At. Data Nucl. Data Tables* **14**, 479.
DESCLAUX, J. P., 1973, *At. Data Nucl. Data Tables* **12**, 311.
DESCLAUX, J. P., 1975, *Comput. Phys. Commun.* **9**, 31.
DESIDERIO, A. M., AND W. R. JOHNSON, 1971, *Phys. Rev. A* **3**, 1267.
DESLATTES, R. D., 1969, *Phys. Rev.* **186**, 1.
DESLATTES, R. D., AND E. G. KESSLER, JR., 1980, in Proceedings of the Workshop on Foundations of the Relativistic Theory of Atomic Structure, Argonne National Laboratory Report No. ANL-80-126, p. 215.
DESLATTES, R. D., E. G. KESSLER, JR., L. JACOBS, AND W. SCHWITZ, 1979, *Phys. Lett.* **71A**, 411.
DOYLE, H. T., 1969, *Adv. At. Mol. Phys.* **5**, 337.
ELTON, L. R. B., 1967, Nuclear charge distribution, in *Landolt–Börnstein Numerical Data and Functional Relationships in Science and Technology*, edited by K.-H. Hellwege and H. Schopper, New Series, Group I, Vol. 2 (Springer-Verlag, Berlin), p. 1.
ENGFER, R., H. SCHNEUWLY, J. L. VUILLEUMIER, H. K. WALTER, AND A. ZEHNDER, 1974, *At. Data Nucl. Data Tables* **14**, 509.
ERICKSON, G. W., 1971, *Phys. Rev. Lett.* **27**, 780.
ERICKSON, G. W., 1977, *J. Phys. Chem. Ref. Data* **6**, 831.
ERICKSON, G. W., AND D. R. YENNIE, 1965, *Ann. Phys. (N.Y.)* **35**, 271.
FREEDMAN, M. S., F. T. PORTER, AND J. B. MANN, 1972, *Phys. Rev. Lett.* **28**, 711.
FRICKE, B., J. P. DESCLAUX, AND J. T. WABER, 1972, *Phys. Rev. Lett.* **28**, 714.
FROESE FISCHER, C., 1977, *The Hartree–Fock Method for Atoms* (Wiley, New York).
GAUNT, J. A., 1929, *Proc. R. Soc. (London)* **A122**, 513.

GRANT, I. P., 1961, *Proc. R. Soc. (London)* **A262**, 555.
GRANT, I. P., 1965, *Proc. Phys. Soc.* **86**, 523.
GRANT, I. P., 1970, *Adv. Phys.* **19**, 747.
GRANT, I. P., AND B. J. MCKENZIE, 1980, *J. Phys. B* **13**, 2671.
GRANT, I. P., AND N. C. PYPER, 1976, *J. Phys. B* **9**, 761.
GRANT, I. P., B. J. MCKENZIE, P. H. NORRINGTON, D. F. MAYERS, AND N. C. PYPER, 1980, *Comput. Phys. Commun.* **21**, 207.
GYULASSY, M., 1974, *Phys. Rev. Lett.* **33**, 921.
HUANG, K.-N., 1976, *Phys. Rev. A* **14**, 1311.
HUANG, K.-N., M. AOYAGI, M. H. CHEN, B. CRASEMANN, AND H. MARK, 1976, *At. Data Nucl. Data Tables* **18**, 243.
JOHNSON, W. R., AND K. T. CHENG, 1976, *Phys. Rev. A* **13**, 692.
KESSLER, E. G., JR., R. D. DESLATTES, D. GIRARD, W. SCHWITZ, L. JACOBS, AND O. RENNER, 1982, *Phys. Rev. A* **26**, 2696.
KIM, Y.-K., 1967, *Phys. Rev.* **154**, 17.
KLARSFELD, S., AND A. MAQUET, 1973, *Phys. Lett.* **43B**, 201.
KOOPMANS, T. H., 1933, *Physica* **1**, 104.
LABZOVSKII, L. N., 1971, *Sov. Phys. JETP* **32**, 1171.
LIBERMAN, D., J. T. WABER, AND D. T. CROMER, 1965, *Phys. Rev.* **137**, A27.
LU, C. C., T. A. CARLSON, F. B. MALIK, T. C. TUCKER, AND C. W. NESTOR, JR., 1971, *At. Data* **3**, 1; 1974, *At. Data Nucl. Data Tables* **14**, 89 (Erratum).
MANN, J. B., AND W. R. JOHNSON, 1971, *Phys. Rev. A* **4**, 41.
MANN, J. B., AND J. T. WABER, 1973, *At. Data* **5**, 201.
MAYERS, D. F., 1957, *Proc. R. Soc. (London)* **A241**, 93.
MAYERS, D. F., G. E. BROWN, AND E. A. SANDERSON, 1959, *Phys. Rev. Lett.* **3**, 90.
MCKENZIE, B. J., I. P. GRANT, AND P. H. NORRINGTON, 1980, *Comput. Phys. Commun.* **21**, 233.
MITTLEMAN, M. H., 1980, in Proceedings of the Workshop on Foundations of the Relativistic Theory of Atomic Structure, Argonne National Laboratory Report No. ANL-80-126, p. 27.
MITTLEMAN, M. H., 1981, *Phys. Rev. A* **24**, 1167.
MOHR, P. J., 1974, *Ann. Phys. (N.Y.)* **88**, 26; **88**, 52.
MOHR, P. J., 1975, *Phys. Rev. Lett.* **34**, 1050.
MOHR, P. J., 1982, *Phys. Rev. A* **26**, 2338.
MOHR, P. J., 1983, Lamb shift in high-Z atoms, in *Relativistic Effects in Atoms, Molecules, and Solids*, edited by G. L. Malli (Plenum Press, New York), p. 145.
PICKUP, B. T., AND O. GOSCINSKI, 1973, *Mol. Phys.* **26**, 1013.
PIEPER, W., AND W. GREINER, 1969, *Z. Phys.* **218**, 327.
ROOTHAAN, C. C. J., 1951, *Rev. Mod. Phys.* **23**, 69.
ROSÉN, A., AND I. LINDGREN, 1968, *Phys. Rev.* **176**, 114.
SAPIRSTEIN, J., 1981, *Phys. Rev. Lett.* **47**, 1723.
SCOFIELD, J. H., 1973, Lawrence Livermore Laboratory Report No. UCRL-51326.
SLATER, J. C., 1951, *Phys. Rev.* **81**, 385.
SMITH, F. C., AND W. R. JOHNSON, 1967, *Phys. Rev.* **160**, 136.
SOFF, G., P. SCHLÜTER, B. MÜLLER, AND W. GREINER, 1982, *Phys. Rev. Lett.* **48**, 1465.
SUCHER, J., 1980, *Phys. Rev. A* **22**, 348.
SUCHER, J., 1983, Foundations of the relativistic theory of many-electron systems, in *Relativistic Effects in Atoms, Molecules, and Solids*, edited by G. L. Malli (Plenum Press, New York), p. 1.
SWIRLES, B., 1935, *Proc. Roy. Soc. (London)* **152**, 625.
THOMAS, T. D., AND R. W. SHAW, JR., 1974, *J. Electron Spectrosc.* **5**, 1081.
UEHLING, E. A., 1935, *Phys. Rev.* **48**, 55.
WICHMANN, E. H., AND N. M. KROLL, 1956, *Phys. Rev.* **101**, 843.
WILLIAMS, A. O., JR., 1940, *Phys. Rev.* **58**, 723.

CHAPTER 2

Relativistic Calculation of Atomic Transition Probabilities

Mau Hsiung Chen

1. INTRODUCTION

Most atomic inner-shell transitions are characterized by high transition energies and involve tightly bound electrons. The effects of relativity on these energetic transitions are, in general, more important than those due to electron–electron Coulomb correlation. Hence, independent-particle models are used in most of the atomic inner-shell transition calculations. In this chapter, we will review the theoretical calculations of atomic inner-shell transition probabilities from independent-particle models and explore some factors affecting the inner-shell transition rates, with special emphasis on the effects of relativity.

The effects of relativity on atomic structure have been thoroughly studied (Desclaux, 1973; Desclaux and Kim, 1975; Rose *et al.*, 1978). The ratios of expectation values of r from Dirac–Fock and Hartree–Fock calculations for several orbitals (Desclaux, 1973) are compared in Fig. 1. The main feature of the structure effects due to relativity is that the inner electrons, being dominated by the direct relativistic effect, are drawn closer to the nucleus as a result of the relativistic electron-mass increase. The relativistic contraction of the inner shells produces a considerable indirect relativistic effect on the outer electrons. In the valence shells, the relativistic modification depends on the angular momentum characterizing the orbital, because of the competition between two opposite effects, namely, direct and indirect relativistic effects. The relativistic orbital is normally more tightly bound than the corresponding nonrelativistic one for s and $p_{1/2}$ electrons, while for $p_{3/2}$ electrons the effect of relativity is quite small. The d or f elec-

Mau Hsiung Chen • Lawrence Livermore Laboratory, Livermore, California 94550.

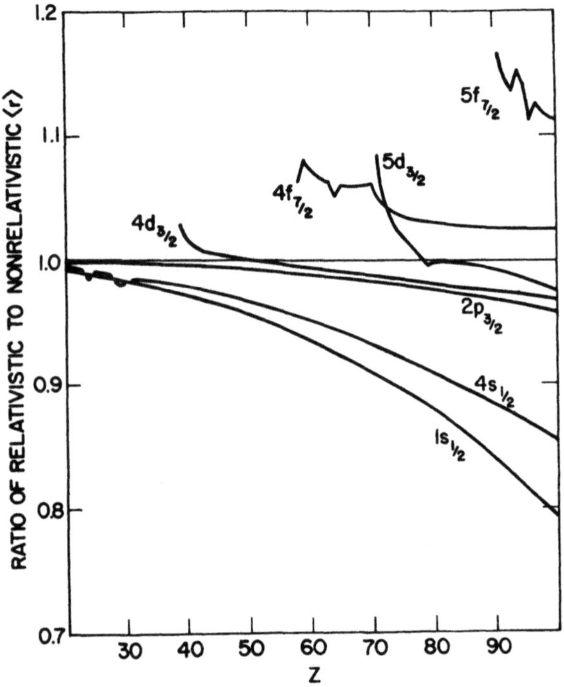

FIGURE 1. Ratios of expectation values of r from a relativistic Hartree–Fock calculation to results from a Hartree–Fock calculation, as functions of atomic number Z. After Desclaux (1973).

trons move out farther from the nucleus than in the nonrelativistic approximation at the beginning of the d or f series, due to the increased screening caused by the inner orbital contraction (i.e., indirect effect). As Z increases, the direct effect becomes larger, which partially cancels the indirect relativistic effect (Fig. 1). These structure changes will in turn affect the transition rates.

The effects of relativity on atomic transition probabilities can arise from several different factors: (1) changes in energies, (ii) relativistic orbital effects caused by the inclusion of the mass–velocity correction, the Darwin term, and spin–orbit interaction in the Dirac Hamiltonian and the indirect relativistic effect; (iii) relativistic aspects of the pertinent operators, viz., the magnetic interaction and retardation correction in the two-electron operator, due to the electron–photon coupling. The net effect depends on the relative strengths and phases of these factors.

The effects of relativity on the transition rates of atoms with a single inner-shell vacancy can be seen more vividly in heavy elements. However, the effects of relativity can sometimes constitute a dominant mechanism in the decay of few-electron ions in the low-Z region, especially for high-spin metastable states.

We will discuss the cases of atoms with a few inner-shell holes and of few-electron ions. Through extensive comparison between relativistic and nonrelativistic Hartree–Fock-type calculations, we will establish a few guidelines as to whether and when full relativistic calculations are required and under what circumstances a nonrelativistic calculation including certain relativistic corrections would be appropriate. From the comparisons between the theoretical results from the independent-particle model and experiment, we will try to determine the reliability of the independent-particle model in inner-shell transition-rate calculations.

2. THE RELATIVISTIC THEORY OF MANY-ELECTRON ATOMS

2.1. *Relativistic Hamiltonian from Quantum Electrodynamics*

All the relativistic self-consistent-field calculations (Grant, 1970; Mann and Waber, 1973; Lindgren and Rosén, 1974; Desclaux, 1973) of atomic structure performed to date are based on a Hamiltonian of the form

$$H = H_{DC} + \sum_{j>i=1}^{N} B(i,j), \tag{1}$$

where

$$H_{DC} = \sum_{i=1}^{N} h_D(i) + \sum_{i<j} \frac{1}{r_{ij}} \tag{2}$$

and $h_D(i)$ is the Dirac single-particle Hamiltonian,

$$h_D(i) = c\boldsymbol{\alpha}_i \cdot \mathbf{p}_i + \beta_i c^2 + V_n(r_i). \tag{3}$$

Here, $\boldsymbol{\alpha}$ and β are Dirac matrices and $V_n(r)$ is the interaction of the electron with the nucleus. The finite nuclear size effect may also be included. Atomic units are used hereafter unless stated otherwise.

$B(i,j)$ in Eq. (1) is the $\sim(\alpha Z)^2$ correction to the Coulomb interaction; in the most advanced calculations (Mann and Johnson, 1971) it is taken as

$$B(i,j) = -\frac{\boldsymbol{\alpha}_i \cdot \boldsymbol{\alpha}_j}{r_{ij}} + (\boldsymbol{\alpha}_i \cdot \boldsymbol{\nabla}_i)(\boldsymbol{\alpha}_j \cdot \boldsymbol{\nabla}_j) \frac{\cos(\omega r_{ij}) - 1}{\omega^2 r_{ij}}, \tag{4}$$

where ω is the wave number of the exchanged virtual photon. In the low-energy limit $\omega \to 0$, Eq. (4) reduces to the Breit interaction (Breit, 1932):

$$B(i, j) = -\frac{1}{2r_{ij}}\left[\boldsymbol{\alpha}_i \cdot \boldsymbol{\alpha}_j + \frac{(\boldsymbol{\alpha}_i \cdot \mathbf{r}_{ij})(\boldsymbol{\alpha}_j \cdot \mathbf{r}_{ij})}{r_{ij}^2}\right]. \tag{5}$$

To make a Hartree–Fock reduction of the Dirac–Coulomb Hamiltonian H_{DC} [Eq. (2)], fully antisymmetrized wave functions are constructed from j–j coupled products of Dirac central-field spinors (Grant, 1970).

$$\psi_{n\kappa m} = \frac{1}{r}\begin{pmatrix} G_{n\kappa}(r)\,\Omega_{\kappa m}(\theta, \phi) \\ iF_{n\kappa}(r)\,\Omega_{-\kappa m}(\theta, \phi)\end{pmatrix}. \tag{6}$$

Here, $G_{n\kappa}$ and $F_{n\kappa}$ are the large and small components of the radial wave functions, respectively, and n is the principal quantum number. The relativistic quantum number is $\kappa = l$ if $j = l - 1/2$ and $\kappa = -(l+1)$ if $j = l + 1/2$. The wave function $\Omega_{\kappa m}$ is the j–j coupled wave function of orbital angular momentum and spin.

The Dirac–Fock equations are obtained by making the total energy stationary with respect to the variations of the radial wave functions. The self-consistent field method is then employed to solve the Dirac–Fock equations. The numerical approach with the implicit constraint of using only positive-energy Hartree–Fock orbitals can avoid the variational collapse frequently suffered by the finite-basis-set technique (Schwarz *et. al.*, 1982; Mark *et al.*, 1982). The Breit interaction and other radiative corrections such as self-energy and vacuum polarization are then included by perturbation theory.

The binding-energy calculations from the Dirac–Fock model described above give results which show good agreement with experimental values. However, Brown and Ravenhall (1951) pointed out that Eq. (2) for a two-electron system yields no normalizable bound-state solutions. The Dirac–Fock approach starting out from an incorrect Hamiltonian [Eq. (2)] still gives good results. This requires some explanation. The difficulty of the Dirac–Coulomb Hamiltonian [Eq. (2)] can be removed by using quantum electrodynamics (Brown *et al.*, 1951; Sucher, 1980; Mittleman, 1981). The modified Hamiltonian can be written as

$$H_+ = \sum_{i=1}^{N} h_D(i) + \sum_{j>i=1}^{N} \Lambda_+(i)\,\Lambda_+(j)\,V(i, j)\,\Lambda_+(i)\,\Lambda_+(j). \tag{7}$$

Here, the projection operator Λ_+ projects onto the positive-energy states of h_D. The modified Hamiltonian H_+ now has stable bound states. The difference between Eqs. (7) and (2) is the presence of the positive-energy state projection operator in Eq. (7) which prevents mixing from the negative-energy continuum into the bound-state solutions.

The unrestricted single-configuration Dirac–Fock equation corresponding to Eq. (7) has been obtained by Sucher (1980),

$$h_D(1)\,\psi_i(1) + \Lambda_+(1)\,Y_i(1)\,\Lambda_+^{(1)}\psi_i(1) - \sum_{j\neq i} \Lambda_+(1)\,X_{ji}(1)\,\Lambda_+(1)\,\psi_j(1) = \sum_j \lambda_{ij}\psi_j(1), \tag{8}$$

where

$$X_{ij}(1) = \langle \psi_i(2)|\, V(1,2)\,|\psi_j(2)\rangle, \tag{9}$$

and

$$Y_i(1) = \sum_{j\neq i} X_{jj}(1). \tag{10}$$

Here, the external-field positive-energy Dirac orbitals ψ_i satisfy

$$\langle \psi_i \mid \psi_j \rangle = \delta_{ij} \tag{11}$$

and

$$\Lambda_+(1)\,\psi_i(1) = \psi_i(1). \tag{12}$$

Equation (8) differs from the Dirac–Fock equation corresponding to the Dirac–Coulomb Hamiltonian [Eq. (2)] only by the presence of the positive-energy projection operator surrounding the potential-energy terms arising from the electron–electron interaction.

2.2. *Interpretation of the Dirac–Fock Approach*

We now turn to the problem of interpretation of numerical Dirac–Fock calculations based on the Dirac–Coulomb Hamiltonian [Eq. (2)]. The numerical Dirac–Fock solutions are obtained with the implicit constraint that only positive-energy single-particle states will be retained. This numerical procedure is equivalent to the Hartree–Fock treatment of the Hamiltonian [Eq. (7)] which contains the positive-energy projection operator explicitly. Therefore, the Dirac–Fock wave function based on the Hamiltonian of Eq. (2) can be viewed as equivalent to the Hartree–Fock solutions based on the more correct Hamiltonian of Eq. (7) (Mittleman, 1981). In light of this interpretation, the results from Dirac–Fock calculations based on Eq. (2) can be used with confidence. However, the

Dirac–Fock approach with only the implicit appearance of Λ_+ should not be used in second-order calculations, since negative-energy states can then appear as intermediate states which can give incorrect results (Bethe and Salpeter, 1959). For a systematic improvement over the relativistic Hartree–Fock approximation, one should use the Hamiltonian explicitly containing the positive-energy projection operator derived from quantum electrodynamics (Sucher, 1980).

3. RELATIVISTIC TRANSITION ENERGIES

Knowledge of accurate transition energies is very important in the analysis of X-ray and Auger spectra. Extensive tabulations of experimental binding energies derived from X-ray (Bearden and Burr, 1967), photoelectron (Siegbahn *et al.*, 1967, 1969), and optical data (Moore, 1949) became available more than a decade ago. Historically, Auger transition energies were estimated by semiempirical (Asaad and Burhop, 1958) and empirical procedures (Bergström and Hill, 1954). With the advance of computers, nonempirical procedures have been developed with varied degrees of sophistication. For inner-shell transitions, the most useful nonempirical approximate method is the Hartree–Fock method (e.g., Hartree, 1927; Froese-Fischer, 1977; Grant, 1970). The early semiempirical and empirical methods and the basic formulation of the Hartree–Fock method for transition-energy calculations have been reviewed by Larkins (1975) and Beck and Nicolaides (1978). In this chapter, we concentrate on the recent development of accurate relativistic calculations of transition energies including relaxation and quantum electrodynamic corrections. We will give a critical appraisal of the results and discuss the needs for further improvement.

3.1. *Atomic Binding Energies and X-ray Transition Energies*

The electron binding energy is a very useful atomic property. A systematic comparison between theory and precision measurements of atomic binding energies can provide sensitive tests of quantum electrodynamic theory and of calculations of electron–electron Coulomb correlation. The knowledge of binding energies can be used to identify species by observing their characteristic X rays. The differences in binding energies among atoms residing in different chemical environments or between gas and solid phases can provide a useful probe to study chemical and solid-state effects. Here we will be concerned mainly with binding energies of free atoms.

Accurate binding-energy calculations should include the effects of relativity (one-electron and Breit interactions), relaxation, quantum electrodynamics, finite nuclear size, and correlation. Their relative importance is illustrated in Fig. 2 for K and L_1 shells. The one-electron relativistic effect [E(REL)] in Fig. 2 is found by taking the difference between Hartree–Fock and Dirac–Fock eigenvalues (Froese-Fischer, 1977; Desclaux, 1973). We will discuss below the methods to obtain the other corrections. At present, there exists no unified theory capable of treating these effects simultaneously. Thus, they are usually calculated additively.

We define the theoretical binding energy as

$$E_B = E_{\text{ion}}(\mathcal{N}-1) - E_G(\mathcal{N}). \tag{13}$$

Here, $E_{\text{ion}}(\mathcal{N}-1)$ and $E_G(\mathcal{N})$ are the total energies of the ion with $(\mathcal{N}-1)$ electrons and the ground state with $\mathcal{N}$ electrons, respectively. The experimental binding energy from photoelectron spectroscopy can be con-

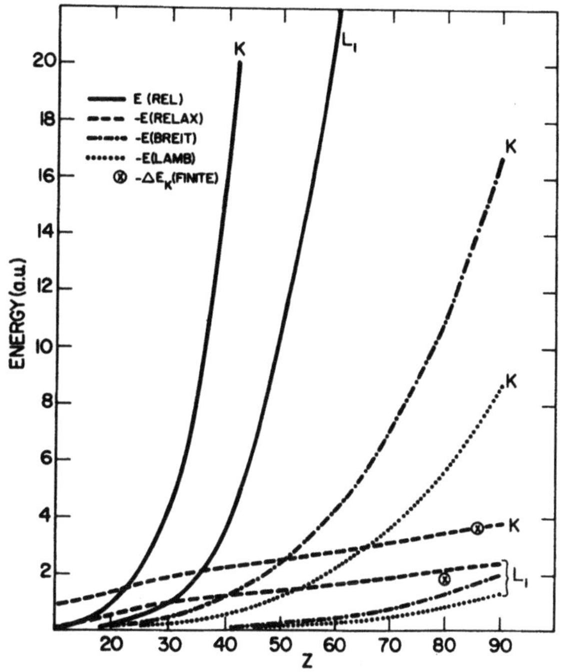

FIGURE 2. Contributions of the one-electron relativistic effect [E(REL)] (Desclaux, 1973; Froese-Fischer, 1977), relaxation [E(RELAX)] (Desclaux, 1973; Huang *et al.*, 1976), Breit interaction [E(BREIT)], Lamb shift [E(LAMB)] (Huang *et al.*, 1976), and the finite-nuclear-size effect [ΔE(Finite)] (Grant, 1970) to atomic electron binding energies, as functions of atomic number Z.

sidered to be equivalent to the binding energy obtained from Eq. (13) if the photon energy is far above ionization threshold.

Since the effect of relativity on inner-shell binding energies becomes important for $Z > 10$ and depends very strongly on atomic number Z [$\sim (\alpha Z)^4$] (see Fig. 2), it is advantageous to start from a relativistic model which includes the relativistic effect *ab initio*. In terms of a relativistic model, the binding energy can be decomposed as

$$E_B = \varepsilon(\mathrm{DF}) + E(\mathrm{RELAX}) + E(\mathrm{MAGR}) + E(\mathrm{LAMB}) + E(\mathrm{CORR}). \quad (14)$$

Here, ε(DF) is the eigenvalue from a Dirac–Fock calculation and is equal to the result from Eq. (13) in the frozen-orbital approximation (Koopmans' theorem). E(RELAX) is the relativistic relaxation energy. The third term, E(MAGR), is the contribution from magnetic and retardation corrections which becomes the Breit interaction in the long-wavelength approximation. The next term, E(LAMB), is the radiative correction including self-energy and vacuum polarization, and the last term E(CORR) is the correlation contribution to the binding energy. We proceed to discuss these terms one by one.

3.1.1. The Relativistic Independent-Particle Model. The effects of relativity on inner-shell binding energies are important even for light atoms. Although for light atoms the effects of relativity can be treated by perturbation theory (Hartmann *et al.*, 1964), the most useful approach which can be applied to all atoms is the relativistic SCF method (e.g., Dirac–Fock or Dirac–Hartree–Slater). Swirles (1935, 1936) was the first to discuss the basic relativistic SCF problem. Grant (1961) reformulated the problem using powerful tensor-operator techniques (Racah, 1942, 1943). Since then relativistic SCF calculations have been routinely performed in many laboratories. Several review articles on this subject have become available (Grant, 1970; Lindgren and Rosén, 1974). Recently, complete packages of Dirac–Fock (Desclaux, 1975; Grant *et al.*, 1980; McKenzie *et al.*, 1980) and Dirac–Hartree–Slater computer codes (Huang *et al.*, 1976; Chen *et al.*, 1981e) capable of calculating total energy including finite nuclear size, Breit interaction, and QED corrections have been developed. Desclaux (1973) has carried out extensive Dirac–Fock calculations for the ground states of all atoms throughout the Periodic Table. A systematic study of atomic binding energies including relaxation effects has also been performed for $2 \leqslant Z \leqslant 106$ using Dirac–Hartree–Slater wave function (Huang *et al.*, 1976; Chen *et al.*, 1981e).

All of these existing calculations are based on the Dirac–Coulomb Hamiltonian [Eq. (2)] and follow the prescription described in Sec. 2.1.

The nuclear potential can be expressed as a point potential (Desclaux *et al.*, 1971). However, for medium-heavy and heavy elements, an extended nucleus has to be used for accurate calculations. The nucleus may be taken as a homogeneously charged sphere (Hofstadter and Collard, 1967) with a radius $R_n = 1.20 \times 10^{-13} A^{1/3}$ cm $= 2.2677 \times 10^{-5} A^{1/3}$ atomic units, where A is the mass number. The nuclear potential experienced by the electron then is

$$V_n(r) = \begin{cases} -\dfrac{Z}{R_n}\left[\dfrac{3}{2} - \dfrac{1}{2}\left(\dfrac{r}{R_n}\right)^2\right], & r < R_n \\ -\dfrac{Z}{r}, & r \geqslant R_n. \end{cases} \tag{15}$$

The nuclear charge can also be taken as the two-parameter Fermi distribution,

$$\rho(r) = \frac{\rho_0}{1 + e^{(r-c)a}}. \tag{16}$$

In accordance with high-energy electron scattering results (Collard *et al.*, 1967), the half-density radius c and parameter a related to skin thickness can be chosen as

$$c = (2.2291 \times 10^{-5} A^{1/3} - 0.90676 \times 10^{-5}) \text{ a.u.}$$

and

$$a = 1.039 \times 10^{-5} \text{ a.u.}$$

The effect of finite nuclear size is only important for inner shells of heavy elements. It reduces the K-shell binding energy with a point nucleus by ~ 50 eV for $_{80}$Hg (Grant, 1970). The increase in binding energy with a Fermi distribution [Eq. (16)] over that with a uniform charge distribution [Eq. (15)] is only significant for the K shells of very heavy atoms (~ 10 eV at $Z = 100$) (Chen *et al.*, 1981e).

Relativistic self-consistent-field calculations of binding energies are usually implemented within the restricted Hartree–Fock (Desclaux, 1973; Grant, 1970) or Hartree–Slater methods. The latter is based on average energy and also uses a local exchange approximation (Lindgren and Rosén, 1974; Liberman *et al.*, 1965, 1971). It has been pointed out by Rosén and Lindgren (1968) that the exchange energy calculated by the Hartree–Slater method introduces nonnegligible errors in the total energy. To overcome this deficiency, a first-order correction to the local approximation can be made by computing the expectation value of the total Hamiltonian using zeroth-order Hartree–Slater eigenfunctions. The total energies from the corrected relativistic Hartree–Slater method agree with those from the corresponding Dirac–Fock model to better than ~1 eV (Huang *et al.*, 1976; Chen *et al.*, 1981e).

For atoms, the effect of relaxation can easily be taken into account by performing separate SCF calculations for ground and hole states. The binding energy is then obtained by taking the total energy difference (ΔSCF method). This zero-order DF or DHS energy is equivalent to the sum of the first two terms in Eq. (14). For extended systems such as molecules and solids, relaxation effects are rather difficult to handle in this way. Methods based on applying the relaxation correction to the ground-state orbital eigenvalue have to be used (Beck and Nicolaides, 1978).

Although *j–j* coupling is the natural coupling scheme for relativistic calculations, LS coupling is more suitable for most atoms with open outer shells, because the electrostatic interaction dominates here over the spin–orbit interaction. To avoid spurious contributions to the binding energies in ΔSCF single *j–j* configuration calculations, one can use the relativistic LS average scheme to treat the open outer shells (Lindgren and Rosén, 1974; Desclaux, 1973)

3.1.2. Magnetic and Retardation Contribution $E(MAGR)$. The exact relativistic two-electron interaction is not known, but the effective two-electron interaction can be obtained from quantum electrodynamics. The effective two-electron potential from QED is gauge dependent. The dynamic corrections to the electrostatic Coulomb interaction due to the exchange of a single transverse photon can be included through the use of the transverse operator [Eq. (4)] or Breit operator [Eq. (5)] from the Coulomb gauge.

In the Lorentz gauge, the transverse operator can be expressed as (Mann and Johnson, 1971)

$$B(i, j) = -\frac{1}{r_{ij}} [\boldsymbol{\alpha}_i \cdot \boldsymbol{\alpha}_j \cos \omega r_{ij} + (1 - \cos \omega r_{ij})], \tag{17}$$

where ω is the wave number of the exchanged virtual photon. The matrix elements of the transverse operators from the Coulomb gauge [Eq. (4)] and Lorentz gauge [Eq. (17)] are identical whenever the unperturbed electron orbitals satisfy the Dirac equations in a local potential (Mann and Johnson, 1971). They are not necessarily identical when evaluated using Dirac–Fock electron orbitals.

Recent work of Sucher (1980) and Mittleman (1981) has indicated that one could, in principle, include the Breit interaction at the SCF level. In most recent calculations the Breit energies, however, are obtained by computing the expectation values of the transverse or Breit operator using SCF wave functions from the Hamiltonian H_{DC} [Eq. (2)] (Desclaux, 1973; Mann and Johnson, 1971; Huang *et al.*, 1976; Chen *et al.*, 1981e).

The reduction of binding energies of core electrons due to the inclusion of the Breit interaction [Eq. (5)] in the ΔSCF calculations is quite significant ($\sim$300 eV for the K shell of $Z=80$) (Fig. 2). The differences between transverse energies [Eq. (4)] and Breit energies [Eq. (5)] are not negligible for K- and L-shell binding energies. The results from ΔSCF calculations according to the definition of binding energy in Eq. (13) using Dirac–Hartree–Slater wave functions are shown in Fig. 3 (Huang *et al.*, 1976; Chen *et al.*, 1981e). The errors in the binding energies due to the long-wavelength approximation are only $\sim$0.3 and $\sim$0.7 eV for M_2 and M_3 levels, respectively, at $Z=86$ (Table I). Hence, for K and L shells of heavy elements, the transverse operator should be used.

3.1.3. Radiative Corrections E(LAMB). The radiative corrections (Lamb shift) of energy levels were discovered by Lamb and Retherford (1947), who found that the $2s_{1/2}$ level of hydrogen was 1058 MHz higher than the $2p_{1/2}$ level, in contradiction to the Dirac one-electron theory. This discrepancy was first explained by Bethe (1947).

There are two main radiative corrections: self-energy and vacuum polarization. The former arises from the electron interaction with the radiation field which leads to mean-square fluctuation in the electron

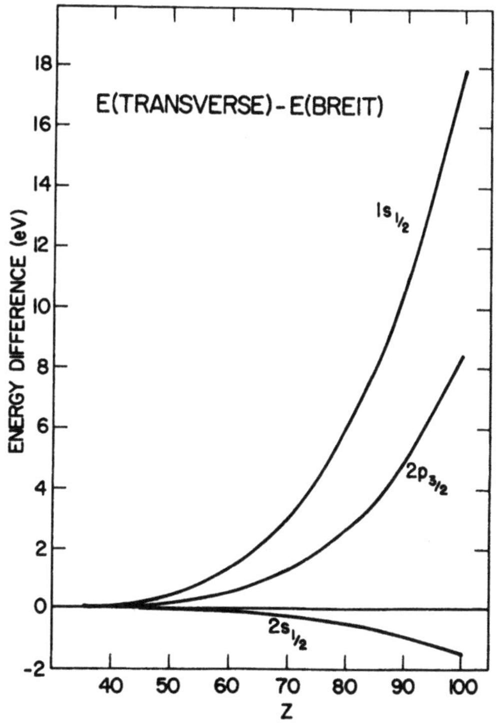

FIGURE 3. Energy differences between the magnetic and retardation corrections from the transverse operator [Eq. (4)] (Chen *et al.*, 1981e) and those from the Breit operator [Eq. (5)] (Huang *et al.*, 1976), as functions of atomic number Z.

TABLE I

Comparison of Dirac–Hartree–Slater (DHS) and Dirac–Fock (DF) calculations of atomic binding energies (in eV).

Contribution	$1s_{1/2}$		$2p_{1/2}$		$2p_{3/2}$		$3p_{1/2}$		$3p_{3/2}$	
	DHS[a]	DF[b]	DHS[a]	DF[b]	DHS[a]	DF[b]	DHS[a]	DF[b]	DHS[a]	DF[b]
					$_{80}$Hg					
Zero-order energy	−83560.04	−83560.44	−14273.13	−14273.52	−12325.97	−12326.45	−3302.43	−3302.55	−2866.53	−2866.77
Breit interaction	297.35	297.19	57.02	56.75	33.26	33.22	10.65	10.86	6.35	5.85
Vacuum polarization	−43.21	−43.13	−0.40	−0.39	0.21	0.21	−0.09	−0.10	0.06	0.06
Higher-order vacuum polarization	1.67		0.03							
Self-energy	198.67	195.47	2.38	−0.15	3.34	−1.77		−0.25		−0.42
Total	−83105.57	−83110.91	−14214.11	−14217.31	−12289.17	−12294.79	−3291.87	−3292.04	−2860.12	−2861.28
					$_{86}$Rn					
Zero-order energy	−98983.55	−98983.80	−17412.86	−17413.19	−14663.17	−14663.63	−4179.31	−4179.38	−3551.56	−3551.78
Breit interaction	380.19	379.90	75.42	75.06	42.17	42.15	14.37	14.70	8.35	7.61
Vacuum polarization	−62.81	−62.70	−0.85	−0.83	0.30	0.30	−0.21	−0.21	0.09	0.09
Higher-order vacuum polarization	2.79		0.06		−0.01		0.01			
Self-energy	262.59	259.55	4.25	0.80	4.76	−2.31		−0.18		−0.57
Total	−98400.79	−98407.05	−17333.99	−17338.17	−14615.96	−14623.50	−4165.13	−4165.07	−3543.12	−3544.66

[a] Chen *et al.* (1981e).
[b] Calculated by the author using Dirac–Fock package (Grant *et al.*, 1980; McKenzie *et al.*, 1980).
[c] Huang *et al.* (1976).

position. The jiggling, bound electron sees a smeared-out nuclear potential (Welton, 1948). To lowest order, the change in electron energy due to this effect is proportional to the square of the electron density at the nucleus. Hence, the effect is more important for inner than for outer levels, and more important for s than p states. The $s_{1/2}$ levels are thus pushed higher than $p_{1/2}$ levels in hydrogenic atoms.

The self-energy shift can be expressed as

$$\Delta E = \left(\frac{1}{\alpha\pi}\right)\frac{(Z\alpha)^4}{n^3}F(Z\alpha). \tag{18}$$

Here, Z is the atomic number, α is the fine-structure constant, and n is the principal quantum number. The function $F(Z\alpha)$ has been found to depend slightly on n for levels with the same l (Erickson, 1971).

Much work has been done on the calculation of the $1s$ and $2s$ self-energy of hydrogenic atoms (Erickson and Yennie, 1965; Erickson, 1971; Lautrup *et al.*, 1972; Brown *et al.*, 1959a, b; Mohr, 1974). Desiderio and Johnson (1971) calculated the K-shell self-energy including the effect of screening for $70 < Z < 90$ using Dirac–Slater wave functions. A systematic calculation of self-energy has been carried out by Mohr (1974, 1975, 1982) for $1s$, $2s$, $2p_{1/2}$, and $2p_{3/2}$ levels of atoms with $Z = 10N$, $1 < N < 11$, for a point Coulomb potential.

Mohr's systematic calculations have been used in recent binding-energy calculations. The effects of screening on the Coulomb self-energies were taken into account by an effective-charge approach (Chen *et al.*, 1981e). The self-energy correction to the binding energy in the Dirac–Fock program of Grant *et al.* (1980) and McKenzie *et al.* (1980) is evaluated through a ΔSCF procedure using Mohr's point Coulomb values for K and L shells and a scaling rule for higher-n levels, with an effective-charge approach to handle the screening effect. Hence, the procedure of Grant *et al.* (1980) and McKenzie *et al.* (1980) also includes a relaxation effect.

The second important radiative correction is the vacuum polarization, which arises because the effective potential seen by the electron is modified by the polarized vacuum charge distribution due to virtual electron–positron pairs. This effect is important for penetrating orbitals which can bring electrons close to the nucleus. The effect of vacuum polarization reduces the self-energy shift slightly. The leading term of the vacuum-polarization potential is known as the Uehling potential (Uehling, 1935), which gives an energy shift for a hydrogenic atom (Bjorken and Drell, 1964) of

$$\Delta E_{\mathrm{vp}} = \frac{-4(Z\alpha)^4}{15\pi n^3\alpha}. \tag{19}$$

The Uehling potential and higher-order terms have also been used in first-order perturbation theory to calculate the vacuum-polarization correction to the binding energy in the ΔSCF scheme (Huang *et al.*, 1976; Chen *et al.*, 1981e).

Radiative corrections to the binding energy grow as $\alpha(Z\alpha)^4 \ln(Z\alpha)$ (Mohr, 1974), much faster than nonrelativistic effects ($\sim Z^2$) and only $\alpha \ln(\alpha Z)$ smaller than relativistic effects (Fig. 2). They must be included in inner-shell binding-energy calculations.

3.1.4. Correlation Effects. The effects of correlation on atomic binding energies are quite important for some cases. These effects are usually estimated from nonrelativistic theory because of the lack of an efficient relativistic correlation theory. Although there exists some evidence that the relativistic effect on atomic correlation energy is quite small (Ramana *et al.*, 1982), the relativistic correlation contribution to binding energies can be quite significant ($\sim$40% for the 1*s* level of $Z=80$; 26% for the 6*s* level at $Z=79$) (Ramana *et al.*, 1982; Savin *et al.* 1983). Since the nonrelativistic correlation theories (Sinanoğlu and Brueckner, 1970) are more suitable for small atoms ($Z<18$), there is a lack of information on correlation contributions to the binding energies for medium and heavy elements. The effect of correlation on binding energies has recently been reviewed by Beck and Nicolaides (1978).

In first approximation, the correlation energies of the passive electrons from ground and hole states can be assumed to cancel each other completely. Hence, we only concern ourselves here with two important correlation contributions to the binding energy, namely, the ground-state correlation correction and the energy shift due to the Coster–Kronig or super-Coster–Kronig fluctuation of the hole states. The ground-state correlation correction arises because of the broken pairs in the hole state. This effect always increases the binding energy since all pair energies are negative and there are more pairs in the ground state than in the hole state. The ground-state correlation effect is the dominant correlation contribution to the binding energy for an outermost shell and for inner shells with orbital angular momentum $l=n-1$, where n is the principal quantum number. The virtual pair correlation energy has been found to be transferable (Öksüz and Sinanoğlu, 1969). Hence, the ground-state correlation corrections to inner-shell binding energies are roughly independent of Z. For the 1*s* level of $_{10}$Ne this correction is $\sim$1 eV, and for the 2*p* level of $_{19}$K it is $\sim$1.5 eV (Beck and Nicolaides, 1978). For outer shells (e.g., 3*d* of Zn), the correction can be as large as 3 eV (Ohno and Wendin, 1979).

Another important correlation contribution to the binding energy is caused by the dynamic relaxation processes in which the core hole fluctuates to Coster–Kronig type or super-Coster–Kronig type intermediate levels in addition to creating electron–hole pair excitations (Ohno and Wendin,

1978; 1979). These effects are important for hole states that can decay by Coster–Kronig or super-Coster–Kronig transitions (e.g., 2*s*, 3*s*, 3*p*, 4*s*, and 4*p* hole states). Some of the important cases are $[2s] \leftrightarrow [2p3d]\,nf$ for $21 < Z < 60$, $[3p] \leftrightarrow [3d^2]\,nf$ for $Z \sim 30$, $[4p] \leftrightarrow [4d^2]\,nf$ for $Z \sim 54$, and $[3s] \leftrightarrow [3p4f]\,ng$ for $Z > 70$.

When the hole state is embedded in the (super) Coster–Kronig continua, Fano's approach (Fano, 1961) of configuration interaction with the continuum can be used to calculate the energy shift, which may be positive or negative. The [2*s*] hole states for $30 < Z < 50$ have been found to shift down by ~4 eV due to the inclusion of the interaction with the $[2p3d]\,\varepsilon f$ Coster–Kronig continuum. This correction removes the discrepancy between the relativistic ΔSCF results and experiments (Chen *et al.*, 1981c).

The effect of (super) Coster–Kronig fluctuations on *M*- and *N*-shell binding energies has been the subject of a random-phase-approximation study (Wendin and Ohno, 1976; Ohno and Wendin, 1977, 1979). They found a significant reduction in binding energy for the 3*p* shell (~3 eV) of Kr and Zn and the 4*s*, 4*p* shells (~10 eV) of Xe. The effect on 3*p* shells of heavy elements has been found to be a 4.6-, 5.4-, and 6.1-eV reduction for Hg, U, and Am, respectively (Chen, 1983).

Although relativistic many-body perturbation theory capable of treating all correlation energies is not yet available due to the complexity and difficulty of dealing with the negative-energy continuum, relativistic random-phase approximation (Johnson and Lin, 1976; Huang and Johnson, 1982) and multiconfiguration Dirac–Fock approaches (Desclaux, 1975; Grant *et al.*, 1980) can be used to tackle certain relativistic correlation problems.

3.1.5. Extra-atomic Relaxation. A systematic comparison between theoretical binding energies of free atoms and experimental results is hampered by the fact that most experiments are done in the solid phase. A meaningful conclusion can be reached only if one takes account of extraatomic relaxation. The binding energy of atoms in a solid is usually smaller than that of free atoms by ~5 eV for 1*s* levels in the range $11 < Z < 14$, and by ~3–5 eV for $2p_{3/2}$ levels of atoms with $29 < Z < 35$ (Aksela *et al.*, 1982).

A simple extraatomic relaxation model for metals was proposed by Shirley (1972) and Ley *et al.* (1973). In this theory, the hole state is assumed to be screened by the conduction electrons forming a semilocalized exciton with the symmetry of the lowest unbound state in the conduction band. Since atomic wave functions, which are used in the calculations, are more compact than in a metal, this model tends to overestimate solid-state shifts. A thermochemical model based on the Born–Haber cycle and the equivalent core approximation has also been applied successfully to calculate the binding-energy shifts between free atoms and corresponding solids (Johansson and Mårtensson, 1980; Jen and Thomas, 1976; Mårtensson and Johansson, 1979). Experimental binding energies from the gas phase and standard state have been tabulated recently (Sevier, 1979; Shirley *et al.*, 1977).

Since the ΔSCF method with Dirac–Hartree–Slater wave functions is much easier to handle, both technically and economically, than the corresponding Dirac–Fock model, it is worthwhile to establish the applicability of the Dirac–Hartree–Slater model in binding-energy calculations. In Table I, we compare the binding energies of $1s$, $2p$, and $3p$ levels of Hg and Rn from the Dirac–Hartree–Slater model including the first-order correction to the local-potential approximation (Huang *et al.*, 1976; Chen *et al.*, 1981e) with results from the Dirac–Fock model, calculated by the author using the program developed by Grant and coworkers (Grant *et al.*, 1980; McKenzie *et al.*, 1980). For $1s$ and $2p$ binding energies, a Fermi distribution of the nuclear charge and the transverse operator for the magnetic interaction and retardation correction were used in both calculations. For $3p$ binding energies, the Breit operator [Eq. (5)] was used in the Dirac–Hartree–Slater calculations. For $1s$ and $2p$ shells, the energy terms calculated by both approaches, except for the self-energy, agree to better than 0.4 eV. For the $3p$ shell, the long-wavelength approximation of the Breit interaction in the DHS calculations introduces only an <0.7-eV error. The higher vacuum-polarization term included in the DHS calculations (Huang *et al.*, 1976; Chen *et al.*, 1981e) accounts for part of the discrepancy in the $1s_{1/2}$ binding energy between these two models. The significant difference in self-energy is due to the fact the relaxation effect is included in the Dirac–Fock calculations but not in the DHS calculations. The question whether the relaxation effect on the self-energy is so important can only be answered by *ab initio* calculations. From the comparison in Table I, we conclude that the corrected DHS method indeed produces reliable binding energies.

The atomic binding energies from a relativistic independent-particle model including relaxation and QED corrections (Huang *et al.*, 1976; Chen *et al.*, 1981e) can be improved by including the following corrections:

(i) Accurate screening and relaxation corrections to K- and L-shell self-energies for heavy elements.

(ii) Self-energy shifts for M_1, $M_{2,3}$, and N_1 shells of heavy elements.

(iii) Ground-state correlation corrections for all shells, especially for $3d$ levels at $Z \sim 30$, $4d$ levels at $Z \sim 46$, and $4f$ levels at $Z \sim 70$.

(iv) For M_1, $M_{2,3}$, N_1, and $N_{2,3}$ levels which are not the outermost shells, the energy shifts due to (super) Coster–Kronig fluctuations has to be included.

The atomic binding energies from the relativistic ΔSCF model including QED corrections can be used to calculate X-ray transition energies. Thus, the calculated X-ray energies will suffer similar deficiencies as those discussed for the atomic binding energies. Although the total binding energies for $1s$ and $2p$ shells of Hg and Rn from DHS and DF calculations (Table I) differ by ~4-7 eV, the $K\alpha_{1,2}$ energies from these two calculations agree within ~2 eV because of partial cancellation between the discrepancies in the $1s$ and $2p$ self-energies. The discrepancies between DHS and DF results of $K\beta_{1,3}$ energies are due to the differences in K-shell self-energies (see Table I).

For $K\alpha$ energies, the theory and experiment agree within 1–2 eV for $Z < 85$ and ~5 eV for $Z > 90$. For $K\beta_{1,3}$ energies, the discrepancy is larger ranging from 5 to 10 eV (Kessler *et al.*, 1982). The larger errors in $K\beta_{1,3}$ energies can be traced to the neglect of energy shifts due to the interaction with the (super) Coster–Kronig continua of $M_{2,3}$ levels (Chen, 1983).

Experimental methods to measure X-ray energies are reviewed by Deslattes and Kessler in Chap. 5 of this volume. A compilation of recent accurate experimental $K\alpha_{1,2}$ and $K\beta_{1,3}$ X-ray energies and comparisons with theories are also presented in their chapter.

3.2. *Auger Energies*

The energy of an Auger transition $A^+(X) \rightarrow A^{2+}(YZ) + e^-$ is given by

$$E(X - YZ) = E(A^+(X)) - E(A^{2+}(YZ)). \tag{20}$$

The earliest method to estimate Auger energies is based on a semiempirical approach. In order to use known experimental binding energies, Eq. (20) is separated into three intermediate steps (Asaad and Burhop, 1958), which yields

$$E(X - YZ) = \mathscr{E}(X) - \mathscr{E}(Y) - \mathscr{E}^*(Z, Y), \tag{21}$$

where $\mathscr{E}^*(Z, Y)$ is the binding energy of the Z subshell in the presence of a Y vacancy. The quantity $\mathscr{E}^*(Z, Y)$ is estimated by using the binding energy of the next-higher element ("$Z + 1$ rule" or one of its variations) (Burhop, 1952; Bergström and Hill, 1954; Coghlan and Clausing, 1973; Erman and Sujkowski, 1961; Albridge and Hollander, 1961; Chung and Jenkins, 1970). This empirical rule has been found to lead to significant errors in low-energy Coster–Kronig or super-Coster–Kronig transitions (McGuire, 1974; Larkins, 1974; Chen *et al.*, 1977a).

A more sophisticated semiempirical model was developed by Asaad and Burhop (1958). The energies of the final two-hole states are expressed in the intermediate-coupling scheme in terms of binding energies, appropriate Slater integrals, and spin–orbit constants. However, the effect of relaxation due to the response of the spectator orbitals to the creation of vacancies is partially ignored in the approach. In order to correct this defect, Shirley (1972, 1973) employed the Hedin–Johansson (1969) formalism to estimate the static relaxation energy. The effect of relativity on the Slater integrals is either ignored or estimated by a scaling procedure.

Recently, Larkins (1976) has carried out a more rigorous semiempirical treatment of Auger energies including relaxation and relativistic corrections. He followed a similar procedure as used by Shirley (1973). The transition energy including multiplet splitting is written as

$$E(X-YZ;{}^{2S+1}L_J)=\mathscr{E}(X)-\mathscr{E}(Y)-\mathscr{E}(Z)-\langle YZ\rangle\,{}^{2S+1}L_J+R[YZ]. \tag{22}$$

The term $\langle YZ\rangle\,{}^{2S+1}L_J$ is the interaction energy between the Y hole and the Z hole, coupled in the multiplet state ${}^{2S+1}L_J$. This term in intermediate coupling can be expressed as a combination of appropriate Slater F^k and G^k integrals and spin–orbit parameters (Asaad and Burhop, 1958). It is assumed that the static relaxation term $R[YZ]$ depends only on the principal quantum numbers of the shells but not on the orbital symmetry and multiplet structure. The relaxation energy and necessary Slater integrals in the energy expression of $\langle YZ\rangle^{2s+1}L_J$ are calculated using the Dirac–Fock model. This semiempirical theory has been used to produce a comprehensive tabulation of *K*-, *L*-, *M*-, and *N*-shell Auger energies (Larkins, 1977a). Except for the rare gases, the energies are given for the solid phase by including extraatomic relaxation energy (Larkins, 1977b, c). The accuracy of the model is estimated to be 1 to 5 eV without including the errors from experimental binding energies.

Although all of the semiempirical models rely on the availability of accurate experimental binding energies they do have some advantage. By virtue of the use of experimental binding energies, they automatically include the Lamb shift, most of the relativistic effects, and some of the correlation and solid-state corrections.

Aside from these semiempirical approaches, Auger transition energies are nowadays calculated by the ΔSCF method. The ΔSCF method can be implemented either by using nonrelativistic SCF approaches: Hartree–Fock (Froese-Fischer, 1969, 1972, 1979) and Hartree–Fock–Slater (Herman and Skillman, 1963), or by using relativistic SCF methods: Dirac–Fock (Grant, 1970; Grant *et al.*, 1980; Desclaux, 1975) and Dirac–Hartree–Slater (Huang *et al.*, 1976; Chen *et al.*, 1977a). The basic Hartree–Fock procedure applied to Auger-energy calculations has been reviewed by Larkins (1975) and Åberg and Howat (1983).

The effect of relativity on *K*-shell Auger energies is quite significant. For example, it increases the *K–LL* Auger energy by ~15 eV for ${}_{20}$Ca and by 245 eV for ${}_{40}$Zr. The effects on *L*, *M*, and *N* shells are much smaller.

However, they are critically important for Coster–Kronig and super-Coster–Kronig transitions: they provide most or all of the (super) Coster–Kronig energies. For L_1–$L_{2,3}M_{4,5}$ Coster–Kronig transitions, the relativistic effect shifts the cutoff point from $Z=40$ to $Z=50$ and increases the L_1–$L_3M_{4,5}$ energy by a factor of 2 for $30<Z<50$ (Chen and Crasemann, 1983).

The nonrelativistic ΔSCF method has been used to calculate *K–LL* energies of Ne, L_1–$L_{2,3}M$ and *L–MM* energies of Ar, and also in the study of the effects of relaxation and defect configurations on Auger transition energies (Larkins, 1975). Chen *et al.* (1977a) have calculated *j–j* configuration-averaged *L*-shell Coster–Kronig energies for $11<Z<103$ using ΔSCF techniques with Dirac–Hartree–Slater wave functions, including QED corrections.

The natural coupling schemes for the nonrelativistic Hartree–Fock and Dirac–Fock approaches are LS and *j–j* coupling, respectively. Intermediate coupling is, however, more appropriate for the final two-hole states in most cases. In the nonrelativistic SCF model, intermediate coupling can be carried out by diagonalizing the appropriate energy matrix of the Coulomb and spin–orbit interactions, obtained by using fixed LS-coupled basis states. For the relativistic SCF approach, intermediate coupling can be implemented either by diagonalizing the energy matrix of Coulomb and Breit interactions with respect to fixed *j–j* coupled basis states (Chen *et al.*, 1980b) or by using the multiconfiguration Dirac–Fock approach, in which the radial wave functions are optimized simultaneously with the diagonalization of the Hamiltonian matrix (Briançon and Desclaux, 1976). For accurate calculations of transition energies, the QED corrections discussed in Sec. 3.1 must also be included.

In Tables II and III, the theoretical *K–LL* Auger energies of Ar and U from relativistic intermediate coupling with configuration interaction (DHS-ICCI) (Chen *et al.*, 1980b) and from a multiconfiguration Dirac–Fock calculation (Briançon and Desclaux, 1976) are compared with results from a semiempirical approach (Larkins, 1977) and experimental data (Krause, 1976; Briançon, 1970). In Table IV, some selected L_3–*MM* Auger energies from semiempirical results, recalculated following Larkins' procedure (Larkins, 1977) using binding energies from ESCA measurements (Siegbahn *et al.*, 1969), are compared with results from a DHS–ICCI calculation (Chen, 1983). The effect of electron–electron Coulomb correlation on Auger energies is largely ignored in most existing calculations. As in the calculation of binding energies, complete cancellation between correlation energies of the passive electrons from initial and final states is assumed. The calculations can then be tailored to include the dominant surviving correlation effects, such as ground-state correlation, (super) Coster–Kronig fluctuations (Sec. 3.1), and final-ionic-state configuration

TABLE II

K–LL Auger energies of argon (in eV).

Transition	Semiempirical[a]	DHS–ICCI[b]	MCDF[c]	Experiment[d]
K–$L_1L_1(^1S_0)$	2510.6	2507.3	2511.6	2508.9 ± 0.6
K–$L_1L_2(^1P_1)$	2576.0	2574.4	2575.6	2575.8 ± 0.4
K–$L_1L_2(^3P_0)$	2599.3	2598.3	2599.2	
K–$L_1L_3(^3P_1)$	2600.1	2599.1	2600.0	2599.4 ± 0.4
K–$L_1L_3(^3P_2)$	2601.4	2600.6	2601.4	
K–$L_2L_2(^1S_0)$	2652.2	2651.9	2649.9	2650.6 ± 0.4
K–$L_2L_3(^1D_2)$	2660.2	2660.3	2661.8	2660.6 ± 0.2
K–$L_3L_3(^3P_0)$	2666.7	2667.4	2668.4	2666.8 ± 0.5
K–$L_3L_3(^3P_2)$	2668.6	2669.8	2670.6	2669.1 ± 0.4

[a] Larkins (1977).
[b] Chen *et al.* (1980b). These energies include the Lamb shift.
[c] Briançon and Desclaux (1976).
[d] Krause (1976).

interaction. The only correlation effect routinely included in the theoretical calculations is the final-ionic-state configuration interaction among double-hole states with the same symmetry which lie close together. For *K–LL* transitions, the configuration interaction among $L_1L_1(^1S_0)$, $L_2L_2(^1S_0)$, and $L_3L_3(^3P_0)$ states has been found to lower the K–L_1L_1 energy by 5–6 eV and increase the K–$L_2L_2(^1S_0)$ and K–$L_3L_3(^3P_0)$ energies by 3–4 eV and 1–2 eV, respectively, for medium and heavy elements and change the tran-

TABLE III

K–LL Auger energies for uranium (in eV).

Transition	Semiempirical[a]	DHS–ICCI[b]	MCDF[c]	Experiment[d]
K–$L_1L_1(^1S_0)$	71776	71739	71749	71745 ± 20
K–$L_1L_2(^1P_1)$	72706	72559	72560	72560 ± 20
K–$L_1L_2(^3P_0)$	72662	72703	72614	72620 ± 20
K–$L_1L_3(^3P_1)$	76334	76318	76332	76320 ± 20
K–$L_1L_3(^3P_2)$	76442	76434	76430	76430 ± 20
K–$L_2L_2(^1S_0)$	73373	73278	73272	73320 ± 40
K–$L_2L_3(^1D_2)$	77157	77143	77125	77130 ± 20
K–$L_3L_3(^3P_0)$	80891	80886	80889	
K–$L_3L_3(^3P_2)$	80954	80970	80955	

[a] Larkins (1977).
[b] Chen *et al.* (1980b).
[c] Briançon and Desclaux (1976).
[d] Briançon (1970).

TABLE IV
Selected L_3–MM Auger energies of krypton (in eV).

Transition	Semiempirical[a]	DHS–ICCI[b]
L_3–$M_1M_1(^1S_0)$	1062.9	1055.4
L_3–$M_1M_2(^1P_1)$	1130.1	1121.7
L_3–$M_2M_2(^1S_0)$	1205.5	1197.9
L_3–$M_2M_3(^3P_1)$	1219.4	1212.8
L_3–$M_3M_3(^3P_2)$	1226.0	1219.3
L_3–$M_2M_5(^1F_3)$	1327.8	1325.7
L_3–$M_2M_4(^1P_1)$	1328.5	1327.8
L_3–$M_3M_4(^3D_3)$	1342.5	1340.5
L_3–$M_4M_4(^1S_0)$	1353.2	1456.3
L_3–$M_4M_5(^1G_4)$	1461.6	1463.7
L_3–$M_5M_5(^3F_4)$	1468.1	1470.4

[a] Larkins (1977).
[b] Chen (1983).

sition energies by 1–2 eV for light atoms (Larkins, 1977a). The contributions from ground-state correlation to the K-shell and double L-shell hole-state binding energies are only a few electron volts. Their contributions to the transition energies are reduced even further because of the partial cancellation between the initial and final states. The effect due to the interaction with the Auger continuum on the K- and L-shell hole states, except for L_1 hole states, has been shown to be negligibly small (Chen *et al.*, 1981c). Hence, the theoretical K–LL Auger energies from relativistic independent-particle intermediate-coupling calculations show generally good agreement with experiment (Tables II and III). However, for Auger transitions involving 2*s*, 3*s*, 3*p*, 4*s*, and 4*p* subshells if they are not the outermost shells (e.g., L_3–M_1M_1 of Kr and $M_{4,5}$–$N_1N_{2,3}$ of Xe), the energy shifts due to the (super) Coster–Kronig fluctuations can be as large as 10–20 eV (see Sec. 3.1). Neglect of this effect could lead to significant errors in transition energies. This is the reason why the L_3–M_1M_1, L_3–$M_{2,3}M_{2,3}$, and L_3–$M_1M_{2,3}$ Auger energies of Kr from DHS–ICCI calculations differ from semiempirical results by ~8 eV (Table IV).

Many-body perturbation theory is very powerful in the calculation of all of the correlation energy. Kelly (1975) used this technique to calculate the K–LL Auger energy of Ne and obtained excellent agreement with experiment. However, little work has been reported on applications to larger atoms, due to the complexity and difficulty of handling negative-energy states in the relativistic formulation of many-body perturbation theory.

In order to compare the theoretical Auger energies calculated for free atoms with experimental data obtained from solids, one must consider extra-atomic relaxation. The extra-atomic relaxation due to the screening response of conduction or valence electrons to the creation of localized holes in the electron emission process always tends to lower the total energy of the system, and thus, to increase the kinetic energy of the emitted electron. The extra-atomic relaxation effect on some of the Auger energies has been estimated by using a semilocalized exciton model (Shirley, 1972; Ley *et al.*, 1973) and a thermochemical model (Mårtensson and Johansson, 1980; Aksela *et al.*, 1982). The solid-state shifts have been found to be ~14–17 eV for K–$L_{2,3}L_{2,3}$ transitions in the range $11 < Z < 15$, ~13–16 eV for L_3–$M_{4,5}M_{4,5}$ transitions for $29 < Z < 34$, and ~12–14 eV for M_5–$N_{4,5}N_{4,5}$ transitions for $46 < Z < 52$ (Aksela *et al.*, 1982).

4. RADIATIVE TRANSITIONS

4.1. *Introduction*

There have been numerous calculations of radiative transition rates. Nonrelativistic calculations have been carried out by Metchnik and Tomlin (1963), Tomlin (1964), Rubinowicz (1960), Tietz (1963), Ramberg and Richtmyer (1937), and Raffray (1966). Relativistic calculations have been performed by Massey and Burhop (1936a, b), Laskar (1955, 1958), Payne and Levinger (1956), Asaad (1959), Taylor and Payne (1960), and Bubushkin (1964, 1965, 1967). In most of the early calculations, hydrogenic wave functions were used. For $E1$ transitions, the nonrelativistic f values for the Coulomb potential are independent of atomic number.

Recently, several systematic studies with Hartree–Fock or Hartree–Slater wave functions have been performed. Manson and Kennedy (1974) calculated the X-ray emission rates for all shells using nonrelativistic Hartree–Slater wave functions. Relativistic Hartree–Slater frozen-core calculations have been carried out by Scofield (1969, 1974a), Rosner and Bhalla (1970), Bhalla (1970a, b), and Lu, Malik, and Carlson (1971). Anholt and Rasmussen (1974) used relativistic Hartree–Fock wave functions. The work of Lu, Malik, and Carlson (1971) and Anholt and Rasmussen (1974) mainly deals with X-ray emission from superheavy elements. Scofield (1974b, c) performed the only relativistic Hartree–Fock calculations in the Coulomb gauge including relaxation effects.

The subject of relativistic radiative transition rates has recently been reviewed by Garstang (1971) and Scofield (1975). In this chapter, we will focus on the question of how relativity affects the transition rates. We explore the effects of relativistic transition energies, wave functions, and retardation on transition rates, and the role played by higher multipole radiation. We also discuss relativistic effects on radiative transitions in multiply ionized atoms.

4.2. *Formulation of Relativistic Radiative Transitions*

In the theory of quantum electrodynamics, the interaction potential between the electromagnetic field and the electron–positron field depends on the choice of gauge (Akhiezer and Berestetskii, 1965). The transition matrix in turn also becomes gauge dependent. The spontaneous emission rate from time-dependent perturbation theory for a transition $i \to f$ is

$$A_{i\to f} = \frac{k}{2\pi} \sum_{\text{pol}} \int d\Omega_k \, |\langle f | \boldsymbol{\alpha} \cdot \hat{\varepsilon} e^{i\mathbf{k}\cdot\mathbf{r}} | i \rangle|^2, \tag{23}$$

where $\mathbf{k}$ is the photon momentum, of magnitude $k = (E_i - E_f)/c$. We follow the procedure of Grant (1974) in the multipole expansion and use the restricted Hartree–Fock scheme with Dirac central-field spinors given by Eq. (6). The transition rate for an atom with an inner-shell vacancy and otherwise closed-shell configuration is

$$\mathrm{A}_{i\to f} = 2k \sum_L \frac{(2j_f+1)}{(2L+1)} \begin{pmatrix} j_i & L & j_f \\ 1/2 & 0 & -1/2 \end{pmatrix}^2 \times (|\bar{M}_{fi}^{(e)}(g_L)|^2 + |\bar{M}_{fi}^{(m)}|^2), \tag{24}$$

where

$$\bar{M}_{fi}^{(e)}(g_L) = \bar{M}_{fi}^{(e)}(0) + g_L \bar{M}_{fi}^{(l)}, \tag{25}$$

$$\bar{M}_{fi}^{(e)}(0) = i^L \left\{ \left(\frac{L}{L+1}\right)^{1/2} [(\kappa_f - \kappa_i)\, I_{L+1}^+(k) + (L+1)\, I_{L+1}^-(k)] - \left(\frac{L+1}{L}\right)^{1/2} [(\kappa_f - \kappa_i)\, I_{L-1}^+(k) - L I_{L-1}^-(k)] \right\}, \tag{26}$$

$$\bar{M}_{fi}^{(l)} = i^L \{(2L+1)\, J^{(L)}(k) + (\kappa_f - \kappa_i)[I_{L+1}^+(k) + I_{L-1}^+(k)] - L I_{L-1}^-(k) + (L+1)\, I_{L+1}^-(k)\}, \tag{27}$$

and

$$\bar{M}_{fi}^{(m)} = i^{L+1} \frac{2L+1}{[L(L+1)]^{1/2}} (\kappa_f + \kappa_i)\, I_L^+(k). \tag{28}$$

The integrals $I_L^{\pm}(k)$ and $\mathcal{J}^{(L)}(k)$ are defined as follows (Rosner and Bhalla, 1970; Grant, 1974):

$$I_L^{\pm}(k)=\int_0^{\infty}(G_fF_i \pm F_fG_i)\,j_L(kr)\,dr, \tag{29}$$

$$\mathcal{J}^{(L)}(k)=\int_0^{\infty}(G_fG_i + F_fF_i)\,j_L(kr)\,dr. \tag{30}$$

Here, j_i, κ_i, and E_i are, respectively, the total angular momentum, relativistic quantum number, and energy of the initial state. The corresponding quantities for the final state are j_f, κ_f, and E_f.

The Coulomb gauge and the length gauge correspond to the choices $g_L=0$ and $g_L=[(L+1)/L]^{1/2}$, respectively (Grant, 1974). The necessary and sufficient condition for the transition matrices for all multipoles to be gauge-invariant is that the transition matrix of electric multipoles for longitudinal photons vanish identically. This condition is automatically satisfied for exact wave functions or for a single-particle atomic model, but it may not necessarily hold for a Dirac–Fock model due to the effect of the nonlocal potential (Grant, 1974). It has been shown for electric dipole transitions that the Coulomb gauge leads to the dipole velocity form in the nonrelativistic limit, while the length gauge leads to the dipole length form (Grant, 1974).

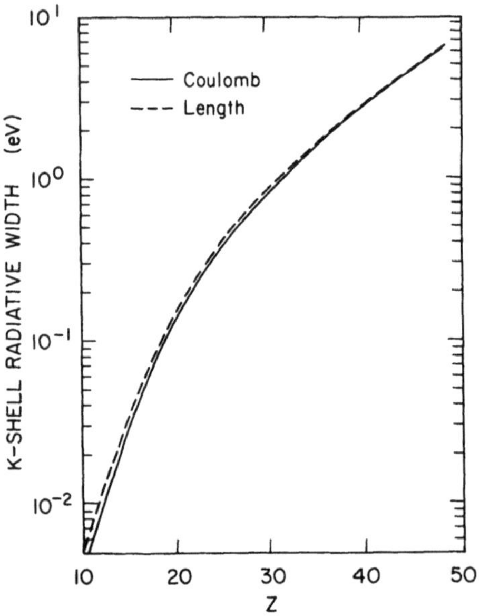

FIGURE 4. Radiative widths (in eV) of atomic [1s] hole states, as functions of atomic number Z. Dirac–Fock calculations were performed in the Coulomb gauge (solid curve) and in the length gauge (dashed curve). From Chen and Crasemann (1983).

Since the Dirac–Fock model provides the most efficient approach to date for the calculation of atomic inner-shell transition rates for medium and heavy elements, a systematic comparison between Dirac–Fock results calculated in the Coulomb gauge and length gauge is desirable. Such a comparison might shed some light on the uncertainty of independent-particle Dirac–Fock calculations. In general, the results from the length gauge are larger than those from the Coulomb gauge. For K-shell radiative widths, the difference is 20% at $Z=10$ and falls to $\sim$4% at $Z=30$ (Fig. 4). For L_2 radiative widths, the difference is approximately a factor of 2 at $Z=18$ and reduces to $\sim$7% at $Z=48$ (Fig. 5). For individual transitions with $\Delta n=0$, the discrepancy is large and persists to medium-heavy elements (e.g., 20% at $Z=48$ for $2s$–$2p$ transitions) (Chen and Crasemann, 1983). Hence, for the calculations of total transition rates of K and L shells and rates of strong individual transitions, the relativistic SCF models (e.g., DF or DHS) are quite adequate for medium and heavy elements.

4.3. *Effects of Relativity, Retardation, and Higher Multipoles*

We now proceed to discuss the difference between the usual nonrelativistic dipole calculation without retardation and the full relativistic multipole calculation with retardation. The contribution of higher multipoles to the transition rates can be estimated easily because there is no interference between the multipole matrix elements in a transition rate. For bound–bound transitions, only a few multipoles of the radiation field are allowed by the selection rules for transitions between states of definite quantum numbers n, l, and j. For K-shell transitions, only the $E1$, $E2$, $M1$, and

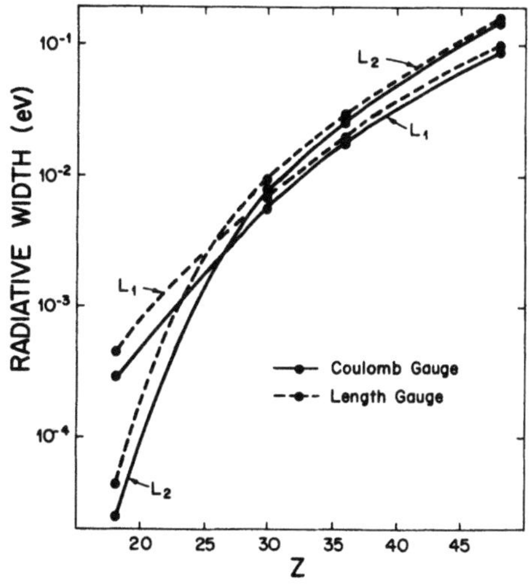

FIGURE 5. L_1 and L_2 radiative widths (in eV), as functions of atomic number Z. The solid curves represent Dirac–Fock results in the Coulomb gauge; the broken curves indicate Dirac–Fock results in the length gauge. From Chen and Crasemann (1983).

$M2$ multipoles give any significant contribution to the total decay rates. Even for uranium ($Z = 92$), the contributions from $E2$, $M1$, and $M2$ are relatively small ($\sim 0.56\%$ $E2$, $\sim 0.44\%$ $M2$, and $\sim 0.17\%$ $M1$) (Scofield, 1969). Hence, higher multipoles make only minor contributions to the radiative transitions, except in the case of forbidden transitions.

The KL_1 transition, being of the $M1$ type, is optically forbidden in the nonrelativistic limit (Breit and Teller, 1940). The KL_1 transition does not vanish in the relativistic multipole calculation, due to the effects of relativity and retardation (Scofield, 1969). The correct leading nonvanishing term in an αZ expansion can only be obtained by considering the higher-order terms in both $j_1(kr)$ and the Dirac wave functions (Gabriel *et al.*, 1981). For the point-Coulomb case, the leading nonvanishing term (Gabriel *et al.*, 1981; Johnson, 1972; Marrus and Schmieder, 1972; Prior, 1972; Kocher *et al.*, 1972) is

$$A(KL_1) = \alpha^9 Z^{10}/972. \tag{31}$$

The effects of relativity and retardation on radiative transition rates are of the same order in an αZ expansion, and are connected through gauge invariance. The first corrections due to the combined effects of relativity and retardation to the nonrelativistic decay rates are of order $(\alpha Z)^2$ (Grant, 1974; Gabriel *et al.*, 1981) and are given in the Coulomb case for $K\alpha$ transitions (Gabriel *et al.*, 1981) by

$$A(K\alpha_1) = (2^9\alpha^3 Z^4/3^8)\left[1 - (\alpha Z)^2\left(\frac{7}{24} + \ln 32 - \ln 27\right)\Big/2\right], \tag{32}$$

and

$$A(K\alpha_2) = (2^8\alpha^3 Z^4/3^8)\left[1 + (\alpha Z)^2\left(\frac{1}{64} + \ln 9 - \ln 8\right)\right]. \tag{33}$$

The coefficients are rather small, so that the nonrelativistic dipole approximation remains fairly good even for heavy atoms (within 8% for $K\alpha_1$ at $Z = 70$). The shift in total transition rates is further reduced because of the partial cancellation in the summing of various individual transition rates (Fig. 6). This same feature, that the combined effects of relativity and retardation on multipole matrix elements remain small, has also been observed in s-state photoeffect (Oh *et al.*, 1976).

To look in more detail at the effects of relativity on radiative transition rates, we explore the individual factors that affect the transition rates. The effects of relativity can arise from changes in transition energy and from the relativistic orbital effect. These factors can be studied by systematically comparing the line strengths and transition rates from nonrelativistic and

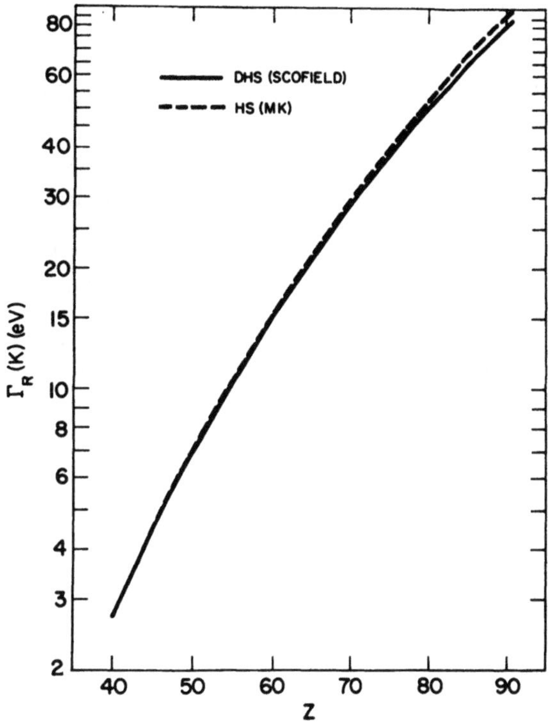

FIGURE 6. K-shell radiative widths (in eV), as functions of atomic number Z. The solid curve represents the results from Dirac–Hartree–Slater calculations (Scofield, 1969, 1974a). The dashed curve indicates results from Hartree–Slater calculations (Manson and Kennedy, 1974).

relativistic calculations. The line strength $S(l_1 j_1; l_2 j_2)$ for an $E1$ transition in the length gauge from $l_1 j_1$ to $l_2 j_2$ can be defined by means of the relation (Garstang, 1978)

$$A(j_1 \to j_2) = \frac{1}{2j_1+1} \frac{4}{3} k^3 S(l_1 j_1; l_2 j_2), \tag{34}$$

where

$$S(l_1 j_1; l_2 j_2) = (2j_1+1)(2j_2+1)(2l_1+1)(2l_2+1) \begin{pmatrix} l_2 & 1 & l_1 \\ 0 & 0 & 0 \end{pmatrix}^2 \times \begin{Bmatrix} j_1 & 1 & j_2 \\ l_2 & \frac{1}{2} & l_1 \end{Bmatrix}^2 I^2, \tag{35}$$

and

$$I = \begin{cases} \int_0^\infty R_{n_1 l_1} r R_{n_2 l_2} r^2 \, dr & \text{nonrelativistic} \quad (36a) \\ \int_0^\infty r[F_1 F_2 + G_1 G_2] \, dr & \text{relativistic without retardation} \quad (36b) \\ \frac{3}{2}\frac{1}{k}[2J^{(1)}(k) + 2I_2^-(k) + (\kappa_2 - \kappa_1) I_2^+(k)] & \text{relativistic with retardation.} \quad (36c) \end{cases}$$

For relativistic $E1$ transitions, the reduction to radial integrals without retardation [Eq. (36b)] was obtained by Babushkin (1964). A slightly different formula was given by Blin-Stoyle (1953). Without retardation, the line strength does not depend on transition energy. A comparison between relativistic and nonrelativistic line strengths without retardation will therefore show the effect produced by the change in wave functions (relativistic orbital effect). A comparison between the transition rates from relativistic calculations with retardation and nonrelativistic dipole rates without retardation will reveal the combined effects of energy, wave functions, and retardation. The ratios of line strengths and transition rates for $1s$–$2p$, $2s$–$2p$, and $2p$–$3s$ transitions between Hartree–Slater results (Manson and Kennedy, 1974) and Dirac–Hartree–Slater results with retardation (Scofield, 1969, 1974a) are shown in Figs. 7–9. The results from Dirac–Hartree–Slater calculations without retardation [Eq. (36b)] (Chen, 1983) are also shown in Figs. 7–9.

In general, relativity causes the transition energy to increase, leading to an increase in transition rates ($A \propto E^3$ for $E1$ transitions without retardation). The dipole moment is reduced by the relativistic orbital contraction; this effect leads to a decrease in transition rates (Weiss, 1976). These relativistic orbital effects are very small for light atoms and reach about 10% at $Z = 55$ for $K\alpha$ transitions (Figs. 7, 8); they are even smaller for L-shell transitions, ($\sim 5\%$ for $2p_{1/2}$–$3s$ transition at $Z = 55$, Fig. 9).

For the $K\alpha_1$ transition rates, the retardation effects are of the same order of magnitude as the relativistic effects, and the effect of the increase in $K\alpha_1$ energy is $\sim 15\%$ at $Z = 55$. The reduction of nonrelativistic $K\alpha_1$ dipole rates is only $\sim 8\%$ at $Z = 70$, because of the partial cancellation between orbital effects, retardation correction, and change in transition energy (Fig. 8). For $K\alpha_2$ transitions, the retardation effect is small ($\sim 5\%$ at $Z = 70$) and out of phase with the relativistic orbital effect; the increase in transition rates due to the energy effect outweighs the reduction due to orbital contraction, so that the relativistic transition rate shows a slight increase over the nonrelativistic rate (Fig. 7). These observations are consistent with the analytic predictions for the Coulomb case [Eqs. (32)–(33)].

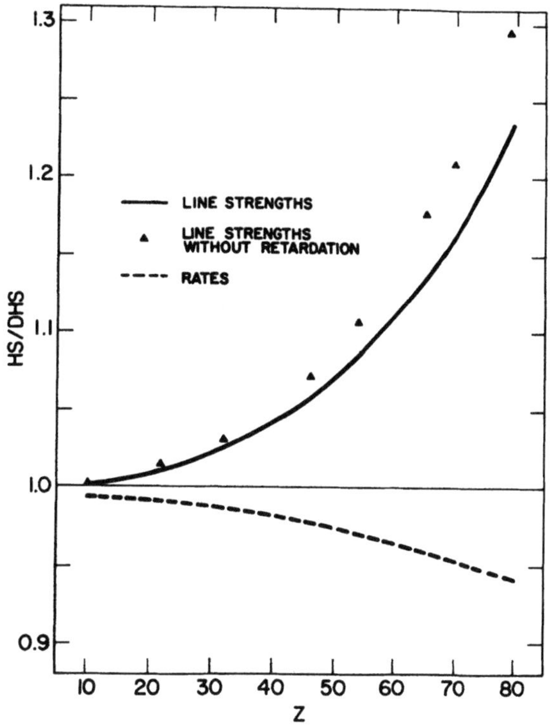

FIGURE 7. Ratios of line strengths and radiative transition rates between Hartree–Slater (HS) (Manson and Kennedy, 1974) and Dirac–Hartree–Slater (DHS) results (Scofield, 1969; 1974a; Chen, 1983), as functions of atomic number Z. The solid curve indicates the line-strength ratios that include the retardation effect in DHS calculations. The triangles represent the line-strength ratios without retardation corrections. The dashed curve represents the ratios of transition rates.

The relativistic effect on the $2s$–$2p_{3/2}$ transition rates is dominated by the change of transition energy. The effect due to shifts in wave functions is only 10% at $Z = 70$ and the retardation effect is negligible because of the small transition energy, while the rate is increased by a factor of 20 due to the increase in transition energy (Fig. 8). Hence, the nonrelativistic line strength for this transition remains fairly good even for high-Z elements.

The relativistic line strength of $2p_{3/2}$–$3s$ transitions, instead of decreasing slowly as a function of Z, increases quite rapidly (Fig. 9). The reason for this strange behavior is related to the fact that these transitions involve strong cancellations in the nonrelativistic radial matrix elements, and the differential orbital contraction tends to remove these cancellations (Weiss, 1976). The energy effects on these $2p_{1/2,3/2}$–$3s$ transitions are quite small due

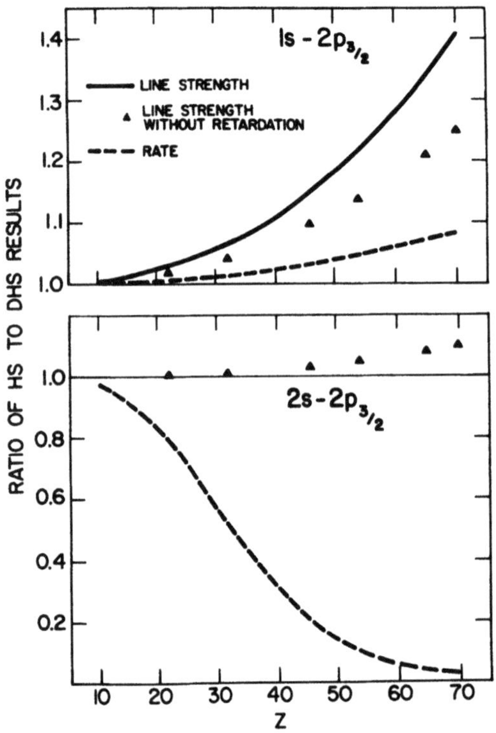

FIGURE 8. Ratios of line strengths and radiative transition rates between Hartree–Slater (Manson and Kennedy, 1974) and Dirac–Hartree–Slater results (Scofield, 1969, 1974a; Chen, 1983), as functions of atomic number Z. The notation is the same as in Fig. 7.

to the near-cancellation in the binding-energy increase. The retardation effects are also small ($<2\%$ at $Z=70$) because of the moderate transition energies (~ 8.7 keV at $Z=80$).

In Fig. 10, doublet transition-rate ratios $K\alpha_2/K\alpha_1$ and $L\beta_4/L\beta_3$ from experiment (Salem *et al.*, 1974) are compared with theoretical results from relativistic Dirac–Hartree–Slater calculations without retardation [Eq. (36b)] (Chen, 1983), and Dirac–Hartree–Slater calculations with retardation (Scofield, 1974a). The deviations from the nonrelativistic value of 0.5 are almost entirely due to retardation in the case of $K\alpha_2/K\alpha_1$ ratios, and mostly due to relativistic effects for $L\beta_4/L\beta_3$ ratios (Fig. 10). The agreement between experimental ratios and the results from Dirac–Hartree–Slater calculations with retardation is quite satisfactory.

4.4. Exchange and Overlap Corrections

In radiative decay, an outer electron filling an inner-shell vacancy causes a change in the potential. This change in potential introduces nonorthogonality between initial- and final-state wave functions. The nonorthogonality will affect the transition rates, especially the branching ratios.

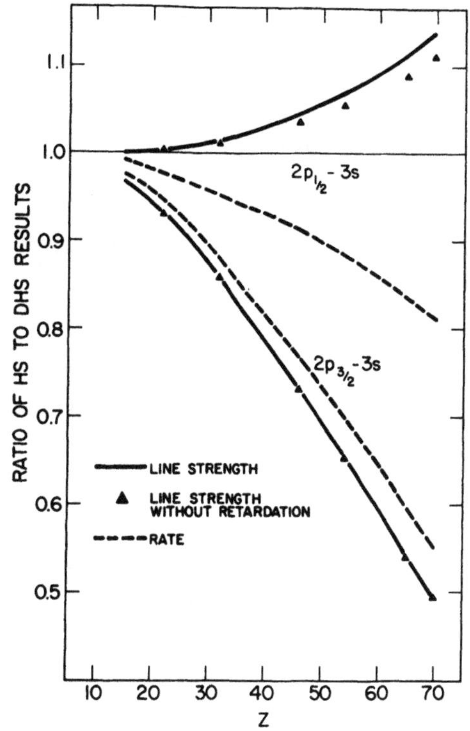

FIGURE 9. Ratios of line strengths and radiative transition rates between Hartree–Slater (Manson and Kennedy, 1974) and Dirac–Hartree–Slater results (Scofield, 1969, 1974a; Chen, 1983), as functions of atomic number Z. The notation is the same as in Fig. 7.

To go beyond the frozen-core or single-potential approximation, separate relativistic Hartree–Fock solutions need to be found for the initial and final states (Scofield, 1974b, c). The transition rates are then calculated including exchange and overlap corrections. A systematic discrepancy between theoretical and experimental intensity ratios, especially for $K\beta/K\alpha$, has been removed by inclusion of the exchange and overlap corrections (Scofield, 1974b, c).

4.5. K X-Ray Hypersatellites

Radiative decay of multiply ionized atoms produces X-ray satellites with energies slightly shifted from those of the corresponding diagram lines. The energies of L X-ray satellites that arise from electric dipole and quadrupole transitions in the presence of one spectator hole in the M or N shell have been calculated by Parente *et al.* (1981) in j–j coupling for atomic numbers $65 < Z < 95$. The computations are relativistic and include relaxation and quantum-electrodynamic corrections (Sec. 3.1).

The Breit interaction is found to substantially affect the predicted multiplet splittings of double-inner-shell-vacancy j–j configurations, such as

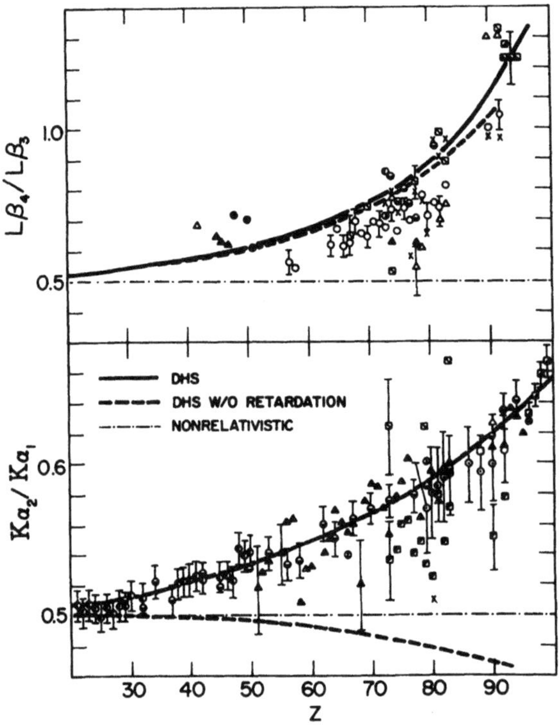

FIGURE 10. Relative X-ray intensities as functions of atomic number Z. Theoretical results from Dirac–Hartree–Slater calculations (solid curve) (Scofield, 1974a), Dirac–Hartree–Slater calculations without retardation corrections (dashed curve) (Chen, 1983), and nonrelativistic calculations (dot-dashed curve) are compared with experiment. Experimental data are taken from the review by Salem *et al.* (1974).

$[1s2p_{1/2}]$ and $[2p_{3/2}3p_{1/2}]$. Not only can the magnitudes of the splittings be greatly changed in some cases, but the level order can also be altered with respect to that predicted from the Coulomb interaction alone (Chen *et al.*, 1982a).

Recently there has been much interest in the decay of K-shell double-hole states (Briand *et al.*, 1971; Åberg and Suvanen, 1980; Chen *et al.*, 1982a). The $K\alpha$ hypersatellites $K\alpha_1^h$ and $K\alpha_2^h$ arise from the double-hole-state transitions $[1s^2] \rightarrow [1s2p_{3/2}]\ ^3P_1$ and $[1s^2] \rightarrow [1s2p_{1/2}]\ ^1P_1$, respectively. Although the contribution of the Breit interaction to the $K\alpha$ X-ray energy is only $\sim 0.4\%$ at $Z = 80$, the contribution to the energy shift of the $K\alpha$ hypersatellite is 16%. Precision measurements of the energy shift can therefore provide a sensitive test of the magnetic and retardation corrections (Desclaux *et al.*, 1974). The energy shift of the $K\alpha$ hypersatellites with respect to the diagram lines has been calculated with a relativistic ΔSCF approach in intermediate coupling including QED corrections (Desclaux *et al.*, 1974; Schreckenbach *et al.*, 1977; Åberg and Suvanen, 1980; Desclaux, 1980; Beatham *et al.*, 1980; Grant and McKenzie, 1980; Chen *et al.*, 1982a). The $K\beta^h_{1,3}$ energy shift has also been calculated (Chen *et al.*, 1982a). Treating the Breit interaction by configuration average has been found to introduce nonnegligible errors. Inclusion of the Breit interaction in the splitting calculation of $[KL]$ hole states reduces the $K\alpha_1$ hypersatellite energy by ~ 10 eV and increases the $K\alpha^h_2$ shift by ~ 16 eV at $Z=80$ (Chen *et al.*, 1982a). The major deficiency in the theory comes from the lack of screening and relaxation corrections of the self-energy and the neglect of electron–electron correlation. As of today, the agreement between experimental and theoretical energy shifts is good within the uncertainties, except for one precision measurement at $Z=80$ where the theoretical prediction exceeds the measured shift by ~ 32 eV (Chen *et al.*, 1982a; Schreckenbach *et al.*, 1977). The size of the discrepancy cannot be accounted for by the screening and relaxation effects on the self-energy and electron–electron correlation. More accurate measurements are needed at high Z in order to determine whether there is a real discrepancy.

The $K\alpha_{1,2}$ hypersatellite intensity ratio has been analyzed in terms of the intermediate-coupling scheme (Åberg *et al.*, 1976; Åberg and Suvanen, 1980; Chen *et al.*, 1982a). For low-Z elements, the $K\alpha^h_1$ X-ray transition $[1s^2] \rightarrow [1s2p]\ ^3P_1$ is dipole-forbidden in LS coupling. X-ray emission then arises from mixing between $^3P_1$ and $^1P_1$ through the spin–orbit interaction. In nonrelativistic computations, intermediate coupling is usually calculated by starting from a Hamiltonian matrix including Coulomb and spin–orbit interactions, with respect to the LS-coupled wave functions. The relativistic analog would then be to consider the interaction between various j–j-coupled states with the same total angular momentum J and the same nl configuration (Desclaux, 1972). For $J=1$ of the $[1s\,2p]$ configuration, the relativistic intermediate-coupling wave functions $\psi_1^{(JM)}$ and $\psi_2^{(JM)}$ can be written as linear superpositions of the wave functions $|1s^{-1}_{1/2}\,2p^{-1}_{1/2}\,JM\rangle$ and $|1s^{-1}_{1/2}\,2p^{-1}_{3/2}\,JM\rangle$ (Åberg and Suvanen, 1980):

$$\psi_1(JM) = -\sin\phi\,|1s^{-1}_{1/2}\,2p^{-1}_{1/2}\,JM\rangle + \cos\phi\,|1s^{-1}_{1/2}\,2p^{-1}_{3/2}\,JM\rangle$$
$$\psi_2(JM) = \cos\phi\,|1s^{-1}_{1/2}\,2p^{-1}_{1/2}\,JM\rangle + \sin\phi\,|1s^{-1}_{1/2}\,2p^{-1}_{3/2}\,JM\rangle. \tag{37}$$

In the nonrelativistic limit with vanishing spin–orbit interaction, the coefficients $\sin\phi = \sqrt{1/3}$ and $\cos\phi = \sqrt{2/3}$ relate ψ_1 to $^1P_1$ and ψ_2 to $^3P_1$, respectively. In the j–j coupling limit, we find $\sin\phi = 1$ and $\cos\phi = 0$.

The relativistic Hamiltonian submatrix for $J=1$ can be evaluated using tensor-operator techniques (Racah, 1942, 1943; Grant, 1970) for the Dirac–Coulomb Hamiltonian [Eq. (2)] (Åberg and Suvanen, 1980) or Dirac–Breit Hamiltonian [Eq. (1)] (Chen *et al.*, 1982a).

The intensity ratios $K\alpha_1^h/K\alpha_2^h$, neglecting the effects of relativity and retardation on the X-ray matrix elements, can be written (Åberg and Suvanen, 1980)

$$R=\left(\frac{\sqrt{2}\sin\phi-\cos\phi}{\sin\phi+\sqrt{2}\cos\phi}\right)^2. \tag{38}$$

The ratios R as a function of Z increase from zero to a j–j coupling limit of 2; they are mainly determined by the spin–orbit-to-exchange-interaction ratios $\xi_{2p}/G^1(1s2p)$ (Åberg *et al.*, 1976; Åberg and Suvanen, 1980).

The mixing coefficients sin ϕ and cos ϕ in Eq. (38) have been calculated through a semiempirical method (Åberg *et al.*, 1976) and a MCDF method with a Dirac–Coulomb Hamiltonian (Åberg and Suvanen, 1980). The intensity ratios R were found from Eq. (38). These intensity ratios have also been studied by Chen, Crasemann, and Mark (1982a) using a fixed j–j basis set. In the latter work, the relativistic intermediate-coupling wave functions were calculated by diagonalizing the Hamiltonian matrix which included the Coulomb and transverse Breit interactions [Eq. (1)]. The relativistic and retardation effects were also included in the calculation of X-ray matrix elements under the frozen-orbital approximation, with the potential corresponding to the initial double K-hole configuration. The K X-ray emission rates, per K hole, are larger for double-hole states than for single-hole states (by $\sim$38% at $Z=18$ and by 9% at $Z=60$) (Chen *et al.*, 1982a). Inclusion of the Breit interaction in the intermediate-coupling calculations reduces the mixing between $[KL_2]J=1$ and $[KL_3]J=1$ states, which in turn causes the $K\alpha_1^h/K\alpha_2^h$ ratios to decrease at low Z ($\sim$25% at $Z=18$) and to increase for medium-Z and heavier elements ($\sim$7% at $Z=60$) (Chen *et al.*, 1982a). The $K\alpha^h$ hypersatellite intensity ratio does not increase monotonically toward the j–j coupling limit of 2 because of the retardation effect. Instead, the $K\alpha^h$ ratio reaches a maximum at $Z\simeq 70$ and then decreases with increasing Z. Similar deviations from nonrelativistic values for the ordinary $K\alpha_1/K\alpha_2$ X-ray ratio caused by the retardation effect have also been noted (Fig. 10) (Salem *et al.*, 1974; Scofield, 1974a). The experimental results of $K\alpha^h$ ratios have been shown to follow the trend obtained by intermediate coupling (Åberg *et al.*, 1976; Åberg and Suvanen, 1980; Chen *et al.*, 1982a). The experimental results are, however, not sufficiently accurate or even lacking for heavy elements to reveal the effects of wave functions, Breit interaction, and retardation correction.

5. RADIATIONLESS TRANSITIONS

5.1. Introduction

Atomic inner-shell vacancies are filled by two major decay processes: photon emission and electron emission. The decay of inner-shell hole states by electron emission is an autoionization process known as the Auger effect. Although Auger transitions, with rare exceptions, are the dominant decay mode of inner-shell vacancies, early attention was largely given to X-ray emission. The Auger effect was predicted by Rosseland (1923) and observed by Auger (1925).

Auger transitions are usually denoted by X–YZ, where X refers to the initial vacancy and YZ represents the final two vacancies. Several specific names are used for the various radiationless processes. *Auger transitions* are those in which none of the final vacancies are in the same principal shell as the initial vacancy. In *Coster–Kronig transitions* (Coster and Kronig, 1935), one of the two final vacancies is in a different subshell of the same principal shell that contained the initial vacancy. In *super-Coster–Kronig* transitions (McGuire, 1972), both final vacancies have the same principal quantum number as the initial vacancy. The term *Auger effect* is also used to describe the entire class of radiationless transitions to inner-shell vacancies.

In 1927, Wentzel formulated the basic ansatz for the calculation of Auger rates. Since then, much theoretical work has been done on the subject. With few exceptions, early theoretical work on Auger transitions has been confined to nonrelativistic calculations. Some used screened hydrogenic wave functions (Burhop, 1935; Callan, 1961; Kostroun *et al.*, 1971) and others used more realistic wave functions (Walters and Bhalla, 1971; Chen and Crasemann, 1973; McGuire, 1969). The simple screened hydrogenic calculations can give quite good total Auger rates. However, the branching ratios from the hydrogenic model are often significantly different from those calculated by more accurate methods, such as the Hartree–Slater model. An extensive list of early Auger transition probability calculations is given by Bambynek *et al.* (1972).

The earliest attempt at a relativistic treatment is that of Massey and Burhop (1936a), who used screened relativistic hydrogenic wave functions to deduce the K–LL rates of gold. Asaad (1959) calculated relativistic K–LL rates of mercury with numerical wave functions in a Hartree potential, and Listengarten (1961, 1962) used the Thomas–Fermi–Dirac potential. Talukdar and Chattarji (1970) computed K–L_1L_1 rates, and Chattarji and Talukdar (1968) evaluated L_1 Coster–Kronig rates with screened relativistic hydrogenic wave functions. Dirac–Hartree–Slater wave functions have also been used to calculate selected K-shell transition rates and the first six of the L_1–MM transition rates (Bhalla, 1970a-c; Bhalla and Ramsdale, 1970a, b;

Bhalla *et al.*, 1970a, b). Recently, rather extensive, systematic Dirac–Hartree–Slater calculations of Auger rates have been carried out for the *K*-shell (Chen *et al.*, 1980a, b; 1983a), *L*-shell (Chen *et al.*, 1979a-c, 1981a), and *M*-shell (Chen *et al.*, 1980c, 1983b) transitions to study the effects of relativity on Auger rates, level widths, and fluorescence yields. The same authors have also calculated relativistic Auger and X-ray emission rates for highly ionized atoms (Chen *et al.*, 1981b, d, 1982b, 1983c) for the purpose of studying relativistic effects on the decay of high-spin states.

In the last decade, several reviews have been written on the subject of the Auger effect (Bambynek *et al.*, 1972; Burhop and Asaad, 1972; McGuire, 1975; Åberg and Howat, 1983; Chattarji, 1976). Bambynek *et al.* (1972) reviewed the subject of X-ray fluorescence yields and basic aspects of Auger and Coster–Kronig transitions, including theory and experiment. McGuire (1975) emphasized the formulation of Auger transitions from multiply ionized atoms. Åberg and Howat (1983) tried to go beyond the present perturbation-theory treatment of the Auger effect, based on Wentzel's ansatz (1927). These authors have treated the decay of autoionizing states and inner-shell hole states by electron and photon emission in the context of the theory of resonance scattering from a unified point of view (Åberg, 1980). In this formalism, the effect of relaxation, final-state Auger channel interaction (Howat *et al.*, 1978), and interference between photon and electron emission (Armstrong *et al.*, 1978) can be taken into account.

Although the effect of relativity on Auger transitions has been mentioned in some of the reviews, there is a need for an in-depth discussion. In this chapter, we will concentrate on the relativistic effect on Auger transitions in atoms with an inner-shell vacancy and in few-electron ions.

5.2. Relativistic Theory of Auger Transitions

The production and deexcitation of inner-shell hole states has traditionally been treated as a two-step process: the decay is separated from the excitation. Breakdown of the two-step model can, however, occur in threshold excitation. For threshold phenomena such as resonant Raman scattering (Sparks, 1974), resonant Raman Auger emission (Brown *et al.*, 1980), and postcollision interaction (Schmidt *et al.*, 1977), a one-step model that unifies atomic excitation and deexcitation is needed (Åberg, 1980; Ohno and Wendin, 1979; Gunnarsson and Schönhammer, 1982; Eisenberger *et al.*, 1976).

In this work, the inner-shell hole state is assumed to be created by an ionization process with energy well above the ionization threshold. The two-step model then becomes a very good approximation to describe the inner-shell decay processes. The Auger transition probabilities can be calculated from perturbation theory. The decay rate is

$$T = \frac{2\pi}{\hbar} \mid \langle \psi_f \mid H - E \mid \psi_i \rangle \mid^2 \rho(\varepsilon). \tag{39}$$

Here, ψ_i and ψ_f are the antiymmetrized many-electron wave functions of the initial and final states of the ion, respectively, and $\rho(\varepsilon)$ is the energy density of final states.

In the frozen-core approximation, only the two-electron operator in Eq. (39) contributes to the transition rate, and the many-electron wave functions ψ_f, ψ_i can be replaced by the two-hole coupled wave functions with the appropriate phase factor for an atom with an initial inner-shell vacancy. The total Auger transition probability for a transition $n_1'\kappa_1' \rightarrow n_1\kappa_1, n_2\kappa_2$, in *j–j* coupling, is (Chen *et al.*, 1979b)

$$T = \tau \frac{1}{2j_1' + 1} \sum_{\substack{J,J' \\ M,M'}} \sum_{\kappa_2'} |D - E|^2, \tag{40}$$

where

$$D = \langle j_1'(1)\, j_2'(2)\, J'M' \mid V_{12} \mid j_1(1)\, j_2(2)\, JM \rangle, \tag{41}$$

$$E = \langle j_1'(1)\, j_2'(2)\, J'M' \mid V_{12} \mid j_1(2)\, j_2(1)\, JM \rangle, \tag{42}$$

and

$$\tau = \begin{cases} 1/2 & \text{if } n_1\kappa_1 = n_2\kappa_2, \\ 1 & \text{otherwise.} \end{cases} \tag{43}$$

The initial hole is characterized by the quantum numbers n_1', κ_1', j_1', the continuum electron by n_2', κ_2', j_2', and the final holes by n_1, κ_1, j_1 and n_2, κ_2, j_2. In Eqs. (41) and (42), $|j_1(1)\, j_2(2)\, JM\rangle$ denotes the (not antisymmetrized) *j–j* coupled two-hole wave function. The continuum wave function is assumed to be normalized so as to represent one electron ejected per unit time (Asaad, 1959).

In the nonrelativistic theory, the two-electron operator represents just the Coulomb repulsion between the two electrons. In the relativistic theory, the dynamic correction due to photon–electron coupling to the Coulomb interaction is included in the two-electron operator. From quantum electrodynamics, the electron–electron interaction operator is gauge dependent.

In the currently available relativistic Auger calculations, the two-electron operator is chosen to be Møller operator, which is based on the Lorentz gauge:

$$V_{12}^M = (1 - \boldsymbol{\alpha}_1 \cdot \boldsymbol{\alpha}_2)\, e^{i\omega r_{12}}/r_{12}. \tag{44}$$

In the Coulomb gauge, the two-electron operator (Huang, 1978) is

$$V_{12}^{c}=\frac{1}{r_{12}}-(\boldsymbol{\alpha}_1\cdot\boldsymbol{\alpha}_2)\frac{e^{i\omega r_{12}}}{r_{12}}+(\boldsymbol{\alpha}_1\cdot\boldsymbol{\nabla}_1)(\boldsymbol{\alpha}_2\cdot\boldsymbol{\nabla}_2)\frac{e^{i\omega r_{12}}-1}{\omega^2 r_{12}}. \tag{45}$$

Here, the $\boldsymbol{\alpha}_i$ are Dirac matrices, and ω is the wave number of the virtual photon.

The Auger matrix elements can be separated into angular and radial parts with the aid of Racah algebra (Racah, 1942, 1943). The direct matrix element of the two-electron operator in the Coulomb gauge (Huang, 1978; Chen and Crasemann, 1983) is

$$\begin{aligned}D=\sum_{\lambda=0}^{\infty}A_\lambda\Bigg\{\Bigg[&\langle W_{11}\gamma_\lambda^C W_{22}\rangle+\Bigg(\frac{\lambda}{2\lambda-1}\langle P_{11}^\lambda\gamma_{\lambda-1}P_{22}^\lambda\rangle\\&+\frac{\lambda+1}{2\lambda+3}\langle Q_{11}^\lambda\gamma_{\lambda+1}Q_{22}^\lambda\rangle\Bigg)\Bigg]\pi(l_1'\lambda l_1)\,\pi(l_2'\lambda l_2)-(1-\delta_{\lambda 0})\\&\times\frac{(\kappa_1'+\kappa_1)(\kappa_2'+\kappa_2)}{\lambda(\lambda+1)}\langle U_{11}\gamma_\lambda U_{22}\rangle\,\pi(l_1'\lambda+1l_1)\,\pi(l_2'\lambda+1l_2)\\&+\Bigg[-\frac{\lambda}{2\lambda-1}\langle V_{11}\gamma_{\lambda-1}P_{22}\rangle+\frac{\lambda+1}{2\lambda+3}\langle V_{11}\gamma_{\lambda+1}Q_{22}\rangle\\&+\lambda(\lambda+1)(\langle P_{11}S_\lambda Q_{22}\rangle+\langle Q_{11}t_\lambda P_{22}\rangle)\Bigg]\pi(l_1'\lambda l_1)\,\pi(l_2'\lambda l_2)\Bigg\},\end{aligned} \tag{46}$$

where

$$\begin{aligned}A_\lambda=(-1)^{j_2+j_2'+J}[j_1,j_1',j_2,j_2']^{1/2}&\begin{Bmatrix}j_1' & j_2' & J\\ j_2 & j_1 & \lambda\end{Bmatrix}\\ \times&\begin{pmatrix}j_1' & j_1 & \lambda\\ -1/2 & 1/2 & 0\end{pmatrix}\begin{pmatrix}j_2' & j_2 & \lambda\\ -1/2 & 1/2 & 0\end{pmatrix},\end{aligned} \tag{47}$$

$$P_{ij}^\lambda(r)=[(\kappa_j-\kappa_i')/\lambda]\,U_{ij}(r)+V_{ij}(r), \tag{48}$$

$$Q_{ij}^\lambda(r)=[(\kappa_j-\kappa_i')/(\lambda+1)]\,U_{ij}(r)-V_{ij}(r), \tag{49}$$

$$W_{ij}(r)=G_i'(r)\,G_j(r)+F_i'(r)\,F_j(r), \tag{50}$$

$$U_{ij}(r)=G_i'(r)\,F_j(r)+F_i'(r)\,G_j(r), \tag{51}$$

$$V_{ij}(r)=G_i'(r)\,F_j(r)-F_i'(r)\,G_j(r), \tag{52}$$

$$\gamma_\lambda^C(r_1,r_2)=r_<^\lambda/r_>^{\lambda+1}, \tag{53}$$

$$\gamma_\lambda(r_1,r_2)=-(2\lambda+1)\,\omega j_\lambda(\omega r_<)\,\mathscr{Y}_\lambda(\omega r_>)+i(2\lambda+1)\,\omega j_\lambda(\omega r_<)\,j_\lambda(\omega r_>), \tag{54}$$

$$S_\lambda = \begin{cases} \dfrac{1}{r_1} j_{\lambda+1}(\omega r_2)\,\mathscr{Y}_\lambda(\omega r_1) - \dfrac{i}{r_1} j_{\lambda+1}(\omega r_2)\, j_\lambda(\omega r_1), & r_1 \geqslant r_2 \\ \dfrac{r_1^{\lambda-1}}{\omega^2 r_2^{\lambda+2}} + \dfrac{1}{r_1} j_\lambda(\omega r_1)\,\mathscr{Y}_{\lambda+1}(\omega r_2) - \dfrac{i}{r_1} j_\lambda(\omega r_1)\, j_{\lambda+1}(\omega r_2), & r_1 < r_2 \end{cases} \quad (55)$$

$$t_\lambda = \begin{cases} \dfrac{r_2^{\lambda-1}}{\omega^2 r_1^{\lambda+2}} + \dfrac{1}{r_1} j_{\lambda-1}(\omega r_2)\,\mathscr{Y}_\lambda(\omega r_1) - \dfrac{i}{r_1} j_{\lambda-1}(\omega r_2)\, j_\lambda(\omega r_1), & r_1 \geqslant r_2 \\ \dfrac{1}{r_1} j_\lambda(\omega r_1)\,\mathscr{Y}_{\lambda-1}(\omega r_2) - \dfrac{i}{r_1} j_\lambda(\omega r_1)\, j_{\lambda-1}(\omega r_2), & r_1 < r_2, \end{cases} \quad (56)$$

$$\langle W_{11}\gamma_\lambda^C W_{22}\rangle = \int_0^\infty \int_0^\infty W_{11}(r_1)\,\gamma_\lambda^C(r_1, r_2)\, W_{22}(r_2)\, dr_1\, dr_2, \quad (57)$$

and

$$\pi(l'\lambda l) = \begin{cases} 1 & \text{if } l' + \lambda + l \text{ is even,} \\ 0 & \text{otherwise.} \end{cases} \quad (58)$$

The quantity $[j_1, j_2, j_1', j_2', \ldots]$ is defined by

$$[j_1, j_2, j_1', j_2', \ldots]^{1/2} = [(2j_1+1)(2j_2+1)(2j_1'+1)(2j_2'+1)\cdots]^{1/2}. \quad (59)$$

The quantity $r_<(r_>)$ is the smaller (larger) of r_1 and r_2, and j_λ and $\mathscr{Y}_\lambda$ are spherical Bessel functions of the first and second kind, respectively.

The exchange matrix element can be obtained by performing the exchange operation $n_1\kappa_1 \leftrightarrow n_2\kappa_2$ on the direct matrix elements [Eq. (46)] and multiplying by the phase factor $(-1)^{J-j_1-j_2}$.

The direct matrix elements of the Møller operator can be obtained from Eq. (46) by dropping the last bracket and changing $\langle W_{11}\gamma_J^C W_{22}\rangle$ into $\langle W_{11}\gamma_\lambda W_{22}\rangle$ (Chen *et al.*, 1979b).

The two-electron matrix elements of V_{12}^M and V_{12}^C are identical in a local-potential model (Mann and Johnson, 1971). The relativistic Auger rate is therefore gauge invariant in the Dirac–Hartree–Slater model. The matrix elements of V_{12}^M and V_{12}^C may not be identical when they are evaluated with Dirac–Fock electron orbitals. It is known, however, that the on-shell matrix elements of Eqs. (44) and (45) give the same results (Mittleman, 1980).

The Auger transition rates have been calculated both in the Lorentz and Coulomb gauges for K- and L-shell initial vacancies, with the use of Dirac–Fock wave functions (Chen and Crasemann, 1983). The two different gauges are found to lead to radiationless transition rates that agree to better than 1% in all cases. The Auger rate is hence seen to be practically gauge invariant in first-order perturbation theory (Chen, 1982; Chen and Crasemann, 1983).

5.3. *Effects of Relativity on Auger Transitions*

In order to explore the general characteristics of relativistic effects on Auger transitions, Chen and Crasemann (1983) have calculated Auger rates with Dirac–Hartree–Slater wave functions and several different two-electron operators: The Coulomb interaction

$$\frac{1}{r_{12}}, \tag{60a}$$

the Coulomb and magnetic interaction without retardation correction

$$(1-\boldsymbol{\alpha}_1\cdot\boldsymbol{\alpha}_2)\frac{1}{r_{12}}, \tag{60b}$$

and the retarded Coulomb and magnetic interaction

$$(1-\boldsymbol{\alpha}_1\cdot\boldsymbol{\alpha}_2)\frac{e^{i\omega r_{12}}}{r_{12}}. \tag{60c}$$

Results are compared with those from nonrelativistic Hartree–Slater calculations with either relativistic or nonrelativistic Auger transition energies, in order to study the relative importance of the various aspects of the effects of relativity.

Comparison of relativistic individual Auger rates with nonrelativistic results show that the relativistic effects are rather uneven (Chen and Crasemann, 1983). For example, the K–L_1L_1 rates are greatly enhanced by relativity, the K–L_3L_3 rates are somewhat reduced (Fig. 11), and K–L_2L_3 transitions are affected only slightly.

The apparent unevenness of the effects of relativity can be analyzed in terms of the different sources of relativistic effects: (i) changes in transition energy, (ii) differences between relativistic and nonrelativistic wave functions, (iii) the contribution from the magnetic interaction, and (iv) retardation of the Coulomb and magnetic interactions. The exact effect of including relativity in the calculation of Auger rates cannot be predicted in general terms. It depends on the relative strengths and phases of all these factors.

Energy effects are obviously very important for Coster–Kronig and super-Coster–Kronig transitions, which generally have low energies and matrix elements that are quite sensitive to energy, especially near threshold.

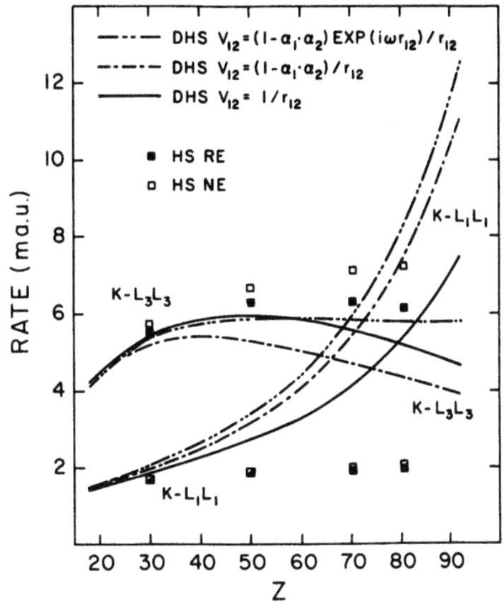

FIGURE 11. Theoretical K–L_1L_1 and K–L_3L_3 Auger transition rates (in milliatomic units) as functions of atomic number Z. Relativistic calculations with Dirac–Hartree–Slater wave functions were performed with the two-electron operators of Eq. (60c) (—-·-·—), Eq. (60b) (—·—), and Eq. (60a) (——). Also shown are nonrelativistic results computed from Hartree–Slater wave functions with relativistic transition energies (solid squares) and with nonrelativistic transition energies (open squares). From Chen and Crasemann (1983).

Without relativity, most of the super-Coster–Kronig and Coster–Kronig transitions would be energetically impossible.

In L_1–$L_{2,3}M_{4,5}$ Coster–Kronig transitions, the L_1–L_3M_5 transition energy is increased by a factor of 2 due to relativity. The large discrepancy between the results for L_1–$L_3M_{4,5}$ transitions from HS and DHS wave functions for $25 < Z < 45$ is almost entirely due to the increase in transition energy (Chen, 1982; Chen and Crasemann, 1983). The L_1–$L_2M_{4,5}$ transition energy from DHS calculations is quite close to that from an HF model because the relativistic effect on the L_1 binding energy is partially canceled by a similar effect on the L_2 binding energy. The L_1–$L_2M_{4,5}$ Coster–Kronig rates for $25 < Z < 45$ from DHS calculations therefore agree quite well with those from an HS model.

The energy effects on ordinary Auger transitions are much smaller, ranging from 2–4% at low to medium Z to 10%–15% for high Z. For transitions involving s or $p_{1/2}$ orbitals in both initial and final states (e.g., K–L_1L_1, K–L_1L_2, and L_2–M_2M_2), the energy effect is small because the relativistic energy shifts largely cancel out between initial and final states.

The relativistic orbital effects due to the inclusion of the mass-velocity term, spin–orbit interaction and the Darwin term in the Dirac Hamiltonian, are typically ~5%–10% on K–LL transitions at medium Z and 25%–30% at high Z (Chen, 1982). The contribution from the magnetic interaction and retardation correction to the Auger transition rates is very small at low Z. It can become as large as that from the relativistic orbital effects for some K- and L-shell transitions (e.g., K–L_1L_1 and L_2–M_4M_5). In general, the contribution from the magnetic interaction and retardation correction is smaller for *L*-shell than for *K*-shell transitions and much smaller for *M*-shell transitions.

For *K–LL* transitions, important effects usually arise from all factors: energy and wave-function changes, magnetic interaction, and retardation correction. Constructive interference can lead to a large increase in transition rates (e.g., K–L_1L_1, K–L_1L_2) (Figs. 11, 12). A near-cancellation can leave the transition rates with little change (e.g., K–L_2L_3).

For the *L*-shell transitions, the interplay of energy, wave-function, magnetic, and retardation effects has also been found (e.g., L_2–M_4M_5) (Chen and Crasemann, 1983). However, for most of the *L–MM* transitions, the relativistic energy changes and the orbital wave-function effect are more important than the current–current interaction term (Fig. 13). This is also true for *M*-shell transitions (Chen *et al.*, 1980c; 1983b). For M–$N_{6,7}N_{6,7}$ transitions, the direct relativistic effects pull the *M* orbital inward, while indirect relativistic effects cause the 4*f* wave function to expand (Fig. 1). Hence, these Auger rates are reduced by as much as 30% in the high-Z region due to the interplay of direct and indirect relativistic effects on wave functions (Chen *et al.*, 1980c, 1983b).

For total Auger level widths, which are sums of a large number of transition rates, the relativistic effects tend to average out to some extent, which is similar to the finding for total radiative widths. Nevertheless, for *K* shells,

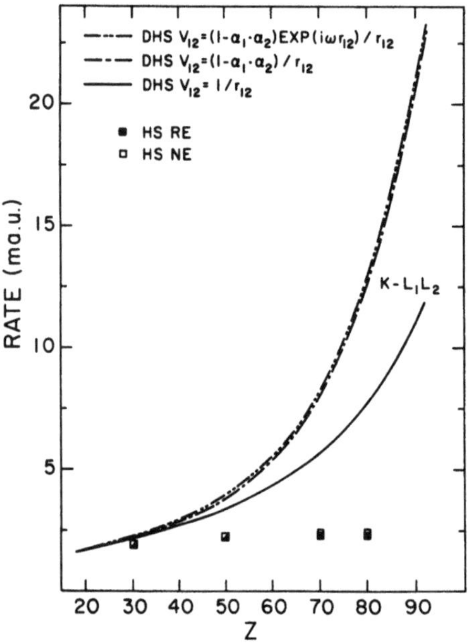

FIGURE 12. Theoretical K–L_1L_2 Auger rates (in milliatomic units), as functions of atomic number Z. The notation is as in Fig. 11. From Chen and Crasemann (1983).

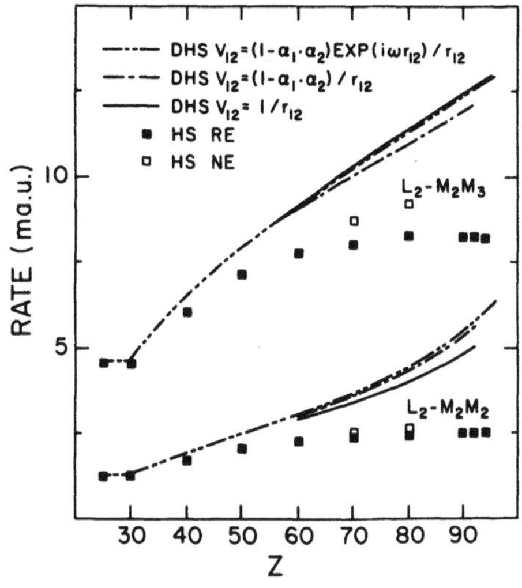

FIGURE 13. Theoretical L_2–M_2M_2 and L_2–M_2M_3 Auger rates (in milliatomic units), as functions of atomic number Z. The notation is as in Fig. 11. From Chen and Crasemann (1983).

a pronounced increase in Auger widths is found, close to a factor of 3 for high-Z atoms (Fig. 14). The effects of relativity reduce the total L_2 Auger rate by ~25% and increase the total L_3 Auger rate by only ~5% at $Z=80$ (Chen and Crasemann, 1983).

For K-shell fluorescence yields, excellent agreement between theoretical results from Dirac–Hartree–Slater calculations and precision measurements has been found (Fig. 15) (Chen *et al.*, 1980a). Good agreement between theory and experiment has also been demonstrated for $L_{2,3}$ and $M_{4,5}$

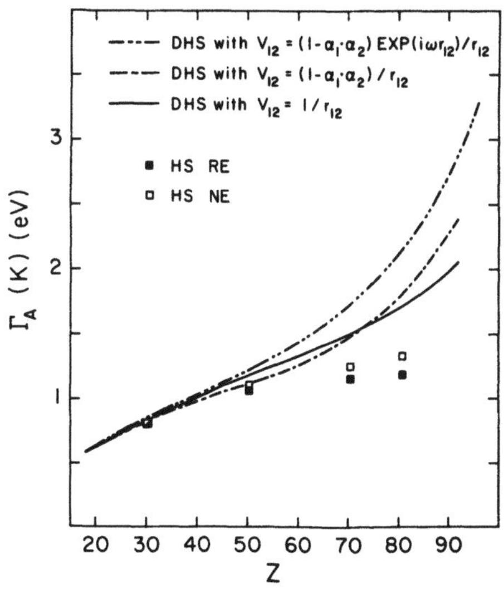

FIGURE 14. Theoretical K-shell Auger widths (in eV), as functions of atomic number Z. The notation is as in Fig. 11. From Chen and Crasemann (1983).

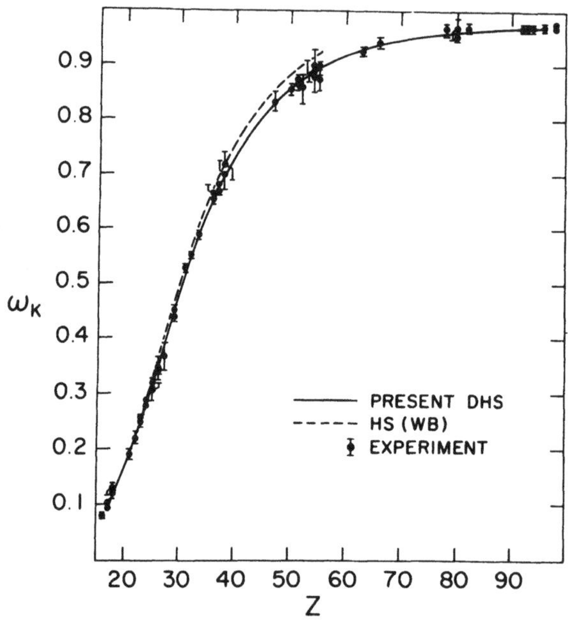

FIGURE 15. The K-shell fluorescence yields ω_K as functions of atomic number Z. The theoretical results from relativistic calculations with Dirac–Hartree–Slater wave functions (solid curve) (Chen *et al.*, 1980a) and nonrelativistic calculations with Hartree–Slater wave functions (broken curve) (Walters and Bhalla, 1971) are compared with experiments (Bambynek *et al.*, 1972). From Chen *et al.* (1980a).

fluorescence yields, Coster–Kronig yields f_{23}, $K\alpha_{1,2}$, $L\alpha_{1,2}$, and $L\beta_1$ X-ray line widths, and the group-intensity ratios $(K\text{–}LX)/(K\text{–}LL)$ (Chen *et al.*, 1980a; 1979b, 1980c, 1981a; Bambynek *et al.*, 1972; McGuire, 1975; Kessler *et al.*, 1982). A slight discrepancy (10%–20%) for $(K\text{–}XY)/(K\text{–}LL)$ ratios might be due to the neglect of the relaxation effect, similar to the case of $K\beta/K\alpha$ ratios in X-ray transitions (Scofield, 1947b).

Although the relativistic calculations improve the theoretical results, a systematic discrepancy between theory and experiment still remains for L_1, M_1, and $M_{2,3}$ level widths (Chen *et al.*, 1979b, 1981a, 1983b; McGuire, 1971, 1972). The independent-particle model has been found to overestimate the Coster–Kronig and super-Coster–Kronig transition rates (Chen *et al.*, 1977b, 1979b, 1981a, 1983b) due to the neglect of the dipole relaxation effects on the core hole. A many-body calculation using the diagrammatic RPAE method has been shown to remove the discrepancy between theory and experiment for $M_{2,3}$ level widths of Zn (Ohno and Wendin, 1979). Similar discrepancies have also been noted for N_1 and $N_{2,3}$ level widths (Fuggle and Alvarado, 1980).

5.4. *Analysis of K–LL and K–MM Auger Spectra*

The *K–LL* Auger spectra have been subject to extensive study both in theory and experiment (Bambynek *et al.*, 1972). For very light elements ($Z < 20$), the electrostatic interaction dominates and LS coupling applies. Effects of electron–electron Coulomb correlation are very important in predicting relative intensities. For elements with atomic numbers in the range $20 < Z < 35$, the spin–orbit interaction is not negligible compared with the electrostatic interaction and nonrelativistic theory including intermediate coupling and configuration interaction is necessary to describe the spectrum (Asaad, 1965; Mehlhorn and Asaad, 1966; Chen and Crasemann, 1973). For $35 < Z < 60$, relativistic effects become quite important, and the *K–LL* spectrum must be calculated relativistically in intermediate coupling with configuration interaction (Asaad and Petrini, 1976; Chen *et al.*, 1980b). For heavy elements ($Z > 60$), relativistic effects play a major role; the results from relativistic DHS calculation in *j–j* coupling agree very well with experiment (Bhalla and Ramsdale, 1970a, b; Chen *et al.*, 1980b).

In the analysis of *K–LL* Auger spectra using relativistic intermediate coupling with configuration interaction, *j–j* coupled basis sets can be used. With such a basis set, intermediate coupling can be treated as configuration interaction. The relativistic Auger matrix elements can be evaluated with the two-hole final state coupled in the *j–j* scheme (Chen *et al.*, 1980a, b). They also have been computed in the *M* scheme (Asaad, 1959; Listengarten, 1961, 1962; Bhalla and Ramsdale, 1970b), which makes it difficult to carry out intermediate coupling and configuration-interaction calculations.

To take into account the final-hole-state correlation and intermediate coupling, a limited configuration-interaction calculation for the final states with the same total angular momentum and parity is performed. For *K–LL* transitions, we have

$$J=0\begin{cases}L_1L_1\\L_2L_2,\\L_3L_3\end{cases}\qquad J=1\begin{cases}L_1L_2\\ \\L_1L_3\end{cases},\qquad\text{and}\quad J=2\begin{cases}L_2L_3\\ \\L_3L_3\end{cases}.$$

The relativistic eigenfunctions in intermediate coupling with configuration interaction can either be calculated by using MCDF which includes the optimization of the orbital wave functions, or by using a fixed *j–j* coupled basis set. In the existing calculations (Asaad and Petrini, 1976; Chen *et al.*, 1980b), the eigenfunctions and eigenvalues are computed by diagonalizing the energy matrix using a fixed basis set.

The relativistic DHS calculations of *K–LL* Auger transition probabilities in intermediate coupling with configuration interaction have been found to agree well with experimental data for a wide range of atomic numbers (Fig. 16) (Chen *et al.*, 1980b; Asaad and Petrini, 1976). The effects of configuration interaction between $L_1L_1(J=0)$ and $L_2L_2(J=0)$ states persist to high-Z atoms. For low-Z elements, some discrepancy between calculated and measured K–LL spectra remains. These differences may be due in part to solid-state effects (Hillig *et al.*, 1974; Breuckmann and Schmidt, 1974). Furthermore, correlation effects other than those represented by the mixing between $L_1L_1(^1S_0)$ and $L_2L_2(^1S_0)$ (e.g., final-state channel–channel coupling) have been shown to be quite important for light atoms (Kelly, 1975; Howat *et al.*, 1978).

Recently, relativistic intermediate coupling with configuration interaction has also been used to treat K–MM Auger spectra for medium and heavy elements (Chen *et al.*, 1983a). Configuration interaction among all the possible final MM double hole states is included in these calculations.

The relativistic effects are quite large on most of the K–MM transitions (e.g., ~a factor of 2 at $Z=80$ for K–M_1M_1). Intermediate coupling

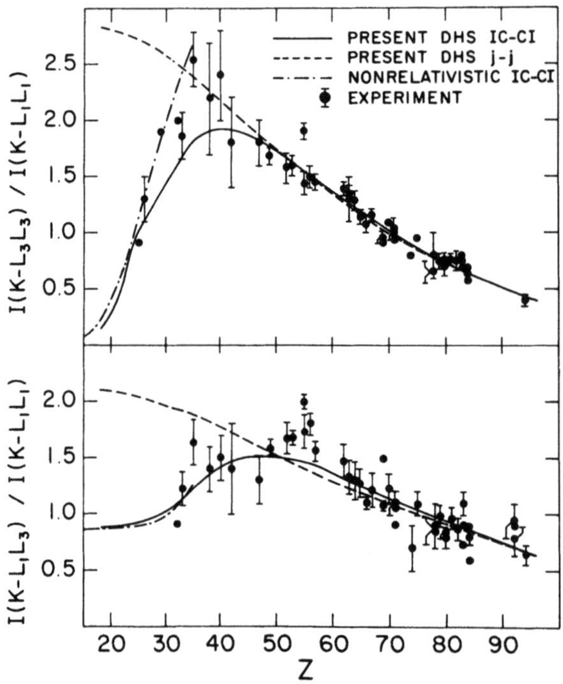

FIGURE 16. K–L_1L_3 and K–L_3L_3 Auger-transition intensities, relative to the K–L_1L_1 intensity, as functions of atomic number Z. DHS results (Chen *et al.*, 1980, b), both in intermediate coupling with configuration interaction (——) and in j–j coupling (- - - -), are compared with nonrelativistic calculations in intermediate coupling with configuration interaction (–·–) (Chen and Crasemann, 1973) and with experimental data (Sevier, 1972). From Chen *et al.* (1980b).

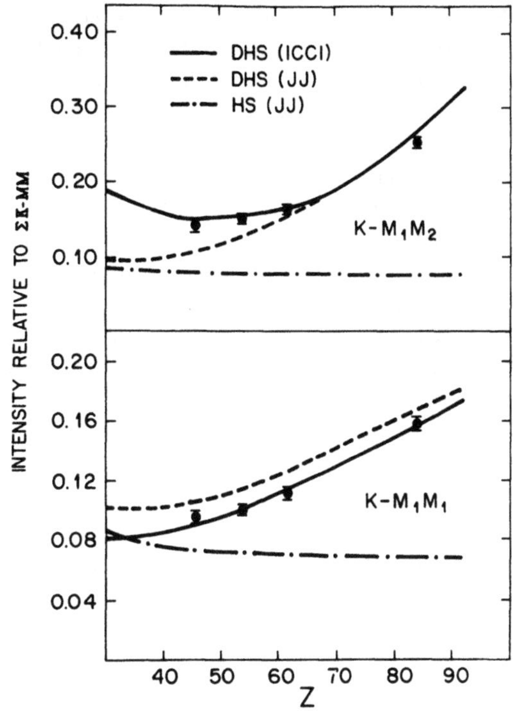

FIGURE 17. Ratio of calculated K–M_1M_2 and K–M_1M_1 Auger-transition rates to the total K–MM transition rate (excluding K–$M_{4,5}M_{4,5}$), as functions of atomic number Z. The theoretical results from Dirac–Hartree–Slater calculations in intermediate coupling with configuration interaction (solid curve), Dirac–Hartree–Slater calculations in j–j coupling (dashed curves), and nonrelativistic Hartree–Slater calculations (dot-dashed curve) are compared with experimental data. From Chen *et al.* (1983a).

drastically improves the agreement between the theoretical and experimental K–M_iM_j intensity ratios (Fig. 17). The effects of configuration interaction among $M_1M_1(J=0)$, $M_2M_2(J=0)$, and $M_3M_3(J=0)$ states persist to the heavy elements, similar to the case of $L_1L_1(J=0)$, $L_2L_2(J=0)$, and $L_3L_3(J=0)$ (Chen *et al.*, 1983a).

The K–$M_1M_{4,5}$ intensity has a strong peak at $Z \approx 63$. This peculiar behavior is caused by level crossing interaction between $M_1M_4(J=2)$, $M_1M_5(J=2)$, and $M_3M_3(J=2)$ (Chen *et al.*, 1983a). The weak K–$M_1M_{4,5}$ transitions pick up some intensity from K–M_3M_3 through the strong final-hole-state configuration interaction. Level crossing is a very common phenomenon in multiply ionized atoms (Weiss, 1976). Great care has to be taken in treating these cases.

6. AUGER AND FLUORESCENCE YIELDS OF MULTIPLY IONIZED ATOMS

X-ray spectra of multiply ionized atoms have been observed in solar flares (Doschek, 1972; Doscheck *et al.*, 1973), in laser-induced plasmas

(Boiko *et al.*, 1976), and in ion–atom collision experiments (Richard, 1975). The Auger spectra from multiply ionized atoms have also been seen in collision experiments (e.g., Mann *et al.*, 1981a, b; Richard, 1975; Schumann *et al.*, 1979; Matthews *et al.*, 1973; Stolterfoht *et al.*, 1973a). The average energy of an Auger spectrum of an atom with a multiple-vacancy configuration has been shown to decrease with increasing number of outer-shell vacancies (Matthews *et al.*, 1974; Stolterfoht *et al.*, 1973b; Stolterfoht *et al.*, 1977; Schneider *et al.*, 1976c) while the average energy of an X-ray spectrum increases with increasing number of outer-shell vacancies (Richard *et al.*, 1969; Richard *et al.*, 1972; Knudson *et al.*, 1971). In order to aid the analysis of the complex X-ray spectra, numerous theoretical calculations of X-ray wavelengths have been performed (Vainshtein and Safronova, 1978; Burkhalter *et al.*, 1975; Doscheck *et al.*, 1973; Ermolaev *et al.*, 1972; Goldsmith, 1974; Hata and Grant, 1982; Cheng *et al.*, 1979; Bely-Dubau *et al.*, 1979). The existing calculations can be classified into four general categories: (1) self-consistent field methods such as Hartree–Fock, Hartree–Slater, and Dirac–Fock (Burkhalter *et al.*, 1975; Doschek *et al.*, 1973; Hata and Grant, 1982; Cheng *et al.*, 1979); (2) Z^{-1}-expansion theory (Vainshtein and Safronova, 1978; Goldsmith, 1974); (3) model potential approaches as in the SUPERSTRUCTURE program of Eissner *et al.* (1974), based on the scaled Thomas–Fermi potential (Bely-Dubau *et al.*, 1979); (4) nonrelativistic variational methods (Accad *et al.*, 1971; Ahlenius and Larsson, 1978; Bunge and Bunge, 1978). The effect of relativity is either included *ab initio* by using a Dirac–Fock model or incorporated through the Breit–Pauli approximation (Vainshtein and Safronova, 1978). As discussed in Sec. 3.1, the contributions of the Breit interaction and QED corrections to the transition wavelengths have been shown to be essential for high-precision work (Hata and Grant, 1982). Systematic calculations of the K Auger transition energies of the multiplet states of the configurations $1s2s^n2p^m$ have also been carried out by Matthews *et al.* (1975) and Groeneveld *et al.* (1975) with the use of the Hartree–Fock model.

The calculations of multiplet Auger and X-ray transition rates and the effect of outer-shell vacancies on fluorescence yields will be discussed in Sec. 6.1. The relativistic effect on transition rates of multiply ionized atoms is reviewed in Sec. 6.2.

6.1. Fluorescence Yields of Atoms with Multiple Vacancies

In heavy-ion–atom collisions, multiple ionization has been found to be more probable than single ionization (Richard, 1975). Most studies concerning multiple vacancies are based on measurements of X-ray spectra. The X-ray fluorescence yields of defect configurations are needed to convert X-ray yields into ionization cross sections.

The early calculations of average fluorescence yields $\bar{\omega}$ of multiple-vacancy atoms (Larkins, 1971; Fortner *et al.*, 1972; Bhalla and Walters, 1973; Bhalla and Hein, 1973) were based on the formula

$$\bar{\omega} = \frac{\bar{\Gamma}_x}{\bar{\Gamma}_x + \bar{\Gamma}_A}. \tag{61}$$

Here, $\bar{\Gamma}_x$ and $\bar{\Gamma}_A$ are the statistical-average X-ray and Auger rates for the relevant configuration. Larkins (1971) and Fortner *et al.* (1972) applied a scaling procedure to the single-vacancy Auger and X-ray rates to obtain $\bar{\Gamma}_A$ and $\bar{\Gamma}_x$ for multiple-vacancy atoms. Bhalla and Walters (1973) and Bhalla and Hein (1973) performed direct calculations of $\bar{\Gamma}_A$ and $\bar{\Gamma}_X$ with Hartree–Slater wave functions for appropriate electronic configurations. The effects on transition rates due to changes of transition energies and wave functions because of the multiple vacancies have been found to be significant, rendering Larkins' simple scaling rule inadequate (Bhalla and Hein, 1973). Furthermore, Eq. (61) has been shown to underestimate the average fluorescence yields, especially for few-electron ions (Chen and Crasemann, 1974a, 1974b; Bhalla, 1974, 1975a). The reason for this deficiency can be traced to the fact that Eq. (61) ignores the possible closure of Auger or X-ray channels due to the selection rules of angular momentum and parity. As an example, the $1s2p^2(^3P)^2P$ state of Li-like ions has unit fluorescence yield in LS coupling because the Auger channel is forbidden due to failure to conserve angular momentum and, simultaneously, parity. This large fluorescence yield which does not follow from Eq. (61) boosts the average fluorescence yield of the $1s\,2p^2$ configuration. To remedy this deficiency, the effective fluorescence yield for an electronic configuration should be evaluated as a weighted sum of multiplet fluorescence yields:

$$\bar{\omega} = \frac{\sum_i P(\alpha_i S_i L_i)\, \omega(\alpha_i S_i L_i)}{\sum_i P(\alpha_i S_i L_i)}, \tag{62}$$

where $P(\alpha_i S_i L_i)$ is the relative population of the initial level, and

$$\omega(\alpha_i S_i L_i) = \frac{\Gamma_x(\alpha_i S_i L_i)}{\Gamma_x(\alpha_i S_i L_i) + \Gamma_A(\alpha_i S_i L_i)}. \tag{63}$$

Here, $\Gamma_x(\alpha_i S_i L_i)$ and $\Gamma_A(\alpha_i S_i L_i)$ are the total X-ray and Auger rates of the multiplet state $\alpha_i S_i L_i$, respectively. The effective fluorescence yield of a charge state can also be calculated as a weighted average of relevant configurations. If the initial states are statistically populated, then we have $P(\alpha_i S_i L_i) = (2L_i + 1)(2S_i + 1)$.

The procedure to obtain theoretical expressions for multiplet X-ray emission rates $\Gamma_x(\alpha_i S_i L_i J_i \rightarrow \alpha_f S_f L_f J_f)$ are given in standard texts on atomic spectra (Shore and Menzel, 1968; Condon and Shortley, 1959; Sobel'man, 1972). Some rather general expressions for multiplet Auger rates of multiple-vacancy atoms have been derived by McGuire (1975) using the hole representation. Fano (1965) developed a general recipe to evaluate the Coulomb matrix elements for atoms with arbitrary numbers of open shells.

Several nonrelativistic calculations using Hartree–Slater wave functions to predict multiplet fluorescence yields [Eq. (63)] and effective fluorescence yields [Eq. (62)] have been carried out: $_{10}$Ne $1s2s^n2p^m$ (Chen and Crasemann, 1975b; Chen *et al.*, 1975; Bhalla, 1975a–c), $_7$N (Bhalla, 1975d), $_9$F (Tunnell *et al.*, 1979), average K satellite and hypersatellite fluorescence yields for $7 < Z < 14$ (Bhalla, 1975e; Tunnel and Bhalla, 1981), and L fluorescence yields of $_{18}$Ar, $_{17}$Cl, and $_{16}$S with multiple M-shell vacancies (Chen and Crasemann, 1974a, b, 1975a, 1977). In general, the statistically averaged fluorescence yields increase as a function of numbers of outer-shell vacancies. The Auger and X-ray emission rates for different multiplet states from the same electronic configuration can differ from each other by as much as a few orders of magnitude. The effective fluorescence yields from Eq. (62) are therefore much larger than those predicted by Eq. (61) for high charge states (Fig. 18). The average fluorescence yields for various configurations belonging to the same charge state are also quite different from each other. Thus, the effective fluorescence yields of each charge state depend quite sensitively on the stripping mechanism (Chen and Crasemann, 1975a). The effective K-shell fluorescence yields of Ne KL^n charge states from collision experiments have been found to agree very well with theoretical results obtained from Eq. (62) with statistical populations (Fig. 18) (Chen *et al.*, 1975). Certainly, relativity can also affect the multiplet fluorescence yields. The effect of the spin–orbit interaction on the Auger transition probabilities and X-ray fluorescence yields of S and $Cl\,(2p)^{-1}(3p)^{-n}$ configurations has been found to be quite significant for lower stages of ionization, and this effect decreases with increasing stages of ionization. It reduces the statistically averaged $L_{2,3}$ fluorescence yields for $(2p)^{-1}(3p)^{-1}$ of Cl and for $(2p)^{-1}(3p)^{-2}$ of S by $\sim 43\%$ and $\sim 28\%$, respectively (Chen and Crasemann, 1977). A similar effect has also been found by McGuire (1976) for Na and Mg.

6.2. *Effects of Relativity on the Decay of Few-Electron Ions*

The transition energies and decay rates of few-electron ions are very important in connection with laboratory and astrophysical plasmas and collision physics. The spectra of heliumlike ions and their satellites due to transitions of the type $1s^2nl$–$1s2pnl$ with $n \geqslant 2$ have received much attention (Boiko *et al.*, 1976; Gabriel, 1972; Goldsmith, 1974; Bhalla *et al.*, 1975; Bely-Dubau *et al.*, 1979; Hata and Grant, 1981, 1982). Precision

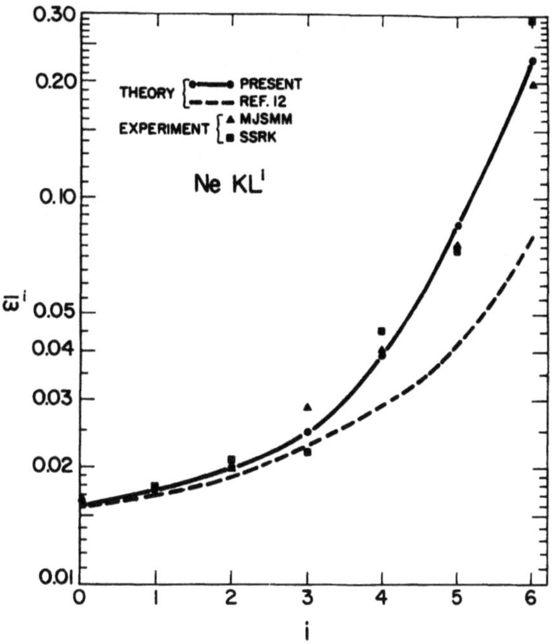

FIGURE 18. Theoretical K-shell fluorescence yields for Ne KL^i charge states, from Eq. (62) (solid curve) (Chen and Crasemann, 1975b), and from Eq. (61) (dashed curve) (Bhalla and Hein, 1973) are compared with measurements. For references to the experimental points, see the former. From Chen *et al.* (1975).

measurements of fine structure in heliumlike ($1s2p$) and lithiumlike ($1s2p2l$) ions can provide a sensitive test of the validity of relativistic atomic-structure theories and quantum-electrodynamic corrections (Livingston *et al.*, 1980; Hata and Grant, 1983a, b; Ermolaev and Swainson, 1983). Furthermore, helium satellite spectra can be a useful tool in plasma diagnostics (Gabriel, 1972; Bhalla *et al.*, 1975). The Auger decay of lithiumlike ions also is of considerable interest because dielectric recombination is the dominant mechanism to populate helium satellite states.

The importance of relativistic effects in the autoionization rates of metastable states has long been recognized in the study of $1s2s2p\ ^4P$ states of doubly excited lithium and N^{4+} and O^{5+} lithiumlike ions (Manson, 1966a, b, 1971). In Manson's calculations, Hartree–Fock wave functions were used. The effects of relativity were incorporated through the Pauli approximation. Bhalla and Gabriel (1976) and Bhalla *et al.* (1975) extended the investigation of $1s2l2l'$ configurations of lithiumlike ions to higher atomic numbers with the use of Hartree–Slater wave functions. The only relativistic effect included in these calculations is the spin–orbit interaction through intermediate coupling. Vainshtein and Safronova (1978) carried out a systematic study of wavelengths and transition probabilities of satellites and resonance lines of H- and He-like ions with $4 < Z < 30$ using perturbation theory in $1/Z$ and the Pauli approximation to treat the effects of relativity. The Auger rates in this work were calculated using unscreened Coulomb wave functions in intermediate coupling.

Recently, Chen *et al.* (1981b, d, 1982b, 1983c) performed systematic relativistic calculations of Auger and X-ray transition rates of highly ionized atoms with an inner-shell vacancy in an otherwise closed-shell configuration (e.g., the $1s2s^2 2p^6$ F-like isoelectronic sequence) and Li-like $1s2s^2$, $1s2p^2$, and $1s2s2p$ isoelectronic sequences, in relativistic intermediate coupling with configuration interaction. The Breit interaction was included both in the Auger matrix elements and in the energy matrix of the intermediate-coupling calculations. Similar relativistic calculations for $1s2s2p\ ^4P$ of the Li-like isoelectric sequence have also been carried out by Cheng *et al.* (1974) and by Bhalla and Tunnell (1981a, b).

For highly ionized atoms with an inner-shell vacancy in an otherwise closed shell (e.g., the $1s2s^2 2p^6$ F-like isoelectronic sequence), relativistic effects on transition rates have been found that are similar to those in normal atoms but slightly more pronounced (Chen *et al.*, 1981d). Transition rates and fluorescence yields are seen to increase with the degree of ionization; removing electrons from outer shells has a much smaller effect than creation of an empty inner shell of spectator holes.

The expressions for relativistic multiplet Auger and X-ray emission rates can be derived by following Fano's procedure (Fano, 1965). For three-electron j–j coupled states, they are given by Chen *et al.* (1981b). The relativistic intermediate-coupling theory described in Sec. 5.4 has been used to treat the Li-like ions with $1s2p2l'$ configurations. The relativistic effects on the decay of these multiplet states are discussed below.

For several open-shell atoms, the effects of relativity can become important very early in the isoelectronic sequence in an intermediate-coupling situation (e.g., for $1s2p^2$ configurations). The Auger rate of the $^2D_{3/2}$ state of the $1s2p^2$ configuration is reduced by 12% already at $Z = 20$ (Chen, 1982), and the radiative lifetime of the $^2D_{5/2}$ state becomes twice that of the $^2D_{3/2}$ at $Z = 30$ (Chen *et al.*, 1982b).

For strong allowed dipole radiative transitions, all theoretical results (Chen *et al.*, 1981b, 1982b, 1983c; Vainshtein and Safronova, 1978; Hata and Grant, 1982; Bhalla and Gabriel, 1976) agree within 5% at $Z \cong 26$ and within 15% at $Z \cong 10$. For allowed Auger transitions, the results from relativistic intermediate coupling (Chen *et al.*, 1981b, 1982b, 1983c) agree within 20% with nonrelativistic Hartree–Slater calculations including spin–orbit mixing for $Z < 26$. These SCF results are quite different from those based on Coulomb wave functions (Vainshtein and Safronova, 1978).

The effect of including the Breit interaction in intermediate coupling is only a few percent for strong transitions (Chen *et al.*, 1981b).

For weak allowed transitions with strong accidental cancellations in the transition matrix elements (e.g., the Auger decay of $1s2s2p\ ^2P^{(-)}_{3/2}$ and radiative decay of $1s2s2p\ ^2P^{(+)}_{3/2}$ at $Z \simeq 30$), the transition rates are very sensitive to the differences between various atomic models, and to relativistic effects (Figs. 19–20). Inclusion of the Breit interaction in the intermediate-coupling calculations can drastically change the transition rates (Chen *et al.*, 1981b). Hence, the fluorescence yields of these states are strongly model dependent (Chen *et al.*, 1981b).

The ^4P states of the $1s2p^2$ and $1s2s2p$ configurations are all Auger-for-

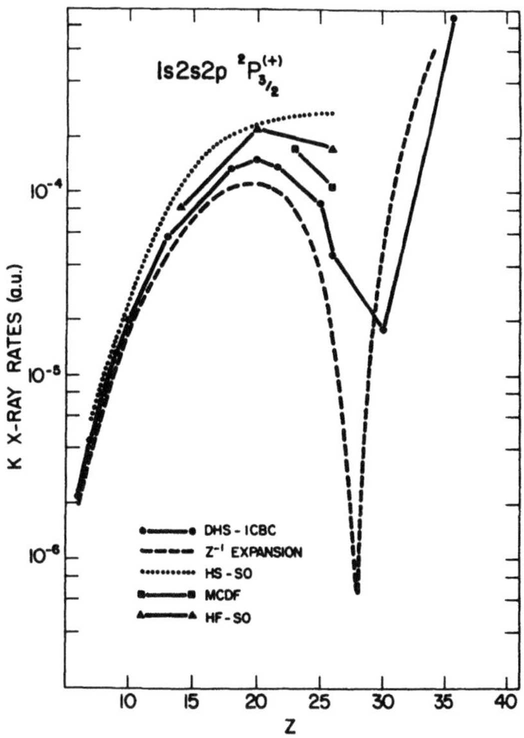

FIGURE 19. Theoretical K X-ray emission rates (in atomic units) for the $1s2s2p\ ^2P^{(+1)}_{3/2}$ state, as functions of atomic number Z: Dirac–Hartree–Slater calculations in intermediate coupling including Breit and Coulomb interaction (—●—) (Chen *et al.*, 1983c, 1981b); Z^{-1} expansion theory (- - - -) (Vainshtein and Safronova, 1978); Hartree–Slater calculations, in intermediate-coupling including spin–orbit interaction only (–·–) (Bhalla and Gabriel, 1976); multiconfiguration Dirac–Fock calculations (—■—) (Hata and Grant, 1982); Hartree–Fock results including spin–orbit interaction (—▲—) (Gabriel, 1972).

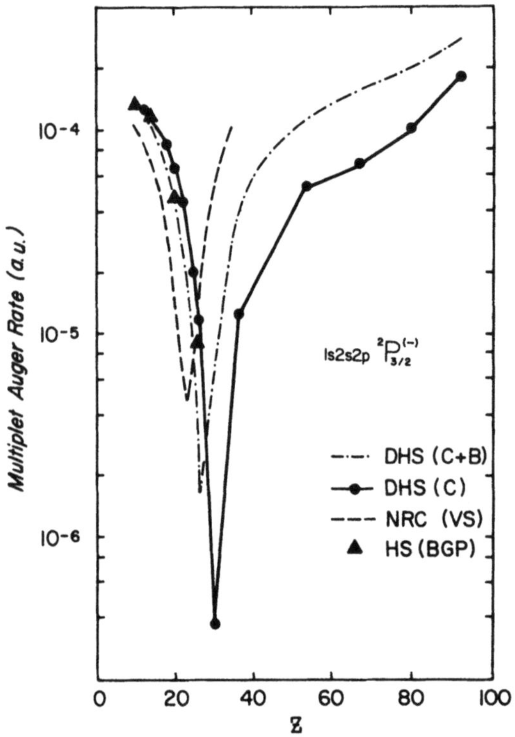

FIGURE 20. Multiplet Auger decay rates (in atomic units) of the $1s2s2p^{(-)}_{3/2}$ state calculated in intermediate coupling as functions of atomic number. The curve labeled DHS (C+B) represents the DHS results computed including both Breit and Coulomb interactions in the energy matrices that determine the intermediate-coupling coefficients (Chen *et al.*, 1981b); the curve labeled DHS (C) was computed including only the Coulomb interaction in these matrices (Chen *et al.*, 1981b); the curve labeled NRC(VS) is from Vainshtein and Safronova (1978) based on Coulomb wave functions, and the curve labelled HS(BGP) indicates the results from Bhalla *et al.* (1975) based on Hartree–Slater wave functions. From Chen *et al.* (1981b).

bidden in the nonrelativistic limit and dipole-forbidden for K X-ray emission (Chen *et al.*, 1981b, 1982b, 1983c). The K-shell Auger and X-ray transition rates of these quartet states are thus completely determined by the effects of relativity. The $^4P_{5/2}$ state of the $1s2s2p$ configuration can decay through Auger transitions thanks to the magnetic spin–spin interaction, or it can decay radiatively by magnetic-quadrupole emission (Manson, 1966b; Cheng *et al.*, 1974; Chen *et al.*, 1981b, 1983c). The Auger decay of ^4P_J states of the $1s2p^2$ configuration and $^4P_{1/2,3/2}$ states of the $1s2s2p$ configuration can occur through mixing with doublet states or by the magnetic interaction. These direct and indirect couplings to the continuum by the fine-structure Hamiltonian are of order α^2 (Manson, 1971) and have been found to be equally important for $^4P_{1/2,3/2}$ states in the low-Z region (Manson, 1971; Chen *et al.*, 1981b, 1982b, 1983c; Bhalla and Tunnell, 1981b). The Auger decay of $^4P_{1/2,3/2}$ states of $1s2p^2$ and $1s2s2p$ configurations are very sensitive to the details of the atomic model because of the strong cancellation between these two effects. Inclusion of the Breit interaction in the intermediate-coupling calculations can change the Auger rates by as much as an order of magnitude (Chen *et al.*, 1981b, 1982b, 1983c). Various theoretical results for Auger rates of $^4P_{1/2,3/2}$ states of the $1s2s2p$ configuration are compared in

Fig. 21. A nonrelativistic intermediate-coupling calculation neglecting the contribution from the magnetic interaction would be in serious error. For the *K* X-ray emission rates of these quartet states the Breit interaction plays a similarly important role (Chen *et al.*, 1981b). Including the Breit interaction in the determination of intermediate-coupling wave functions can affect the X-ray rates by as much as a factor of 2 (Fig. 22). Consequently, the *K*-shell fluorescence yields are also significantly modified by the effects of relativity (Chen *et al.*, 1981b). Similar relativistic effects can be expected for other high-spin states, such as $1s2s2p^2\ ^5P$ and $1s2p^3\ ^5S$.

The $^2P_{1/2}$ and $^2P_{3/2}$ states of the $1s2p^2$ configuration, being Auger forbidden in the nonrelativistic limit, acquire their Auger strengths from $^2S_{1/2}$ and $^2D_{3/2}$ states, respectively, through the spin–orbit interaction. The Auger

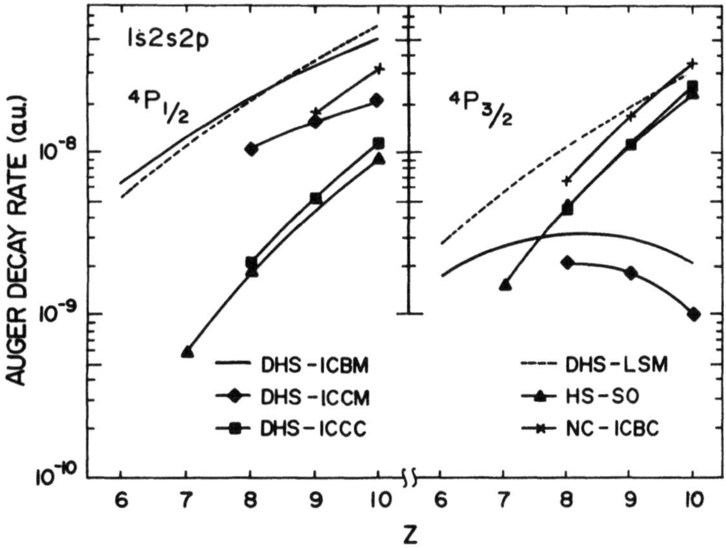

FIGURE 21. Auger decay rates of the $^4P_{1/2,3/2}$ states of the $1s2s2p$ configuration of Li-like ions of atomic number Z. The Dirac–Hartree–Slater calculations in intermediate coupling, including Coulomb and magnetic interaction in both the energy matrices and the Auger transition matrices (DHS–ICBM), are compared with the following: DHS calculations with only the Coulomb interaction in the energy matrix but also the magnetic interaction in the Auger matrix (DHS–ICCM); DHS calculations with only the Coulomb interaction in both the energy and Auger matrices (DHS–ICCC), and DHS computations in LS coupling, including contributions from the magnetic interaction in the Auger matrix elements (DHS–LSM) (Chen *et al.*, 1983c); nonrelativistic Hartree–Slater calculations in intermediate coupling with spin–orbit mixing and Coulomb interaction in the Auger matrix (Bhalla and Gabriel, 1976) (HS–SO); results of $1/Z$ expansion theory based on Coulomb wave functions (Vainshtein and Safronova, 1978). From Chen *et al.* (1983c).

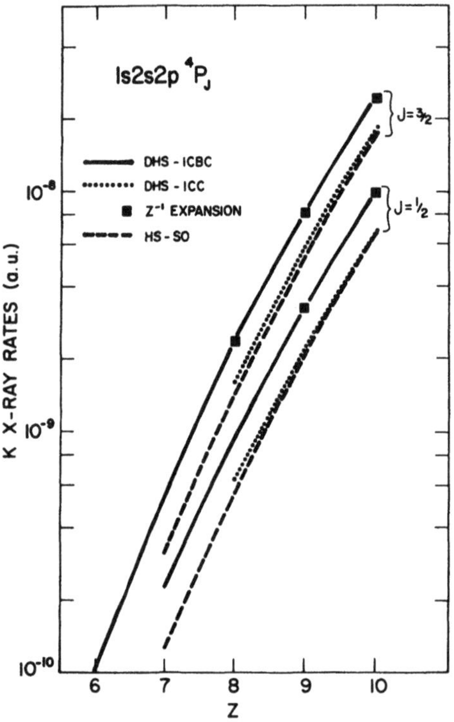

FIGURE 22. K X-ray transition rates (in atomic units), as functions of atomic number. Dirac–Hartree–Slater calculations in intermediate coupling including both Coulomb and Breit interactions in the energy matrices that determine intermediate-coupling coefficients (Chen *et al.*, 1983c) (DHS–ICBC) are compared with: Dirac–Hartree–Slater computations in intermediate coupling including only the Coulomb interaction in the energy matrices (DHS–ICC) (Chen *et al.*, 1983c); Z^{-1} expansion-theory results of Vainshtein and Safronova (1978); Hartree–Slater calculations in intermediate coupling including spin–orbit mixing (HS–SO) (Bhalla and Gabriel, 1976).

rates of ^2P states of the $1s2p^2$ configuration increase over four orders of magnitude from $Z=6$ to $Z=30$ (Chen *et al.*, 1982b).

There exists strong configuration interaction between the $1s2p^2\ {}^2S$ and $1s2s^2\ {}^2S$ states: the Auger decay of the former is reduced by a factor of 3 and the fluorescence yield of the latter is increased by six orders of magnitude at low and medium Z due to this effect (Chen *et al.*, 1981d, 1982b).

For ^4P states of the $1s2p^2$ configuration, the $E1$ transition $1s2p^2 \rightarrow 1s2s2p$ filling the $2s$ vacancy is a very important decay route for low-Z atoms. This radiative transition has been observed in the study of fine structure and lifetimes of the quartet states (Livingston and Berry, 1978). Accurately to predict the K-shell fluorescence yields and lifetime of these quartet states, one must include this $E1$ radiative transition in the theoretical calculations for $Z < 14$ (Chen *et al.*, 1982b).

Comparison between theoretical and experimental lifetimes of ^4P states of $1s2l2l'$ Li-like ions reveals the generally good agreement between relativistic theory and experiment (Fig. 23), except for $^4P_{1/2,3/2}$ states of the $1s2s2p$ configuration (Chen *et al.*, 1981b, 1982b, 1983c).

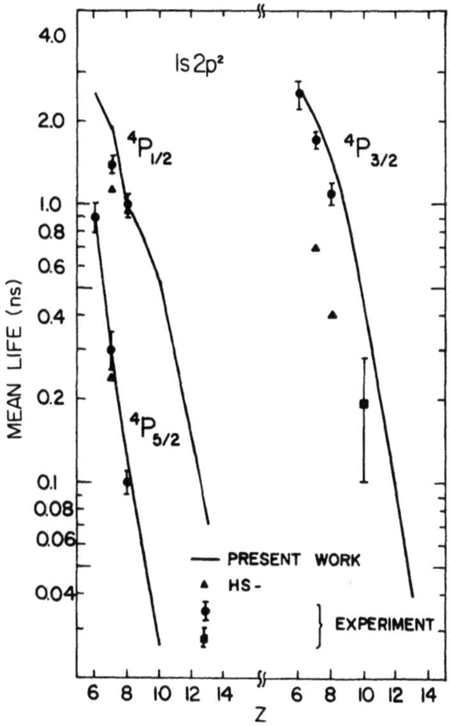

FIGURE 23. Lifetimes of the $1s2p^2\ ^4P$ states. The Dirac–Hartree–Slater results (Chen *et al.*, 1982b) (solid curves) are compared with nonrelativistic HS calculations (triangles) (calculated by C. P. Bhalla, quoted in Livingston and Berry, 1978) and with experimental data. For reference to experimental points, see Chen *et al.* (1982b). From Chen *et al.* (1982b).

7. SUMMARY

The relativistic independent-particle relaxed-orbital Dirac–Fock calculations of radiative transitions for atoms with an inner-shell vacancy are in quite good agreement with experiments. The gauge dependence of total radiative rates from Dirac–Fock calculations is small ($\lesssim 3\%$ for K shells for $Z \gtrsim 30$, and $\lesssim 7\%$ for L shells for $Z \gtrsim 48$). Some general conclusions concerning the effects of relativity on radiative transitions can be drawn:

1. The relativistic line strength is reduced due to relativistic orbital contraction.
2. For strong transitions (e.g., $1s$–$2p$, $2p$–$3d$), the combined effects of relativity and retardation are small because of near-cancellation between energy, wave function, and retardation effects.
3. For weak transitions ($\Delta n = 0$) or cases with large cancellations in nonrelativistic dipole matrix elements, the relativistic effect is large.
4. The effects of relativity and retardation can, in some cases, strongly affect the relative intensity (e.g., $K\alpha_2/K\alpha_1$ and $L\beta_4/L\beta_3$).

5. The combined effects of relativity and retardation on total radiative rates are quite small because of partial cancellation among the effects on various individual transitions.
6. In intermediate coupling of multiple-vacancy atoms (e.g., $K\alpha^h$ hypersatellites and $1s2p^2$ configurations), the effect of relativity can be important even for light atoms, and the Breit interaction should be included in the determination of the intermediate-coupling wave functions.

Relativistic DHS calculations of Auger transitions in intermediate coupling with limited final-hole-state configuration interaction have been found to be in reasonable agreement with experiments for $Z \gtrsim 18$. For low-Z atoms, however, other aspects of correlation effects, such as initial-state configuration interaction and final continuum state channel–channel interaction, must be included. Furthermore, the independent-particle model is inadequate to treat low-energy Coster–Kronig and super-Coster–Kronig transitions.

The effect of relativity on radiationless transitions has been studied in a systematic manner for atoms with an inner-shell vacancy and for few-electron ions. Some general conclusions can be drawn:

1. The relativistic effects on Auger transitions can arise from changes in energies, shifts in wave functions, and magnetic and retardation corrections in the two-electron operator. The net effect depends on the relative strengths and phases of these factors. Thus, the effect of relativity on individual transitions is very uneven.
2. For Coster–Kronig and super-Coster–Kronig transitions, the effect due to the relativistic change in transition energy is the dominant factor.
3. For normal medium-Z and heavy atoms, the magnetic interaction and retardation corrections are as important as the relativistic orbital effect in inner shells, while at low Z, the orbital effect is more important.
4. For transitions involving outer shells, the wave-function effect is the most important.
5. The relativistic effect on the total Auger rates is much reduced due to the partial cancellation among the individual transitions.
6. For highly ionized atoms, the effect of relativity can be very severe in the intermediate-coupling situation.
7. The effect of relativity is very critical in the decay of metastable states (e.g., ^4P of $1s2p^2$ and $1s2s2p$ configurations); a consistent treatment including all aspects of relativistic effects is necessary.

ACKNOWLEDGMENTS

This chapter was written while the author was a member of the Physics Department and Chemical Physics Institute of the University of Oregon, Eugene, Oregon. The work was supported in part by the U. S. Air Force Office of Scientific Research through contract No. F49620-83-K0020.

REFERENCES

ÅBERG, T., 1980, *Phys. Scr.* **21**, 495.
ÅBERG, T., AND G. HOWAT, 1983, in *Handbuch der Physik*, Vol. 31, edited by W. Mehlhorn (Springer, Berlin).
ÅBERG, T., J. P. BRIAND, P. CHEVALLIER, A. CHETIOUI, J. P. ROZET, M. TAVERNIER, AND A. TOUATI, 1976, *J. Phys. B* **9**, 2815.
ÅBERG, T., AND M. SUVANEN, 1980, in *Advances in X-Ray Spectroscopy*, edited by C. Bonnelle and C. Mandé (Pergamon, New York).
ACCAD, Y., C. L. PEKERIS, AND B. SCHIFF, 1971, *Phys. Rev. A* **4**, 516.
AHLENIUS, T., AND S. LARSSEN, 1978, *Phys. Rev. A* **18**, 1329.
AKHIEZER, A. I., AND V. B. BERESTETSKII, 1965, *Quantum Electrodynamics.* (Interscience, New York).
AKSELA, S., R. KUMPULA, H. AKSELA, AND J. VÄYRYNEN, 1982, *Phys. Scr.* **25**, 45.
ALBRIDGE, R. G., AND J. M. HOLLANDER, 1961, *Nucl. Phys.* **27**, 554.
ANHOLT, R. AND J. O. RASMUSSEN, 1974, *Phys. Rev. A* **9**, 585.
ARMSTRONG, L. JR., C. E. THEODOSIOU, AND M. J. WALL, 1978, *Phys. Rev. A* **18**, 2538.
ASAAD, W. N., 1959, *Proc. R. Soc. London, Ser. A* **249**, 555.
ASAAD, W. N., 1965, *Nucl. Phys.* **66**, 494.
ASAAD, W. N. AND E. H. S. BURHOP, 1958, *Proc. Phys. Soc. London* **71**, 369.
ASAAD, W. N., AND D. PETRINI, 1976, *Proc. R. Soc. London Ser. A* **350**, 381.
AUGER, P., 1925, *J. Phys. Radium* **6**, 205.
BABUSHKIN, F. A., 1964, *Acta Phys. Pol.* **25**, 749.
BABUSHKIN, F. A., 1965, *Opt. Spectrosk.* **19**, 3 [Opt. Spectrosc. (USSR) **19**, 1].
BABUSHKIN, F. A., 1967, *Acta. Phys. Pol.* **31**, 459.
BAMBYNEK, W., B. CRASEMANN, R. W. FINK, H.-U. FREUND, H. MARK, C. D. SWIFT, R. E. PRICE, AND P. V. RAO, 1972, *Rev. Mod. Phys.* **44**, 716.
BEARDEN, J. A., AND A. F. BURR, 1967, *Rev. Mod. Phys.* **39**, 125.
BEATHAM, N., I. P. GRANT, B. J. MCKENZIE, AND S. J. ROSE, 1980, *Phys. Scr.* **21**, 423.
BECK, D. R., AND C. A. NICOLAIDES, 1978, in *Excited States in Quantum Chemistry*, edited by C. A. Nicolaides and D. R. Beck (Reidel, Dordrecht).
BELY-DUBAU, F., A. H. GABRIEL, AND S. VOLONTÉ, 1979, *Mon. Not. R. Astron. Soc.* **186**, 405.
BERGSTRÖM, I., AND R. D. HILL, 1954. *Ark. Fys.* **8**, 21.
BETHE, H. A., 1947, *Phys. Rev.* **72**, 339.
BETHE, H. A., AND E. E. SALPETER, 1959, *Quantum Mechanics of One- and Two-Electron Atoms* (Academic, New York), p. 198.
BHALLA, C. P., 1970a, *J. Phys. B* **3**, L9.
BHALLA, C. P., 1970b, *Phys. Rev. A* **2**, 722.
BHALLA, C. P., 1970c, *J. Phys. B* **3**, 916.
BHALLA, C. P., 1970d, *Phys. Rev. A* **2**, 2575.
BHALLA, C. P., 1974, *Phys. Fenn. Suppl. S1* **9**, 435.

BHALLA, C. P., 1975a, *Phys. Rev. A* **12**, 122.
BHALLA, C. P., 1975b, *J. Electron Spectrosc. Relat. Phenom.* **7**, 287.
BHALLA, C. P., 1975c, *J. Phys. B* **8**, 1200.
BHALLA, C. P., 1975d, *J. Phys. B* **8**, 2792.
BHALLA, C. P., 1975e, *J. Phys. B* **8**, 2787.
BHALLA, C. P., AND A. H. GABRIEL, 1976, in *Beam-Foil Spectroscopy*, edited by I. A. Sellin and D. J. Pegg (Plenum, New York), Vol. I, p. 121.
BHALLA, C. P., AND M. HEIN, 1973, *Phys. Rev. Lett.* **30**, 39.
BHALLA, C. P., AND D. J. RAMSDALE, 1970a, *J. Phys. B* **3**, L 14.
BHALLA, C. P. AND D. J. RAMSDALE, 1970b, *Z. Phys.* **239**, 95.
BHALLA, C. P., AND T. W. TUNNELL, 1981a, *Z. Phys. A* **303**, 199.
BHALLA, C. P., AND T. W. TUNNELL, 1981b, in *Inner-Shell and X-ray Physics of Atoms and Solids*, edited by D. J. Fabian, H. Kleinpoppen, and L. M. Watson (Plenum, New York, 1981), p. 285.
BHALLA, C. P., AND D. L. WALTERS, 1973. Proc. Int. Conf. Inner Shell Ionization Phenomena Future Appl. Atlanta, edited by R. W. Fink, S. T. Manson, J. M. Palms, and R. V. Rao, U.S. At. Energy Comm. report No. 720404, Oak Ridge, Tennessee.
BHALLA, C. P., D. J. RAMSDALE, AND H. R. ROSNER, 1970a, *Phys. Lett.* **31A**, 122.
BHALLA, C. P., H. R. ROSNER, AND D. J. RAMSDALE, 1970b, *J. Phys. B* **3**, 1232.
BHALLA, C. P., A. H. GABRIEL, AND L. P. PRESNYAKOV, 1975, *Mon. Not. R. Astron. Soc.* **172**, 359.
BJORKEN, J. D., AND S. D. DRELL, 1964, *Relativistic Quantum Mechanics* (McGraw-Hill, New York), p. 158.
BLIN-STOYLE, R. J., 1953, *Proc. Phys. Soc. A* **66**, 729.
BOIKO, V. A., A. YA. FAENOV, AND S. A. PIKUZ, 1976, *J. Quant. Spectrosc. Radiat. Transfer* **19**, 11.
BREIT, G., 1932, *Phys. Rev.* **39**, 616.
BREIT, G., AND E. TELLER, 1940, *Astrophys. J.* **91**, 215.
BREUCKMANN, B., AND V. SCHMIDT, 1974, *Z. Phys.* **268**, 235.
BRIANÇON, C., 1970, *Ann. Phys. (Paris)* **5**, 151.
BRIANÇON, C., AND J. P. DESCLAUX, 1976, *Phys. Rev. A* **13**, 2157.
BRIAND, J. P., P. CHEVALLIER, M. TAVERNIER, AND J. P. ROZET, 1971, *Phys. Rev. Lett.* **27**, 777.
BROWN, G. E., AND D. G. RAVENHALL, 1951, *Proc. R. Soc. London, Ser. A* **208**, 532.
BROWN, G. E., J. S. LANGER, AND G. W. SCHAEFER, 1959a, *Proc. R. Soc. London, Ser. A* **251**, 92.
BROWN, G. E., AND D. F. MAYERS, 1959b, *Proc. R. Soc. London, Ser. A* **251**, 105.
BROWN, G. S., M. H. CHEN, B. CRASEMANN, AND G. E. ICE, 1980, *Phys. Rev. Lett.* **45**, 1937.
BUNGE, C. F. AND A. V. BUNGE, 1978, *Phys. Rev. A* **17**, 816.
BURHOP, E. H. S., 1935, *Proc. R. Soc. London, Ser. A* **148**, 272.
BURHOP, E. H. S., 1952, *The Auger Effect and Other Radiationless Transitions* (Cambridge U. P., Cambridge).
BURHOP, E. H. S., AND W. N. ASAAD, 1972, *Adv. At. Mol. Phys.* **8**, 163.
BURKHALTER, P. G., D. J. NAGEL, AND R. D. COWAN, 1975, *Phys. Rev. A* **11**, 782.
CALLAN, E. J., 1961, *Phys. Rev.* **124**, 793.
CHATTARJI, D., 1976, *The Theory of Auger Transitions* (Academic, London).
CHATTARJI, D., AND B. TALUKDAR, 1968, *Phys. Rev.* **174**, 44.
CHEN, M. H., in *X-Ray and Atomic Inner-Shell Physics—1982* edited by B. Crasemann, AIP Conference Proceeding No. 94 (American Institute of Physics, New York), p. 331.
CHEN, M. H., 1983, unpublished.
CHEN, M. H., AND B. CRASEMANN, 1973, *Phys. Rev. A* **8**, 7.
CHEN, M. H., AND B. CRASEMANN, 1974a, *Phys. Fenn. Suppl. S1*, **9**, 250.
CHEN, M. H., AND B. CRASEMANN, 1974b, *Phys. Rev. A* **10**, 2232.
CHEN, M. H., AND B. CRASEMANN, 1975a, *Phys. Rev. A* **12**, 710.
CHEN, M. H., AND B. CRASEMANN, 1975b, *Phys. Rev. A* **12**, 959.

CHEN, M. H., AND B. CRASEMANN, 1977, *Phys. Rev. A* **16**, 1495.
CHEN, M. H., AND B. CRASEMANN, 1983, *Phys. Rev. A* **28**, 2829.
CHEN, M. H., B. CRASEMANN, AND D. L. MATTHEWS, 1975, *Phys. Rev. Lett.* **34**, 1309.
CHEN, M. H., B. CRASEMANN, K. N. HUANG, M. AOYAGI, AND H. MARK, 1977a, *At. Data Nucl. Data Tables* **19**, 97.
CHEN, M. H., B. CRASEMANN, M. AOYAGI, AND H. MARK, 1977b, *Phys. Rev. A* **15**, 2312.
CHEN, M. H., B. CRASEMANN, AND H. MARK, 1979a, *At. Data Nucl. Data Tables* **24**, 13.
CHEN, M. H., E. LAIMAN, B. CRASEMANN, M. AOYAGI, AND H. MARK, 1979b, *Phys. Rev. A* **19**, 2253.
CHEN, M. H., B. CRASEMANN, M. AOYAGI, AND H. MARK, 1979c, *Phys. Rev. A* **20**, 385.
CHEN, M. H., B. CRASEMANN, AND H. MARK, 1980a, *Phys. Rev. A* **21**, 436.
CHEN, M. H., B. CRASEMANN, AND H. MARK, 1980b, *Phys. Rev. A* **21**, 442, and unpublished results.
CHEN, M. H., B. CRASEMANN, AND H. MARK, 1980c, *Phys. Rev. A* **21**, 449.
CHEN, M. H., B. CRASEMANN, AND H. MARK, 1981a, *Phys. Rev. A* **24**, 177.
CHEN, M. H., B. CRASEMANN, AND H. MARK, 1981b, *Phys. Rev. A* **24**, 1852.
CHEN, M. H., B. CRASEMANN, AND H. MARK, 1981c, *Phys. Rev. A* **24**, 1158.
CHEN, M. H., B. CRASEMANN, K. R. KARIM, AND H. MARK, 1981d, *Phys. Rev. A* **24**, 1845.
CHEN, M. H., B. CRASEMANN, M. AOYAGI, K. N. HUANG, AND H. MARK, 1981e, *At. Data Nucl. Data Tables* **26**, 561.
CHEN, M. H., B. CRASEMANN, AND H. MARK, 1982a, *Phys. Rev. A* **25**, 391.
CHEN, M. H., B. CRASEMANN, AND H. MARK, 1982b, *Phys. Rev. A* **26**, 1441.
CHEN, M. H., B. CRASEMANN, AND H. MARK, 1983a, *Phys. Rev. A* **27**, 1213.
CHEN, M. H., B. CRASEMANN, AND H. MARK, 1983b, *Phys. Rev. A* **27**, 2989.
CHEN, M. H., B. CRASEMANN, AND H. MARK, 1983c, *Phys. Rev. A* **27**, 544.
CHENG, K. T., C. P. LIN, AND W. R. JOHNSON, 1974, *Phys. Lett.* **48A**, 437.
CHENG, K. T., Y. K. KIM, AND J. P. DESCLAUX, 1979, *At. Data Nucl. Data Tables* **24**, 111.
CHUNG, M. F., AND L. H. JENKINS, 1970, *Surf. Sci.* **22**, 479.
COLLARD, H. R., L. R. B. ELTON, AND R. HOFSTADTER, 1967, *Nuclear Radii*, edited by H. Schopper (Springer, Berlin), Landolt-Börnstein, New Series, Group I, Vol. 2.
COGHLAN, W. A., AND R. E. CLAUSING, 1973, *At. Data* **5**, 317.
CONDON, E. U., AND G. H. SHORTLEY, 1959, *The Theory of Atomic Spectra* (Cambridge U. P., Cambridge).
COSTER, D. AND R. DE L. KRONIG, 1935, *Physica* **2**, 13.
DESCLAUX, J. P., 1972, *Int. J. Quant. Chem.* **6**, 25.
DESCLAUX, J. P., 1973, *At. Data Nucl. Data Tables* **12**, 311.
DESCLAUX, J. P., 1975, *Comp. Phys. Commun.* **9**, 31.
DESCLAUX, J. P., 1980, *Phys. Scr.* **21**, 436.
DESCLAUX, J. P., AND Y. K. KIM, 1975, *J. Phys. B* **8**, 1177.
DESCLAUX, J. P., D. F. MAYERS, AND F. O'BRIEN, 1971, *J. Phys. B* **4**, 631.
DESCLAUX, J. P., CH. BRIANÇON, J. P. THIBAUD, AND R. J. WALEN, 1974, *Phys. Rev. Lett.* **32**, 447.
DESLATTES, R. D., AND E. G. KESSLER, JR., 1985, Chapter 5 in this volume.
DESIDERIO, A. M., AND W. R. JOHNSON, 1971, *Phys. Rev. A* **3**, 1267.
DOSCHEK, G. A., 1972, *Space Sci. Rev.* **13**, 765.
DOSCHEK, G. A., J. F. MEEKINS, AND R. D. COWAN, 1973, *Solar Phys.* **29**, 125.
EISENBERGER, P., P. M. PLATZMAN, AND H. WINICK, 1976, *Phys. Rev. Lett.* **36**, 623.
EISSNER, W., M. JONES, AND H. NUSSBAUMER, 1974, *Comp. Phys. Commun.* **8**, 270.
ERICKSON, G. W., 1971, *Phys. Rev. Lett.* **27**, 780.
ERICKSON, G. W., AND O. R. YENNIE, 1965, *Ann. Phys. (N.Y.)* **35**, 271.
ERMAN, P., AND Z. SUJKOWSKI, 1961, *Ark. Fys.* **20**, 209.
ERMOLAEV, A. M., M. JONES, AND K. J. H. PHILLIPS, 1972, *Astrophys. Lett.* **12**, 53.

ERMOLAEV, A. M., AND R. A. SWAINSON, 1983, *J. Phys. B* **16**, L35.
FANO, U., 1961, *Phys. Rev.* **124**, 1866.
FANO, U., 1965, *Phys. Rev.* **140**, A67.
FORTNER, R. J., R. C. DER, T. M. KAVANAGH, AND J. D. GARCIA, 1972, *J. Phys. B* **5**, L73.
FROESE-FISCHER, C., 1969, *Comp. Phys. Commun.* **1**, 151.
FROESE-FISCHER, C., 1972, *Comp. Phys. Commun.* **4**, 107.
FROESE-FISCHER, C., 1977, *The Hartree–Fock Method for Atoms* (Wiley, New York).
FUGGLE, J. C., AND S. F. ALVARADO, 1980, *Phys. Rev. A* **22**, 1615.
GABRIEL, A. H., 1972, *Mon. Not. R. Astron. Soc.* **160**, 99.
GABRIEL, O. V., S. CHAUDHURI, AND R. H. PRATT, 1981, *Phys. Rev. A* **24**, 3088.
GARSTANG, R. H., 1971, in *Topics in Modern Physics*, edited by W. E. Brittin and H. Odabasi (Colorado Associated University Press, Boulder).
GOLDSMITH, S., 1974, *J. Phys. B* **7**, 2315.
GRANT, I. P., 1961, *Proc. R. Soc. London, Ser. A* **262**, 555.
GRANT, I. P., 1970, *Adv. Phys.* **19**, 747.
GRANT, I. P., 1974, *J. Phys. B* **12**, 1458.
GRANT, I. P., D. F. MAYERS, AND N. C. PYPER, 1976, *J. Phys. B* **9**, 2777.
GRANT, I. P., AND B. J. MCKENZIE, 1980, *J. Phys. B* **13**, 2671.
GRANT, I. P., B. J. MCKENZIE, P. H. NORRINGTON, D. F. MAYERS, AND N. C. PYPER, 1980, *Comp. Phys. Commun.* **21**, 207.
GROENEVELD, K. O., R. MANN, G. NOLTE, S. SCHUMANN, R. SPOHR, AND B. FRICKE, 1975, *Z. Phys. A* **294**, 191.
GUNNARSSON, O., AND K. SCHÖNHAMMER, 1982, in *X-Ray and Atomic Inner-Shell Physics—1982*, edited by B. Crasemann, AIP Conference Proceedings No. 94 (American Institute of Physics, New York), p. 517.
HARTMANN, H., AND E. CLEMENTI, 1964, *Phys. Rev.* **133**, A1295.
HARTREE, D. R., 1927, *Proc. Cambridge Phil. Soc.* **24**, 89.
HATA, J., AND I. P. GRANT, 1981, *J. Phys. B* **14**, 2111.
HATA, J., AND I. P. GRANT, 1982, *Mon. Not. R. Astron. Soc.* **198**, 1081.
HATA, J., AND I. P. GRANT, 1983a, *J. Phys. B* **16**, 507.
HATA, J., AND I. P. GRANT, 1983b, *J. Phys. B* **16**, 523.
HEDIN, L., AND G. JOHANSSON, 1969, *J. Phys. B* **2**, 1336.
HERMAN, F., AND S. SKILLMAN, 1963, *Atomic Structure Calculations*, (Prentice-Hall, Englewood cliffs, New Jersey).
HILLIG, H., B. CLEFF, W. MEHLHORN, AND W. SCHMITZ, 1974, *Z. Phys.* **268**, 225.
HOFSTADTER, R., AND H. R. COLLARD, 1967, Nuclear Radii Determined by Electron Scattering, in *Numerical Data and Functional Relationships in Science and Technology* (Springer, Berlin), Group I, Vol. 2, p. 21.
HOWAT, G., T. ÅBERG, AND O. GOSCINSKI, 1978, *J. Phys. B* **11**, 1575.
HUANG, K.-N., 1976, *Phys. Rev. A* **14**, 1311.
HUANG, K.-N., 1978, *J. Phys. B* **11**, 787.
HUANG, K.-N., M. AOYAGI, M. H. CHEN, B. CRASEMANN, AND H. MARK, 1976, *At. Data Nucl. Data Tables* **18**, 243.
HUANG, K.-N., AND W. R. JOHNSON, 1982, *Phys. Rev. A* **25**, 634.
JEN, J. S., AND T. D. THOMAS, 1976, *Phys. Rev. B* **13**, 5284.
JOHANSSON, B., AND N. MÅRTENSSON, 1980, *Phys. Rev. B* **21**, 4427.
JOHNSON, W. R., 1972, *Phys. Rev. Lett.* **29**, 1123.
JOHNSON, W. R., AND C. D. LIN, 1976, *Phys. Rev. A* **14**, 565.
KELLY, H. P., 1975, *Phys. Rev. A* **11**, 556.
KESSLER JR., E. G., R. D. DESLATTES, D. GIRARD, W. SCHWITZ, L. JACOBS, AND O. RENNER, 1982, *Phys. Rev. A* **26**, 2696.

KNUDSON, A. R., D. J. NAGEL, P. G. BURKHALTER, AND K. L. DUNNING, 1971, *Phys. Rev. Lett.* **26**, 1149.

KOCHER, C. A., AND J. E. CLENDENIN, AND R. NOVICK, 1972, *Phys. Rev. Lett.* **29**, 615.

KOSTROUN, V. O., M. H. CHEN, AND B. CRASEMANN, 1971, *Phys. Rev. A* **3**, 533.

KRAUSE, M. O., 1976, in Proc. Soc. Int. Conf. Inner Shell Ioniz. Phen, edited by W. Mehlhorn and R. Brenn, Freiburg 1976, Invited papers, p. 184.

LAMB, W. E., JR., AND R. C. RETHERFORD, 1947, *Phys. Rev.* **72**, 241.

LARKINS, F. P., 1971, *J. Phys. B* **4**, L29.

LARKINS, F. P., 1974, *J. Phys. B* **7**, 37.

LARKINS, F. P., 1975, in *Atomic Inner Shell Processes*, edited by B. Crasemann (Academic, New York), p. 377.

LARKINS, F. P., 1976, *J. Phys. B* **9**, 47.

LARKINS, F. P., 1977a, *At. Data Nucl. Data Tables* **20**, 311.

LARKINS, F. P., 1977b, *J. Phys. C* **10**, 2453.

LARKINS, F. P., 1977c, *J. Phys. C* **10**, 2461.

LASKAR, W., 1955, *J. Phys. Radium* **16**, 644.

LASKAR, W., 1958, *Ann. Phys. (Paris)* **13**, 258.

LAUTRUP, B. E., A. PETERMANN, AND E. DE RAFAEL, 1972, *Phys. Rep.* **3**, 193.

LEY, L., S. P. KOWALCZYK, F. R. MCFEELY, R. A. POLLAK, AND D. A. SHIRLEY, 1973, *Phys. Rev. B* **8**, 2392.

LIBERMAN, D. A., J. T. WABER, AND D. T. CROMER, 1965, *Phys. Rev.* **137**, A27.

LIBERMAN, D. A., J. T. WABER, AND D. T. CROMER, 1971, *Comp. Phys. Commun.* **2**, 107.

LINDENBERG, J., AND Y. OHRN, 1973, *Propagators in Quantum Chemistry* (Academic, New York).

LINDGREN, I., AND A. ROSÉN, 1974, *Case Stud. At. Phys.* **4**, 93.

LISTENGARTEN, M. A., 1961, *Izv. Akad. Nauk SSR, Ser. Fiz.* **25**, 803 [*Bull. Acad. Sci. USSR, Phys. Ser.* **25**, 803].

LISTENGARTEN, M. A., 1962, *Izv. Akad. Nauk SSR, Ser. Fiz.* **26**, 182 [*Bull. Acad. Sci. USSR, Phys. Ser.* **26**, 182].

LIVINGSTON, A. E., AND H. G. BERRY, 1978, *Phys. Rev. A* **17**, 1966.

LIVINGSTON, A. E., S. J. HINTERLONG, J. A. POIRIER, R. DE SERIO, H. G. BERRY, 1980, *J. Phys. B* **13**, L139.

LU, C. C., F. B. MALIK, AND T. A. CARLSON, 1971, *Nucl. Phys. A* **175**, 289.

MANN, J. B., AND W. R. JOHNSON, 1971, *Phys. Rev. A* **4**, 41.

MANN, J. B., AND J. T. WABER, 1973, *At. Data* **5**, 202.

MANN, R., F. FOLKMANN, AND H. F. BEYER, 1981a, *J. Phys. B* **14**, 1161.

MANN, R., H. F. BEYER, AND F. FOLKMANN, 1981b, *Phys. Rev. Lett.* **46**, 646.

MANSON, S. T., 1966a, *Phys. Rev.* **145**, 35.

MANSON, S. T., 1966b, *Phys. Lett.* **23**, 315.

MANSON, S. T., 1971, *Phys. Rev. A* **3**, 147.

MANSON, S. T., AND D. J. KENNEDY, 1974, *At. Data Nucl. Data Tables* **14**, 111.

MARK, F., AND W. H. E. SCHWARZ, 1982, *Phys. Rev. Lett.* **48**, 673.

MARRUS, R., AND R. W. SCHMIEDER, 1972, *Phys. Rev. A* **5**, 1160.

MÅRTENSSON, N., AND B. JOHANSSON, 1979, *Solid State Commun.* **32**, 791.

MÅRTENSSON, N., AND B. JOHANSSON, 1980, *Phys. Rev. Lett.* **45**, 482.

MASSEY, H. S. W., AND E. H. S. BURHOP, 1936a, *Proc. R. Soc. London Ser. A* **153**, 661.

MASSEY, H. S. W., AND E. H. S. BURHOP, 1936b, *Proc. Cambridge Phil. Soc.* **32**, 461.

MATTHEWS, D. L., B. M. JOHNSON, J. J. MACKEY, AND C. F. MOORE, 1973, *Phys. Rev. Lett.* **31**, 1331.

MATTHEWS, D. L., B. M. JOHNSON, J. J. MACKEY, L. E. SMITH, W. HODGE, AND C. F. MOORE, 1974, *Phys. Rev. A* **10**, 1177.

MATTHEWS, D. L., B. M. JOHNSON, AND C. F. MOORE, 1975, *At. Data Nucl. Data Tables* **15**, 40.

McGUIRE, E. J., 1969, *Phys. Rev.* **185**, 1.
McGUIRE, E. J., 1971, *Phys. Rev. A* **3**, 587.
McGUIRE, E. J., 1972, *Phys. Rev. A* **5**, 1043.
McGUIRE, E. J., 1974, *Phys. Rev. A* **9**, 1840.
McGUIRE, E. J., 1975, in *Atomic Inner-Shell Processes*, edited by B. Crasemann (Academic, New York), p. 293.
McGUIRE, E. J., 1976, *Phys. Rev. A* **14**, 1402.
McKENZIE, B. J., I. P. GRANT, AND P. H. NORRINGTON, 1980, *Comput. Phys. Commun.* **21**, 233.
MEHLHORN, W., AND W. N. ASAAD, 1966, *Z. Phys.* **191**, 231.
METCHNIK, V., AND S. G. TOMLIN, 1963, *Proc. Phys. Soc. London* **81**, 956.
MITTLEMAN, M. H., 1980, in Proc. Workshop on Foundations of the Relativistic Theory of Atomic Structure, Argonne National Laboratory report No. ANL-80-128, p. 27.
MITTLEMAN, M. H., 1981, *Phys. Rev. A* **24**, 1167.
MOHR, P. J., 1974, *Ann. Phys. (N. Y.)* **88**, 52.
MOHR, P. J., 1975, *Phys. Rev. Lett.* **34**, 1050.
MOHR, P. J., 1982, *Phys. Rev. A* **26**, 2338.
MOORE, C. E., 1949, in *Atomic Energy Levels*, Vol. 2; (1952), Vol. 2; (1958), Vol. 3. Nat. Bur. Stand. US Circ. 467.
OH, S. D., J. McENNAN, AND R. H. PRATT, 1976, *Phys. Rev. A* **14**, 1428.
OHNO, M., AND G. WENDIN, 1977, *Solid State Commun.* **24**, 75.
OHNO, M., AND G. WENDIN, 1978, *J. Phys. B* **11**, 1557.
OHNO, M., AND G. WENDIN, 1979, *J. Phys. B* **12**, 1305.
ÖKSÜZ, I., AND O. SINANOĞLU, 1969, *Phys. Rev.* **181**, 54.
PARENTE, F., M. H. CHEN, B. CRASEMANN, AND H. MARK, 1981, *At. Data Nucl. Data Tables* **26**, 383.
PAYNE, W. B., AND J. S. LEVINGER, 1956, *Phys. Rev.* **101**, 1020.
PRIOR, M. H., 1972, *Phys. Rev. Lett.* **29**, 611.
RACAH, G., 1942, *Phys. Rev.* **62**, 438.
RACAH, G., 1943, *Phys. Rev.* **63**, 367.
RAFFRAY, Y., 1966, *J. Phys. (Paris)* **27**, 657.
RAMANA, M. V., A. K. RAJAGOPAL, AND W. R. JOHNSON, 1982, *Phys. Rev. A* **25**, 96.
RAMBERG, E. G., AND F. K. RICHTMYER, 1937, *Phys. Rev.* **51**, 913.
RICHARD, P., 1975, in *Atomic Inner-Shell Processes*, edited by B. Crasemann (Academic, New York), p. 73.
RICHARD P., I. L. MORGAN, T. FURUTA, AND D. BURCH, 1969, *Phys. Rev. Lett.* **23**, 1009.
RICHARD, P., W. HODGE, AND C. F. MOORE, 1972, *Phys. Rev. Lett.* **29**, 393.
ROSE, S. J., I. P. GRANT, AND N. C. PYPER, 1978, *J. Phys. B* **11**, 1171.
ROSÉN, A., AND I. LINDGREN, 1968, *Phys. Rev.* **176**, 114.
ROSNER, H. R., AND C. P. BHALLA, 1970, *Z. Phys.* **231**, 347.
ROSSELAND, S. 1923, *Z. Phys.* **14**, 1972.
RUBINOWICZ, A., 1960, *Acta Physiol. Polon.* **19**, 21.
SALEM, S. I., S. L. PANOSSIAN, AND R. A. KRAUSE, 1974, *At. Data Nucl. Data Tables* **14**, 91, and references therein.
SAVIN, A., P. SCHWERDTFEGER, H. PREUSS, H. SILBERBACH, AND H. STOLL, 1983, *Chem. Phys. Lett.* **98**, 226.
SCHMIDT, V., N. SANDNER, W. MEHLHORN, M. Y. ADAM, AND F. WUILLEUMIER, 1977, *Phys. Rev. Lett.* **38**, 63.
SCHNEIDER, D., C. F. MOORE, AND B. M. JOHNSON, 1976, *J. Phys. B* **9**, L153.
SCHRECKENBACH, K., H. G. BÖRNER, AND J. P. DESCLAUX, 1977, *Phys. Lett. A* **63**, 330.
SCHUMANN, S., K. O. GROENEVELD, G. NOLTE, AND B. FRICKE, 1979, *Z. Phys. A* **289**, 245.
SCHWARZ, W. H. E., AND E. WECHSEL-TRAKOWSKI, 1982, *Chem. Phys. Lett.* **85**, 94.

SCOFIELD, J. H., 1969, *Phys. Rev.* **179**, 9.
SCOFIELD, J. H., 1974a, *At. Data Nucl. Data Tables* **14**, 121.
SCOFIELD, J. H., 1974b, *Phys. Rev. A* **9**, 1041.
SCOFIELD, J. H., 1974c, *Phys. Rev. A* **10**, 1507.
SCOFIELD, J. H., 1975, in *Atomic Inner-Shell Processes*, edited by B. Crasemann (Academic, New York), Vol. I.
SEVIER, K. D., 1972, *Low Energy Electron Spectroscopy* (Wiley, New York).
SEVIER, K. D., 1979, *At. Data Nucl. Data Tables* **24**, 323.
SHIRLEY, D. A., 1972, *Chem. Phys. Lett.* **17**, 312.
SHIRLEY, D. A., 1973, *Phys. Rev. A* **7**, 1520.
SHIRLEY, D. A., R. L. MARTIN, S. P. KOWALCZYK, F. R. MCFEELY, AND L. LEY, 1977, *Phys. Rev. B* **15**, 544.
SHORE, B. W., AND D. H. MENZEL, 1968, *Principles of Atomic Spectra* (Wiley, Nex York).
SIEGBAHN, K., C. NORDLING, A. FAHLMAN, R. NORBERG, K. HAMRIN, J. HEDMAN, G. JOHANSSON, T. BERGMARK, S. KARLSSON, I. LINDGREN, AND B. LINDBERG, 1967, ESCA-Atomic, Molecular and Solid State Structure Studied by Means of Electron Spectroscopy, *Nova Acta Regiae Soc. Sci Ups.* **4**, 20.
SIEGBAHN, K., C. NORDLING, G. JOHANSSON, J. HEDMAN, P. F. HEDÉN, K. HAMRIN, U. GELIUS, T. BERGMARK, L. O. WERME, R. MANNE, AND Y. BAER, 1969, *ESCA Applied to Free Molecules* (North-Holland, Amsterdam).
SINANOĞLU, O., AND K. A. BRUECKNER, 1970, *Three Approaches to Electron Correlation in Atoms* (Yale U. P., New Haven).
SOBEL'MAN, I. I., 1972, *Introduction to the Theory of Atomic Spectra* (Pergamon, Oxford).
SPARKS, C. J., 1974, *Phys. Rev. Lett.* **33**, 262.
STOLTERFOHT, N., F. J. DE HEER, AND J. VAN ECK, 1973a, *Phys. Rev. Lett.* **30**, 1159.
STOLTERFOHT, N., H. GOBLER, AND U. LEITHÄUSER, 1973b, *Phys. Lett* **45A**, 351.
STOLTERFOHT, N., D. SCHNEIDER, R. MANN, AND F. FOLKMANN, 1977, *J. Phys. B* **10**, L281.
SUCHER, J., 1980, *Phys. Rev. A* **22**, 348.
SWIRLES, B., 1935, *Proc. R. Soc. London, Ser. A* **152**, 625.
SWIRLES, B., 1936, *Proc. R. Soc. London, Ser. A* **157**, 680.
TALUKDAR, B., AND D. CHATTARJI, 1970, *Phys. Rev. A* **1**, 33.
TAYLOR, G. R., AND W. B. PAYNE, 1960, *Phys. Rev.* **118**, 1549.
TIETZ, F., 1963, *Ann. Phys. (Leipzig)* **10**, 340.
TOMLIN, S. G., 1964, *Australian J. Phys.* **17**, 452.
TUNNEL, T. W., C. CAN, AND C. P. BHALLA, 1979, *IEEE Trans. Nucl. Sci.* **NS-26**, 1124.
TUNNELL, T. W., AND C. P. BHALLA, 1981, *Phys. Lett.* **86A**, 13.
UEHLING, E. A., 1935, *Phys. Rev.* **48**, 55.
VAINSHTEIN, L. A., AND U. I. SAFRONOVA, 1978, *At. Data Nucl. Data Tables* **21**, 49.
WALTERS, D. L., AND C. P. BHALLA, 1971, *Phys. Rev. A* **3**, 519.
WEISS, A. W., 1976, *Beam-Foil Spectroscopy*, edited by I. A. Sellin and D. J. Pegg (Plenum Press, New York).
WELTON, T. A., 1948. *Phys. Rev.* **74**, 1157.
WENDIN, G., AND M. OHNO, 1976, *Phys. Scr.* **14**, 148.
WENTZEL, G., 1927, *Z. Phys.* **43**, 524.

CHAPTER 3

Many-Body Effects in Energetic Atomic Transitions

T. N. Chang

1. INTRODUCTION

When a target atom is bombarded with a beam of energetic photons, electrons, heavy charged particles, or other atoms, one or more of the atomic electrons from the target atom may be ejected or excited to different orbits. In many of these atomic processes, the quantitative study of the transition based on the straightforward independent-particle calculation, when compared with the experimental measurement, fails to provide a satisfactory physical interpretation. The failure of the independent-particle calculation suggests that the interaction to be considered should not be limited to the one between the incident particle and the ejected or excited electron alone. A more comprehensive quantitative estimation should also include interactions between (i) the ejected or the excited electron and the remaining electrons of the target atom, (ii) the incident particle and the remaining electrons of the target atom, and (iii) electrons within the target atom at various stages of the transition.

As the energy of the incident particle E_i exceeds the ionization threshold ε_I of the atomic innermost shell, the ejection of any inner-subshell electron will leave the atom in an unstable state with one (or more) inner-shell vacancies. The unstable atom can then decay through a radiationless process involving electron ejection, such as an Auger transition, or a radiative process involving emission of a photon. For heavy atoms, the decay of a K-shell vacancy can lead to new vacancies in the L, M, or higher shells which may decay by further radiative or radiationless processes. Following the ejection of electrons or the emission of photons, all vacancies will eventually reach the outermost subshell.

T. N. Chang • Physics Department, University of Southern California, Los Angeles, California 90089-0484.

For most atoms, during this relaxation process, the excessive energy released via photon emission is usually less (much less for lighter atoms) than that carried away by the ejected electrons. However, since the photons have a much longer mean free path than electrons in most media, it is a lot easier experimentally to detect photon emission than electron ejection. Most of the earlier work on inner-shell processes was indeed centered around the determination of the fluorescence yield ω_x which measures the probability that a vacancy in a specific atomic shell x decays through a radiative process. On the other hand, the radiationless process usually involves the change of electronic orbits of at least two electrons whereas photon emission can be interpreted approximately as the transition of a single electron, whence the study of the radiationless process almost always provides more detailed dynamical information on the atomic transition in a many-electron atom.

In this chapter, instead of presenting a historical survey or a comprehensive review, we will concentrate on a few selected works which have successfully led to the understanding of the role played by many-electron interactions in energetic atomic transitions. We will also limit most of our discussion to the nonrelativistic processes and leave those dominated by relativistic effects to other chapters in this volume. More comprehensive reviews of some of the subjects included in this chapter have been given by Bambynek *et al.* (1972) McGuire (1975), Krause (1975), and Mehlhorn (1976).

In Sec. 2, we discuss high-energy processes involving electrons from deep inner shells. A brief discussion of the postcollison interaction is presented. Examples are given to identify the important many-electron interactions which are responsible for the differences between experimental results and the independent-particle estimation for inner-shell binding energies, Auger rates, and the relative intensities for satellite lines. In Sec. 3, low-energy electron transitions from the first or second subshell next to the outermost (or valence) subshell are discussed. Quantitative assessment of the relative importance of the final-state inner-shell interaction and the initial-state configuration interaction is presented. Finally, in Sec. 4, we conclude with a brief discussion of some of the existing difficulties which will require further theoretical and experimental investigation.

2. HIGH-ENERGY PROCESSES: TRANSITIONS FROM INNER SHELLS

2.1. *Introduction*

When the energy of the incident particle is sufficiently high to ionize one of the deep inner-shell electrons, the observed ejected-electron spectrum is usually characterized by a primary electron with kinetic energy that varies with incident-particle energy, and subsequent Auger electrons with fixed energy. To illustrate the effect of the many-electron interaction in deep inner-shell processes, we limit our discussion to the simplest process with a photon as the incident particle.

Schematically, such an inner-shell process is usually dominated by two successive transitions,

$$h\nu + A \rightarrow A^{+}([nl]^{-1}) + e^{-}_{\text{ph}}(\varepsilon_{\text{ph}}) \quad \text{photoelectron emitted,} \tag{1}$$

$$A^{+}([nl]^{-1}) \rightarrow A^{++}([n_1 l_1, n_2 l_2]^{-1}) + e^{-}_{A}(\varepsilon_A) \quad \text{Auger electron emitted.} \tag{2}$$

The kinetic energy of the ejected photoelectron varies with photon energy, i.e.,

$$\varepsilon_{\text{ph}} = h\nu - \varepsilon_I(nl), \tag{3}$$

where $\varepsilon_I(nl)$ is the binding energy (or ionization energy) of the inner-shell nl electron. The energy of the Auger electron is independent of the photon energy and given by the difference between the initial ionic state and the final ionic state, i.e.,

$$\varepsilon_A = E(A^{+}([nl]^{-1})) - E(A^{++}([n_1 l_1, n_2 l_2]^{-1})). \tag{4}$$

These two transitions are known as the diagram transition (i.e., the normal transition). In an independent-particle description, the first transition is the promotion of one inner-shell electron to the continuum with transition amplitude

$$\langle \psi_{\text{ph}}(\mathbf{r}) | D(\mathbf{r}) | \chi_{nl}(\mathbf{r}) \rangle, \tag{5}$$

where ψ_{ph} and χ_{nl} are the independent-particle wavefunctions of the photoelectron and the initial orbital electron in the nl subshell. The photon interaction is approximated by the dipole operator $D(\mathbf{r})$. The Auger transition can be estimated by the Coulomb matrix

$$\langle \psi_a(n_1 l_1, n_2 l_2) | 1/r_{12} | \psi_b(nl, \varepsilon_A l') \rangle, \tag{6}$$

where ψ_a represents the two-particle wave function for the final two-hole state and ψ_b, the wave function for the combined state of inner-shell nl electron and ejected $\varepsilon_A l'$ Auger electron. Both ψ_a and ψ_b are antisymmetrized and usually given by the Slater-determinant wave function in an independent-particle approximation.

When the photon energy decreases to a value close to the ionization threshold of the inner-shell nl electron, the Auger decay can be significantly influenced by the presence of the slowly escaping primary photoelectron.

The transitions given by Eqs. (1) and (2) can no longer be considered as a two-step process. This is known as the postcollision interaction (see also Chaps. 4 and 9). One of the most important consequences of the postcollision interaction is the observation of the energy exchange between the slow primary photoelectron and the fast secondary Auger electron. Since the time of interaction is very short, the postcollision interaction can be treated by a sudden approximation such that the slow electron experiences a sudden transition from a potential due to the initial A^+ ionic field to a potential due to the final A^{++} ionic field. This sudden change in potential leads to a net energy loss for the slow electron and at the same time the same amount of energy gain by the Auger electron. The postcollision interaction can also lead to an exchange of angular momentum. This is most likely to be observed in electron-impact ionization (Heideman *et al.*, 1980), where the scattered electron and the ejected primary electron are both moving slowly and therefore are highly correlated. A more detailed discussion of the dynamics of the postcollision interaction can be found elsewhere (Schmidt, 1982).

In general, the effect to the many-electron interaction is measured by the deviation of the experimentally observed ionization energy (or binding energy), the cross section (or intensity), and the linewidth from the calculated values in the independent-particle approximation. It can also be examined by more complicated transitions such as

$$h\nu + \mathrm{A} \rightarrow \mathrm{A}^{+*}([n_1 l_1, n_2 l_2]^{-1}, n'l') + e^-, \tag{7a}$$

$$h\nu + \mathrm{A} \rightarrow \mathrm{A}^{++*}([n_1 l_1, n_2 l_2]^{-1}) + e_1^- + e_2^-, \tag{7b}$$

where two of the atomic electrons are either excited or ionized. The remaining ions can further decay via Auger transitions such as

$$\mathrm{A}^{+*}([n_1 l_1, n_2 l_2]^{-1}, n'l') \rightarrow \mathrm{A}^{++}([n_2 l_2, n_3 l_3]^{-1}) + e^-, \tag{8a}$$

$$\mathrm{A}^{++*}([n_1 l_1, n_2 l_2]^{-1}) \rightarrow \mathrm{A}^{3+}([n_1 l_1, n_3 l_3, n_4 l_4]^{-1}) + e^-. \tag{8b}$$

Experimentally, transitions (7a) and (8a), (8b) are characterized by the presence of satellite lines. Process (7b) usually leads to a continuous energy distribution between two simultaneously ionized electrons. Additional electrons may be ejected in transitions similar to (8a), (8b), which will lead to higher multiply charged ions, until all the vacancies eventually reach the outermost subshell.

2.2. *Binding Energy*

We first illustrate the importance of the many-electron interaction by examining the binding energy of the inner-shell electron nl_j as measured in the photoelectron spectrum (see also Chaps. 1, 2, and 5). For example, some of the measured binding energies of inner-shell s electrons are listed in Table I. These experimentally observed values are compared with theoretical energies from the best independent-particle relativistic Dirac–Hartree–Slater (or Dirac–Fock) self-consistent-field calculation (Huang *et al.*, 1976) which includes contributions from vacuum polarization, Breit energy, and the relaxation effect. The calculated values are seen to be systematically larger than the experimental ones.

To remove this discrepancy, Chen *et al.* (1981) suggested that an energy shift due to the interaction between the inner-shell hole state and the corresponding radiationless continua should be included in the quantitative estimation of the binding energy of the inner-shell electron. This energy shift of the initial hole state $|i\rangle$ is approximately given by the principal-value integral of the second-order perturbation energy correction from the interaction due to the final continuum $|f\rangle$, i.e.,

$$\Delta E = \sum_f p \int \frac{|V_{if(\varepsilon)}|^2}{E_i - (E_f + \varepsilon)}\, d\varepsilon, \tag{9}$$

where V_{if} is the Coulomb matrix similar to the one given in Eq. (6), and E_f is the energy threshold of the final-state continuum.

As demonstrated quantitatively by Chen *et al.* (1981), the level shifts for the $2s$ electron for an atom with $Z = 18$–92 is dominated by the contribution from the interaction with the $[LM]^{-1}\varepsilon l$ Coster–Kronig continua, because the $2s$ hole state is close to the thresholds of the Coster–Kronig continua with relatively large Coulomb matrix elements which are sensitive to variation of the energy. On the contrary, the contribution from the

TABLE I

Comparison of inner ns subshell binding energy (in eV) between experiments and the best independent-particle calculation of Huang *et al.* (1976).

Atom	Subshell	Theory (Huang *et al.*, 1976)	Experiment	Δ (theory − exp.)
Hg	$4s$	817.6	809.4[a]	8.2
Xe	$4s$	222.67	213.32[b]	9.25
Xe	$3s$	1152.4	1148.7[b]	3.7
Kr	$3s$	295.85	292.8[b]	3.05
Mg	$2s$	97.59	96.5[c]	1.09

[a] Svensson *et al.* (1976a).
[b] Svensson *et al.* (1976b).
[c] Mehlhorn *et al.* (1977).

$[M]^{-1}\varepsilon l$ Auger continua to the $2s$ shift is negligible as the $2s$-hole state is well separated from the continuum threshold and the Coulomb matrix elements are relatively small and not particularly sensitive to energy variation.

A more dramatic example is found in hole states which decay through super-Coster–Kronig transitions. In a qualitative theoretical study of the photoelectron spectrum of Xe in the $4s$–$4p$ region, Wendin and Ohno (1976) have pointed out that the interaction between the $[4p]^{-1}$ hole state and the $[d^2]^{-1}\varepsilon f$ super-Coster–Kronig continuum and its corresponding $[4d^2]^{-1}nf$ Rydberg state has led to complete smearing of the $[4p]^{-1}$ levels, with most of the $4p_{3/2}$ transition intensity picked up by a localized $[4d^2]^{-1}4f$ state.

2.3. *Intensity*

We now turn our attention to the effect of the many-electron interaction on the intensity of inner-shell transitions. We first examine the K-shell Auger process. For heavy atoms, the K-shell Auger transition is dominated by the relativistic effect. Systematic quantitative calculations based on the relativistic independent-particle (Dirac–Hartree–Slater) approximation in j–j coupling have been carried out by Chen *et al.* (1980a). Qualitatively, when only deep inner shells are involved in the transition, the calculated K-shell fluorescence yields and the intensity ratios $(K\text{–}LX)/(K\text{–}LL)$ and $(K\text{–}XY)/(K\text{–}LL)$ are either in excellent or good agreement with experimental values (Sevier, 1972). The $(K\text{–}L_3M_1)/(K\text{–}L_1M_1)$ intensity ratios for heavy atoms are also found to be in satisfactory agreement with the limited available experimental data (Sevier, 1972).

2.3.1. *Final-Ionic-State Configuration Interaction.*

For medium or light atoms, the $(K\text{–}LM)/(K-L_1M_1)$ intensity ratios [normalized to $I(K\text{–}L_1M_1)=1$] obtained in the same calculation by Chen *et al.* (1980a) are found to be consistently smaller than observed ratios. Similarly, in the K–LL transitions, the relative intensities for the $K\text{–}L_1L_2$, $K\text{–}L_2L_2$, $K\text{–}L_3L_3$, $K-L_1L_3$, and $K\text{–}L_3L_3$ $(^3P_0)$ transitions with respect to the $K\text{–}L_1L_1$ transition are also found to deviate from the experimental results for medium to light atoms ($Z<50$).

For the K–LL transitions, this difficulty was first considered by Asaad (1965) who showed that mixing between the $1s^22p^6(^1S)$ and $1s^22s^22p^4(^1S)$ configurations in the final ionic state, when included in the calculation, could considerably improve the agreement between calculated relative intensities and experiment for atoms ranging from $Z=12$ to 36. Using a similar theoretical approach, Chen *et al.* (1980b) have recently carried out a more comprehensive relativistic intermediate-coupling calculation including the $2s_{1/2}2s_{1/2}$, $2s_{1/2}2p_{1/2}$, $2s_{1/2}2p_{3/2}$, $2p_{1/2}\,2p_{3/2}$, and $2p_{3/2}2p_{3/2}$ configuration in the $J=0$, 1, and 2 final two-hole ionic states. Their calculated relative intensities for atoms ranging from $Z=18$ to 96 compare well with experimental results, with the exception of K–$L_2L_2(^1S_0)/K$–$L_1L_1(^1S_0)$ for light atoms ($Z \lesssim 30$). (See also Chap 2, Sec. 5, and Chap. 4, Sec. 5.)

2.3.2. Final-State Interchannel Interaction. For K–LM transitions in medium to light atoms and K–LL transitions in light atoms, the outermost subshell or the subvalence subshells (i.e., the first or second next to the outermost subshell) are involved in the Auger process. The final-ionic-state configuration interaction discussed earlier can only account for part of the difference between the quantitative estimation from an independent-particle calculation and the observed data. Unlike in high-Z atoms, where the final-state holes are screened by the outer-shell electrons and therefore interact weakly with the ejected Auger electron, for medium-to-light atoms the interaction between the K–LL and K–LM final-state holes and the outgoing Auger electron is subject only to little screening from electrons in the outermost subshell. Consequently, the final-state interaction between continuum channels is expected to influence the relative intensity, particularly for relatively weak transitions. This is best illustrated in the study of the K–LL spectrum of Ne and the K–LX spectra of Mg.

We first consider the K–LL transition for the Ne atom. In LS coupling, there are five different K–LL lines corresponding to the $1s^22p^6(^1S)$, $1s^22s2p^5(^1P)$, $1s^22sp^5(^3P)$, $1s^22s^22p^4(^1S)$, and $1s^22s^22p^4(^1D)$ final states. In the earlier nonrelativistic calculations (Asaad, 1965; Bhalla, 1973), only $1s^22p^6(^1S)$ and $1s^22s^22p^4(^1S)$ were included in the final-ionic-state configuration-interaction calculation, and therefore only the intensities of these two lines were affected. To study the effect due to other many-electron interactions, Kelly (1975) carried out a comprehensive many-body perturbation calculation (MBPT) which included all non-negligibly contributing terms up to second order. Kelly also estimated the contribution from some higher-order terms and included them in the calculated Auger rates for all five transitions.

The important final-state configurations included in Kelly's calculation are (i) $1s^22s^22p^4(^1D)\,kd$, (ii) $1s^22s^22p^4(^1S)\,ks$, (iii) $1s^22s2p^5(^{1,3}P)\,kp$, (iv) $1s^22p^6(^1S)\,ks$, and (v) and $1s2s^22p^5(^{1,3}P)\,kp$. The corresponding initial-state configurations included are (i) $1s2s^22p^4ll$ ($l=0$–3), (ii) $1s2s2p^5pd$, (iii) $1s2s2p^5sp$, (iv) $1s2p^6ll$ ($l=0$–2), and (v) $2s2p^6ll$ ($l=0$–2). In Table II we list all the important many-body interactions which Kelly found to contribute significantly to the correction of the Auger rate from the independent-particle Hartree–Fock calculation. For the strongest transition, K–$L_{2,3}L_{2,3}(^1D)$, the change of the Auger rate from the Hartree–Fock result is small. The main corrections are found to come from the final-state interchannel interaction due to the $1s^22s2p^5(^{1,3}P)\,kp$ configuration and the initial-state configuration interaction due to the $1s2s^22p^4ll$ configuration. For the K–$L_1L_1(^1S)$ transition, in addition to the final-ionic-state configuration

TABLE II
The important contributing many-electron interactions for K–LL transitions included in the MBPT calculation of Kelly (1975).

	Transition	Final state	Initial state
K–$L_{2,3}L_{2,3}(^1D)$	$(1s2s^22p^6 \rightarrow 1s2s^22p^4(^1D)\,kd)$	$1s^22s2p^5(^{1,3}p)\,kp$	$1s2s^22p^4ll$
K–$L_1L_1(^1S)$	$(1s2s^22p^6 \rightarrow 1s^22p^6(^1S)\,ks)$	$1s^22s^22p^4(^1S)\,ks$	
		$1s^22s2p^5(^{1,3}P)\,kp$	
K–$L_1L_{2,3}(^1P)$	$(1s2s^22p^6 \rightarrow 1s^22s2p^5(^1P)\,kp)$	$1s^22s^22p^4(^1D)\,kd$	
		$1s^22p^6(^1S)\,ks$	
K–$L_1L_{2,3}(^3P)$	$(1s2s^22p^6 \rightarrow 1s^22s2p^5(^3P)\,kp)$	$1s2s^22p^5(^{1,3}P)\,kp$	$1s2s2p^5ll'$
K–$L_{2,3}L_{2,3}(^1S)$	$(1s2s^22p^6 \rightarrow 1s^22s^22p^4(^1S)\,ks)$	$1s^22p^6(^1S)\,ks$	$1s2s^22p^4ll$
		$1s^22s^22p^4(^1D)\,kd$	
		$1s^22s2p^5(^{1,3}P)\,kp$	

interaction due to the $1s^22s^22p^4(^1S)$ ionic state considered in the earlier nonrelativistic calculation (Asaad, 1965; Bhalla, 1973), the final-state interchannel interaction due to the K–$L_1L_{2,3}(^{1,3}P)$ transitions also contribute strongly. The influence from the strong transition K–$L_{2,3}L_{2,3}(^1D)$ is not expected to be large as there is no second-order contributing term representing such final-state interchannel interaction. For the K–$L_1L_{2,3}(^1P)$ transition, the calculated Auger rate is strongly affected by the final-state interchannel interaction due to the K–$L_1L_1(^1S)$ and strong K–$L_{2,3}L_{2,3}(^1D)$ transitions. For the K–$L_1L_{2,3}(^3P)$ transition, unlike the ^1P transition, the corrections from the final-state interchannel interactions corresponding to the K–$L_1L_1(^1S)$ and K–$L_{2,3}L_{2,3}(^1D)$ transitions are close to zero from the near-complete cancellation due to symmetry. The main many-body corrections, instead, come from the final-ionic-state configuration interaction due to the mixing with the $1s2s^22p^5(^3P)$ configuration and the initial-state configuration interaction due to mixing with the $1s2s2p^5pd$ and $1s2s2p^5sp$ configurations. For the K–$L_{2,3}L_{2,3}(^1S)$ transition, the final-state interactions corresponding to all four other transitions are strong and contribute significantly to the Auger-rate calculation. In addition, significant correction is also found from initial-state mixing with the $1s2s^22p^4ll$ configuration. In addition to the work of Kelly, Howat *et al.* (1978) have developed an alternative approach which is slightly more convenient in its numerical application but includes only the final-state interactions, without an explicit procedure to estimate the many-electron contribution from the initial-state configuration interaction.

In Table III we summarize the theoretical calculated relative intensities of the $K-L_{2,3}L_{2,3}(^1D,\ ^1S)$ and K–$L_1L_{2,3}(^{1,3}P)$ transitions with respect

TABLE III

Theoretical and experimental relative and total K–LL Auger rates of Ne. The relative rates are normalized to the K–$L_1L_1(^1S)$ rate. The total rates are in units of 10^{-3} a.u. The experimental relative rates are from Krause *et al.* (1971); the experimental total rate is from Svensson *et al.* (1976b).

Transition	Kelly (1975) HF	Kelly (1975) MBPT	Howat *et al.* (1978)	Experiment
K–$L_1L_{2,3}(^1P)$	2.14	2.79	3.24	2.83
(^3P)	0.83	1.00	1.69	1.03
K–$L_{2,3}L_{2,3}(^1S)$	0.48	1.57	1.59	1.57
(^1D)	5.98	10.07	10.10	10.05
K–LL(Total)	9.91	8.05	8.98	9.9 ± 0.8

to the K–$L_1L_1(^1S)$ transition. The relative intensities calculated by Kelly (1975) are in excellent agreement with the measured values of Krause *et al.* (1971). The effect of the many-electron interaction to the K–LL transition is clearly demonstrated by the large differences found in the relative intensities between the Hartree–Fock and MBPT results. The difference between the results of Kelly (1975) and Howat *et al.* (1978) can be attributed to the lack of the initial-state configuration interaction in the later calculation. For example, in the K–$L_1L_{2,3}(^3P)$ transition, the correction due to $1s2s2p^5ll$ initial-state configuration mixing is comparable to the final-ionic-state configuration interaction due to the $1s2s^22p^5(^{1,3}P)\,kp$ continuum. Consequently, the relative intensity calculated by Howat *et al.* is significantly different from that of Kelly. On the other hand, although the initial-state configuration mixing from $1s2s^22p^4ll$ is not negligible in the K–$L_{2,3}L_{2,3}(^1S,\ ^1D)$ transitions, the many-body correction is dominated by the final-state interactions and therefore the difference in Auger rates between these two calculations is small. Despite the excellent agreement between theory and experiment for the relative intensities, the calculated total K–LL Auger rate is still found to be smaller than the experimental value (Svensson *et al.*, 1976b).

With the same theoretical procedure, Howat (1978) has extended his calculation to the K–LL and K–LM transitions in Mg. In Table IV we list his calculated relative intensities of various K–LL and K–LM transitions with respect to the dominant K–$L_{2,3}L_{2,3}(^1D)$ transition in three different stages of approximation: (A) independent-particle Hartree–Fock; (B) with all final-ionic-state configuration interactions included; (C) with all final-state interchannel interactions included. Again, the corrections due to

TABLE IV

Comparison between theoretical (Howat, 1978) and experimental (Breuckmann, 1979) relative K–LL and K–LM Auger rates of Mg. The relative rates are normalized to the K–$L_{2,3}L_{2,3}(^1D)$ rate. A: HF; B: HF + all final ionic state configuration interactions; C: HF + all final state interchannel interactions.

Transition	Theory (Howat, 1978) A	B	C	Experiment (Breuckmann, 1979)
K–$L_1L_1(^1S)$	0.183	0.132	0.091	0.092
L–$L_1L_{2,3}(^1P)$	0.400	0.396	0.315	0.295
(^3P)	0.161	0.158	0.137	0.134
K–$L_{2,3}L_{2,3}(^1D)$	0.080	0.111	0.131	0.153
K–L_1M_1	0.021	0.037	0.035	0.030
K–$L_{2,3}M_1$	0.028	0.034	0.035	0.036

the final-state interchannel interaction discussed earlier are responsible for the good agreement between theory and experiment in the quantitative study of the relative intensities of various K–LL and K–LM transitions.

2.4. *Satellites*

Study of the satellites in photoelectron and Auger-electron spectra should lead in principle to a more detailed understanding of the many-electron interaction in inner-shell atomic processes. However, only a few theoretical attempts have been made in this area in recent years. We limit our discussion to a few examples which may identify the direction for future theoretical effort.

2.4.1. Satellites in Photoelectron Spectra. The best-studied satellite spectrum associated with the ejection of an inner-shell photoelectron is the $1s$ spectrum of Ne. Experimentally, the photoelectron lines are represented by processes similar to those indicated by Eqs. (1) and (7a), i.e.,

$$h\nu + \mathrm{Ne}(1s^22s^22p^6) \rightarrow \begin{bmatrix} \mathrm{Ne}^+(1s2s^22p^6;\ ^2S) + e^-(kp); \quad \text{diagram line}, & (10a) \\ \mathrm{Ne}^{+*}(1s2s^22p^5(^{3,1}P)\, np;\ ^2S) + e^-(kp), & (10b) \\ \mathrm{Ne}^{+*}(1s2s2p^6(^{3,1}S)\, ns;\ ^2S) + e^-(kp). & (10c) \end{bmatrix}$$

The satellite intensity I_j for a specific final ionic state $|j\rangle$ is usually calculated by using the sudden approximation, i.e.,

$$I_j \sim |\langle \psi_i([1s]^{-1};\ ^2S) | \psi_j([1s, nl]^{-1} n'l';\ ^2S) \rangle|^2, \tag{11}$$

where ψ_i represents the wave function for the final ionic state $1s2s^2 2p^6(^2S)$ corresponding to the diagram line (10a), and ψ_j represents the wave function for the final ionic state with one of the $2s$ or $2p$ electrons promoted to the $n's$ or $n'p$ excited orbit. In an independent-particle calculation, ψ_i and ψ_j are given by single Slater determinants. The intensity ratio between the $(^3P)n'p$ and $(^1P)n'P$ satellite lines should be 3:1, which disagrees strongly with the experimental ratio of approximately 1 (Gelius, 1974).

To resolve this discrepancy, Krause *et al.* (1968) were first to suggest that the final-ionic-state configuration interaction between the $1s2s^2 2p^5(^3P)\,np(^2S)$ and $1s2s2p^5(^1P)\,np(^2S)$ configurations might be responsible for the deviation from the statistical ratio of 3. A more elaborate calculation was subsequently carried out by Martin and Shirley (1976) to examine the effect of various initial- and final-state configuration interactions to the relative intensities of Ne $1s$ satellites. In their calculation, Martin and Shirley first employed optimized multiconfiguration Hartree–Fock (MCHF) wave functions, as suggested by Krause *et al.* (1968), for each state in the intensity estimation. They also removed the nonorthogonality effect by successively applying the Schmidt orthogonalization procedure to the MCHF wave function for each state. As listed in Table V (second and third column), the calculated intensities relative to the $1s$ diagram line are still substantially different from the experimental values. Having failed to achieve a satisfactory theoretical interpretation of the experimental results, Martin and Shirley suggested an alternative approach, expanding the final ionic state to include configurations that correspond to all excited orbits of interest. As for the initial state, in addition to the $1s^2 2s^2 2p^6$ Hartree–Fock

TABLE V

Relative intensity (in %) of the Ne $1s2s^2 2p^5(^{3,1}P)\,np$ satellite lines with respect to the $1s$ diagram line. The theoretical results are from Martin and Shirley (1976).

	Theory			
State	MCHF	Orthogonalized MCHF	Configuration interaction	Experiment (Gelius, 1974)
$(^3P)\,3p$	9.25	3.35	2.47	3.15
$(^1P)\,3p$	5.45	1.97	2.60	3.13
$(^3P)\,4p$	2.31	0.74	1.48	2.02
$(^1P)\,4p$	0.89	0.36	0.70	0.90
$(^3P)\,4p$	0.95	0.29	0.43	0.42
$(^1P)\,5p$	0.31	0.13	0.11	0.17

configuration, they included both singly excited $1s^22s^22p^5np$ and doubly excited $1s^22s^22p^4np^2$ and $1s^22p^6ns^2$ configurations in their wave functions. With such extended configuration mixing, both in the initial and final states, they were able to obtain about 80% of the experimental relative intensities for $1s2s^22p^5(^{1,3}P)np$ satellite lines (cf. Table V, column 4).

A qualitative study of the satellite intensities associated with the subvalence subshells, as in the $2s$ and $2p$ photoelectron spectra of Ar, has also been carried out by Dyall *et al.* (1981) and Dyall and Larkins (1982).

2.4.2. Satellites in the Auger Spectrum. The satellites associated with the Auger process are more complicated to analyze than those in the photoelectron spectrum. Most of the existing theoretical studies are more qualitative than quantitative in nature. Here we only discuss briefly the study of the satellites associated with the $L_{2,3}$–MM Auger transitions. Schematically, the $L_{2,3}$–MM diagram lines and their associated satellites are represented by the transitions

$$A^+(1s^22s^22p^53s^23p^6\cdots) \rightarrow A^{++*}(1s^22s^22p^63s3p^5(^{1,3}P)\cdots) + e^-, \tag{12a}$$

$$A^+(1s^22s^22p^53s^23p^6\cdots) \rightarrow A^{++*}(1s^22s^22p^63s^23p^3{}^{3nd}_{ns}(^{1,3}P)\cdots) + e^-. \tag{12b}$$

For light atoms, when the final-state configuration interaction between the $3s3p^5(^{1,3}P)$ and $3s^23p^3nd/ns(^{1,3}P)$ is strong, a noticeable amount of the intensity from the diagram lines can be transferred to the satellites which are identified by final states with $3s^23p^3nd(^{1,3}P)$ and $3s^23p^3ns(^{1,3}P)$ configurations. A theoretical attempt to identify the line positions and estimate the intensities has been carried out by Dyall and Larkins (1981) in a final-state multiconfiguration calculation.

3. LOW-ENERGY PROCESSES: TRANSITIONS FROM SUBVALENCE SUBSHELLS

In this section, we discuss the effect of many-electron interactions on atomic processes involving electrons from the first or second inner subshell from the valence subshell. We first consider weak inner-subshell transitions in the presence of strong outer-shell transitions. Then we discuss strong inner-subshell transitions accompanied by weak outer-shell transitions.

3.1. Weak Inner-Shell Transitions in the Presence of Strong Outer-Shell Transitions

To illustrate the importance of many-electron interactions in a weak inner-shell transition in the presence of strong outer-shell transitions, we examine one of the simplest cases, e.g., the photoionization of the first inner subshell $2s$ of Ne,

$$Ne(1s^2 2s^2 2p^6) + h\nu \rightarrow Ne^+(1s^2 2s 2p^6) + e^-. \quad (13)$$

At photon energy above the $2s$ ionization threshold, three transitions, $2s \rightarrow k_{2s}p$, $2p \rightarrow k_{2p}d$, and $2p \rightarrow k_{2p}s$, contribute to the total photoionization cross section. Near the $2s$ ionization threshold, the observed partial cross section for the strong outer-shell transition $2p \rightarrow k_{2p}d$ is about one order of magnitude greater than that of the weak transitions $2p \rightarrow k_{2p}s$ and $2s \rightarrow k_{2s}p$. The partial cross section for the $2s \rightarrow k_{2s}p$ transition calculated in the independent-particle approximation with single-configuration Hartree–Fock initial- and final-state wave functions is about twice the experimental observed value. A substantial difference between the dipole-length and dipole-velocity calculations is also found in the independent-particle calculation, as shown in Fig. 1.

As pointed out earlier, the outer-shell $2p \rightarrow kd$ transition is the dominant process in this spectral region; it is therefore expected that an accurate theoretical estimation for the $2s \rightarrow kp$ partial cross section should also include the effect due to the $2p \rightarrow kd$ transition. Quantitatively, this effect can be estimated by including in the final-state wave function an additional con-

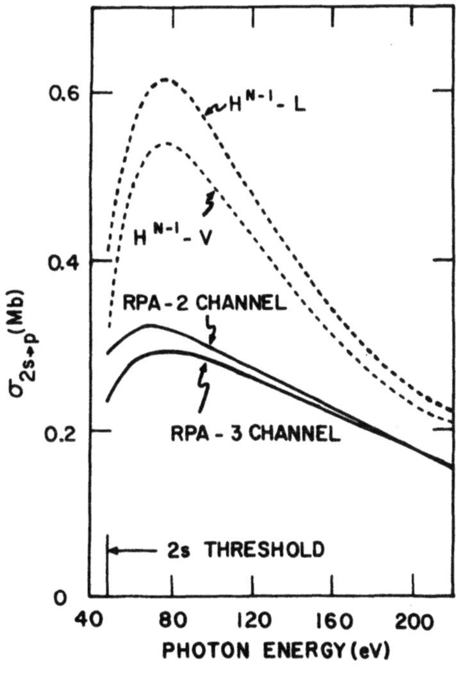

FIGURE 1. Partial photoionization cross section of the $2s \rightarrow k_{2s}p$ transition in Ne (Chang and Olsen, 1981).

figuration $1s^22s^22p^5k_{2p}d$ to the Hartree–Fock $1s^22s2p^6k_{2s}p$ configuration. Similarly, the effect due to the weak outer-shell $2p \rightarrow ks$ transition can also be included by adding the $1s^22s^22p^5k_{2p}s$ configuration in the final-state wave function.

Numerically, these final-state intershell interactions together with the pertaining initial-state configuration interactions can be estimated by solving a set of coupled integrodifferential equations, i.e.,

$$[H_{l=1}^{N-1}(r) - (\varepsilon_{2s} + E)]\ \psi_p(r) = f_p(\phi_p, \psi_d, \phi_d, \psi_s, \phi_s; r) \tag{14a}$$

$$[H_{l=1}^{N-1}(r) - (\varepsilon_{2s} - E)]\ \phi_p(r) = f_p(\psi_p, \phi_d, \psi_d, \phi_s, \psi_s; r) \tag{14b}$$

$$[H_{l=2}^{N-1}(r) - (\varepsilon_{2p} + E)]\ \psi_d(r) = f_d(\phi_d, \psi_p, \phi_p, \psi_s, \phi_s; r) \tag{14c}$$

$$[H_{l=2}^{N-1}(r) - (\varepsilon_{2p} - E)]\ \phi_d(r) = f_d(\psi_d, \phi_p, \psi_p, \phi_s, \psi_s; r) \tag{14d}$$

$$[H_{l=0}^{N-1}(r) - (\varepsilon_{2p} + E)]\ \psi_s(r) = f_s(\phi_s, \psi_p, \phi_p, \psi_d, \phi_d; r) \tag{14e}$$

$$[H_{l=0}^{N-1}(r) - (\varepsilon_{2p} - E)]\ \phi_s(r) = f_s(\psi_s, \phi_p, \psi_p, \phi_d, \psi_d; r), \tag{14f}$$

where ε_{2s} and ε_{2p} are the Hartree–Fock orbital energy eigenvalues and E is the photon energy. The effective single-particle radial Hamiltonian H_l^{N-1} for the ^1P dipole transition from nl_i to $k_{nl_i}l$ is given by

$$H_l^{N-1}(r)\, h_l(r) = H_l^N(r)\, h_l(r) - (1 - P_l)\Bigg[\tfrac{2}{3}(l \,\|\, V^l(\chi_{nl_i}, h_l; r) \,\|\, l_i)\, \chi_{nl_i}(r)$$

$$+ \sum_k (-1)^k \begin{Bmatrix} l & k & l \\ l_i & 1 & l_i \end{Bmatrix} (l \,\|\, V^k(\chi_{nl_i}, \chi_{nl_i}; r) \,\|\, l)\, h_l(r)\Bigg], \tag{15}$$

where

$$(l \,\|\, V^k(a, b; r) \,\|\, l') = (l \,\|\, C^{[k]} \,\|\, l')(l_a \,\|\, C^{[k]} \,\|\, l_b) \int_0^\infty ds\, a(s)\, b(s)\, \frac{r_<^k}{r_>^{k+1}}. \tag{16}$$

The Hartree–Fock radial Hamiltonian H_l^N is defined elsewhere (Chang and Fano, 1976) and the matrix elements of $C^{[k]}$ follow the convention of Edmonds (1957). The detailed expression for the coupling terms which represent the final-state intershell interactions and the ground-state configuration interactions are given in our previous work (Chang, 1978).

The effective wave functions thus generated are orthogonal to the occupied Hartree–Fock ground-state orbital wave functions, with the help of the projection operator P_l. The energy values $\varepsilon_{2s} + E$ and $\varepsilon_{2p} + E$ can be replaced by k_{2s}^2 and k_{2p}^2, respectively. The photoelectron momenta k_{2s} and k_{2p} of the ionized electrons from the $2s$ and $2p$ subshells are related by the equation

$$k_{2s}^2 - k_{2p}^2 = \varepsilon_{2s} - \varepsilon_{2p}. \tag{17}$$

The coupled integrodifferential Eqs. (14) are solved with the boundary conditions

$$\psi_l(r) \xrightarrow[r\to\infty]{} A_l \sin\left[kr + \frac{1}{k}\ln(2kr) - \frac{l\pi}{2} + \delta_l\right], \tag{18a}$$

$$\phi_l(r) \xrightarrow[r\to\infty]{} 0, \tag{18b}$$

$$\psi_l(r) \xrightarrow[r\to 0]{} r^{l+1}, \tag{18c}$$

and

$$\phi_l(r) \xrightarrow[r\to 0]{} r^{l+1}. \tag{18d}$$

The solution is normalized so that the coefficient A_l equals unity for the transition $nl_i \to k_{nl_i} l$.

Physically, the effective wave functions ϕ_l and ψ_l can be identified with various initial- and final-state configurations. As illustrated in Table VI, ψ_p, ψ_d, and ψ_s represent the wave functions corresponding to the ionized electron in the $k_{2s}p$, $k_{2p}d$, and $k_{2p}s$ states, respectively, and ϕ_p, ϕ_d, and ϕ_s represent the initial-state wave function correction corresponding to the $1s^2 2p^6 pp$, $1s^2 2s^2 2p^4 dd$, and $1s^2 2s^2 2p^4 ss$ configurations, respectively. The many-electron interactions such as the initial-state configuration interactions and the final-state inner-shell interactions are included explicitly through the coupling terms in Eqs. (14) as discussed earlier. The effective wave functions obtained in this procedure can then be used to evaluate the partial photoionization cross section. A more detailed description has been given elsewhere (Chang, 1978). The theoretical procedure outlined above is

TABLE VI

Identification of the effective wave function with various initial- and final-state configurations.

	Configuration	Effective wave function
Initial state (^1S)	$1s^2 2p^6 pp$	ϕ_p
	$1s^2 2s^2 2p^4 dd$	ϕ_d
	$1s^2 2s^2 2p^4 ss$	ϕ_d
Final state (^1P)	$1s^2 2s 2p^6 k_{2s} p$	ψ_p
	$1s^2 2s^2 2p^5 k_{2p} d$	ψ_d
	$1s^2 2s^2 2p^5 k_{2p} s$	ψ_s

equivalent to the usual RPAE (random phase approximation with exchange) calculation (Chang and Fano, 1976).

To examine the effect of the final-state intershell interaction due to the strong outer-shell $2p \rightarrow k_{2p}d$ transition, one needs only to solve the first four of Eqs. (14) by setting ψ_s and ϕ_s equal to zero. (This will be referred to as two-channel RPA calculation.) As shown in Fig. 1, the final-state intershell interaction due to the strong $2p \rightarrow k_{2p}d$ transition results in a substantial reduction of the $2s$ partial cross section. Also, the difference between the dipole-length and dipole-velocity calculation is reduced to about 2%–3%. The ground-state configuration interaction which dominates the outer-shell $2p \rightarrow k_{2p}d$ transition (Chang, 1978) plays a secondary role in this process. The effect due to the weaker outer-shell transition $2p \rightarrow ks$ can be estimated by expanding the numerical calculation to include all six of Eqs. (14). As shown in Fig. 1, such a calculation, referred to as three-channel RPA calculation, yields a smaller but still noticeable reduction of the $2s$ partial cross section. Direct comparison of the ratio between the partial cross sections of $2s$ and $2p$ ionization shows good agreement between the theoretical result and experimental measurements (Chang and Olsen, 1981). Similar but more dramatic effects are found for heavier rare-gas atoms (Chang, 1978; Johnson and Cheng, 1979).

3.2. Strong Inner-Shell Transitions with Weak Outer-Shell Transitions

We now turn our attention to a different class of inner-shell processes in which the outer-shell transition is weak compared with the inner-shell transition. A systematic study of this type of process is made possible by the recent experimental study (Bizau *et al.*, 1981; Wuilleumier, 1982) of the photoelectron spectrum of the inner subshells of Na with its valence electron excited by a strong laser field into various higher orbits. Again, we examine one of the simplest processes which involves the photoionization of the inner $2p$ electron of Na with its outer-shell valence electron at various stages of excitation, i.e.,

$$\mathrm{Na}^*(1s^2 2s^2 2p^6 n_e l_e) + h\nu \rightarrow \mathrm{Na}^+(1s^2 2s^2 2p^5 n_e l_e) + e^-, \tag{19}$$

where $n_e l_e$ denotes the outer-shell electron orbit. Since the $2s \rightarrow k_{2s}p$ transition is weak compared with the $2p \rightarrow k_{2p}d$ transition, we shall limit out discussion to the $2p$ ionization only.

With few minor modifications, a quantitative procedure similar to that outlined in the first part of this section can be employed to analyze the effect of the many-electron interaction on the ionization cross section. First, the Hartree–Fock radial Hamiltonian H_l^N is modified by adding a static-with-exchange potential term due to the excited outer-shell $n_e l_e$ electron, i.e.,

$$V_{ex}(r) f_l(r) = [l_e, l]^{-1/2} (l \| V^0(\chi_{n_e l_e}, \chi_{n_e l_e}; r) \| l) f_l(r) - [l]^{-1} \sum_k (-1)^k (l \| V^k(\chi_{n_e l_e}, f_l; r) \| l_e) \chi_{n_e l_e}(r), \quad (20)$$

where $[l] = 2l + 1$. With photon energy limited to the region between $2s$ and $2p$ ionization thresholds, the effective wave functions ψ_p and ϕ_p in Eqs. (14) can be set to zero in the numerical calculation. Second, since the outer-shell $n_e l_e \to k(l_e \pm 1)$ transition is at least one order of magnitude weaker than the inner-shell $2p \to k_{2p} d$ transition, it is more convenient to estimate the final-state intershell interaction between these transitions in a perturbation calculation. A more detailed discussion of the calculation is given elsewhere (Chang and Kim, 1982a).

Several conclusions can be drawn from the quantitative calculation outlined above for $2p$ ionization. (i) For the $2p \to k_{2p} d$ transition, the independent-particle Hartree–Fock calculation with ψ_s, ϕ_s, and ϕ_d set equal to zero yields a partial cross section about 15%–25% below the value calculated with many-electron interactions included. (ii) For the $2p \to k_{2p} s$ transition, the independent-particle Hartree–Fock calculation yields a partial cross section about 25%–35% below the calculated value with many-electron interactions included. (iii) As in the outer-shell $np \to kd$ transition in rare-gas atoms, the initial-state configuration interaction due to the

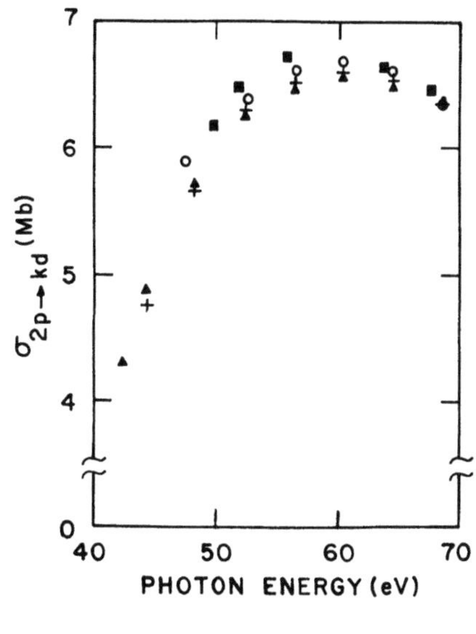

FIGURE 2. Calculated photoionization cross section of the $2p \to kd$ transition in Na initially in (i) the ground state (▲), (ii) the excited state with valence electron in $3p$ orbit (+), (iii) the excited state with valence electron in $3d$ orbit (○), and (iv) the ground ionic state (■) (Chang and Kim, 1982b).

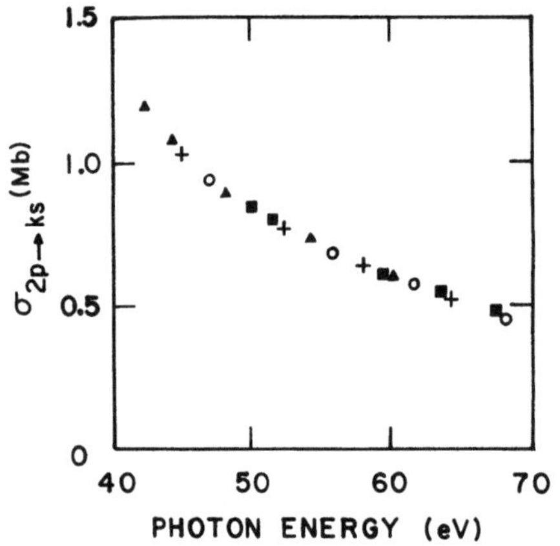

FIGURE 3. Calculated photoionization cross section for the $2p \rightarrow ks$ transition for Na initially in (i) the ground state (▲), (ii) the excited state with valence electron in $4p$ orbit (○), (iii) the excited state with valence electron in $3d$ orbit (+), and (iv) the ground ionic state (■) (Chang and Kim, 1982b).

(np^4dd) configuration dominates the partial cross section correction to the independent-particle result for the inner-shell $2p \rightarrow k_{2p}d$ transition in the present case. The contribution from the final-state intershell interaction due to the weak $2p \rightarrow k_{2p}s$ transition is negligible. (iv) For the weak $2p \rightarrow k_{2p}s$ transition, the final-state intershell interaction due to the $2p \rightarrow k_{2p}d$ strong transition is large and comparable to the initial-state configuration interaction due to the $(2p^4ss)$ configuration. (v) The final-state intershell interaction due to the outer-shell $n_el_e \rightarrow k(l_e \pm 1)$ transition yields a correction of less than 2% to the partial cross section for the inner-shell $2p$ ionization.

In addition, the calculated partial cross section for $2p$ ionization at fixed photon energy is found to be practically independent of the excited orbit occupied by the valence electron. This is clearly demonstrated in Fig. 2 and 3, where the partial cross sections for the $2p \rightarrow k_{2p}d$ and $2p \rightarrow k_{2p}s$ transitions for Na atom initially in (a) the ground state, (b) the excited state with valence electron in $3p$ orbit, (c) the excited state with valence electron in $3d$ orbit, and (d) the ground ionic state are found to be within 5% of each other. Again, similar conclusions can also be drawn for heavier alkali atoms (Chang and Kim, 1982b).

4. CONCLUDING REMARKS

With specific examples, we have examined in detail the importance of many-electron interactions in atomic inner-shell transitions. It becomes clear that for medium to light atoms, many-body effect play an extremely important part, both in the physical interpretation and the quantitative estimation of inner-shell atomic processes. Several important aspects have emerged from the discussion.

First, it is necessary to treat both the initial- and final-state many-electron interactions in a consistent theoretical procedure if an accurate physical interpretation is to be established. This is particularly evident in the *K–LL* Auger transition rate and the Ne 1*s* satellite-intensity calculation. Second, the intensity transfer from the strong transitions often dominates the transition amplitude of the weak transitions if configuration mixing is allowed. The only exception is when symmetry considerations limit the mixing, as seen in the Ne $K\text{–}L_1L_1(^1S)$ and $K\text{–}L_1L_{2,3}(^3P)$ intensity calculations. Third, the final-state interactions between continuum channels are important when the outermost or near-outermost subshell electrons are involved in the transition. This effect is significant even when the transitions corresponding to both continua are relatively weak. Fourth, the main theoretical difficulty remains in the final-state coupling between configurations corresponding to various continuum channels, although the procedure outlined in Sec. 3 seems to be effective in dealing with such final-state interchannel interactions. The relativistic counterpart (i.e., the relativistic random-phase approximation or RRPA) developed by Johnson and Lin (1979) and Johnson and Cheng (1979) is also found to be effective as discussed by Johnson and Cheng in Chap. 1 of this volume.

We conclude with a brief discussion of the recent investigation of the photoionization of 4*f* electrons in Hg, where a surprisingly large difference is found to exist between the experimental measurement and the theoretical calculation. At photon energies above 200 eV, the partial photoionization cross sections for all three subshells, 5*d*, 5*p*, and 4*f*, are relatively small and comparable in magnitude. On the theoretical side (Johnson and Radojevic, 1982), the calculation was carried out in the relativistic random phase approximation (an approach found to be extremely successful in many other applications), which includes explicitly both initial- and final-state configurations corresponding to the 5*d*, 5*p*, and 4*f* ionization channels. On the experimental side (Kobrin *et al.*, 1983), the partial cross sections are determined to within 30% by normalizing the photoelectron intensity to the 5*d* partial cross section of 8 Mb at 70 eV. Despite the good agreement between theory and experiment in the spin–orbit branching ratios and the asymmetry parameters for the 5*d*, 5*p*, and 4*f* photoelectrons at photon energy below 200 eV, the experimentally determined 4*f* partial cross section is found to be 3 to 4 times larger than the theoretical estimation from the RRPA calculation at energies above 200 eV. More theoretical and experimental investigations in the intermediate-energy region are clearly needed for a better understanding of the many-electron interactions in inner-shell atomic processes.

ACKNOWLEDGMENTS

This work is supported in part by NSF under grants No. PHY80-09146 and PHY84-08333. The author would also like to acknowledge support from the National Science Council during his sabbatical leave at the National Central University in Taiwan in 1982–1983.

REFERENCES

ASAAD, W. N., 1965, *Nucl. Phys.* **66**, 494.

BAMBYNEK, W., B. CRASEMANN, R. W. FINK, H.-U. FREUND, H. MARK, C. D. SWIFT, R. E. PRICE, AND P. V. RAO, 1972, *Rev. Mod. Phys.* **44**, 716.

BHALLA, C. P., 1973, *Phys. Lett.* **A44**, 103.

BIZAU, J. M., J. L. LEGOUËT, D. EDERER, P. KOCH, F. WUILLEUMIER, J. L. PICQUÉ, AND P. DHEZ, 1981, in XIIth Int. Conf. Phys. Electronic Atomic Collisions, Abs. Post Deadline Papers, edited by S. Datz, p. 1.

BREUCKMANN, B., 1979, *J. Phys. B* **12**, L609.

CHANG, T. N., 1978, *Phys. Rev. A* **18**, 1448.

CHANG, T. N., AND U. FANO, 1976, *Phys. Rev. A* **13**, 263.

CHANG, T. N., AND T. OLSEN, 1981, *Phys. Eev. A* **23**, 2394.

CHANG, T. N., AND Y. S. KIM, 1982a, *J. Phys. B* **15**, L835.

CHANG, T. N., AND Y. S. KIM, 1982b, in *X-Ray and Atomic Inner-Shell Physics—1982*, edited by B. Crasemann, AIP Conf. Proc. No. 94 (American Institute of Physics, New York), p. 633.

CHEN, M. H., B. CRASEMANN, AND H. MARK, 1980a, *Phys. Rev. A* **21**, 436.

CHEN, M. H., B. CRASEMANN, AND H. MARK, 1980b, *Phys. Rev. A* **21**, 442.

CHEN, M. H., B. CRASEMANN, AND H. MARK, 1981, *Phys. Rev. A* **24**, 1158.

DYALL, K. G., AND F. P. LARKINS, 1981, in *Inner-Shell and X-Ray Physics of Atoms and Solids*, edited by D. J. Fabian, H. Kleinpoppen, and L. M. Watson (Plenum Press, New York), p 319.

DYALL, K. G., AND F. P. LARKINS, 1982, *J. Phys. B* **15**, 1021.

DYALL, K. G., F. P. LARKINS, K. D. BOMBEN, AND T. D. THOMAS, 1981, *J. Phys. B* **14**, 2551.

EDMONDS, A. R., 1957, *Angular Momentum in Quantum Mechanics* (Princeton U. P., New Jersey), p. 76.

GELIS, U., 1974, *J. Electron. Spectrosc.* **5**, 985.

HEIDEMAN, H. G. M., W. VAN DER WATER, AND L. J. M. VON MOERGESTEL, 1980, *J. Phys. B* **13**, 2801.

HOWAT, G., 1978, *J. Phys. B* **11**, 1589.

HOWAT, G., T. ÅBERG, AND O. GOSCINSKI, 1978, *J. Phys. B* **11**, 1575.

HUANG, K.-N., M. AOYAGI, M. H. CHEN, B. CRASEMANN, AND H. MARK, 1976, *At. Data Nucl. Data Tables* **18**, 243.

JOHNSON, W. R., AND K. T. CHENG, 1979, *Phys. Rev. A* **20**, 978.

JOHNSON, W. R., AND C. D. LIN, 1979, *Phys. Rev. A* **20**, 964.

JOHNSON, W. R., AND V. RADOJEVIC, 1982, *Phys. Lett.* **92A**, 75.

KELLY, H. P., 1975, *Phys. Rev. A* **11**, 556.

KOBRIN, P. H., P. A. HEIMANN, H. G. KERKHOFF, D. W. LINDLE, C. M. TRUESDALE, T. A. FERRETT, U. BECKER, AND D. A. SHIRLEY, 1983, *Phys. Rev. A* **27**, 3031.

KRAUSE, M. O., 1975, in *Atomic Inner-Shell Process* edited by B. Crasemann (Academic, New York), Vol. II, p. 33.

KRAUSE, M. O., T. A. CARLSON, AND R. D. DISMUKES, 1968, *Phys. Rev.* **170**, 37.

KRAUSE, M. O., T. A. CARLSON, AND W. E. MODDEMAN, 1971, *J. Phys. (Paris)* **32**, C4-139.

MARTIN, R. L., AND D. A. SHIRLEY, 1976, *Phys. Rev. A* **13**, 1475.

MCGUIRE, E. J., 1975, in *Atomic Inner-Shell Processes*, edited by B. Crasemann (Academic, New York), Vol. I, p. 293.

MEHLHORN, W., 1976, in *Photoionization and Other Probes of Many-Electron Interactions*, edited by F. J. Wuilleumier (Plenum Press, New York), p. 309.

MEHLHORN, W., B. BREUCKMANN, AND D. HAUSAMANN, 1977, *Phys. Scr.* **16**, 177.

SCHMIDT, V., 1982, in *X-Ray and Atomic Inner-Shell Physics—1982*, edited by B. Crasemann AIP Conf. Proc. No. 94 (American Institute of Physics, New York), p. 544.

SEVIER, K. D., 1972, *Low-Energy Electron Spectroscopy* (Wiley, New York).

SVENSSON, S., M. MÅRTENSSON, E. BASILIER, P. A. MALMQUIST, U. GELIUS, AND K. SIEGBAHN, 1976a, *J. Electron. Spectrosc.* **9**, 51.

SVENSSON, S., M. MÅRTENSSON, E. BASILIER, P. A. MALMQUIST, U. GELIUS, AND K. SIEGBAHN, 1976b, *Phys. Scr.* **14**, 141.

WENDIN, G., AND M. OHNO, 1976, *Phys. Scr.* **14**, 148.

WUILLEUMIER, F. J., 1982, in *X-Ray and Atomic Inner-Shell Physics*—1982, edited by B. Crasemann, AIP Conf. Proc. No. 94 (American Institute of Physics, New York), p. 615.

CHAPTER 4

Auger-Electron Spectrometry of Core Levels of Atoms

W. Mehlhorn

1. INTRODUCTION

Although the Auger effect was discovered experimentally by P. Auger in 1925 (Auger, 1925) and has been theoretically interpreted by G. Wentzel (1927) as a nonradiative transition, the first systematic experimental and theoretical investigations started only in the mid-1950s. On the experimental side this became possible due to the development of high-resolution β-ray spectrometers which, besides for β-ray spectroscopy, were also used for the study of Auger transitions following internal conversion or electron capture in radioactive decay. On the theoretical side the development was mainly due to the availability of fast electronic computers (for a historic sketch see Crasemann, 1973).

Later in the mid 1960s, high-resolution electrostatic energy analyzers and a new technique, the use of gaseous targets and the ionization by an external beam of electrons or photons, were introduced to the field of Auger-electron spectrometry (AES). The application of AES to surface physics and the study of ion–atom collisions by means of AES brought again strong impetus to this field. Today AES is a powerful tool for studying the structure and the dynamics (e.g., energies, transition probabilities, and electron correlation) of atoms and molecules and investigating collision processes where the degree of inner-shell ionization plus simultaneous outer-shell ionization is probed via the Auger transitions.

In the present chapter AES will be discussed only in connection with *free* atoms and with main emphasis on the study of the structure and the dynamics of the atoms. In Sec. 2 the perturbation-theoretical treatment of

W. Mehlhorn • Fakultät für Physik, Universität Freiburg, D-7800 Freiburg, Federal Republic of Germany.

the Auger effect and an important extension beyond it (Åberg, 1976; Åberg and Howat, 1982) is sketched. In Sec. 3 the experimental arrangement for AES of free atoms is briefly discussed. Next, we discuss diagram Auger transitions (Sec. 4) and satellite Auger transitions (Sec. 5), the latter being caused by electron correlation. In Secs. 6 and 7 we present examples of satellite Auger spectra of multiply ionized target and projectile atoms, respectively. Finally, in Sec. 8 we discuss the nonisotropic emission of Auger electrons and in Sec. 9 the effects of the postcollision interaction in inner-shell ionization on the Auger spectrum. The spin polarization of Auger electrons of free atoms recently predicted by Klar (1980), Kabachnik (1981), and Huang (1982) is not treated in this contribution.

Several monographs (Burhop, 1952; Burhop and Asaad, 1972; Carlson, 1975; Chattarji, 1976; Åberg and Howat, 1982) and review articles (Åberg, 1976; Mehlhorn, 1976, 1978, 1982, 1983a, 1983b; Weightman, 1982) have been devoted to various aspects of the Auger effect, and most recently Åberg and Howat (1982) went beyond Wentzel's original perturbation-theory approach by treating the Auger effect in the broader context of resonance scattering theory.

2. THEORY OF AUGER TRANSITIONS AND BASIC CONSIDERATIONS

2.1. *Definitions and Notation*

A quasidiscrete state embedded in the continuum of the next-higher charge state of the system is called an Auger or autoionizing state. It can decay either by electron emission (nonradiative transition) or by emission of radiation. The former decay mode is by far the dominant one unless the transition energy is large (>5 keV) or selection rules hinder the nonradiative transition. An Auger state is created by inner-shell ionization, an autoionizing state by excitation of one (inner-shell) electron or of several electrons. By inner-shell (core) electrons all electrons except the outermost are meant. Although there is no fundamental difference between the decay of an Auger and an autoionizing state, in the following only Auger transitions will be discussed.

To be more specific, we consider the Ne atom as example. Here only $1s$ ionization leads to an Auger state, whereas $2s$ ionization leads to an excited Ne^+ atom the energy of which is still below the continuum threshold of Ne^{++}, whence, it decays only by emission of radiation. The Auger state $1s^{-1}$ (or $\underline{1s}$, an underline under the orbital denoting a vacancy) decays to final ionic states $2s^{-2}$, $2s^{-1}2p^{-1}$, or $2p^{-2}$ giving rise to normal or diagram Auger transitions. To describe Auger transitions either the X-ray or the orbital notation for vacancies is used, e.g., $K\text{–}L_1L_{2,3}$, $1s^{-1}\text{–}2s^{-1}2p^{-1}$, or $\underline{1s}-\underline{2s2p}$.

In an energy diagram in which the ejected electron is included in the final state (Fig. 1), the diagram Auger transition

$$A^+(\text{inner-shell}) \rightarrow A^{++}(E_\beta) + e^-(\varepsilon_{\text{Auger},\beta}) \tag{1}$$

is indicated by a horizontal arrow connecting initial and final states of equal energy, with

$$\varepsilon_{\text{Auger},\beta} = E(A^+) - E_\beta(A^{++}). \tag{2}$$

Here $E(A^+)$ and $E_\beta(A^{++})$ are the positive energies above the ground state of the neutral atom A.

If one or both of the final vacancies $n_1 l_1^{-1}$ and $n_2 l_2^{-1}$ occur in the same principal shell but different subshells than the initial vacancy nl^{-1}, the corresponding transitions $nl^{-1}\text{–}nl_1^{-1}nl_2^{-1}$ are called Coster–Kronig (CK) transitions (Coster and Kronig, 1935) and super-Coster–Kronig (sCK) transitions (McGuire, 1972), respectively. These differ from ordinary Auger transitions by their much smaller energies ($E_A > E_{\text{CK}} > E_{\text{sCK}}$) and much larger transition probabilities.

Starting with more than one vacancy, e.g., $1s^{-1}2p^{-1}$, the corresponding Auger transitions are called (multiple-ionization) satellite transitions. Since Auger transitions are due to the Coulomb interaction between two electrons, transitions in which actually three electrons are involved can occur only through electron correlation; they are called double Auger transitions. In a double Auger transition, either two electrons can be ejected simultaneously,

$$A^+(\text{inner shell}) \rightarrow A^{+++} + 2e^-_{\text{Auger}}, \tag{3}$$

leading to a continuous energy distribution of the two electrons, or one electron is ejected and the other electron is excited:

$$A^+(\text{inner shell}) \rightarrow A^{++*} + e^-_{\text{Auger}}. \tag{4}$$

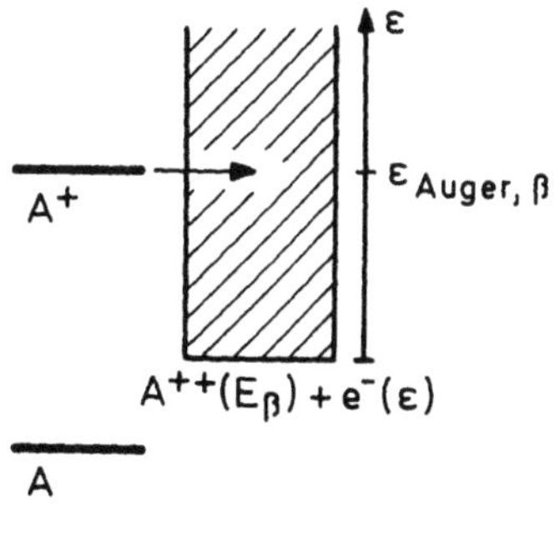

FIGURE 1. Auger transition in an energy diagram for the total final system of ion A^{++} and electron.

In the latter case the electrons e^-_{Auger} have discrete energies and lead to (electron correlation) satellite lines in the Auger spectrum (see Sec. 5.1). The former case of Eq. (3), in which two electrons are ejected simultaneously, has been studied mostly by means of ion-charge spectrometry (see, e.g., Åberg, 1975, and references therein).

2.2. *Theory of Auger Transitions; the Wentzel Ansatz*

The process of inner-shell ionization followed by the emission of an Auger electron is usually interpreted as a two-step process, in which the decay is treated independently from the primary ionization. This approach is correct only if the following conditions are fulfilled: First, the primary ejected electron has large enough energy so that it does not interact with the Auger electron, i.e., we neglect what is known as postcollision interaction (see Sec. 9). Second, the interaction of the core hole state with the Auger continuum is weak, in which case the Auger decay can be treated by perturbation theory giving a level width to the core hole state. However, if the interaction is very strong (e.g., in the case of super-Coster–Kronig transitions), then both the core hole state and the continuum are strongly modified. In this case the Auger decay and the inner-shell ionization must be treated in a unified manner (Wendin and Ohno, 1976; Ohno and Wendin, 1979; Wendin, 1981, 1982; Yafet, 1980; Combet Farnoux and Ben Amar, 1980; Åberg, 1981; Combet Farnoux 1982; Gunnarsson and Schönhammer, 1982).

In the following we treat the Auger transition as an independent process. According to Wentzel (1927), the Auger transition in the nonrelativistic theory is caused by the Coulomb interaction e^2/r_{12} between the two electrons which participate directly in the transition. For example, let us consider an Auger transition for which the initial Auger state is an atom with one inner-shell vacancy nl^{-1} but otherwise closed shells. The final state is an atom with two outer-shell vacancies $(n_1 l_1 n_2 l_2)^{-1}$ coupled to S and L (we consider for the moment *pure LS* coupling) and with energy E_β plus the ejected Auger electron $\varepsilon_A l_A$, where ε_A and l_A are the energy and the orbital angular momentum of the Auger electron. Then the initial and final $(\mathcal{N}-1)$-electron-state wave functions are (Cleff and Mehlhorn, 1974a)

$$\Psi_i^{(\mathcal{N}-1)} = \Psi^{(\mathcal{N}-1)}(nl^{-1}, S_i L_i M_{S_i} M_{L_i}, \Pi_i), \tag{5}$$

$$\Psi_{f,l_A}^{(\mathcal{N}-1)} = \Psi^{(\mathcal{N}-1)}(n_1 l_1^{-1} n_2 l_2^{-1}(SLM_S M_L)\, \varepsilon_A l_A m_{l_A} m_{s_A}, \Pi_f), \tag{6}$$

where Π_i and Π_f are the parities.

Then the probability for the Auger transition $|i\rangle \rightarrow |f, l_A\rangle$ is given by (Åberg and Howat, 1982)

$$P_{i \rightarrow f,l_A} = \frac{2\pi}{\hbar} \frac{1}{(2S_i+1)(2L_i+1)} \times \sum_{M_{L_i}M_{S_i}M_L M_S m_{l_A} m_{s_A}} |\langle \Psi_{f,l_A}^{(N-1)} | H - E | \Psi_i^{(N-1)} \rangle|^2 \rho(\varepsilon_A), \tag{7a}$$

where H is the total Hamiltonian of the $(N-1)$-electron ion and E its total energy. In Eq. (7a) we have summed over the final and averaged over the initial states; $\rho(\varepsilon_A)$ is the density of continuum states and depends on the normalization of the continuum orbital. If we normalize to one electron per unit of energy range, then $\rho(\varepsilon_A) = 1$.

For the wave functions $\Psi_i^{(N-1)}$ and $\Psi_f^{(N-1)}$ we now use the independent-particle frozen-core approximation, i.e., we use Slater determinants of one-electron spin orbitals and the same potential for the initial and final bound and continuum spin orbitals. Since $\Psi_f^{(N-1)}$ and $\Psi_i^{(N-1)}$ differ by two spin orbitals, the matrix element of Eq. (7a) reduces in the independent-particle frozen-core approximation to

$$\langle \Psi_{f,l_A}^{(N-1)} | H - E | \Psi_i^{(N-1)} \rangle = \langle \Psi_{f,l_A}^{(N-1)} | \sum_{\nu > \mu} \frac{e^2}{r_{\mu\nu}} | \Psi_i^{(N-1)} \rangle. \tag{7b}$$

Equation (7a) together with (7b) yields Wentzel's ansatz for the nonradiative transition probability derived by perturbation theory (Wentzel, 1927).

The matrix element in Eq. (7b) is nonzero only if the selection rules for the Coulomb interaction between $\Psi_i^{(N-1)}$ and $\Psi_{f,l_A}^{(N-1)}$ are fulfilled. These are, for *pure* LS-coupling,

$$\Delta L = \Delta S = \Delta M_L = \Delta M_S = 0, \tag{8a}$$

$$\Delta J = \Delta M = 0, \qquad \Pi_i = \Pi_f. \tag{8b}$$

In order to apply these selection rules we expand $\Psi_{f,l_A}^{(N-1)}$ in terms of wave functions $\Phi_f^{(N-1)}(S_f L_f)$ having a definite symmetry $S_f L_f$. This yields (Cleff and Mehlhorn, 1974a)

$$\Psi_{f,l_A}^{(N-1)} = \sum_{L_f M_{L_f}} \sum_{S_f M_{S_f}} (-1)^{l_A - L - M_{L_f} + 1/2 - S - M_{S_f}} [(2L_f+1)(2S_f+1)]^{1/2} \times \begin{pmatrix} L & l_A & L_f \\ M_L & m_{l_A} & -M_{L_f} \end{pmatrix} \begin{pmatrix} S & 1/2 & S_f \\ M_S & m_{s_A} & -M_{S_f} \end{pmatrix} \times \Phi_f^{(N-1)}(n_1 l_1^{-1} n_2 l_2^{-1}(SL)\, \varepsilon_A l_A, S_f L_f M_{S_f} M_{L_f}, \Pi_f), \tag{9}$$

where

$$\begin{pmatrix} l & l' & L \\ m_l & m_l' & -M_L \end{pmatrix}$$

are the 3-j symbols.

As was mentioned above, $\Psi_i^{(\mathcal{N}-1)}$ and $\Phi_f^{(\mathcal{N}-1)}$ in Eqs. (7b) and (9) differ in exactly two spin orbitals, viz., $n_1l_1n_2l_2$ in $\Psi_i^{(\mathcal{N}-1)}$ and $nl\varepsilon_A l_A$ in $\Phi_f^{(\mathcal{N}-1)}$. Since the Coulomb interaction e^2/r_{12} is a two-particle operator, it causes the two-electron transition $n_1l_1n_2l_2 \rightarrow nl\varepsilon_A l_A$. Then the $(\mathcal{N}-1)$-electron matrix element of Eq. (7b) reduces exactly to a two-electron matrix element describing the transition $n_1l_1n_2l_2 \rightarrow nl\varepsilon_A l_A$; all other electrons do not participate in the transition. In this approximation the transition probability $P(nl^{-1} \rightarrow n_1l_1^{-1}n_2l_2^{-1}, SL)$ becomes (see, e.g., Walters and Bhalla, 1971a; Åberg and Howat, 1982, p. 517)

$$P(nl^{-1} \rightarrow n_1l_1^{-1}n_2l_2^{-1}, SL)$$
$$= \frac{2\pi}{\hbar}\frac{(2L+1)(2S+1)}{2(2l+1)}\sum_{l_A}\left|\langle nl\varepsilon_A l_A, SLM_SM_L|\frac{e^2}{r_{12}}|n_1l_1n_2l_2, SLM_SM_L\rangle\right|^2. \tag{10}$$

In Eq. (10) the selection rules (8) have been used as well as the fact that the Coulomb matrix element is independent of M_L and M_S. Since several partial waves l_A of the Auger electron can couple with l of the initial vacancy to give L, we must sum over l_A. In Eq. (10) the initial- and final-state wave functions are two-particle Slater determinants. The evaluation of the Coulomb matrix element leads to (Shore and Menzel, 1968)

$$\langle nl\varepsilon_A l_A, SLM_SM_L|\frac{e^2}{r_{12}}|n_1l_1n_2l_2, SLM_SM_L\rangle = \tau(-1)^{l_1+l_A+L}$$
$$\times \sum_{k=0}^{\infty}\left[R^k(n_1l_1n_2l_2, nl\varepsilon_A l_A)\langle l\| C^k \|l_1\rangle\langle l_A\| C^k \|l_2\rangle \begin{Bmatrix} l & l_A & L \\ l_2 & l_1 & k \end{Bmatrix}\right.$$
$$\left. + (-1)^{L+S} R^k(n_1l_1n_2l_2, \varepsilon_A l_A nl)\langle l\| C^k \|l_2\rangle\langle l_A\| C^k \|l_1\rangle \begin{Bmatrix} l & l_A & L \\ l_1 & l_2 & k \end{Bmatrix}\right], \tag{11}$$

with

$$\langle l\| C^k \|l'\rangle = (-1)^l[(2l+1)(2l'+1)]^{1/2}\begin{pmatrix} l' & k & l \\ 0 & 0 & 0 \end{pmatrix}.$$

The

$$\begin{pmatrix} l' & k & l \\ 0 & 0 & 0 \end{pmatrix} \quad \text{and} \quad \begin{Bmatrix} l & l_A & L \\ l_2 & l_1 & k \end{Bmatrix}$$

are $3j$ and $6j$ symbols. The factor τ is 1 or 1/2 depending on whether $n_1 l_1$ and $n_2 l_2$ are nonequivalent or equivalent electrons, respectively. The $R^k(n_1 l_1 n_2 l_2, nl\varepsilon_A l_A)$ and $R^k(n_1 l_1 n_2 l_2, \varepsilon_A l_A nl)$, the radial parts of the matrix elements, are the direct and exchange Slater integrals defined by

$$R^k(n_1 l_1 n_2 l_2, n_3 l_3 n_4 l_4)$$

$$= e^2 \iint R_{n_1 l_1}(r_1)\, R_{n_2 l_2}(r_2)\, \frac{r_<^k}{r_>^{k+1}}\, R_{n_3 l_3}(r_1)\, R_{n_4 l_4}(r_2)\, r_1^2 r_2^2\, dr_1\, dr_2, \quad (12)$$

where $r_<$ and $r_>$ are the lesser and the greater of r_1 and r_2, respectively.

In the sum over k in Eq. (11) actually only a few of the angular factors $\langle l \| C^k \| l' \rangle$ are $\neq 0$ due to the condition that the angular momenta k, l, and l' have to fulfill the triangular relation

$$|l - l'| \leqslant k \leqslant l + l' \quad (13a)$$

and the relation

$$k + l + l' = \text{even number.} \quad (13b)$$

So far we have assumed that the two final vacancies $n_1 l_1^{-1}$, $n_2 l_2^{-1}$ and the initial vacancy nl^{-1} can be described in pure LS coupling. Except for $l = 0$, the initial vacancy is better described by nlj^{-1}, so that j–j coupling of the nlj electron and the Auger electron $\varepsilon_A l_A j_A$ in Eq. (10) is appropriate. General expressions for transition probabilities, equivalent to Eqs. (10) and (11), have been given for j–j coupling in the initial and final states (Asaad, 1963; Åberg and Howat, 1982; p. 518). The more general case is that the two final vacancies couple to SLJ in intermediate coupling. Asaad and Burhop (1958) originally introduced intermediate coupling to the K–LL Auger transitions, starting from transition probabilities in pure LS coupling [Eq. (10)]. Later, Asaad and Mehlhorn (1968) introduced the mixed coupling scheme, in which the initial vacancy and the Auger electron are treated in j–j coupling and the final vacancies either in LS or in intermediate coupling (Mehlhorn and Stalherm, 1968). General expressions for transition probabilities in mixed coupling with pure LS coupling or intermediate coupling in the final state have been given by El Ibyari *et al.* (1972) and by McGuire (1975b, 1978), respectively (see also Åberg and Howat, 1982,

p. 520). In the case of intermediate coupling the selection rules (8a) must be replaced by

$$\Delta S = 0, \pm 1, \qquad \Delta L = 0, \pm 1, \tag{8c}$$

whereas the selection rules (8b) are always valid.

For the case of atoms that have open shells before inner-shell ionization, general expressions for the nonradiative transition probability in *j–j* or *LS* coupling for the initial state and *LS* coupling for the final state have been worked our by McGuire (1975a). Even the most general case, in which both initial and final states are in intermediate coupling, has been treated (McGuire, 1975b). Calculations of Auger transition probabilities for open-shell atoms have been carried out for various problems (Schmidt, 1973; Hillig *et al.*, 1974; McGuire, 1974a, 1974b, 1975c, 1975d, 1975e, 1976, 1977) but especially in connection with the fluorescence yields of multiply ionized atoms, e.g., *K*-shell Auger rates of neon (Chen and Crasemann, 1975; Chen *et al.*, 1975; Bhalla *et al.*, 1973a; Bhalla, 1975a–d), fluorine (Can *et al.*, 1982) and argon (Bhalla, 1973b) and 2*p*-shell rates of argon (Chen and Crasemann, 1974), chlorine and sulfur (Chen and Crasemann, 1977).

As shown in the preceding section, the Auger effect in the non-relativistic limit is a consequence of the electron–electron interaction. In the relativistic regime the Auger effect is treated as due to the exchange of virtual photons between the pair of interacting electrons. For details see, e.g., Chattarji (1976), Åberg and Howat (1982), and Chen (1985).

2.3. *The Auger Effect Treated beyond the Wentzel Ansatz*

Åberg (1976) and Åberg and Howat (1982) have treated the Auger effect in the broader context of resonance scattering and have thus established the close relationship between autoionization and the Auger effect. The main extension beyond the Wentzel approach is made by the explicit inclusion of configuration interaction between the quasidiscrete Auger state and the final continua, originally developed by Fano (1961) and Fano and Prats (1963) for the autoionization process, into the theory of the Auger effect. Since the excitation followed by autoionization is generally treated as a one-step process, we will adopt this view for the moment also for the process of ionization and Auger decay. We further assume—for simplicity—that the primary ionization is produced by photons $\hbar\omega$. Then Fig. 1 must be modified as indicated in Fig. 2. The two transition paths (a) and (b) in Fig. 2 correspond to the following processes:

$$\text{(a)}\quad \hbar\omega + \mathrm{A} \rightarrow \mathrm{A}^{+}(E_{\Phi}) + e^{-}(\tau) \rightarrow \mathrm{A}^{++}(E_{\beta}) + e^{-}(\varepsilon_1) + e^{-}(\varepsilon_2), \tag{14a}$$

$$\text{(b)}\quad \hbar\omega + \mathrm{A} \xrightarrow{\hspace{6em}} \mathrm{A}^{++}(E_{\beta}) + e^{-}(\varepsilon_1) + e^{-}(\varepsilon_2). \tag{14b}$$

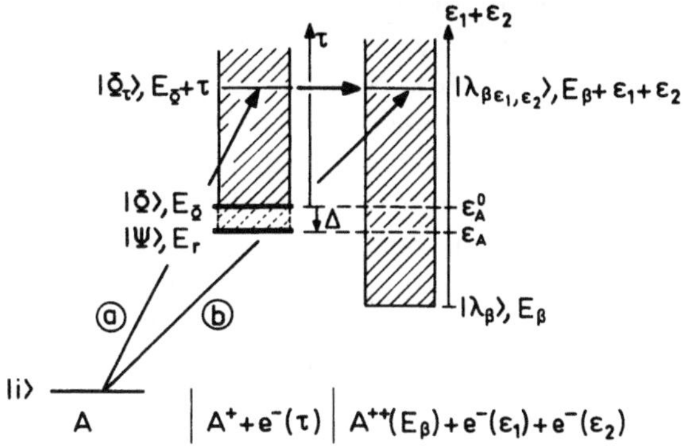

FIGURE 2. Inner-shell ionization and Auger decay as viewed in the resonance scattering theory. The wave function $|\Phi\rangle$ with energy E_Φ of the inner-shell ionized atom A^+ is without the interaction with the continuum $|\lambda_{\beta\varepsilon_1}\rangle = |\lambda_\beta\rangle\,|\varepsilon_1\rangle$; inclusion of the interaction yields $|\Psi\rangle$ with energy E_r [Eq. (19)].

Following Åberg and Howat (1982, p. 488), we have for the amplitude of the transition from the initial state $|i\rangle$ of the neutral $\mathcal{N}$-electron atom to the final state $|\lambda_{\beta\varepsilon_1,\varepsilon_2}\rangle$ with two outgoing electrons, in which the primary ejected electron (ε_2) and the Auger electron (ε_1) are coupled to a discrete ($\mathcal{N}-2$)-electron state λ_β (with energy E_β):

$$T = \langle \lambda_{\beta\varepsilon_1,\varepsilon_2}, 0|\, H_{\text{int}}\, |i, \omega\rangle + \int_0^\infty \frac{\langle \lambda_{\beta\varepsilon_1,\varepsilon_2}, 0|\, H_{\text{el}} - E\, |\Phi_\tau, 0\rangle\langle \Phi_\tau, 0|\, H_{\text{int}}\, |i, \omega\rangle\, d\tau}{\varepsilon_1 + \varepsilon_2 - \varepsilon_A - \tau + i\Gamma/2}. \tag{15}$$

Atomic units are used throughout in this section. In Eq. (15) we have dropped the superscripts +, − used in the original Eq. (4.8) of Åberg and Howat (1982). The wave function Φ_τ stands for the ($\mathcal{N}-1$)-electron Auger state Φ and the outgoing photoelectron with energy τ. The operator H_{el} is the full $\mathcal{N}$-electron Hamiltonian, and H_{int} stands for the interaction of the photon with the atom. The energy $\varepsilon_A = E_r - E_\beta$ is the true kinetic energy of the Auger electron, including the level shift $\Delta = E_r - E_\Phi$ due to the interaction of Φ with $\lambda_{\beta\varepsilon_1}$ [see Eq. (19) and Fig. 2]. The total width of the Auger state Φ is denoted by Γ.

The first part of the transition amplitude T of Eq. (15) is due to the direct transition (path b in Fig. 2), whereas the second part (resonance amplitude) corresponds to path a. In the case of excitation followed by autoionization, both parts of the corresponding transition amplitude, the direct and the resonance amplitudes are in general nonzero. This leads to interference effects (Fano profile) in the cross section of photoionization or inelastically scattered electrons (Fano, 1961; Fano and Cooper, 1965) and of ejected electrons (Balashov *et al.*, 1973; Lipovetsky and Senashenko, 1974; see also Mehlhorn, 1975, 1978). Now in the case of Eq. (15) the direct amplitude, corresponding to two-electron ionization, is negligibly small; in the independent-particle frozen-core approximation it would even vanish, since H_{int} is a one-electron operator. This is also true for primary ionization by particle impact if H_{int} is the Born operator $\exp(i\mathbf{K}\mathbf{r})$.

We now assume that τ is large enough so that the interaction between $e^-(\tau)$ and the residual ion A^+ can be neglected; then $|\Phi_\tau\rangle = |\Phi\rangle|\tau\rangle$. We approximate also the final wave function in the same way, $|\lambda_{\beta\varepsilon_1,\varepsilon_2}\rangle = |\lambda_{\beta\varepsilon_1}\rangle|\varepsilon_2\rangle$. Then we have

$$\langle\lambda_{\beta\varepsilon_1,\varepsilon_2}, 0|\, H_{\text{el}} - E\,|\Phi_\tau, 0\rangle = \langle\varepsilon_2|\tau\rangle\langle\lambda_{\beta\varepsilon_1}|\, H_{\text{el}}(N-1) - E(N-1)\,|\Phi\rangle. \quad (16)$$

Since $|\varepsilon_2\rangle$ should be evaluated in the field of the residual doubly ionized ion and $|\tau\rangle$ in that of the ion with one inner-shell vacancy, it follows that $\langle\varepsilon_2|\tau\rangle \neq \delta(\varepsilon_2 - \tau)$. Only in the case of $\tau \gg 0$, which implies $\hbar\omega \gg E_\Phi$ and that postcollision interaction can be neglected, are $|\tau\rangle$ and $|\varepsilon_2\rangle$ orthogonal and $\langle\varepsilon_2|\tau\rangle = \delta(\varepsilon_2 - \tau)$ (Åberg, 1981). Assuming the latter condition to be fulfilled the cross section for the emission of an electron with kinetic energy ε_1 is given by Eq. (4.10) of Åberg and Howat (1982):

$$\frac{d\sigma}{d\varepsilon_1} = \frac{\Gamma_\beta(\varepsilon_A)/2\pi}{(\varepsilon_1 - \varepsilon_A)^2 + \Gamma^2/4}\,\sigma(\tau). \quad (17)$$

Here, $\sigma(\tau)$ is the photoionization cross section, and

$$\Gamma_\beta(\varepsilon_A) = 2\pi\,|\langle\lambda_{\beta\varepsilon_A}|\, H_{\text{el}}(N-1) - E(N-1)\,|\Phi\rangle|^2 \quad (18)$$

is the partial level width due to the nonradiative transition probability of the Auger state Φ into the continuum state $\lambda_{\beta\varepsilon_A}$, as given by Eq. (7a). In case of several continua $\lambda_{\beta\varepsilon_A}$, the total width Γ in the denominator of Eq. (17) is given by $\Gamma = \sum \Gamma_\beta$. (If the subsequent decay of the final state had also been included in the above theory, then the total width Γ would be the sum of the widths pertaining to the initial and final states.)

The main feature of Eq. (17) is that the differential cross section is separated into two parts: the photoionization cross section $\sigma(\tau)$ and the Auger decay probability. The part describing the decay yields a symmetrical Lorentzian line shape, provided the widths Γ_β and Γ are constant and so is the shift Δ contained in $\varepsilon_A = E_\Phi + \Delta - E_\beta$ [Eq. (19)]. For Auger transitions the widths Γ_β and Γ are small (<1 eV) and essentially constant over this range, and the shift Δ is even negligibly small (see next paragraph). Experimentally, all Auger lines have been found so far to possess symmetrical line shapes, provided the high-energy condition (neglect of postcollision interaction) is fulfilled. This can be taken as evidence that the direct amplitude (path *b* of Fig. 2) is indeed negligibly small. On the other hand, a strong variation of the widths Γ_β and Γ and the shift Δ, especially likely for super-Coster–Kronig transitions in which these quantities are large, would exclude a symmetrical Lorentzian distribution. For that case, ionization and decay have been treated as a one-step process (Ohno and Wendin, 1979; Wendin, 1981).

In Eq. (17) the true kinetic energy ε_A of the Auger electron is related to the energy $\varepsilon_A^0 = E_\Phi - E_\beta$ by $\varepsilon_A = \varepsilon_A^0 + \Delta$ (see Fig. 2), where $\Delta = E_r - E_\Phi$ is caused by the interaction of the vacancy state Φ with the Auger continuum $\lambda_{\beta\varepsilon_1}$ and is given by Eq. (3.8) of Åberg and Howat (1982):

$$\begin{aligned}\Delta &= E_r - E_\Phi \\ &= \mathscr{P}\int_0^\infty \frac{|\langle\lambda_{\beta\varepsilon_1}|\, H_{\text{el}} - E\,|\Phi\rangle|^2\, d\varepsilon_1}{(E-E_\beta)-\varepsilon_1} \cong \mathscr{P}\int_0^\infty \frac{|V_{\Phi,\beta\varepsilon_1}|^2\, d\varepsilon_1}{(E_\Phi - E_\beta)-\varepsilon_1}.\end{aligned} \tag{19}$$

The symbol $\mathscr{P}$ indicates the principal value of the integral, and we have

$$V_{\Phi,\beta\varepsilon_1} = \langle\lambda_{\beta\varepsilon_1}|\sum e^2/r_{\mu\nu}\,|\Phi\rangle = \langle\lambda_{\beta\varepsilon_1}|\, H_{\text{el}} - E\,|\Phi\rangle, \tag{20}$$

where the second equality in Eq. (20) follows if Φ and $\lambda_{\beta\varepsilon_1}$ are constructed in the independent-particle frozen-core approximation. The energy E in Eq. (19) has been taken to be the energy E_Φ of the inner-shell vacancy state.

In an Auger transition, in which the initial and the two final vacancies are in different principal shells, the interaction $|V_{\Phi,\beta\varepsilon_1}|^2_{\text{Auger}}$ is of the order of 10^{-4} to 10^{-3} a.u. and the energy difference $\varepsilon_A^0 = E_\Phi - E_\beta$ is large ($\gg 1$ a.u.). Since $|V_{\Phi,\beta\varepsilon_1}|^2_{\text{Auger}}$ depends only weakly on ε in the neighborhood of ε_A^0, the positive and negative contributions in Eq. (19) cancel almost completely, which results in $\Delta \cong 0$. In case of a Coster–Kronig transition or a super-Coster–Kronig transition, the transition energies are much smaller and $|V_{\Phi,\beta\varepsilon_1}|^2$ is roughly 10 times or even up to 10^2 times larger than $|V_{\Phi,\beta\varepsilon_1}|^2_{\text{Auger}}$. This may lead to a shift Δ of several eV for Coster–Kronig transitions (Chen *et al.*, 1981) and even up to about 10 eV for super-Coster–Kronig transitions, e.g., $\underline{4p} - \underline{4d}^2\varepsilon f$ (Wendin, 1981; Wendin and Ohno, 1976). In addition, due to the very strong interaction in a super-Coster–Kronig transition, the initial vacancy can be completely diluted into the continuum, as has been found, e.g., for the $\underline{4p}$ states for atomic numbers $48 \leqslant Z \leqslant 58$. In these cases, inner-shell ionization and Auger transition cannot be treated as two independent processes, but rather they must be treated in a unified manner (Wendin, 1981).

3. EXPERIMENTAL ARRANGEMENTS

In Fig. 3, the general scheme of an apparatus for electron spectrometry is shown. As primary beams either electrons, ions, or photons are used. The target atoms are either in gaseous or in vapor form (in case of metals). The ejected electrons are energy-analyzed by means of an electrostatic electron spectrometer (such as the 30° parallel-plate spectrometer in Fig. 3); an energy resolution of 1×10^{-3} is almost standard, and resolution of 2×10^{-4} can be achieved in special cases. Electrons are detected by means of a channeltron or, better still, by means of a multichannel detection system. One of the spectrometers in Fig. 3 can be rotated relative to the primary beam direction for angular-distribution measurements; the other spectrometer serves as a monitor. For comprehensive articles on techniques of electron spectrometry see, e.g., Siegbahn and Karlsson (1982) and Stolterfoht (1983).

The progress made in AES during the last two decades can be appreciated, for example, in the $L_{2,3}$–MM Auger spectrum of Ar. In Fig. 4a the first Auger spectrum of Ar, measured with a gaseous target and a resolution of 2.4×10^{-2} is shown (Mehlhorn, 1960). Figure 4b shows the same spectrum, measured with resolution 5×10^{-4} (Werme *et al.*, 1973); the high-energy group, measured by Huster (1981) with resolution 3×10^{-4}, is displayed in Fig. 4c.

Contrary to ionization by photons and electrons, ionization by ion impact introduces kinematic effects in the Auger spectra. In ion–atom collisions the inner-shell vacancy is created either in the target atom or in the projectile ion. In both cases the Auger electron is influenced by

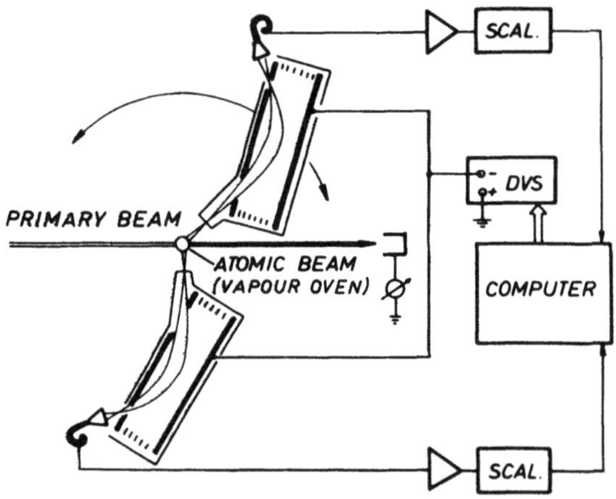

FIGURE 3. Principal experimental arrangement for AES. From Sandner and Schmitt (1978).

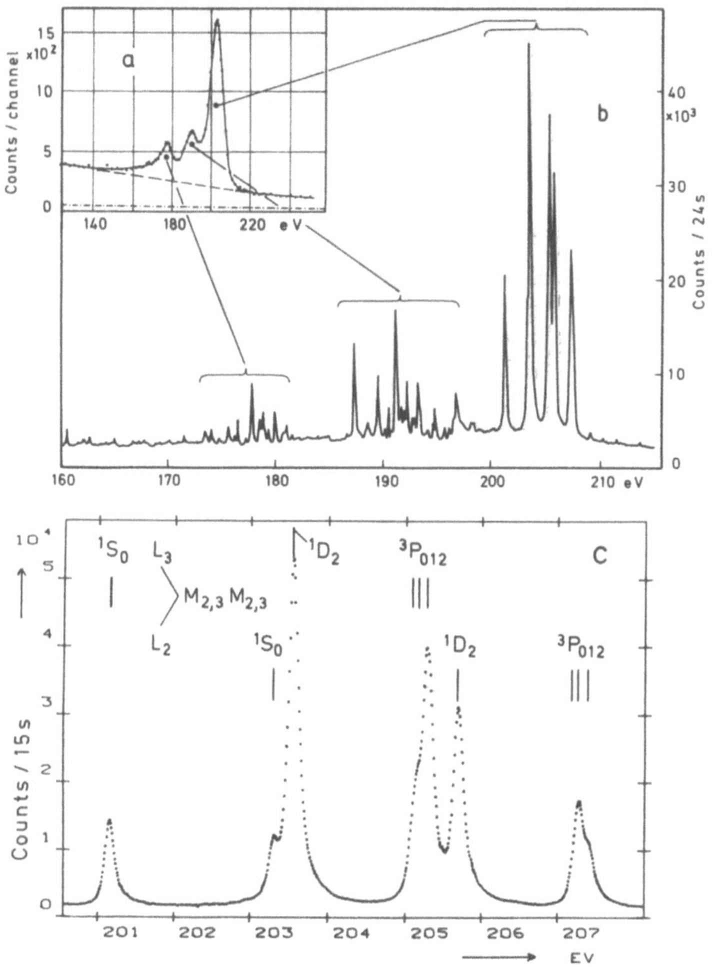

FIGURE 4. Progress in AES demonstrated by at $L_{2,3}$–MM Auger spectrum of Ar. (a) First spectrum of gaseous Ar excited by X-ray bremsstrahlung, at 2.4% spectrometer resolution (Mehlhorn, 1960); (b) spectrum excited by 3-keV electrons, taken at 5×10^{-4} resolution (Werme *et al.*, 1973); (c) $L_{2,3}$–$M_{2,3}M_{2,3}$ group of the spectrum excited by 2-keV electrons, at 3×10^{-4} resolution (Huster, 1981).

kinematic effects because it is emitted from a moving particle, viz., the recoil target atom or the fast-moving projectile ion. A discussion of these effects has been provided by Stolterfoht (1983). In the case of Auger emission from the projectile ion this leads (a) to a transformation of the energy (Doppler shift) and the intensity of the Auger electrons from the emitter frame to the laboratory frame (see, e.g., Dahl *et al.*, 1976a); and (b) to a kinematic broadening of Auger lines observed in the laboratory frame (see, e.g., Dahl *et al.*, 1976a; Bachmann *et al.*, 1982). The kinematic broadening of Auger lines in projectile Auger-electron spectrometry is considered in Sec. 7. In the case of Auger emission from the recoil target atom, only line broadening is observed which decreases with increasing projectile energy (Stolterfoht, 1983).

4. DIAGRAM AUGER TRANSITIONS

4.1. *Energies*

The energy of an Auger transition $\underline{nl}$–$\underline{n}_1\underline{l}_1\underline{n}_2\underline{l}_2(SLJ)$ is given according to Eq. (2) by

$$\varepsilon_{\text{Auger}} = E(A^+(\underline{nl})) - E(A^{++}(\underline{n}_1\underline{l}_1\underline{n}_2\underline{l}_2, SLJ)), \tag{21}$$

where the two vacancies in the final state couple to a specific SLJ. The transition energies are nowadays usually computed by performing separate SCF calculations for the total energies of the initial and final states (ΔSCF method), which includes automatically the relaxation of the initial and final states. The approximations used are either nonrelativistic Hartree–Fock (HF) (Froese-Fischer, 1969, 1972) or relativistic Dirac–Fock (DF) (Grant, 1970; Grant *et al.*, 1980; Desclaux, 1975) and Dirac–Slater (DS) (Huang *et al.*, 1976). Since the basic SCF calculations are performed either in LS coupling (HF) or j–j coupling (DF, DS), the intermediate coupling SLJ of final states has to be carried out either separately by diagonalizing the appropriate matrix of spin–orbit and Coulomb interactions or by using a relativistic multiconfiguration approach, in which the intermediate coupling can be treated completely *ab initio* (Briançon and Desclaux, 1976). Earlier calculations of transition energies were mostly semiempirical and were mainly based on the expressions given by Asaad and Burhop (1958) for the intermediate coupling scheme for two-hole states. Inclusion of a relaxation correction (Shirley, 1973; Larkins, 1976) considerably improved the Asaad–Burhop expressions; a comprehensive tabulation of Auger transition energies was furnished by Larkins (1977).

In order to illustrate the power of such calculations, we compare in Tables I and II the experimental energies of the K–LL transitions of neon and uranium with theoretical HF (Kelly, 1975) and DF results (Briançon and Desclaux, 1976) and with semiempirical values of Larkins (1977). In the case of U, the theoretical DF values agree with the experimental results within experimental errors. For Ne, the HF energies differ from experiment by about 1 eV, whereas the results of a many-body perturbation theory

TABLE I
Ne K–LL Auger energies (in eV).

Transition		Semiempirical[a]	HF[b]	MBPT[b]	Experimental[c]
K–L_1L_1	$^1S_0$	747.0	747.0	748.2	748.3(1)
K–$L_1L_{2,3}$	$^1P_1$	770.9	770.9	771.7	771.8(1)
	^3P	781.8	783.0	782.4	782.3(1)
K–$L_{2,3}L_{2,3}$	$^1S_0$	800.9	801.0	801.3	800.8(1)
	$^1D_2$	803.4	806.0	804.5	804.5(1)
	$^3P_{0,2}$	806.2			807.6(1)

[a] Larkins (1977), with configuration interaction between $^1S_0$ of $\underline{2s}^2$ and $\underline{2p}^2$.
[b] Kelly (1975).
[c] Krause *et al.* (1971).

calculation by Kelly (1975) (MBPT in Table I) agree perfectly with experiment.

These two examples seem to demonstrate that electron correlation plays a minor role in the case of transition energies. The reason is that for deep core transitions the electron-correlation effects on the initial- and final-state energies are usually in the eV range and partially cancel in the transition energy. For final states in the outermost or penultimate shell, electron correlation is generally larger, e.g., if excited states involving d orbitals are close. Examples of strong correlation effects on the energies of final Auger states are $\underline{3s3p}\,^1P$ and $\underline{3s}^2\,^1S$ of Ar (McGuire, 1975d; Dyall and Larkins,

TABLE II
U K–LL Auger energies (in eV).

Transition	Semiempirical[a]	DF[b]	Experimental[c]
K–L_1L_1 $^1S_0$	71776	71749	71745(20)
K–L_1L_2 $^1P_1$	72606	72560	72560(20)
$^3P_0$	72662	72614	72620(20)
K–L_1L_3 $^3P_1$	76334	76332	76320(20)
$^3P_2$	76442	76430	76430(20)
K–L_2L_2 $^1S_0$	73373	73272	73320(40)
K–L_2L_3 $^1D_2$	77157	77125	77130(20)
K–L_3L_3 $^3P_0$	80891	80889	
$^3P_2$	80954	80955	

[a] Larkins (1977), with configuration interaction between $^1S_0$ of L_1L_1 and L_2L_2.
[b] Briançon and Desclaux (1976).
[c] Briançon (1970).

1982b) and $4s4p\,{}^1P$ and $4s^2\,{}^1S$ of Kr (McGuire, 1975c) and Sr (Mehlhorn *et al.*, 1977).

Another source of electron correlation is the interaction of the initial Auger state (and also of the final Auger state if this decays by electron emission) with the Auger continua causing a shift Δ [see Eq. (19)]. As stated in Sec. 2.3, the shift Δ is usually small in the case of an Auger state and can thus be neglected. The situation is different for Coster–Kronig and especially for super-Coster–Kronig states. Here the shift Δ can amount to several eV and even up to 10 eV and the transition energies will be correspondingly smaller compared with pure ΔSCF values (Chen *et al.*, 1981; Wendin, 1981; see also Mehlhorn, 1983a).

4.2. *Intensities*

For a long time, the *K–LL* Auger spectra (Ne up to U) and the outer-core level spectra of the rare-gas atoms (e.g., *L–MM* of Ar, *M–NN* of Kr, and *N–OO* and $M_{4,5}$–*NN* of Xe) were studied most thoroughly. From about 1974 on, high-resolution Auger spectra of core levels of free metal atoms have been measured (Mehlhorn *et al.*, 1977). In most cases, closed-shell atoms have been investigated, e.g., Mg (Breuckmann and Schmidt, 1974; Väyrynen *et al.*, 1977; Breuckmann, 1979), Ca (Breuckmann, 1978), Sr (Mehlhorn *et al.*, 1977), Ba (Mehlhorn *et al.*, 1977; Breuckmann, 1978), Zn (Aksela *et al.*, 1974; Mehlhorn *et al.*, 1977), Cd (Aksela and Aksela, 1974, 1977b), Hg (Aksela *et al.*, 1977a), but also a few open-shell atoms were studied, e.g., Na (Hillig *et al.*, 1974; Breuckmann, 1978), K (Breuckmann, 1978; Aksela *et al.*, 1981), Rb (Menzel and Mehlhorn, 1981), Ag (Väyrynen *et al.*, 1980), Mn (Väyrynen, 1981; Väyrynen and Aksela, 1981), and Cu (Aksela and Sivonen, 1982).

On the theoretical side, the radial integrals R^k must be calculated numerically. They depend on the coice of the central-field approximation of the atom used for the calculation of the bound and continuum wave functions. In order to ensure orthogonality of the initial and final wave functions, *all* bound orbitals *and* the continuum orbital are calculated for the same atomic potential; usually the $V(nl^{-1})$ potential is taken. This procedure, the frozen-core approximation, neglects the relaxation effect of the orbitals in the final state. In a few cases, the effect of relaxation on the transition probabilities has been studied (Howat *et al.*, 1978; Chen *et al.*, 1978).

Systematic and comprehensive *ab initio* calculations of Auger transition rates for different shells throughout the Periodic Table have been performed by three groups: McGuire at Sandia Laboratories [Hartree–Slater (HS) approximation], Bhalla and co-workers at Kansas State University (HS and

DS approximations), and Chen, Crasemann, and co-workers at the University of Oregon (HS and DS approximations).

In Table III, experimental relative intensities of K–LL transitions of Ne, the best-studied case so far (Körber and Mehlhorn, 1966; Mehlhorn *et al.*, 1968; Krause *et al.*, 1971), are compared with theoretical values calculated for a HS potential (Walters and Bhalla, 1971a) as well as for a HF potential (Kelly, 1975). The total K–LL transition probabilities in atomic units are also given. The experimental total transition probability $P_{K\text{-}LL}$ can be evaluated from the experimentally known K-shell width $\Gamma(K) = (270 \pm 20)$ meV (Svensson *et al.*, 1976) using the relation

$$P_{K\text{-}LL} = P_K - P_{\gamma K} - P_{K\text{-}LLL} = \frac{\Gamma(K)}{\hbar} \frac{1-\omega_K}{1+r}. \tag{22}$$

Here, $P_{\gamma K}$ and $P_{K\text{-}LLL}$ are the probabilities for radiative transitions $K-L$ and for double Auger transitions K–LLL, ω_K is the K-shell fluorescence yield, and r is the ratio $P_{K\text{-}LLL}/P_{K\text{-}LL}$. With $\omega_K = 0.025$ and $r = 0.08$ (Carlson and Krause, 1965), one obtains $P_{K\text{-}LL} = (90 \pm 8) \times 10^{-4}$ a.u.

Although the total K–LL transition probabilities in Table III agree within 10% with the experimental values, the relative probabilities of individual transitions disagree especially for the $^1S_0$ and $^1P_1$ components. The main reason for this discrepancy is the complete neglect of many-electron effects, viz., electron-correlation and relaxation effects. For example, the

TABLE III

Comparison of theoretical relative and total absolute K–LL transition rates of Ne with experiment. The relative rates are normalized to the ^1D rate; the total rates are in units of 10^{-4} a.u.

Transition		HF[a]	HS[b]	HS[b,c] with FISCI	Experiment[d]
K–L_1L_1	^1S	16.7	15.7	11.8	10.0(1)
K–$L_1L_{2,3}$	^1P	35.8	35.9	35.9	28.2(4)
	^3P	8.8	11.0	11.0	10.2(4)
K–$L_{2,3}L_{2,3}$	^1S	8.0	7.7	11.6	15.6(2)
	^1D	100.0	100.0	100.0	100.0
K–LL (total)		99.1	88.3	88.3	90 ± 8[e]

[a] Kelly (1975).
[b] Walters and Bhalla (1971b).
[c] Bhalla (1973a), corrected values.
[d] Krause *et al.* (1971). These values are in good agreement with those obtained by Mehlhorn *et al.* (1968).
[e] Svensson *et al.* (1976).

initial- and final-state wave functions in Eq. (7a) have been approximated by independent-particle functions. The relaxation of the final state constitutes a many-electron effect in the sense that those electrons not directly involved in the Auger transition are coupled to the active electrons via the relaxed average potential. The neglect of relaxation (frozen-core approximation) is therefore another possible source for the discrepancy between theoretical and experimental transition probabilities.

Howat *et al.* (1978) were the first to go beyond the frozen-core approximation. They showed that the transition probability then is no longer given by Eq. (7b) with $\sum e^2/r_{\mu\nu}$ as operator, but by Eq. (7a) (see also Åberg and Howat, 1982). If Ψ_f and Ψ_i are again approximated by Slater determinants, where relaxed orbitals are taken for Ψ_f, then the reduction of Eq. (7a) to Eq. (10) leads to additional terms, which are overlap integrals and Slater integrals between initial orbitals and relaxed final orbitals (for details see Howat *et al.*, 1978).

We now discuss the effects of electron correlation on the transition probabilities. Electron correlation (EC) is usually treated through configuration interaction (CI), and the term CI will be used as equivalent to EC in the following. There are several channels through which electron correlation can enter into the transition matrix elements. As indicated in Fig. 5, electron correlation is usually divided into initial-state configuration interaction (ISCI) and final-state configuration interaction (FSCI); the final-state CI is conveniently subdivided into final-ionic-state CI (FISCI) and final-continuum-state CI (FCSCI). ISCI, FISCI, and FCSCI not only alter the relative intensities and widths of diagram Auger lines but also lead to additional initial and final Auger states which give rise to satellite transitions (Sec. 5).

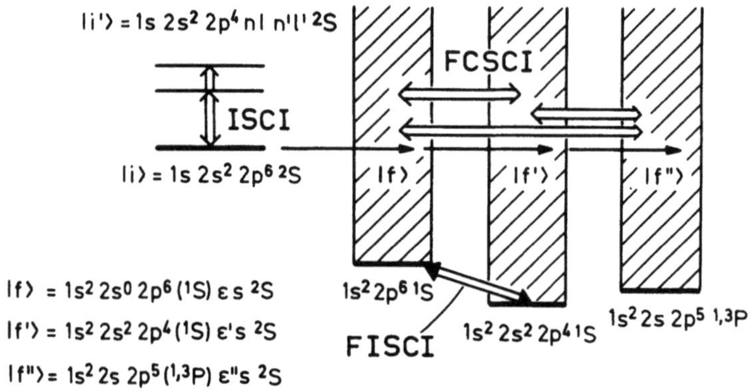

FIGURE 5. Possibilities of electron correlation entering into the Auger matrix element in the case of *K–LL* transitions of Ne. From Mehlhorn (1983a).

Asaad (1965) was first to show that inclusion of CI between the final ionic states $\underline{2s}^2\ {}^1S_0$ and $\underline{2p}^2\ {}^1S_0$ (FISCI) considerably improves agreement between theoretical and experimental *K–LL* intensities, especially for low Z. But even with FISCI, agreement of the intensities of the *K–LL* spectrum of neon was not satisfactory, and deviations as large as 20% between theory and experiment persisted (cf. third column of Table III). Kelly (1975), using many-body perturbation theory to calculate the *K–LL* rates of neon, included not only FISCI but also FCSCI and ISCI and obtained perfect agreement regarding the relative intensities (see Table IV). Examples of final continuum states $|f\rangle$, $|f'\rangle$, and $|f''\rangle$ and initial states $|i\rangle$ and $|i'\rangle$ which couple through CI and can alter the relative transition probabilities of diagram transitions are given in Fig. 5. Although Kelly also corrected in his calculation for the relaxation of final orbitals, the total transition probability was still smaller than the experimental value by about 10%. Kelly showed that FCSI is as important as FISCI in the calculation of transition rates (at least for neon), whereas ISCI is less important but non-negligible.

Howat *et al.* (1978) as well calculated *K–LL* transition probabilities using their new Auger theory in which FCSCI between the various final continuum channels is explicitly included. These authors also included the orbital relaxation effect. For the continuum orbitals they used either a potential generated by restricted HF initial $\underline{1s}\ {}^2S$ state orbitals (*i* in Table IV), or by restricted transition-operator orbitals based on $\underline{1s}$–$\underline{2s2p}$ configurations (*ii* in Table IV). Agreement with experimental rates for neon is much better than that for calculations that include only FISCI (although

TABLE IV

Comparison of theoretical relative and total absolute *K–LL* transition rates of Ne (including electron correlation) with experiment. The relative rates are normalized to the K–$L_{2,3}L_{2,3}({}^1D)$ rate, the total rates are in units of 10^{-4} a.u.

Transition		Kelly (1975) HF	Kelly (1975) MBPT	Howat *et al.* (1978) (i)	Howat *et al.* (1978) (ii)	Petrini (1982) CC	Experiment Krause *et al.* (1971)
K–L_1L_1	${}^1S$	16.7	9.9	9.9	8.2	9.7	10.0(1)
K–$L_1L_{2,3}$	${}^1P$	35.8	27.8	32.1	27.3	27.2	28.2(4)
	${}^3P$	8.8	9.9	16.7	12.7	10.7	10.2(3)
K–$L_{2,3}L_{2,3}$	${}^1S$	8.0	15.6	15.7	15.1	15.2	15.6(4)
	${}^1D$	100.0	100.0	100.0	100.0	100.0	100.0
K–LL (total)		99.1	80.5	105.7	89.8	85.9	90 ± 8[a]

[a] Svensson *et al.* (1976).

not as good as Kelly's values), which again demonstrates that ISCI is not negligible. On the other hand, the total *K–LL* rate calculated by Howat *et al.* (1978) is in better agreement with the experimental result.

Very recently, Petrini (1982) has calculated *K–LL* transition rates of neon by analyzing the resonance behavior of the scattering matrix for the $e^- + Ne^{2+}$ collision system in the close-coupling approximation, the results (CC in Table IV) agree well with experiment.

Howat (1978) has included FISCI and FCSCI as well for the *K* Auger rates of Mg and obtained results that are in excellent agreement with experiment for free Mg atoms (Breuckmann, 1979); here too, FCSCI proved as important as FISCI.

The foregoing examples demonstrate that full inclusion of electron correlation (at least FISCI and FCSCI) brings theory into good accord with experiment for Auger transitions in which the outermost or next to outermost shell is involved in the final state. For *K–LL* transitions and higher atomic numbers, one would expect that FCSCI becomes smaller because of the smaller interaction between the final ionic core and the Auger electron, the latter having increasingly larger energies. Thus, the experimental relative intensities of *K–LL* lines of argon (Krause, 1975; Schneider *et al.*, 1976a) are in satisfactory agreement with theoretical HS and DS values for which only FISCI is included (Fig. 6). On the other hand, more recent experimental results by Asplund *et al.* (1977) deviate from the earlier experimental values and are thus in less good agreement with theory. For

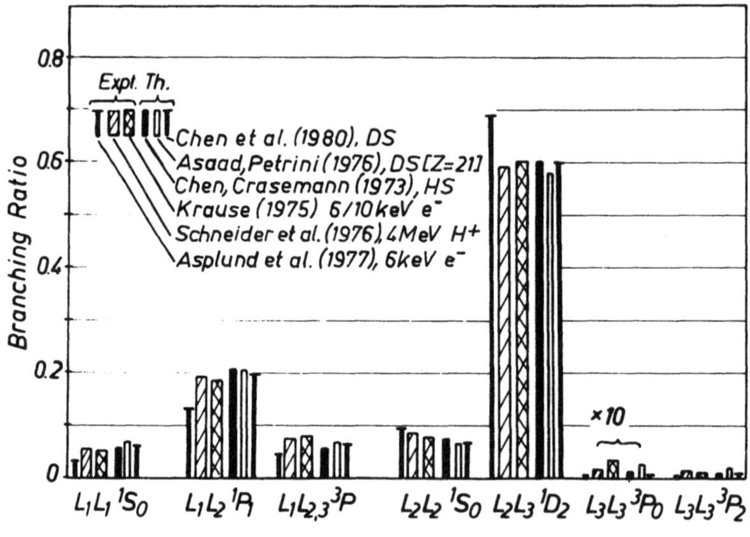

FIGURE 6. Bar diagram of the *K–LL* spectrum of Ar; comparison between experiment and theory.

$Z > 18$ relativistic calculations of the K–LL rates in the intermediate coupling scheme with FISCI (Asaad and Petrini, 1976; Chen *et al.*, 1980) agree reasonably well with experiment (see also Krause, 1976).

For other than K Auger spectra and for which the final vacancies are in the outermost or next to outermost shells, very strong FISCI has been found if excited d electrons are involved. For example, the final Auger states $sp^5(^{1,3}P)$ (diagram states) interact with the excited states $s^2p^3\{^{d}_{s'}\}(^{1,3}P)$ according to

$$\begin{array}{ll} sp^5(^{1,3}P) & \\ s^2p^3d(^{1,3}P) & \text{strong CI} \\ s^2p^3s'(^{1,3}P). & \text{weak CI} \end{array} \qquad (23)$$

An energy shift of the diagram lines results (McGuire, 1975c,d) and intensity is transferred from the diagram lines to the FISCI satellite lines that have $s^2p^3\{^{d}_{s'}\}(^{1,3}P)$ as final states (Mehlhorn, 1976; see also Sec. 5).

Strong FISCI effects of the kind mentioned before have been found in the $3d$–$4s4p^5(^{1,3}P)\,5s^m$ diagram spectra of Kr ($m = 0$) (Mehlhorn *et al.*, 1972; Werme *et al.*, 1972; McGuire, 1975d), Rb ($m = 1$) (Menzel and Mehlhorn, 1981), and Sr ($m = 2$) (Mehlhorn *et al.*, 1977) and in the $2p$–$3s3p^5(^{1,3}P)\,4s^m$ spectra of Ar ($m = 0$) (Werme *et al.*, 1973; McGuire, 1975c; Dyall and Larkins, 1982b), K ($m = 1$) (Breuckmann, 1978), and Ca ($m = 2$) (Breuckmann, 1978). For example, the full $L_{2,3}$ Auger spectrum of Ar is displayed in Fig. 7 (see also Fig. 4b). The $L_{2,3}$–MM diagram lines are the peaks indicated in black in the spectrum. Most of the other lines between 185 and 200 eV are due to (multiple-ionization) satellite transitions of the type $L_{2,3}M$–MMM. The initial double ionization $L_{2,3}M$ is either due to shakeoff of an M electron during $L_{2,3}$ ionization or produced in a preceding L_1–$L_{2,3}M$ Coster–Kronig transition. These $L_{2,3}M$–MMM transitions have been discussed by McGuire (1975c). Here we are interested in the diagram lines $L_{2,3}$–$3s3p^5\ ^{1,3}P$ (lines 45, 48, 52, and 60 of Fig. 7). In a multi-CI calculation of configurations given in Eq. (23) Dyall and Larkins (1982b) obtained the positions and intensities of diagram as well as FISCI satellite lines (bar spectrum in Fig. 7). The 26% and 54% intensity transfer from ^3P and ^1P $L_{2,3}$–$3s3p^5$ diagram lines, respectively, to the corresponding satellite lines brings the experimental intensities of $L_{2,3}$–$3s3p^5$ diagram lines into much better agreement with theory (see, e.g., Mehlhorn and Stalherm, 1968). A similar calculation was done earlier by McGuire (1975c); in particular, McGuire reassigned lines No. 24/29 to the diagram transitions $L_{2,3}$–$3s^2$, instead of lines No. 32/37 assigned by Mehlhorn and Stalherm (1968), improving also here the agreement with theoretical intensities.

For spectra for which the final states are inner-shell states it has been

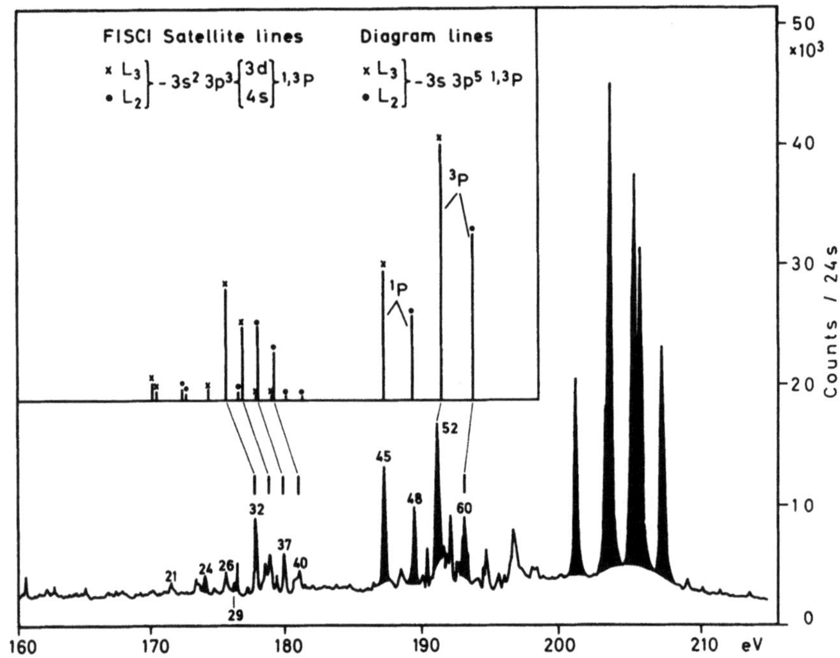

FIGURE 7. $L_{2,3}$–MM spectrum of Ar, excited by 3-keV electrons (from Werme *et al.*, 1973). The diagram lines are the black solid peaks. The FISCI satellite lines and their parent diagram lines are indicated by the bar spectrum. Their intensities and energies are relative to the experimental $L_{2,3}$–$3s3p^5\ ^1P$ diagram lines 45 and 48, taken from Dyall and Larkins (1982b).

found that FISCI plays only a minor role if relative intensities are considered, e.g., $M_{4,5}$–$N_{4,5}N_{4,5}$ of Xe (Hagmann *et al.*, 1974; Aksela *et al.*, 1979), Cd (Aksela and Aksela, 1974), and Ag (Väyrynen *et al.*, 1980).

4.3. *Linewidths*

The natural width Γ of the Auger line of a transition $\underline{n}\underline{l}$–$\underline{n}_1\underline{l}_1\underline{n}_2\underline{l}_2$, SLJ is

$$\Gamma = \Gamma_i(\underline{n}\underline{l}) + \Gamma_f(\underline{n}_1\underline{l}_1\underline{n}_2\underline{l}_2, SLJ), \tag{24}$$

where Γ_i and Γ_f are linked to the total decay probabilities (nonradiative and radiative) P_i and P_f of the initial and final Auger states, respectively, by $\Gamma_{i,f} = \hbar P_{i,f}$. In order to compare with theory or with experimental level widths, the approximation

$$\Gamma = \Gamma_i(\underline{n}\underline{l}) + \Gamma_f(\underline{n}_1\underline{l}_1) + \Gamma_f(\underline{n}_2\underline{l}_2) \tag{25}$$

is usually employed in place of Eq. (24), where the width $\Gamma_f(\underline{n}_1\underline{l}_1\underline{n}\,\underline{l}_2)$ of the two-vacancy state is decomposed into the individual level widths. Natural widths of Auger lines have been measured only in a few cases of high-resolution spectra (Table V). A compilation of experimental *K–LL* Auger linewidths and comparison with semiempirical level widths based on Eq. (25) has been compiled by Krause and Oliver (1979).

Caution must be taken when Eq. (25) is applied to transitions in which one (or both) of the final one-vacancy states $\underline{n}_1\underline{l}_1$ and $\underline{n}_2\underline{l}_2$ decay through CK or even sCK transitions. Since the energy of these transitions is usually small, the presence of the second vacancy in the final state in Eq. (24) introduces a relatively large decrease of the transition energy of the final-state decay process. This in turn leads to a substantial change of the

TABLE V

Experimental natural widths (FWHM) of Auger lines in free atoms[a] (Krause, 1980).

Element	Line	Γ (eV)	Ref.
$_{10}$Ne	K–$L_{2,3}L_{2,3}$	0.23(3)	Ridder *et al.* (1976)[b]
$_{11}$Na	K–$L_{2,3}L_{2,3}$	0.30(5)	Breuckmann (1978)
	L_1–$L_{2,3}M$	0.24(2)	Breuckmann *et al.* (1977)
$_{12}$Mg	K–$L_{2,3}L_{2,3}$	0.35(5)	Breuckmann (1979)
		0.25	Väyrynen *et al.* (1977)
$_{13}$Ar	K–$L_{2,3}L_{2,3}$	0.94(3)	Krause (1975)
			Asplund *et al.* (1977)
	L_3–$M_{2,3}M_{2,3}$	0.12(2)	Ridder *et al.* (1976)[b]
			Väyrynen and Aksela (1979)
	L_1–$L_{2,3}M_{2,3}$	1.9(2)	Mehlhorn (1968b)
$_{19}$K	L_2–$M_{2,3}M_{2,3}$	0.20(2)	Breuckmann (1978)
$_{20}$Ca	L_2–$M_{2,3}M_{2,3}$	0.37(3)	Breuckmann (1978)
$_{30}$Zn	L_3–$M_{4,5}M_{4,5}$	0.50(15)	Aksela *et al.* (1974)
	M_2–$N_{4,5}N_{4,5}$	2.1(2)	Mehlhorn *et al.* (1977)
	M_3–$N_{4,5}N_{4,5}$	2.1(2)	Mehlhorn *et al.* (1977)
$_{36}$Kr	L_3–$M_{4,5}M_{4,5}$	0.82	Aksela *et al.* (1980a)
	M_5–$N_{2,3}N_{2,3}$	0.10(1)	Svensson *et al.* (1976)
$_{47}$Ag	M_5–$N_{4,5}N_{4,5}$	0.21	Väyrynen *et al.* (1980)
$_{48}$Cd	M_5–$N_{4,5}N_{4,5}$	0.32(5)	Aksela *et al.* (1974, 1977b)
$_{49}$In	M_5–$N_{4,5}N_{4,5}$	0.4(1)	Aksela *et al.* (1980b)
$_{54}$Xe	M_5–$N_{4,5}N_{4,5}$	0.64(4)	Svensson *et al.* (1976)
			Hagmann *et al.* (1974)
			Aksela *et al.* (1979)
	N_5–$O_{2,3}O_{2,3}$	0.10(1)	Svensson *et al.* (1976)
$_{56}$Ba	N_5–$O_{2,3}O_{2,3}$	0.16(2)	Breuckmann (1978)
$_{80}$Hg	N_7–$O_{4,5}O_{4,5}$	0.22	Aksela *et al.* (1977a)

[a] Widths are listed only for the major component of a group.
[b] Values quoted are revised.

transition probability P_f (and level width Γ_f) and may even forbid a super-Coster–Kronig transition on energetic grounds. Examples are the $\underline{3d}$–$\underline{4p4d}$ transitions of Xe, where the single $\underline{4p}$ vacancy state is strongly shifted and broadened by sCK transitions (Wendin, 1981; Wendin and Ohno, 1976) but the $\underline{4p4d}$ double-vacancy state is not (Aksela *et al.*, 1979).

An interesting case, showing the importance of many-body effects in the calculation of level widths (and thus of linewidths) is constituted by the $\underline{3p}$–$\underline{3d}^2$ sCK and $\underline{3p}$–$\underline{3d4s}$ CK transitions of Zn. From the spectrum shown in Fig. 8, a linewidth Γ of 2.1(2) eV had been determined (Mehlhorn *et al.*, 1977). Since both final states $\underline{3d}^2$ and $\underline{3d4s}$ decay only radiatively, the linewidth Γ is identical with the level width Γ_i of the initial states, yielding $\Gamma(\underline{3p}_{1/2}) = \Gamma(\underline{3p}_{3/2}) = (2.1 \pm 0.2)$ eV. These values are smaller by a factor of 2 than those found by McGuire (1972) using a frozen $V(3p^{-1})$ potential and HS approximation, but they accidentally agreed with those calculated by Yin *et al.* (1974), who used a frozen-neutral-atom potential. Later, Chen *et al.* (1978) investigated the effect produced by relaxation of final-state orbitals and by the exchange between the continuum electron and bound electrons on the $\underline{3p}$ sCK and CK transition rates of Zn. Taking both into account reduced, but did not eliminate, the discrepancies between theoretical and experimental rates. Only in a many-body calculation (random-phase

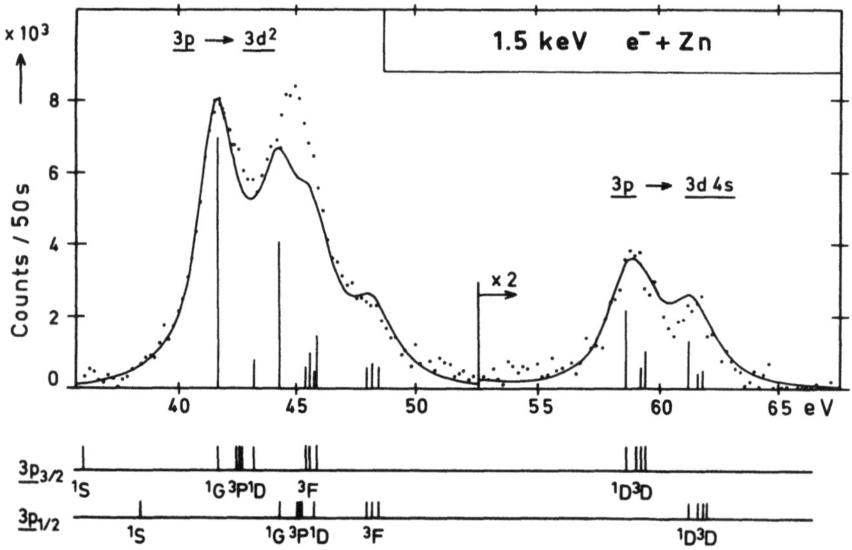

FIGURE 8. $\underline{3p}$–$\underline{3d}^2$ sCK transitions and $\underline{3p}$–$\underline{3d4s}$ CK transitions of free Zn atoms. The bar spectrum indicates the theoretical relative intensities calculated in j–j coupling for the $3p$ vacancy, and in intermediate coupling for the final vacancies, using McGuire's (1972) matrix elements. The solid line is the theoretical spectrum based on a linewidth $\Gamma = (2.1 + 0.2)$ eV. From Mehlhorn *et al.* (1977).

approximation), in which inner-shell ionization and Auger emission are treated in a unified manner (Ohno and Wendin, 1979), level widths were obtained as $\Gamma(\underline{3p}_{1/2}) = 2.2$ eV and $\Gamma(\underline{3p}_{3/2}) = 2.1$ eV, in excellent agreement with experiment. A similar discrepancy existed for the $2s$-level width of Ar (Mehlhorn, 1968b; McGuire, 1971a; Dyall and Larkins, 1982c) which could partly be resolved by most recent investigations (Karim *et al.*, 1984; Karim and Crasemann, 1984, 1985).

5. AUGER SATELLITE TRANSITIONS DUE TO MANY-ELECTRON EFFECTS

Many-electron effects consisting of electron correlation and relaxation not only alter the intensities and energies of diagram transitions, as pointed out in Sec. 4.2, but also cause satellite transitions which would be entirely forbidden in the independent-particle frozen-core model. Figure 9 shows a level diagram similar to that in Fig. 5, except that here the ground state (GS) is also included. Due to CI in the initial Auger state (ISCI), additional excited states can be reached in the primary ionization process (e.g., $\Psi^{(N-1)}_{2,\mathrm{IS}}$ in Fig. 9). Their decay leads to satellite lines (broken arrow in Fig. 9). Final-ionic-state CI (FISCI) and final-continuum CI (FCSCI) open new continuum channels (e.g., $\Psi^{(N-2)}_{2,\mathrm{FIS}} + e^-$ in Fig. 9), and the decay of an Auger diagram state into these continua leads again to satellite lines (broken arrow in Fig. 9).

If the ionic final state $\Psi^{(N-2)}_{2,\mathrm{FIS}}$ is connected via CI to $\Psi^{(N-2)}_{1,\mathrm{FIS}}$, then both states must have equal parity and symmetry. This is not necessarily the case when the final ionic states $\Psi^{(N-2)}$ are coupled through FCSCI. For example, in Sec. 4.2 we have seen that FCSCI couples the final ionic states $2s^0 2p^6\ {}^1S$ and $2s2p^5\ {}^{1,3}P$ via continuum channels $|f\rangle = 1s^2 2s^0 2p^6({}^1S)\varepsilon s\ {}^2S$ and $|f''\rangle = 1s^2 2s 2p^5({}^{1,3}P)\varepsilon'' p\ {}^2S$ (Fig. 5), which leads to a change of relative intensities of the transitions $\underline{1s}$–$\underline{2s}^2\ {}^1S$ and $\underline{1s}$–$\underline{2s2p}\ {}^{1,3}P$. Although FCSCI is

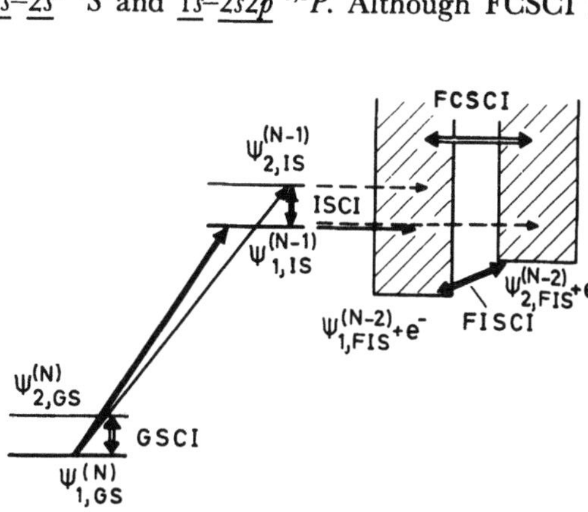

FIGURE 9. Various channels of CI causing satellite Auger transition: GSCI, ISCI, FISCI, and FCSCI. Diagram and CI satellite Auger transitions are indicated by the horizontal solid and broken lines, respectively.

rather important in the quantitative understanding of low-energy Auger diagram spectra (Sec. 4.2), it has not yet been applied to the case of Auger satellites.

If a substantial amount of excited-state configuration is admixed to the ground-state wave function $\Psi^{(N)}_{1,\mathrm{GS}}$, then an excited ionic state $\Psi^{(N-1)}_{2,\mathrm{IS}}$ can even be reached directly without CI in the initial ionic state. Since ground-state correlation of closed-shells atoms is usually small (see, e.g., Süzer *et al.*, 1976), it will not be further considered here.

Excited states, $\Psi^{(N-1)}_{k,\mathrm{IS}}$ and $\Psi^{(N-2)}_{k,\mathrm{FIS}}$, which can be reached by shakeup transitions in the conventional sense in the primary ionization and in the Auger process, respectively, can also give rise to satellite lines. Here the shakeup transitions are due to the relaxation which final orbitals experience when an inner-shell electron is removed or when the Auger transition occurs. In this view, the two-electron process (ionization and simultaneous shakeup) can be treated as a two-step process: first, ionization of an inner-shell electron, and second, excitation (shakeup) of an outer-shell electron. By conventional shake theory we mean the approach in which the initial- and final-state wave functions of the second step are approximated by Slater determinants of spin orbitals and relaxed spin orbitals are taken for the final state (Åberg, 1969). The shakeup probability is then basically given by a product of one-electron overlap matrix elements, $|\langle\overline{n_2 l}|n_1 l\rangle|^2$, the bar indicating a relaxed orbital. From these overlap matrix elements the selection rule $\Delta l=0$ follows for shake transitions. Clearly, a general theory in which the initial- and final-state wave functions are given by *relaxed* CI wave functions would automatically include also the $\Delta l=0$ excitations (see, e.g., Martin and Shirley, 1977). We therefore subsume the corresponding satellite transitions under the wider class of CI Auger satellite lines. Only in cases in which Auger satellites due to $\Delta l=0$ excitations can be explained quantitatively by a pure relaxation process of a central-field potential is the term "shakeup satellites" justified. Such cases include, e.g., the Auger satellites due to $3s\rightarrow 4s$ shakeup associated either with $1s$ ionization or $\underline{1s}$–LL transitions in Na (Hillig *et al.*, 1974; Breuckmann, 1978) and Mg (Breuckmann, 1979).

5.1. *Satellite Transitions Due to Final-Ionic-State Configuration Interaction (FISCI)*

Transitions of this kind are frequently also called discrete double Auger transitions [cf. Eq. (4)]. Because electron correlation is largest in the outermost shell we expect large satellite intensities if the final Auger state involves electrons from this shell. If, in addition, the energies of the final Auger diagram state and of a CI-excited state are close, then we expect very intense FISCI satellites. It has been found that this is especially the case when *d* electrons are involved in the final excited states, e.g., for configurations described in Eq. (23). Let us consider, for example, the diagram

(D) and satellite (S) transitions of the $L_{2,3}$–MM spectrum of Ar with the final-ionic-state configurations

$$D = 2p^6 3s 3p^5\ {}^1P \quad \text{and} \quad S_j = 2p^6 3s^2 3p^3 \begin{Bmatrix} ns \\ nd \end{Bmatrix} {}^1P. \tag{26}$$

In the CI model, the diagram and satellite final-ionic-state wave functions are (Mehlhorn, 1976)

$$\Phi_D = c_{00}\Phi_D^0 + \sum_j c_{0j}\Phi_{S_j}^0$$

and (27)

$$\Phi_{S_k} = c_{k0}\Phi_D^0 + \sum_j c_{kj}\Phi_{S_j}^0,$$

where Φ_D^0 and $\Phi_{S_j}^0$ are determinantal wave functions. The initial-state wave function Ψ_i^0 and the total-final-state wave functions Ψ_f, including the ejected Auger electron εl, are

$$\Psi_i^0 = |2p^5 3s^2 3p^6\ {}^2P\rangle,$$
$$\Psi_{f,D} = |\Phi_D({}^1P)\,\varepsilon_0 l\ {}^2P\rangle, \qquad \Psi_{f,S_k} = |\Phi_{S_k}({}^1P)\,\varepsilon_k l\ {}^2P\rangle. \tag{28}$$

Then the diagram and associated satellite intensities are

$$I_D = K\,|c_{00}\langle\Phi_D^0\varepsilon_0 l|\,V\,|\Psi_i^0\rangle + \sum_j c_{0j}\langle\Phi_{S_j}^0\varepsilon_0 l|\,V\,|\Psi_i^0\rangle|^2,$$
$$I_{S_k} = K\,|c_{k0}\langle\Phi_D^0\varepsilon_k l|\,V\,|\Psi_i^0\rangle + \sum_j c_{kj}\langle\Phi_{S_j}^0\varepsilon_k l|\,V\,|\Psi_i^0\rangle|^2, \tag{29}$$

with $V = \sum e^2/r_{\mu\nu}$. The matrix elements $\langle\Phi_{S_j}^0\varepsilon l|\,V\,|\Psi_i^0\rangle$ vanish identically because Ψ_i^0 and $\Phi_{S_j}^0\varepsilon l$ differ by more than two spin orbitals. If $I_D^0 = K\,|\langle\Phi_D^0\varepsilon_0 l|\,V\,|\Psi_i^0\rangle|^2$ is the diagram-line intensity without FISCI, then we find from Eq. (29)

$$I_D = |c_{00}|^2\, I_D^0 \quad \text{and} \quad I_{S_k} = |c_{k0}|^2\, I_D^0, \tag{30}$$

with

$$I_D + \sum_k I_{S_k} = I_D^0. \tag{31}$$

In principle one should also include in Eq. (26) the continuum channels $2p^6 3s^2 3p^3 \{{}^{\varepsilon s}_{\varepsilon d}\}\ {}^1P$, which are the corresponding final states of continuum double Auger transitions in which two electrons are ejected simultaneously [Eq. (3)]. The intensity of these continuum double Auger transitions is not at all negligible (see, e.g., Åberg, 1975), but they have been omitted here for clarity of presentation. Any electron correlation in the initial-state wave function Ψ_i^0 has also been neglected for the same reason.

From Eq. (30) it follows that the intensity of satellite lines is proportional to the admixture $|c_{k0}|^2$ of the diagram configuration to the excited-state configuration, i.e., the intensity loss of the diagram line reappears as intensity of associated satellite lines.

The first clear identification of FISCI satellite lines was made for the satellite lines associated with the $\underline{3d}$–$4s4p^5\ ^{1,3}P$ diagram transitions of Kr (Mehlhorn *et al.*, 1972; Werme *et al.*, 1972; McGuire, 1975c). More recently, exactly the same type of FISCI satellites has also been found for Rb (Menzel and Mehlhorn, 1981) and for Sr (Mehlhorn *et al.*, 1977). FISCI satellite transitions associated with the $\underline{2p}$–$3s3p^5(^{1,3}P)\,4s^m$ diagram transitions have been discussed for Ar ($m=0$) (Mehlhorn *et al.*, 1972; Werme *et al.*, 1973; McGuire, 1975d; Dyall and Larkins, 1982b), K ($m=1$) (Breuckmann, 1978), and Ca ($m=2$) (Breuckmann, 1978), and the satellite spectrum connected with the K–LM transitions of Ar was treated by Darko *et al.* (1981).

For example, the FISCI satellite spectrum associated with the $\underline{2p}$–$3s3p^5\ ^{1,3}P$ diagram spectrum of Ar has been computed by Dyall and Larkins (1982b) in a multi-CI calculation and is displayed as a bar spectrum in Fig. 7. These authors also included CI in the initial $\underline{2p}$ vacancy state and found a total intensity transfer of 26% and 54% from the ^1P and ^3P diagram lines, respectively, to the satellite lines. The satellite spectrum agrees reasonably well with experiment (lines No. 21 to 40 between 172 and 182 eV in Fig. 7). The satellite lines associated with the $\underline{2p}$–$\underline{3s}^2\ ^1S$ diagram transitions, not displayed in Fig. 7, were also calculated by Dyall and Larkins (1982b).

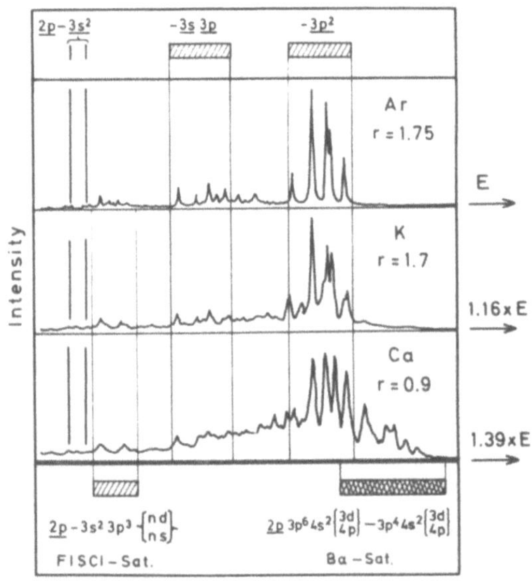

FIGURE 10. Comparison of experimental $\underline{2p}$–MM spectra of Ar, K, and Ca excited by 2-keV electrons (Breuckmann, 1978). The energy scales of the K and Ca spectra have been changed in order to let the Auger groups $\underline{3s}^2$, $\underline{3s3p}$, and $\underline{3p}^2$ fall on top of each other. The energy regions of FISCI satellites in Ar, K, and Ca and of strong Bα satellites in Ca are indicated by the hatched and cross-hatched areas below the Ca spectrum. The ratio r of diagram intensities $I(L_3$–$MM)/I(L_2$–$MM)$ is also given.

It has been found (Breuckmann, 1978) that the total relative intensity of FISCI satellites $\underline{2p}$–$3s^2 3p^3 \{ {nd \atop ns} \}$ remains approximately unchanged from Ar to Ca (Fig. 10). It therefore appears that this kind of electron correlation is still strong when the final diagram state is not in the outermost but in the penultimate shell. The same has also been found for the $\underline{3d}$–$4s^2 4p^3 \{ {nd \atop ns} \}$ satellites in the $\underline{3d}$–NN spectra from Kr to Sr (Mehlhorn *et al.*, 1977).

5.2. *Satellite Transitions Due to Initial-State Configuration Interaction (ISCI)*

The initial states of ISCI transitions are excited ($N-1$)-electron states (Fig. 9) which in addition to the inner-shell vacancy have at least one additional excited electron, the excitation of which is either through ground-state (GSCI) or initial-ionic-state CI (ISCI) or both CI together. In Fig. 11 we illustrate the essential features of these excitation processes, associated with the $2p$ ionization of Ca as example. A quantitative treatment would correspond to what has been outlined for FISCI in Eqs. (26)–(30).

The excited states $\Psi_{k,\mathrm{IS}}^{(N-1)}$ associated with the $\underline{nl}$ vacancy state also lead to satellites in the nl photoelectron spectrum; calculations of the ground- and ionic-state CI have been performed, e.g., for Ne($\underline{1s}$) (Martin and Shirley, 1976), Ar($\underline{2s}$) (Dyall and Larkins, 1982a), Ar($\underline{2p}$) (Dyall *et al.*, 1981), Xe($\underline{5s}$) (Adam *et al.*, 1978), Ba($\underline{5p}$) (Rose *et al.*, 1980). The general theory is outlined by Martin and Shirley (1977).

Assuming the ground-state CI for closed-shell atoms to be small ($c_{10}^2 \cong 1$; $c_{1j}^2 \cong 0$ for $j \neq 0$), the excitation probability of vacancy state $\Psi_{k,\mathrm{IS}}$ (Fig. 11) is proportional to a_{k0}^2. There is always a certain admixture a_{k0}^2 of ionic diagram configuration into excited ionic states $\Psi_{k,\mathrm{IS}}$ leading to ISCI satellites of

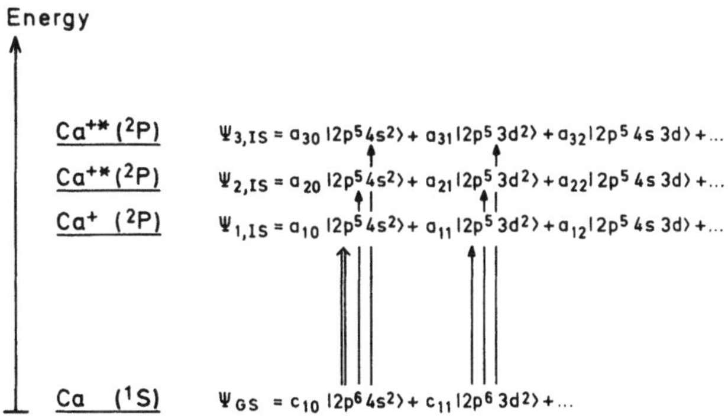

FIGURE 11. Model to illustrate the various paths to the excited inner-shell vacancy state $\Psi_{k,\mathrm{IS}}$, either by GSCI or by ISCI. The zero-order CI transition amplitude (⇑) leads only to state $\Psi_{1,\mathrm{IS}}$ and is proportional to $c_{10}a_{10}$. The first-order CI transition amplitudes (↑) exciting $\Psi_{2,\mathrm{IS}}$ e.g., are proportional to $c_{10}a_{20}$ and $c_{11}a_{21}$.

generally low intensity. But there are special cases in which a large fraction or even the larger part of the initial ionic diagram configuration is mixed into ISCI states, resulting in a complete redistribution of diagram-line intensity into the satellite intensity.

For example, in Fig. 10 the ratio of diagram intensities $I(\underline{2p}_{3/2}\text{–}MM)/I(\underline{2p}_{1/2}\text{–}MM)$ varies from $r = 1.75$ for Ar and 1.7 for K to $r = 0.9$ for Ca. At the same time, the intensity at the high-energy side of the $\underline{2p}\text{–}\underline{3p}^2$ Auger group increases dramatically from K to Ca. These two facts have been interpreted as due to a very strong CI in the initial ionic state $\underline{2p}_{3/2}$ of Ca (Breuckmann, 1978).

In Fig. 12, the measured $\underline{2p}\text{–}\mathcal{N}_1\mathcal{N}_1$ spectrum of Ca is shown (Weber, 1984). In this case the final state is a closed-shell ion, $3s^2 3p^6\ {}^1S$, and neglecting CI we expect exactly one doublet according to the $\underline{2p}_{1/2,3/2}$ splitting. At least four strong lines have been found, where the lines a_1/b_1 and a_2/b_2 have the 3.6-eV fine-structure splitting of the $\underline{2p}_{1/2,3/2}$ vacancy states. Again, the additional satellite lines can only be explained through very strong CI in the initial $\underline{2p}$ vacancy state. Shakeup in the conventional sense, i.e., $4s \rightarrow 5s$ excitation accompanying the $2p$ ionization, can be ruled out since the measured 1.2-eV energy difference between the lines a_1/a_2 and b_1/b_2 is much smaller than the excitation energy $\Delta E(4s \rightarrow 5s)$ of ~ 5 eV.

According to Fig. 11, the Ca^{+*} state $\Psi_{2,\mathrm{IS}}(^2P)$ can be reached from the

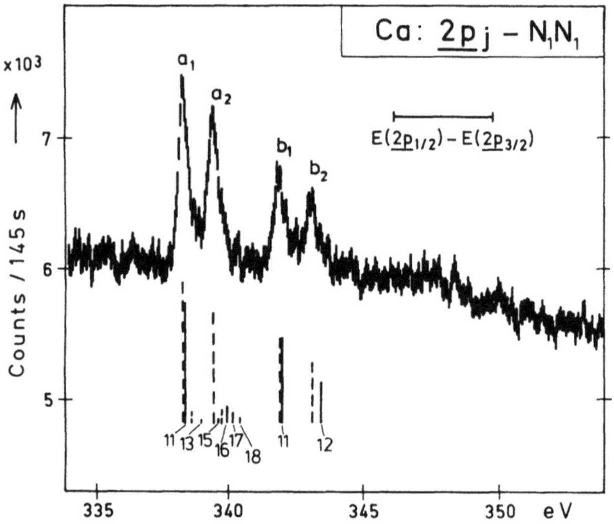

FIGURE 12. $\underline{2p}_j\text{–}\mathcal{N}_1\mathcal{N}_1$ Auger spectrum of Ca excited by 2-keV electrons (Weber, 1984). The energy splitting of the doublets a_1/b_1 and a_2/b_2 correspond to the fine-structure splitting of the initial $\underline{2p}_{1/2}$ and $\underline{2p}_{3/2}$ levels. The computer fit to the experimental spectrum yields the broken bars. The theoretical relative intensities and energies are given by the solid bars; they have been normalized to line b_1. The numbers at the solid bars refer to k of Fig. 13.

ground state via two paths, the admixture of $2p^5 4s^2$ into $\Psi_{2,\mathrm{IS}}(^2P)$ and the admixture of $2p^5 3d^2(^1S)$ into the ground state. A MCHF calculation (Breuckmann, 1978) of the two configurations $4s^2$ and $3d^2(^1S)$ with either a ground-state core or a $\underline{2p}$ vacancy core yields as admixtures of the $3d^2(^1S)$ configuration into Ψ_{GS} and $\Psi_{1,\mathrm{IS}}$ the values $c_{11}^2 = 0.02$ and $a_{11}^2 = 0.43$, respectively. This clearly shows that for a $\underline{2p}$ vacancy in Ca ISCI dominates by far over GSCI.

In order to obtain quantitative results for the initial fine-structure vacancy states $\underline{2p}_j$ as well, multiconfiguration Dirac–Fock calculations have been performed (Breuckmann *et al.*, 1983; Weber, 1984), using the MCDF program of Grant *et al.* (1980). For the $\underline{2p}_j$ ionic states the configurations $4s^2$, $3d^2$, and $3d4s$ were included, yielding a total of 12 and 19 states with $J = 1/2$ and $J = 3/2$, respectively. For the $J = 0$ neutral atomic states, the CI of $4s^2$, $3d_{5/2}^2$, and $3d_{3/2}^2$ was considered and admixtures of $c_{10}^2(4s^2) = 0.9916$, $c_{11}^2(3d_{5/2}^2) < 1 \times 10^{-4}$, and $c_{12}^2(3d_{3/2}^2) = 0.0084$ to the ground state Ψ_{GS} were found, in agreement with the MCHF calculation. The MCDF energies for the neutral Ca atom and for the $J = 1/2$ and $J = 3/2$ manifold of $\underline{2p}_j$ vacancy states are plotted in Fig. 13 on a relative energy scale. The squared values of the mixing coefficients $a_{k0}^2(4s^2)$ of $\underline{2p}_j 4s^2$ diagram configuration into the various $\underline{2p}_j$ vacancy states are also listed if they exceed 0.1. As can be seen by the strength of the mixing coefficients of Fig. 13 ISCI is exceedingly strong in the case of a $\underline{2p}_j$ state in Ca. In addition, the MCDF calculation clearly shows that ISCI gives different results for the two fine-structure vacancy states. Assuming that the excitation probability of an initial vacancy state is proportional to $a_{k0}^2(4s^2)$ (the small admixture of $3d_{3/2}^2$ configuration into the ground state being neglected), then in case of $\underline{2p}_{1/2}$ ionization one expects two initial ionic states with about equal probability whereas in the case of $\underline{2p}_{3/2}$ ionization at least four initial ionic states with comparable strengths are excited.

Although a unique distinction between diagram and satellite $\underline{2p}_j$ ionic states of Ca is not possible, the $J = 1/2$ state with largest probability ($k = 12$) and that $J = 3/2$ state having the observed fine structure splitting of 3.6 eV ($k = 16$) are taken as "diagram" states. Assuming now that the ratio of ionization probabilities of the $2p_{3/2}$ and the $2p_{1/2}$ electrons by 2-keV incident electrons has the statistical value of 2, the ratio r' of excitation of initial $\underline{2p}_j$ "diagram" Auger states ($k = 12$ for $J = 1/2$ and $k = 16$ for $J = 3/2$, see Fig. 13) is

$$r' = \frac{2\,|c_{1,0}(4s^2)\,a_{16,0}(4s^2) + c_{1,2}(3d_{3/2}^2)\,a_{16,2}(3d_{3/2}^2)|_{J=3/2}^2}{|c_{1,0}(4s^2)\,a_{12,0}(4s^2) + c_{1,2}(3d_{3/2}^2)\,a_{12,2}(3d_{3/2}^2)|_{J=1/2}^2} = 1.03, \qquad (32)$$

where the indices have been separated by commas to avoid ambiguity. Although the excitation of the initial $\underline{2p}_j$ Auger states is almost only via one channel, namely, through admixture of the $4s^2$ configuration into the $\underline{2p}_j$

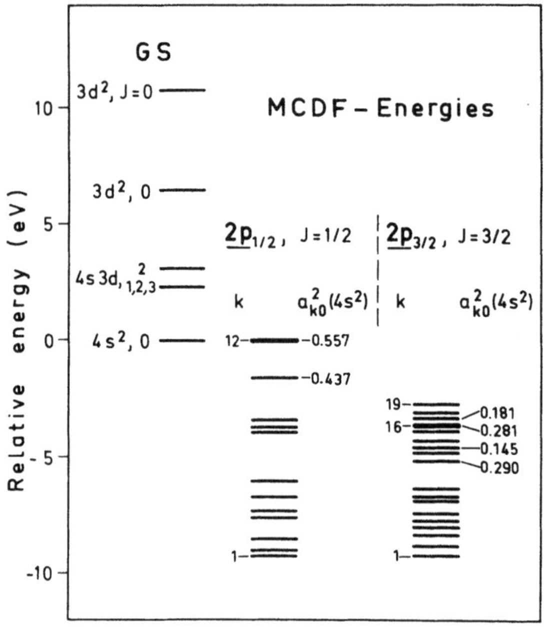

FIGURE 13. MCDF energies of outer-shell configurations $4s^2$, $4s3d$, and $3d^2$ for the neutral Ca atom (GS) and for the $J = 1/2$ and $J = 3/2$ levels of $\underline{2p}_j$ vacancy configurations given on a relative energy scale. The energies of the $4s^2$ ground state and the $\underline{2p}_{1/2}$ ($k = 12$) vacancy state are set to zero. The numbering of levels (12 for $J = 1/2$ and 19 for $J = 3/2$) is indicated by k; the admixture $a_{k0}^2(4s^2)$ of configuration $4s^2$ into states $\Psi_{k,\mathrm{IS}}(\underline{2p}_j)$ is also given if it exceeds 0.1. Levels $k = 12$ ($J = 1/2$) and $k = 16$ ($J = 3/2$) with fine-structure splitting 3.6 eV are taken as initial "diagram" states.

vacancy state, the decay of a $\underline{2p}_j$ Auger state to one final MM state will occur through several coherent channels, whereby according to MCDF calculations the outermost electrons stay as spectators in one of the configurations $4s^2$, $4s3d$, or $3d^2$. A quantitative calculation of the ratio r of line intensities must therefore also include FISCI of $4s^2 + 4s3d + 3d^2$ for the MM Auger states and the coherent decay via the various CI channels. For example, FISCI of the final $M_{2,3}M_{2,3}$ states $2p^6 3s^2 3p^4(4s^2 + 4s3d + 3d^2)\,J_f$ yields a total of MCDF states of 18, 32, and 47 for $J_f = 0$, 1, and 2, respectively. The calculation of relative Auger intensities of $\underline{2p}_j - MM$ transitions is therefore very extensive and has not yet been performed. In any case, the smaller value of the ratio r' of ionization probabilities [Eq. (32)] by a factor of 2 compared to the statistical value indicates, at least qualitatively, that also a much smaller value than 2 for the ratio r of corresponding line intensities can be expected, which would agree with the experimental value $r = 0.9(1)$.

The situation is much simpler for the calculation of relative Auger intensities of $\underline{2p}_j - N_1N_1$ transitions, where the final state of the Ca^{++} ion has the closed shell configuration $2p^6 3s^2 3p^6$ and FISCI is very small. Also here, an initial ionic MCDF state $\underline{2p}_j$ decays coherently via all single configuration states, but there is only *one* final single configuration state. In this case the relative Auger line intensities have been evaluated (Weber, 1984) using the $\underline{2p}_j(k)$–$\mathcal{N}_1\mathcal{N}_1$ transition probabilities calculated by Chen (1984). The theoretical relative intensities and relative energies are plotted in Fig. 12 as bar spectrum (solid lines), they have been normalized to the experimental bar spectrum (broken lines), at line b_1. As can be seen, the theoretical relative intensities and energies agree rather well with the experimental values of lines a_1 and b_2. In the case of line a_2 the theory predicts four lines, their total intensity is too small compared to the experimental value. One possible reason for this discrepancy is that the number of configurations ($4s^2$, $4s3d$, and $3d^2$) used in the ISCI calculation is too small and inclusion of more configurations would improve the result.

As it was mentioned above, another interesting feature of the $\underline{2p}$–MM spectrum of Ca, in contrast to the spectra of Ar and K, is the large intensity on the high energy side of the $\underline{2p}$–$\underline{3p}^2$ Auger group (see Fig. 10). Experimental investigation of the intensity I_α of this part of the spectrum relative to the intensity $I(\underline{2p}$–$\underline{3p}^2)$ of the $\underline{2p}$–$\underline{3p}^2$ group as function of energy E_0 of incident electrons yielded a strong increase of $I_\alpha/I(\underline{2p}$–$\underline{3p}^2)$ when E_0 was lowered close to threshold (Weber, 1984; Aksela, 1982). The same intensity dependence was found for the $B\alpha$ satellites in the $\underline{1s}$ Auger spectrum of neon (Kihara *et al.*, 1983). It was therefore concluded (Weber, 1984) that the intensity I_α in the Ca spectrum has its origin from $B\alpha$ transitions, where initially the $2p$ electron is *excited* to an nl level (e.g., $3d$, $4p$) and the $\underline{2p}$–$\underline{3p}^2$ Auger transition occurs within the core with the excited electron as a spectator [for the notation $B\alpha$ satellite see Krause *et al.* (1971)]. An MCDF calculation of the transition energies of $B\alpha$ satellites with $nl = 3d$ or $4p$ yielded an energy range which is given by the cross-hatched area in Fig. 10 (Weber, 1984).

Another important point in Fig. 13 is that for the Ca atom with a $2p$ vacancy the energy of the $3d$ orbital has changed relative to that of the $4s$ orbital compared to the neutral Ca atom. This is known as collapse of the $3d$ orbital (Connerade, 1976; Mansfield, 1976), which occurs for the *neutral*-atom $3d$ orbital for Z around 21 but for the $\underline{2p}$-vacancy atom already for $Z = 20$. A similar collapse occurs also for the $4d$ and $5d$ electrons between $Z = 38$–40 and $Z = 56$–58, whence one would expect similar strong effects of ISCI for the inner-shell vacancy states of Sr and Ba. Such has indeed been found for the $5p$ ionization of Ba (Connerade *et al.*, 1979; Rose *et al.*, 1980) and to a lesser extend also for the $4d$ ionization of Ba (Mehlhorn *et al.*, 1977).

As an example we show in Fig. 14 the $\underline{5p}_{3/2,1/2}$–$\underline{6s}^2(^1S_0)$ Auger spectrum of Ba (Mehlhorn *et al.*, 1977). Although one expects only one Auger doublet, the spectrum is rich in satellite lines. The ratio r of the diagram doublet lines $I(\underline{5p}_{2/2})/I(\underline{5p}_{1/2})$ is 0.8(1), much smaller than the expected statistical value of 2. The quantitative interpretation of this spectrum was given by Con-

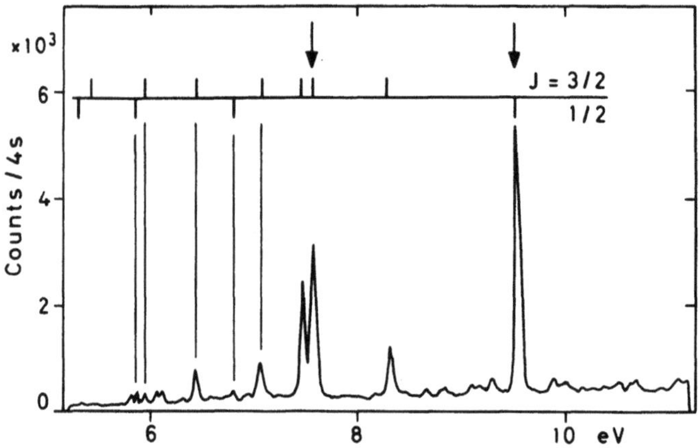

FIGURE 14. Experimental $\underline{5p}$–$\underline{6s}^2$ Auger spectrum of Ba excited by 2-keV electrons (Mehlhorn *et al.*, 1977). The diagram lines $\underline{5p_j}$–$\underline{6s}^2$ are indicated by arrows. The experimental Ba I $5p$ photoabsorption series limits are indicated as tick marks; the assignment of series limits to $J = 1/2$ or $3/2$ levels based on a MCDF calculation is also indicated (Connerade *et al.*, 1979; Rose *et al.*, 1980).

nerade *et al.* (1979) and Rose *et al.* (1980). On the basis of a MCDF calculation for the $J = 1/2$ and $J = 3/2$ levels of the configurations $5p^5(6s^2 + 6s5d + 5d^2)$, these authors calculated the positions of the $J = 1/2$ and $J = 3/2$ levels and could assign these levels to the series limits found in the $5p$ absorption spectrum of Ba (Connerade *et al.*, 1979). These series limits are identical with the initial states of the $\underline{5p_j}$ Auger spectrum and are indicated in the upper part of Fig. 14 as tick marks. We note that strong CI occurs only for the $J = 3/2$ levels. Their total intensity is distributed into five intense components, whereas for the $J = 1/2$ levels only two small satellites exist. If one now adds the intensities of the $J = 3/2$ and the $J = 1/2$ components, then the ratio r becomes 1.8(1), close to 2.

6. AUGER SPECTRA OF MULTIPLY IONIZED ATOMS

Here we discuss Auger spectra for which the initial state of the atom has in addition to an inner-shell vacancy, one or several outer-shell vacancies. For example, in the case of Ne, we have the initial configurations $1s2s^{2-n}2p^{6-m}$ with $n + m \leqslant 6$. The degree of multiple ionization of the target atom formed in a collision depends strongly on the projectile and its energy. For photons, electrons, and high-energy protons usually one additional outer-shell vacancy (with 10%–20% probability) is created with an inner-shell ionization. In the corresponding Auger spectrum, the additional lines associated with an initial multiple ionization therefore have a much smaller

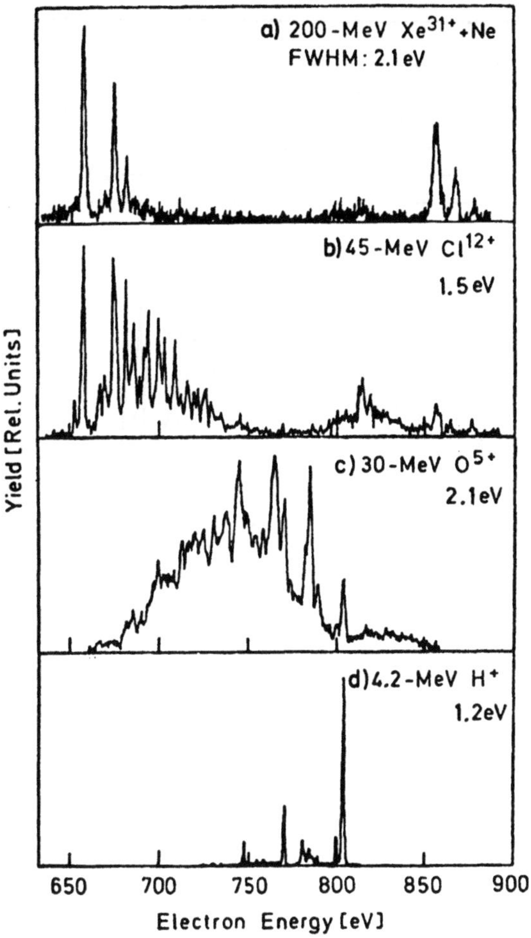

FIGURE 15. Comparison of Ne K Auger spectra excited by various projectiles. The resolution (FWHM) is given for each spectrum. (a) 200-MeV Xe^{31+} + Ne (Stolterfoht *et al.*, 1977), (b) 45-MeV Cl^{12+} + Ne (Schneider *et al.*, 1976b), (c) 30-MeV O^{5+} + Ne (Burch *et al.*, 1975), (d) 4.2-MeV H^+ + Ne (Stolterfoht *et al.*, 1973). From Stolterfoht *et al.* (1977).

intensity than the diagram lines (Figs. 15d, 17) and are called (multiple-ionization) satellite lines.

For impact by highly charged, heavy ions the average degree of outer-shell ionization increases with q/v_{ion}, where q is the charge and v_{ion} is the velocity of the ion (Mann *et al.*, 1982). Here the intensity of the diagram lines is in general much smaller than the intensity of satellite lines or even vanishes (Figs. 15, 16); then the term "multiple-ionization Auger spectrum" is appropriate.

In ion–atom collisions the inner-shell ionization plus simultaneous

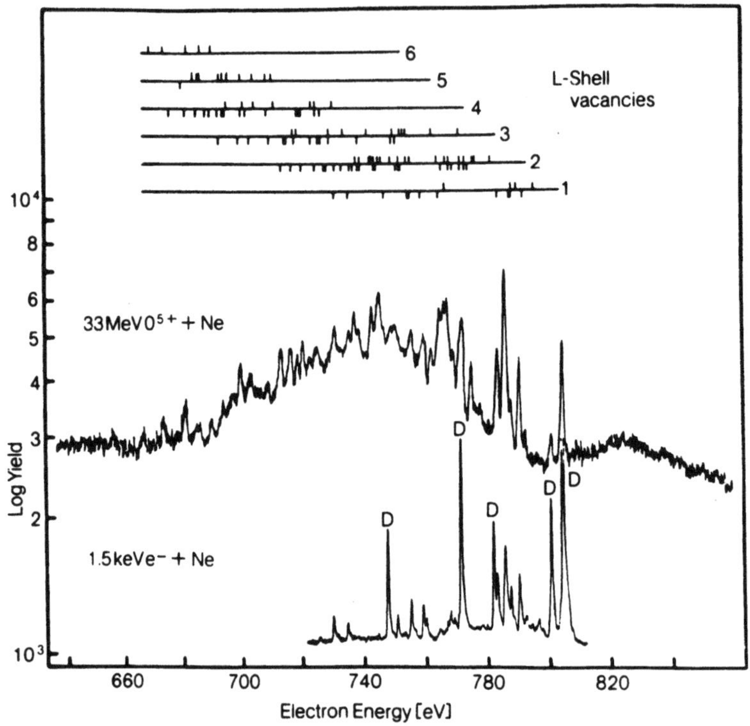

FIGURE 16. Neon K Auger spectra produced in 33-MeV O^{5+} and 1.5-keV e^- collisions. Diagram lines are labeled D. The calculated satellite transition energies are shown as tick marks on the lines labeled with the numbers of L-shell vacancies. From Matthews *et al.* (1974).

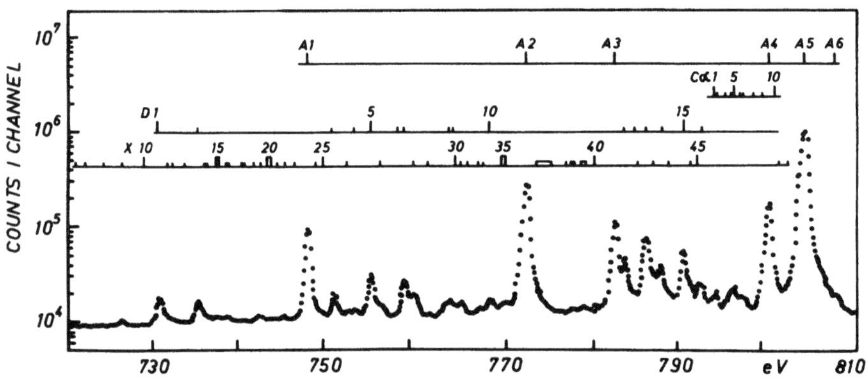

FIGURE 17. Logarithmic plot of the Ne K Auger spectrum excited by 4.5-keV electrons. Lines A1 to A6 are K–LL diagram lines, "D" lines are KL–LLL satellite lines; other lines are identified in the text. From Krause *et al.* (1971).

outer-shell ionization can occur either in the target or in the projectile atom. Study of the inner-shell ionized projectile atoms has opened a new field of research, projectile Auger-electron spectrometry, which is discussed separately in Sec. 7.

The general characteristics of multiple-ionization Auger spectra are as follows:

(a) The *average* energy of an Auger spectrum of a multiple-vacancy configuration decreases with increasing number of outer-shell vacancies. This has been shown in many experiments (Figs. 15, 16). Detailed calculations for K Auger transition energies of the multiplet states of configurations $1s2s^{2-n}2p^{6-m}$ have been carried out by Matthews *et al.* (1975) (see tick-mark spectra in Fig. 16) and Groeneveld *et al.* (1975).

(b) The number of Auger components increases with increasing number of vacancies (for Ne see Fig. 16). Only for an almost empty outer shell does the number of components decrease. Since the total energy spread of the Auger spectrum of one configuration is quite large (≈ 50 eV for Ne) and generally more than one multiple-vacancy configuration is excited, many Auger components overlap in a more or less structured intensity distribution (Figs. 15, 16). It is therefore rather difficult to obtain quantitative information from an Auger spectrum as shown, e.g., in Fig. 15c. For this reason, we discuss in detail only the Auger spectra from configurations with one additional outer-shell vacancy, e.g., the $(1s2p)^{-1}$ spectrum of Ne, or with only two electrons left in the L shell, e.g., the $1s2l2l'$ spectra of Li-like atoms. Detailed calculations and discussions of other two-vacancy spectra have been carried out, e.g., for $L_{2,3}M$–MMM of Ar (McGuire, 1975d) and $M_{4,5}N$–NNN of Kr (McGuire, 1975c).

6.1. *The* $(1s2p)^{-1}$ *Auger Spectrum of Ne*

The initial $1s2s^22p^5$ configuration of Ne has the multiplets ^1P and ^3P and we expect a total of 13 allowed $KL_{2,3}$–$LLL_{2,3}$ transitions with final configurations and terms

$$1s^22p^5(^2P), \qquad 1s^22s2p^4(^2S, {}^2P, {}^2D, {}^4P), \quad \text{and} \quad 1s^22s^22p^3(^2P, {}^2D),$$

according to the selection rules of Eqs. (8a), (8b) (Körber and Mehlhorn, 1966). Figure 17 shows the K Auger spectrum of Ne excited by 4.5-keV electrons and plotted on a logarithmic intensity scale (Krause *et al.*, 1971). In addition to the 5 normal K–LL Auger transitions (lines A1 to A5; line A6 has zero intensity in pure LS coupling), 16 KL–LLL satellite lines have been identified (lines D1 to D16 in Fig. 17); 13 of them are due to $KL_{2,3}$–$LLL_{2,3}$ transitions, while three lines (D8, D12 and D16) are due to KL_1–LLL_1 transitions. Other lines are due to transitions following $1s$ ionization accompanied by shakeup of an $n=2$ electron (Cα in Fig. 17) or due to discrete double Auger transitions.

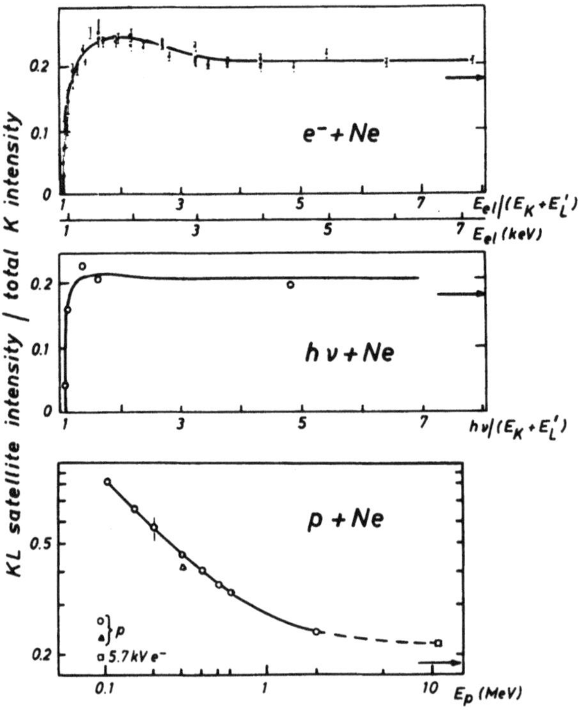

FIGURE 18. Ratio of KL double ionization to all K ionization in Ne as a function of energy of incident electrons or photons (Carlson *et al.*, 1970) or protons (Schneider *et al.*, 1973). The KL/K ratio is taken to be the ratio of KL Auger satellites (i.e., KL_1 and $KL_{2,3}$ satellites) to all K Auger transitions. E_K and E_L' are the binding energies of a K-shell electron and the shakeoff electron in the presence of a K vacancy, respectively. The arrow indicates the result of the conventional shakeoff theory (Carlson and Nestor, 1973). From Mehlhorn (1976).

In the case of inner-shell ionization plus additional outer-shell ionization (e.g., $1s2p$ in Ne), the observed double ionization rate for $h\nu$, e^-, and fast H^+ impact reasonably agrees with conventional shake theory, provided the sudden limit is reached in these experiments. The basis and the limits of the shakeoff concept have been discussed by Åberg (1969), and extensive calculations of shakeoff probabilities have been performed by Carlson and Nestor (1973). In Fig. 18 the experimental ratio of KL satellite intensity to total K Auger intensity for ionization by photons and electrons (Carlson *et al.*, 1970) and protons (Schneider *et al.*, 1973) is shown as function of primary energy. The result of the shakeoff theory in the limit of the sudden approximation is indicated by the arrow (Carlson and Nestor, 1973); this accounts for ~85% of the observed ratio. From Fig. 18 it can be concluded that, e.g., for electron-impact ionization, the sudden-approximation limit seems to be reached already for an incident-electron energy $E_{el} > 4(E_K + E_{L'})$. In contrast, it has been found recently (Genz,

1982; Löw *et al.*, 1983) that the ratio KL/K of ionization by electron impact of, e.g., Mg still decreases for energies $E_{el} > 50E_K$ and reaches a constant value in agreement with the shakeoff calculation only at very large impact energies. This striking difference of the energy dependence of the KL/K ratio between Ne and Mg (and a variety of other elements, cf. Löw *et al.*, 1983) calls for further investigations, especially to measure this ratio for Ne also at much larger energies E_{el} than given in Fig. 18.

Schmidt (1973) has calculated the transition probabilities of $KL_{2,3}$–$LLL_{2,3}$ satellite lines of Ne; his results are compared in Fig. 19 with the relative experimental satellite intensities (lines D1–D7, D9–D11, and D13–D15 of Fig. 17). In order to obtain good agreement between theoretical and experimental relative intensities, Schmidt (1973) had to assume that the formation of multiplets ^3P and ^1P in the $1s2p$ double ionization by 4.5 keV electron impact is in the ratio 1.36 instead of the expected (statistical) ratio 3. The latter value of 3 follows from the conventional shake theory. Similar results have been found also in the $KL_{2,3}$ Auger spectrum excited by proton impact (Bhalla *et al.*, 1973b) and in the $KL_{2,3}$ X-ray spectrum excited by electron impact (Ågren *et al.*, 1978). The reason for the deviation from the statistical value is the FCSCI between the two continua $1s2s^22p^5(^1P)\varepsilon p\ ^2S$ and $1s2s^22p^5(^3P)\varepsilon' p\ ^2S$. Application of FCSCI to the problem of $2p$ shakeoff in Ne (Chattarji *et al.*, 1978) yielded a ratio of 2.18.

The observation of an intensity ratio less than 3 could also be due to different nonisotropic distributions of Auger electrons corresponding to different alignments of the ^1P and ^3P multiplet states of the $1s2p$ vacancy con-

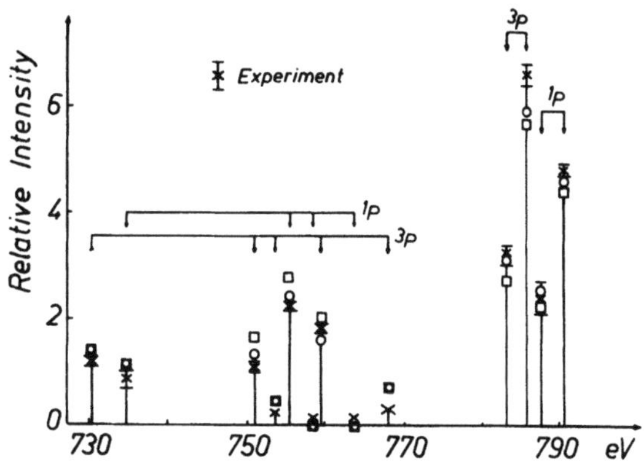

FIGURE 19. Relative intensities of $KL_{2,3}$–$LLL_{2,3}$ satellite transitions of Ne excited by 4.5-keV electrons (Krause *et al.*, 1971) (Ī) and comparison with theory (Schmidt, 1973). The initial state of the satellite transitions (either ^3P or ^1P) is indicated. The theoretical values have been obtained by using either the radial integrals R^k calculated by McGuire (1969) (□) or those extracted from the K–LL intensities (Krause *et al.*, 1971) (○). The excitation ratio of initial multiplet states $^3P/^1P$ must be assumed to be 1.36 in order to achieve good agreement with the experimental intensities. From Mehlhorn (1976).

figuration, the Auger spectra having been measured at a fixed angle of 90°. Within the shakeoff model, however, there is no mechanism for an alignment of the multiplet states of the $1s2p$ configuration. On the other hand, Jamison *et al.* (1977, 1978) and Ost *et al.* (1982) found rather strong alignment in the $1s2p$ vacancy production of Al by 1–3 MeV proton impact, indicating that for protons in this energy range double collision processes still contribute substantially to the $1s2p$ ionization. But even in case of alignment it has been shown (Mehlhorn, 1978, p. 125) that the ^1P and the ^3P multiplet states have the same alignment (here it was assumed that the cross sections $Q(SLM_SM_L)$ of magnetic substates are independent of SM_S), and corresponding transitions to the same final state have the same angular distributions. The intensity *ratio* of corresponding lines should therefore be independent of an alignment.

6.2. *Auger Spectra of Li-like Target Ions*

In a collision of a highly charged, heavy ion with a target atom, e.g., 200-MeV Xe^{31+} on Ne, where a *K*-shell electron is ionized, most or even all of the *L*-shell electrons are removed simultaneously. If only one electron remains in the $1s$ shell, only an ion having at least three electrons (Li-like configuration) can decay by electron emission. The initial Li-like configurations lead to an electron spectrum with only a few lines. In Fig. 15a the lines at energies between 650 and 700 eV are due to the decay of the three-electron configurations $1s2l2l'$ and the lines above 800 eV are due to initial configurations $1s2lnl'$ with $n=3$, 4, 5, 6. The latter Li-like configurations are formed by the capture of an electron from the residual target atoms by target recoil ions having a He-like configuration in a metastable state (Mann *et al.*, 1981a,b 1982). From the observed small kinematic

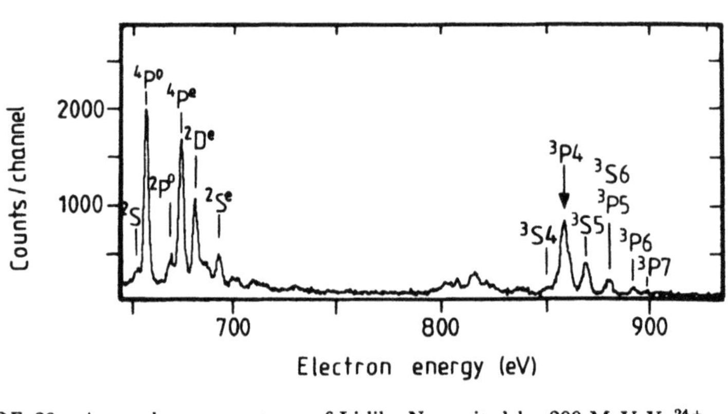

FIGURE 20. Auger-electron spectrum of Li-like Ne excited by 200-MeV Xe^{24+} projectiles. The $1s2l2l' \rightarrow 1s^2\varepsilon l$ transitions give rise to the lines below 700 eV; for explanation of lines at energies $\geqslant 850$ eV see Fig. 21. From Mann *et al.* (1981b).

broadening of the Auger lines a very low projectile-induced recoil energy of the target ion of $\lesssim 10$ eV has been calculated (Mann *et al.*, 1981b). The high degree of target ionization in a single collision, the small recoil energy of the ionized target atoms, and the well-defined recoil of about 90° with respect to the incident projectile favor these recoil ions as systems for high-resolution spectrometry. So far, the following Li-like recoil target ions have been studied: N^{4+}, O^{5+}, F^{6+}, Ne^{7+} (Mann *et al.*, 1981a,b 1982), and Ar^{15+} (Beyer *et al.*, 1982).

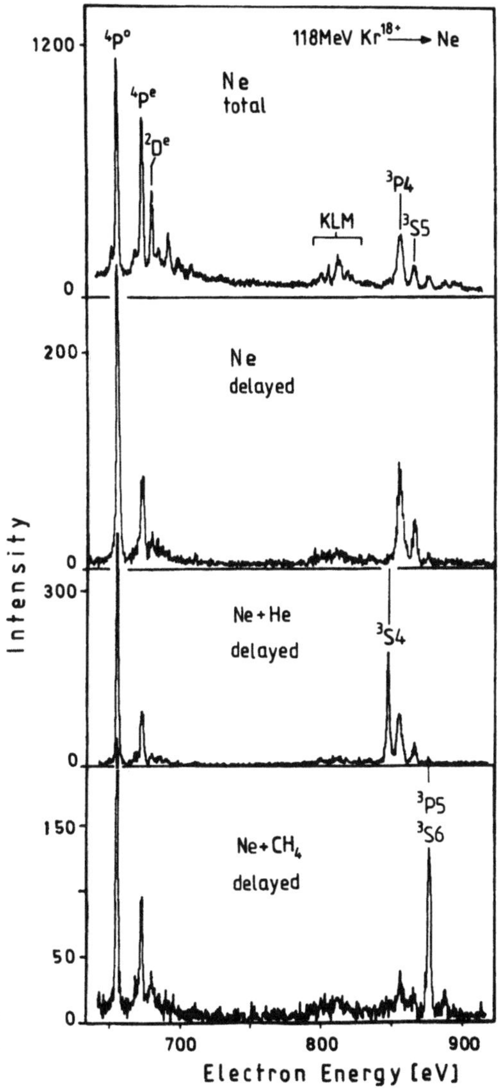

FIGURE 21. Total and delayed (>10 ns) electron spectra of $1s2lnl'$ configurations of Ne excited in 118-MeV $Kr^{18+}+Ne$ collisions in various gas mixtures. The lines $^3S4$, $^3S5$, $^3S6$, $^3P4$, $^3P5$ are due to the decay of states formed by electron capture into main shells $n=4, 5, 6$ of metastable $(1s2s)\,^3S$ and $(1s2p)\,^3P_{0,2}$ core ions. From Mann *et al.* (1982).

For example, in Fig. 20 the electron spectrum of Li-like Ne excited by 200-MeV Xe^{24+} projectiles is shown (Mann *et al.*, 1981b). The Auger lines due to initial configurations $1s2s^2\ {}^2S$, $1s2s2p\ {}^4P^0$, ${}^2P^0$ and $1s2p^2\ {}^4P^e$, ${}^2D^e$, ${}^2S^e$, decaying to the common final ionic state $1s^2\ {}^1S$ have been identified. The experimental energies of these transitions agree within the experimental error of ± 1 eV with calculated energies sing a MCDF program by Desclaux (1975). The lines in the energy range 800 to 900 eV are due to the decay of configurations $1s2lnl'$ which are discussed below.

Another interesting aspect of the slowly moving recoil ions is the possibility for the study of electron-capture processes in the collision of highly charged, low-energy ions with surrounding neutral target atoms. The basis for these capture processes is the formation of metastable states of He-like recoil ions in the collision with the projectile, states which live long enough so that electron capture occurs before they decay radiatively. It has been found that this electron capture occurs predominantly into selective outer shells n depending on the binding energy I_P of the electron in the donating target atom (Mann *et al.*, 1981a,b). Consequently, different electron lines became prominent in the Auger spectra when different gases (with different I_P) were mixed to the target gas under investigation. For example, in Fig. 21 the total and time-delayed ($t > 10$ ns) $1s2lnl'$ spectra of Ne from 118-MeV Kr^{18+} impact on pure Ne target and mixtures with He and CH_4, respectively, are compared. The strong lines in the delayed spectra at energies $\gtrsim 840$ eV are attributed to the decay of three-electron states having metastable cores $1s2s\ {}^3S$ and $1s2p\ {}^3P_{0,2}$ and an additional electron in an outer shell of $n = 4$ to 6. The different principal quantum number n of the outer-shell electrons demonstrate the selectivity of the capture process depending on the I_P of the admixed atoms to the target gas (Mann *et al.*, 1981b, 1982).

7. PROJECTILE AUGER-ELECTRON SPECTROMETRY

Earlier spectroscopic measurements employing projectile Auger-electron spectrometry have dealt primarily with the study of metastable Li-like $1s2s2p\ {}^4P_{5/2}$ states of various Z (Pegg *et al.*, 1973; Sellin, 1973; Sellin, 1978). In these experiments the lifetime of the decaying projectile state was determined by measuring the decay length along the beam and the velocity of the projectile. Early attempts to measure the prompt projectile Auger spectra as well by viewing the back of the exciting foil were not successful, mainly due to line-broadening effects. Real progress towards high-resolution projectile Auger-electron spectrometry was made when the projectiles were excited in single gas collisions and the electron spectrum was recorded at an extreme forward direction relative to the beam of projectiles (Rødbro *et al.*, 1979). Earlier work on projectile Auger-electron spectrometry in single gas collisions is described by Dahl *et al.* (1976a,b).

The broadening of Auger lines in projectile-electron spectrometry is a kinematic effect. Neglecting a velocity spread of projectile ions, the kinematic broadening is due to the scattering of projectile ions in the excitation process into a cone of half-angle β and due to the finite angular acceptance $\Delta\theta$ of the electron spectrometer. The broadenings ΔE_β and $\Delta E_{\Delta\theta}$ due only to scattering of the projectiles (Dahl *et al.*, 1976a) and due only to the acceptance $\Delta\theta$ of the electron spectrometer (Schneider *et al.*, 1977), respectively, have been shown to be

$$\Delta E_\beta = 4\sqrt{\varepsilon E}\sin\theta\sin\beta, \qquad \Delta E_{\Delta\theta} = 2\sqrt{\varepsilon E}\sin\theta\cos\beta\,\Delta\theta. \tag{33}$$

Here, we have $\varepsilon = \frac{1}{2}mu^2$, where u is the velocity of the projectile ion; $E = \frac{1}{2}mv^2$ is the energy of the Auger electron in the laboratory frame, and θ is the angle of observation of the Auger electron relative to the primary beam direction. Since ΔE_β and $\Delta E_{\Delta\theta}$ depend on θ through $\sin\theta$, an observation at

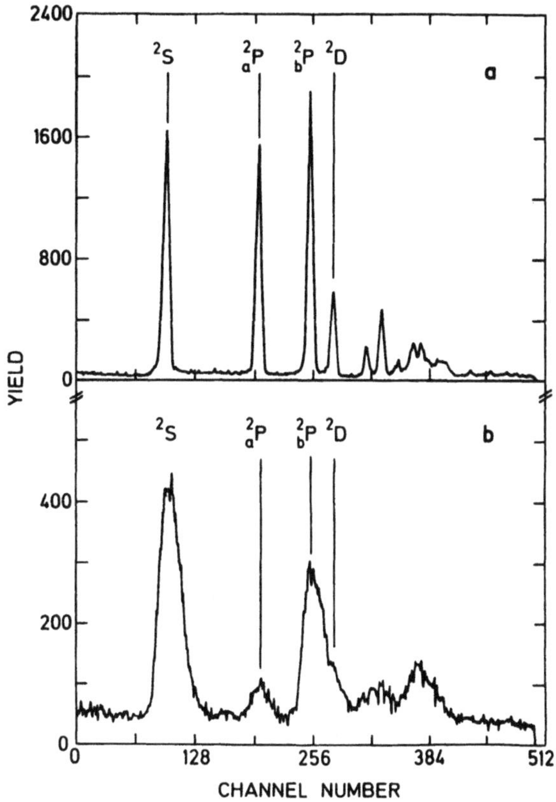

FIGURE 22. Lithium autoionization electron spectra for 250-keV $Li^+ + He$, recorded at (a) $\theta = 5°$ and (b) $\theta = 60°$. From Bisgaard *et al.* (1981).

$\theta = 0°$ or 180° would eliminate the broadening completely. In Fig. 22 the effect of θ on the linewidth can be seen for the $1s2l2l' \rightarrow 1s^2 + e_A^-$ transitions of Li in $Li^+ + He$ collisions. Thus, high-resolution electron spectrometry should be performed at as small (or as large) angles θ as possible. For example, Rødbro *et al.* (1979) studied the electron spectra of singly and doubly core-excited three- and four-electron systems of Li, B, Be, and C by means of projectile-electron spectrometry at an angle $\theta = 6.4°$. In Fig. 23 their high-resolution spectrum of Li-like Be^+ excited in 500-keV $Be^+ + CH_4$ collisions is shown. Numerous states, previously not seen, have been resolved in this spectrum and identified by comparison with theoretical results (see also Chung, 1982; Davis and Chung, 1982).

In order to determine the alignment as well of projectile Auger states one must measure the angular distribution of Auger electrons (Sec. 8). Since at least one spectrum must be measured at large angles θ, one again runs into problems with line broadening. In order to reduce the kinematic line broadening $\Delta E_{\Delta\theta}$, Bachmann *et al.* (1982) have proposed a refocusing method which indeed has been applied successfully: the broadening of Li lines excited in 280-keV $Li^+ + CH_4$ collisions and observed at $\theta = 20°$ was reduced by a factor of 2 (Bruch *et al.*, 1983).

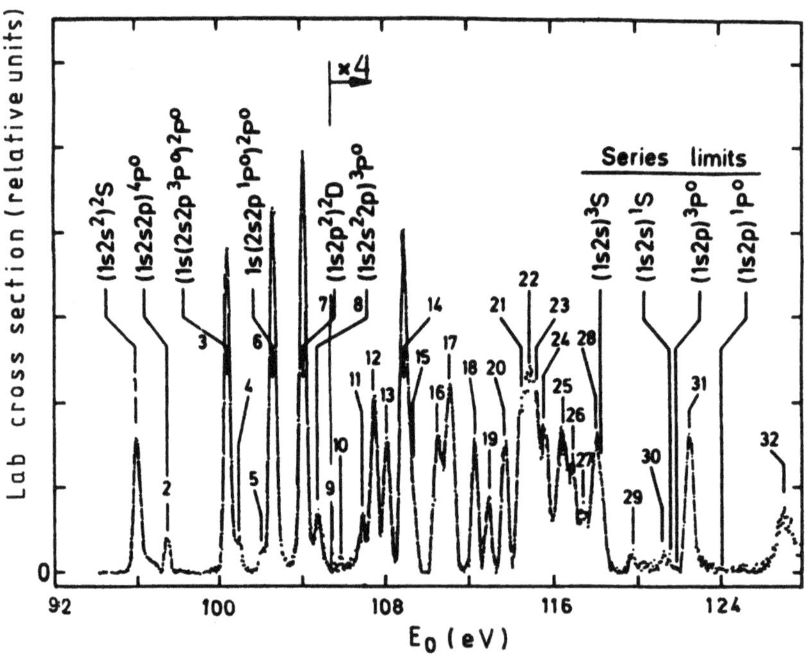

FIGURE 23. Beryllium Auger spectrum of singly core-excited three- and four-electron states in 500-keV $Be^+ + CH_4$ collisions. From Rødbro *et al.* (1979).

8. ANISOTROPIC ANGULAR DISTRIBUTION OF AUGER ELECTRONS

The anisotropic angular distribution of Auger electrons is uniquely related to the alignment of the initial Auger state $A^+(JM)$. A state is said to be aligned if the population probabilities $P(JM)$ of magnetic substates M are unequal, but with $P(JM)=P(J, -M)$.

The question of an anisotropic angular distribution of Auger electrons came up first in connection with the new technique of Auger-electron spectrometry, i.e., the use of gaseous targets and the ionization by an external directed beam of photons or particles (Mehlhorn, 1960). Later it was predicted (Mehlhorn, 1968a) that ionization of an inner-shell electron with $j \geqslant 3/2$ by a directed unpolarized particle beam would lead to alignment of the ionic state. Auger electrons emitted in the subsequent decay manifest this alignment through their anisotropic angular distribution. The first experimental evidence for an anisotropic angular distribution of Auger electrons was provided for the L_3–$M_{2,3}M_{2,3}(^1S_0)$ electrons of Ar from electron-impact ionization (Cleff and Mehlhorn, 1971), and the theory of the anisotropic angular distribution was first formulated by Cleff and Mehlhorn (1974a). Alignment in inner-shell photoionization was predicted in 1972 (Flügge *et al.*, 1972; Jacobs, 1972); the first experimental evidence was provided by Caldwell and Zare (1977). Since 1977 the interest in investigations, experimental as well as theoretical, of alignment in inner-shell ionization has been rapidly increasing (for recent review articles see, e.g., Mehlhorn, 1982; Jitschin, 1982).

We now consider the inner-shell ionization process

$$\begin{Bmatrix} P \\ \gamma \end{Bmatrix} + A(J_0) \rightarrow A^+(JM) + \begin{Bmatrix} P+e^- \\ e^-_{ph} \end{Bmatrix}, \tag{34a}$$

where the ion $A^+(JM)$ decays in a second step

$$A^+(JM) \rightarrow A^{++}(J_f) + e^-_{\text{Auger}}(l'j'). \tag{34b}$$

As projectiles P, either electrons or protons will be considered. We shall assume that the target atoms are randomly oriented and that the beam of projectiles P is unpolarized. The photons γ are either unpolarized or linearly or circularly polarized. We shall further assume that $SLJM$ coupling is valid, i.e.,

$$\begin{aligned} &\text{fine structure splitting} \gg \text{level width } \Gamma \\ &\qquad \gg \text{hyperfine structure splitting;} \end{aligned} \tag{35}$$

this inequality certainly holds for inner-shell vacancy states.

In almost all experimental alignment studies, axial symmetry has been used (noncoincidence experiments). Only very recently in coincidence experiments (Sewell and Crowe, 1982; Völkel and Sandner, 1983) has the symmetry of the experiment been determined by the plane of the incident and scattered particle P.

The ionic state formed by impact ionization or by photoionization is in general anisotropic (aligned and oriented). Theoretically, the anisotropy of a state is described by a set of anisotropy parameters $\mathcal{A}_{k\kappa}$ which are uniquely related to the matrix elements of the density matrix of the relevant ionization process (here and in the following we will use the notation of Berezhko and Kabachnik, 1977, and of Berezhko *et al.*, 1978a,b). On the basic assumption that ionization and decay can be treated as a two-step process, the general angular correlation of emitted Auger electrons can be expressed in terms of the parameters $\mathcal{A}_{k\kappa}$ and of coefficients α_k, where the latter depend only on the decay process. Auger decay, in which parity is conserved, restricts k to only even values. Hence it follows, that an orientation ($\mathcal{A}_{1\kappa} \neq 0$) of the initial Auger state cannot be observed via the angular distribution of Auger electrons. The number of independent nonzero parameters $\mathcal{A}_{k\kappa}$ ($\kappa = -k + \cdots + k$) which enter into the angular distribution of Auger electrons is determined by the symmetry of the experiment.

In the following we will discuss the angular distribution of Auger electrons for experiments with axial symmetry in particle and photon ionization and with plane symmetry in particle ionization (coincidence experiments). For simplicity we shall only consider closed-shell atoms A having $J_0 = 0$ before inner-shell ionization.

8.1. Particle-Impact Experiments with Axial Symmetry

We assume that an unpolarized beam of projectiles P defines the z axis, and that the target atoms are randomly oriented. We also assume that neither the scattered projectile nor the ionized electron is detected. Consequently, the state of the ion is axially symmetric about z and symmetric with respect to reflection in the plane perpendicular to z. This restricts the anisotropy parameters $\mathcal{A}_{k\kappa}$ of the ionic state $A^+(JM)$ to $\mathcal{A}_{k0}$, where k can take on only even values $2, 4, \ldots, 2J - 1$. It follows that only inner-shell vacancy states with $J \geqslant 3/2$ are aligned. For a given values of J, the $\mathcal{A}_{k0}$ can be expressed in terms of the occupation probabilities $P(J\,|M|)$. For example, for $J = 3/2$ we have

$$\mathcal{A}_{20} = \frac{P(3/2\,|3/2|) - P(3/2\,|1/2|)}{P(3/2\,|3/2|) + P(3/2\,|1/2|)}. \tag{36}$$

The angular distribution of Auger electrons is then given by

$$I(\theta) = \frac{I_0}{4\pi}\left\{1 + \sum_{k=2}^{k_{\max}} \alpha_k \mathcal{A}_{k0} P_k(\cos\theta)\right\}, \tag{37}$$

where the $P_k(\cos\theta)$ are Legendre polynomials, θ is the angle relative to the incident direction, and $k_{\max}$ is determined by (Berezhko and Kabachnik, 1977)

$$k_{\max} \leqslant 2\min(\mathcal{J}, l', j'). \tag{38}$$

The quantum numbers $\mathcal{J}$, l', and j' are given in Eqs. (34). The coefficients α_k in Eq. (37) depend on the Auger decay amplitudes and relative phases between different Auger partial waves. Only in the special case of $\mathcal{J}_f = 0$, where a single Auger wave is emitted, are the α_k independent of the decay amplitude. This special case is therefore particularly suited to an experimental determination of the parameters $\mathcal{A}_{k0}$. For example, Auger electrons of the transition

$$\mathrm{A}^+(np_{3/2}^{-1}\ {}^2P_{3/2}) \to \mathrm{A}^{++}(\mathcal{J}_f = 0) + e^-_{\mathrm{Auger}}(\varepsilon p_{3/2})$$

have $k = 2$ and $\alpha_2 = -1$.

Calculations of alignment parameters $\mathcal{A}_{20}$ for impact ionization of different subshells nlj by electrons or protons have been performed by Berezhko and Kabachnik (1977) and Sizov and Kabachnik (1980) using the plane-wave Born approximation (PWBA) and various approximations for the atomic wave functions (hydrogenic, screened hydrogenic, HS, and HF).

On the experimental side, the alignment $\mathcal{A}_{20}$ has been determined by means of the angular distribution of Auger electrons for the following cases: (a) electron impact ionization: Mg L_3 (DuBois *et al.*, 1981; Rødbro *et al.*, 1978), Ar L_3 (Cleff and Mehlhorn, 1974b; Sandner and Schmitt, 1978; DuBois and Rødbro, 1980), and Kr M_4, M_5 (Sandner *et al.*, 1980); (b) proton impact ionization: Mg L_3 (DuBois *et al.*, 1981; Rødbro *et al.*, 1978). Alignment measurements by means of anisotropic emission or polarization of X-radiation are described by Mehlhorn (1982), Cleff (1982), and Jitschin (1982).

As an example of alignment by electron-impact ionization, the experimental $\mathcal{A}_{20}$ values of Ar L_3 are compared in Fig. 24 with theoretical results. The experimental $\mathcal{A}_{20}$ agree well with theoretical PWBA values for HS wave functions (Berezhko and Kabachnik, 1977; Kabachnik, 1978) in the high-energy range $[E/E(L_3) > 10]$. The change of sign for $E/E(L_3) > 100$ has been explained as due to relativistic effects of the incident electrons (Sandner and Schmitt, 1978). For the near-threshold range $[E/E(L_3) < 4]$, experimental values deviate systematically from the theoretical HS values in PWBA, indicating the invalidity of the plane-wave Born approximation in this region. Recently Berezhko and Kabachnik (1980, 1982), by using distorted-wave Born approximation, obtained agreement with the latest experiment by DuBois and Rødbro (1980) down to $E/E(L_3) = 1.4$. This result contradicts the earlier experimental findings of Sandner and Schmitt (1978).

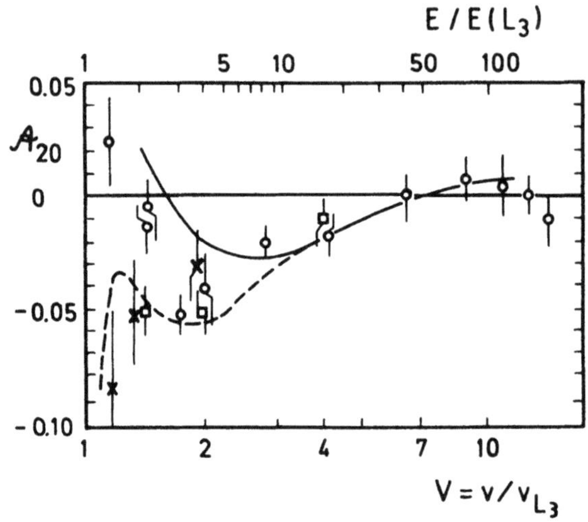

FIGURE 24. Alignment of the L_3-subshell of Ar due to electron-impact ionization, as function of the reduced velocity $V = v/v_{L_3}$ or reduced energy $E/E(L_3)$ of incident electrons. Experiment: □, Cleff and Mehlhorn (1974b); ○, Sandner and Schmitt (1978); ×, (DuBois and Rødbro (1980). Theory: —, PWBA and HS model (Berezhko and Kabachnik, 1977; Kabachnik, 1978); ---, DWBA and HS model (Berezhko and Kabachnik, 1980, 1982).

As can be seen from Fig. 24, the alignment parameter $\mathscr{A}_{20}$ is small in magnitude ($|\mathscr{A}_{20}| \lesssim 0.05$), resulting in only a small anisotropy of the angular distribution of Auger electrons. This is generally true for electron-impact ionization in an axially symmetric experiment. For proton impact and small reduced velocity $V = v/v_{nl} < 0.3$ of the projectile, where v is the proton velocity and v_{nl} the velocity of the electron being ionized, the alignment parameter $\mathscr{A}_{20}$ can be much larger in magnitude (Sizov and Kabachnik, 1980; Cleff, 1982; Jitschin, 1982).

8.2. *Photon-Impact Experiments with Axial Symmetry*

The angular distribution of Auger electrons following photoionization (by electric-dipole transitions) is (Berezhko *et al.*, 1978a)

$$I(\theta) = \frac{I_0}{4\pi}\,[1 + \alpha_2 \mathscr{A}_{20}(\gamma)\, P_2(\cos\theta)]. \tag{39}$$

Equation (39) follows from Eq. (37) and the fact that the electric-dipole operator leads to only $k \leqslant 2$ of $\mathscr{A}_{k\kappa}$. The angle θ in Eq. (39) is relative to the axis of symmetry (direction of incident photons for unpolarized, circularly and elliptically polarized light, direction of the electric vector for linearly polarized light). It is interesting to note that the alignment parameter

$\mathscr{A}_{20}(\gamma)$ depends only on the squares of the radial integrals $R_{nl,\varepsilon l\pm 1}$ of electric-dipole matrix elements and the polarization characteristics of incident photons. General expressions for $\mathscr{A}_{20}(\gamma)$ and numerical values for various subshells have been given by Berezhko *et al.* (1978a). The case of elliptically polarized light has been treated recently also by Klar (1982).

So far, the only measured angular distribution of Auger electrons following photoionization is that for $Xe(\underline{4d}_{5/2})$ (Southworth *et al.*, 1981, 1983), who used linearly polarized synchrotron radiation:

$$Xe^{+}(\underline{4d}_{5/2}) \rightarrow Xe^{++}(\underline{5s}^{2}\ {}^{1}S_{0}) + e^{-}_{A1} \tag{40a}$$

$$\rightarrow Xe^{++}(\underline{5s5p}\ {}^{1}P_{1}) + e^{-}_{A2}. \tag{40b}$$

For Auger transitrons leading to a final ${}^{1}S_{0}$ state one obtains (Berezhko *et al.*, 1978a) $\alpha_2 = \sqrt{8/7}$, and the angular anisotropy of Eq. (39) is given by

$$\beta_2 = \alpha_2 \mathscr{A}_{20}(\gamma_{\text{lin}}) = (8/7)^{1/2}\, \mathscr{A}_{20}(\gamma_{\text{lin}}). \tag{41}$$

In Fig. 25 the experimental values of β_2 for Auger electrons e^{-}_{A1} and e^{-}_{A2} are compared with theoretical values $(8/7)^{1/2}\,\mathscr{A}_{20}(\gamma_{\text{lin}})$. The theoretical values of $\mathscr{A}_{20}(\gamma_{\text{lin}})$ as functions of photoelectron energy have been calculated (Southworth *et al.*, 1983) by using HF photoionization radial matrix

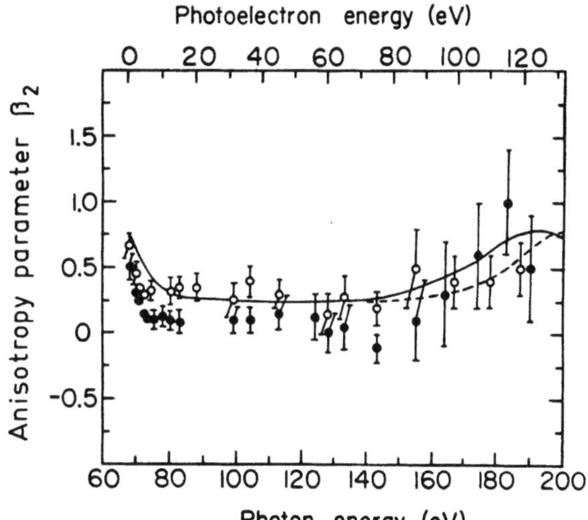

FIGURE 25. Anisotropy parameter β_2 for $\underline{4d}_{5/2}$–OO Auger transitions of Xe following photoionization. Experiment: ○, $\underline{4d}_{5/2}$–$\underline{5s}^{2}({}^{1}S_{0})$ transition; ●, $\underline{4d}_{5/2}$–$\underline{5s5p}({}^{1}P_{1})$ transition. The theoretical curves, given by $\beta_2 = \sqrt{8/7}\,\mathscr{A}_{20}(\gamma_{\text{lin}})$, are for a transition of the general type $\underline{4d}_{5/2}$–OO$({}^{1}S_{0})$. Dashed curve: HF and length form of dipole matrix element; solid curve: HF and velocity form. From Southworth *et al.* (1983).

elements. The experimental anisotropy β_2 of Auger electrons e^-_{A1} agrees well with the general shape of the theoretical curve, although it is somewhat smaller near threshold. The coefficient α_2 of the Auger transition of Eq. (40b) depends on three Auger radial matrix elements and their phases; nevertheless it has a constant value. Consequently, the anisotropies β_2 of electrons e^-_{A1} and e^-_{A2} should only differ by a constant factor, which is qualitatively the case (Fig. 25).

8.3. *Experiments with Plane Symmetry*

The symmetry plane (zx) is defined by the incident and the scattered particle (Fig. 26). We deal with unpolarized incident electrons and an unpolarized target. The state of the ion formed is then symmetric about reflection in the reaction plane (zx), and the anisotropy parameters are interrelated (Berezhko *et al.*, 1978b):

$$\mathcal{A}_{k\kappa} = (-1)^{k+\kappa}\, \mathcal{A}_{k,-\kappa}. \tag{42}$$

We consider only the special case which has already been investigated: the ionization of Ar L_3 and the measurement of the angular distribution of L_3–$M_{2,3}M_{2,3}(^1S_0)$ Auger electrons in coincidence with the scattered incident electrons (Sewell and Crowe, 1982, 1984a,b; Völkel and Sandner, 1983; Sandner and Völkel, 1984; Sandner, 1985). Here, according to Eq. (42), the number of independent and observable anisotropy parameters is three, namely, $\mathcal{A}_{20}$, $\mathcal{A}_{21}$, and $\mathcal{A}_{22}$, and the angular distribution of Auger electrons is (Berezhko *et al.*, 1978b):

$$I(\theta,\phi) = \frac{I_0}{4\pi}\left[1 + \alpha_2 \sum_{\kappa=0}^{2} \mathcal{A}_{2\kappa}\left(\frac{4\pi}{2k+1}\right)^{1/2} Y_{2\kappa}(\theta,\phi)\right]. \tag{43}$$

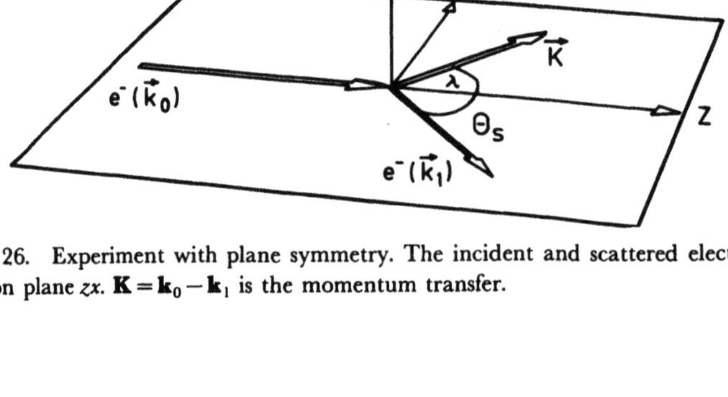

FIGURE 26. Experiment with plane symmetry. The incident and scattered electrons define the reaction plane zx. $\mathbf{K} = \mathbf{k}_0 - \mathbf{k}_1$ is the momentum transfer.

The angle θ is taken relative to the incident-beam direction, the plane $\phi = 0$ is the scattering plane, and the $Y_{2\kappa}$ are spherical harmonics.

In the experiment by Sewell and Crowe (1982), the angular distribution of Auger electrons was measured in the scattering plane, whereas Völkel and Sandner (1983) measured the Auger electrons simultaneously in 12 directions (θ, ϕ), the 12 directions lying on a cone with axis in the x direction (Fig. 26) and with a 39.2° half opening angle. Neither experiment alone could determine all three parameters $\mathcal{A}_{2\kappa}$; only the combined evaluation of both experiments (Völkel and Sandner, 1983) leads to a determination of $\mathcal{A}_{20}$, $\mathcal{A}_{21}$, and $\mathcal{A}_{22}$. First results for incident electrons of 1 keV and scattered electrons of 747 eV (the ionized electron has an energy $E_1 = 5$ eV) and scattering angle $\theta_s = 15°$, compared to theoretical values obtained in PWBA (Berezhko *et al.*, 1978b), seemed to indicate the complete breakdown of the PWBA. Later, in more extensive experiments (Sewell and Crowe, 1984a; Sandner and Völkel, 1984; Sandner, 1985) it was found that shifts and distortions of the shape of the Auger line under investigation occur. These are due to postcollision interaction (Sec. 9) and interference effects (Fano profile, Sec. 2.3). Taking these effects into account one of the alignment experiments (Sandner and Völkel, 1984) now yields values that are not in contradiction with PWBA.

9. POSTCOLLISION INTRACTION EFFECTS IN AUGER SPECTRA

In Sec. 2.2 it was assumed that Auger decay can be treated independently from the primary ionization process. One necessary condition for this assumption is that the interaction between the primary ejected inner-shell electron and the Auger electron can be neglected. Now we take this interaction explicitly into account and discuss the corresponding effects seen in Auger spectra.

We consider an inner-shell photoionization process followed by an Auger transition

$$\begin{array}{lll} \gamma(h\nu) + \mathrm{A} \to \mathrm{A}^{+}(E_B) & + e^{-}(E_1) & \qquad e^{-}(E_1 - \Delta\varepsilon) \\ \qquad\qquad\quad \big\downarrow & \qquad\qquad \longrightarrow & \\ \qquad\qquad\quad \longrightarrow & \mathrm{A}^{++} + e_A^{-}(E_A^0) & \qquad \mathrm{A}^{++} + e_A^{-}(E_A^0 + \Delta\varepsilon). \end{array} \tag{44}$$

As the excess energy $E_1 = h\nu - E_B$ decreases, the interaction between the slow photoelectron and the fast Auger electron causes an exchange of energy and angular momentum. This interaction of particles in the final state is called postcollision interaction, PCI, and leads to an energy loss $\Delta\varepsilon$ of the photoelectron and a corresponding energy gain $\Delta\varepsilon$ of the Auger electron. Experimentally, the effect is seen as a shift and a distortion of the shape of the corresponding lines in the electron spectrum. The PCI effect in inner-shell ionization processes has found great experimental and theoretical interest in the last few years (for references see, e.g., Schmidt *et al.*, 1981; Schmidt, 1982; Huster and Mehlhorn, 1982; Niehaus, Chap. 9, this volume).

For inner-shell *photoionization*, Niehaus (1977) has predicted within a semiclassical framework the PCI shifts $\Delta\varepsilon_m$ and shapes of Auger lines. Both shift and lineshape depend (apart from the lifetime τ or width Γ of the inner-shell vacancy) only on the excess energy E_1. An improved semiclassical model for PCI in photoexcited Auger decay was recently given by Helenelund *et al.* (1983) and Russek and Mehlhorn (1985). In this model the correct asymptotic line shape, a Lorentz distribution with width Γ_{Auger}, is obtained for $E_1 \to \infty$ (in the earlier Niehaus model, a δ function was

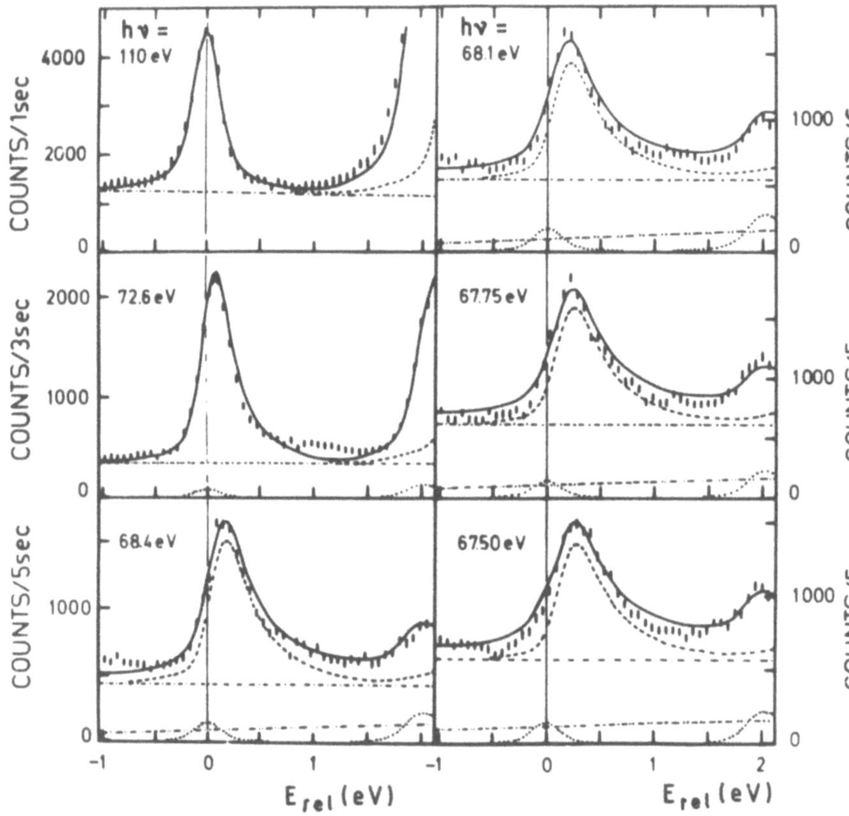

FIGURE 27. Experimental PCI effect (shift and shape) of the Xe $\underline{4d}_{5/2}$–$\underline{5p}^2(^1S_0)$ Auger line following the photoionization of a $\underline{4d}_{5/2}$ electron (with binding energy $E_B = 67.55$ eV) for various excess energies $E_1 = h\nu - E_B$. The solid line is the calculated PCI intensity distribution. From Schmidt *et al.* (1981).

obtained as asymptotic line shape). In both models only the energy exchange between the slow photoelectron and the fast Auger electron was considered. Recently, Niehaus (1983) improved his earlier model by including also angular-momentum exchange and showed that this may have an observable effect on the Auger line shape. Quantum-mechanical treatments have also been formulated (Amusia *et al.*, 1977, 1980; Åberg, 1980, 1981), but to date neither shifts nor line shapes have been calculated quantitatively.

As an example, in Fig. 27 the experimental Xe $\underline{4d}_{5/2}$–$\underline{5p}^2\ {}^1S_0$ Auger peak following the photoionization of a $4d_{5/2}$ electron by monochromatized synchrotron radiation for different excess energies E_1 is compared with the theoretical line shape (solid lines in Fig. 27). In Fig. 28 the corresponding experimental PCI shifts $\Delta\varepsilon_m$ and those obtained by Southworth *et al.* (1983) are compared with theory. As can be seen from Figs. 27 and 28, the theoretical shapes and shifts agree well with the experimental results.

In the case of inner-shell *ionization* by *electron impact* we have two slowly moving electrons e_1 and e_2 (sharing the excess energy E_1) which interact with the Auger electron. No general theoretical treatment of the PCI effect for inner-shell ionization by electron impact has yet emerged, i.e., an adequate formulation of the interaction of an electron with two slowly moving and correlated electrons is still missing. Only recently the PCI shifts and shapes of Auger lines were calculated with a classical model for the following two limiting cases of E_1: (a) the excess energy E_1 is large enough for the motion of electrons e_1 and e_2 to be treated as uncorrelated (Huster and

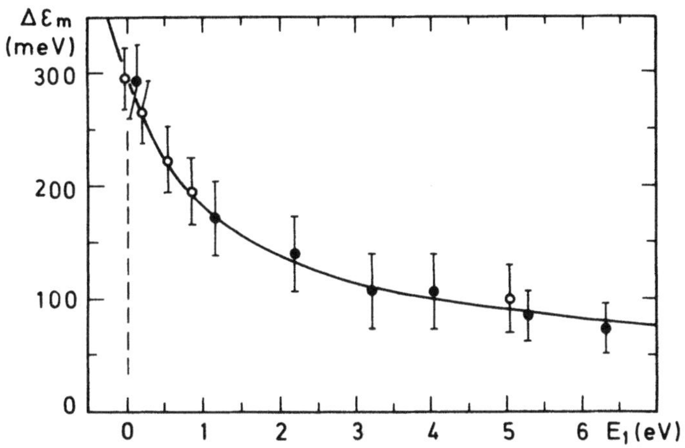

FIGURE 28. PCI shifts $\Delta\varepsilon_m$ of the Xe $\underline{4d}_{5/2}$–$\underline{5p}^2({}^1S_0)$ Auger line after photoionization, as function of the excess energy E_1. The theoretical curve has been obtained according to Niehaus (1977) with $\Gamma = 110$ meV. The experimental data are from Schmidt *et al.* (1981) (○) and from Southworth *et al.* (1983) (●).

Mehlhorn, 1982); (b) $E_1 \to 0$; the motion is highly correlated and well established (Huster and Mehlhorn, 1982; Watanabe *et al.*, 1983).

As example for PCI phenomena observed in Auger spectra following the ionization by electron impact, the Ar $L_{2,3}$–$M_{2,3}M_{2,3}$ Auger spectrum (lines 2 to 7) and a reference line, M_5–$4p^2\ {}^1S_0$ of Kr (line 1) are shown in Fig. 29 for various excess energies E_1. The solid-line spectra are based on theoretical PCI line shapes given by Niehaus (1977). For the PCI shifts $\overline{\Delta\varepsilon}_m(E_1)$ of Ar $L_{2,3}$–$M_{2,3}M_{2,3}$ Auger lines, the following dependence on the excess energy E_1 ($E_1 > 10$ eV) has been found (Huster and Mehlhorn, 1982):

$$\overline{\Delta\varepsilon}_m(E_1) = (5.3 \pm 0.4)\, \Gamma E_1^{-(0.45 \pm 0.04)}, \tag{45}$$

with Γ, E_1, and $\overline{\Delta\varepsilon}_m$ in eV. Helenelund *et al.* (1983) and Hedman *et al.* (1982) found in their investigations of the PCI shifts of Ar $L_{2,3}$–$M_{2,3}M_{2,3}$

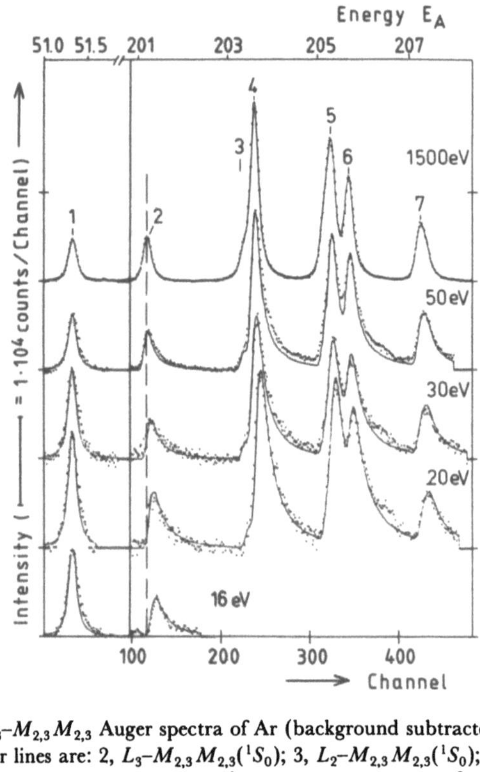

FIGURE 29. $L_{2,3}$–$M_{2,3}M_{2,3}$ Auger spectra of Ar (background subtracted) for different excess energies E_1. The Ar lines are: 2, L_3–$M_{2,3}M_{2,3}({}^1S_0)$; 3, L_2–$M_{2,3}M_{2,3}({}^1S_0)$; 4, L_3–$M_{2,3}M_{2,3}({}^1D_2)$; 5, L_3–$M_{2,3}M_{2,3}({}^3P_{0,1,2})$; 6, L_2–$M_{2,3}M_{2,3}({}^1D_2)$; 7, L_2–$M_{2,3}M_{2,3}({}^3P_{0,1,2})$. Line 1 is the Kr M_5–$N_{2,3}N_{2,3}({}^1S_0)$ reference line with excess energy $(E_1 + 155)$ eV. From Huster and Mehlhorn (1982).

Auger lines close agreement with Eq. (45). Relation (45) should hold for any inner-shell vacancy, provided the corresponding width Γ is used. In Fig. 30 the experimental PCI shifts $\overline{\Delta\varepsilon_m}$ of Ne K and Ar L_3 Auger electrons are compared with those obtained by means of Eq. (45) with $\Gamma(\text{Ne } K) = 0.27$ eV and $\Gamma(\text{Ar } L_3) = 0.13$ eV. As can be seen, the data agree reasonably well with the predictions of Eq. (45), although the shifts measured by Wakiya *et al.* (1981) follow more closely the power laws $E_1^{-0.72}$ for Ar L_3 and $E_1^{-0.65}$ for Ne K ionization. The shift $\overline{\Delta\varepsilon_m}(E_1 = 0)$ for $E_1 = 0$ has been derived as (Huster and Mehlhorn, 1982; Watenabe *et al.*, 1983)

$$\overline{\Delta\varepsilon_m}(E_1 = 0) = (8\sqrt{3}\,\Gamma/15)^{2/3}, \tag{46}$$

where Γ and $\overline{\Delta\varepsilon_m}$ are in atomic units. Equation (46) yields $\overline{\Delta\varepsilon_m}(E_1 = 0)$ as 1.19 eV and 0.84 eV for Ne K and Ar L_3 ionization, respectively. These results are indicated as arrows in Fig. 30; they are larger by a factor $(32/3)^{1/3} = 2.20$ than the threshold value $\Delta\varepsilon_m(E_1 = 0)$ for photoionization (Niehaus, 1977).

Another effect which changes the energy and width of an Auger line excited by photoionization near threshold is the resonant Raman Auger (RRA) effect (Brown *et al.*, 1980; Ice *et al.*, 1982; Crasemann and Wuilleumier, Chap. 7, this volume). This effect occurs mainly for photoionization below threshold, where the shift of an Auger line due to RRA effect is negative, opposite to that of PCI.

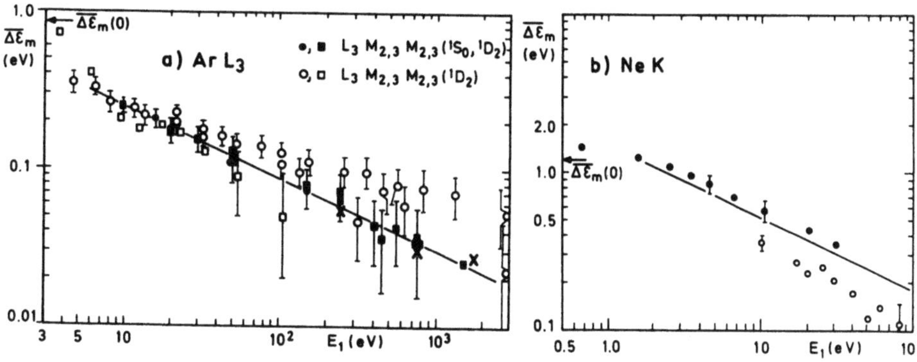

FIGURE 30. PCI shifts $\overline{\Delta\varepsilon_m}$ of Auger lines after inner-shell ionization by electron impact, as function of excess energy E_1. (a) Shift of L_3–$M_{2,3}M_{2,3}(^1S_0, ^1D_2)$ lines of Ar. Experiment: ●, ■, Huster and Mehlhorn (1982); □, Wakiya *et al.* (1981); ○, Helenelund *et al.* (1983). The solid line represents Eq. (45) with $\Gamma = 130$ meV; the crosses (×) are theoretical values of the PCI shifts (Huster and Mehlhorn, 1982); the arrow indicates the predicted shift for $E_1 = 0$. (b) Shift of the K–$L_{2,3}L_{2,3}(^1D_2)$ line of Ne. Experiment: ●, Hink *et al.* (1981); ○, Wakiya *et al.* (1981). The solid line is calculated from Eq. (45) with $\Gamma = 270$ meV. The predicted shift for $E_1 = 0$ is indicated by the arrow.

ACKNOWLEDGMENTS

I am very grateful to T. Åberg for valuable comments on the manuscript and to V. Schmidt for several helpful discussions. I also thank M. H. Chen for calculating the Ca $\underline{2p}$–N_1N_1 transition probabilities and W. Sandner, M. Völkel, and M. Weber for providing me with their results prior to publication.

REFERENCES

Åberg, T., 1969, *Ann. Acad. Sci. Fenn. A* VI **308**, 1.

Åberg, T., 1975, Two-photon emission, the radiative Auger effect and the double Auger process, in *Atomic Inner-Shell Processes*, vol. I, edited by B. Crasemann (Academic, New York), pp. 353–375.

Åberg, T., 1976, Theory of atomic decay following inner-shell ionization, in *Photoionization and Other Probes of Many-Electron Interactions*, edited by F. Wuilleumier, NATO ASI Series (Plenum Press, New York), pp. 273–285.

Åberg, T., 1980, *Phys. Scripta* **21**, 495.

Åberg, T., 1981, The creation and decay of inner-shell vacancies as a single scattering process, in *Inner-Shell and X-ray Physics of Atoms and Solids*, edited by D. J. Fabian, H. Kleinpoppen, L. M. Watson (Plenum, New York), pp. 251–260.

Åberg, T., and Howat, G., 1982, Theory of the Auger effect, in *Handbuch der Physik*, vol. XXXI, edited by W. Mehlhorn (Springer Verlag, Heidelberg), pp. 469–619.

Adam, M. Y., Wuilleumier, F., Sandner, N., Schmidt, V., and Wendin, G., 1978, *J. Phys. (Paris)* **39**, 129.

Ågren, H., Nordgren, J., Selander, L., Nordling, C., and Siegbahn, K., 1978, *J. Electron. Spectrosc.* **14**, 27.

Aksela, S., 1982, University of Oulu, private communication.

Aksela, H., and Aksela, S., 1974, *J. Phys. B* **7**, 1262.

Aksela, S., and Sivonen, J., 1982, *Phys. Rev. A* **25**, 1243.

Aksela, S., Väyrynen, J., and Aksela, H., 1974, *Phys. Rev. Lett.* **33**, 999.

Aksela, H., S. Aksela, J. S. Jen, and T. D. Thomas, 1977a, *Phys. Rev. A* **15**, 985.

Aksela, S., H. Aksela, M. Vuontisjärvi, J. Väyrynen, and E. Lähteenkorva, 1977b, *J. Electron. Spectrosc.* **11**, 137.

Aksela, S., H. Aksela, and T. D. Thomas, 1979, *Phys. Rev. A* **19**, 721.

Aksela, H., S. Aksela, and J. Väyrynen, 1980a, *Phys. Rev. A* **22**, 1116.

Aksela, S., J. Väyrynen, H. Aksela, and S. Pennanen, 1980b, *J. Phys. B* **13**, 3745.

Aksela, S., M. Kellokumpu, H. Aksela, and J. Väyrynen, 1981, *Phys. Rev. A* **23**, 2374.

Amusia, M. Ya., M. Yu. Kuchiev, S. A. Sheinerman, and S. I. Sheftel, 1977, *J. Phys. B* **10**, L535.

Amusia, M. Ya., M. Yu. Kuchiev, and S. A. Sheinerman. 1980, The Postcollision Interaction as a Manifestation of Many-Electron Correlations, in *Coherence and Correlation in Atomic Collisions*, edited by H. Kleinpoppen and J. F. Williams (Plenum Press, New York), pp. 297–319.

Asaad, W. N., 1963, *Nucl. Phys.* **44**, 415.

Asaad, W. N., 1965, *Nucl. Phys.* **66**, 494.

Asaad, W. N., and E. H. S. Burhop, 1958, *Proc. Phys. Soc. London* **71**, 369.

Asaad, W. N., and W. Mehlhorn, 1968, *Z. Phys.* **217**, 304.

ASAAD, W. N., AND D. PETRINI, 1976, *Proc. R. Soc. London Ser. A* **350**, 381.
ASPLUND, L., P. KELVE, B. BLOMSTER, H. SIEGBAHN, AND K. SIEGBAHN, 1977, *Phys. Scr.* **16**, 268.
AUGER, P., 1925, *J. Phys. Radium* **6**, 205.
BACHMANN, P., A. EBERLEIN, AND R. BRUCH, 1982, *J. Phys. E* **15**, 207.
BALASHOV, V. V., S. S. LIPOVETSKY, AND V. S. SENASCHENKO, 1973, *Sov. Phys. JETP* **36**, 858.
BEREZHKO, E. G., AND N. M. KABACHNIK, 1977, *J. Phys. B* **10**, 2467.
BEREZHKO, E. G., AND N. M. KABACHNIK, 1980, *J. Phys. B* **13**, L445.
BEREZHKO, E. G., AND N. M. KABACHNIK, 1982, *J. Phys. B* **15**, 2075.
BEREZHKO, E. G., N. M. KABACHNIK, AND V. S. ROSTOVSKY, 1978a, *J. Phys. B* **11**, 1749.
BEREZHKO, E. G., N. M. KABACHNIK, AND V. V. SIZOV, 1978b, *J. Phys. B* **11**, 1819.
BEYER, H. F., R. MANN, F. FOLKMAN, AND P. H. MOKLER, 1982, *J. Phys. B* **15**, 3853.
BHALLA, C. P., 1973a, *Phys. Lett.* **44a**, 103.
BHALLA, C. P., 1973b, *Phys. Rev. A* **8**, 2877.
BHALLA, C. P., 1975a, *J. Phys. B* **8**, 1200.
BHALLA, C. P., 1975b, *J. Phys. B* **8**, 2787.
BHALLA, C. P., 1975c, *J. Electron Spectrosc.* **7**, 287.
BHALLA, C. P., 1975d, *Phys. Rev. A* **12**, 122.
BHALLA, C. P., N. O. FOLLAND, AND M. A. HEIN, 1973a, *Phys. Rev. A* **8**, 649.
BHALLA, C. P., D. L. MATTHEWS, AND C. F. MOORE, 1973b, *Phys. Lett.* **46a**, 336.
BISGAARD, P., P. DAHL, B. FASTRUP, AND W. MEHLHORN, 1981, *J. Phys. B* **14**, 2023.
BREUCKMAN, B., 1978, Ph.D. thesis, Universität Freiburg, unpublished.
BREUCKMAN, B., 1979, *J. Phys. B* **12**, L609.
BREUCKMAN, B., AND V. SCHMIDT, 1974, *Z. Phys.* **268**, 235.
BREUCKMAN, E., B. BREUCKMAN, W. MEHLHORN, AND W. SCHMITZ, 1977, *J. Phys. B* **10**, 3135.
BREUCKMAN, B., R. HUSTER, W. MENZEL, W. WEBER, W. MEHLHORN, AND K. G. DYALL, 1983, in *Abstracts of Papers, XIII ICPEAC, Berlin 1983*, p. 168.
BRIANÇON, CH., 1970, *Ann. Phys. (Paris)* **5**, 151.
BRIANÇON, CH., AND J. P. DESCLAUX, 1976, *Phys. Rev. A* **13**, 2157.
BROWN, G. S., M. H. CHEN, B. CRASEMANN, AND G. E. ICE, 1980, *Phys. Rev. Lett.* **45**, 1937.
BRUCH, R., A. GUTENKUNST, AND K. H. MÜLLER, 1983, *J. Phys. E*, to be published.
BURCH, D., N. STOLTERFOHT, D. SCHNEIDER, H. WIEMAN, AND J. S. RISLEY, 1975, Nuclear Physics Laboratory Annual Report, University of Washington, Seattle.
BURHOP, E. H. S., 1952, *The Auger Effect and Other Radiationless Transitions* (Cambridge U.P., Cambridge).
BURHOP, E. H. S., AND W. N. ASAAD, 1972, The Auger effect, in *Advances in Atomic and Moleclar Physics*, Vol. 8, edited by D. Bates and I. Esterman (Academic, New York), pp. 164–284.
CALDWELL, C. D., AND R. N. ZARE, 1977, *Phys. Rev. A* **16**, 255.
CAN, C., T. W. TUNNEL, AND C. P. BHALLA, 1982, *J. Electron. Spectrosc.* **27**, 75.
CARLSON, T. A., 1975, *Photoelectron and Auger Spectroscopy* (Plenum Press, New York).
CARLSON, T. A., AND M. O. KRAUSE, 1965, *Phys. Rev. Lett.* **14**, 390.
CARLSON, T. A., AND C. W. NESTOR, JR., 1973, *Phys. Rev. A* **8**, 2887.
CARLSON, T. A., W. E. MODDEMAN, AND M. O. KRAUSE, 1970, *Phys. Rev. A* **1**, 1406.
CHATTARJI, D., 1976, *The Theory of Auger Transitions* (Academic, New York).
CHATTARJI, B., W. MEHLHORN, AND V. SCHMIDT, 1978, *J. Electron Spectroscopy* **13**, 97.
CHEN, M. H., 1984, private communication.
CHEN, M. H., 1985, Chapter 2 of this volume.
CHEN, M. H., AND B. CRASEMANN, 1974, *Phys. Rev. A* **10**, 2232.
CHEN, M. H., AND B. CRASEMANN, 1975, *Phys. Rev. A* **12**, 959.
CHEN, M. H., AND B. CRASEMANN, 1977, *Phys. Rev. A* **16**, 1495.
CHEN, M. H., B. CRASEMANN, AND D. L. MATTHEWS, 1975, *Phys. Rev. Lett.* **34**, 1309.
CHEN, M. H., B. CRASEMANN, M. AOYAGI, AND H. MARK, 1978, *Phys. Rev. A* **18**, 802.

CHEN, M. H., B. CRASEMANN, AND H. MARK, 1980, *Phys. Rev. A* **21**, 442.
CHEN, M. H., B. CRASEMANN, AND H. MARK, 1981, *Phys. Rev. A* **24**, 1158.
CHUNG, K. T., 1982, *Phys. Rev. A* **25**, 1596.
CLEFF, B., 1982, *Acta Phys. Pol. A* **61**, 285.
CLEFF, B., AND W. MEHLHORN, 1971, *Phys. Lett.* **37A**, 3.
CLEFF, B., AND W. MEHLHORN, 1974a, *J. Phys. B* **7**, 593.
CLEFF, B., AND W. MEHLHORN, 1974b, *J. Phys. B* **7**, 605.
COMBET FARNOUX, F., 1982, *Phys. Rev. A* **25**, 287.
COMBET FARNOUX, F., AND M. BEN AMAR, 1980, *Phys. Rev. A* **21**, 1975.
CONNERADE, J. P., 1976, *Proc. R. Soc. London Ser. A* **347**, 575.
CONNERADE, J. P., M. W. D. MANSFIELD, D. A. NEWSOM, D. A. TRACY, M. A. BAIG, AND K. THIMM, 1979, *Phil. Trans. R. Soc. London* **290**, 327.
COSTER, D., AND R. DE L. KRONIG, 1935, *Physica*, **2**, 13.
CRASEMANN, B. 1973, Auger and Radiative Transition Probabilities, in Proc. Int. Conf. Inner-Shell Ionization Processes and Future Applications, Atlanta 1972, edited by R. W. Fink, S. T. Manson, J. M. Palms, and P. V. Rao, report No. CONF-720404, USAEC Technical Information Center, Oak Ridge, pp. 9–42.
CRASEMANN, B., AND F. WUILLEUMIER, 1985, Chapter 7 of this volume.
DARKO, T. H. SIEGBAHN, AND P. KELFVE, 1981, *Chem. Phys. Lett.* **81**, 475.
DAHL, P., M. RØDBRO, B. FASTRUP, AND M. E. RUDD, 1976a, *J. Phys. B* **9**, 1567.
DAHL, P., M. RØDBRO, G. HERMANN, B. FASTRUP, AND M. E. RUDD, 1976b, *J. Phys. B* **9**, 1581.
DAVIS, B. F., AND K. T. CHUNG, 1982, *J. Phys. B* **15**, 3113.
DESCLAUX, J. P., 1975, *Comp. Phys. Commun.* **9**, 31.
DUBOIS, R. D., AND M. RØDBRO, 1980, *J. Phys. B* **13**, 3739.
DUBOIS, R. D., L. MORTENSON, AND M. RØDBRO, 1981, *J. Phys. B* **14**, 1613.
DYALL, K. G., AND F. P. LARKINS, 1982a, *J. Phys. B* **15**, 1021.
DYALL, K. G., AND F. P. LARKINS, 1982b, *J. Phys. B* **15**, 2793.
DYALL, K. G., AND F. P. LARKINS, 1982c, *J. Phys. B* **15**, 4103.
DYALL, K. G., F. P. LARKINS, K. D. BOMBEN, AND T. D. THOMAS, 1981, *J. Phys. B* **14**, 2551.
EL IBYARI, S. N., W. N. ASAAD, AND E. J. MCGUIRE, 1972, *Phys. Rev. A* **5**, 1048.
FANO, U., 1961, *Phys. Rev.* **124**, 1866.
FANO, U., AND J. W. COOPER, 1965, *Phys. Rev.* **137A**, 1364.
FANO, U., AND F. PRATS, 1963, *Proc. Natl. Acad. Sci. India A* **33**, 553.
FLÜGGE, S., W. MEHLHORN, AND V. SCHMIDT, 1972, *Phys. Rev. Lett.* **29**, 7.
FROESE-FISCHER, C., 1969, *Comp. Phys. Commun.* **1**, 151.
FROESE-FISCHER, C., 1972, *Comp. Phys. Commun.* **4**, 107.
GENZ, H., 1982, Inner-Shell Ionization by Relativistic Electron Impact, in *X-Ray and Atomic Inner-Shell Physics—1982*, edited by B. Crasemann, AIP Conference Proceedings No. 94 (American Institute of Physics, New York), p. 85–99.
GRANT, I. P., 1970, *Adv. Phys.* **19**, 747.
GRANT, I. P., B. J. MCKENZIE, P. H. NORRINGTON, D. F. MAYERS, AND N. C. PYPER, 1980, *Comp. Phys. Commun.* **21**, 207.
GROENEVELD, K. O., R. MANN, G. NOLTE, S. SCHUMANN, R. SPOHR, AND B. FRICKE, 1975, *Z. Phys. A* **274**, 191.
GUNNARSSON, O., AND K. SCHÖNHAMMER, 1982, One-Step Model of X-Ray Photoemission and Auger Processes, in *X-Ray and Atomic Inner-Shell Physics, 1982*, AIP Conference Proceedings No. 94, edited by B. Crasemann (American Institute of Physics, New York), pp. 517–529.
HAGMANN, S. G. HERMAN, AND W. MEHLHORN, 1974, *Z. Phys.* **266**, 189.
HEDMAN, S., K. HELENELUND, L. ASPLUND, U. GELIUS, AND K. SIEGBAHN, 1982, *J. Phys. B* **15**, L799.
HELENELUND, K., HEDMAN, L. ASPLUND, U. GELIUS, AND K. SIEGBAHN, 1983, *Phys. Scr.* **27**, 245.

HILLIG, H., B. CLEFF, W. MEHLHORN, AND W. SCHMITZ, 1974, *Z. Phys.* **268**, 225.

HINK, W., L. KEES, H. P. SCHMITT, AND A. WOLF, 1981, Near *K*-Ionization Threshold Auger-Electron Measurements for Neon Under Electron Impact, in *Inner-Shell and X-Ray Physics of Atoms and Solids*, edited by D. J. Fabian, H. Kleinpoppen, and L. M. Watson, (Plenum Press, New York), pp. 327–330.

HOWAT, G., 1978, *J. Phys. B* **11**, 1589.

HOWAT, G., T. ÅBERG, AND O. GOSCINSKI, 1978, *J. Phys. B* **11**, 1575.

HUANG, K.-N., 1982, *Phys. Rev. A* **26**, 2274.

HUANG, K.-N., M. AOYAGI, M. H. CHEN, B. CRASEMANN, AND H. MARK, 1976, *At. Data Nucl. Data Tables* **18**, 243.

HUSTER, R., 1981, private communication.

HUSTER, R., AND W. MEHLHORN, 1982, *Z. Phys.* **307**, 67.

ICE, G. E., G. S. BROWN, G. B. ARMEN, M. H. CHEN, B. CRASEMANN, J. LEVIN, AND D. MITCHELL, 1982, Atomic Inner-Shell Threshold Excitation with Synchrotron Radiation, in *X-Ray and Atomic Inner-Shell Physics—1982*, AIP Conference Proceedings No. 94, edited by B. Crasemann (American Institute of Physics, New York), pp. 105–113.

JACOBS, V. L., 1972, *J. Phys. B* **5**, 2257.

JAMISON, K. A., AND P. RICHARD, 1977, *Phys. Rev. Lett.* **38**, 484.

JAMISON, K. A., P. RICHARD, F. HOPKINS, AND D. L. MATTHEWS, 1978, *Phys. Rev. A* **17**, 1642.

JITSCHIN, W., 1982, Alignment Studies in Inner-Shell Processes, in *X-Ray and Atomic Inner-Shell Physics—1982*, AIP Conference Proceedings No. 94, edited by B. Crasemann (American Institute of Physics, New York), pp. 74–84.

KABACHNIK, N. M., 1978, private communication.

KABACHNIK, N. M., 1981, *J. Phys. B* **14**, L337.

KARIM, K. R., AND CRASEMANN, B., 1984, *Phys. Rev. A* **30**, 1107.

KARIM, K. R., AND CRASEMANN, B., 1985, *Phys. Rev. A* **31**, 709.

KARIM, K. R., CHEN M. H., AND CRASEMANN, B., 1984, *Phys. Rev. A* **29**, 2605.

KELLY, H. P., 1975, *Phys. Rev. A* **11**, 556.

KIHARA, S., T. TAKAYANAGI, K. WAKIYA, AND H. SUZUKI, 1983, in *Abstracts of Papers, XIII ICPEAC, Berlin 1983*, p. 155.

KLAR, H., 1980, *J. Phys. B* **13**, 4741.

KLAR, H., 1982, *J. Phys. B* **15**, 4535.

KÖRBER, H., AND W. MEHLHORN, 1966, *Z. Phys.* **191**, 217.

KRAUSE, M. O., 1975, *Phys. Rev. Lett.* **34**, 633.

KRAUSE, M. O., 1976, The *K* Auger Spectrum: New Developments, in *Proc. Intern. Conf. on Inner-Shell Ionization Phenomena, Freiburg 1976*, edited by W. Mehlhorn and R. Brenn, (Universität Freiburg, FRG), pp. 184–197.

KRAUSE, M. O., 1980, private communication. Most of the data of Table V are taken from his compilation.

KRAUSE, M. O., AND J. H. OLIVER, 1979, *J. Phys. Chem. Ref. Data* **8**, 329.

KRAUSE, M. O., T. A. CARLSON, AND W. E. MODDEMAN, 1971, *J. Phys. (Paris)* **32**, C4-139.

LARKINS, F. P., 1976, *J. Phys. B* **9**, 47.

LARKINS, F. P., 1977, *At. Data Nucl. Data Tables* **20**, 311.

LIPOVETSKY, S. S., AND V. S. SENASCHENKO, 1974, *J. Phys. B* **7**, 693.

LÖW, W., H. GENZ, AND A. RICHTER, 1983, in *Abstracts of Papers, XIII ICPEAC, Berlin 1983*, p. 167.

MANN, R., H. F. BEYER, AND F. FOLKMANN, 1981a, *Phys. Rev. Lett.* **46**, 646.

MANN, R., F. FOLKMANN, AND H. F. BEYER, 1981b, *J. Phys. B* **14**, 1161.

MANN, R., H. F. BEYER, AND F. FOLKMANN, 1982, Excited States Formed by Electron Capture in Highly Charged Recoiled Ions, in *Physics of Electronic and Atomic Collisions*, edited by S. Datz (North-Holland, Amsterdam), pp. 683–695.

MANSFIELD, M. W. D., 1976, *Proc. R. Soc. London Ser. A* **348**, 143.
MARTIN, R. L., AND D. A. SHIRLEY, 1976, *Phys. Rev. A* **13**, 1475.
MARTIN, R. L., AND D. A. SHIRLEY, 1977, Many-Electron Theory of Photoemission, in *Electron Spectroscopy: Theory, Techniques and Applications*, edited by C. R. Brundle and A. D. Baker (Academic, London), Vol. I, pp. 75–117.
MATTHEWS, D. L., B. M. JOHNSON, J. J. MACKEY, L. E. SMITH, W. HODGE, AND C. F. MOORE, 1974, *Phys. Rev.* **10**, 1177.
MATTHEWS, D. L., B. M. JOHNSON, AND C. F. MOORE, 1975, *At. Data. Nucl. Data Tables* **15**, 40.
MCGUIRE, E. J., 1969, *Phys. Rev.* **185**, 1.
MCGUIRE, E. J., 1971a, *Phys. Rev. A* **3**, 1801.
MCGUIRE, E. J., 1971b, Research Report SC-RR-710075, Sandia Laboratories.
MCGUIRE, E. J., 1972, *Phys. Rev. A* **5**, 1043.
MCGUIRE, E. J., 1974a, *Phys. Rev. A* **10**, 13.
MCGUIRE, E. J., 1974b, *Phys. Rev. A* **10**, 32.
MCGUIRE, E. J., 1975a, Auger and Coster–Kronig transitions, in *Atomic Inner-Shell Processes*, edited by B. Crasemann (Academic, New York), Vol. I, pp. 293–330.
MCGUIRE, E. J., 1975b, *Phys. Rev. A* **11**, 10.
MCGUIRE, E. J., 1975c, *Phys. Rev. A* **11**, 17.
MCGUIRE, E. J., 1975d, *Phys. Rev. A* **11**, 1880.
MCGUIRE, E. J., 1975e, *Phys. Rev. A* **12**, 330.
MCGUIRE, E. J., 1976, *Phys. Rev. A* **14**, 1402.
MCGUIRE, E. J., 1977, *Phys. Rev. A* **16**, 2365.
MCGUIRE, E. J., 1978, *Phys. Rev. A* **17**, 182.
MEHLHORN, W., 1960, *Z. Phys.* **160**, 247.
MEHLHORN, W., 1968a, *Phys. Lett.* **26A**, 166.
MEHLHORN, W., 1968b, *Z. Phys.* **208**, 1.
MEHLHORN, W., 1975, Resonances Seen in Secondary Electron Spectra, in *The Physics of Electronic and Atomic Collisions, Proceedings of IX ICPEAC, Seattle*, edited by J. S. Risley and R. Geballe (University of Washington Press, Seattle), pp. 756–765.
MEHLHORN, W., 1976, Correlation Effects in Auger-Electron Spectroscopy, in *Photoionization and other Probes of Many-Electron Interactions*, edited by F. Wuilleumier, NATO ASI Series (Plenum Press, New York), pp. 309–330.
MEHLHORN, W., 1978, Electron Spectroscopy of Auger and Autoionizing States: Experiment and Theory, Lecture Notes, University of Aarhus (unpublished).
MEHLHORN, W., 1982, Alignment in Ionization by Ion, Electron and Photon Impact, in *X-Ray and Atomic Inner-Shell Physics—1982*, AIP Conference Proceedings No. 94, edited by B. Crasemann (American Institute of Physics, New York), pp. 55–73.
MEHLHORN, W., 1983a, Structure and Dynamics of Atoms Probed by Inner-Shell Ionization, in *Atomic Physics 8*, Proc. of the VIII Intern. Conf. on Atomic Physics, Göteborg 1982, edited by I. Lindgren (Plenum Press, New York), pp. 213–241.
MEHLHORN, W., 1983b, Electron Decay Processes, in *Fundamental Processes in Energetic Atomic Collisions*, edited by J. S. Briggs, H. Kleinpoppen, and H. O. Lutz, NATO ASI Series (Plenum Press, New York), pp. 253–293.
MEHLHORN, W., AND D. STALHERM, 1968, *Z. Phys.* **217**, 294.
MEHLHORN, W., D. STALHERM, AND H. VERBEEK, 1968, *Z. Naturforsch A* **23**, 287.
MEHLHORN, W., W. SCHMITZ, AND D. STALHERM, 1972, *Z. Phys.* **252**, 399.
MEHLHORN, W., B. BREUCKMANN, AND D. HAUSAMANN, 1977, *Phys. Scr.* **16**, 177.
MENZEL, W., AND W. MEHLBORN, 1981, 3*d* Auger and 3*p* Coster–Kronig Spectrum of Free Rubidium Atoms, in *Inner-Shell and X-Ray Physics of Atoms and Solids*, edited by D. J. Fabian, H. Kleinpoppen, and L. M. Watson (Plenum Press, New York), pp. 319–322.
NIEHAUS, A., 1977, *J. Phys. B* **10**, 1845.

Niehaus, A., 1983, *J. Phys. B* **16**, L135.

Ohno, M., and G. Wendin, 1979, *J. Phys. B* **12**, 1305.

Ost, S., M. Brüssermann, J. Wigger, G. Richter, and B. Cleff, 1982, Abstract of Contr. Papers, Intern. Conf. on X-Ray and Atomic Inner-Shell Physics, Eugene, Oregon, p. 290.

Pegg, D. J., I. A. Sellin, R. Peterson, J. R. Mowat, W. W. Smith, M. D. Brown, and J. R. MacDonald, 1973, *Phys. Rev. A* **8**, 1350.

Petrini, D., 1982, *Can. J. Phys.* **60**, 644.

Rødbro, M., R. D. DuBois, and V. Schmidt, 1978, *J. Phys. B* **11**, L551.

Rødbro, M., R. Bruch, and P. Bisgaard, 1979, *J. Phys. B* **12**, 2413.

Ridder, D., J. Dieringer, and N. Stolterfoht, 1976, *J. Phys. B* **9**, L307.

Rose, S. J., I. P. Grant, and J. P. Connerade, 1980, *Phil. Trans. R. Soc. London* **296**, 527.

Russek, A., and W. Mehlhorn, 1985, *J. Phys. B* **18**, (to be published).

Sandner, W., 1985, New applications of electron–electron coincidences in *Fundamental Processes in Atomic Collision Physics*, edited by H. Kleinpoppen, J. S. Briggs, and H. O. Lutz, NATO ASI Series (Plenum Press, New York).

Sandner, W., and W. Schmitt, 1978, *J. Phys. B* **11**, 1833.

Sandner, W., and M. Völkel, 1984, *J. Phys. B* **17**, L597.

Sandner, W., M. Weber, and W. Mehlhorn, 1980, Energy Dependence of the Alignment in Inner-Shell Ionization by Electron Impact, in *Coherence and Correlation in Atomic Collisions*, edited by H. Kleinpoppen and J. F. Williams (Plenum Press, New York), pp. 215–222.

Schmidt, V., 1973, Transition Probabilities of *KL–LLL* Auger Satellites in Neon, in Proc. Intern. Conf. on Inner-Shell Ionization Processes and Future Applications, Atlanta 1972, edited by R. W. Fink, S. T. Manson, J. M. Palms, and P. V. Rao, report No. Conf-720404, USEAC Technical Information Center, Oak Ridge, Tenn., pp. 548–558.

Schmidt, V., 1982, Post-Collision Interaction in Inner-Shell Ionization, in *X-Ray and Atomic Inner-Shell Physics—1982* AIP Conference Proceedings No. 94, edited by B. Crasemann (American Institute of Physics, New York), pp. 544–558.

Schmidt, V., S. Krummacher, F. Wuilleumier, and P. Dhez, 1981, *Phys. Rev. A* **24**, 1803.

Schneider, D., D. F. Burch, and N. Stolterfoht, 1973, in Abstracts of Papers, VIII ICPEAC, Beograd, p. 729.

Schneider, D., K. Roberts, W. Hodge, and C. F. Moore, 1976a, *Phys. Rev. Lett.* **36**, 1065.

Schneider, D., C. F. Moore, and B. M. Johnson, 1976b, *J. Phys. B* **9**, L153.

Schneider, D., R. Bruch, W. H. E. Schwarz, T. C. Chang, and C. F. Moore, 1977, *Phys. Rev. A* **15**, 926.

Sellin, I. A., 1973, *Nucl. Instrum. Methods* **110**, 477.

Sellin, I. A., 1978, Extensions of Beam Foil Spectroscopy, in *Structure and Collisions of Ions and Atoms*, edited by I. A. Sellin, Topics in Current Physics, Vol. 5 (Springer, Berlin) pp. 273–308.

Sewell, E. C., and A. Crowe, 1982, *J. Phys. B* **15**, L357.

Sewell, E. C., and A. Crowe, 1984a, *J. Phys. B* **17**, L547.

Sewell, E. C., and A. Crowe, 1984b, *J. Phys. B* **17**, 2913.

Shirley, D. A., 1973, *Phys. Rev. A* **7**, 1520.

Shore, B. W., and D. H. Menzel, 1968, *Principles of Atomic Spectra* (Wiley, New York).

Siegbahn, H., and L. Karlsson, 1982, Photoelectron Spectroscopy, in *Handbuch der Physik*, Vol. XXXI, edited by W. Mehlhorn (Springer, Heidelberg), pp. 215–446.

Sizov, V. V., and N. M. Kabachnik, 1980, *J. Phys. B* **13**, 1601.

Southworth, S. H., P. H. Kobrin, C. M. Truesdale, D. Lindle, S. Owaki, and D. Shirley, 1981, *Phys. Rev. A* **24**, 2257.

Southworth, S. H., U. Becker, C. M. Truesdale, P. H. Kobrin, D. W. Lindle, S. Owaki, and D. A. Shirley, 1983, *Phys. Rev. A* **28**, 261.

Stolterfoht, N., 1983, Techniques of High-Resolution Auger-Electron and X-Ray Spectroscopy in Energetic Ion–Atom Collisions, in *Fundamental Processes in Energetic Atomic Collisions*, edited by J. S. Briggs, H. Kleinpoppen, and H. O. Lutz, NATO ASI Series (Plenum Press, New York), 295–318.

Stolterfoht, N., H. Gabler, and U. Leithäuser, 1973, *Phys. Lett.* **45A**, 351.

Stolterfoht, N., D. Schneider, R. Mann, and F. Folkmann, 1977, *J. Phys. B* **10**, L281.

Süzer, S., S.-T. Lee, and D. A. Shirley, 1976, *Phys. Rev. A* **13**, 1842.

Svensson, S., N. Mårtensson, E. Basilier, P. A. Malmquist, U. Gelius, and K. Siegbahn, 1976, *Phys. Scr.* **14**, 141.

Väyrynen, J., 1981, *J. Electron. Spectrosc.* **22**, 27.

Väyrynen, J., and S. Aksela, 1979, *J. Electron. Spectrosc.* **16**, 423.

Väyrynen, J., and S. Aksela, 1981, *J. Electron. Spectrosc.* **23**, 119.

Väyrynen, J., S. Aksela, and H. Aksela, 1977, *Phys. Scr.* **16**, 452.

Väyrynen, J., S. Aksela, M. Kellokumpu, and H. Aksela, 1980, *Phys. Rev. A* **22**, 1610.

Völkel, M., and W. Sandner, 1983, in Abstracts of Papers, XIII ICPEAC, Berlin, p. 142.

Wakiya, K., H. Suzuki, T. Takayangai, M. Muto, S. Ito, V. Iketaki, and S. Ohtani, 1981, in *Abstracts of Papers, XII ICPEAC*, Gatlinburg, p. 247.

Walters, D. L., and C. P. Bhalla, 1971a, *Phys. Rev. A* **3**, 19.

Walters, D. L., and C. P. Bhalla, 1971b, *At. Data* **3**, 301.

Watanabe, T., T. Ishihara, and J. Mizuno, 1983, *J. Phys. B* **16**, L107.

Weber, W., 1984, Diplom Thesis, Universität Freiburg.

Weightman, P., 1982, *Rep. Prog. Phys.* **45**, 753.

Wendin, G., 1981, Breakdown of One-Electron Pictures in Photoelectron Spectra, in *Structure and Bonding*, Vol. 45 (Springer, Heidelberg).

Wendin, G., 1982, Dynamic effects in core spectra, in *X-Ray and Atomic Inner-Shells Physics—1982*, AIP Conference Proceedings No. 94, edited by B. Crasemann (American Institute of Physics, New York), pp. 495–516.

Wendin, G., and M. Ohno, 1976, *Phys. Scr.* **14**, 148.

Wentzel, G., 1927, *Z. Phys.* **43**, 524.

Werme, L. O., T. Bergmark, and K. Siegbahn, 1972, *Phys. Scr.* **6**, 141.

Werme, L. O., T. Bergmark, and K. Siegbahn, 1973, *Phys. Scr.* **8**, 149.

Yafet, Y., 1980, *Phys. Rev. B* **21**, 5023.

Yin, L. I., I. Adler, T. Tsang, M. H. Chen, D. Ringers, and B. Crasemann, 1974, *Phys. Rev. A* **9**, 1070.

CHAPTER 5

Experimental Evaluation of Inner-Vacancy Level Energies for Comparison with Theory

Richard D. Deslattes and Ernest G. Kessler, Jr.

1. INTRODUCTION AND OVERVIEW

Evidence can be found elsewhere in this volume of considerable progress on the theoretical side in calculations of atomic inner-shell energy levels. Especially as regards the deeper lying levels of heavier atoms one is obliged to use a relativistic formulation which entails a number of questions. Firstly any structure which emerges from the Dirac one-electron problem shares with it questions about proper accounting of the role of negative-energy states. This seems tractable via projection operators (Sucher, 1980; Mittleman, 1981) and will not be further considered here. Secondly, in the case of a many-electron atom, one does not have a closed-form relativistic Hamiltonian; the usual approach is to construct a relativistic analog of the nonrelativistic Hamiltonian (Weiss, 1981; Desclaux, 1979) with the static electron–electron interaction included in the Hamiltonian while transverse photon exchange (the Breit interaction), self-energy, and vacuum polarization are treated perturbatively. Ultimately such an ad hoc procedure tends to be judged retrospectively by its ability to account for experimental results. Naturally, precise comparisons with accurate experimental data are sensitive to the distribution of charge within the nucleus (its size and shape) as well as to (unevaluated) higher-order quantum electrodynamic terms. It is precisely in the hope of identifying the best available procedures and data for the purpose of such a comparison that the present study as well as our previous work have been undertaken.

Richard D. Deslattes and Ernest G. Kessler, Jr. • Quantum Metrology Group, National Bureau of Standards, Washington, D.C. 20234.

The energy-level values emerging from theory are differences between self-consistent-field calculations (ΔSCF) for fully or partially relaxed electronic configurations for atoms with inner-shell vacancies present and the neutral-atom ground state. Under these circumstances it is natural to attempt to extract from experiments energy term values with which such results might be compared. As will be described somewhat more carefully below, experimental algorithms for term-value estimation, especially of deep vacancy levels, have associated with them appreciable difficulties even in favorable cases and, at least for some methods, a disturbing lack of generality.

On the other hand, one can imagine directly applying the ΔSCF approach to configurations in which a single vacancy appears in two distinct shells connected by an allowed radiative transition. In this case one has the possibility of a direct ΔSCF estimate of the transition energy, i.e., the wavelength or energy which would be expected for the corresponding emission line. From the experimental side, emission spectra are generally available and can be measured with reasonably high precision and accuracy. In addition, because of the rather similar effect of external electrons in shifting energy levels of deep core states, it is expected that differences between solids and gases should not be very large and, as will be seen below, this has indeed proven to be the case.

Of course there are problems in obtaining spectra which have adequate precision and accuracy as well as problems associated with the fact that one never observes a pure single-vacancy intershell transition. To attend to the last point first, the reader need only be reminded that the production of a single vacancy with no other consequence is extremely unlikely and additional vacancies and excitations generally accompany each primary ionizing event. Emission processes then take place in an atom with one or more spectator vacancies. If the spectator vacancies are sufficiently close to the inner shells involved in the primary transition, then these so-called satellite lines are reasonably well resolved from the parent and are merely a nuisance. On the other hand, when the parasitic excitation or ionization occurs in a more outer region of the atom, perturbations of the emission spectra are small and the corresponding satellites are inexorably tangled with the parent line. The result of this awkward situation is that the lines are shifted about and contain almost unmodelable fine detail. Up to a certain point, however, one can simply ignore this problem and accept the fact that there will be an effective limit on the interpretation of experimental results regardless of the precision and accuracy with which the line profile is delineated. Identifying this limiting region is perhaps one of the main burdens of our contribution to this volume.

Although less interesting physically, the technical problems of precise and accurate spectroscopy do require some measure of attention. Spectroscopic resolving powers of the order of a few to 20,000 are appropriate since the linewidths in the X-ray region yield Q ($Q = E/\Delta E = \lambda/\Delta\lambda$) values which are typically 1000 to 3000. Although vast quantities of X-ray data were accumulated long ago, few of these early instances approach in quality the best among the more modern results. It transpires also that the interesting level of confrontation between theory and experiment which is now possible lies just beyond the boundary between what has been possible recently and what was accumulated in the past. Our emphasis will therefore be largely on the results of most recent work especially as regards the heavier elements. Certain earlier wavelength ratios in the low- to mid-Z region have been verified, at least by sampling, and remain useful for the purpose of comparisons with theory.

Aside from adequate resolution and scale linearity, there is one other consideration needed to produce an ensemble of experimental results appropriate for theoretical consideration. This last point has to do with setting the magnitude of the scale. Some readers may recall that not too long ago wavelengths in the X-ray region were determined on a conventional scale based on an effectively arbitrary definition of an X-unit (approximately 0.1 pm or 1.0 m-Å). One then had the problem of determining the ratio between the X-unit and the milliangstrom. In the past this conversion factor had to either be evaluated from general knowledge of the physical constants or by means of a ruled-grating measurement. These approaches are not adequate for today's task. An alternative became available several years ago whereby one can now measure directly the interplanar spacing of a crystal by combined optical and X-ray interferometry and use this result, together with absolute angular measurement technique, to determine X-ray wavelengths directly in terms of the optical secondary standard used in the interferometric measurement. When this optical standard is, in effect, the same stabilized laser used to determine the Rydberg constant, the resulting X-ray wavelengths emerge in units exactly appropriate for theoretical comparison. A more complex but equally satisfactory chain can be made through the use of gamma-ray secondary standards. This latter possibility arises because the same technology, namely, X-ray interferometry plus absolute goniometry, has been used for the determination of a number of gamma-ray lines in the region from somewhat below 100 keV to nearly 1 MeV. These gamma-ray secondary standards can be used in high orders with bent-crystal instruments to determine X-ray wavelengths on a scale that is as practical as that available from direct diffraction measurements.

Therefore in reaching what we consider to be the best available body of experimental information about inner-shell energy-level differences, we will be making use of three types of input. First will be those lines which have been directly measured with high-resolution double-diffraction instruments.

Second will be those obtained with high-resolution curved-crystal optics relative to gamma-ray standards, and third will be those (low-energy) lines whose wavelength ratios with respect to directly measured X-ray lines have been taken from a very restricted set of earlier measurements.

2. METHODS FOR DETERMINING ENERGY LEVELS AND LEVEL DIFFERENCES

We shall be considering application of X-ray absorption spectroscopy (XAS), X-ray photoelectron spectroscopy (XPS), Auger electron spectroscopy (AES), appearance-potential spectroscopy (APS), and X-ray emission spectroscopy (XES) to the problem of single-vacancy energy level or, in the last-mentioned case, energy-level difference determination. Our aim is to identify a suitable experimental data base for comparison with first-principles calculations. This objective leads to a pattern of choices tending to avoid problems associated with effects of the state of aggregation or chemical combination and to neglect, where possible, the disturbing effects of the inevitable multivacancy contamination of predominantly single-vacancy processes.

There are, of course, instrumental problems associated with each of the methods noted above as will be discussed in more detail below. Aside from these instrumental problems and the complications noted above, there are intrinsic limitations arising from Heisenberg broadening of the inner levels. The corresponding level widths range from 300 ppm for the sharpest levels (in the mid-Z region) to more than 1000 ppm in the high-Z region. Considering this width and complications such as those arising from multivacancy effects, it is generally not possible to hope to obtain energy-level values (or their differences) to any better than 1% of the linewidth. Naturally, as one goes toward lower Z or less deep final hole states, even this level of "pointing precision" is not interpretable while from the standpoint of measurement technique, an order of magnitude greater refinement is readily attained.

2.1. *Absorption Spectroscopy*

Analyses of phenomenological absorption structures near edges to obtain estimates of binding energies differ according to chemical combination and state of aggregation. We consider first the case of rare gases which is, in most respects, the simplest. Some remarks attend the case of molecular gases and compounds, which are, however, not generally important for the purposes considered here. Finally, we discuss absorption spectra of metals including difficulties of interpretation and conventions which have been used to extract numerical estimates of atomic energy levels.

The near-edge region in rare-gas spectra has long been modeled, following Parratt (1939), as a Rydberg series plus an arctangent continuum. Most work of this sort was carried out on Ar (Watanabe, 1965; Nakamura *et al.*, 1968) and to a lesser extent on Ne and certain edges of the other rare gases (Brogren, 1949). Most recently Breinig *et al.* (1980) made extensive redeterminations in Ar, Kr, and Xe using synchrotron radiation and *a priori* theory. The intense synchrotron radiation continuum gave more precise experimental data than were available previously while the absence of excessive numbers of free parameters gave more robust threshold estimates than would be obtained if the Rydberg series parameters were also extracted from the data. The reason for using *a priori* theory is evident from the spectra shown in Fig. 1a, 1b since, except in the case of the Ne (not shown) and Ar *K* edges, the series, though surely present, are not so evident in the data as typified by the Xe L_{II} edge shown in Fig. 1b. As a cautionary input, we include also, in Fig. 1c, a still more refined view of the Ar *K*-edge region (Deslattes *et al.*, 1983) where it is clear that the continuum oscillator strength drops much more rapidly above the threshold than the normally assumed λ^3 variation would suggest. Thus, while a λ^3 term has in the past either been used to modulate the arctangent function or considered to be substantially constant in the fitted region, the more rapid decrease shown in Fig. 1c has not been allowed for as yet and its presence will surely "pull" the threshold estimate extracted from the data. At the same time, the arctangent model itself, which follows from the Weisskopf–Wigner result for transitions into a uniformly dense continuum of states characterized by a common (Lorentzian) lifetime broadening, is at least suspect in the region near the ionization threshold. The resulting threshold estimates are given in Table I, where the indicated uncertainties are those given by Breinig *et al.* (1980). Also included in Table I are threshold energies obtained by photoelectron spectroscopy (Siegbahn *et al.*, 1969) and from the tabulation by Sevier (1979) obtained by adjusting available absorption, emission, and photoelectron data. The recent absorption and photoelectron data agree very well while the tabulated results are systematically smaller. Although the absorption results in Table I are encouraging as a testing ground for theoretical calculations, they are very sparse and are not likely to be improved because of the model dependence of the analysis.

In the cases of molecular gases and of chemical compounds, there is generally appreciable migration of charge from some toward other atoms participating in binding. The processes involved, though complex, lead to shifts in inner-vacancy energy levels whose systematic study is informative in regard to the nature and strengths of the chemical bonding process. Although much of the discussion of these shifts is expressed in terms of atomic reference states, there is little in the way of an empirical data base which would permit inversion of these discussions to yield atomic term

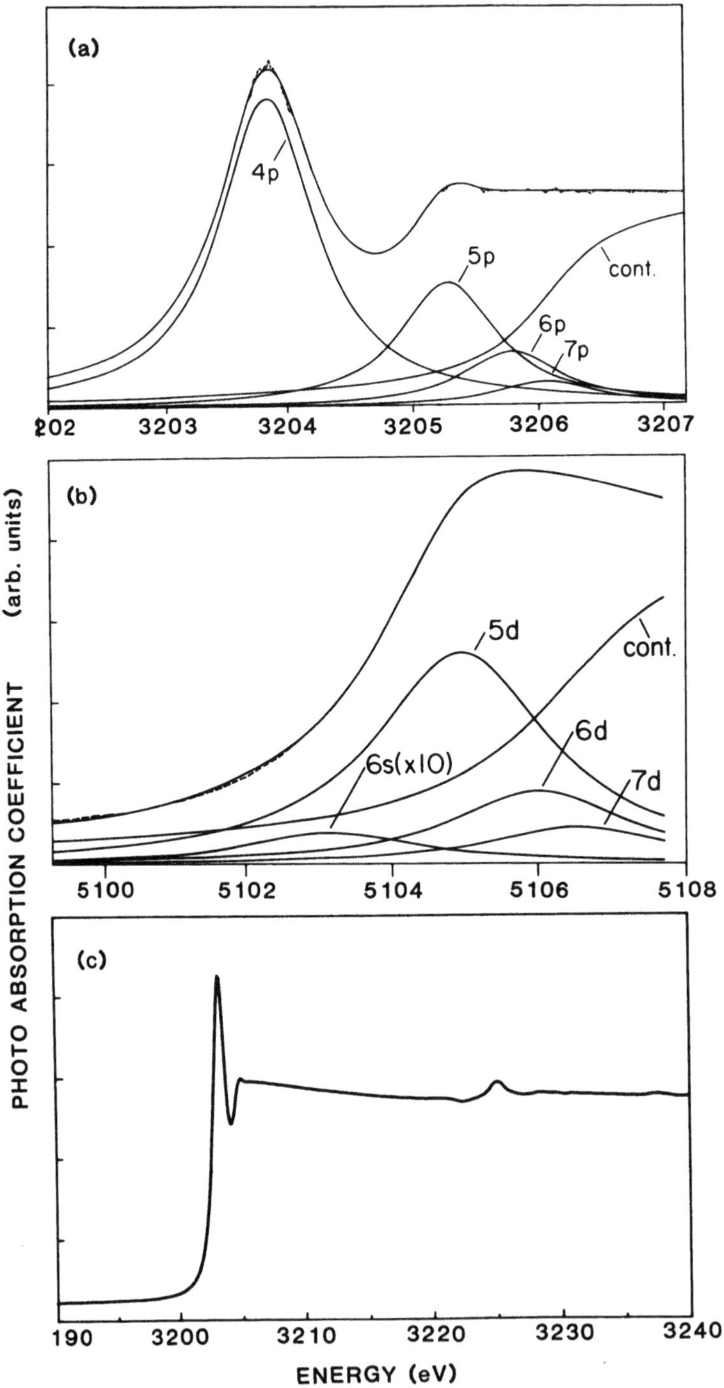

FIGURE 1. Measured absorption spectra and theoretical decomposition near the (a) Ar *K* edge, and (b) Xe L_2 edge (Breinig *et al.*, 1980). A recent trace of the Ar *K* edge over a wider energy range is shown in (c) (Deslattes *et al.*, 1983).

TABLE I

Atomic electron binding energies for rare gases.

Z	Element	Shell	XAS (eV)[a]	XPS (eV)[a]	Sevier (1979)(eV)
18	Ar	$1s$	3206.3(3)	3205.9(5)	3202.9(3)
36	Kr	$1s$	14327.2(8)		14325.6(8)
54	Xe	$2s$	5452.9(5)	5453.2(4)	5452.8(4)
		$2p_{1/2}$	5107.0(5)	5107.2(4)	5103.7(4)
		$2p_{3/2}$	4786.3(6)	4787.3(4)	4782.2(4)

[a] XAS, X-ray absorption spectroscopy; XPS, X-ray photoelectron spectroscopy.

energies. Besides the fact that such an inversion is difficult in principle, there is a much larger practical obstacle. That obstacle is that most chemical-shift measurements have been made using precise instruments that are generally not capable of absolute energy measurements whether from elements or their compounds. The instrumentation and procedures capable of yielding accurate wavelength or energy results, on the other hand, tend to be somewhat inflexible and are usually inefficient; they are most assuredly not widely diffused among research groups working in X-ray spectroscopy. For these reasons, we shall not discuss estimates of atomic term values which might be extracted from spectra of gaseous molecules and chemical compounds even though such inferences are, in principle, possible.

The situation of elemental metals is rather different than those described above for free atoms, polyatomic molecules, and chemical compounds. The customary assumption dating back to the first decades of X-ray spectroscopy and made explicit in a review by Skinner (1940) is that "onset" of X-ray absorption corresponds to the beginning of the conduction continuum. Somewhat on the basis of physical modeling and partly as a convention, this "onset" has been associated with the "first" or lowest energy inflection point in the absorption profile of the near-edge region. This same "onset" or first inflection may likewise be associated with the metal's Fermi energy. [Detailed and critical tests of this association in both emission and absorption spctra have only recently been reported (Callcott *et al.*, 1978).]

The intrinsic complexity involved in realistic calculations of photoabsorption threshold profiles has been widely recognized. In fact computer codes have been generated and made generally available (Durham *et al.*, 1982) for dealing with the near-edge region. Detailed studies relating the condensed-phase threshold behavior to that of free atoms have been reported in only a few cases (Soules and Shaw, 1959; Malzfeldt *et al.*, 1983). On the other hand, Bearden and Burr (1967) noted a general concordance between photoelectron thresholds (relative to the respective Fermi levels) and crudely analyzed absorption spectra. Their conclusion, namely, that "with sufficient care the X-ray absorption measurements could be made competitive with the photoelectron method" can still be supported in general. It reflects an appropriate deference to the results of photoelectron spectroscopy, to which we now turn; after that it will be appropriate to indicate the results of a recent comparison exercise.

2.2. *Photoelectron and Auger-Electron Spectroscopies*

Among all techniques available for experimental determination of inner-vacancy energy levels, X-ray photoelectron spectroscopy (XPS) would appear to be the most sure-footed. If one assumes a well-known monochromatic photon probe, electron groups liberated from various inner shells give separation energies which are, in fact, rigorous estimators of term values. This rigor notwithstanding, separation energies obtained by XPS are sensitive to the physical and chemical state of the probed atom. Indeed this particular sensitivity is widely exploited in chemistry and materials science.

If, however, one's focus is, as ours is now, to find experimental results suited for comparison with *atomic* theory, the above-mentioned sensitivity indicates a need for adjustment of the raw data to obtain free-atom term estimates. To first order, one goes from a Fermi-level referenced binding energy to one referenced to the vacuum by simply adding the work function to the result obtained from an experiment on a metallic solid (Lotz, 1968, 1970). Two difficulties intervene: firstly the work function itself is not terribly well defined for a polycrystalline sample, especially without full surface characterization. Secondly, there is the important problem of extra-atomic relexation, which, inter alia, is operative in aggregated systems but not in free atoms (Shirley *et al.*, 1977; Jen and Thomas, 1976; Mehlhorn *et al.*, 1977).

Gas-phase photoelectron spectroscopy is naturally immune to such ambiguities as those noted above; however, it is not available in all cases, especially toward deeper levels of higher-Z systems. To gain an impression of the utility of experiments involving solids, we can make comparisons between them and gas-phase XPS. The good agreement for XAS and XPS illustrated in Table I suggests that these two experimental methods are both rigorous for gases. Likewise, one would expect good agreement for XAS and XPS applied to metals where the reference is now the Fermi level instead of the vacuum. Table II brings together the two types of comparisons mentioned above: solid-phase vs. gas-phase XPS and solid-phase XPS vs. solid-phase XAS. XPS data are from the recent summary by Sevier (1979), where references to the original works may be found, while X-ray absorption results are those summarized by Bearden and Burr (1967) and Mehlhorn *et*

TABLE II
Atomic electron binding energies for solids.

Z	Element	Shell	XPS (eV)[a] (gas)	XPS (eV)[a] (solid)	AES (eV)[a] (gas)	XAS (eV)[a] (solid)
11	Na	1s		1074.0(1)	1079.1(3)	1072
		2s	70.84(10)	65.7(1)		
		3s	5.14	3.5		
12	Mg	1s	1311.2(3)	1306.7(1)	1311.3(3)	1303
		2s	94.0(5)	92.25(10)	96.5(3)	63
		$2p_{1/2}$	55.0(5)	53.2(1)	57.85(10)	57.83
		$2p_{3/2}$	54.8(5)	52.9(1)	57.57(10)	57.54
21	Sc	2s	503.2(10)	501.5(3)		
		$2p_{1/2}$	408.4(10)	407.08(10)		
		$2p_{3/2}$	403.9(10)	402.15(10)		
		3s	56.4(10)	54.64(10)		
		$3p_{1/2}$, $3p_{3/2}$	33.6(10)	31.84(10)		
26	Fe	2s	857(2)	853(2)		
		$2p_{1/2}$	733(1)	725.15(20)		720.8
		$2p_{3/2}$	720(1)	712.05(20)		707.4
		3s	104(1)	96.06(20)		
		$3p_{1/2}$, $3p_{3/2}$	66(1)	57.54(20)		54
30	Zn	2s	1203(1)	1200.46(25)		
		$2p_{1/2}$	1052.0(3)	1049.39(15)		1045
		$2p_{3/2}$	1029.1(3)	1026.26(15)		1022
		3s	145.0(1)	144.18(15)		
		$3p_{1/2}$		95.61(15)	98.7(7)	86
		$3p_{3/2}$		93.00(15)	96.1(5)	
82	Pb	$4f_{7/2}$	144.0(5)	104.5		143.6(1)

[a] XPS, X-ray photoelectron spectroscopy; AES, Auger electron spectroscopy; XAS, X-ray absorption spectroscopy.

al. (1977). Also included in the table are results from gas-phase Auger-electron spectroscopy (Mehlhorn *et al.*, 1977). In general binding-energy measurements using AES are not as numerous or precise as those from XPS. In a recent commentary (Powell *et al.*, 1982), the AES imprecision is estimated to exceed that of XPS by about a factor of 10. However, the gas-phase AES results presented here have a precision comparable to and agree well with the XPS results, thus increasing confidence in the gas-phase values.

The data selected for inclusion in Table II are intended to be representative rather than exhaustive. The most pervasive feature of the comparison is the vapor-solid shift (after work-function adjustment) due to extra-atomic relaxation. We feel that the differences appearing in Table II are indicative of the limiting precision of all approaches to term-value estimates from solid samples unless more detailed models are used.

2.3. *Appearance-Potential Spectroscopy*

Appearance potential spectroscopy (APS) is capable of yielding numerical results of considerable precision, although details of technique and interpretation remain significant. In the earliest studies from which precise energy thresholds were extracted, APS experiments employed crystal spectrometers to isolate the center of specific emission lines (Ulmer, 1982). The resulting intensity was then monitored as a function of the exciting voltage and a certain feature of this excitation function associated with the "threshold." In more recent years the monochromator has been eliminated and various modulations introduced both to sharpen the (pseudo-) resolution of this process and to alleviate the lack of sensitivity of the monochromatized approach. Where good technique has been coupled with sanitary vacuum conditions and adequate specimen characterization, results of quite excellent precision have emerged.

In a recent study results from carefully executed APS measurements were compared with those obtained by photoelectron spectroscopy. Results of several such comparisons have been presented by Anderson *et al.* (1982), with particular emphasis on the appropriateness of Fermi-energy referenced XPS measurements. These authors obtained fairly consistent offsets of the order of 1 ± 0.5 eV for the $2p_{1/2}$ core levels of first transition group elements. The joint uncertainty indicated reflects contributions of 0.2 eV from XPS and 0.35 eV from APS. Without belaboring details, we believe it evident that APS spectra are more critically dependent on surface conditions and vacuum sanitation than are at least some alternatives. A corollary to the last remark is that there are many elements in the Periodic Table for which such methods are less useful, e.g., those difficult to bring into an "ultrahigh vacuum" and those lacking a sufficiently high degree of metallic conductivity. According to the above discussion, we are of the opinion that APS, though precise when it is appropriate, is not a generally useful tool in estimating atomic energy levels.

2.4. *X-Ray Emission Spectroscopy*

In the single-vacancy, active-electron approximation X-ray emission processes connect the levels determined by the above-mentioned methods. Although there is an evident loss in the directness of level determinations, there are certain compensations which, as will be shown, make emission spectroscopy particularly attractive for the purposes intended here, namely, as a testing ground for atomic calculations. Among the characteristics which lead to this attractiveness, four stand out: their interpretation (neglecting satellite problems which will be discussed below) is largely model independent; their chemical and physical state shifts are small for the case that neither vacancy is in the valence shell; they are numerous so that significant overdetermination within a given element is, in principle, available; they are available from many elements spaced throughout the Periodic Table including the high-Z region. The importance of this last characteristic needs to be emphasized. None of the spectroscopies considered above have been applied to core levels in the mid- to high-Z region. Even in the $Z=20$ to 30 region, only *outer*-level energies are measured; deeper levels are then estimated by using X-ray emission line data together with the outer-level XPS base value. As $Z=90$ is approached, energies of several emission lines are needed, some of which may be weak and not well known.

Of course emission spectra are not without their own problems, to which we draw some attention at this point. Most serious among the problems are those associated with satellite processes. There are more than a few mechanisms which lead to emission processes not represented on a single-vacancy energy-level diagram. These processes are too numerous and complex to be adequately discussed in this review. However, their main influence on precise emission-line measurements can be understood by considering the class of satellites involving "spectator holes." More adequate discussions have been given by Larkins (1982) and by Åberg and Suvanen (1982).

Satellites due to the presence of spectator holes have been long known and are referred to as Wentzel–Druyvestein satellites. These arise with considerable integrated strength because the sudden change in atomic potential occasioned by the primary photoionization event allows the remaining electrons to find themselves in various eigenstates of the final potential most especially in those corresponding to extra inner- (or outer-) shell vacancies. Neglecting the resulting changes in symmetry (which must be included in a realistic calculation) one then considers that a previously identified single-vacancy transition, the parent, will take place in a potential modified by the presence of the extra, spectator vacancies. If a spectator hole is sufficiently interior to the atom, the resulting satellite is well resolved and its presence no more than a small nuisance in emission spectroscopy. If, however, the spectator hole is in an outer shell the resulting Coulomb perturbation is small and the shifts, though significant, are comparable in magnitude to the linewidth. Thus we arrive at the situation generally encountered, namely, that any measured emission line (except those obtained by threshold excitation) must be assumed to be contaminated by numerous satellites whose presence may or may not be obvious from "lumpiness" in the profile but which nevertheless have the potential to shift and distort the line.

Actually the problem is both more difficult and less serious than these initial considerations might suggest. An example of considerable difficulty is the problem of multiplet splitting mentioned above. Here we emphasize that it is both significant in size and ubiquitous in occurrence. Consider the simple case of an extra $3p$ or $3s$ vacancy in atomic Ar. As suggested in Fig. 2, these shift the single-vacancy levels by the direct part of the Coulomb interactions and split them by the exchange. The array of double-vacancy initial states shown then couples to the associated complex of $3s$ and $3p$ vacancy final states by the transition array, β_x. One component of this, γ, lies within the diagram line and must be considered when studying double-vacancy excitation functions. The corresponding experimental data from the $K\beta$ region are shown in Fig. 2. We note that β'', though clearly a satellite, has not yet been well understood; it probably arises from the presence of a spectator $3s$ vacancy (Deslattes *et al.*, 1983).

Discussion thus far was restricted to a simple transition in a closed-shell atom. In the case of atoms with incomplete outer shells the situation is much more complex. Even the single-vacancy initial state has splittings arising from angular-momentum coupling between it and the incomplete outer shell. Introduction of still other vacancies (spectators) complicates the initial-state array even further. Naturally the final states accessible from this initial manifold themselves form a complex manifold. In spite of this evident complexity such calculations are nowadays practical, at least for free atoms. What we hope to show below, however, is that they are not necessary in order to make at least some progress.

Before moving to the empirical situation we should mention in passing the complications besides those of incomplete shells which have been purposely neglected in the above discussion. Perhaps most serious from the point of view of distortion of the diagram lines, is the problem of multiple spectator vacancies. As one might expect, this leads to progressive asymmetric distortion as is frequently encountered. We have also neglected specific many-electron effects which occur even in atoms and have remarkable influence on both line shape and position. These last-mentioned effects are especially prominent in the $2s$ level where Coster–Kronig and super-Coster–Kronig processes are possible. We have also neglected satellite processes involving two-electron transitions and combined radiative and Auger processes, both of which are also possible in free atoms. In addition to those processes (which are possible in free atoms) there are others specifically arising from matter in its condensed state such as extra-atomic relaxation, plasmon satellites, and phonon coupling.

If all of the above-mentioned effects were simultaneously operative and conspicuous, it would be very hard to imagine extracting any information

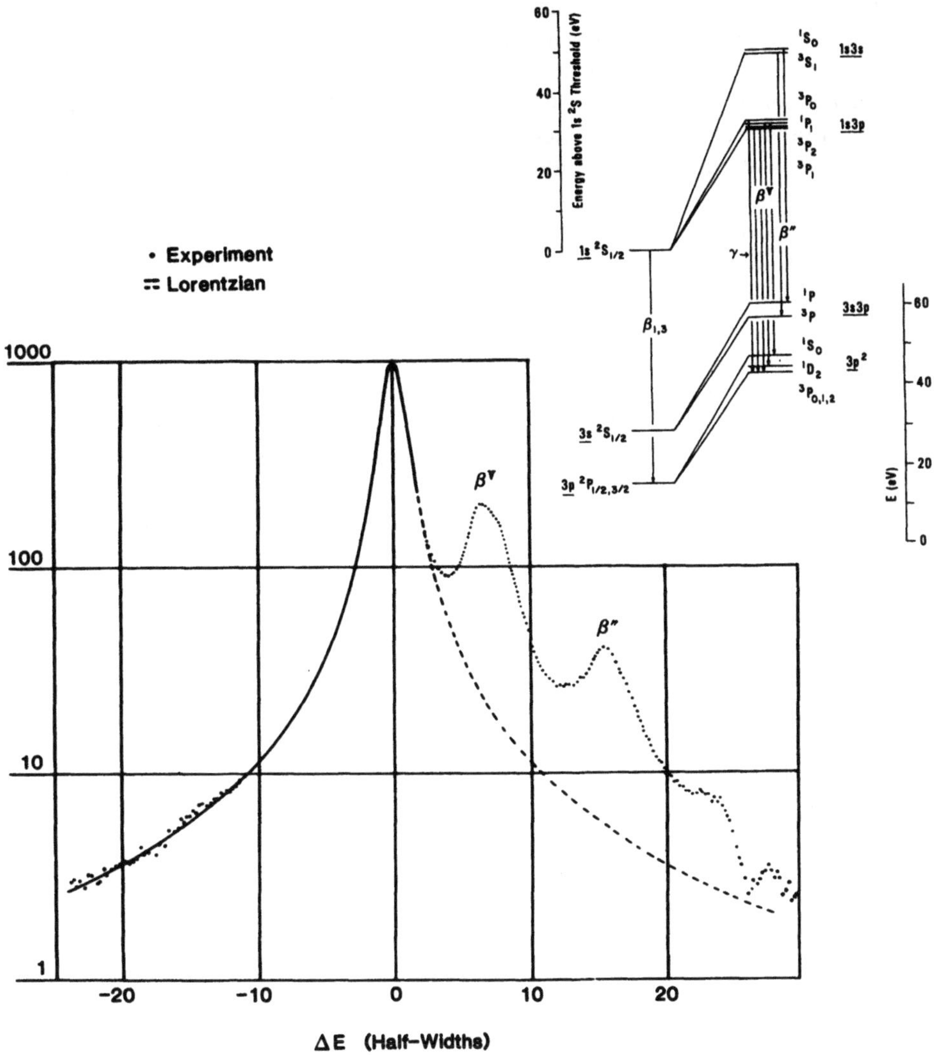

FIGURE 2. Argon $K\beta$ region. The level diagram insert shows the principal single- (left) and double- (right) vacancy transitions which are expected to produce X-ray emission near the $K\beta$ energy. The spectrum is the Ar $K\beta$ emission following 10-keV electron bombardment. The dashed line is a Lorentzian fit to the data.

from an X-ray emission line regarding calculated single-vacancy atomic levels. Fortunately the coupling constants are small enough that the location of the maximum (and to a somewhat lesser extent the centroid) of an X-ray emission line is relatively insensitive to the operation of these mechanisms. We offer two pieces of evidence in this regard. First of all we reproduce in Fig. 3 a somewhat controversial delineation of the spectral profile of Cu $K\alpha_{1,2}$ (Sauder *et al.*, 1977). These data show what appear to be convincing suggestions of the kind of added detail expected on account of multiple-vacancy processes. In passing we note that the same line has been studied by Deutsch and Hart (1982), who are convinced that this detail is spurious and who have fitted the line assuming many-body perturbations in the solid sample are responsible for the "smooth" asymmetry seen in their data. Regardless of the outcome of this question we note that this line pair has been measured over a very large number of years by a variety of people who find that its peak position (relative to other prominent X-ray lines) agrees probably to a few ppm if we consider only the more recent measurements and almost certainly within about 10 ppm if we consider a broader class of measurements. This level of agreement corresponds to an energy spread of no more than 0.1 eV. Chemical shifts associated with this and similar lines are sufficiently small that their detection is an exercise of considerable difficulty. Finally as will be seen in later sections where recent data are compared with theory, systematic studies of the wavelengths of prominent characteristic X-ray lines covering a range of atomic numbers can be considered even with great refinement whether or not the studied range contains gas-phase species.

The approach which we are thus suggesting, namely, to neglect these complications, is not the only one available. Specifically, Borchert *et al.* (1978) have calculated explicitly the positions and relative intensities of the most prominent components due to double-vacancy processes in certain heavy-element spectra. This model "stick" spectrum was then folded with an estimated lifetime plus instrument-broadening function and the result fitted to the experimental data. The shift in peak position between the simple model ignoring satellites and the more complex (though necessary incomplete) model amounted to approximately 0.2 eV for the prominent diagram lines of Au. We consider that the large amount of extra effort entailed and the uncertainty of this result makes it inappropriate to insist on applying the more complex approach in every case. Instead we treat this shift as indicating one of several considerations which limit the validity of this kind of study.

In spite of the above-described important limitations on X-ray emission spectroscopy, we nonetheless feel it to be a most desirable testing ground for atomic calculations. In particular, as long as we stay with relatively deep inner shells it is possible to use samples of any physical form which are

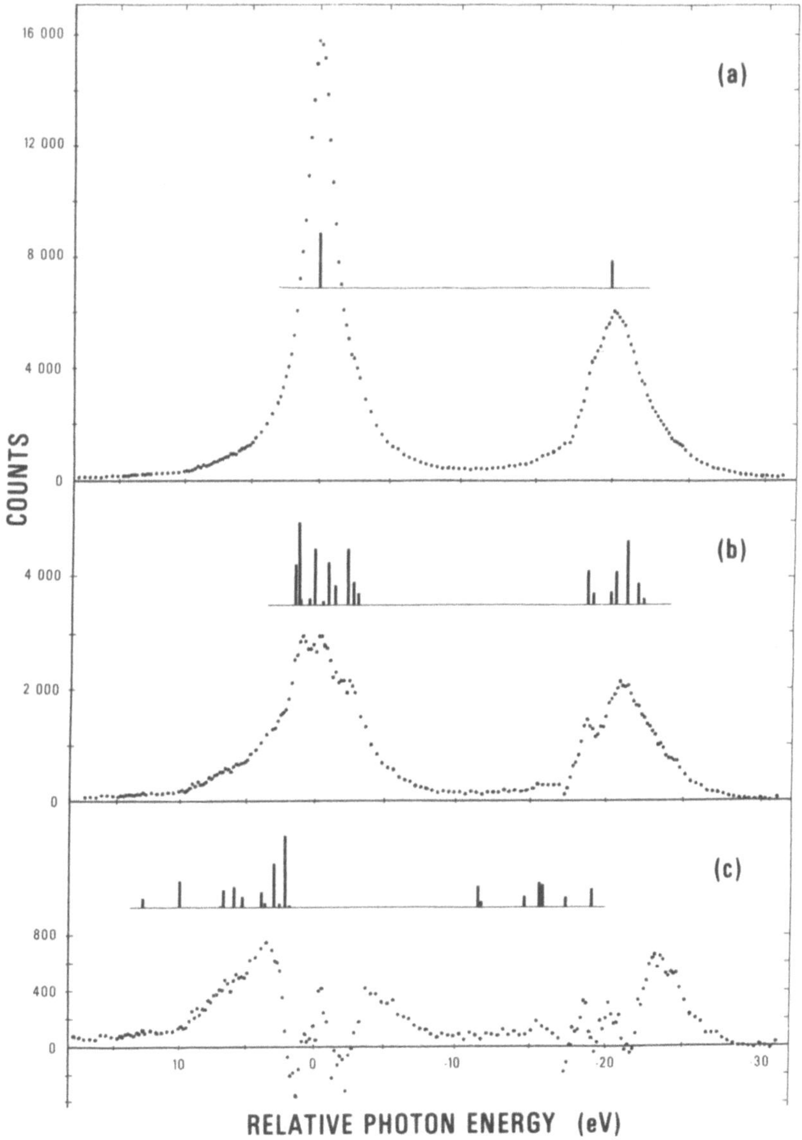

FIGURE 3. The Cu $K\alpha_{1,2}$ spectrum showing (a) the complete observed spectrum; (b) spectrum (a) minus the single vacancy model spectrum of (a); (c) spectrum (b) minus the 3*d* spectator model vacancy spectrum of (b). The model spectrum in (c) is the 3*p* spectator vacancy spectrum.

available in sufficient quantities to serve as targets and thus access all of the (stable) elements in the Periodic Table. Secondly, although the ultimate refinement of our interpretation is limited by the presence of complex processes which we do not feel able to model in detail, the result of this limited interpretation is, as far as it goes, substantially model independent. Lastly, the complex array of emission lines connecting single-vacancy energy levels not only permits an overdetermination, but perhaps more importantly, permits us to study their systematics with respect to theory so as to identify those parts of the theoretical estimates which are most likely to be in difficulty. For example, it is clear that *K*-series transitions to *L*-shell final states are sensitive to the interior of the wave functions while *L–M* transitions explore outer regions and *K–M* series connect the two.

3. EXPERIMENTAL TECHNIQUES FOR HIGH-ACCURACY SPECTROSCOPY

From this point onward our focus is on photon spectroscopy with a strong emphasis on emission spectroscopy for reasons discussed above. The same questions of technique and scale arise equally in the case of absorption spectroscopy. Furthermore, a careful treatment of X-ray emission wavelengths (or energies) is not without relevance to photoelectron spectroscopy.

Dispersive elements required in an overall study of inner-shell transitions include ruled gratings at both normal and grazing incidence as well as, of course, crystals. Normal-incidence gratings cover the range of valence ionization energies and detailed spectroscopy in this region yields very precise estimates of outer-shell vacancy levels. Also in the region of effective normal-incidence spectroscopy, one finds the intrashell or $\Delta n = 0$ transitions for a large portion of the Periodic Table. Grazing-incidence gratings go beyond the normal-incidence region to reach transitions among the outermost inner levels and the innermost levels of at least the first-row elements.

To reach the deeper levels of mid- to high-Z atoms, there is no choice but to employ crystal-diffraction spectroscopy. Available instrumental strategies involve use of flat crystals and curved crystals in both transmission and reflection geometry (Knowles, 1979). Flat crystals can be employed either individually (with appropriate mechanical collimation) or in pairs in double-crystal spectrometers. Such instruments allow one, in principle, to obtain spectral wavelengths from careful angle measurement (goniometry) if the lattice period is known independently. If the lattice period is not otherwise known such instruments can be thought of as broad-range interpolators or extrapolators. Specifically, diffraction study of any one line profile (whose wavelength is known) permits normalization of all others similarly obtained. Alternatively, crystals can be curved (singly or doubly) and possibly also ground to obtain focusing instruments. Generally speaking curved-crystal spectrometers offer relatively high efficiency but need wavelength markers established otherwise. Satisfactory normalization of the wavelength scale over an appreciable range requires, in general, a number of reference wavelengths spaced throughout the range except in the case that there is a very close coincidence between the reference wavelength and the spectral feature under investigation. This last-mentioned case is particularly important in studies of high-Z one-electron spectra using a bent crystal in combination with a position-sensitive detector as discussed below (Sec. 4.4).

3.1. *Wavelength Determination in the Grating Region*

The issues suggested by the above remarks take somewhat different forms according to the wavelength region and type of spectrometer. In the case of grating instruments excellent wavelength reference values are available by means of the so-called Ritz principle standards (Edlén, 1963). These arise in cases where uv or vuv transitions connect states which are also connected by a group of lower-energy transitions (three is usually the minimum number) which can, in turn, be related to interferometrically determined secondary standards in the visible. These latter are perforce known in terms of hydrogen Balmer α to much greater precision and accuracy than is required for the purposes of the present discussion.

Although the network of Ritz-type reference wavelengths can, in principle, be extended to very short wavelengths, this has not been done as yet. Because of limitations in the performance of grazing-incidence instruments in the nanometer wavelength region and the many steps required in a chain to this or shorter wavelengths, we believe that this route to the X-ray region will not likely become available. On the other hand, with diligence and use of well-chosen normal-incidence measurements whose summation would yield markers in the region accessible to crystal diffraction, such methodology is rigorous. Were it not for the possibility of using X-ray interferometry (see Sec. 3.4), this would clearly be the route of choice in contrast to those approaches (Tyrén, 1940; Edlén and Svensson, 1965; Kirkpatrick *et al.*, 1967) which attempted to use calculable spectral references.

3.2. *Wavelength Problems in Crystal-Diffraction Spectroscopy*

Beyond first-row elements, one is inevitably concerned with crystal-diffraction spectroscopy. This region has always had certain peculiar difficulties when pressed toward high accuracies. Until very recent times, the accuracy level at which difficulties were encountered was well beyond what could be thought of as meaningful for comparison with atomic calculations. As will be seen when we discuss such comparisons, the effect of X-ray wavelength scale uncertainties is marginally perceptible even at the present time. Thus we feel it useful to present a very brief indication of the current situation and even some words about the past since data in the earlier literature need to be reinterpreted to be of maximum value.

The basic problem was already evident in the earliest days of X-ray spectroscopy when questions about the scale of X-ray wavelengths were answered with uncertainties in order of magnitude. Very quickly this situation improved via Bragg's (1913, 1914a, b) evaluation of the NaCl unit-cell dimension from combining knowledge of the oil-drop e and the electrochemical Faraday with density and atomic weight (molar mass). Even so, it was soon clear that relative determinations of X-ray wavelengths could be carried out with much greater precision than could be obtained in reckoning crystal grating spacings.

This last-mentioned situation led Manne Siegbahn (1919) to the first of several proposals for defining a local scale. He proposed simply to take the NaCl result then available and treat it as exact in terms of a new unit, xu, which initially was assumed to be near 1 mÅ. It was soon realized that NaCl was an unfortunate choice from the standpoint of crystallographic perfection. Since much successful spectroscopy was by then being carried out using calcite (Iceland spar), the xu was redefined so that the (211) interplanar distance of "the purest calcite" should be (after correction for index of refraction) 3029.04 xu at 18°C (Siegbahn, 1925; Siegbahn, 1943).

Discovery of X-ray diffraction by ruled gratings by Compton and Doan (1925) and its rapid perfection to an advanced precision measurement tool by Bäcklin (1935) and by Bearden (1931) led to the canonical formulation of the X-ray wavelength problem: Namely, given a local scale presumably well defined and internally consistent (but see below), it should be used to establish the wavelength of a relatively long-wavelength X-ray line which could then be diffracted by a ruled grating and thereby connected to the visible reference scale. The output of such an exercise was usually referred to as $\Lambda = \lambda_g/\lambda_S$, where λ_g and λ_S were, respectively, the grating and Siegbahn (xu) values for the transfer wavelength which had no special significance.

Over most of the period when this formulation of the wavelength problem was appropriate one had $\Lambda \sim 1.00202$ (Bragg, 1947; Wood, 1947) with an uncertainty which varied widely but in the best of times was near 20 ppm. Thus, except for inserting the dominant 0.2% offset, numerical results for comparison with atomic theory could well afford to neglect the more arcane aspects of the X-ray scale problem. As suggested above, we are now moving toward a condition where the luxury of such neglect may no longer be affordable. Accordingly certain further details need to be indicated to make the literature accessible to the extent it deserves and is needed.

3.3. Local Scales and Conversion Factors

Although "official" definitions of the xu tended to emphasize crystal "standards," these seldom intervened in actual practice. Partly this reflected the fact the interplanar spacings of available calcite were noticeably variable (Bearden, 1965a), but mostly it was simply more convenient to use X-ray emission lines. Two of the lines most frequently used were Cu $K\alpha_1$ ($\lambda =$ 1537.400 xu) and Mo $K\alpha_1$ ($\lambda = 707.831$ xu) (Thomsen and Burr, 1968). For reasons not entirely clear, the Cu reference was most often used in the U.S. while the Mo reference seemed to dominate European work.

The above numbers were assumed to be consistent on the basis of two theses carried out in Siegbahn's laboratory which gave the same result for their ratio (Bearden, 1965b). As occasionally suspected by crystallographers (Cooper, 1965), this ratio was shown by Bearden *et al.* (1964) in the early sixties to be inconsistent by about 40 ppm.

What happened next is well documented and forms the basis of Bearden's widely used compilation (Bearden, 1967). The network of lines among which accurate ratios were directly measured was expanded to five with extensions toward both longer wavelengths, Cr $K\alpha_2$, and toward shorter wavelengths, Ag $K\alpha_1$ and W $K\alpha_1$. To avoid continued ambiguity in the xu and to get rid of the 0.2% offset, Bearden (1965b) proposed a new scale unit, namely Å*, defined so that $\lambda(\mathrm{W}\ K\alpha_1) = 0.2090100$ Å* exactly. The scale was so chosen that the Å*/Å ratio was unity on the basis of density and isotopic abundances of a group of silicon crystals combined with Avogadro's constant from the 1963 Cohen and DuMond adjustment of the fundamental physical constants (Cohen and DuMond, 1964).

In the following years several developments suggested that the Å*/Å ratio was appreciably different from unity and that precision X-ray wavelength measurements could be routinely made on the Ångstrom scale. The first event was a revision of the 1963 adjustment which moved N_A by 30 ppm with a corresponding 10 ppm shift in the Å* scale (Cohen and DuMond, 1965). Secondly, as will be described below, we began to make direct X-ray wavelength measurements using crystals calibrated by X-ray/optical interferometry. The wavelength values obtained this way depend on a crystal spacing measured directly in terms of an optical reference and thus appear more robust than do calculated spacings as used in the work of Henins and Bearden (1964). Lastly, as also described below, the more recent reference wavelength ratios obtained by us display a satisfactory agreement with the ratios obtained by Bearden *et al.* (1964) except for those involving W $K\alpha_1$.

To summarize this somewhat complex story we note in particular that recourse to the literature still needs some attention to the reference wavelengths used. The excellent compilation of Bearden (1967) set out to take care of this problem once and for all, and to a large extent has done so except for the 10–20 ppm offset of the Å* scale. There are, however, several areas of the Periodic Table (i.e., high Z) where the literature values which served as input to Bearden's compilation can be significantly improved.

3.4. Wavelengths Based on X-Ray Interferometry

Achromatic X-ray Moiré interference was demonstrated by Bonse and Hart in the 1960s (Bonse and Hart, 1965). By 1970 it was clear that addition of suitable optical elements and a visible reference wavelength source would permit direct determination of an interplanar spacing. Such a measurement can be combined with density and isotopic abundance measurements to determine Avogadro's constant. Alternatively, in order to obtain wavelengths in the X-ray and γ-ray region we need a comparison scheme for relating lattice periods determined in the interferometer crystal to other species and specimens as required by the various wavelength regions of interest. With appropriate crystals, wavelength determination then reduces to the problem of accurate angle measurement for what we refer to as the class of "direct-reading" instruments. Focusing instruments require reference wavelengths obtained from "direct-reading" systems as will be discussed below (Sec. 3.5).

Our version of a three-step process is shown in Fig. 4. The first step, Fig. 4a, used an I_2 stabilized NeHe laser similar to that used in a then recent (0.01 ppm) determination of the components of Balmer α to obtain the Rydberg constant, R_∞. Thus all subsequent measurements are immediately in units appropriate for comparison with theory. In a second step, Fig. 4b, a sibling (adjacent) crystal sample was used to determine repeat distances in other specimens and species. These determinations, if they are not to be limited by X-ray line shapes, need to use nearly equal crystal repeat distances as may be contrived by exploring reciprocal space, e.g., d(Si 440) differs only slightly ($\sim$200 pm) from d(Ge 800) as pointed out by Baker and Hart (1975). Thus only the small difference yields a wavelength proportional signal.

These first two steps have been carried out in our laboratory and at the Physikalisch-Technische Bundesanstalt (PTB) (Becker *et al.*, 1981). Claimed accuracies of 0.15 ppm and 0.05 ppm were reported in the two cases, and after sample interchange and preliminary evaluation there appears to remain a significant difference near 1.8 ± 0.5 ppm (Becker *et al.*, 1982) which is currectly under study. On the other hand the accuracy levels we finally

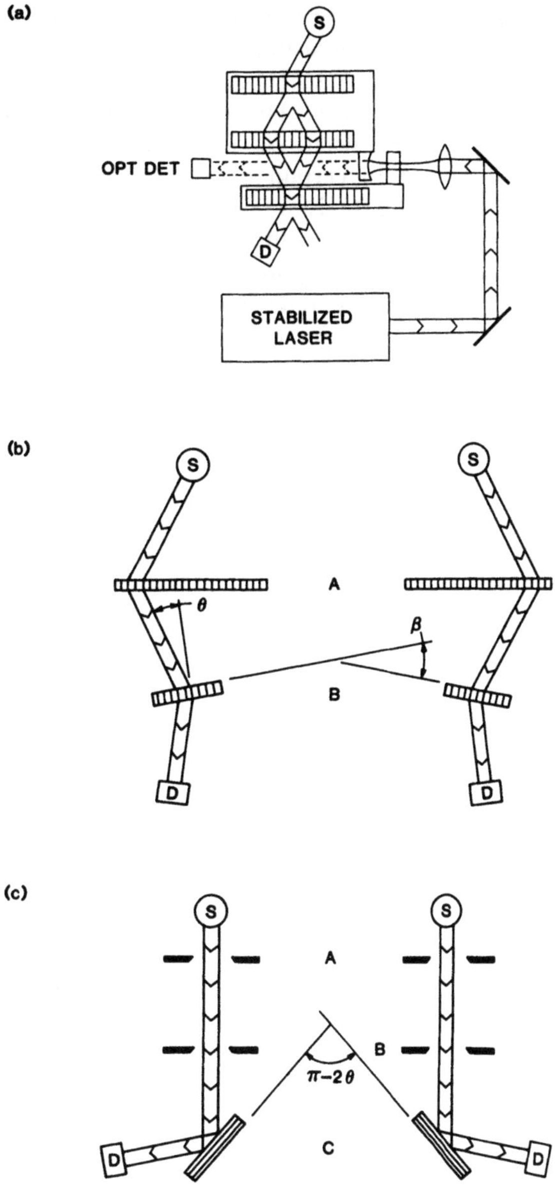

FIGURE 4. NBS three-step X-ray wavelength measurement scheme. The X-ray source is indicated by S and the X-ray detector by D. (a) Schematic of crystal lattice spacing measurement using separated crystal interferometer; (b) schematic of a quasinondispersive comparison of nearly equal lattice spacings with long crystal at position A and standard or unknown crystal at position B; (c) schematic of crystal diffraction angle measuring spectrometer with collimating slits at A and B and crystal at C.

reach in the comparison with theory are such that 1 ppm uncertainties on the experimental side are not significant.

The third step suggested very schematically in Fig. 4c has thus far been carried out only in our laboratory. The wavelength emerges in this case from measurement of $\pi - 2\theta_B$, where θ_B is determined by the Bragg–Laue relation $\lambda = 2d \sin \theta$ subject to certain corrections noted below. The slit collimation suggested by Fig. 4c is impractical especially toward shorter wavelengths, and the needed collimation is obtained by various arrangements of two crystals as suggested in Fig. 5. At longer wavelengths, crystals must be used in surface reflection as shown in Fig. 5a, thereby entailing a (large) index of refraction correction, viz.,

$$\Delta\lambda/\lambda = -4d^2(\delta/\lambda^2), \quad (1)$$

where $\delta = 1 - n$ is the decrement in refractive index. For wavelengths near and below 0.1 nm (1 Å), the mixed geometry of Fig. 5b is convenient and operation of the second crystal in the Laue condition provides complete freedom from index-of-refraction effects. Angular measurements in the regions appropriate to both (a) and (b) geometries are of modest difficulty but within the range of conventional metrological practice. For instance at a Bragg–Laue angle of 45°, angular accuracy of 0.1 arcsec yields 1 ppm wavelength values.

For high-Z K-series spectra as well as for γ-rays the angles became rather small and the direct-reading instrument configuration of choice becomes that shown in Fig. 5c. In order to reach the 1 ppm level interferometric angle measurements calibrated from first principles are needed. The main reason for mentioning γ-rays here is that they provide convenient reference wavelengths for curved-crystal instruments used in the study of high-Z K-series X-ray lines. As described elsewhere in detail (Deslattes *et al.*, 1980) we have developed a form of polarization-encoded angle interferometry which is effective below the 0.1 milliarcsec level and has an on-scale range of $\pm 15°$. For calibration there are null-pointing photoelectric autocollimators whose noise level is significantly below 1 milliarcsec. By using such a system to measure each of the external interfacial angles of, for example, a 24-sided optical polygon, the closure condition provides a first-principles measure of the interferometer constant K such that an interferometer reading, η, yields an angular position via $\theta = \sin^{-1}(\eta/K)$. Accuracy levels available for K, namely, ~ 0.03 ppm, are well beyond the needs of X-ray spectroscopy.

A number of new results have been obtained over the past several years in our laboratory using direct-reading instruments operating in all configurations shown in Fig. 5 (Kessler *et al.*, 1982). The accuracy levels

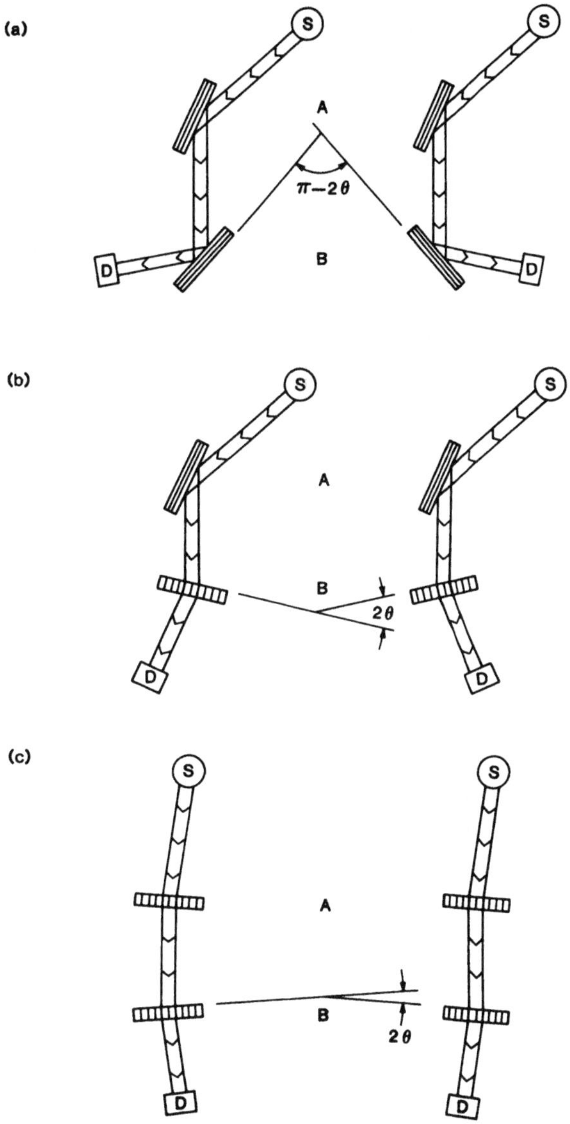

FIGURE 5. Two-crystal diffraction angle measuring spectrometer using: (a) surface reflection on crystals A and B, (b) surface reflection and transmission on crystal A and crystal B, respectively, (c) transmission on crystals A and B. The X-ray source is indicated by S and the X-ray detector by D.

obtained when measuring X-ray emission lines are always limited by the unmodeled characteristics of the line profiles themselves. This limitation is typically near one hundredth part of the width or in exceptional cases maybe half that. Since X-ray lines have widths (FWHM) ranging from 300 to 1000 ppm we obtain results with claimed accuracies (estimate of 1σ) in the range 2–5 ppm. In the case of γ-ray lines we are limited by counting statistics and instrumental resolution but generally have obtained precisions below 1 ppm, often near 0.3 ppm. Results at this level are naturally sensitive to the resolution of the apparent discrepancy between our work and that of PTB. On the other hand, conclusions suggested below regarding X-ray lines are, however, effectively independent of the outcome of the lattice-parameter conflict.

3.5. Wavelength Measurements with Focusing Instruments

For the most part curved-crystal instruments are used in a one-sided "fixed" geometry mode that is not well suited to precision spectroscopy except when nearly degenerate calibration lines are available (there are special circumstances where this is both possible and rather important as discussed below in Sec. 4.4). In general, precise focusing-crystal spectroscopy has used one of the configurations shown in Fig. 6, although other con-

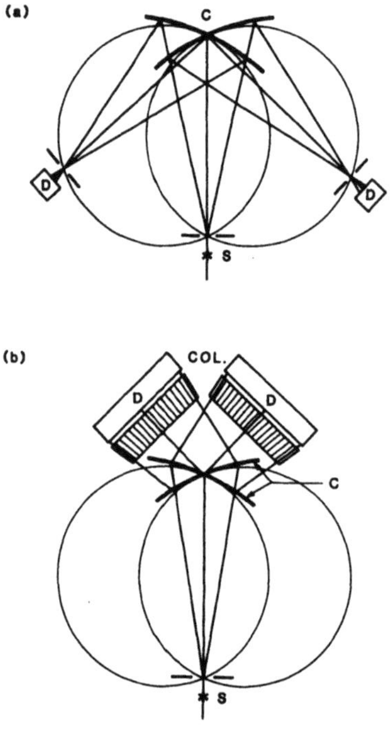

FIGURE 6. Curved-crystal spectrometers of the DuMond geometry employing (a) surface reflection, and (b) transmission. The X-ray source is labeled by S, the crystal by C, the detector by D, and the collimator COL. Both the left and right reflections are shown.

figurations are possible (Blokhin, 1965). In each case, spectra are recorded with both a left-handed and a right-handed disposition of the source and detector. Examples of curved-crystal spectrometers are the reflection instrument of Sandström (1952) and the DuMond transmission instruments of Seppi *et al.* (1962), Borchert *et al.* (1975), Piller *et al.* (1973), Smirnov *et al.* (1968), and Koch *et al.* (1980) to name a few. Significant improvements in the angle-measuring capability of curved-crystal spectrometers have been made over the past 20 years. The more sophisticated instruments now employ angle interferometers having a precision of a few milliarcseconds.

At first glance it would apear possible that measurement configurations illustrated in Fig. 6 should be useful as direct-reading instruments. In principle one could know the lattice parameter and even allow for the distortion incidental to bending. It would then simply be a matter of reading $2\theta_B$ by some adequate means and obtaining wavelength results directly. Such thoughts have motivated construction of several significant instruments including systems for both reflection and transmission geometry. However, none of these instruments has been productive in this suggested direct mode of operation.

Instead, each has operated primarily as a relative instrument employing (usually many) calibration lines and using multiple diffraction orders to achieve a measure of overdetermination in each datum; the practice of recording left and right reflections has generally been retained even in the relative mode. While most precision curved-crystal work has used X-ray lines (determined by two-crystal spectroscopy) as references, an important group of such measurements have used γ-ray lines as references.

This last-mentioned (and at first sight unusual) situation arose from historical circumstances that are easy to understand. Specifically, for a long time high-Z X-ray lines were valuable anchor points for γ-ray measurements. This reflected, to some extent, the use of ionization spectrometers but in the more important case of crystal spectrometers it arose from the limited narrowing of crystal-diffraction profiles (from natural crystals) at high energies. In principle the intrinsic resolution (resolving power) of a diffraction instrument is largely independent of wavelength. The dispersion, however, varies inversely with energy so that, for crystals characterized by an effective mosaic spread (the situation of most natural crystals), there is an energy above which performance deteriorates, i.e., the resolving power instead of remaining constant decreases as E^{-1}. Thus a peculiar phenomenology arises where spectrally narrower γ-ray lines of higher energy appear broader than X-ray lines. This effect can be seen very clearly in the flat-crystal work of Knowles (1962).

In any case, several very careful studies were undertaken in the past ten years using good curve-crystal instruments to record both γ-ray lines and X-ray lines. The intent of these efforts was to determine γ-ray wavelengths in terms of the then better known X-ray wavelengths. However, when our work was able to establish the γ-ray lines directly, these same curved-crystal studies could be inverted, becoming excellent sources of information about the X-ray lines. Since the possibility was realized, it has been consciously invoked in several significant studies. Most prominent among these are the continuing efforts at the Institut Laue-Langevin using the 5.7-m bent-crystal instrument GAMS 1 (Barreau *et al.*, 1982). This work using the general β and γ excitation inside the high-flux reactor and the earlier measurements of Borchert (1976) are unique sources of information in the transuranic region up to the present time.

4. SELECTED EXPERIMENTAL RESULTS

It is our purpose in this section to bring together the main body of experimental X-ray wavelength measurements which can contribute in a critical way to tests of self-consistent-field calculations. In two regions, $20 < Z < 30$ and $40 < Z < 50$, our report is selective but otherwise complete. Data included were chosen on the basis of (1) small uncertainty, and (2) traceability to the Rydberg energy scale which is used in calculations.

Over the past ten years a significant number of measurements have been reported with an uncertainty less than 10 ppm. Thus measurements having an uncertainty significantly greater than 10 ppm contribute little to the tests of theory. All the measurements that are used in the experiment–theory comparison have an accuracy <25 ppm except for a few measurements at very high Z.

The experimental X-ray wavelengths in our selection were either derived from direct-reading instruments or from bent-crystal instruments with γ-ray reference lines or, less frequently, from X-ray to X-ray ratios. As described above, the primary optical reference is well connected to the Rydberg constant and thus the wavelengths are well suited for comparison with theoretical calculations.

Tables presented in the following sections are given in energy units, namely, eV. This choice reflects recent trends and is harmless even though one, in fact, measures wavelengths. Measured wavelengths were simply converted to energies using the energy–wavelength conversion factor from the most recent adjustment of the fundamental constants (Cohen and Taylor, 1973). The theoretical energy scale is established by expressing the Rydberg in energy units with the conversion factor taken from the same adjustment. Any change in the energy–wavelength conversion factor will shift both the experimental and theoretical energy scales and thus will leave the comparisons intact.

4.1. *Measurements from Direct-Reading Instruments*

Absolute angle-measuring spectrometers using interferometrically calibrated crystals are referred to as direct-reading instruments. The only external reference is an optical wavelength which is well related to the Rydberg constant. The measurements reported here have been made on four different spectrometers, three of which are equipped with angle interferometers while one uses angle encoders.

The elements which were measured were chosen for their availability, their ease of handling, their distribution in the Periodic Table, or because they had traditionally served as wavelength standards. Excitation was by electron bombardment of solid targets except for Xe which was studied as a gas. Fluorescent excitation was used for gaseous Ar and for K (in KCl) (LaVilla, 1983). The measured results are summarized in Table III. For the most part these results have been reported previously (Kessler *et al.*, 1979; Kessler *et al.*, 1982) and further details can be found in these references.

4.2. *Measurements Referred to Directly Measured γ-Ray Lines*

Borchert (1976), Barreau *et al.* (1982), and Hungerford *et al.* (1981) have used curved-crystal spectrometers to make high-precision X-ray measurements relative to the ^{198}Au 411-keV γ-ray line. This standard line was measured by the authors and collaborators (Kessler *et al.*, 1978) using a direct reading spectrometer (Deslattes *et al.*, 1980) and is the most carefully measured γ-ray wavelength standard in this low-energy region. A provisional value for the energy of the ^{198}Au standard was used by Borchert so the published values were corrected. The calibration lines used by Barreau *et al.* and Hungerford *et al.* were fission-product lines from ^{235}U and ^{238}U and the 517-keV line from ^{36}Cl, all of which have been measured relative to the 411-keV line of ^{198}Au. All of the curved-crystal measurements involved the use of multiple orders and radioactive sources. The results are presented in Table IV.

4.3. *Measurements Referred to Directly Measured X-Ray Lines*

Bearden and his collaborators have made careful X-ray wavelength-ratio measurements of the $K\alpha_{1,2}$ lines of a number of elements (Bearden *et al.*, 1964). Some of these elements have also been measured by direct-reading spectrometers. Thus it is possible to check some of the published ratios and to deduce a few new wavelengths.

In Table V ratios from Bearden *et al.* involving Cu, Mo, Ag, and W $K\alpha_1$ are compared with ratios of directly measured wavelengths. Agreement appears to be satisfactory except for ratios involving W, and even for W the

TABLE III

Measurements using direct-reading spectrometers.

Z	Element	X-ray line	Energy (eV)	Uncertainty (eV)
18	Ar	$K\alpha_2$	2955.684	0.009
		$K\alpha_1$	2957.813	0.008
19	K	$K\alpha_2$	3311.217	0.006
		$K\alpha_1$	3313.969	0.005
29	Cu	$K\alpha_1$	8047.865	0.008
42	Mo	$K\alpha_1$	17479.485	0.010
47	Ag	$K\alpha_2$	21990.44	0.10
		$K\alpha_1$	22163.06	0.03
54	Xe	$K\alpha_2$	29458.44	0.05
		$K\alpha_1$	29778.97	0.10
		$K\beta_3$	33563.42	0.12
		$K\beta_1$	33624.45	0.12
56	Ba	$K\alpha_2$	31816.82	0.06
		$K\alpha_1$	32193.47	0.07
		$K\beta_3$	36303.58	0.12
		$K\beta_1$	36377.68	0.08
60	Nd	$K\alpha_2$	36847.74	0.08
		$K\alpha_1$	37360.98	0.07
62	Sm	$K\alpha_2$	39523.64	0.10
		$K\alpha_1$	40118.74	0.06
68	Er	$K\alpha_2$	48221.92	0.20
		$K\alpha_1$	49127.56	0.12
		$K\beta_3$	55480.08	0.35
		$K\beta_1$	55673.88	0.18
74	W	$K\alpha_1$	59319.23	0.05
79	Au	$K\alpha_2$	66991.16	0.22
		$K\alpha_1$	68804.94	0.18
		$K\beta_3$	77575.51	0.61
		$K\beta_1$	77980.30	0.38
82	Pb	$K\alpha_2$	72805.89	0.24
		$K\alpha_1$	74970.59	0.17
		$K\beta_3$	84450.99	0.60
		$K\beta_1$	84939.63	0.34
90	Th	$K\alpha_2$	89957.62	0.20
		$K\alpha_1$	93347.98	0.25
		$K\beta_3$	104817.21	0.69
		$K\beta_1$	105602.19	0.53
92	U	$K\alpha_2$	94651.45	0.56
		$K\alpha_1$	98432.21	0.28
		$K\beta_3$	110416.38	0.65
		$K\beta_1$	111295.80	0.65

disagreement is small compared to the experiment–theory differences presented in Sec. 5.3. Table VI contains an energy value for Cr $K\alpha_2$ obtained from a Cr $K\alpha_2$/Mo $K\alpha_1$ ratio from Bearden *et al.* (1964) and the directly measured Mo $K\alpha_1$ energy.

Bearden (1967) also reported K-series measurements for elements with $44 \leqslant Z \leqslant 51$ with respect to Ag $K\alpha_1$. By using the directly measured Ag $K\alpha_1$ energy, K-series energies for elements $44 \leqslant Z \leqslant 51$ are available. Energies for Ru ($Z=44$), In ($Z=49$), and Sb ($Z=51$) are included in Table VI. For low Z Henins (1971) measured the ratio of the unresolved Al $K\alpha$ line to the Cu $K\alpha_1$ line. By using the Al $K\alpha_1/K\alpha_2$ ratio from Bearden (1967) and assuming equal widths and an intensity ratio of 2:1, energy values for Al $K\alpha_{1,2}$ were obtained.

Borchert *et al.* (1980) have precisely measured energy ratios of K-series X rays from high- and mid-Z elements. The high-Z X rays were diffracted in second or third order while the low-Z X rays were diffracted in first order. The elements and lines were chosen so that the two profiles nearly overlapped and were essentially recorded simultaneously, thus eliminating uncertainty caused by reproducibility of the angular setting and nonlinearity of the angular scale. The Bi $K\alpha_2$ energy is available by combining the ratio of Bi $K\alpha_2$ (third order)/Ag $K\beta_1$ (first order) and the Ag $K\beta_1$ energy given in Table VI. The Ho $K\alpha_2$ energy is available by combining the ratio of Th $K\alpha_2$ (second order)/Ho $K\alpha_2$ (first order) and the directly measured Th $K\alpha_2$ given in Table III. The U $K\beta_1$ (second order)/Er $K\beta_1$ (first order) ratio (1.999120) can be compared with the ratio obtained by direct measurement (1.999067) in order to demonstrate the level of agreement of the two experimental measurements. Although the ~27 ppm difference between these ratios is a cause for some concern, we have included the Bi $K\alpha_2$ and Ho $K\alpha_2$ measurements in Table VI.

Although it is our desire to compare experimental and theoretical values for $K\alpha_2$, $K\alpha_1$, $K\beta_3$, and $K\beta_1$ transitions for as many elements as possible, not all four lines have been measured for all elements. In order to get experimental values for all the lines, it was assumed that the ratios $K\alpha_1/K\alpha_2$, $K\alpha_1/K\beta_3$, and $K\alpha_1/K\beta_1$ from Bearden (1967) are valid. Energies obtained in this way are also included in Table VI.

4.4. *One-Electron and Few-Electron Spectra*

Within the past year we have begun to see important results coming from spectroscopy of one-electron systems having $Z \gg 1$. Although long known in astronomy and frequently observed in plasmas, somewhat cleaner (and unfortunately weaker) sources are needed to get spectra for critical comparison with theory. Such spectra have by now been produced and studied with relatively high accuracy both under beam–foil excitation

(Fe^{+25} and Cl^{+16}) (Briand *et al.*, 1983; Stockli *et al.*, 1983) and as excited recoils from collisions with very heavy, highly charged ions (Ar^{+17}) (Beyer *et al.*, 1983).

The weakness and importance of these spectra have led to the use of fixed-geometry crystal spectrometers with linear position-sensitive proportional counters as detectors. *In situ* calibration has been carried out with respect to nearly degenerate $K\alpha$ lines of the element $Z+1$. In the cases of Cl^{+16} and Ar^{+17} new measurements of Ar $K\alpha_{1,2}$ and K $K\alpha_{1,2}$ were carried out with the cooperation of R. E. LaVilla (1983). In the case of Fe^{+25} an

TABLE IV
Measurements referred to directly measured γ-ray lines.

Z	Element	X-ray line	Reference	Energy (eV)	Uncertainty (eV)
69	Tm	$K\alpha_2$	*a*	49772.99	0.12
		$K\alpha_1$	*a*	50741.81	0.09
		$K\beta_1$	*a*	57509.14	0.15
82	Pb	$K\alpha_2$	*b*	72807.7	10
		$K\alpha_1$	*b*	74972.0	10
90	Th	$K\alpha_2$	*a*	89957.5	0.7
			c	89957	2
		$K\alpha_1$	*a*	93348.3	0.6
			c	93348	2
		$K\beta_3$	*a*	104820.4	1.6
			c	104822	2
		$K\beta_1$	*c*	105606	2
91	Pa	$K\alpha_2$	*c*	92284	2
		$K\alpha_1$	*c*	95867	2
		$K\beta_3$	*c*	107586	2
		$K\beta_1$	*c*	108418	2
92	U	$K\alpha_2$	*a*	94653.0	0.5
			c	94656	2
		$K\alpha_1$	*a*	98434.1	0.5
			c	98435	2
		$K\beta_3$	*c*	110416	3
		$K\beta_1$	*c*	111300	2

[a] Borchert (1976).
[b] Hungerford *et al.* (1981).
[c] Barreau *et al.* (1982).

Table continued

TABLE IV *(Continued)*

Z	Element	X-ray line	Reference	Energy (eV)	Uncertainty (eV)
93	Np	$K\alpha_2$	*c*	97069	3
		$K\alpha_1$	*c*	101057	3
		$K\beta_3$	*c*	113308	4
		$K\beta_1$	*c*	114244	3
94	Pu	$K\alpha_2$	*a*	99523.8	1.2
			c	99530	2
		$K\alpha_1$	*a*	103734.7	0.6
			c	103741	2
		$K\beta_3$	*c*	116242	2
		$K\beta_1$	*c*	117233	2
95	Am	$K\alpha_2$	*c*	102032	3
		$K\alpha_1$	*c*	106474	3
		$K\beta_3$	*c*	119240	2
		$K\beta_1$	*c*	120280	2
96	Cm	$K\alpha_2$	*c*	104591	2
		$K\alpha_1$	*c*	109273	2
		$K\beta_3$	*c*	122303	2
		$K\beta_1$	*c*	123404	2
97	Bk	$K\alpha_2$	*c*	107195	5
		$K\alpha_1$	*c*	112128	5
		$K\beta_3$	*c*	125415	7
		$K\beta_1$	*c*	126578	7
98	Cf	$K\alpha_2$	*c*	109838	8
		$K\alpha_1$	*c*	115036	8

TABLE V

Comparison of X-ray wavelength ratios.

Wavelength ratio	NBS	Bearden *et al.*	Differences (NBS-Bearden *et al.*, ppm)
$\lambda(\text{Cu } K\alpha_1)/\lambda(\text{Mo } K\alpha_1)$	2.1719406 (1.2 ppm)	2.171945 (1.5 ppm)	−2.0
$\lambda(\text{Cu } K\alpha_1)/\lambda(\text{Ag } K\alpha_1)$	2.7539058 (1.6 ppm)	2.753917 (1.6 ppm)	−4.1
$\lambda(\text{Mo } K\alpha_1)/\lambda(\text{Ag } K\alpha_1)$	1.2679471 (1.4 ppm)	1.2679495 (1.6 ppm)	−1.9
$\lambda(\text{Cu } K\alpha_1)/\lambda(\text{W } K\alpha_1)$	7.3708037 (1.3 ppm)	7.370757 (1.8 ppm)	6.3
$\lambda(\text{Mo } K\alpha_1)/\lambda(\text{W } K\alpha_1)$	3.3936489 (1.1 ppm)	3.393620 (1.9 ppm)	8.5
$\lambda(\text{Ag } K\alpha_1)/\lambda(\text{W } K\alpha_1)$	2.6764909 (1.6 ppm)	2.676463 (1.9 ppm)	10.4

TABLE VI
Measurements referred to directly measured X rays.

Z	Element	X-ray line	Energy (eV)	Uncertainty (eV)
13	Al	$K\alpha_2$	1486.269	0.032
		$K\alpha_1$	1486.695	0.006
24	Cr	$K\alpha_2$	5405.565	0.010
		$K\alpha_1$	5414.787	0.071
29	Cu	$K\alpha_2$	8027.917	0.023
42	Mo	$K\alpha_2$	17374.40	0.22
		$K\beta_3$	19590.37	0.42
		$K\beta_1$	19608.47	0.42
44	Ru	$K\alpha_2$	19150.61	0.18
		$K\alpha_1$	19279.29	0.18
		$K\beta_3$	21634.79	0.23
		$K\beta_1$	21656.89	0.23
47	Ag	$K\beta_3$	24911.71	0.30
		$K\beta_1$	24942.58	0.30
49	In	$K\alpha_2$	24002.18	0.28
		$K\alpha_1$	24209.90	0.28
		$K\beta_3$	27237.68	0.36
		$K\beta_1$	27275.73	0.36
51	Sb	$K\alpha_2$	26110.95	0.25
		$K\alpha_1$	26359.03	0.25
		$K\beta_3$	29679.40	0.42
		$K\beta_1$	29725.72	0.32
60	Nd	$K\beta_3$	42166.51	0.74
		$K\beta_1$	42271.17	0.57
67	Ho	$K\alpha_2$	46700.28	0.26
		$K\alpha_1$	47547.40	0.58
74	W	$K\alpha_2$	57982.64	0.81
		$K\beta_3$	66952.40	1.07
		$K\beta_1$	67245.45	1.08
83	Bi	$K\alpha_2$	74816.70	0.31
		$K\alpha_1$	77109.75	1.37
		$K\beta_1$	87344.65	2.16

available estimate of Co $K\alpha_{1,2}$ sufficed for the initial measurement. The results obtained for Ar and K are included in the data summary presented in the previous sections.

If one considers the results to date, both published and preliminary, they are not critically significant. Specifically, while they are consistent with theoretical estimates of, for example, the Lamb shift, they are not sufficiently accurate to challenge these estimates. The potential of such measurements is, however, very high indeed. One notes, in particular, that in all experimental profiles the X-ray reference lines and the Lyman α doublet components have about the same widths. Yet in all cases, especially Cl^{+16} and Ar^{+17}, the one-electron line should be much narrower than the $Z+1$ X-ray line due to the absence of (fast) nonradiative decay channels. Thus simple instrumental improvements would already be significant.

Further discussion of the details of these experiments would take us too far afield. Let the following few surmises stand in lieu of fuller discussion. Although the recoil experiments are most promising for alleviating Doppler-effect troubles, beam–foil experiments can be improved greatly. For instance, use of a two-dimensional position-sensitive detector permits correction for first-order Doppler shift for each photon, preserving high rates and retaining narrow lines. Secondly, as shown below (Sec. 6.1), the effective sensitivity of a Lyman α $1s$ Lamb-shift measurement near $Z=18$ is greater than for a similar measurement near $Z=36$. On the other hand, by the time one reaches $Z=36$ the transition rate for $2s\rightarrow 1s$ is large enough ($\tau\sim 10^{-11}$ sec) that this radiative decay is readily observable but very much narrower than $2p\rightarrow 1s$. Thus a measurement of $\lambda(2s\rightarrow 1s)$ with sufficient resolution and referenced to sharp (γ-ray) lines in high order would yield a test of QED with sensitivity available in no other way, provided Doppler problems could be eliminated by use of recoils or treated by a sufficiently ingenious procedure that their effect would be suitably limited.

Besides the one-electron case, considerable importance attaches to other few-electron spectra and their Z dependences. Although the particular case of He-like spectra is already widely exploited in plasma diagnostics, there are fewer systematic data available for relatively unperturbed sources which might be used as tests for calculations regarding He-like ions. Similar remarks apply to other few-electron spectra where at present there are only first-principles calculations and semiempirical "series formulas" as discussed, for example, by Martin (1981). Evidently the same kinds of experimental technology (source, spectrometer, and detector) used in the case of one-electron spectra can be applied also in these cases. Since theoretical estimates are in these cases less precise, one can well imagine relaxing the requirement of near degeneracy for the reference spectrum. There appears to be a considerable range of few-electron spectra in the X-ray region that is ripe for study at the present time.

5. THEORETICAL CALCULATIONS AND COMPARISON WITH EXPERIMENT

Theoretical estimates of single-vacancy energy levels depend both on properties of the fundamental interaction as one encounters also in one-electron spectra and on the adequacy of treatment of the many-electron aspects of the problem. On the fundamental side there are questions arising from the essential incompleteness of Dirac's equation. Specifically, it must be supplemented by a verbal statement to the effect that the vacuum state corresponds to all negative energy states being filled. Issues of this kind were extensively discussed in a recent workshop (Berry *et al.*, 1981). The essence of the situation as revealed in these discussions can be found in several articles which have appeared in more generally accessible media (Mittleman, 1981; Sucher, 1980). The following discussions are at best only suggestive regarding the present state of understanding of the fundamental interactions versus many-electron aspects of this theoretical problem.

Even when fundamental interactions are well understood, there remain several choices as to how to acknowledge the simultaneous presence of many electrons in a neutral heavy atom even in its ground state but more emphatically when it contains one (or more) inner-shell vacancies. At any level of current interest all approaches have self-consistent-field characteristics although they span a wide range of complexity and rigor. Although nonrelativistic Schrödinger calculations are highly developed and quite useful for low-Z inner-shell calculations, we neglect them entirely. Instead, we focus only on calculations in which relativistic effects are explicitly included so that discussion of the entire range of Z can proceed on a common footing.

Even with the above restrictions, several calculational procedures have been described in the literature. In the following Sec. 5.1, we inventory those procedures which have been pressed to high accuracy if not over a broad range of Z at least in more than a few cases. In the end it proved desirable for us to install the relativistic self-consistent-field codes developed by Grant and collaborators (Sec. 5.2) on our local computer and directly produce a theoretical reference data base. This procedure is described in Sec. 5.2 for the programs we used and compared in a few cases with results from the more rigorous approach of Desclaux and the less rigorous but all-Z results of Huang *et al.* (Sec. 5.2). After these comparisons among theoretical results, we turn finally in Sec. 5.3 to comparisons between the locally generated theoretical and experimental data sets.

5.1. Relativistic Self-Consistent-Field Calculations

The simplest and most widely used approximation which acknowledges the multielectron character of the problem is the relaxed-orbital relativistic

Hartree–Fock–Slater (HFS) approximation. The effectiveness of using this local exchange approximation in lieu of the more complicated (relativistic) Hartree–Fock procedure (nonlocal exchange) was demonstrated by Rosén and Lindgren (1968). These authors showed that the improvement in total energy estimates by use of an "optimized" exchange potential previously realized in the nonrelativistic case (Lindgren, 1965; 1966) carried over as well to the relativistic case.

Further development of this approach led to the extensive calculations by Huang *et al.* (1976) which went beyond the Rosén and Lindgren calculation by using the HFS wave functions as zeroth-order eigenfunctions to compute total energies thereby including a first-order correction to the local approximation. The Huang *et al.* calculations were important since they were carried out with high numerical precision and covered a wider range of Z ($2 \leqslant Z \leqslant 106$) than any other calculation then available. Unfortunately this monumental effort contained two features which have impeded the more critical comparisons with experiment that became possible over the past several years. Firstly, the nucleus was taken to be a uniformly charged sphere with radius proportional to $A^{1/3}$ with A assigned the molar mass value of the most abundant isotope of each element. Secondly, the Breit interaction was represented by its long-wavelength limit. Subsequently Chen *et al.* (1981a) have given some results obtained with more accurate estimates of the nuclear charge distribution and the Breit interaction. Unfortunately these efforts, though helpful as far as they go, have not yet led to a full revision of the original compilation.

In many ways the numerical calculations of J. P. Desclaux have served as benchmarks for both experimental comparisons and other theoretical efforts. The central basis is a multiconfiguration Dirac–Fock program (Desclaux, 1975) which permits choice of a nuclear charge model at least up to the level of a Fermi distribution. Although Desclaux has not generated an "all-Z" table, his work has extended to a sufficient number of experimentally accessible situations that quite general conclusions are possible. The generally rigorous approach available in this work together with the explicit results for physically distinct perturbations make its generalization quite desirable. Desclaux has, in fact, made his "code" available through the CPC Library, Queens University, Belfast, Northern Ireland. Unfortunately, full exploitation of its results requires mastery of the experimental universe and a ready theoretical wit.

The physical basis of Desclaux's calculation is shared by another extensive computational resource which has been developed by I. Grant and collaborators (Grant *et al.*, 1980; McKenzie *et al.*, 1980). This effort has likewise been made available in a variety of ways including through the above-mentioned CPC library of the Queens University, Belfast, Northern Ireland. Experience using the above-mentioned computational schemes has indicated an impressive level of agreement (Kim, 1983). Through the kindness of Professor Grant and the specific efforts of Dr. K. Dyall we obtained a copy of this program. With the help of local experts, especially Dr. A. W. Weiss, we have been able to install this program on the local NBS computer and use it to generate a theoretical data base specifically tailored for comparison with our experimental results.

5.2. *Theoretical Relativistic SCF Estimates*

The Oxford relativistic SCF program was installed on the NBS Univac 1108 computer from a magnetic tape and calculations were attempted for all elements and levels for which experimental values are reported in Table III, IV, and VI.

This program requires choosing a number of options for each level. For each hole-state calculation all possible terms permitted in the j–j coupling scheme were included. For valence electrons we required that the lower j states for a given n, l level be filled before electrons were permitted in a higher j state. To illustrate this point, consider Pb, which has two $6p$ valence electrons. These two electrons were required to be in the $6p_{1/2}$ level with a total $j=0$ rather than occupy the $6p_{1/2}$ level and/or the $6p_{3/2}$ level. According to these rules the $2p_{3/2}$ hole state in Pb has only one term rather than 13 terms. The program returns an energy value for each term and these were averaged using a weighting of $2j+1$ to obtain a final energy for each hole state. There is some ambiguity in the weighting because the energy levels are not described exactly by single-term wave functions.

The importance of using a Breit interaction which takes into account that electron velocities may be large was pointed out by Mann and Johnson (1971) and Chen *et al.* (1981a). For inner-electron transitions in heavy atoms, the difference between using a velocity-dependent or independent Breit interaction is more than 10 eV. For all calculations a velocity-dependent Breit interaction was used.

The nuclear charge density was assumed to be a two-parameter Fermi distribution having the form

$$\rho(r) = \rho_0/(1 + e^{(r-c)/a}) \tag{2}$$

where $c = (2.2291 \times 10^{-5}\, A^{1/3} - 0.90676 \times 10^{-5})$ a.u. and $a = 1.039 \times 10^{-5}$ a.u. The molar mass, A, was chosen to be the molar mass of a naturally occurring sample of the element except for high-Z elements which do not occur naturally in which case the molar mass of the most abundant long-lived isotope was used.

There was some difficulty in obtaining convergence when valence f

electrons were present. This was particularly true for $67 \leqslant Z < 70$ and $Z > 93$. Using the Cowan–Mann modification to the exchange potential (Cowan and Mann, 1975), adjusting the grid of the calculation, and using initial wave functions obtained by artificially increasing Z by one usually resulted in convergence of the self-consistent-field calculation.

Typical computation times ranged from near 40 min for the U $2p_{3/2}$ hole state (43 terms) to a few minutes for one of the hole states in Xe (1 term).

The Oxford program does not include higher-order vacuum polarization terms. These terms are of order $\alpha^2(Z\alpha)$, $\alpha(Z\alpha)^3$, $\alpha(Z\alpha)^5$, and $\alpha(Z\alpha)^7$ and shift inner-electron transition energies by 0.1 and 5 eV at $Z = 55$ and 95, respectively. Higher-order vacuum polarization corrections are tabulated for all Z by Huang *et al.* (1976) and this tabulation was used to correct the transition energies calculated by the Oxford program. In this and the following sections all of the theoretical calculations labeled as "Oxford calculations" are corrected for higher-order vacuum polarization. The one exception is in the immediate following comparison of the Oxford calculations with those of J. P. Desclaux, who does not include higher-order vacuum polarization.

Results of calculations made with the Oxford program are compared with calculations by Desclaux in Table VII. The Hg $1s_{1/2}$ binding energy is taken from Desclaux (1981) and the differences between high-Z and low-Z X rays are from Borchert *et al.* (1980). The latter calculations were made in conjunction with a series of experimental measurements of high- and low-Z X rays in different orders which were chosen because the crystal reflections occurred at approximately the same angular setting. For the Hg $1s_{1/2}$ binding energy the two calculations show discrepancies of the order of 1.5 eV for the zero-order energy, vacuum polarization, and self-energy with a total discrepancy of 2 eV. Similar discrepancies exist between the two calculations for comparing high- and low-Z X-ray energies.

In Fig. 7, the Oxford calculations are compared with those of Huang *et al.* (1976) and Chen *et al.* (1981a) as a function of Z. The calculations differ by approximately 0.5 eV and -2 eV in the $Z = 90$ range for $K\alpha_2$ and $K\alpha_1$, respectively. Thus the Dirac–Hartree–Slater approach appears not to differ from the Dirac–Fock approach by much more than 2 eV. For $K\beta_1$ and $K\beta_3$, the discrepancy in the calculations increases monotonically from -1 eV at $Z = 50$ to -7 eV at $Z = 95$. The larger discrepancies for the $K\beta_{1,3}$ transitions cannot be explained by the fact that the Chen *et al.* corrections to the Huang *et al.* calculations were not extended to the $M_{II,III}$ levels. The binding energies of the $3p$ levels shift only by about 0.5 eV and 1.5 eV for $Z = 80$ and 100, respectively, when the Chen corrections are applied to the Huang calculations. Chen (1983) has pointed out to us that the main difference between the Huang–Chen and Oxford calculation is in the treat-

TABLE VII
Comparison of Desclaux and Oxford calculations.

Hg $1s_{1/2}$ binding energy		
Contribution	Desclaux (eV)	Oxford (eV)
DF point nucleus	−83613.5	
Finite nucleus	54.4	
Zero-order energy	−83559.1	−83560.44
Breit interaction	297.4	297.19
Vacuum polarization	−44.2	−43.13
Self-energy	197.0	195.47
	−83108.9	−83110.91

High/low Z X-ray comparison

El.	Trans.	Ord.	Δ Desclaux (eV)	Δ Oxford (eV)
Bi	$K\alpha_2$	2	−4.08	−2.03
Ag	$K\beta_1$	1		
Th	$K\alpha_1$	2	−26.37	
Ho	$K\alpha_2$	1		
U	$K\beta_1$	2	−23.49	
Er	$K\beta_1$	1		

ment of self-energy. The Oxford calculation includes relaxation effects on self-energy while the Huang–Chen calculation does not. The Oxford calculation of the self-energy correction is numerically smaller than the Huang–Chen calculation for the K, L_{II}, and L_{III} levels by similar amounts (2 to 4 eV at $Z \approx 80$). Thus the $K\alpha_{1,2}$ transition is somewhat insensitive to the choice of self-energy calculation. However, the M_{II} and M_{III} self-energy corrections are small in either calculation so that the self-energy term for the $K\beta_{1,3}$ transitions is essentially that of the K level. A large part of the discrepancy in Fig. 7 for $K\beta_{1,3}$ is due to the different methods for calculating self-energy. Chen (1983) further suggests that the importance of relaxation effects on self-energy can only be answered by *ab initio* calculations.

In summary, the above comparisons of calculated inner-shell transition energies suggest that all the calculations which have been considered are in agreement to a few tenths of eV at $Z = 50$ and a few eV at $Z = 90$. The contributions to the calculated energies differ somewhat in form but included a zero-order energy including a finite-nucleus correction; a velocity-dependent

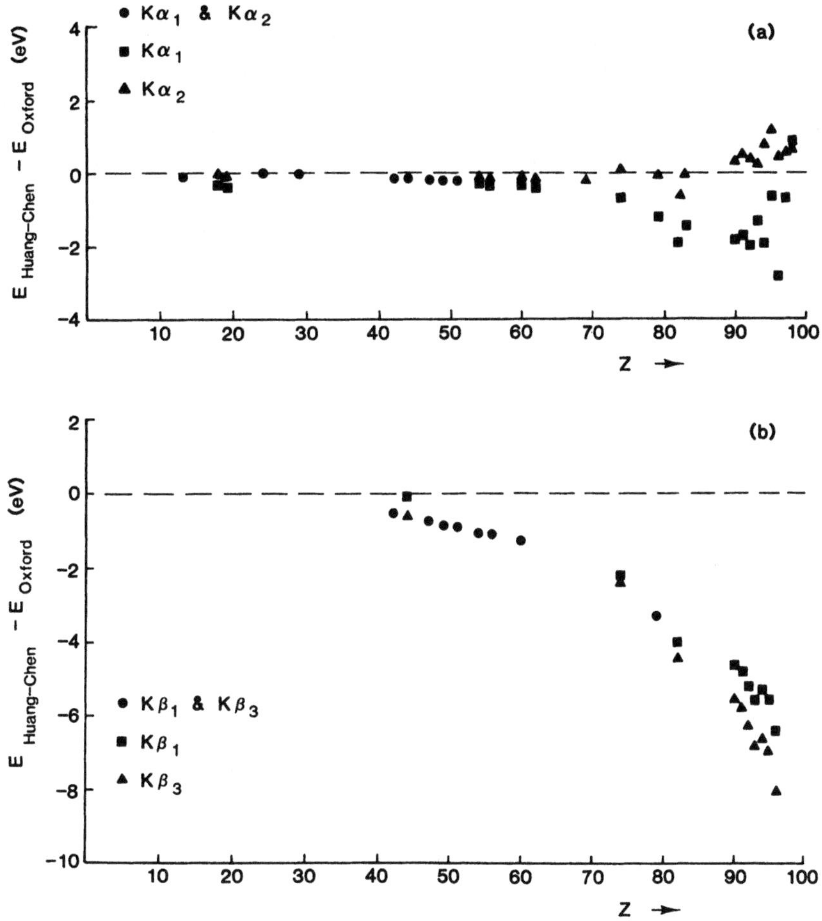

FIGURE 7. Energy differences between theoretical calculations of Huang-Chen and Oxford for (a) $K\alpha_1$ and $K\alpha_2$, and (b) $K\beta_1$ and $K\beta_3$. The symbols are ●: (a) $K\alpha_1$ and $K\alpha_2$, (b) $K\beta_1$ and $K\beta_3$ (differences too small to differentiate on plot); ▲: (a) $K\alpha_2$, (b) $K\beta_3$; ■: (a) $K\alpha_1$, (b) $K\beta_1$.

Breit interaction; first- and higher-order vacuum polarization; and self-energy. All of the calculations may suffer from the omission of important contributions such as configuration interaction with nearby bound excited states and with radiative and radiationless continua as suggested by Chen *et al.* (1981b).

5.3. Comparison with Experiment

We turn now to the comparison of the experimental data base which was assembled in Sec. 4 with theoretical calculations made with the Oxford program. The experimental and theoretical values are compiled in Table VIII along with the uncertainties and the experiment–theory differences. In Figs. 8 and 9 these differences are presented in graphical form.

The $K\alpha_{1,2}$ transitions shown in Fig. 8a,b exhibit two more-or-less distinct regions. For $Z < 85$, the experiment–theory discrepancies increase slightly with Z and have values of ~ 1.5 and 2.5 eV at $Z \sim 45$ and 85, respectively. In this region, the experimental uncertainties are typically $\leqslant 1$ eV and the corrections (vacuum polarization, velocity dependent Breit, nuclear shape, and self-energy) to the theoretical calculations are small. For $Z > 85$ the discrepancy no longer is a smooth function of Z but oscillates crossing zero at $Z \sim 90$ and 94 and possibly 98. In this region the experimental uncertainties are typically > 2 eV and the theoretical corrections approach 10 eV in some instances.

The $K\beta_{1,3}$ transitions shown in Fig. 9a,b also exhibit regions and behavior similar to the $K\alpha_{1,2}$ transitions but the regions are less distinct and the behavior is exaggerated. For $Z < 85$ the discrepancies are larger and increase more rapidly with Z, having values of ~ 3.5 and 6.5 eV at $Z \sim 45$ and 85, respectively. For $Z > 85$ the oscillatory behavior is again in evidence but the amplitude of the oscillations is larger and the only zero crossing is at $Z \sim 91$.

5.4. *Conclusions Derived from Comparison*

In the experiment–theory comparison of the previous section systematic discrepancies were noted. Our discussion in this section focuses on possible explanation for these discrepancies.

The almost constant 1.5-eV discrepancy for $K\alpha_1$ and $K\alpha_2$ over a wide range of Z may result from uncalculated configuration interactions as suggested by Chen *et al.* (1981b). All the calculations assume a single configuration and ignore interaction between configurations. The inner-level hole states may interact with nearby bound excited states and with radiation and radiationless continua to produce such shifts.

The same or similar mechanism has been suggested as an explanation of the larger $K\beta_{1,3}$ discrepancies. Here the interaction of the $M_{\mathrm{II,III}}$ levels with the decay continua is expected to produce larger shifts than the interactions of the $L_{\mathrm{II,III}}$ levels produce. Preliminary calculations by Chen (1982) predict shifts on the order of 6 eV for the M levels due to coupling to the continua. Such a shift has the possibility of making the experiment–theory discrepancies of the $K\alpha$ and $K\beta$ transitions more nearly equal.

The self-energy term is more significant at high Z and requires the choice of an effective atomic number to account for electron screening. This could possibly account for the larger discrepancies at high Z but probably cannot explain the oscillatory behavior.

TABLE VIII

Comparison of experimental and theoretical energies.

Z	Element	X-ray line	Experimental energy (eV)	Theoretical energy (eV)	Energy diff. expt.−theor. (eV)
13	Al	$K\alpha_2$	1486.269 ± 0.032	1486.893	−0.624
		$K\alpha_1$	1486.695 ± 0.006	1487.612	−0.917
18	Ar	$K\alpha_2$	2955.684 ± 0.009	2955.564	0.120
		$K\alpha_1$	2957.813 ± 0.008	2957.703	0.110
19	K	$K\alpha_2$	3311.217 ± 0.006	3310.964	0.253
		$K\alpha_1$	3313.969 ± 0.005	3313.746	0.223
24	Cr	$K\alpha_2$	5405.565 ± 0.010	5404.444	1.212
		$K\alpha_1$	5414.787 ± 0.071	5414.507	0.280
29	Cu	$K\alpha_2$	8027.917 ± 0.023	8027.307	0.610
		$K\alpha_1$	8047.865 ± 0.008	8047.365	0.500
42	Mo	$K\alpha_2$	17374.40 ± 0.22	17373.063	1.337
		$K\alpha_1$	17479.485 ± 0.010	17478.476	1.009
		$K\beta_3$	19590.37 ± 0.42	19586.004	4.366
		$K\beta_1$	19608.44 ± 0.42	19603.956	4.484
44	Ru	$K\alpha_2$	19150.61 ± 0.18	19149.160	1.450
		$K\alpha_1$	19279.29 ± 0.18	19278.467	.823
		$K\beta_3$	21634.79 ± 0.23	21631.484	3.306
		$K\beta_1$	21656.89 ± 0.23	21653.520	3.370
47	Ag	$K\alpha_2$	21990.44 ± 0.10	21989.090	1.350
		$K\alpha_1$	22163.06 ± 0.03	22161.961	1.099
		$K\beta_3$	24911.71 ± 0.30	24908.498	3.212
		$K\beta_1$	24942.58 ± 0.30	24939.758	2.822
49	In	$K\alpha_2$	24002.18 ± 0.28	24000.681	1.499
		$K\alpha_1$	24209.90 ± 0.28	24208.714	1.186
		$K\beta_3$	27237.68 ± 0.36	27234.285	3.395
		$K\beta_1$	27275.73 ± 0.36	27272.797	2.933
51	Sb	$K\alpha_2$	26110.95 ± 0.25	26109.054	1.896
		$K\alpha_1$	26359.03 ± 0.25	26357.602	1.428
		$K\beta_3$	29679.40 ± 0.42	29675.628	3.772
		$K\beta_1$	29725.72 ± 0.32	29722.638	3.082
54	Xe	$K\alpha_2$	29458.44 ± 0.05	29456.833	1.607
		$K\alpha_1$	29778.97 ± 0.10	29777.266	1.704
		$K\beta_3$	33563.42 ± 0.12	33559.172	4.248
		$K\beta_1$	33624.45 ± 0.12	33621.349	3.101

Table continued

TABLE VIII *(Continued)*

Z	Element	X-ray line	Experimental energy (eV)	Theoretical energy (eV)	Energy diff. expt.−theor. (eV)
56	Ba	$K\alpha_2$	31816.82 ±0.06	31814.773	2.047
		$K\alpha_1$	32193.47 ±0.07	32191.720	1.750
		$K\beta_3$	36303.58 ±0.12	36299.368	4.212
		$K\beta_1$	36377.68 ±0.08	36373.736	3.944
60	Nd	$K\alpha_2$	36847.74 ±0.08	36846.181	1.559
		$K\alpha_1$	37360.98 ±0.07	37360.057	0.923
		$K\beta_3$	42166.51 ±0.74	42163.632	2.878
		$K\beta_1$	42271.17 ±0.57	42269.236	1.934
62	Sm	$K\alpha_2$	39523.64 ±0.10	39522.278	1.362
		$K\alpha_1$	40118.74 ±0.06	40118.105	0.635
67	Ho	$K\alpha_2$	46700.28 ±0.26		
		$K\alpha_1$	47547.40 ±0.58		
68	Er	$K\alpha_2$	48221.92 ±0.20		
		$K\alpha_1$	49127.56 ±0.12		
		$K\beta_3$	55480.08 ±0.35		
		$K\beta_1$	55673.88 ±0.18		
69	Tm	$K\alpha_2$	49772.99 ±0.12	49771.590	1.400
		$K\alpha_1$	50741.81 ±0.09		
		$K\beta_1$	57509.14 ±0.15		
74	W	$K\alpha_2$	57982.64 ±0.81	57979.792	2.848
		$K\alpha_1$	59319.23 ±0.05	59317.754	1.476
		$K\beta_3$	66952.40 ±1.07	66946.315	6.085
		$K\beta_1$	67245.45 ±1.08	67239.316	6.134
79	Au	$K\alpha_2$	66991.16 ±0.22	66988.551	2.609
		$K\alpha_1$	68804.94 ±0.18	68803.838	1.102
		$K\beta_3$	77575.51 ±0.62	77568.769	6.741
		$K\beta_1$	77980.30 ±0.38	77973.898	6.402
82	Pb	$K\alpha_2$	72805.89 ±0.24	72803.448	2.442
		$K\alpha_1$	74970.59 ±0.17	74968.492	2.098
		$K\beta_3$	84450.99 ±0.60	84444.190	6.800
		$K\beta_1$	84939.63 ±0.34	84932.515	7.115
83	Bi	$K\alpha_2$	74816.70 ±0.31	74813.180	3.520
		$K\alpha_1$	77109.75 ±1.37	77106.863	2.887
		$K\beta_1$	87344.65 ±2.16	87343.035	1.615

Table continued

TABLE VIII *(Continued)*

Z	Element	X-ray line	Experimental energy (eV)	Theoretical energy (eV)	Energy diff. expt. − theor. (eV)
90	Th	$K\alpha_2$	89957.62 ± 0.20	89956.827	0.793
		$K\alpha_1$	93347.98 ± 0.25	93350.663	−2.683
		$K\beta_3$	104817.21 ± 0.69	104813.328	3.882
		$K\beta_1$	105602.19 ± 0.53	105598.045	4.145
91	Pa	$K\alpha_2$	92284 ± 2	92284.663	−0.663
		$K\alpha_1$	95867 ± 2	95868.592	−1.592
		$K\beta_3$	107586 ± 2	107588.280	−2.280
		$K\beta_1$	108418 ± 2	108419.492	−1.492
92	U	$K\alpha_2$	94651.45 ± 0.56	94652.946	−1.496
		$K\alpha_1$	98432.21 ± 0.28	98436.228	−4.018
		$K\beta_3$	110416.38 ± 0.65	110414.006	2.374
		$K\beta_1$	111295.80 ± 0.65	111294.105	1.695
93	Np	$K\alpha_2$	97069 ± 3	97067.826	1.174
		$K\alpha_1$	101057 ± 3	101060.380	−3.380
		$K\beta_3$	113308 ± 4	113297.726	10.274
		$K\beta_1$	114244 ± 3	114229.073	14.927
94	Pu	$K\alpha_2$	99525.4 ± 1	99525.073	0.327
		$K\alpha_1$	103735.2 ± 0.6	103737.239	−2.039
		$K\beta_3$	116242 ± 2	116235.658	6.342
		$K\beta_1$	117233 ± 2	117221.070	11.930
95	Am	$K\alpha_2$	102032 ± 3	102031.073	0.927
		$K\alpha_1$	106474 ± 3	106472.625	1.375
		$K\beta_3$	119240 ± 2	119234.213	5.787
		$K\beta_1$	120280 ± 2	120276.313	3.687
96	Cm	$K\alpha_2$	104591 ± 2	104582.309	8.691
		$K\alpha_1$	109273 ± 2	109266.650	6.350
		$K\beta_3$	122303 ± 2	122289.386	13.614
		$K\beta_1$	123404 ± 2	123390.960	13.040
97	Bk	$K\alpha_2$	107195 ± 5	107184.455	10.545
		$K\alpha_1$	112128 ± 5	112122.635	5.365
		$K\beta_3$	125415 ± 7		
		$K\beta_1$	126578 ± 7		
98	Cf	$K\alpha_2$	109838 ± 8	109833.368	4.632
		$K\alpha_1$	115036 ± 8	115037.847	−1.847

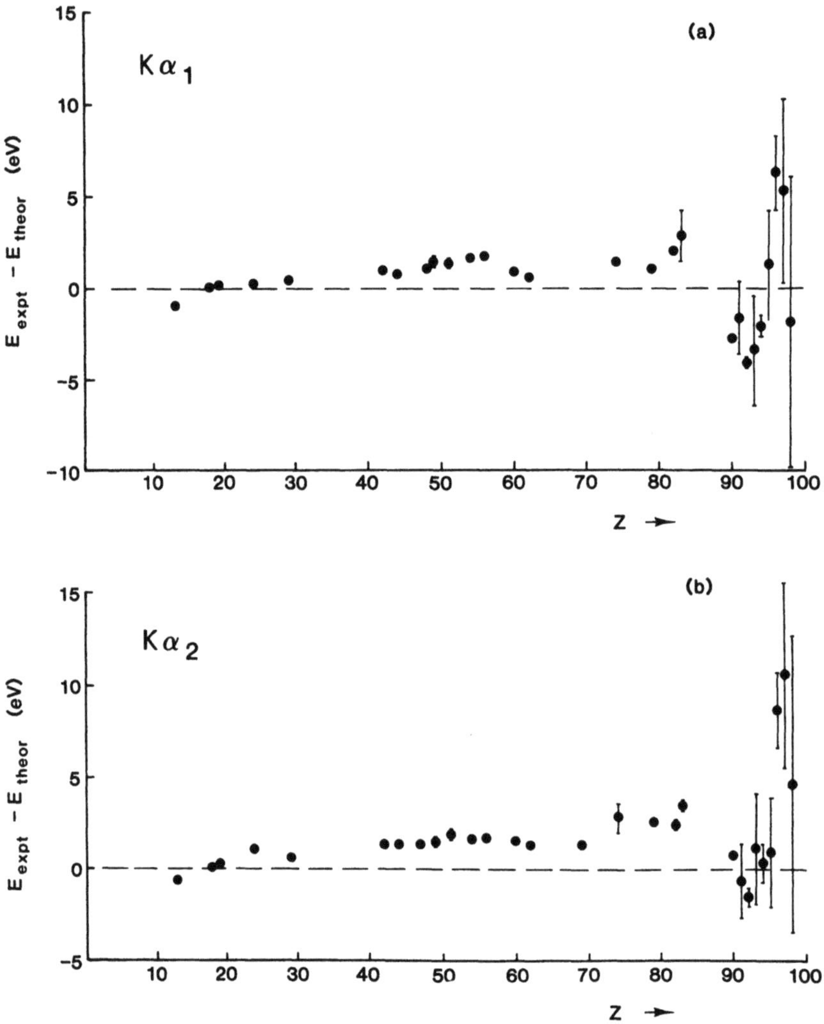

FIGURE 8. Energy differences between experimental values and theoretical estimates of Oxford for (a) $K\alpha_1$ and (b) $K\alpha_2$.

The effect of the nuclear charge distribution has been discussed by Chen *et al.* (1981a) for a Fermi nuclear charge distribution and a uniform sphere. The effect of a deformed nuclear charge distribution on the calculated energies has not been reported, but possibly should be investigated.

Finally, more precise experimental measurements are needed especially for $Z > 90$ and may make a significant contribution to our understanding of these complex atomic systems.

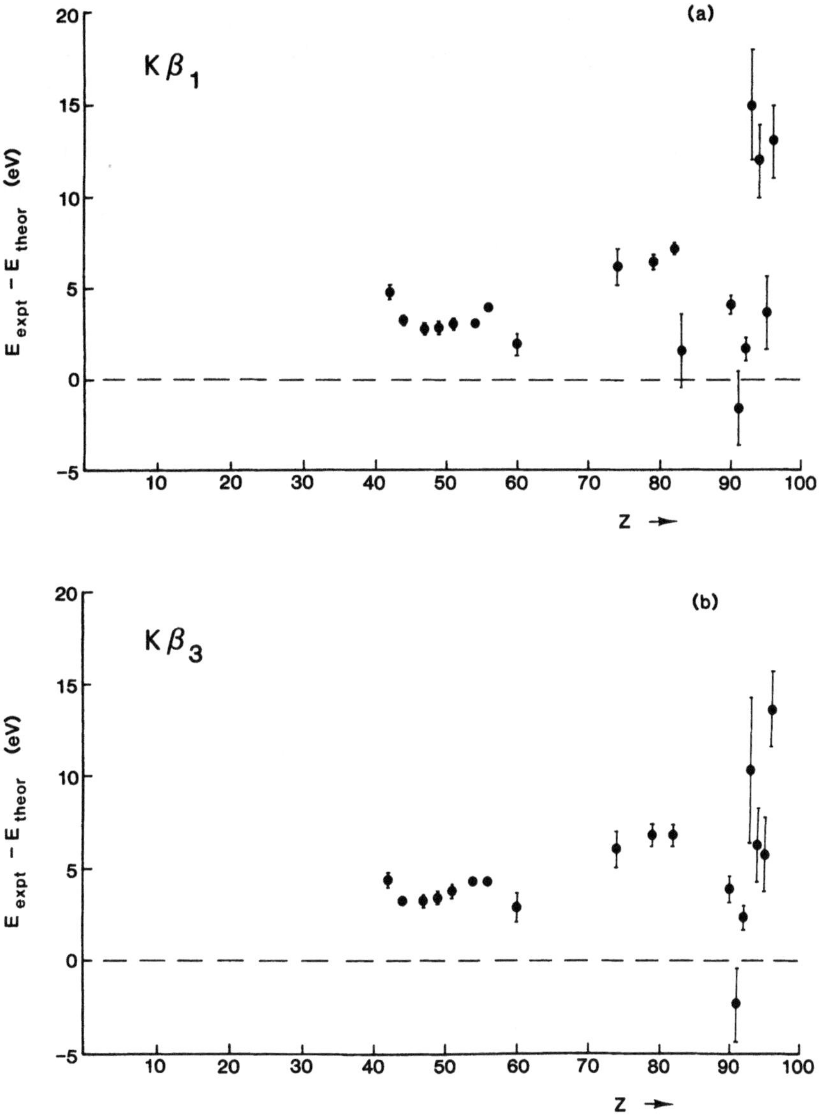

FIGURE 9. Energy differences between experimental values and theoretical estimates of Oxford for (a) $K\beta_1$ and (b) $K\beta_3$.

6. SUMMARY AND OUTLOOK

The review of experimental data on X-ray emission lines along with the comparison with relativistic SCF theory given above is, we believe, representative of the state of the art in the summer of 1983. The important trends in the comparisons reported are not yet understood in detail but do appear to suggest that the theory is rather more competent in dealing with some classes of electron configurations than with others. We need next to consider how what has been learned in this way connects with other experiments on one- (or few-) electron systems. We then turn to some speculation about further X-ray measurements which might be undertaken and where limitations regarding their interpretation are likely to intervene. Finally we attempt to deduce from the data sample what could be projected for elements and lines not directly studied.

6.1. Relation between Single-Electron and X-Ray Spectra

As indicated earlier (Sec. 4.4), new results are becoming available on one-electron spectra of systems in the range $16 < Z < 40$. Since very accurate and relatively simple theoretical estimates are at hand for such spectra, why should they not be considered as the sole (or at least principal) arena for tests of the fundamental interaction? In this way one could avoid the relatively untidy combination of fundamental interaction and description of complex many-electron systems which is required for estimating level and transition energies for X-ray processes. Not only that, but since the one-electron spectra are immune to Auger processes, the lines are appreciably narrower than are corresponding X-ray lines and, to a great extent, devoid of satellite processes (but see below). The answer to such questions which we offer is rather complex but it is an answer which we feel provides a reasonable guide for our own efforts which include work on both types of spectra and may perhaps be of interest to others.

For the one-electron systems, Kugel and Murnick (1977) have reviewed both the theoretical situation and experimental results available at that time. We focus on the theoretical QED corrections to the energies of $1s$ and $2s$ electrons and their dependence on nuclear charge. Our discussion is in terms of hypothetical "Lamb-shift" experiments where the general inverse cubic dependence on principal quantum number, n, and on Z^4 appear to favor experiments on $1s$ levels in as high-Z systems as are tractable. A more discerning estimate of the potential sensitivity of such experiments considers the ratio, Q, of shift, S, to linewidth, Γ, which also varies as Z^4 leading to a constant value for Q. Actually the elementary expression for Q needs modification by the structure function $F(Z\alpha)$ given by Cheng and Johnson

(1976). The more complete result, which follows easily from the summary by Kugel and Murnick, is

$$Q = S/\Gamma = 1.3F(Z\alpha). \tag{3}$$

The function $F(Z\alpha)$ is in the range of 2–3 for Z's 16–18 and drops to a broad minimum in the region $50 < Z < 120$ with values near 0.3. In the region of still higher Z, $F(Z\alpha)$ rises progressively, attaining the value 3 once more near $Z = 150$.

Assuming adequate experimental resources it would appear that the experimentally attractive region near $Z = 18$ is probably superior in sensitivity for study of QED effects to the often targeted region near $Z = 36$. This is particularly the case when experimental practicalities are noted. Specifically, while the entire range $Z < 50$ is accessible to present-day accelerator technology by beam–foil techniques, only the region near $Z = 18$ is thus far available by means of the recoil stripping process. Since Doppler linewidths and shifts dominate the uncertainties associated with beam–foil experiments, the recoil method (where these effects are negligible) appears preferable. These arguments will need significant revision when adequate observational technique can be applied to magnetic dipole transitions $2s \rightarrow 1s$ in the region near $Z = 36$ (and above) where their study becomes practical. Such transitions are already available in beam–foil experiments but their potentially very small linewidths cannot be fully exploited because of Doppler broadening and uncertainty in Doppler shifts. Should such species become available as recoils in the future their study would be very attractive indeed. The main technical problem will center about how to deliver an adequate optically based (probably γ-ray) secondary standard to the immediate vicinity of the $2s \rightarrow 1s$ emission line.

Actually all the above stripped-ion spectra share some problems that are only now being addressed as experimental results start to become available. These problems are to be found both in beam–foil and recoil experiments although there are likely to be significant differences in the two cases. One sees in the data so far obtained for example of Cl^{+16} in a beam–foil experiment (Stockli *et al.*, 1983) and a recoil experiment on Ar^{+17} (Beyer *et al.*, 1983) that there are significant line-shape anomalies especially in the way of high tails on the low-energy side of line center. It is likely that these effects arise from the presence of one or more spectator electrons in outer orbits when the atom radiates by a transition of the active electron from the $n = 2$ state.

The present-day situation of experiments in one-electron systems aimed at tests of the QED corrections is typified by the sample of recent results indicated in Table IX. Although we have listed both $n = 2$ and $n = 1$ results

TABLE IX
Lamb-shift measurements of hydrogenic atoms.

Element	Level	Theoretical Lamb shift	Experimental Lamb shift	Percent difference
H	1s	8149.43 MHz	8151 ± 30 MHz[a]	0.02 ± 0.37
H	2s	1057.884 MHz	$1057.845 \pm .009$ MHz[b]	$0.0037 \pm .0009$
O^{+7}	2s	2196.21 GHz	2192 ± 15 GHz[c]	0.19 ± 0.68
F^{+9}	2s	3360 GHz	3339 ± 35 GHz[d]	0.6 ± 1
P^{+14}	2s	20.254 THz	20.18 ± 0.25 THz[e]	0.37 ± 1.23
Cl^{+16}	2s	31.35 THz	31.19 ± 0.22 THz[f]	0.5 ± 0.7
Ar^{+17}	2s	38.25 THz	38.0 ± 0.6 THz[g]	0.7 ± 1.6
Fe^{+25}	1s	3.9 eV	3.4 ± 0.6 eV[h]	13 ± 15

[a] Weiman and Hänsch (1980).
[b] Lundeen and Pipkin (1981).
[c] Curnutte *et al.* (1982).
[d] Kugel *et al.* (1975).
[e] Pellegrin *et al.* (1982).
[f] Wood *et al.* (1982).
[g] Gould and Marrus (1978).
[h] Briand *et al.* (1983).

for each Z together quite different technologies are required in these experiments. The reader should also note that the units used in each of the referenced studies have been retained so that the units appearing in columns 3 and 4 are not consistent. On the other hand, the results given in column 5 indicate the fractional size (in percent) of the deviation of the experimental result from its QED estimate and the uncertainty therein. This mode of expression gives (to a first order) a meaningful basis for comparison regarding implication for QED of various experimental results. [Actually the situation is more complex since such features as the structure function, $F(Z\alpha)$, suggest that studies should be carried out over a range of nuclear charge even if this did not lead, for example, to a more sensitive test of QED.] The pattern of results in Table IX invites several observations concerning the present state of affairs. First of all, studies of atomic hydrogen remain the most highly developed and continue to be the most informative both in $n=2$ and $n=1$. For $n=2$, these studies appear to be nearing a stage at which only marginal improvements are possible while for $n=1$, an order of magnitude improvement seems likely. Secondly, results on higher-Z one-electron spectra available to date are less instructive in either $n=1$ or $n=2$ experiments. We note, however, that the $n=1$ experiments in $Z \gg 1$ systems are less difficult to realize and have only been recently undertaken so that it is reasonable to expect some improvement in the quality of these results over the next few years. In fact, efforts are currently underway aimed at improved measurement of Lyman α in one-electron argon recoils. It appears that such measurements could be fully competitive with the far more precise results available from the lower-energy spectra of atomic H. Evidently, successful investigation of the spontaneous $2s \rightarrow 1s$ emission mentioned above would change this picture quite appreciably.

With the above background, we can address the question of whether X-ray experiments on complex neutral atoms have anything to add to the above discussions. First of all, even in the relatively low region of Z where one-electron spectra have also been obtained, the precision and accuracy available for X-ray lines (due to their larger intensity under laboratory electron bombardment) permit effective cancellation of the linewidth advantage afforded by investigation of one-electron systems. The real import of inner-shell X-ray spectra lies, however, in our ability to pursue such studies to higher regions of $Z\alpha$ than are presently regarded as practical in one-electron species. One gains a sense of this application by noting that at $Z=90$ the radiative correction amounts to about 300 eV and agreement with theory is within ± 2 eV. Thus, if one had reasonable confidence in the many-body formulation, this agreement could be taken as a verification of the QED calculation within $\pm 0.6\%$. There are, of course, considerable uncertainties associated with the various formulations of many-electron theory especially for the case of inner-shell vacancies. On the other hand, where several of these have demonstrated a reasonable level of concordance in detail and with regard to Z-dependent trends, it is probably safe to assert their significance with respect to the fundamental interactions.

In considering the roles of spectroscopy of one-electron systems and spectroscopy of inner-vacancy transitions the following surmises seem evident: In the region of Z where both approaches are possible, one-electron systems are more discerning except for the (important) problem of attendant spectators. X-ray lines, though broader, are available with higher intensities over a broad range of Z, thus mitigating, to some extent, the advantage associated with narrower profiles in the one-electron case. Inner-shell vacancy spectra at the present time and for the foreseeable future are accessible over a much wider range of $Z\alpha$ than are one-electron spectra. X-ray spectra are contaminated by satellite structures to a likely greater extent (Hansen *et al.*, 1985) than one-electron spectra are perturbed by spectators. In an asymptotic state where all techniques have been developed to their intrinsic limits, there will remain valid roles for both approaches to put to the test not only our ideas about fundamental interaction but also our methodologies for treatment of the simultaneous presence of many electrons in the system. This last-mentioned point may ultimately be the more significant since the atomic domain may well be an appropriate testing area for methods which would appear capable of predictions for actual polyatomic systems including complex molecules and condensed systems.

6.2. Future Measurements and Applications

The X-ray experimental results we have reported above are rather near to what can be hoped for in the way of interpretable accuracy. What is meant by this statement is that the approach which we have used is limited by failure to include details of multivacancy satellites as discussed by Borchert *et al.* (1978). Also, elsewhere in this volume, Hansen *et al.* (1985) discuss many mechanisms which can lead to shifts in X-ray line positions of the same order as those arising from neglecting the presence of unresolved satellite components within the line profile. Although these effects can be studied by intensive observation, it is difficult to think of such studies being carried forward over a very wide range of atomic number. Similarly, although readily available atomic theory can be modified to take account of these effects, the labor and computational cost appear to be sufficiently large as to militate against considering such an exercise over a wide range of Z for a relatively great number of cases.

In view of the above considerations we chose to restrict the level of precision at which interpretation is attempted to one where such complications may be neglected. It turns out that this level, near a few percent of the linewidth, is comparable with the statistical precision obtainable with levels of effort that can reasonably be extended to a large number of cases. It also turns out, as seen above, that for inner-shell transitions, the comparisons with atomic SCF theory make reasonable sense even though solid samples are used in the majority of cases. Our conclusion is thus that X-ray spectroscopy has approached a limiting utility in the work reported, and we are accordingly not enthusiastic to press harder to attempt to obtain and interpret results of higher accuracy.

Besides the question of possibly more intensive study of spectra from tractable materials, there arise inevitably questions about the spectra of less tractable systems. Such questions often emerge in practical situations where there is an externally generated need to know something that is not known directly and would also be difficult or hazardous or simply inconvenient to obtain directly. In view of these considerations several tables covering all Z have been attempted in the past and have proven to be widely useful.

Looking back over these efforts we were led to consider how the results summarized here might be used to reach an improved version of such a table. In this regard it is helpful to consider how at least some of the available examples were assembled and then offer an alternative suggestion below. At the earliest time people used graphical plots, often linearized for example via Moseley's law, i.e., E vs. Z^2. Missing elemental data were interpolated or extrapolated with reasonable success. Subsequent efforts have been either predominantly experimental such as in the work of Bearden (1967) or have attempted to arrive at an overall set of "recommended" energy levels by a combination of X-ray and electron spectroscopies along with certain measures of theoretical guidance. Examples can be found in the widely used tables of Lotz (1970), Carlson *et al.* (1969), and Sevier (1979). Expecially for the region of high Z there are several tabulations among which that due to Porter and Friedman (1978) is noteworthy.

To our knowledge, Porter and Friedman (1978) were the first to emphasize the potential of using a detailed SCF reference calculation and studying the Z dependence of the residual differences. These residuals are reasonably expected to vary more smoothly with Z than the raw data or the data scaled against any reference theory itself smooth in Z. Any reasonable SCF theory contains some representation of shell effects and angular-momentum coupling. A decent theoretical basis set such as we used above contains rather adequate representations of these effects and many others including such matters as nuclear size effects and QED corrections. Residuals in this case should indeed be smooth in Z (or negligible). The comparisons given above verify this situation to a higher level of accuracy than has been seen previously.

We therefore suggest that an all-Z table of X-ray emission energies could be obtained from data on residuals as shown in Figs. 8 and 9. One would interpolate or extrapolate the residual values for missing elements and combine these with reference theoretical calculations to obtain estimates for lines not otherwise observed. To assure ready distinction between observed and estimated values, the latter should appear either in a distinct (and weaker) typeface or in braces. Assuming such were available, it could lead (when combined with outer-shell XPS results) to a further table of estimated term values provided the important distinction between Fermi-level reference and vacuum level reference was maintained as discussed in Sec. 2.2 above.

ACKNOWLEDGMENTS

In preparing this chapter we were helped by a number of people whose assistance we wish to acknowledge. J. P. Desclaux and Y.-K. Kim shared with us their broadly informed outlook on the general theoretical situation. In addition, Kim helped us verify proper operation of the Oxford program and also furnished numerical comparisons between the earlier Huang–Chen calculations and results obtained from the Oxford program. M. H. Chen and B. Crasemann provided us with prepublication results in the $50 < Z < 70$ region and estimates of shifts due to coupling of inner vacancy states with adjacent decay continua.

We are particularly indebted to Ken Dyall of Oxford, who arranged for us (the present authors and R. E. LaVilla) to receive an updated program tape containing the computational codes developed by I. P. Grant and collaborators. Dyall was also most helpful when difficulties intervened in application of these programs. At NBS we obtained much help and guidance from A. W. Weiss both as regards computational specifics and the general physical situation. Weiss and Charles Clark (also of NBS) provided us with comparisons among results obtained by different approaches.

Last of all, we appreciate the continued interest of W. C. Martin (NBS) in our work in this area as well as his careful reading of this manuscript.

REFERENCES

ÅBERG, T., AND M. SUVANEN, 1982, Relativistic Theory of X-Ray Satellites, in *Advances in X-Ray Spectroscopy*, edited by C. Bonnelle and C. Mandé (Pergamon, Oxford), p. 1.

ANDERSON, C. R., R. N. LEE, J. F. MORAR, AND R. L. PARK, 1982, *J. Vac. Sci. Technol.* **20**, 617.

BÄCKLIN, E., 1935, *Z. Phys.* **93**, 450.

BAKER, J. F. C., AND M. HART, 1975, *Acta Crystallogr.* **A31**, 364.

BARREAU, G., H. G. BÖRNER, T. VON EGIDY, AND R. W. HOFF, 1982, *Z. Phys. A* **308**, 209.

BEARDEN, J. A., 1931, *Phys. Rev.* **37**, 1210.

BEARDEN, J. A., 1965a, *Phys. Rev.* **137**, B181.

BEARDEN, J. A., 1965b, *Phys. Rev.* **137**, B455.

BEARDEN, J. A., 1967, *Rev. Mod. Phys.* **39**, 78.

BEARDEN, J. A., AND A. F. BURR, 1967, *Rev. Mod. Phys.* **39**, 125.

BEARDEN, J. A., A. HENINS, J. G. MARZOLF, W. C. SAUDER, AND J. S. THOMSEN, 1964, *Phys. Rev.* **135**, A899.

BECKER, P., K. DORENWENDT, G. EBELERG, R. LAUER, W. LUCAS, R. PROBST, H.-J. RADEMACKER, G. REIM, P. SEYFRIED, AND H. SIEGERT, 1981, *Phys. Rev. Lett.* **46**, 1540.

BECKER, P., P. SEYFRIED, AND H. SIEGERT, 1982, *Z. Phys. B* **48**, 17.

BERRY, H. G., K. T. CHENG, W. R. JOHNSON, AND Y.-K. KIM, editors, 1981, Proceedings of the Workshop on Foundations of the Relativistic Theory of Atomic Structure, Argonne National Laboratory report No. ANL-80-126.

BEYER, H., R. D. DESLATTES, F. FOLKMANN, AND R. E. LAVILLA, 1985, *J. Phys. B*, **18**, 207.

BLOKHIN, M. A., 1965, *Methods of X-Ray Spectroscopic Research* (Pergamon, Oxford).

BONSE, U., AND M. HART, 1965, *Appl. Phys. Lett.* **6**, 155.

BORCHERT, G. L., 1976, *Z. Naturforsch.* **31a**, 102.

BORCHERT, G. L., W. SCHECK, AND O. W. B. SCHULT, 1975, *Nucl. Instrum. Methods* **124**, 107.

BORCHERT, G. L., P. G. HANSEN, B. JONSON, H. L. RAVN, AND J. P. DESCLAUX, 1980, in *Atomic Masses and Fundamental Constants 6*, edited by J. A. Nolen, Jr., and W. Benenson (Plenum Press, New York), p. 189.

BORCHERT, G. L., P. G. HANSEN, B. JONSON, I. LINDGREN, H. L. RAVN, O. W. B. SCHULT, AND P. TIDEMAND-PETERSON, 1978, *Phys. Lett.* **65A**, 297.

BRAGG, W. H., 1913, *Proc. R. Soc. London Ser. A* **88**, 428.

BRAGG, W. H., 1914a, *Proc. R. Soc. London Ser. A* **89**, 246.

BRAGG, W. H., 1914b, *Proc. R. Soc. London Ser. A* **89**, 430.

BRAGG, W. L., 1947, *J. Sci. Instrum.* **24**, 27.

BREINIG, M., M. H. CHEN, G. E. ICE, F. PARENTE, B. CRASEMANN, AND G. S. BROWN, 1980, *Phys. Rev. A* **22**, 520.

BRIAND, J. P., M. TAVERNIER, P. INDELICATO, R. MARRUS, AND H. GOULD, 1983, *Phys. Rev. Lett.* **50**, 832.

BROGREN, G., 1949, *Nova Acta Regiae Soc. Sci. Ups.* **14**, No. 4.

CALLCOTT, T. A., E. T. ARAKAWA, AND D. L. EDERER, 1978, *Jpn. J. Appl. Phys. Suppl.* **17-2**, 149.

CARLSON, T. A., C. W. NESTOR, JR., F. G. MALIK, AND T. C. TUCKER, 1969, *Nucl. Phys. A* **135**, 57.

CHEN, M. H., 1982, preliminary calculations.

CHEN, M. H., 1983, private communication.

CHEN, M. H., B. CRASEMANN, M. AOYAGI, K.-N. HUANG, AND H. MARK, 1981a, *At. Data Nucl. Data Tables* **26**, 561.

CHEN, M. H., B. CRASEMANN, AND H. MARK, 1981b, *Phys. Rev. A* **24**, 1158.

CHENG, K. T., AND W. R. JOHNSON, 1976, *Phys. Rev. A* **14**, 1943.

COHEN, E. R., AND J. W. M. DUMOND, 1964, in *Proc. 2nd Int. Conf. Nuclidic Masses 1963*, edited by W. H. Johnson, Jr. (Springer, New York), p. 152.

COHEN, E. R., AND J. W. M. DUMOND, 1965, *Rev. Mod. Phys.* **37**, 537.

COHEN, E. R., AND B. N. TAYLOR, 1973, *J. Phys. Chem. Ref. Data* **2**, 663.

COMPTON, A. H., AND R. L. DOAN, 1925, *Proc. Natl. Acad. Sci. U.S.* **11**, 598.

COOPER, A., 1965, *Acta Crystallogr.* **18**, 1078.

COWAN, R. D., AND J. B. MANN, 1975, *J. Comput. Phys.* **16**, 160.

CURNUTTE, B., C. L. COCKE, AND R. D. DUBOIS, 1982, *Nucl. Instrum. Methods* **202**, 119.

DESCLAUX, J. P., 1975, *Comput. Phys. Commun.* **9**, 31.

DESCLAUX, J. P., 1979, *Phys. Scr.* **21**, 436.

DESCLAUX, J. P., 1981, State-of-the-Art for Multiconfiguration Dirac–Fock Calculations, in Proceedings of the Workshop on Foundations of the Relativistic Theory of Atomic Structure, edited by H. G. Berry, K. T. Cheng, W. R. Johnson, and Y.-K. Kim, Argonne National Laboratory report No. ANL-80-126, p. 78.

DESLATTES, R. D., E. G. KESSLER, W. C. SAUDER, AND A. HENINS, 1980, *Ann. Phys. (N.Y.)* **129**, 378.

DESLATTES, R. D., R. E. LAVILLA, P. L. COWAN, AND A. HENINS, 1983, *Phys. Rev. A* **27**, 923.

DEUTSCH, M., AND M. HART, 1982, *Phys. Rev. B* **26**, 5558.

DURHAM, P. J., J. B. PENDRY, AND C. H. HODGES, 1982, *Comput. Phys. Commun.* **25**, 193.

EDLÉN, B., 1963, *Rept. Prog. Phys.* **XXVI**, 181.

EDLÉN, B., AND L. Å. SVENSSON, 1965, *Ark. Fys.* **28**, 427.

GOULD, H., AND R. MARRUS, 1981, *Phys. Rev. Lett.* **41**, 1457.

GRANT, I. P., B. J. MCKENZIE, P. H. NORRINGTON, D. F. MAYERS, AND N. C. PYPER, 1980, *Comput. Phys. Commun.* **21**, 207.

HANSEN, P. G., B. JONSEN, G. L. BORCHERT, AND O. W. B. SCHULT, 1985, Chapter 6 of this volume.

HENINS, A., 1971, in *Precision Measurements and Fundamental Constants*, edited by D. N. Langenberg and B. N. Taylor, National Bureau of Standards Special Publication No. 343 (GPO, Washington, D.C.), p. 255.

HENINS, I., AND J. A. BEARDEN, 1964, *Phys. Rev.* **135**, A890.

HUANG, K.-N., M. AOYAGI, M. H. CHEN, B. CRASEMANN, AND H. MARK, 1976, *At. Data Nucl. Data Tables* **18**, 243.

HUNGERFORD, P., H. H. SCHMIDT, R. BRISSOT, G. BARREAU, AND H. G. BÖRNER, 1981, *Z. Naturforsch. Teil A* **36**, 919.

JEN, J. S., AND T. D. THOMAS, 1976, *Phys. Rev. B* **13**, 5284.

KESSLER, E. G., JR., R. D. DESLATTES, A. HENINS, AND W. C. SAUDER, 1978, *Phys. Rev. Lett.* **40**, 171.

KESSLER, E. G., JR., R. D. DESLATTES, AND A. HENNINS, 1979, *Phys. Rev. A* **19**, 215.

KESSLER, E. G., JR., R. D. DESLATTES, D. GIRARD, W. SCHWITZ, L. JACOBS, AND O. RENNER, 1982, *Phys. Rev. A* **26**, 2696.

KIM, Y.-K., 1983, private communication.

KIRKPATRICK, H. A., J. W. M. DUMOND, AND E. R. COHEN, 1967, Remeasurement of the Conversion Constant, *A*, in *Proceedings of the Third International Conference on Atomic Masses*, edited by R. C. Baker (University of Manitoba Press, Winnipeg, 1965), p. 347.

KNOWLES, J. W., 1962, *Can. J. Phys.* **40**, 237.

KNOWLES, J. W., 1979, *Nucl. Instrum. Methods* **162**, 677.

KOCH, H. R., H. G. BÖRNER, J. A. PINSTON, W. F. DAVIDSON, J. FOUDOU, R. ROUSSILLE, AND O. W. B. SCHULT, 1980, *Nucl. Instrum. Methods* **175**, 401.

KUGEL, H. W., AND D. E. MURNICK, 1977, *Rep. Prog. Phys.* **40**, 297.

KUGEL, H. W., M. LEVENTHAL, D. E. MURNICK, C. K. N. PATEL, AND O. R. WOOD, II., 1975, *Phys. Rev. Lett.* **35**, 647.

LARKINS, F. P., 1982, Electron Correlations Manifested by Satellites in Atomic Spectra, in *X-Ray and Atomic Inner-Shell Physics—1982*, AIP Conference Proceedings No. 94, edited by B. Crasemann (American Institute of Physics, New York), p. 530.

LAVILLA, R. E., 1983, private communication.

LINDGREN, I., 1965, *Phys. Lett.* **19**, 382.

LINDGREN, I., 1966, *Ark. Fys.* **31**, 59.

LOTZ, W., 1968, *J. Opt. Soc. Am.* **58**, 915.

LOTZ, W., 1970, *J. Opt. Soc. Am.* **60**, 206.

LUNDEEN, S. R., AND F. M. PIPKIN, 1981, *Phys. Rev. Lett.* **46**, 232.

MALZFELDT, W., W. NIEMANN, P. RABE, AND N. SCHWENTNER, 1983, in *EXAFS and Near Edge Structure*, edited by A. Bianconi, L. Incoccia, and S. Stipcich (Springer-Verlag, New York), p. 203.

MANN, J. B., AND W. R. JOHNSON, 1971, *Phys. Rev. A* **4**, 41.

MARTIN, W. C., 1981, Energy Levels of Highly Ionized Atoms, in Proceedings of the Workshop on Foundations of the Relativistic Theory of Atomic Structure, edited by H. G. Berry, K. T. Cheng, W. R. Johnson, and Y.-K. Kim, Argonne National Laboratory Report No. ANL-80-126, p. 195.

MCKENZIE, B. J., I. P. GRANT, AND P. H. NORRINGTON, 1980, *Comput. Phys. Commun.* **21**, 233.

MEHLHORN, W., B. BREUKMANN, AND D. HAUSAMANN, 1977, *Phys. Scr.* **16**, 177.

MITTLEMAN, M. H., 1981, *Phys. Rev. A* **24**, 1167.

NAKAMURA, M., M. SASANUMA, S. SATO, M. WATANABE, H. YAMASHITA, Y. IGUCHI, A. EJIRI, S. NAKAI, S. YAMAGUCHI, T. SAGAWA, Y. NAKAI, AND T. OSHIO, 1968, *Phys. Rev. Lett.* **21**, 1303.

PARRATT, L. G., 1939, *Phys. Rev.* **56**, 295.

PELLEGRIN, P., Y. EL MOSRI, L. PALFFY, AND R. PRIEELS, 1982, *Phys. Rev. Lett.* **49**, 1762.

PILLER, O., W. BEER, AND J. KERN, 1973, *Nucl. Instrum. Methods* **107**, 61.

PORTER, F. T., AND M. S. FREEDMAN, 1978, *J. Phys. Chem. Ref. Data* **7**, 1267.

POWELL, C. J., N. E. ERIKSON, AND T. JACH, 1982, *J. Vac. Sci. Technol.* **20**, 625.

ROSÉN, A., AND I. LINDGREN, 1968, *Phys. Rev.* **176**, 114.

SANDSTRÖM, A. E., 1952, *Ark. Fys.* **4**, 517.

SAUDER, W. C., J. R. HUDDLE, J. D. WILSON, AND R. E. LAVILLA, 1977, *Phys. Lett. A* **63**, 313.

SEPPI, E. J., H. HENRIKSON, F. BOEHM, AND J. W. M. DUMOND, 1962, *Nucl. Instrum. Methods* **16**, 17.

SEVIER, K. D., 1979, *At. Data Nucl. Data Tables* **24**, 323.

SHIRLEY, D. A., R. L. MARTIN, S. P. KOWALCZYK, F. R. MCFEELY, AND L. LEY, 1977, *Phys. Rev. B* **15**, 544.

SIEGBAHN, M., 1919, *Ark. Mat. Astron. Fys.* **14**, 1.

SIEGBAHN, M., 1925, *The Spectroscopy of X Rays* (Oxford U. P., London), p. 84.

Siegbahn, M., 1943, *Nature* **151**, 502.

Siegbahn, K., C. Nordling, G. Johansson, J. Hedman, P. F. Hédén, K. Hamrin, U. Gelius, T. Bergmark, L. O. Werme, R. Manne, and Y. Baer, 1969, *ESCA Applied to Free Molecules* (North-Holland, Amsterdam).

Skinner, H. W. B., 1940, *Phil. Trans. R. Soc. London Ser. A* **239**, 95.

Smirnov, A. I., V. A. Shaburov, V. L. Alexeyev, D. M. Kaminker, A. S. Rylnikov, and O. I. Sumbaev, 1968, *Nucl. Instrum. Methods* **60**, 103.

Soules, J. A., and C. H. Shaw, 1959, *Phys. Rev.* **113**, 470.

Stockli, M. P., P. Richard, R. Mann, R. D. Deslattes, P. L. Cowan, R. E. LaVilla, K. Jones, M. Meron, and K.-H. Schartner, 1983, *Bull. Am. Phys. Soc.* **28**, 781.

Sucher, J., 1980, *Phys. Rev. A* **22**, 348.

Thomsen, J. S., and A. F. Burr, 1968, *Am. J. Phys.* **36**, 803.

Tyrén, F., 1940, *Nova Acta Reg. Soc. Sci. Upsal. (IV)* **12**, No. 1.

Ulmer, K., 1982, Isochromat Spectroscopy, in *Advances in X-Ray Spectroscopy*, edited by C. Bonnelle and C. Mandé (Pergamon, Oxford), p. 451.

Watanabe, T., 1965, *Phys. Rev.* **139**, A1747.

Weiman, C., and T. W. Hänsch, 1980, *Phys. Rev. A* **22**, 192.

Weiss, A., 1981, Review of Atomic Structure Calculations, in Proceedings of the Workshop on Foundations of the Relativistic Theory of Atomic Structure, edited by H. G. Berry, K. T. Cheng, W. R. Johnson, and Y.-K. Kim, Argonne National Laboratory Report No. ANL-80-126, p. 272.

Wood, E. A., 1947, *Phys. Rev.* **72**, 436.

Wood, O. R., II, C. K. N. Patel, D. E. Murnick, E. T. Nelson, M. Leventhal, H. W. Kugel, and Y. Niv, 1982, *Phys. Rev. Lett.* **48**, 398.

CHAPTER 6

Mechanisms for Energy Shifts of Atomic *K* X Rays

P. G. Hansen, B. Jonson, G. L. Borchert, and O. W. B. Schult

1. INTRODUCTION

The energies of the *K*-series X rays (Fig. 1) are for a given element almost independent of the source and the mode of excitation. Nevertheless, by means of suitably designed spectrometers, to be discussed in Sec. 2, it is possible to observe minute energy shifts. The main part of the present Chapter is devoted to experiments in which such shifts are observed, and to their interpretation.

There is a double aim behind the study of X-ray energy shifts. First of all the experiments can be designed to exclude certain mechanisms and to be sensitive to others. In this way it is possible to test our understanding of the processes involved; several contributions that had not been considered in earlier work have in the last few years been recognized. A second reason for investigating shifts is that they provide new probes for studying nuclei, atoms, and solids.

We give a short account of the mechanisms to be discussed. The first two have been known for a long time and have been reviewed in detail elsewhere; we nevertheless mention them briefly, both for the sake of completeness and because they give additional contributions when some of the other mechanisms are studied.

1. *Isotope shifts.* They originate from the effect of finite nuclear size on the 1*s* binding energy (Sec. 3).
2. *Chemical shifts.* They originate from changes in the atomic screening.

P. G. Hansen • Institute of Physics, University of Aarhus, Aarhus, Denmark. B. Jonson • CERN-ISOLDE, CERN, Geneva, Switzerland. G. L. Borchert and O. W. B. Schult • Institut für Kernphysik, Kernforschungsanlage, Jülich, Federal Republic of Germany.

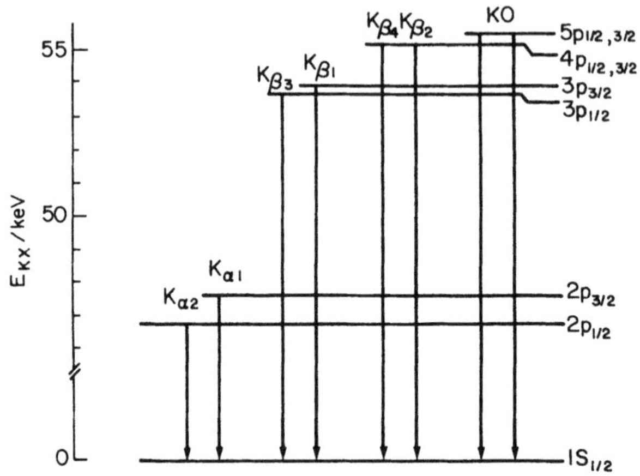

FIGURE 1. Origin and nomenclature for the K-series X rays. The energy scale is for the case of holmium ($Z=67$).

These changes are caused by outer electrons participating in the formation of chemical bonds (Sec. 4).

3. *Hyperfine shifts.* For $1s^{-1}$ excitations caused by radioactive decays, intensity rules related to the total atomic angular momentum F lead to shifts that may reach several eV. These shifts furnish a method for measuring nuclear magnetic moments (Sec. 5) and possibly also for determining multipole mixing of beta transitions.
4. *Dynamic shifts.* Outer-shell excitations associated with the excitation mechanism modify temporarily the atomic screening and give rise to small shifts (Sec. 6).
5. *Atomic-structure shifts.* Nuclear beta decay transforms an element into a neighboring one but preserves temporarily the electronic structure of the mother atom. Structural differences influence (via the screening) the X-ray energies in a characteristic way that is not yet fully understood (Sec. 7).
6. Other contributions (Sec. 8).

Finally, we remind the reader that there has been remarkable progress in the understanding of absolute X-ray energies, and that contributions at the eV level can no longer be disregarded entirely. Improved experimental methods (Borchert, 1976; Borchert *et al.*, 1980c; Deslattes and Kessler, Chap. 5 of this volume) and more refined theoretical calculations (Chen *et al.*, 1981) now give agreement to within 1–2 eV for the K X-ray energies of heavier elements corresponding to a precision of $(1–2)\times 10^{-5}$.

2. THE EXPERIMENTAL METHOD

The energy shifts of *K* X-ray lines dealt with here are very small in comparison with their natural widths Γ_{nat} (Nelson *et al.*, 1969, 1970; Krause and Oliver, 1979). Diffraction spectrometers which have the highest resolving power are therefore used for X-ray shift measurements. Improvements in crystal diffraction spectroscopy during the last two decades have been described by Knowles (1979). Since then, progress has continued, and very high resolving powers have been obtained not only with a double-flat crystal spectrometer (Deslattes and Kessler, 1979) but also with curved-crystal diffractometers for which either a favorable lattice plane has been chosen (Koch *et al.*, 1980) or the crystal lamina were cut out of the crystal block in the optimum way so that bending of the lamina did not distort the lattice planes (Sumbaev, 1957, 1959; Sumbaev and Lapin, 1980).

The use of curved crystals in focusing diffraction spectrometers (Cauchois, 1932a) results in an increase of the counting rate by about three orders of magnitude. For the spectroscopy of radiation from very small sources it is advisable to employ the DuMond geometry (DuMond, 1947), which has been successfully utilized by various groups (Boehm, 1975; Brockmeier *et al.*, 1965; Van Eijk and Schütte, 1970; Van Eijk and Vissher, 1970; Borchert *et al.*, 1981b) for isotope-shift measurements and for studies of other mechanisms for energy shifts in atomic *K* X-ray spectra (Boehm, 1975; Borchert *et al.*, 1977, 1978a, b; Hansen *et al.*, 1981; Borchert *et al.*, 1980b). Figure 2 shows the X-ray beam geometry of the DuMond-type spectrometer (Borchert *et al.*, 1980a) installed for X-ray energy-shift studies at CERN. In cases where larger sources are available, the Cauchois geometry (Cauchois, 1932a, 1932b) can be used as well. In fact, the X-ray energy-shift measurements by the groups of Boehm (Chesler and Boehm, 1968; Boehm, 1975; Bhattacherjee *et al.*, 1969; Lee and Boehm, 1973; Boehm and Lee, 1974) and Sumbaev (Sumbaev and Mezentsev, 1965a, b; Sumbaev *et al.*, 1969; Smirnov *et al.*, 1969; Ryl'nikov *et al.*, 1972; Sumbaev, 1976, 1978; Molkanov *et al.*, 1976; Egorov *et al.*, 1978) have been carried out with Cauchois spectrometers, where the X-ray beam direction is reversed with respect to the one depicted in Fig. 2. Descriptions of such spectrometers can be found in various publications (Seppi *et al.*, 1962; Boehm, 1975; Chesler and Boehm, 1968; Henrikson and Boehm, 1974; Sumbaev, 1978). For a given curved crystal and optimized geometries, the counting rate in the DuMond-type setup equals that of the Cauchois geometry. The latter even permits a spectrograph mode of operation if a suitable position-sensitive detector is available, so that a segment of the spectrum can be recorded simultaneously. The Cauchois spectrometer allows the use of a very narrow detector which minimizes background, but it needs larger sources which must be prepared so that the required homogeneity is obtained. These sour-

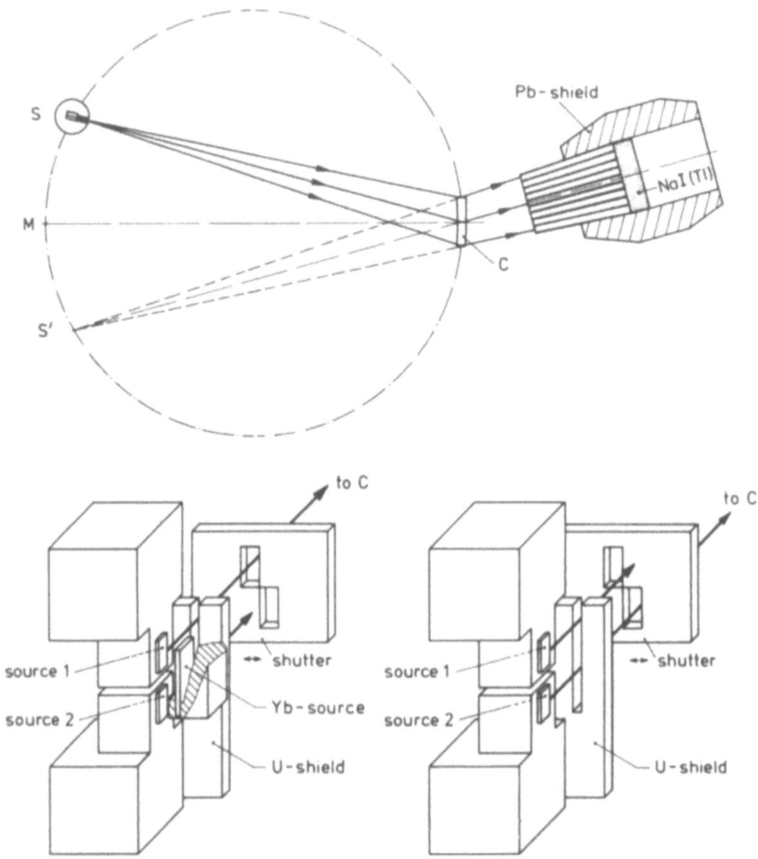

FIGURE 2. Measurements of very small X-ray energy shifts are carried out with high-resolution curved-crystal spectrometers. The upper part of the figure shows the principle of the DuMond geometry (DuMond, 1947) with a narrow source (S) and an extended detector (NaI). Reversed beam direction is characteristic of the Cauchois geometry (Cauchois, 1932a) with an extended source at the NaI position and a narrow detector at S. The lower part of the figure shows more details of the sources used with the DuMond-type spectrometer at CERN. At the left, the photoionization source (^{169}Yb) arrangement is shown as it was employed during isotope-shift measurements (Borchert *et al.*, 1981b). At the right side is the setup as used during the measurements of hyperfine shifts (Borchert *et al.*, 1977). For measurements of other shifts, only one X-ray source was photoionized by a smaller Yb source, the other source emitting X rays excited through its own decay.

ces are 2–3 orders of magnitude more active than the very narrow sources of a DuMond spectrometer. The DuMond geometry is preferable for measurements with rare isotopes which are available only in small quantities, and in cases where very small sources are produced, e.g., through the collection of radionuclei (Borchert *et al.*, 1977) obtained at isotope separators (Ravn, 1979). The larger detector in a DuMond spectrometer gives a higher background and requires better shielding.

In order to obtain data with the highest statistical quality for a given measuring time, the experiment should be carried out under optimized conditions (Schult, 1965). This implies that the source width in a DuMond spectrometer or the detector-slit width in a Cauchois spectrograph are in proper relation to the geometrical width of the X-ray line at the position S of Fig. 2. The reflection power of the crystal can be maximized if the crystal plate is cut out of the block in the right way so that the lattice planes are curved when the crystal is bent (Sumbaev and Smirnov, 1963). Alternatively, the reflectivity can be increased by application of electric fields (Dousse and Kern, 1976). Under optimum conditions the combination of the width of the source (slit) at S of Fig. 2 and of the defocussing width of the curved crystal, together with the width of the crystal diffraction pattern for the X-ray energy, yields a total width for a γ-ray reflection smaller than that for the X-ray line reflection, because of the natural linewidth Γ_{nat} of the X ray (Fig. 3).

The X-ray shifts can be as small as 10^{-3} of the natural linewidth, and their measurement requires great care. Instead of measuring absolute X-ray energies E_x with an accuracy of $10^{-7}E_x$, the X-ray energy shift can be obtained much more easily through a relative measurement, namely, of the difference between the energies of one X-ray line emitted from two or more samples (Lee and Boehm, 1973; Sumbaev, 1978; Borchert *et al.*, 1980a).

For this purpose a measurement strategy is adopted by which systematic errors are efficiently eliminated so that the uncertainty in the shift measurement is essentially due to the statistical error. For shift measurements with Cauchois spectrometers, samples containing 0.1–2 g of material were pressed into pills and fluorescence-excited by means of an X-ray tube or a ^{169}Yb source of 200 Ci (Chesler and Boehm, 1968; Boehm, 1975; Sumbaev, 1978). The samples were carefully controlled for homogeneity and chemical purity (Boehm, 1975), and part of the aberrational shift from the remaining inhomogeneous source distribution was compensated (Sumbaev and Mezentsev, 1965b; Boehm, 1975) by spinning the sample about an axis through its center. The samples were brought into the source position alternatively with a suitable source-changing device (Boehm, 1968; Sumbaev, 1978). Many source changes were carried out with the detector slit fixed at the measuring angle θ, and θ was subsequently changed to the next measuring position. The DuMond geometry also per-

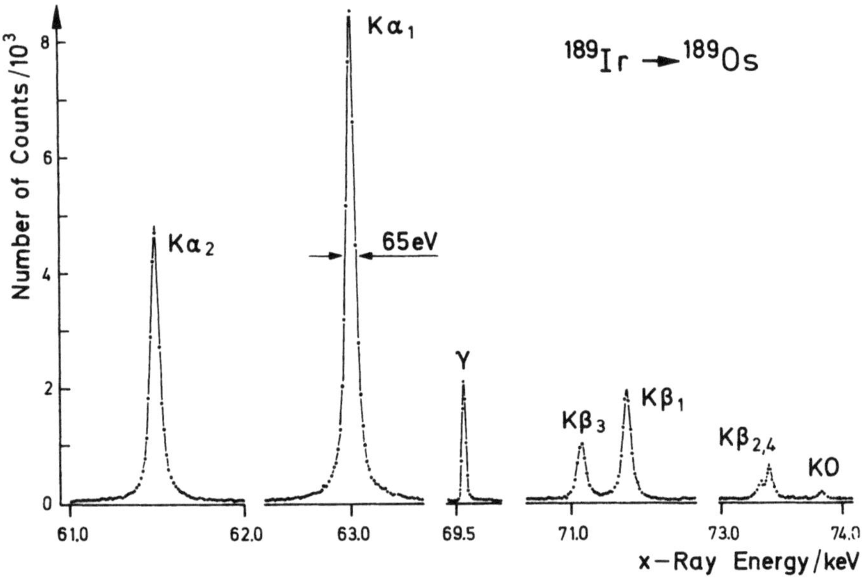

FIGURE 3. Typical K X-ray spectrum as emitted during the electron-capture beta decay of ^{189}Ir. The measured width of the $K\alpha_1$ line is only 18% larger than the natural linewidth (Nelson *et al.*, 1969, 1970). This increase is due to the geometrical width of the source and the defocusing width of the curved crystal. These contributions alone yield the linewidth of the 69.5-keV γ transition. Recent measurements also include the determination of the shift of the KO "line" which constitutes electron transitions from shells with main quantum number $n \geqslant 5$ to the K shell ($n = 1$).

mits the exchange of small sources (Brockmeier *et al.*, 1965) or the installation of a defining slit (Van Eijk and Schütte, 1970) very close to a slightly larger source. The geometrical reproducibility in this case has to be $\sim 10\,\mu$m, which is not easily assured by a mechanical source-changing mechanism.

Another solution, suggested by the smallness of the sources, is to mount the two samples slightly above and below the ideal source point S in the symmetry plane of the spectrometer, as shown in Fig. 2. Aberrations increase quadratically with the distance from S, hence they vanish for geometrically identical sources at equal distances above and below the symmetry plane. Small inhomogeneities lead to aberrations which are negligible in comparison with statistical errors for sources at close distance and, in addition, they can be essentially eliminated by an additional measurement with the sources interchanged (Fig. 2). The symmetry plane is determined experimentally by verifying that identical sources give a null result. Typical sources at the CERN-DuMond spectrometer (Borchert *et al.*, 1980a) were $\lesssim 5$ mm high, $\lesssim 2$ mm in the direction of the X-ray beam, and $\lesssim 0.15$ mm in the direction of the dispersion of the spectrometer. Photoionization (PI) was induced with a ^{169}Yb source of up to 50 Ci. For each measuring point a shutter was moved back and forth many times in the direction of the dispersion; this transmits the beam from only one source at a time to the spectrometer for periods which can be chosen so that the statistical accuracy in the shift measurement is maximal. The rapid cycling at the same time eliminates the effects of drift of the apparatus. The two sources were adjusted to yield well-overlapping X-ray reflections, as shown in Fig. 4. The shift arising from the (inevitable) residual displacement is corrected through the measurement of the reflections at positive (+) and negative (−) Bragg angles, θ. The point is that a mechanical displacement shifts both Bragg reflections by $\delta\theta_{\rm displ}$ in the same direction so that their distance remains constant,

$$(\theta^+ + \delta\theta_{\rm displ}) - (\theta^- + \delta\theta_{\rm displ}) = \theta^+ - \theta^-,$$

while their center of gravity is shifted by $\delta\theta_{\rm displ}$. On the other hand, an energy change shifts the Bragg reflections by $\delta\theta_{\rm energy}$ in opposite directions

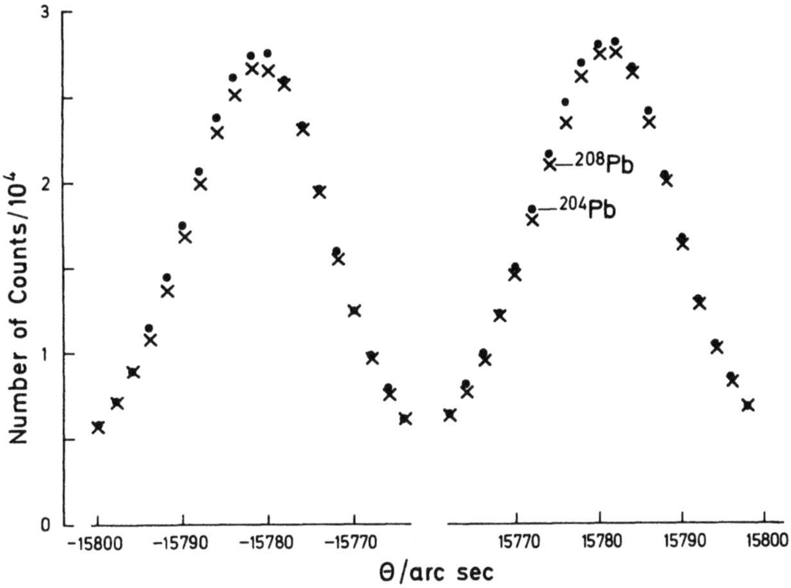

FIGURE 4. Reflections of the $K\alpha_1$ lines of ^{204}Pb and ^{208}Pb recorded at positive and negative Bragg angles with the DuMond-type diffraction spectrometer at CERN. The measuring time per point was 4 min.

so that their center of gravity remains fixed while their distance changes from

$$\theta^{+}-\theta^{-}$$

to

$$(\theta^{+}+\delta\theta_{\text{energy}})-(\theta^{-}-\delta\theta_{\text{energy}})=\theta^{+}-\theta^{-}+2\delta\theta_{\text{energy}}.$$

The measurement of the X-ray energy shift then consists of a series of scans (each containing ~20 measuring points) across the pair of X-ray reflections at $\theta^{+}, \theta^{-}, \theta^{+}, \theta^{-}$,... until the desired statistical accuracy is obtained. In a similar way, shift measurements are performed with a Cauchois spectrometer, where many source changes are carried out at each measuring point.

Because of the smallness of the shifts compared with the experimental linewidths, the method of data reduction must be such that it does not introduce artificial shifts. Various methods have been applied (Boehm, 1975; Sumbaev, 1976, 1978; Hansen, 1978; Borchert *et al.*, 1980a). The best way to be sure of avoiding artificial shifts is to check that different approaches give consistent results (Borchert *et al.*, 1982).

3. THE ISOTOPE SHIFT

The shift in the X-ray energy when δA neutrons are added to an isotope with mass number A is referred to as the isotope shift. For heavier nuclei the energy change is approximately proportional to the variation in the mean-square charge radius $\delta\langle r^2\rangle_{A,A+\delta A}$. The largest shifts are observed for the K X rays, and arise in the $1s_{1/2}$ level while the contributions from the p levels are negligible (Boehm, 1975; Lee and Boehm, 1973). For this reason measurements of isotope shifts concentrate on $K\alpha_1$, the most intense of the K X rays.

K X-ray isotope-. hift measurements (Boehm, 1975) can be carried out in the absence of other shifts. Chemical shifts (Sumbaev, 1976) are avoided if the isotopically enriched samples exist in identical and stable forms with the same structure (Chesler and Boehm, 1968). Shake-off (Borchert *et al.*, 1978a) and other mechanisms (Borchert *et al.*, 1978b) can be eliminated if all samples are excited through photoionization. The full shift is the sum of the Coulomb shift, the mass shift, and the nuclear polarization shift,

$$\delta E(1s_{1/2})=\delta E(K\alpha_1)=\delta E_{\text{Coul}}+\delta E_{\text{mass}}+\delta E_{\text{polar}}.$$

The total mass shift [the sum of the normal and specific mass shifts (Boehm,

1975, 1968)] can be approximated (Chesler and Boehm, 1968; Sumbaev *et al.*, 1969; Boehm and Lee, 1974) by

$$\delta E_{\text{mass},A \to A+\delta A} \approx 2/3(\delta A/1836A^2)\, E(K\alpha_1).$$

The mass shift for heavy elements is very small in comparison with the measurement errors: for Pb, $\delta E_{\text{mass},A \to A+1} \approx 0.7$ meV. The influence of nuclear polarization (Breit, 1957) by the electron is also small (Reiner and Wilets, 1962; Hoffmann *et al.*, 1984).

As discussed in greater detail by Seltzer (1969), it is useful to introduce the rapidly converging series

$$\delta E_{\text{Coul}} = \sum C_n \delta\langle r^{2n} \rangle = C_1 \lambda,$$

where the parameter λ thus defined usually deviates by less than 10% from $\delta\langle r^2 \rangle$.

The advantage of the electronic X-ray isotope shift measurement is its insensitivity to nuclear polarization, in contrast to muonic X-ray isotope-shift measurements in which the nuclear polarization shift (Rinker and Speth, 1978) is not only a large fraction of the measured shift (Kessler *et al.*, 1975) but also is difficult to explain (Yamazaki *et al.*, 1979). Furthermore, the total mass shift is very small and the electronic part of the calculation (Seltzer, 1969) can be carried out with good accuracy. Optical isotope-shift measurements (Manning *et al.*, 1950; Brix *et al.*, 1952; Steudel, 1952; Blaise, 1958; Heilig and Steudel, 1974; Moscatelli *et al.*, 1982) suffer from our insufficient knowledge of the specific mass shift, which is essential here. Furthermore, the electronic calculations of the initial and final states are, in general, not precise enough to give an accurate value for the λ parameter. For these reasons, the K X-ray isotope shifts serve for the normalization of optical data (Van Eijk and Vissher, 1970; Van Eijk and Schütte, 1970; Merton, 1919; Korostyleva, 1957; Brix and Kopfermann, 1958; Stacey, 1966; Golovin and Striganov, 1967; Boehm, 1975; Lee and Boehm, 1971; Heilig and Steudel, 1974; Lee *et al.*, 1978; Borchert *et al.*, 1982). A plot of the shifts of the $K\alpha_1$ line versus the shifts of an optical line (e.g., Fig. 5), yields for negligible mass shift (Stacey, 1966) the electronic factor C_1 for the optical transition. Combining this parameter with optical-shift data for unstable isotopes (Brix *et al.*, 1952; Thompson *et al.*, 1982; Moscatelli *et al.*, 1982), it is possible to determine the nuclear radius also for unstable isotopes, as shown in the example in Fig. 6. Isotope shifts of X rays have been measured in many cases (Brockmeier *et al.*, 1965; Boehm, 1975; Bhattacherjee *et al.*, 1968; Boehm, 1968; Chesler and Boehm, 1968; Bhattacherjee *et al.*, 1969; Lee and Boehm, 1971, 1973; Lee *et al.*, 1978; Boehm and Lee, 1974; Sumbaev and Mezentsev, 1965a; Sumbaev *et al.*, 1969; Ryl'nikov *et al.*, 1972). The K X-

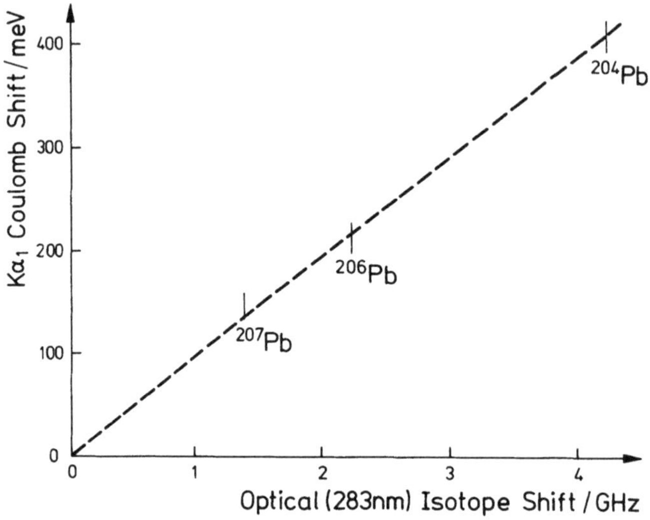

FIGURE 5. Comparison of the Coulomb energy shifts of the $K\alpha_1$ X-ray line with the optical isotope shifts of the 283.3-nm line (Thompson *et al.*, 1982) measured for ^{207}Pb, ^{206}Pb, and ^{204}Pb, each with respect to ^{208}Pb. The Coulomb energy shift allows for the small mass shift of the $K\alpha_1$ line. The mass shift of the optical transition is unknown but is assumed to be small in heavy elements (Stacey, 1966). The figure serves to test this hypothesis: for a negligible mass shift of the 283-nm line theoretical isotope shifts should be proportional to the $K\alpha_1$Coulomb shifts. While the uncertainties of the optical data correspond to the widths of the data bars, their lengths equal twice the errors of the X-ray energy shifts. Unfortunately, the X-ray energy shifts cannot be measured much more accurately. Therefore, the data are only consistent with negligible mass shift in the optical data.

ray data have contributed in an important way to the understanding of nuclear radii (Boehm, 1968, 1975; Lee and Boehm, 1971; Bhattacherjee *et al.*, 1968; Uher and Sorensen, 1966). Space does not permit a detailed discussion of this subject here; we shall simply give some recent references and point to the important interplay of the X-ray data with those from optical spectra, electron scattering, and muonic X rays. Recently, X-ray shift data (Borchert *et al.*, 1981a) of the stable lead isotopes have been compared with the droplet model (Myers, 1969, 1977; Myers and Swiatecki, 1974) and with $\delta\langle r^2\rangle$ values calculated in the framework of the shell model (Speth *et al.*, 1974; Borchert *et al.*, 1982). Although the general agreement is rather good, the odd–even staggering is not explained. McGrory and Brown (1982) have discussed in detail the shell-model calculations of nuclear charge radii. Nuclear charge distributions have been dealt with elsewhere (Malaguti *et al.*, 1982).

The X-ray shift data have also been compared with the results from

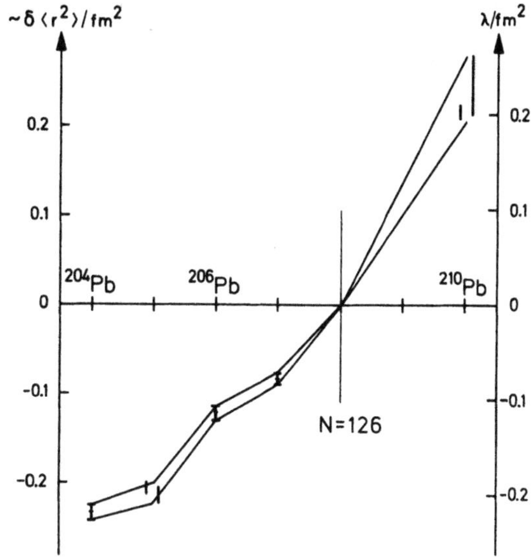

FIGURE 6. The measurements of X-ray isotope shifts or optical isotope shifts yield the nuclear parameter λ (Seltzer, 1969), which is related to the shift δ of the mean square charge radius $\langle r^2 \rangle$. The figure shows these quantities for a few lead isotopes relative to ^{208}Pb. The data for ^{204}Pb, ^{206}Pb, and ^{207}Pb are from the $K\alpha_1$ X-ray shifts (Lee and Boehm, 1973; Borchert *et al.*, 1982) and those for ^{205}Pb and ^{210}Pb are from optical isotope shift measurements of the 283-nm line (Thompson *et al.*, 1982) and the 405-nm line (Brix *et al.*, 1952), respectively. Because of the large uncertainties in the calculation of the electron wave functions at the nuclear site, the optical shifts were normalized with the use of the X-ray shift data. The normalization was accomplished once with the assumption of zero optical mass shift (left data bar for ^{205}Pb and ^{210}Pb) and once for a mass shift as obtained from the fit to the X-ray data (right data bars).

muonic isotope-shift measurements (Boehm, 1975; Lee and Boehm, 1973; Lee *et al.*, 1978; Borchert *et al.*, 1982). For samarium isotopes, the comparison of muonic, optical, and electronic X-ray shifts shows good agreement (Powers *et al.*, 1979). For the mercury isotopes the optical data (Bonn *et al.*, 1976) indicate larger shifts than those given by the X-ray data (Lee *et al.*, 1978). This could be due to insufficient accuracy of the optical screening calculation. The muonic data (Hahn *et al.*, 1979) are consistent with the X-ray shifts except for ^{201}Hg, where the nuclear polarization correction of the muonic data seems to be the major source of error.

For the stable isotopes of lead, the shift between ^{208}Pb and ^{204}Pb, obtained from the muonic data alone, clearly disagrees with the shift derived from combining muonic data (Kessler *et al.*, 1975) with electron scattering data (Emrich *et al.*, 1981; Rychel, 1980; Fricke, private communication, 1981; Borchert *et al.*, 1982) when the nuclear polarization correction was applied as calculated by Skardhamar (1970). The combined $(\mu + e)$ data are consistent with but systematically smaller than the electronic X-ray shift data.

Although most of the isotope shifts can be measured with an accuracy which is sufficient for the comparison with theory, at present all methods suffer from different limitations. The electronic K X-ray shifts, being very small compared with the linewidths, cannot be measured with the desired accuracy. The muonic X-ray shifts are large and can be measured accurately, but the nuclear polarization adds a shift which cannot be calculated reliably. Optical isotope-shift measurements (Heilig and Steudel, 1974), which have been leading in the field for many decades (Brix, 1977, 1981, 1982) and which have opened up the possibility of measuring nuclear charge radii even of radioisotopes (Brix *et al.*, 1952), have benefited enormously from the introduction of laser techniques (Otten, 1981; Rebel, 1980; Rebel and Schatz, 1982; Timmermann, 1980; Lindgren *et al.*, 1981). Thanks to the combination of muonic, electronic X-ray, and optical data it is today possible to obtain a global picture of nuclear charge radii, including even isotopes far from stability (Otten, 1981; Thibault, 1981).

4. THE CHEMICAL SHIFT

Chemical shifts are observed as the differences in energies between X rays from a given element incorporated in different chemical compounds. For the lighter elements, in particular, the bonding effects may change the X-ray spectra in a major way as discussed by Nagel and Baun (1974). For the inner shells of the heavier elements, chemical shifts were believed to be negligible (Sandström, 1957), until Sumbaev and Mezentsev (1965a) reported shifts for the $K\alpha_1$ lines of molybdenum and tin. A large body of data has since accumulated and has been reviewed on several occasions by Sumbaev (1969, 1976, 1978) and by Boehm (1975). As the chemical shift is closely related to several phenomena to be discussed in the following we give here a brief summary of its origin, key data, and applications.

4.1. Origin of the Chemical Shift

The discussion of chemical shifts (see, for example, Sumbaev, 1976) starts from the notion that a valence electron participating in the formation of a chemical bond is removed or "pulled away" from the atom. A decrease in electronic screening ensues and leads to an increase in inner-shell binding energies; the differences in the binding energies are, however, much smaller, but the naively expected trend still is an increase in the X-ray energies. The real situation is more complicated as it is not sufficient to consider the direct interaction between the valence electron and the $1s$ and np states responsible for the K X-ray emission. Just to take an example, the direct effect would be close to zero for the case of a $4f$ electron (illustrated in Fig. 7); the fact that very large negative shifts are seen there arises from a readjustment ("compression") of the electron density distribution in the atom.

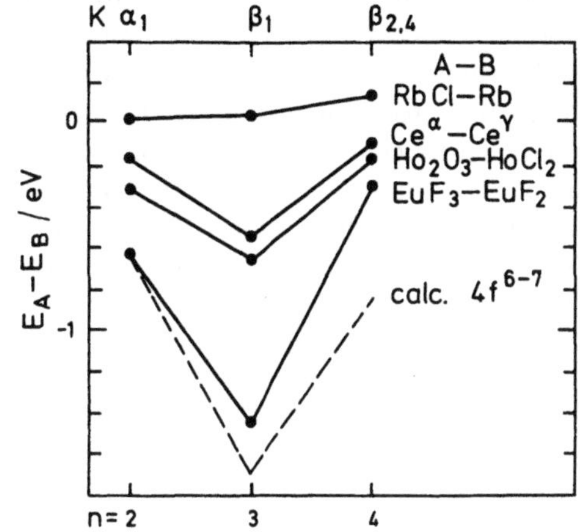

FIGURE 7. Examples of chemical shifts (from Sumbaev, 1976, 1978). The shift due to the removal of a 5s electron (the pair RbCl–Rb) is small and positive, whilst the shifts involving the removal of a 4f electron (rare-earth elements) are large and negative, indicating the compression effect as the main source of the shift. The calculation of the 4f shift shown is based on a self-consistent free-atom–ion calculation.

The chemical shifts can be treated in a somewhat more quantitative fashion by performing self-consistent-field calculations of the energy changes associated with the removal of an electron in a given atomic state (Sumbaev, 1969; Lee *et al.*, 1971, 1973; Petrovich *et al.*, 1971; Alder *et al.*, 1972). These free-atom–ion calculations are still very far from representing a chemical bond, but may be expected to represent an extreme value, which is approached for strongly ionic bonds. It is possible to obtain a convincing semiempirical presentation of the chemical shifts in metal/metal-oxide systems by introducing the concept of bond ionicities related to the difference in electronegativities of the partners in the compound (Sumbaev, 1969, 1976).

The 2p, 3p, 4p, and 5p electrons are situated in different parts of the atom and undergo different chemical shifts. For this reason, the energy shifts of the $K\alpha_{1,2}$, $K\beta_{1,3}$, $K\beta_{2,4}$, and KO lines characterize a valence electron by a "fingerprint" pattern.

4.2. *Examples of Data and Applications*

Examples of measured chemical shifts are shown in Fig. 7. The "fingerprints" are very clear for 4f electrons but relatively uncharacteristic for s, p, and d electrons. For this reason, much work has centered on the rare-earth elements (see Sumbaev, 1978), and additional pertinence is lent to this research by the current interest in valence fluctuations.

The theoretical picture (Johansson and Rosengren, 1975) of valence change in the rare earths involves the transition of a localized (nonbonding) 4f electron into a delocalized (bonding) 5d6s state, i.e., one electron is promoted from the 4f state into the 5d6s conduction band. However, for the case of cerium, which under pressure undergoes a phase transition from the normal trivalent γ phase to the more dense α phase, this explanation seems to be ruled out by thermochemical data (Johansson, 1978). Instead it has been suggested (Johansson, 1974) that the γ–α transformation is due to a fundamental change of the f state itself, a transition sometimes looked upon as a Mott transition. Such itinerant f states are found among the earlier actinide metals (Johansson, 1975). It is therefore interesting that for the Ce α–γ phase transition the experiment of Shaburov *et al.* (1973) shows a "fingerprint" (Fig. 7) that would agree with a decrease in the $4f$ occupancy by 0.5 electron. However, it is not yet clear whether the experiment is at variance with the Mott transition picture. Further experimental data would be welcome.

The problem, however, is of interest not only to solid-state physicists. Inasmuch as the standard state for experimental X-ray energies is the metallic state (see the article by Deslattes and Kessler, this volume) while theoretical calculations are carried out for free atoms (see, for example, Chen *et al.*, 1981), the possible existence of shifts related to the electronic structure of metals could eventually become important for the understanding of absolute X-ray energies. In Section 7.4, we mention data for the $5d$ elements that may point in this direction. The subject of shifts in binding energies of core electrons between free atoms and the atoms in the metallic state is one that has been widely studied, theoretically and also experimentally, especially in photoionization experiments. The shifts dealt with in the present paper correspond to differences in core-level shifts for very deep-lying states. Being small, their existence would not necessarily be negated by the success of a semiempirical theory of core-level shifts, which does not invoke any details of the internal electronic structure (Johansson and Mårtensson, 1980).

5. THE 1s HYPERFINE SHIFT

The interaction between the atomic electrons and the electric and magnetic moments of the nucleus leads to the hyperfine structure of atomic states, which is well-known from optical spectra (Kopfermann, 1958). For K X rays, the splitting is dominated by the contribution from the $1s_{1/2}$ state (Fig. 8), a contribution which is very much larger than that of the p states.

Even the 1s splittings, however, are not resolvable by X-ray spectroscopy, because they amount to no more than a few percent of the natural linewidth. Furthermore, in photoionization the net shift of the unresolved line vanishes because the two line components are populated with intensities that correspond to their statistical weights. The first demonstration of 1s hyperfine structure was based on a determination of a minute difference in

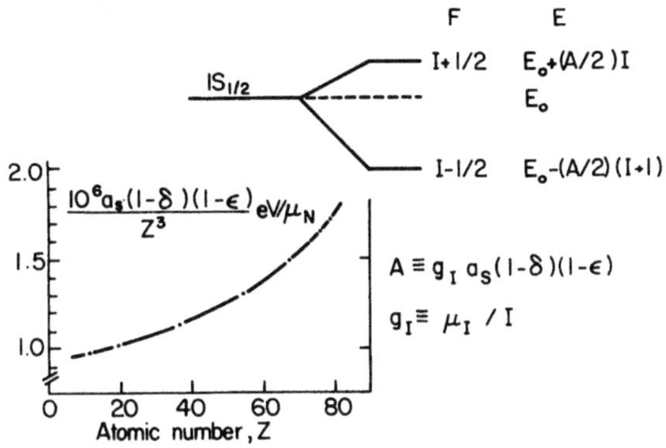

FIGURE 8. Hyperfine splitting of the 1*s* state caused by a nucleus with spin *I* and gyromagnetic ratio $g_I = (\mu_I/I)$ (in nuclear magnetons). The atomic parameter for the 1*s* state a_s (in eV/μ_N) has been calculated with the relativistic Dirac–Fock program of Rosén and Lindgren (1968, 1972). Approximate corrections $[(I-\delta)$ and $(I-\varepsilon)]$ for finite nuclear size and the Bohr–Weisskopf effect have been taken from Kopfermann (1958). The curve allows a fast estimate of the magnetic hyperfine shifts. Note that contrary to normal convention the total angular momentum *F* arising from the coupling of the nuclear and the atomic angular momenta, *I* and *J*, is defined to include only core electrons (here $1s_{1/2}$), the valence electrons having no measurable influence on the hyperfine shifts. The calculated splittings of the atomic *p* states are at least two orders of magnitude smaller than the 1*s* splittings and need not concern us here.

the natural *K* linewidths of the pairs of isotopes $^{151,153}Eu$ and $^{121,123}Sb$, characterized by large differences between the magnetic moments of the two isotopes (Ryl'nikov *et al.*, 1972).

It was demonstrated by Borchert *et al.* (1977) that the 1*s* hyperfine shifts can be observed by using a radioactive X-ray source decaying by electron-capture beta decay (EC). In this situation, the population of the hyperfine levels is no longer statistical, and a net shift is measured. In the following two subsections we outline the theoretical and experimental background for this finding.

5.1. *Population of the Hyperfine-Structure Components in EC Beta Decay and Internal Conversion (IC)*

For allowed Gamow–Teller beta decay there is no exchange of orbital angular momentum, because the corresponding operator involves only the spin (and isospin) components of leptons and nucleons. If the spin change in the decay is ± 1, it can be seen from elementary arguments (Fig. 9) that only one of the two hyperfine components can be formed. Another way of realizing this fact is to consider the total angular momentum F_i of the initial state, defined as the vector sum of the nuclear spin and the angular momentum of the core electron (Fig. 8). Since the only particle leaving the atomic

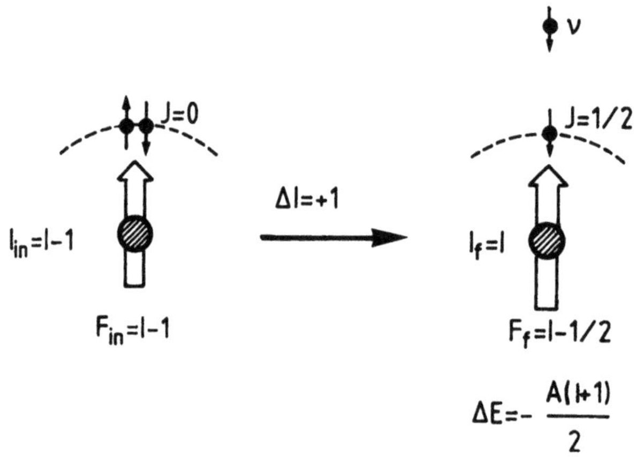

FIGURE 9. The F selection rule for the population of hyperfine levels is demonstrated here for the case in which the nuclear spin increases by one unit. The initial state has a nuclear spin $I-1$ and atomic angular momentum $J=0$. The nucleus undergoes beta decay, with capture of an electron, so that the remaining inner electron can only have its spin antiparallel to the nuclear spin. The hyperfine component with the inner electron spinning parallel to the nucleus is not formed.

system is a neutrino carrying $\frac{1}{2}$ unit of angular momentum, it can immediately be seen that the final state must have $F_f = F_i \pm \frac{1}{2}$. Only one of these two components can be realized, however, if the nuclear spin change is unity. For this reason, we refer to this effect as an "F selection rule" in beta decay. Through the use of standard angular-momentum algebra, the relative populations of the two hyperfine states $I_f + \frac{1}{2}$ and $I_f - \frac{1}{2}$ can easily be extracted also for the case $\Delta I = 0$. The result (Borchert *et al.*, 1977) and the corresponding energy shifts are listed in Table I, which shows that for multipolarity zero (Fermi decay) the population ratio is statistical, $(I+1)/I$, producing no shift, while the intensity ratio for multipolarity one is $I/(I+1)$, leading to a shift of $-\frac{1}{2}A$. Here A is the magnetic hyperfine-interaction parameter, proportional to the magnetic moment (Fig. 8). The intensity rules of Table I apply also to quasiallowed first-forbidden transitions (Borchert *et al.*, 1977, 1978a) and to K excitations following internal conversion of M1 electromagnetic transitions (Egorov *et al.*, 1978).

It is simple to generalize to transitions of multipolarity λ higher than unity. This has been done by Wang *et al.* (1978) and Grushko *et al.* (1981). The latter authors give the following expression for the transition matrix element to the hyperfine state F_f:

$$H(F_f) = (2F_f + 1)^{1/2}(2j_2 + 1)^{1/2}(-1)^{\lambda} \begin{Bmatrix} I_i & I_f & \lambda \\ 1/2 & j_2 & F_f \end{Bmatrix} b_{\kappa}(\tau\lambda).$$

TABLE I
Branching ratios and energy shifts[a] of K X rays from allowed or quasiallowed EC beta decay or from M1 electromagnetic transitions.

Multipole order	ΔI	Relative intensity $F=I+\frac{1}{2}$	Relative intensity $F=I-\frac{1}{2}$	Energy shift[b]
1	-1	1	0	$\frac{1}{2}AI$
1	0	I	$(I+1)$	$-\frac{1}{2}A$
1	1	0	1	$-\frac{1}{2}A(I+1)$
0	0	$(I+1)$	I	0

[a] From Borchert *et al.* (1977).
[b] The quantity A is the usual hyperfine interaction constant proportional to the nuclear g factor (see Fig. 8).

Here, j_2 is the angular momentum of the outgoing conversion electron of amplitude b (characterized by the state κ), and τ denotes the type (electric/magnetic) of transition. The total transition probability to the state F_f is proportional to the square of the transition matrix element, summed over κ. Grushko *et al.* (1981), and Fedorov and Ryl'nikov (1979) have pointed out that for mixed transitions an interference term will arise between the different multipolarities. This interference provides a new method for determining the ratio of Fermi to Gamow–Teller matrix elements in beta decay.

The selection rule for $|\Delta I| = 1$, $\lambda = 1$ (Fig. 9) turns out to be a special case of a more general rule for stretched $M\lambda$ transitions (that is, with $|I_f - I_i| = \lambda$). We note that the free electron may carry angular momentum $j_2 = \lambda \pm \frac{1}{2}$ but that in the nonrelativistic approximation only one of these occurs for magnetic transitions. [As an example, $M2$ internal conversion allows electron emission in a $p_{3/2}$ state ($\kappa = -2$) but not in a $p_{1/2}$ state. Both can contribute to an E1 transition.] By evaluating the $6j$ symbols in the above expression, we find that for $I_f - I_i = \pm\lambda$, only the transition to $F_f = I_f \mp \frac{1}{2}$ is allowed. In the general case, both transitions will occur, but still leading to a nonstatistical population.

We may at this point ask to what extent the F selection rules and nonstatistical populations discussed above constitute a new effect. It is clear that this cannot be the case from the point of view of beta-decay theory: were it so, then the calculated EC/β^+ ratios would change drastically and fail to agree with experiment. The usual derivation of EC decay rates in fact involves an averaging that implicitly takes the hyperfine structure into account (see, for example, Wu and Moszkowski, 1965). For experiments that involve the polarization of nuclei, photons, or leptons, on the other hand, the selection rules imposed by angular momentum conservation have been discussed by Dolginov (1956, 1958) and by Bernstein *et al.* (1958), for the case of muonic atoms. One consequence will be the possibility of directional and polarization correlations involving X rays and nuclear γ rays (Rupnik and Crasemann, 1972). A very clear discussion of hyperfine effects in inner-shell processes has been given in a review by Emery (1975). On the other hand, the influence of the nonstatistical population of the F states on the energies of X rays does not seem to have been noted previously, as can be seen by comparing the work of Hahn *et al.* (1976) with that of Wang *et al.* (1978).

We note finally that the $1s^{-1}$ lifetimes of 10^{-15} to 10^{-17} s for heavier atoms are much shorter than lifetimes of excited nuclear states, and that the hyperfine shifts may therefore serve to measure magnetic moments of very short-lived nuclear states (Borchert *et al.*, 1977; Egorov *et al.*, 1978; Grushko *et al.*, 1981).

5.2. *Experimental Observations of 1s Hyperfine Shifts*

The $1s$ hyperfine shift has been detected in three experiments. Two of these were carried out under conditions in which other contributions (see Sects 6 and 7) were expected to be small:

(i) The first experiment (Borchert *et al.*, 1977) served to compare the energies of K X rays from the EC decays of ^{131,132}Cs, for which the main characteristics are indicated in Fig. 10. The intensity rules of Table I predict

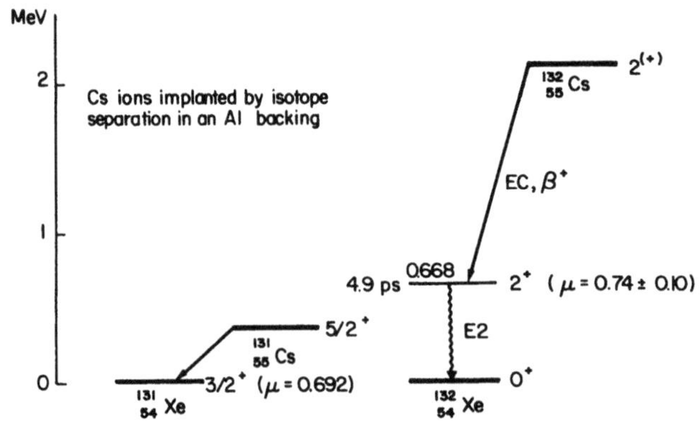

FIGURE 10. The difference between the $1s$ hyperfine structure shifts in $^{131,132(2+)}$Xe were measured from the K X rays emitted following EC decay of the radioisotopes ^{131,132}Cs, with half-lives of 9.7d and 6.5d, respectively (Borchert *et al.*, 1977). The relevant parts of the decay schemes (Lederer *et al.*, 1978) are indicated in the figure; μ denotes the nuclear magnetic moment in units of μ_N.

a hyperfine shift of 113 meV. The only other expected contribution is the nuclear volume effect, which has been estimated to 13 meV, leading to a total predicted shift of 126 meV. This value agrees well with the experimental result for $K\alpha_1$ of 112 ± 11 meV and for $K\beta_1$ of 170 ± 40 meV. The experimental technique eliminates chemical shifts because the two sources were prepared identically by implanting Cs ions into aluminum backings. It also eliminates shake-off and shake-up contributions, which anyway are only 1–2 meV for EC, and which furthermore must cancel in a comparison.

(ii) The experiment on $K\alpha_1$ X rays from internal conversion of a 276-keV M4 transition in ^{133}Ba (Grushko *et al.*, 1981) offers similar advantages. From the information given in Fig. 11 it will be seen that chemical shifts should be absent (identical $BaCO_3$ samples were used). The shifts of the lines due to associated atomic excitations are expected to be much bigger than in the previous case (see Sec. 6), of the order of 60 meV, but they should be close to identical for photoionization and internal conversion. The only contribution is thus a 1.3-meV correction for the nuclear-volume effect, after correction for which the experiment indicated an increase in energy to 56 ± 8 meV for the $K\alpha_1$ X ray arising from internal conversion. This shift corresponds to a magnetic moment $\mu = \pm 0.51 \pm 0.07 \mu_N$ for the $3/2^+$ state. There is, unfortunately, no independent measurement of μ in this case.

(iii) In a third experiment (Egorov *et al.*, 1978), X rays from internal conversion of an $M1$ transition (fed in beta decay) were compared with those from photoionization. The main characteristics are indicated in Fig. 12. Shake-off and shake-up are expected to cancel, leaving as the main problem chemical shifts and atomic-structure shifts (see Secs. 4 and 7). The authors solved this problem by noting that the shifts arising from different occupations of the $4f$ levels are known to be 2.68 times larger for $K\beta_1$ than for $K\alpha_1$. A correction based on this fact led to a hyperfine shift of 560 ± 50 meV, which corresponds to a magnetic moment of $4.1 \pm 0.4 \mu_N$, in good agreement with the lieterature value of $4.09 \pm 0.06 \mu_N$. An interesting result is obtained if we correct the raw data for the chemical shifts

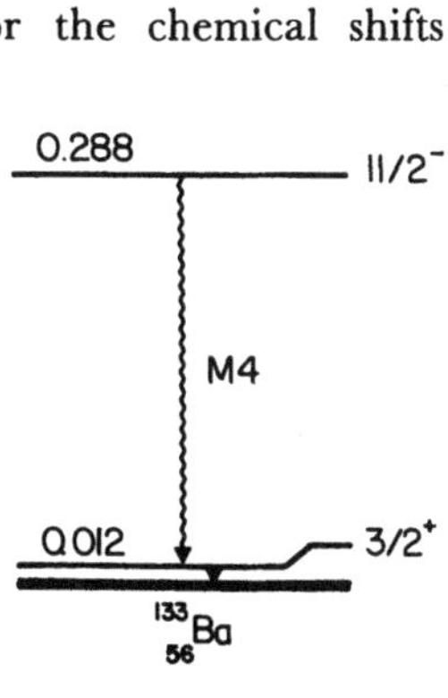

FIGURE 11. The hyperfine shift associated with K X rays from M4 internal conversion in ^{133}Ba (as a carbonate) was measured by making a comparison with a sample of stable $^{135}BaCO_3$ excited by photoionization (Grushko *et al.*, 1981). The decay scheme is from Lederer *et al.* (1978).

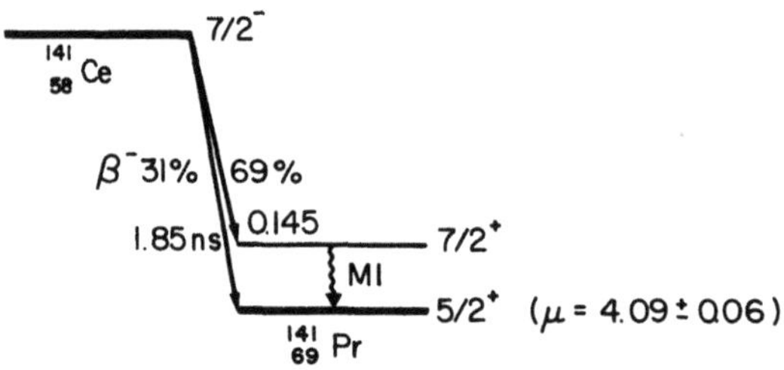

FIGURE 12. The decay scheme (Lederer *et al.*, 1978) of ^{141}Ce, used in the ^{141}Ce–^{141}Pr hyperfine structure experiment (Egorov *et al.*, 1978).

(Table II). The two measurements yield consistent corrected shifts, which must contain a 560-meV contribution from the hyperfine shift, leaving residual shifts of ~260 meV and ~850 meV for $K\alpha_1$ and $K\beta_1$. The positive sign indicates that the decay product of cerium has about $\frac{1}{2}$ $4f$ electron more than normal praseodymium, whereas β^- decay without "healing" would be expected to lead to praseodymium with one electron less. (See the discussion of the atomic structure effect, Sec. 7 and also Fig. 21, below.) The interpretation—in agreement with other evidence to be presented later—seems to be that the $4f$ shell has healed during the lifetime of the 1.85-ns state of ^{141}Pr.

We may conclude that the selection-rule effects associated with radioactive decays have offered the first possibility of observing the $1s$ hyperfine shifts in atoms. It would seem that the theory is sufficiently evident and that the data are sufficiently convincing, so that no great additional effort is required. Still, a few precise experiments under clean conditions [such as (i) and (ii) above] would be most welcome. Calculated hyperfine shifts enter into the analysis of atomic-structure effects, to be discussed in Sec. 7 of this

TABLE II

Shifts (meV) of praseodymium X rays observed in the $^{141}Ce(\beta^-)$–^{141}Pr experiment (Egorov *et al.*, 1978) after correction (Sumbaev, 1978) for chemical shifts of the CeO_2–Ce, PrO_2–Pr, Pr_2O_3–Pr systems.

$K\alpha_1$	$K\beta_1$	
PrO_2 reference	790 ± 25	1465 ± 200
Pr_2O_2 reference	870 ± 40	1350 ± 200

chapter, and the good agreement obtained for the rare-earth experiments (see Fig. 17) provides additional support for the conclusion reached here.

6. THE DYNAMIC SHIFT

This section concerns shifts caused by the excitation mechanism. We shall see that electron-capture beta decay (EC) provides a "clean" way of creating a $1s$ hole, and that a comparison with X rays from photoionization will, at least in principle, render the shifts due to spectator vacancies observable.

The K X rays may be thought of as arising from a single $1s^{-1}$ excitation, but the real situation is more complicated and must include also excited neutral atoms, double-vacancy excitations, etc. Those excitations that involve the inner shells can shift the X-ray energies considerably and give rise to "satellite lines," a phenomenon that has been known since the early days of X-ray spectroscopy (Hedin, 1974). The intensities of satellite lines depend on the excitation mechanism and energy. Sandström (1957) thus notes that these lines "do not appear in the secondary radiation generated through the irradiation of matter with X rays." Another extreme limit is reached when atoms are bombarded with heavy ions of high energy. Under such circumstances, highly charged intermediate states are formed.

In low-energy nuclear transitions and in photoionization, the most easily detectable effects come from an additional vacancy in one of the inner shells. These secondary-ionization effects are weak but resolvable and have attracted the main interest of experimentalists (Freedman, 1974; Walen and Briançon, 1975). The outer-shell ionization is much more abundant, as is known from calculations and from studies of charges of recoil ions from β^- decay (Carlson *et al.*, 1968, and references therein), but the effects tend to be obscured by the natural linewidth. There are nevertheless many cases in which skew or complex line shapes have been observed in the K X rays of light and medium-weight elements. As an example we quote an experiment by Sauder *et al.* (1977), who studied the Cu $K\alpha_{1,2}$ lines with good resolution. Complex line shapes were found and attributed to $3d^{-1}$ and $3p^{-1}$ spectator vacancies. From a calculated multiplet structure of the double-vacancy peaks combined with the single-vacancy line, all broadened by the instrumental resolution and by the natural width, it was possible to fit the complex peak shape. In units of the total $1s^{-1}$ excitations, this fit requires 27% associated $3d^{-1}$ and 2% $3p^{-1}$. There are no calculations of the expected shake-off for copper, but the calculations by Nestor *et al.* (1966) predict 15%–20% shake-off for the outer shells in argon and krypton, so the interpretation by Sauder *et al.* (1977) appears entirely reasonable.

We now turn to the rather limited theoretical and experimental information relating to dynamic energy shifts of K X rays.

6.1. Outer-Shell Shake-Off and Its Effect on K X-Ray Energies

A clear and detailed treatment of atomic excitation processes associated with nuclear transitions has been presented by Freedman (1974). We recall here a few of the essential features and first consider β^- decay, which for our purposes resembles K-shell photoionization.

It is well known that β^- decay proceeds essentially to the ground state of the daughter atom. The gain in binding energy is approximately $\delta B = -\partial E(Z)/\partial Z$, where $E(Z)$ is the total atomic energy of the system with atomic number Z. The sudden approximation leads to an expectation value of the atomic excitation energy of only $\langle \varepsilon^* \rangle = -\frac{1}{2}\partial^2 E(Z)/\partial Z^2$, which can be demonstrated by the usual techniques for deriving energy-weighted sum rules (Serber and Snyder, 1952). Taking the simplest Thomas–Fermi estimate, $E(Z) = -20.9 \times Z^{7/3}$ eV, we find for gold a total electronic energy $E(Z) = -5.6 \times 10^5$ eV, $\delta B = 1.7 \times 10^4$ eV, and a mean excitation energy $\langle \varepsilon^* \rangle = 140$ eV. Thus, more than 99% of the atomic energy release is available in the nuclear process. In the words of Freedman (1974), the transition is a "whole-atom one-step process."

For the EC process it was pointed out already by Primakoff and Porter (1953) that outer-shell excitations are essentially absent in $1s$ capture in a heavy atom. The reason is that the $1s$ orbit is very much bunched about the nucleus; consequently, the simultaneous disappearance of a nuclear charge and a $1s$ electron will affect the potential in the main part of the atom relatively little. (The same argument explains why, in this context, β^- decay and photoionization resemble each other.)

In order to arrive at quantitative estimates of the energy shifts of X rays, it is necessary to turn to detailed atomic models. The usual method for calculating the influence of shake-off processes is to use a cascade picture. In this approximation, the initial process (e.g., $1s$ photoionization) creates a distribution of vacancies, which in a subsequent series of steps (X-ray emission and Auger cascades) lead to the final (observed) states.

Carlson *et al.* (1968) (see also Nestor *et al.*, 1966) used the sudden approximation to calculate initial vacancy distributions for beta-decay processes and photoionization. The use of the sudden approximation is justified, as the passage time for exiting or produced particles is much shorter than the orbital periods involved. In these calculations, a self-consistent-field (Hartree–Fock) calculation was used to obtain the wave functions of the initial and final atomic states. From the single-electron overlap integrals, the excitation probabilities were obtained as unity, minus the probability that a given state is not excited, minus a correction for the exclusion of already filled states. Calculations of the excitation probabilities following $1s$, $2s$, $3s$, and $4s$ electron capture have also been carried out by Crasemann *et al.* (1979), who show that the total excitation probabilities are two orders of magnitude smaller for EC than for PI. This clearly confirms that EC provides an extremely "clean" $1s^{-1}$ excitation, i.e., it has few double-vacancy excitations. As an example, Table III shows the vacancy-production probabilities for xenon for PI and EC.

In order to convert the excitation probabilities into contributions to the X-ray energy shift, it is necessary to calculate the shifts in binding energy in the presence of a spectator hole. For accurate calculations this implies the double difference, for example for $K\alpha_1$ in the presence of a $4f$ hole,

$$\Delta E(K\alpha_1) = \langle 1s^{-1}4f^{-1}|\,V\,|1s^{-1}4f^{-1}\rangle - \langle (2p_{3/2})^{-1}4f^{-1}|\,V\,|(2p_{3/2})^{-1}4f^{-1}\rangle - [\langle 1s^{-1}|\,V\,|1s^{-1}\rangle - \langle (2p_{3/2})^{-1}|\,V\,|(2p_{3/2})^{-1}\rangle],$$

where V denotes the total atomic energy. (This is how the atomic structure

TABLE III

Vacancy probabilities and K X-ray energy shifts for Xe ($Z = 54$), $K\alpha_1$.[a]

	Vacancy probability $P_i \times 10^4$		
Subshell	PI[b]	EC[c]	$\Delta E(K\alpha_1)$[c] (eV)
$1s_{1/2}$	0.18	0.89	643
$2s_{1/2}$	3.19	1.89	86
$2p_{1/2}$	4.18	0.09	105
$2p_{3/2}$	7.73	0.05	86
$3s_{1/2}$	8.80	1.36	7.49
$3p_{1/2}$	13.8	0.10	9.82
$3p_{3/2}$	26.6	0.03	
$3d_{3/2}$	24.1	0.01	−0.398
$3d_{5/2}$	35.4	0.02	
$4s_{1/2}$	26.0	1.71	1.09
$4p_{1/2}$	43.7	0.06	1.43
$4p_{3/2}$	86.6	0.02	
$4d_{3/2}$	187	0.08	−0.491
$4d_{5/2}$	287	0.12	
$5s_{1/2}$	166	2.28	0.059
$5p_{1/2}$	497	0.02	0.089
$5p_{3/2}$	1229	0.07	

[a] Estimated energy shift for $K\alpha_1(E_{K\alpha_1} = 29779$ eV), $\Delta E(K\alpha_1, \text{PI}) = 51$ meV; $\Delta E(K\alpha_1, \text{EC}) = 1.2$ meV.

[b] From Nestor *et al.* (1966).

[c] Relativistic self-consistent field calculation as described by Rosén and Lindgren (1968, 1972). Note that only one calculation has been performed for each spin–orbit pair in the outer shells.

shifts discussed in Sec. 7 were calculated.) In Table III and in the calculations by Borchert *et al.* (1978b), both based on the self-consistent-field techniques of Rosén and Lindgren (1968, 1972), the shifts were approximated by differences in single-electron removal energies for a static Hartree–Fock field so that only one calculation was carried out for each spectator state. [A previous calculation (Borchert *et al.*, 1978a) had considered only the direct Coulomb terms and had overestimated the shifts by a factor of ~2.5; see also the discussion of "direct" and "compression" effects, Sec. 4.1.]

The net shifts obtained by combining excitation probabilities and calculated energy shifts are illustrated in Table III and in Figs. 13 and 14. It will be seen that the dynamic shift is expected to raise the energies of K

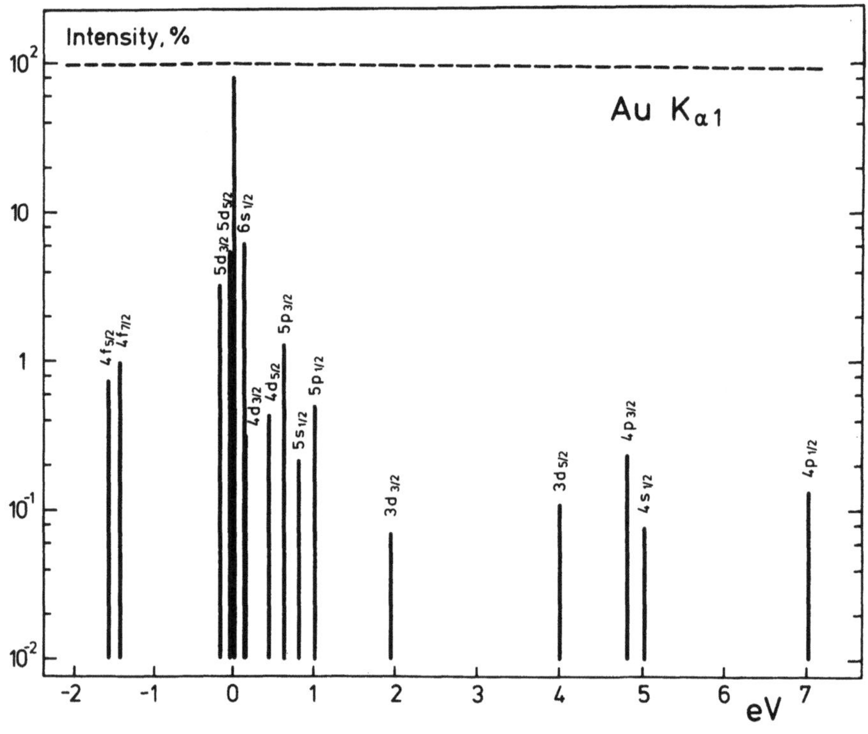

FIGURE 13. The Au $K\alpha_1$ line $[(1s_{1/2})^{-1} \rightarrow (2p_{3/2})^{-1}]$ has a natural width $\Gamma = 54$ eV, and it is probably reasonable to assume the same width for the unresolved satellite lines, here characterized by the quantum numbers nlj of the spectator holes. Hyperfine structure of the single-vacancy X ray (heavy line) and of the satellites is not shown. The dynamic shift results from an asymmetric contribution, hence an accurate calculation must take into account the precise definition of the line position ["the location parameter" (Hansen, 1978)]. The dashed line represents the contour of the complex. (From Borchert *et al.*, 1981b.)

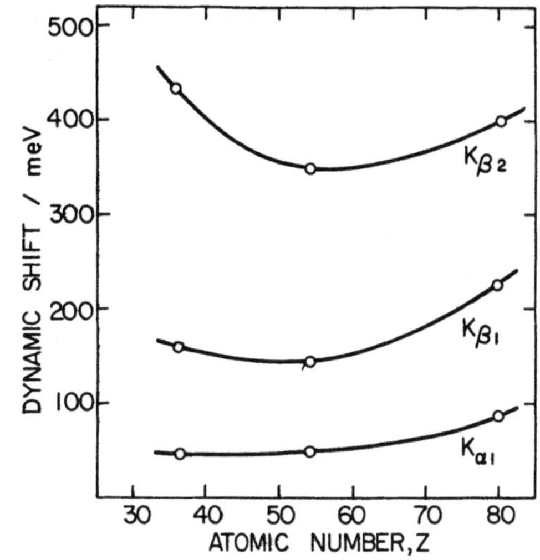

FIGURE 14. Dynamic shift in photoionization calculated as by Borchert *et al.* (1978b). The line position is as defined by Hansen (1978) with a measuring interval $2a = 1.2\Gamma$.

X rays following photoionization by 50–400 meV and to be completely negligible for those following EC. In view of other contributions that are not yet fully understood, it is doubtful whether the differences are observable at our present level of understanding of X rays, although they may have been detected in one experiment, to be discussed in the following.

The discussion of dynamic shifts would not be complete without once again reminding the reader that the cascade picture of inner-shell excitations is only an approximation. A sound quantum-mechanical treatment would have to consider the creation and decay of inner-shell vacancies as a single process leading to the (experimentally observed) final states. The theory of X-ray line shapes, to which we return in Sec. 8, is of this kind, as are also the calculations by Åberg (1981).

Finally, we mention that most experimental and theoretical studies have dealt with shake-off from the 1*s* shell rather than from the outer shells with which we have been concerned here. The 1*s* process clearly serves as an acid test of the theoretical techniques, and for this reason it is relevant for the present chapter. References to much work on this subject can be traced through the book edited by Fabian *et al.* (1981) and the paper of Law and Suzuki (1982).

6.2. *Experimental Detection of the Dynamic Shift*

The first experiment to attempt to detect dynamic effects by comparing electron capture (EC) and photoionization (PI) was carried out by Schnopper (1967), who measured the *K* X-ray spectra from *K* capture in radioactive ^{55}Fe and from electron excitation of manganese metal. The spectra turned out to be essentially identical and the precision on the absolute energies (± 2 eV) was too low to permit conclusions about energy shifts. In a subsequent experiment, Hahn *et al.* (1976) found that for Ta $K\alpha$ the energy for PI exceeded that for EC by about 0.8 eV, but as they believed that shake-off would contribute most strongly for the EC process, they could only understand the sign by postulating a selective multiple shake-off of $4f$ electrons. It was soon pointed out (Freedman, 1977; Borchert *et al.*, 1977) that there are strong theoretical arguments (cited above as well as in Schnopper's paper) for expecting shake-off to be very weak in the EC process, and that the until then unknown $1s$ hyperfine shift would increase the effect to 1.2 eV. The quantitative explanation of the data, which meanwhile have been confirmed in several experiments, still remains an interesting problem, which we shall discuss in Sec. 7.4.

A second attempt to detect the contribution to K X-ray energies from shake-off was made by Borchert *et al.* (1978a), who in a subsequent paper presented a more refined theoretical analysis (1978b). In this experiment, a strong source of 64-h ^{197}Hg was implanted into an aluminum backing, and the energies of the gold X rays were compared with those induced by photoionization in a gold foil. The presence of a 24-h isomer ^{197m}Hg in the radioactive sample complicated the analysis in several ways (Borchert *et al.*, 1978a), but it was possible to confirm these corrections by repeating the Au $K\alpha_1$ measurement 238 h after the start of the experiment. The error due to counting statistics amounted to 3×10^{-4} of the natural line-width; and the total error was 1.5–2 times bigger when possible systematic errors were taken

TABLE IV

Energy shifts[a] of gold K X rays in meV.

Line	Calculated value			Experiment
	Shake-off[b]	"6s"[b]	Total[c]	
$K\alpha_1$	70	83	543	568 ± 36
$K\alpha_2$	70	94	554	567 ± 46
$K\beta_1$	124	95	609	620 ± 50

[a] Defined as the energy for X rays from photoionization of gold minus the energy for X rays from EC of 65-h ^{197}Hg. The experimental data (Borchert *et al.*, 1978a, b) are presented here without contributions from the 23.8-h isomer and from tails of neighboring mercury X rays.

[b] The energy shifts due to shake-off and valence electrons were calculated by the techniques discussed in Secs. 6.1 and 7.1.

[c] The total theoretical shifts include a 378-meV contribution from magnetic hyperfine structure and a 12-meV isomer shift.

into account. The results, recalculated to represent the pure 64-h activity, are listed in Table IV. When the hyperfine shift, a small correction for the isomer shift, the shake-off, and the atomic-structure effects are taken into account, the agreement is excellent. This analysis, however, rests heavily on two assumptions: one from nuclear physics and one from chemistry and solid-state physics. The hyperfine shift in the (first-forbidden, "quasiallowed") EC decay is $-6-392(1+y^2)^{-1}$ meV, where y^2 is the intensity ratio of multipole order zero to multipole order one. From nuclear-structure arguments it seems reasonable to assume $y^2=0.05$; even though this correction is small, it would be desirable to have a measured value.

The second and more serious problem concerns the correction for the atomic-structure shift (Sec. 7). It was assumed in the experiment that in metallic samples the elements gold and mercury (here implanted into aluminum) have electronic structures that can be approximated by the

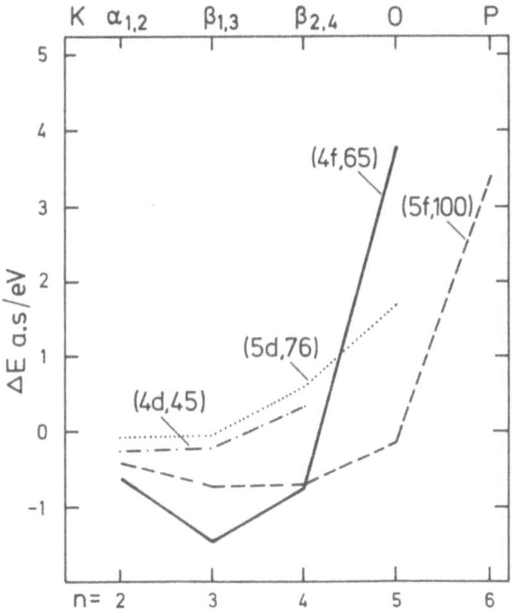

FIGURE 15. Atomic structure shifts ($E_{PI}-E_{EC}$) for the K-series X rays ($np \rightarrow 1s$) as functions of the quantum number n. Each pair of atoms is characterized by the quantum numbers (nl) of the orbital that is being filled at this point and by the atomic number Z of the daughter. These calculations have been carried out with Desclaux's (1975) Dirac–Fock computer program, in which the energy differences are taken between total energies, so that full relaxation is included in the calculation. Note that the shift displayed here is the negative of that for the addition of the indicated electron.

free-atom configurations $6s$ and $6s^2$, outside a platinum core. There is now evidence that such assumptions fail entirely for the $5d$ elements (Fig. 19), and there is no *a priori* reason why we should fare better in the Au–Hg case and in the rare earths (Figs. 16 and 17), as seems to be the case. For the Au–Hg case, the agreement is quantitative: the weighted difference between theory and experiment for all lines is 16 ± 26 meV, but 104 ± 26 meV if the shake-off shift is not included.

In summary we can conclude that the experimental evidence is consistent with a small shake-off shift, of the magnitude shown in Fig. 14. Further experimental work is clearly desirable.

7. THE ATOMIC-STRUCTURE SHIFT FOR TRANSITIONS WITH $\Delta Z = 1$

The process of nuclear electron capture (EC) transforms an element with atomic number Z into one with atomic number $Z-1$. The associated X rays are therefore emitted from an atom containing one electron "too many," in which the outer shells have the electronic structure of the next-heavier element. By comparing the X-ray energies with those obtained from processes with $\Delta Z = 0$ (photoionization, internal conversion), it is possible to observe energy shifts that reflect the difference in atomic structure of two neighboring elements. For this reason, we refer to a shift caused by this mechanism as an atomic-structure shift (Borchert *et al.*, 1978b).

It is clear from the discussion in Sec. 6 that the outer shells are temporarily in a stable state for an atom formed in EC decay. When the same atom is formed in photoionization (PI), it will be able to bind one more electron but the lifetime of a $1s$ hole (10^{-15}–10^{-17} s for one of the heavier atoms) seems somewhat short for a rearrangement to occur. We may therefore expect the atomic-structure shift (PI–EC) to provide a "snapshot"

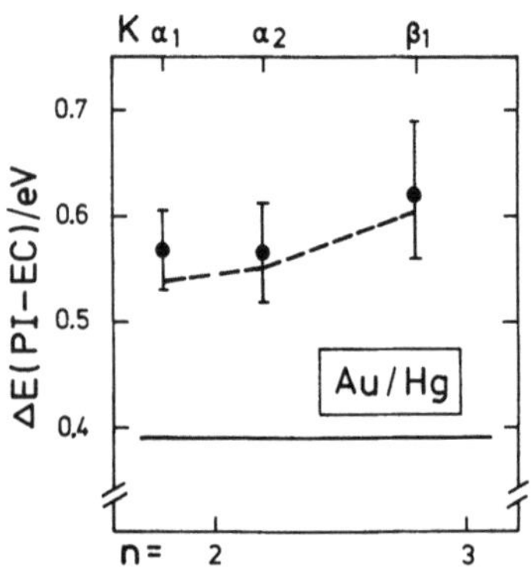

FIGURE 16. The energy shifts $E(\text{PI}, np \to 1s)$–$E(\text{EC}\ np \to 1s)$, with $n = 2,3$, for the pair ^{197}Au(PI)–^{197}Hg(EC)Au. The experimental results of Borchert *et al.* (1978a, b) are denoted by dots, together with their total uncertainties. The horizontal line corresponds to the sum of isomer shift (12 meV) and hyperfine shift (378 meV). It represents the nuclear contribution. The total shifts, given by the dashed line, consist of isomer shift, hyperfine shift, dynamic shift, and atomic-structure shift due to an additional $6s$ electron for the EC case. These data are also presented in Table IV. Note the offset zero of the y axis.

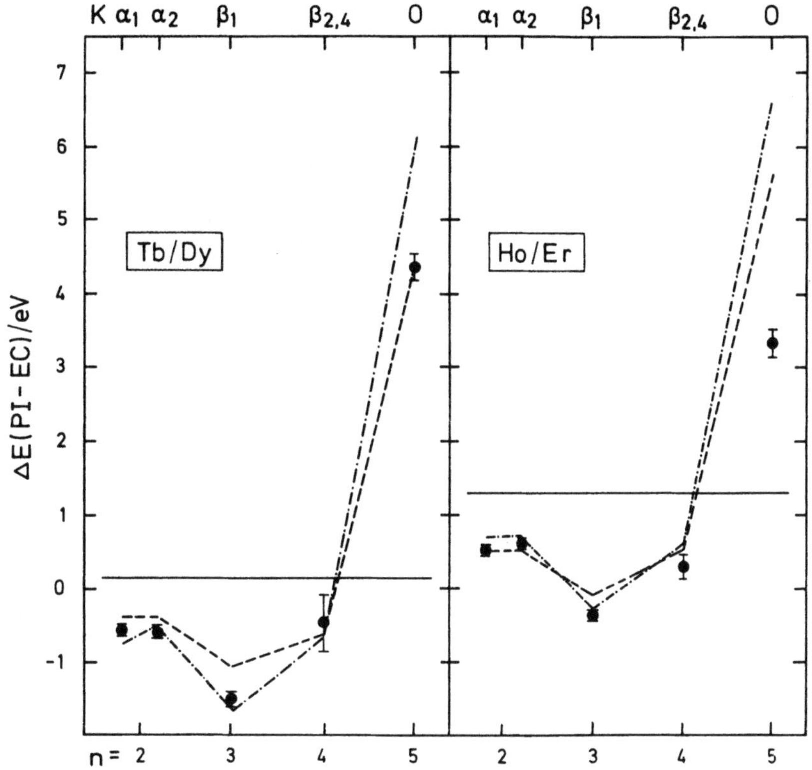

FIGURE 17. The energy shifts $E(\mathrm{PI}, np \rightarrow 1s)-E(\mathrm{EC}, np \rightarrow 1s)$ for X rays from metallic samples of rare-earth elements ($n=2$ to $n=5$). The left-hand side shows data for the pair ^{159}Tb(PI)–^{159}Dy(EC) ^{159}Tb (Hansen *et al.*, 1981) and the right-hand side shows data for the pair ^{165}Ho(PI)–^{165}Er(EC) ^{165}Ho (Borchert *et al.*, 1978b). The experimental results are represented by dots together with the corresponding total uncertainties. In cases where the error bar is not shown, it is smaller than the diameter of the dot. The horizontal line indicates the calculated hyperfine contribution to the shifts (Sec. 4). Relativistic atomic calculations have been carried out by the method of Desclaux (1975). The dot-dashed lines represent the total theoretical shifts for the assumed configuration "Xe core" $6s^2 5d(4f_{5/2})^6 (4f_{7/2})^2$ for terbium, and the configurations $(4f_{7/2})^3$, $(4f_{7/2})^4$, $(4f_{7/2})^5$ for dysprosium, holmium, and erbium, respectively. These are the expected electronic configurations for these (trivalent) metals. The dashed lines are obtained under the assumption of divalent metals, an assumption that is ruled out by simple arguments from solid-state physics (Johansson, 1975).

of the two structures. For the case of β^- decay followed by *K* internal conversion, the time available for rearrangements is given by the nuclear lifetime, and the observed shift may depend upon the life-time of the intermediate state. Such effects very likely play a role in the ^{141}Ce(β^-) experiment discussed in Sec. 5.2 (iii).

It is possible to calculate the influence of different atomic structures by

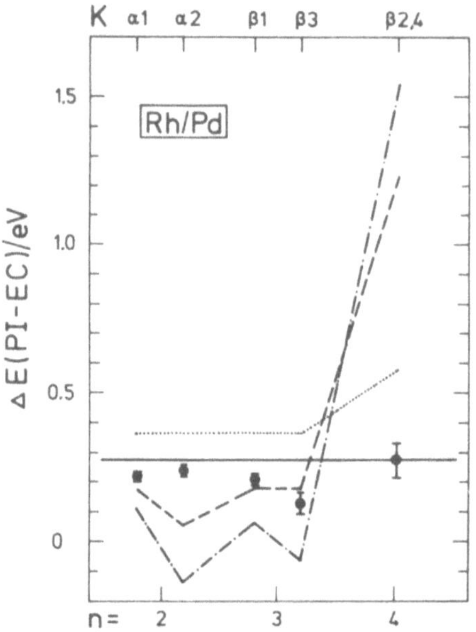

FIGURE 18. The energy shifts $E(\mathrm{PI}, np \to 1s)-E(\mathrm{EC}, np \to 1s)$ as a function of $n(=2\text{–}4)$ for the pair ^{103}Rh(PI)–^{103}Pd(EC) ^{103}Rh. The experimental results of Hansen et al. (1981) are denoted by dots with the corresponding error bars. Again the horizontal line represents the calculated hyperfine shift contribution of 278 meV. The three curves shown correspond to the calculated total shift for three different configurations. ^{103}Rh is assumed to have the configuration "Kr core" $(4d_{3/2})^4$, $(4d_{3/2})^4$, $(5s_{1/2})^1$. The configuration of ^{103}Pd is taken as $(4d_{3/2})^4$ $(4d_{5/2})^4$ $(5s_{1/2})^2$ (dotted line), $(4d_{3/2})^4$ $(4d_{5/2})^5$ $(5s_{1/2})$ (dashed line), and $(4d_{3/2})^4$ $(4d_{5/2})^6$ (dash–dotted line).

the self-consistent-field method used for treating the dynamic shift (Sec. 6). Again it is essential to calculate the shifts as a double difference between the total energies of the four states involved, thus taking into account the full relaxation. To illustrate the expected pattern of atomic-structure shifts, we show in Fig. 15 the results obtained for pairs of elements in which the occupation of the indicated orbital differs by unity. It is characteristic that addition of a typical valence electron (s or p) gives rise to a small negative shift of $K\alpha_1$, $K\beta_1$, $K\beta_2$, whereas the d and, even more so, the f orbitals lead to large positive* shifts corresponding to the compression effect (Table III, Sec. 4). Experiments on transition elements therefore should offer the best possibilities for observing the atomic-structure shift.

The atomic-structure shift is not identical with the chemical shift discussed in Sec. 4: chemical shifts are observed between different chemical states of the same element, while the atomic-structure shift is observed between different elements in the same chemical state, usually the metallic

* The sign conventions for the X-ray energy shifts may lead to confusion, especially since the definition of the signs is not specified in some papers. We have, here and in previous papers, presented the experimental data as "energy in photoionization minus energy in EC" (often abbreviated PI – EC). The atomic-structure shift is thus the negative of the shift due to the addition of an extra electron. In the calculation of the dynamic shift, we are dealing with vacancies (see Table III and Fig. 13). Consequently, the $4f$ shifts appear negative throughout this paper.

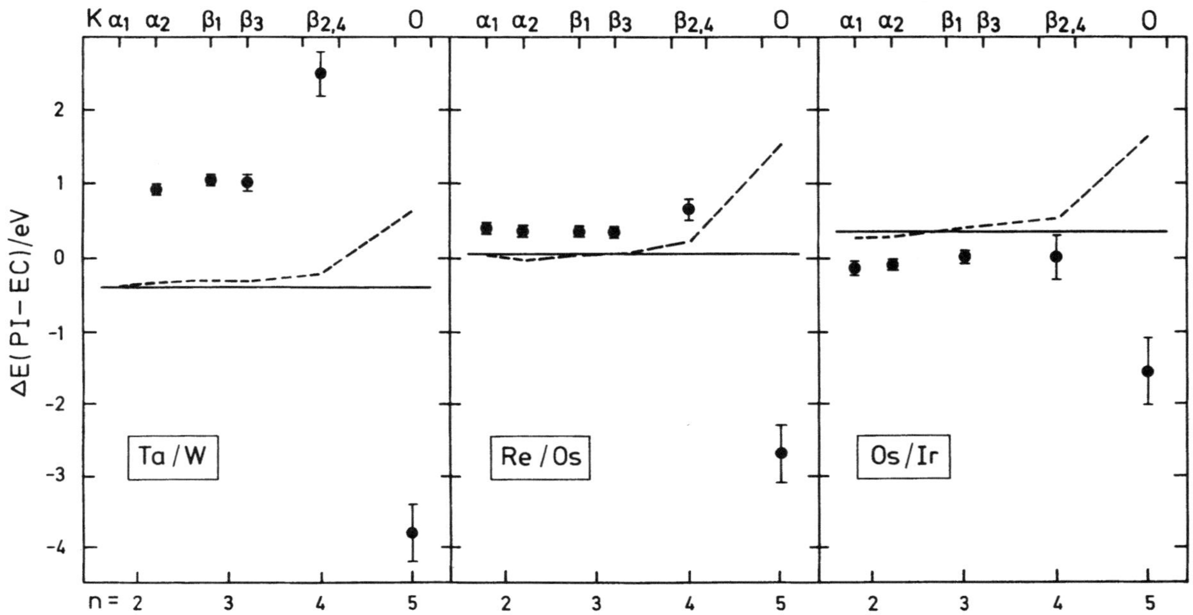

FIGURE 19. The energy shifts $E(\mathrm{PI}, np \rightarrow 1s)-E(\mathrm{EC}, np \rightarrow 1s)$ in the $5d$ elements for X-ray transitions, as functions of n. The experimental results for three different pairs of elements (Hansen *et al.*, 1981) are given by the dots with corresponding error bars. The solid lines denote the calculated hyperfine shift, whereas the dotted lines correspond to the total shifts including the atomic structure shift. The configuration for Hf is "Yb core" $5d^2$; analogous configurations apply for the other elements. The disagreement is general; note especially the drastic jump of the KO shift.

state. The difference is well illustrated by Figs. 7 and 17, which show that the Z, $Z-1$ calculation describes the atomic-structure shift quantitatively while the corresponding free-atom-ion estimate of the chemical shift is an overestimate; a more accurate calculation would have to take the nature of the chemical bond into account.

We now discuss the experimental evidence on atomic-structure shifts, grouped according to the atomic orbitals. The atomic-structure shifts used in this section are based on single-electron configurations calculated in a Dirac–Fock computer program developed by Desclaux (1975). Previous results (Borchert *et al.*, 1978b; Hansen *et al.*, 1981), which were based on calculations developed by Rosén and Lindgren (1968, 1972) agree to within typically 0.1 eV so that the conclusions are not sensitive to details of the atomic model.

7.1. *The 6s Elements*

The only data until now are for the pair ^{197}Au(PI), ^{197}Hg(EC) discussed in Sec. 6.2. The results of Table IV are presented graphically in Fig. 16, which shows excellent agreement with a calculation for which the underlying assumption is that metallic Au and Hg samples can be described by free-atom calculations (cf. Sec. 6.2). The net contribution of the atomic-structure shift is, however, too small in this case to furnish a sensitive test of the theory.

7.2. *The 4f Elements*

The rare-earth elements furnish an ideal testing ground for the atomic-structure shift, not only because the $4f$ electrons give large shifts (Fig. 15), but also because these elements (excluding Eu and Yb) have nearly identical chemical properties. Thus, holmium and erbium (Fig. 17) both crystallize in an *hcp* lattice and differ by less than 1% in the lattice constants. There are therefore strong reasons for believing that chemical shifts should be absent. The good overall agreement between theory and experiment is a striking verification of the existence of an atomic-structure shift.

It is interesting to perform the calculation assuming other configurations than $6s^2 5d 4f^n$, which is known to be the right one for the four elements involved here (see Johansson, 1975). If instead we assume $6s^2 4f^{n+1}$, corresponding to the divalent structure of Eu and Yb metals, we get poorer agreement (Fig. 17), and versions involving either a change in the number of $5d$ electrons or a change of two $4f$ electrons can be completely excluded.

In order to test whether the results are very sensitive to the chemical environment, the Ho–Er experiment was repeated in part with ^{165}Er(EC) implanted in Al metal. The results (for $K\alpha_1$ and $K\beta_1$) were unchanged within the experimental errors (Borchert *et al.*, 1978b).

In two other experiments, energies from photoionization and beta decay of rare-earth elements were compared. The experiment on β^- decay of ^{141}Ce was discussed in Sec. 5.2 (Egorov *et al.*, 1978); in the other experiment, Eu *K* X rays from photoionization of Eu_2O_3 were compared with those from the EC decay of ^{153}Gd (Wang *et al.*, 1978). In both cases, however, the shifts contain also contributions from internal conversion and from (possible) effects of nuclear lifetimes. The latter problem is discussed in Sec. 9.

It is interesting to compare the shift pattern due to the atomic-structure effect with that caused by chemical binding (Sec. 4). The physical difference is evident: the atomic-structure shift in the rare earths arises from the fact that in the EC process there is exactly one electron more which is in a pure and localized state. On the other hand, the chemical shift is caused by the involvement of 4*f* electrons in a chemical bond. Consequently, the amount of "removal" of the 4*f* electron depends sensitively on the ionicity of the bond, and the theoretical value from free-atom-ion calculations is reached only as a limit (Fig. 7).

7.3. *The 4d Elements*

The only case studied involves metallic samples of ^{103}Rh(PI)–^{103}Pd(EC) (Hansen *et al.*, 1981). The shifts do not agree with calculations based on simple assumptions about the electronic structure of the two elements (Fig. 18). The situation seems to be similar to that of the 5*d* elements.

7.4. *The 5d Elements*

A more extensive set of data for the atomic structure effect is available for elements in which the 5*d* shell is being filled. The first measurements were performed with oxide sources on the strong lines in the pair ^{181}Ta(PI)–^{181}W(EC) (Hahn *et al.*, 1976; Wang *et al.*, 1978). A complete set of data with metallic sources was obtained for the pairs ^{181}Ta(PI)–^{181}W(EC), ^{185}Re(PI)–^{185}Os(EC), and ^{189}Os(PI)–^{189}Ir(EC) (Borchert *et al.*, 1980b; Hansen *et al.*, 1981).

In the case of Ta and W the chemical shift between metal and oxide is known (Sumbaev, 1976). The common lines of the different Ta–W experiments can therefore be compared directly, and they agree within the quoted errors. The experimental data for these three pairs together with the results of the corresponding shift calculations for the atomic-structure effect of an additional $5d$ electron are shown in Fig. 19.

It is interesting to note that in these cases the calculation reproduces neither the pattern nor the absolute values of the shifts. The calculation predicts a smooth behavior with an amplitude of about ± 0.5 eV. Even for the transitions of the innermost shells, the experimental data differ from the theoretical predictions by more than 1.5 eV, whereas the shifts of the KO transitions amount to more than 4 eV.

A systematic compilation of the results is provided in Fig. 20, which shows the differences between experimental and calculated shifts as a function of the occupation number of the $5d$ shell. The different scales are denoted by the corresponding transitions. It is striking that all transitions from the L, M, N shells show a linear trend and change sign in the vicinity of osmium.

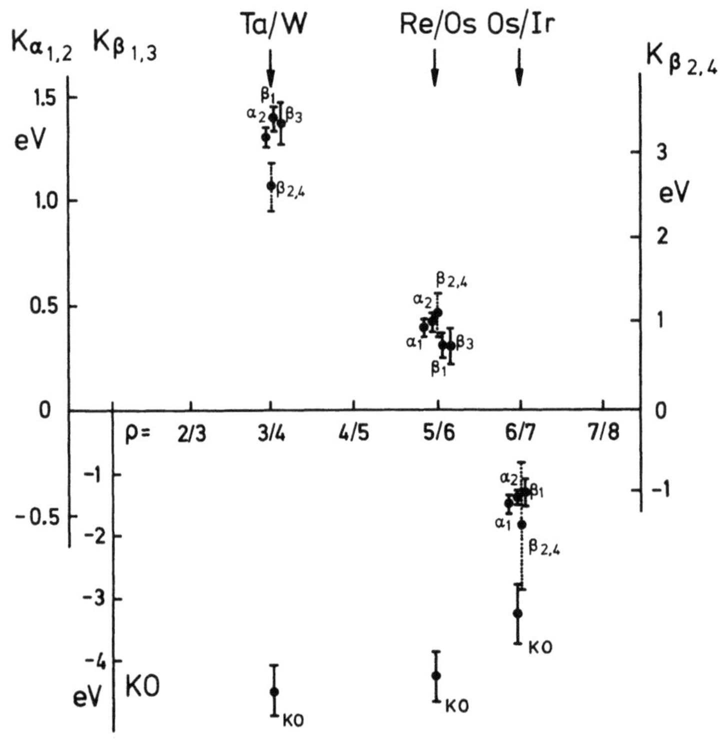

FIGURE 20. The differences between experimental and calculated total shifts of the X-ray transition energies in the $5d$ elements. For each measured pair of elements, the difference $\Delta E_{\text{exp}} - \Delta E_{\text{calc}}$ is shown as a function of the occupation number p of the $5d$ shell. Note that the scale of the shift differences varies. It seems that the shift differences from the L, M, and N shells follow a linear trend with p.

By contrast, the difference between the calculated and the experimental shifts for the *KO* transitions is more or less constant but as large as 4 eV. There is no quantitative explanation for this behavior in the 5*d* elements. It is striking to note, however, that the change of sign for the differences coincides with the minimum of the corresponding atomic radii and the minimum of the 5*d* and 6*s* binding energies (Huang *et al.*, 1976). There might be an indication that one would have to take into account the influence of the metallic lattice (Wang *et al.*, 1978a, b; Borchert *et al.*, 1978b; Hansen *et al.*, 1981). (This interpretation is clearly related to the subject of core-level shifts, referred to in Sec. 4.2.) The interpretation of the anomalous W–Ta shifts as a solid-state effect finds support in an experiment in which the energies of Ta $K\alpha_1$ X rays from a 2% alloy of tantalum in tungsten were compared with those of pure tantalum metal (Wang, 1979). The shift of -0.46 ± 0.14 eV is in the right direction but insufficient to explain the full effect.

8. OTHER CONTRIBUTIONS

Additional small contributions to the X-ray energies are considered in this section. These contributions have not been studied experimentally and, except for special cases, it is safe to neglect them.

8.1. Deviations from the Breit–Wigner Single-Level Line Shape

As described so far in the present chapter, the modifications of X-ray line shapes have had their origin in unresolved multiplet structures such as hyperfine structure (Sec. 5) and satellite lines (Sec. 6). It has been assumed that the basic line shape could be described by the Breit–Wigner formula, which is a good approximation because the natural width of a *K* X ray is at most 10^{-3} of the energy. At a certain level of precision it becomes necessary, however, to consider other contributions to the energy arising from energy conservation (the tails of the line cannot extend to infinity) and nonresonant terms. We now examine these factors.

We are aided in this task by the existence of a complete theory of the line shape of X rays following EC beta decay. This theory (developed by Glauber and Martin, 1956; Martin and Glauber, 1958) predicts that E1 internal bremsstrahlung becomes extremely intense at low energies, where it merges continuously with the characteristic X rays. The theory has been extended to cover the X-ray region by De Rújula (1981), who derived approximate expressions which have been used in the following.

As we shall now be dealing with asymmetric line profiles, it is useful to begin by considering the definition of the line position, or "location parameter," in statistical terminology. Measuring the photon energy k and natural width Γ of the line in units of the transition energy, we need only consider the profile

$$G(x)=\frac{1+2\alpha x}{x^2+\gamma^2/4}, \tag{1}$$

which is sufficiently general for our purposes, and where

$$1+x=k/E[K_x] \text{ and } \gamma=\Gamma/E[K_x]. \tag{2}$$

The resulting shift may be written

$$\Delta x=C\alpha\gamma^2,$$

where the constant C depends on the definition of the location parameter. Taking this to be the mode [i.e., the maximum with $dG(x)/dx=0$], we have $C=0.25$. A more useful definition is to truncate the line and to define the location parameter as the center of gravity after a bias correction (Hansen, 1978). With this definition and a truncation interval of approximately 1.2γ, we obtain $C\cong 0.4$.

For allowed X rays, the transition probability to a given final state nlj may be written (De Rújula, 1981)

$$\frac{d\omega[nlj]}{dk}=\text{const }|Q_{nP}|^2\,\phi(k), \tag{3}$$

where the resonant structure is contained in the matrix element Q_{np}, and where the phase-space factor divided by the photon energy k is

$$\phi(k)=(Q-B[nlj]-k)^2\,k, \tag{4}$$

with Q denoting the energy available and $B[nlj]$ binding energy of the electron in the daughter atom. As the shift arises from contributions inside the natural width $\Gamma(K_x)$ of the X ray, it is sufficient to consider linear terms in the variable x [Eq. (8.2)] and we find the contribution from ϕ

$$\Delta E_{\text{EC},1}=0.4\Gamma^2\left[\frac{0.5}{E}-\frac{1}{Q-B[1s_{1/2}]}\right] \tag{5}$$

which is small in almost all cases. For the decay of ^{235}Np($Q=123.0$ keV, $B[1s_{1/2}]=115.6$ keV, and $\Gamma=110$ eV) we find a resulting shift of -0.63 eV.

The matrix element Q_{nP} contains a sum over intermediate states. At the position of the *K* X ray, the nonresonant contribution can be approximated by a constant β so that the line shape is

$$G_1(x) = |(x + \tfrac{1}{2}i\gamma)^{-1} + \beta|^2, \tag{6}$$

which to the lowest order in x leeds to the shift

$$\Delta E_{\mathrm{EC},2} \cong \frac{0.4\beta\Gamma^2[K_x]}{E[K_x]}. \tag{7}$$

With β calculated from the expressions given by De Rújula (1981), the shifts for uranium X rays are -0.005, $+0.009$, and -0.064 eV for $K\alpha_1$, $K\beta_1$, and $K\beta_1$, respectively. Similar interference effects between resonant and nonresonant contributions have been studied for the $2s2p$ autoionizing state of helium ("Fano resonances"; see Hedin, 1974).

For photoionization there is to our knowledge no equivalent to the Glauber–Martin theory. It is easy to calculate the phase-space contribution in this case for an incident photon energy k_0 and with an outgoing free electron, and one finds, when Coulomb interactions with the outgoing electron are included, a constant shift

$$\Delta E_{\mathrm{PI},1} \cong \frac{0.2\Gamma^2[K_x]}{E[K_x]}. \tag{8}$$

The constant term vanishes in experiments in which EC is compared with PI. The absence (to lowest order) of an energy-dependent term is reassuring for the X-ray experimentalist, since the measured energies will not depend appreciably on the choice of photoexcitation source. This, of course, no longer holds when the excitation line coincides with the *K* edge, a case that has been studied theoretically and experimentally by Grushko *et al.* (1978).

The contribution from nonresonant terms would have to include also the *d* states. This effect has not been evaluated, but it is tempting to speculate that it could be more important here than in EC decay.

8.2. The Coupling of Atomic and Nuclear Excitations

The energy shifts arising from finite nuclear size have been discussed in Sec. 3. The corresponding off-diagonal contribution will lead to mixing of atomic and nuclear excitations and to small energy shifts. Such phenomena are well known in muonic and pionic atoms, but they are essentially negligible in electronic atoms.

It was pointed out by Morita (1973) that such processes might become observable by resonant enhancement in cases where a nuclear excited level with the right quantum numbers coincides in energy with an X-ray transition. In a first experiment of this kind (Otozai *et al.*, 1978), a nuclear isomer ^{189m}Os was formed in bombardments with 70–100-keV electrons. The production threshold agreed with the Os K binding energy. The process, referred to by the authors as "nuclear excitation by electron transition" (NEET), can be viewed as inverse internal conversion. In the ^{189}Os case the probability of nuclear excitation of 1.2×10^{-7} corresponds to a coupling matrix element of 0.89 eV. The calculated shift for this case would be only 0.35 meV.

In a second experiment on ^{237}Np, a marginal effect was observed in direct photon counting for an $E1$–$K\alpha_1$ pair (Saito *et al.*, 1980). The deduced

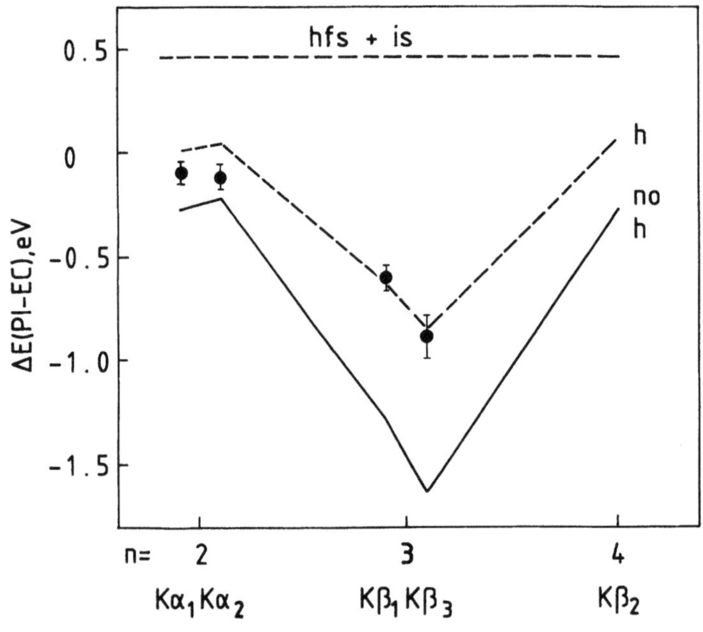

FIGURE 21. Approximately 36% of the X rays emitted in EC decay of ^{153}Gd originate in internal conversion. The nuclear intermediate states (in $^{153}_{63}$Eu) have half-lives of 10^{-9}–10^{-10} s. The analysis shown here aims at detecting whether the Gd atoms have had time to shed the surplus 4*f* electron (to "heal") before the X rays from conversion are emitted. The experimental data (Wang *et al.*, 1978b) compare PI and EC from trioxides of the two elements so that there is no chemical shift. The upper line indicates the contribution from the hyperfine shift and the isotope shift. In addition to the dynamic shift, the theoretical estimates include either an atomic-structure shift only for the X rays associated with EC (dashed line), or the same atomic-structure shift is assumed for all X rays (solid curve). The atomic-structure shift was calculated according to Rosén and Lindgren (1968, 1972). The experimental indication thus is that only the X rays from EC are shifted, so that the Gd structure is lost in no more than 10^{-9}–10^{-10} s (Wang *et al.*, 1978a, b). The ^{141}Ce experiment discussed in Sec. 5.2 (iii) and Table II suggests a similar conclusion.

off-diagonal matrix element was 26 ± 4 eV, leading to a calculated shift in this case of 0.36 eV. Effects of this kind could thus accidentally play a role for X-ray energies.

9. CONCLUDING REMARKS

Although considerable progress has been made in our understanding of the mechanisms for energy shifts in *K* X rays, there still are points that seem to need further clarification, and there are many possibilities for applications.

The dynamic shift has not been well established, and it would be interesting to have more precise measurements for checking the expectations of Fig. 14. The atomic-structure shifts for the 4*d* and 5*d* elements (Figs. 18–20) present a major puzzle: it is not at all clear where the additional shifts have their origin. If, as most authors are inclined to believe (Sec. 7.4), the effect originates in an incomplete description of the electronic structure, then the mechanism could involve either the free atom or, perhaps most likely, the electronic structure of the solid. In the latter case the problem would be intimately linked to the core-level binding-energy shifts of metals. More experiments are clearly needed.

We have here and there touched on applications of the effects discussed in this chapter, and in Sec. 5.2 we have cited the first determinations of nuclear magnetic moments by a new technique. As pointed out in the previous paragraph, the atomic-structure shift leads to investigations of the electronic structure of atoms and solids. A further development (Fig. 21) could be the determination of the healing time of certain electronic defects.

ACKNOWLEDGMENTS

The authors are indebted to Jens Ulrik Andersen, Alvaro De Rújula, and Börje Johansson for reading parts of the manuscript and for helpful comments.

NOTE ADDED IN PROOF. Recent studies by Rose and Borchert (1985) have revealed that the low values obtained for the *KO* lines in Figs. 17, 19, and 20 arise from a simple and purely experimental effect. They point out that in photoionization (but not in EC beta decay) the *KO* line lies so near the *K* edge of the source material that the self-absorption in emission varies appreciably across the line. The resulting skewness toward lower energies manifests itself as a decrease in the energy. This contribution can easily be calculated to vary from −1.8 eV for terbium to −3.2 eV for osmium. With this correction, which had not been considered in earlier work, the behavior of the *KO* components falls in line with that of other members of the *K* series.

REFERENCES

ÅBERG, T., 1981, in Fabian *et al.* (1981), p. 251.

ALDER, K., G. BAUR, AND U. RAFF, 1972, *Helv. Phys. Acta* **45**, 765.

AZÁROFF, L. V., editor, 1974, *X-Ray Spectroscopy* (McGraw-Hill, New York).

BERNSTEIN, J., T. D. LEE, AND C. N. YANG, 1958, *Phys. Rev.* **111**, 313.

BHATTACHERJEE, S. K., F. BOEHM, AND P. L. LEE, 1968, *Phys. Rev. Lett.* **20**, 1295.

BHATTACHERJEE, S. K., F. BOEHM, AND P. L. LEE, 1969, *Phys. Rev.* **188**, 1919.

BLAISE, J., 1958, *Ann. Phys. (Paris)* **3**, 1019.

BOEHM, F., 1968, CALT-63-104 AEC Research and Development Report, presented at the Sommerfeld Memorial Meeting, Munich, September 1968.

BOEHM, F., 1975, in *Atomic Inner-Shell Processes*, edited by B. Crasemann (Academic, New York), p. 411.

BOEHM, F., AND P. L. LEE, 1974, *At. Data Nucl. Data Tables* **14**, 605.

BONN, J., G. HUBER, H.-J. KLUGE, AND E. W. OTTEN, 1976, *Z. Phys.* **A276**, 203, and references therein.

BORCHERT, G. L., 1976, *Z. Naturforsch., Teil A* **31**, 102.

BORCHERT, G. L., P. G. HANSEN, B. JONSON, H. L. RAVN, O. W. B. SCHULT, AND P. TIDEMAND-PETERSSON, 1977, *Phys. Lett* **63A**, 15.

BORCHERT, G. L., P. G. HANSEN, B. JONSON, I. LINDGREN, H. L. RAVN, O. W. B. SCHULT, AND P. TIDEMAND-PETERSSON, 1978a, *Phys. Lett.* **65A**, 297.

BORCHERT, G. L., P. G. HANSEN, B. JONSON, I. LINDGREN, H. L. RAVN, O. W. B. SCHULT, AND P. TIDEMAND-PETERSSON, 1978b, *Phys. Lett.* **66A**, 347.

BORCHERT, G. L., P. G. HANSEN, B. JONSON, H. L. RAVN, O. W. B. SCHULT, AND P. TIDEMAND-PETERSSON, 1980a, *Nucl. Instrum. Methods* **178**, 209.

BORCHERT, G. L., P. G. HANSEN, B. JONSON, H. L. RAVN, AND O. W. B. SCHULT, 1980b, Proc. XVIII Int. Winter Meeting on Nuclear Physics, Bormio, 1980, edited by I. Iori (University of Milan), p. 579.

BORCHERT, G. L., P. G. HANSEN, B. JONSON, H. L. RAVN, AND J. P. DESCLAUX, 1980c, in *Atomic Masses and Fundamental Constants 6*, edited by J. A. Nolen and W. B. Benenson (Plenum, New York), p. 189.

BORCHERT, G. L., O. W. B. SCHULT, J. SPETH, P. G. HANSEN, B. JONSON, H. RAVN, AND J. B. MCGRORY, 1981a, Proc. 4th Int. Conf. on Nuclei far from Stability, Helsingør, 1981 (CERN, Geneva, CERN 81-09), p. 56.

BORCHERT, G. L., P. G. HANSEN, B. JONSON, H. L. RAVN, AND O. W. B. SCHULT, 1981b, *Hyperfine Interact.* **10**, 1115.

BORCHERT, G. L., O. W. B. SCHULT, J. SPETH, P. G. HANSEN, B. JONSON, H. L. RAVN, AND J. B. MCGRORY, 1982, *Nuovo Cimento*, **73A**, 273.

BREIT, G., 1957, *J. Opt. Soc. Am.* **47**, 446.

BRIX, P., 1977, *Naturwissenschaften* **64**, 293.

BRIX, P., 1981, *Phys. Blätter* **37**, 181.

BRIX, P., 1982, Max-Planck-Institut für Kernphysik, Heidelberg Report No. MPI H-1982-V3.

BRIX, P. AND H. KOPFERMANN, 1958, *Rev. Mod. Phys.* **30**, 517.

BRIX, P., H. V. BUTTLAR, F. G. HOUTERMANS, AND H. KOPFERMANN, 1952, *Z. Phys.* **133**, 192.

BROCKMEIER, R. T., F. BOEHM, AND E. N. HATCH, 1965, *Phys. Rev. Lett.* **15**, 132.
CARLSON, T. A., C. W. NESTOR, JR., T. C. TUCKER, AND F. B. MALIK, 1968, *Phys. Rev.* **169**, 27.
CAUCHOIS, Y., 1932a, *C. R. Acad. Sci.* **194**, 362; 1479.
CAUCHOIS, Y., 1932b *J. Phys. Rad.* **3**, 320.
CHEN, M. H., B. CRASEMANN, M. AOYAGI, K.-N. HUANG, AND H. MARK, 1981, *At. Data Nucl. Data Tables* **26**, 561.
CHESLER, R. B., AND F. BOEHM, 1968, *Phys. Rev.* **166**, 1206.
CRASEMANN, B., M. H. CHEN, J. P. BRIAND, P. CHEVALLIER, A. CHETIOUI, AND M. TAVERNIER, 1979, *Phys. Rev. C* **19**, 1042.
DE RÚJULA, A., 1981, *Nucl. Phys.* **B188**, 414.
DESCLAUX, J. P., 1975, *Comput. Phys. Commun.* **9**, 31.
DESLATTES, R. D., AND E. G. KESSLER, JR., 1979, in *Atomic Masses and Fundamental Constants 6*, edited by J. A. Nolen and W. B. Benenson (Plenum Press, New York), p. 203.
DESLATTES, R. D., AND E. G. KESSLER, JR, 1985, Chapter 5 of this volume.
DOLGINOV, A. Z., 1956, *Nucl. Phys.* **2**, 723.
DOLGINOV, A. Z., 1958, *Nucl. Phys.* **6**, 460.
DOUSSE, J.-CL., AND J. KERN 1976, *Phys. Lett.* **A59**, 159.
DUMOND, J. W. M., 1947, *Rev. Sci. Instrum.* **18**, 626.
EGOROV, A. I., A. A. RODIONOV, A. S. RYL'NIKOV, A. E. SOVESTNOV, O. I. SUMBAEV, AND V. A. SHABUROV, 1978, *Zh. Eksp. Teor. Fiz. Pis'ma Red.* **27**, 514 [English transl.: *JETP Lett.* **27**, 483].
EMERY, G. T., 1975, in *Atomic Inner-Shell Processes*, edited by B. Crasemann (Academic, New York), p. 214.
EMRICH, H. J., G. FRICKE, M. HOEHN, K. KÄSER, M. MALLOT, H. MISKA, B. ROBERT-TISSOT, D. RYCHEL, L. SCHALLER, L. SCHELLENBERG, H. SCHNEUWLY, B. SHERA, H. G. SIEBERLING, R. STEFFEN, H. D. WOHLFAHRT, AND Y. YAMAZAKI, 1981, Proc. 4th Int. Conf. on Nuclei far from Stability, Helsingør, 1981 (CERN, Geneva, CERN 81-09), p. 33.
FABIAN, D. J., H. KLEINPOPPEN, AND L. M. WATSON, 1981, editors, *Inner-Shell and X-Ray Physics of Atoms and Solids* (Plenum Press, New York).
FEDOROV, V. V., AND A. S. RYL'NIKOV, 1979, *Zh. Eksp Teor. Fiz.* **76**, 1986 [English transl.: *Sov. Phys. JETP* **49**, 1007].
FREEDMAN, M. S., 1974, *Ann. Rev. Nucl. Sci.* **24**, 209.
FREEDMAN, M. S., 1977, *Phys. Rev. A* **16**, 1315.
GLAUBER, R. J., AND P. C. MARTIN, 1956, *Phys. Rev.* **104**, 158.
GOLOVIN, A. F., AND A. R. STRIGANOV, 1967, *Usp. Fiz. Nauk* **93**, 111 [English transl.: *Sov. Phys. Usp.* **10**, 658 (1968)].
GRUSHKO, A. I., T. B. MEZENTSEVA, N. M. MIFTAKHOV, YU. P. SMIRNOV, A. E. SOVESTNOV, O. I. SUMBAEV, AND V. A. SHABUROV, 1978, *Zh. Eksp. Teor. Fiz.* **74**, 501 [English transl.: *Sov. Phys. JETP* **47**, 263].
GRUSHKO, A. I., K. E. KIR'YANOV, N. M. MIFTAKHOV, A. S. RYL'NIKOV, YU. P. SMIRNOV, AND V. V. FEDOROV, 1981, *Zh. Eksp. Teor. Fiz.* **80**, 120 [English transl.: *Sov. Phys. JETP* **53**, 59].
HAHN, A. A., K.-C. WANG, F. BOEHM, H. E. HENRIKSON, AND P. VOGEL, 1976, *Phys. Rev. Lett.* **37**, 1741.
HAHN, A. A., J. P. MILLER R. J. POWERS, A. ZEHNDER, A. M. RUSHTON, R. E. WELSH, A. R. KUNSELMAN, P. ROBERSON, AND H. K. WALTER, 1979, *Nucl. Phys.* **A314**, 361.
HANSEN, P. G., 1978, *Nucl. Instrum. Methods* **154**, 321.
HANSEN, P. G., G. L. BORCHERT, B. JONSON, H. L. RAVN, AND O. W. B. SCHULT, 1981, in Fabian *et al.* (1981), p. 163.
HEDIN L., 1974, in Azároff (1974), p. 226.
HEILIG, K., AND A. STEUDEL, 1974, *At. Data Nucl. Data Tables* **14**, 613.
HENRIKSON, H. E., AND F. BOEHM, 1974, *Nucl. Instrum. Methods* **121**, 269.

HOFFMANN, B., G. BAUR, AND J. SPETH, 1984, *Z. Phys. A* **315**, 57.
HUANG, K.-N., M. AOYAGI, M. H. CHEN, B. CRASEMANN, AND H. MARK, 1976, *At. Data Nucl. Data Tables* **18**, 243.
JOHANSSON, B., 1974, *Phil. Mag.* **30**, 469.
JOHANSSON, B., 1975, *Phys. Rev. B* **11**, 2740.
JOHANSSON, B., 1978, *J. Phys. Chem. Solids* **39**, 467.
JOHANSSON, B., AND N. MÅRTENSSON, 1980, *Phys. Rev. B* **21**, 4427.
JOHANSSON, B., AND A. ROSENGREN, 1975, *Phys. Rev. B* **11**, 2836.
KESSLER, D., H. MES, A. C. THOMPSON, H. L. ANDERSON, M. S. DIXIT, C. K. HARGROVE, AND R. J. MCKEE, 1975, *Phys. Rev. C* **11**, 1719.
KNOWLES, J. W., 1979, *Nucl. Instrum. Methods* **162**, 677.
KOCH, H. R., H. G. BÖRNER, J. A. PINSTON, W. F. DAVIDSON, J. FANDON, R. ROUSSILLE, AND O. W. B. SCHULT, 1980, *Nucl. Instrum. Methods* **175**, 401.
KOPFERMANN, H., 1958, *Kernmomente* (Academic, New York).
KOROSTYLEVA. L. A., 1957, *Opt. Spektrosk.* **3**, 536.
KRAUSE, M. O., AND J. H. OLIVER, 1979, *J. Phys. Chem. Ref. Data* **8**, 329.
LAW, J., AND A. SUZUKI, 1982, *Phys. Rev. C* **25**, 514.
LEDERER, C. M., V. S. SHIRLEY, E. BROWNE, J. M. DAIRIKI, R. E. DOEBLER, A. A. SHIHAB-ELDIN, L. J. JARDINE, J. K. TULI, AND A. B. BUYRN. 1978, *Table of Isotopes*, Seventh Ed. (Wiley, New York).
LEE, P. L., AND F. BOEHM, 1971, *Phys. Lett.* **35B**, 33.
LEE, P. L., AND F. BOEHM, 1973, *Phys. Rev.* C **8**, 819.
LEE P. L., E. C. SELTZER, AND F. BOEHM, 1972, *Phys. Lett.* **38A**, 29.
LEE, P. L., F. BOEHM, AND P. VOGEL, 1974, *Phys. Rev. A* **9**, 614.
LEE, P. L., F. BOEHM, AND A. A. HAHN, 1978, *Phys. Rev. C* **17**, 1859.
LINDGREN, I., A. TIMMERMANN, AND E. MATTHIAS, 1981, in Book of Abstracts for the European Conf. on Atomic Physics, Heidelberg 1981 (European Physical Society, Heidelberg), Vol. 5A, Part 1, p. 235.
MALAGUTI, F., A. UGUZZONI, E. VERONDINI, AND P. E. HODGSON, 1982, *Riv. Nuovo Cimento* **5**, 1.
MANNING, T. E., G. E. ANDERSON, AND W. W. WATSON, 1950, *Phys. Rev.* **78**, 417.
MARTIN, P. C., AND R. J. GLAUBER, 1958, *Phys. Rev.* **109**, 1307.
MCGRORY, J. B., AND B. A. BROWN, 1982, Proceedings Conf. on Lasers in Nuclear Physics, Oak Ridge, Tenn., 1982 (to be published) and references therein.
MERTON, T. R., 1919. *Nature* **104**, 406.
MOLKANOV, L. I., YU. S. GRUSHKO, I. M. BAND, G. A. SHADRINA, AND M. B. TRZHASKOVSKAYA, 1976, *Zh Eksp. Teor. Fiz.* **70**, 2219 (in Russian).
MORITA, M., 1973, *Prog. Theor. Phys.* **49**, 1574.
MOSCATELLI, F. A. O. REDI, H. H. STROKE, AND R. L. WIGGINS, 1983, *J. Opt. Soc. Am.* **73**, 156.
MYERS. W. D., 1969, *Phys. Lett.* **30B**, 451.
MYERS, W. D., 1977, *Droplet Model of Atomic Nuclei* (Plenum, New York).
MYERS, W. D., AND W. J. SWIATECKI, 1974, *Ann. Phys.* (*N.* Y.) **84**, 186.
NAGEL, D. J., AND W. L. BAUN, 1974, in Azároff (1974), p. 445.
NELSON, G. C., W. JOHN, AND B. G. SAUNDERS, 1969, *Phys. Rev.* **187**, 1; 1970, *Phys. Rev. A* **2**, 542.
NESTOR, C. W., T. C. TUCKER, T. A. CARLSON, F. B. MALIK, AND C. FROESE, 1966, Oak Ridge National Laboratory report No. ORNL-4027.
OTOZAI, K., R. ARAKAWA, AND T. SAITO, 1978, *Nucl. Phys. A* **297**, 97.
OTTEN, E.-W., 1981, Proc. 4th Int. Conf. on Nuclei far from Stability, Helsingør, 1981 (CERN, Geneva CERN 81-09), p. 3.
PETROVICH. E. V., YU. P. SMIRNOV, V. S. ZYKOV, A. I. GRUSHKO, O. I. SUMBAEV, I. M. BAND, AND M. B. TRZHASKOVSKAYA, 1971, *Zh. Eskp. Teor. Fiz.* **61**, 1756 [English transl.: *Sov. Phys. JETP* **34**, 935].

POWERS, R. J., P. BARREAU, B. BIHOREAU, J. MILLER, J. MORGENSTERN, J. PICARD, AND L. ROUSSEL, 1979, *Nucl. Phys. A* **316**, 295.

PRIMAKOFF, H., AND F. T. PORTER, 1953, *Phys. Rev.* **89**, 930.

RAVN, H. L., 1979, *Phys. Rep.* **54**, 201.

REBEL, H., 1980, Kernforschungszentrum Karlsruhe report No. 2974.

REBEL, H., AND SCHATZ, 1982, Kernforschungszentrum Karlsruhe report No. 3344.

REINER, A. S., AND L. WILETS, 1962, *Nucl. Phys.* **36**, 457.

RINKER, G. A., AND J. SPETH, 1978, *Nucl. Phys. A* **306**, 360.

ROSE, TH., AND G. L. BORCHERT, 1985 (to be published).

ROSÉN, A., AND I. LINDGREN, 1968, *Phys. Rev.* **176**, 114.

ROSÉN, A., AND I. LINDGREN, 1972, *Phys. Scr.* **6**, 109.

RUPNIK, T., AND B. CRASEMANN, 1972, *Phys. Rev. C* **6**, 1780.

RYCHEL, D., 1980, Diploma-Thesis, Institute for Nuclear Physics, Johannes Gutenberg University, Mainz.

RYL'NIKOV, A. S., A. I. EGOROV, G. A. IVANOV, V. I. MARUSHENKO, A. F. MEZENTSEV, A. I. SMIRNOV, O. I. SUMBAEV, AND V. V. FEDOROV, 1972, *Zh. Eksp. Teor. Fiz.* **63**, 53 [English transl.: *Sov. Phys. JETP* **36**, 27 (1973)].

SAITO, T., A. SHINOHARA, AND K. OTAZAI, 1980, *Phys. Lett.* **92B**, 293.

SANDSTRÖM, A. E., 1957, in *Handbuch der Physik*, edited by S. Flügge (Springer, Berlin), p. 78.

SAUDER, W. C., J. R. HUDDLE, J. D. WILSON, AND R. E. LAVILLA, 1977, *Phys. Lett.* **63A**, 313.

SCHNOPPER, H. W., 1967, *Phys. Rev.* **154**, 118.

SCHULT, O. W. B., 1965, *Kerntechnik* **7**, 309.

SELTZER, E., 1969, *Phys. Rev.* **188**, 1916.

SEPPI, E. J., H. E. HENRIKSON, F. BOEHM, AND J. W. M. DUMOND, 1962, *Nucl. Instrum. Methods* **16**, 17.

SERBER, R., AND H. S. SNYDER, 1952, *Phys. Rev.* **87**, 152.

SHABUROV, V. A., I. M. BAND, A. I. GRUSHKO, T. B. MEZENTSEVA, YU. P. PETROVICH, A. E. SMIRNOV, A. E. SOVESTNOV, O. I. SUMBAEV, M. B. TRZHASKOVSKAYA, AND I. A. MARKOVA, 1973, *Zh. Eksp. Teor. Fiz.* **65**, 1157 [English transl.: *Sov. Phys. JETP* **38**, 573].

SKARDHAMAR, H. F., 1970, *Nucl. Phys.* **A151**, 154.

SMIRNOV, Y. P., O. I. SUMBAEV, E. V. PETROVICH, V. S. ZYKOV, A. I. EGOROV, AND A. I. GRUSHKO, 1969, *Zh. Eksp. Teor. Fiz.* **57**, 1139 [English transl.: *Sov. Phys. JETP* **30**, 622].

SPETH, J., L. ZAMICK, AND P. RING, 1974, *Nucl. Phys. A* **232**, 1.

STACEY, D. N., 1966, *Rep. Progr. Phys.* **29**, 171.

STEUDEL, A., 1952, *Z. Phys.* **133**, 438.

SUMBAEV, O. I., 1957, *Zh. Eksp. Teor. Fiz.* **32**, 1276 (in Russian).

SUMBAEV, O. I., 1959, *Izv. Akad. Nauk SSSR, Ser. Fiz.* **23**, 880 (in Russian).

SUMBAEV, O. I., 1969, *Zh. Eksp. Teor. Fiz.* **57**, 1716 [English transl.: *Sov. Phys. JETP* **30**, 927 (1970)].

SUMBAEV, O. I., 1976, Chemical Effects in X-Ray Spectroscopy, in *Modern Physics and Chemistry* (Academic, London), Vol. 1, p. 31.

SUMBAEV, O. I., 1978, *Usp. Fiz. Nauk* **124**, 281 [English transl.: *Sov. Phys. Usp.* **21**, 141].

SUMBAEV, O. I., AND E. G. LAPIN, 1980, *Zh. Eksp. Teor. Fiz.* **78**, 803 [English transl.: *Sov. Phys. JETP* **51**, 403].

SUMBAEV, O. I., AND A. F. MEZENTSEV, 1965a, *Zh. Eksp. Teor. Fiz.* **48**, 445 [English transl.: *Sov. Phys. JETP* **21**, 295].

SUMBAEV, O. I., AND A. F. MEZENTSEV, 1965b, *Zh. Eksp. Teor. Fiz.* **49**, 459 [English transl.: *Sov. Phys. JETP* **22**, 323 (1966)].

SUMBAEV, O. I., AND A. I. SMIRNOV, 1963, *Nucl. Instrum. Methods* **22**, 125.

SUMBAEV, O. I., A. F. MEZENTSEV, V. I. MARUSHENKO, A. S. RYL'NIKOV, AND G. A. IVANOV, 1969, *Yad. Fiz.* **9**, 906 [English transl.: *Sov. J. Nucl. Phys.* **9**, 529].

THIBAULT, C., 1981, Proc. 4th Int. Conf. on Nuclei far from Stability, Helsingør, 1981 (CERN, Geneva, CERN 81-09), p. 47.
THOMPSON, R. C., A. HANSER, K. BEKK, G. MEISEL, AND D. FRÖLICH, 1982, *Z. Phys.* **A305**, 89.
TIMMERMANN, A., 1980, *Z. Phys. A* **296**, 93.
UHER, R. A., AND R. A. SORENSEN, 1966, *Nucl. Phys.* **86**, 1.
VAN EIJK, C. W. E., AND F. SCHÜTTE, 1970, *Nucl. Phys. A* **151**, 459.
VAN EIJK, C. W. E., AND M. J. C. VISSHER, 1970, *Phys. Lett.* **33B**, 349.
WALEN, R. J., AND C. BRIANÇON, 1975, in *Atomic Inner-Shell Processes*, edited by B. Crasemann (Academic, New York), Vol. 1, p. 233.
WANG, K. C., 1979, thesis, Dept. of Physics, California Institute of Technology.
WANG, K. C., A. A. HAHN, AND F. BOEHM, 1978a, *Phys. Rev. A* **17**, 1735.
WANG, K. C., A. A. HAHN, F. BOEHM, AND P. VOGEL, 1978b, *Phys. Rev. A* **18**, 2580.
WU, C. S., AND S. A. MOSZKOWSKI, 1965, *Beta Decay* (Wiley-Interscience, New York).
YAMAZAKI, Y., H. D. WOHLFAHRT, E. B. SHERA, M. V. HOEHN, AND R. M. STEFFEN, 1979, *Phys. Rev. Lett.* **42**, 1470.

CHAPTER 7

Atomic Physics Research with Synchrotron Radiation

Bernd Crasemann and François Wuilleumier

1. INTRODUCTION

Two decades after the first use of synchrotron radiation in atomic spectroscopy, and nine after the discovery of X rays, we stand at the threshold of major advances in facilities for the production of synchrotron light that are certain to engender a great increase in its use for the exploration of atomic inner shells. This appears to be a good time, therefore, for some retrospection and a survey, albeit brief and incomplete, of applications of synchrotron light in atomic and molecular physics.

X rays have made a profound impact upon science, medicine, and technology, starting almost immediately upon discovery of the new radiation by Wilhelm Conrad Röntgen during the Christmas holidays of 1895. Applications to military medicine were among the first: for an expedition undertaken by Great Britain in Africa in 1896, the medical department of the War Office ordered two sets of Röntgen-ray apparatus to be sent up the Nile, to be used by the army surgeons in locating bullets in soldiers (Anon., *Electrical World*, 1896; Anon., *Scientific American*, 1896). In the battle of Tsushima during the Russo-Japanese war, the renowned cruiser *Aurora* carried a laboratory in which the naval doctor V. S. Kravchenko is said to have used X rays for diagnosis on 40 wounded sailors (Blokhin, 1957).

The key role of X rays in the development of modern physics is well known. Highlights include the discovery of X-ray diffraction by Max von Laue and its application to the analysis of crystal structure by W. L. Bragg (both in 1912); the unequivocal determination of atomic numbers of the

Bernd Crasemann • Department of Physics and Chemical Physics Institute, University of Oregon, Eugene, Oregon 97403. François Wuilleumier • LURE et Laboratoire de Spectroscopie Atomique et Ionique, 91405 Orsay Cedex, France.

elements through Moseley's law (1913); the discovery of radiationless transitions by Pierre Auger and the conclusive demonstration of the particle nature of photons by A. H. Compton, both in 1923.

The use of X rays from Röntgen tubes as a tool for research in atomic physics has been limited primarily by the fact that appreciable intensity of the radiation is only attained in a few characteristic lines of the target material, of fixed frequencies for any given anode. The potential for the use of photons, from the vacuum ultraviolet to hard X rays, for investigations of atomic structure and transitions has been vastly enhanced in the last few years by the advent of synchrotron radiation facilities (Fig. 1). The characteristics of these powerful new sources and their range of applications have been reviewed in several monographs (Sokolov and Ternov, 1968; Basov, 1976; Kunz, 1979; Winick and Doniach, 1980; Crasemann and Ice, 1985). Here we summarize only some of the basic concepts.

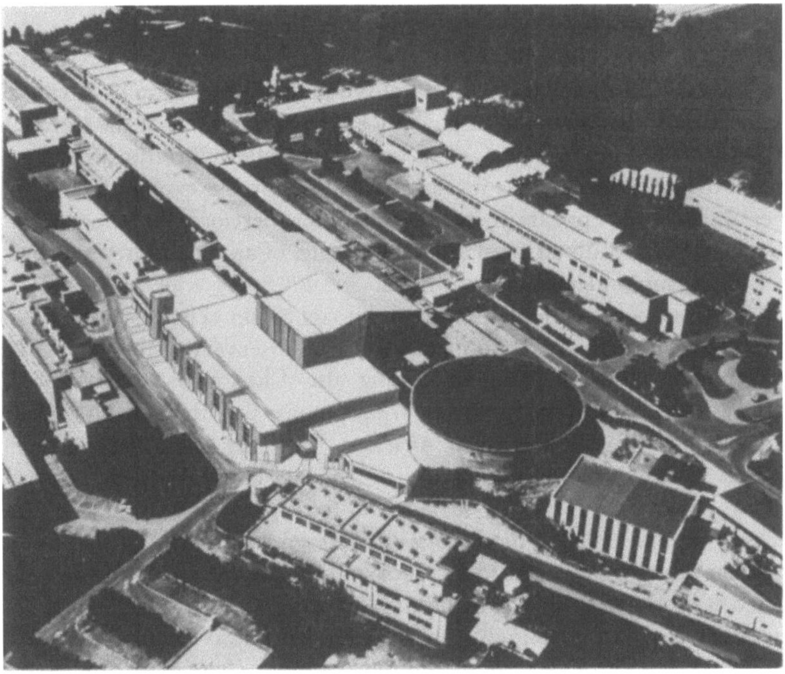

FIGURE 1. Aerial view of the LURE site in Orsay, France, with the electron linear accelerator in the upper left. Two rings currently generate synchrotron radiation: DCI in the "igloo" in the right foreground, and ACO in the tall square building at the target end of the linear accelerator. In front of the ACO structure, an artist's sketch indicates the building that will contain France's newest and most powerful synchrotron-radiation facility, Super-ACO, scheduled for completion in 1986.

2. SYNCHROTRON RADIATION

The electromagnetic radiation emitted by electrons that move at relativistic velocities under transverse acceleration in storage rings has unique properties that make it an ideal probe of atomic and molecular structure. Intensity and tunability are the most important of these properties. Free atoms generally need to be studied in the gas phase, and the tenuity of the target must be offset by a high incident photon flux if significant signal is to be obtained. The power radiated in the bending magnets of storage rings is considerable; it is proportional to the fourth power of the electron energy and inversely proportional to the radius of curvature of the orbit. Thus, 100 mA of 3-GeV electrons confined to a circular path of 10-m radius emit more than 70 kW of electromagnetic radiation. The electrons radiate in a tight cone in the forward direction, of root-mean-square aperture mc^2/E or 0.2 mrad for $E = 3$ GeV. Substantial flux can thus be brought to experimental apparatus well removed from the ring. From a 5-mrad segment of the orbit in the foregoing example (an amount that can be readily used by a single experiment), more than 3×10^{12} photons per second are contained within a 2-eV bandwidth at 1.8 keV. Formidable as this flux is, it can be enhanced yet further by several orders of magnitude through the use of insertion devices, described below.

The spectrum of synchrotron radiation from the bending magnets of a storage ring is produced by the superposition of many harmonics of the fundamental frequency of revolution of the electrons; the individual harmonic lines are smeared out into a continuum that extends from the visible through the vacuum-ultraviolet and soft X-ray range into the hard X-ray regime (Fig. 2). The spectrum is characterized by a "critical photon energy" ε_c: half the radiated power is contained in photons of less than ε_c, and half above. The critical photon energy is proportional to the magnetic field and to the square of the stored-electron energy. The photon flux peaks at $0.4\varepsilon_c$ and falls off nearly exponentially above ε_c.

For virtually all experimental applications, a slice of the synchrotron-radiation spectrum is selected by monochromatization—with diffraction gratings at the longer wavelengths and by Bragg-crystal diffraction in the X-ray range. Resolution better than 1 eV at 5 keV and 0.1 eV at 100 eV is attained, and the photon energy is readily tuned over a wide range of the spectrum by adjusting the monochromators, usually under computer control.

The intensity of synchrotron radiation can be greatly enhanced and its spectral distribution modified through the use of insertion devices—wigglers and undulators—that are now coming into widespread use (Farge, 1980; Spencer and Winick, 1980; Brown *et al.*, 1983). These insertion devices are periodic arrays of dipole magnets or bifilar-wound solenoids placed in a

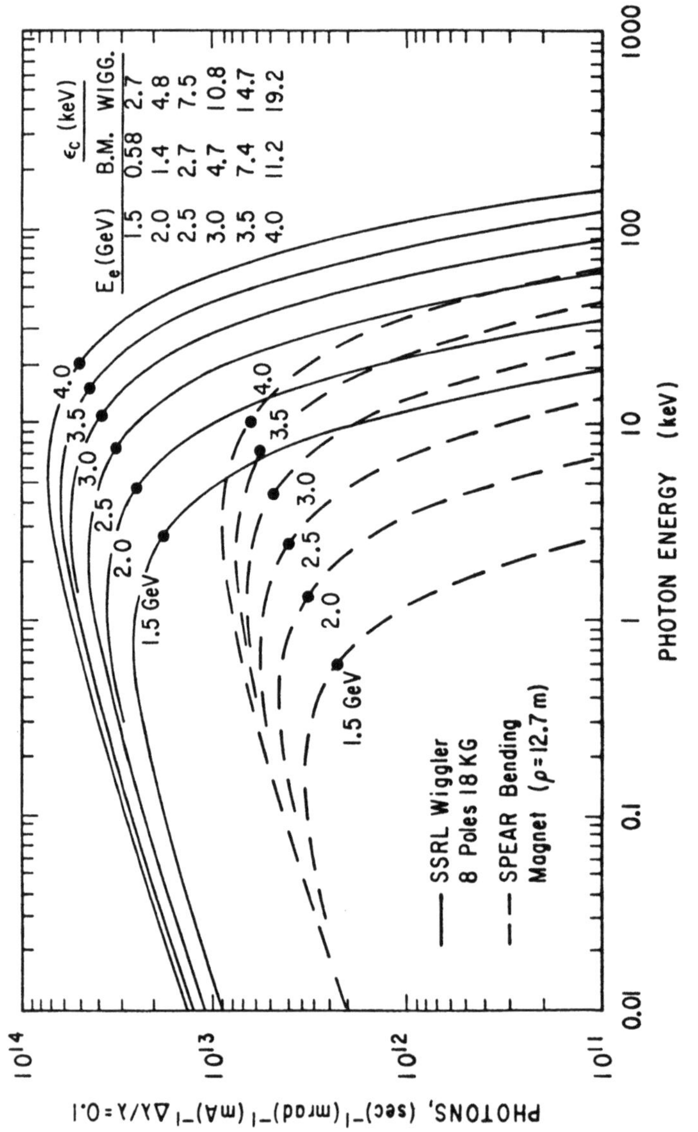

FIGURE 2. Spectral distribution of the synchrotron radiation produced by a SPEAR bending magnet and by an 8-pole wiggler in the Stanford Synchrotron Radiation Laboratory.

straight section of a storage ring. Conventional or superconducting electromagnets or power-saving and compact permanent magnets can be used. The spatially alternating magnetic field causes the electrons to oscillate about their straight path, but the electrons leave the device without net deflection or displacement.

In wigglers, the angular excursions of the electron beam are relatively large; the spectrum of the emitted radiation has the conventional shape but the intensity is enhanced by as much as the number of pole pairs, due to superposition of the radiation from the individual source points. The critical energy is increased by the ratio of the magnetic field in the wiggler to that in the bending magnets, making it possible for a low-energy ring to provide radiation that extends into the X-ray spectrum.

In undulators, the angular excursion of the electron beam is comparable to the natural opening angle of synchrotron-radiation emission, mc^2/E. The radiation from an undulator is therefore concentrated in the smallest possible opening angle. The result is very high brightness. It becomes possible to attain very high resolution upon monochromatization, because the resolution is normally limited by the geometry: source size and angular divergence of the photon beam. At least equally important is the modification of the spectral shape produced by an undulator. With a large number of poles, interference effects in the radiation produced at essentially colinear source points result in a spectrum with quasimonochromatic peaks. In the limit of very small angular excursions and many poles, the radiation is concentrated in the fundamental peak. The suppression of radiation of other energies results in very high brightness, exceeding that produced in ring-bending magnets by nearly the square of the number of undulator pole pairs. Concentration of the formidable intensity of undulator radiation in a narrow frequency band allows exceedingly high flux to be delivered to experimental target areas without placing intolerable heat loads on the optical components of the beam line.

The generation of stimulated radiation from interaction of the circulating electron beam with the spatially oscillating magnetic field of an undulator has recently been achieved, in the first successful operation of a storage-ring free-electron laser (Billardon *et al.*, 1983). It now seems reasonable to predict the realization, within the next few years, of a free-electron laser that is continuously tunable from ~ 1800 Å to the infrared.

Progress with the design of wigglers and undulators has made it highly attractive to develop a next generation of sources that will rely almost exclusively upon insertion devices for the production of synchrotron radiation (*SuperAco Storage Ring, The*, 1983; Thompson *et al.*, 1981; Winick *et al.*, 1983; Sah *et al.*, 1983; Bienenstock *et al.*, 1983). Once such facilities are built and appropriate instrumentation has been developed, the use of synchrotron radiation in all areas of research, but especially in atomic physics, can be expected to take a large quantum jump.

While intensity and tunability are the most important properties of synchrotron radiation for the purposes of atomic investigations, there are other extremely useful features (Bienenstock and Winick, 1983). These include a sharp time structure (typically, nanosecond pulses separated by as much as a microsecond) that makes it possible to conduct time-of-flight electron and ion spectrometry. Synchrotron radiation has a very high degree of polarization, reaching 100% in the orbital plane at the critical energy. This property has become particularly useful for angle-resolved photoelectron spectrometry. It is furthermore suggested that superposition of radiation from two undulators can be used to produce circularly polarized X rays and open intriguing additional possibilities for experimentation.

3. TECHNIQUES

Absorption spectrometry is the simplest and historically oldest method, first used with ultraviolet synchrotron radiation in a pioneering program at the National Bureau of Standards in the early 1960s. This work made great impact as it led to the discovery of a wealth of discrete resonances in the photoionization continua of rare gases. The first atomic absorption spectrum obtained with synchrotron radiation, that of He in the 200-Å region, revealed series of autoionizing states (Fig. 3) (Madden and Codling, 1963; Cooper *et al.*, 1963; Codling, 1979). The discovery was of crucial importance for the understanding of electron correlation effects because it represents the simplest possible example of the highly correlated motion of two excited electrons.

In more recent times, X-ray absorption measurements have been used successfully to analyze the structure of inner-shell edges in rare gases, leading to precise determinations of hole-state energy levels (Fig. 4) (Breinig *et al.*, 1980). Total photoabsorption cross sections exhibit the effects of the deviation of the atomic potential from the Coulomb potential and of the inadequacy of independent-particle central-field models, clearly showing the influence of correlations and extraatomic effects (Fig. 5) (Wolff *et al.*, 1972). The scope of absorption spectrometry is limited, however, in that the results normally reflect the undifferentiated response of the atomic electrons as a whole. Thus, in the photoabsorption spectrum of Xe in the ultraviolet, we note the absence of 5*s*, 4*d*, 4*p*, and 4*s* edges (Fig. 6); a theoretical analysis of contributions from each subshell reveals that the cross section between 70 and 700 eV is dominated by the 4*d* partial cross section which goes through a Cooper minimum near 200 eV and produces a second maximum around 350 eV (Krause, 1981). Clearly, it is important to make measurements that

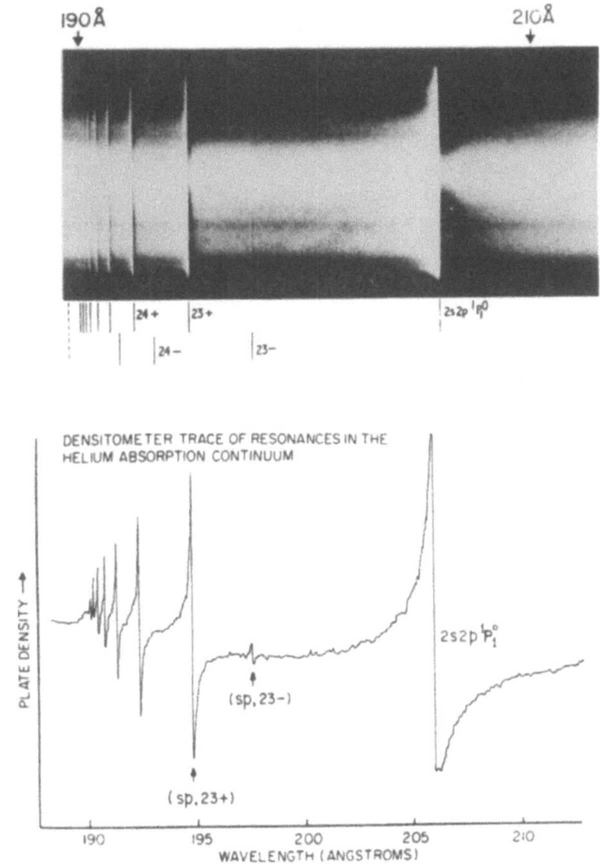

FIGURE 3. Section of the atomic absorption spectrum of He near 60 eV which led to the discovery of autoionizing two-electron excitations (courtesy of Professor K. Codling).

enable one to distinguish between the various exit channels, if the complex gross features are to be understood.

Photoelectron spectrometry provides the necessary discrimination between final states. Combined with tunable incident light, it has proven an extremely valuable tool for the study of electronic structure and dynamics (Wuilleumier, 1981). In order to fully exploit the unique properties of synchrotron radiation, it is important to perform measurements of photoelectron intensity as a function of three independent variables: wavelength of the incident radiation, photoelectron kinetic energy that is characteristic of the atomic level structure, and angle of emission. In addition, the polarization of the incident radiation must be monitored if one wants to

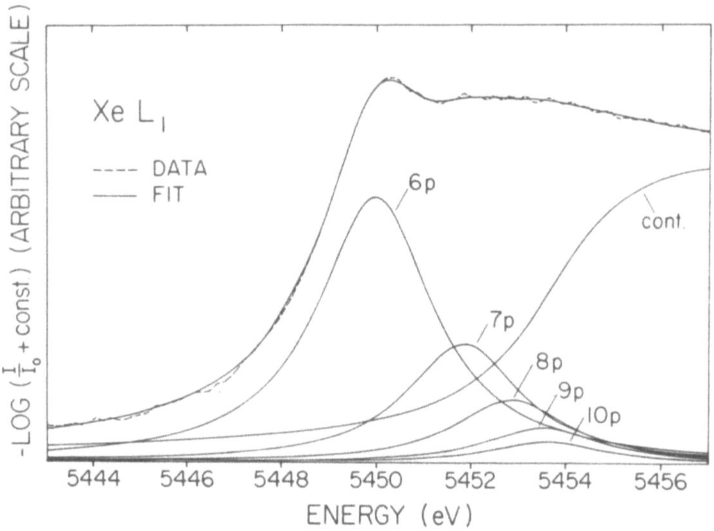

FIGURE 4. The L_{I} absorption edge of Xe, measured under high resolution with synchrotron radiation at SSRL. The analysis shows contributions from partial cross sections for the excitation of 2*s* electrons to *np* states; the limit of the Rydberg series leads to an accurate L_{I} binding energy (from Breinig *et al.*, 1980).

determine the energy-dependent asymmetry parameter $\beta(\varepsilon)$ from the angular distribution. This parameter is important in the complete description of the photoionization process because it is very sensitive to such many-body effects as interchannel coupling. Triply differential photoionization experiments in which all these variables are measured can currently be performed in a number of synchrotron radiation centers in the U.S., France, Germany, England, and Japan; electron spectrometers suitable for this work are cylindrical-mirror or hemispherical electrostatic analyzers or time-of-flight devices (Sonntag and Wuilleumier, 1983).

A last step toward the complete characterization of the final state is taken when the spin polarization of the photoelectron is also determined. With circularly polarized ultraviolet light from the Bonn 2.5-GeV synchrotron, Heinzmann and Schäfers (1980) have shown how one can study the variation of the spin polarization along with the photoionization cross section (Fig. 7); this variation is strong as the incident-photon energy sweeps through autoionization resonances (Fig. 8).

Other techniques applied in connection with synchrotron radiation include Auger and fluorescence spectrometry, which permit study of the details of atomic and molecular deexcitation following photoexcitation, and

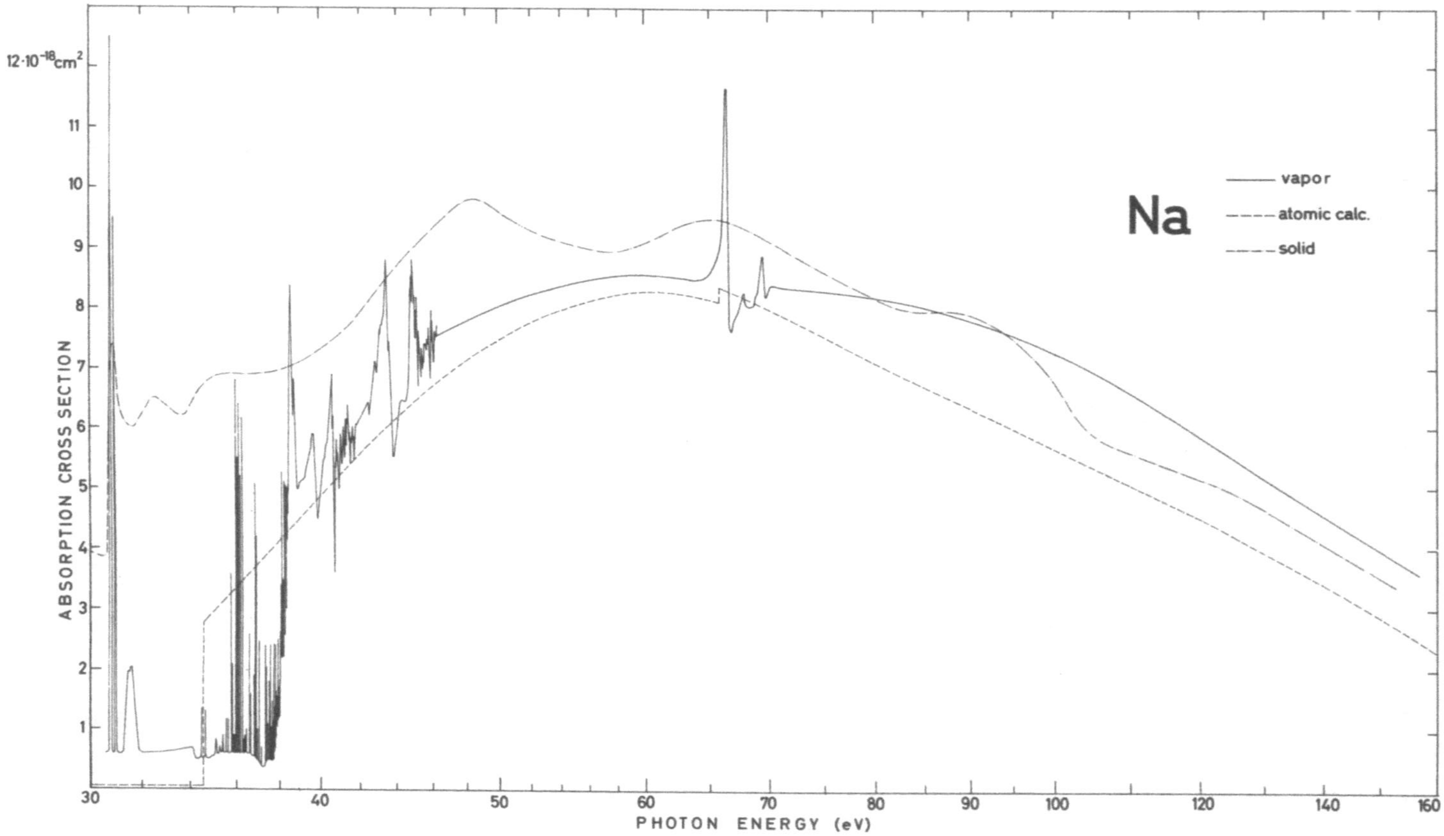

FIGURE 5. Absorption spectrum of Na in solid and vapor forms, measured at DESY (courtesy of Professor Bernd Sonntag).

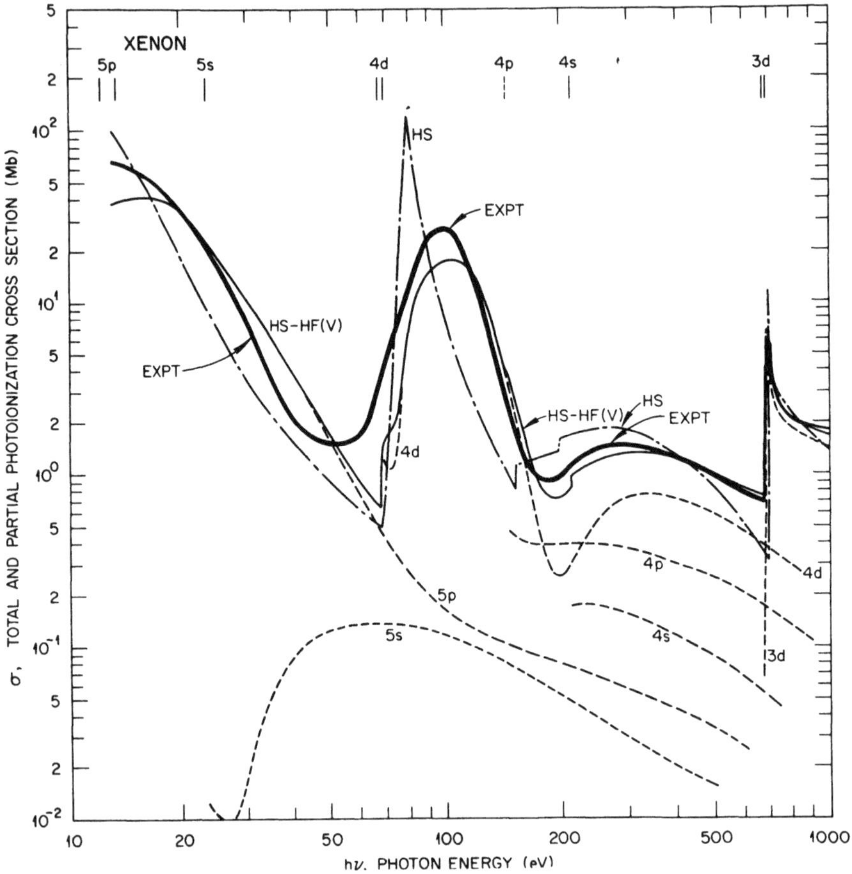

FIGURE 6. The photoabsorption spectrum of Xe. Theoretical results [HS-HF(V)] are partitioned into subshell components (from Krause, 1981).

X-ray scattering, which probes the atomic charge distribution and hence, in its inelastic part, becomes a sensitive indicator of Coulomb correlation in few-electron systems (Fig. 9) (Ice *et al.*, 1978).

4. MANY-ELECTRON EFFECTS

We now turn to some interesting problem areas that are being studied by the techniques which we have mentioned. One prevailing subject is the characterization of many-body effects in atomic structure and transitions, as

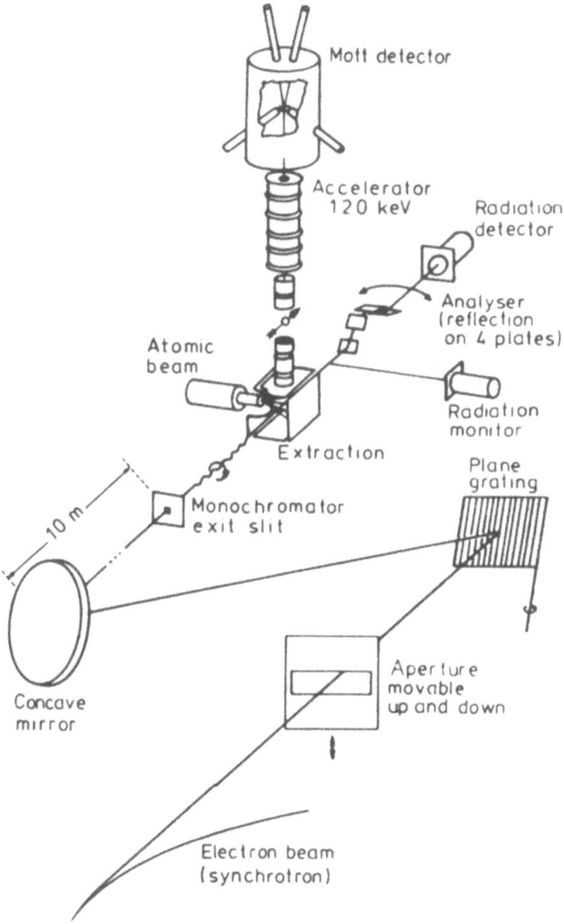

FIGURE 7. Apparatus for measuring the spin polarization of photoelectrons (from Heinzmann and Schäfers, 1980).

a guide for the development of theoretical approaches to describe them. The many-body problem pervades all of physics, to be sure, but what makes atoms an ideal proving ground to pursue it is that here the potential is known, and the dominant charge of the massive nucleus makes the central-field model quite a good first approximation. Many-electron effects can thus be studied and treated selectively—often by perturbative techniques—and insights gained that will carry over into the much more difficult situations encountered in nuclear matter and elsewhere (Wuilleumier, 1976).

Many-body effects often manifest themselves very distinctly in the photoelectron spectra of shallow core levels. For example, the

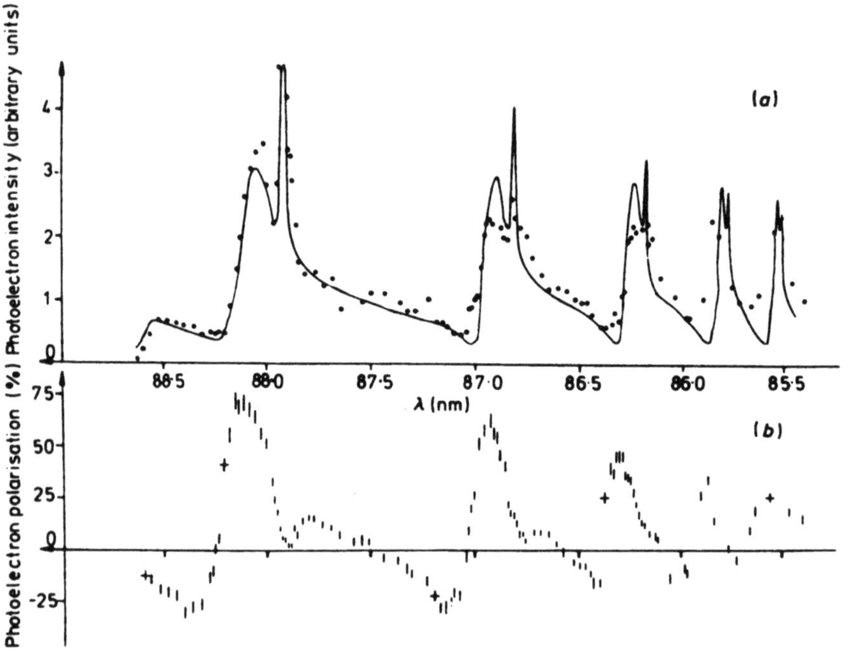

FIGURE 8. Cross section (a) and spin polarization (b) of Kr photoelectrons in the region between the $4p_{3/2}$ and $4p_{1/2}$ ionization thresholds, measured with circularly polarized synchrotron radiation (from Heinzmann and Schäfers, 1980).

photoionization cross sections of transition-metal $3d$ levels go through an enormous excursion near the energy needed to excite an inner $3p$ electron to an empty orbital. This resonant photoemission process was first observed in atomic Cu (Chandesris *et al.*, 1981); it was further investigated systematically in several other elements of the series, viz., Cr, Mn, Fe, Co, and Ni (Bruhn *et al.*, 1982). In Fig. 10, we show the measured intensity of the Cr $3p^6 3d^4 4s\ ^6D$ photoemission line from the work of Bruhn *et al.* (1982). The experimental points are well approximated by a Fano-type line shape. Such a phenomenon is not at all predicted by independent-particle models of the Hartree–Fock type; it arises from interference with the $3p \rightarrow 3d$ excitation, which is activated at this energy and causes an autoionization resonance enhancement in the cross section. A many-body perturbation treatment that includes all of the important excitation and ionization channels correctly reproduces this phenomenon which single-particle calculations are incapable of predicting.

Outer-shell photoionization cross sections can also exhibit strong many-body effects. We have seen in Fig. 6 that $4d$ photoemission dominates

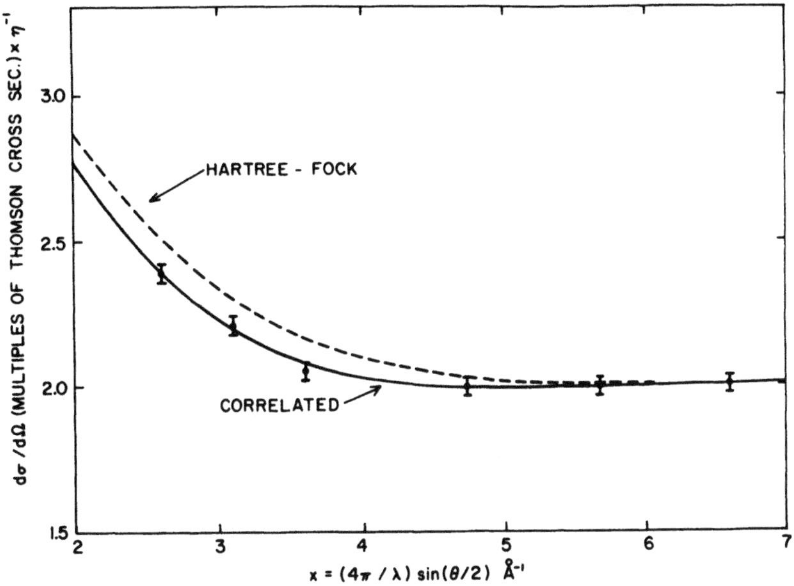

FIGURE 9. X-ray scattering cross section of H_2, showing effect of electron–electron correlation (from Ice *et al.*, 1978).

the total cross section of Xe from the 4*d* ionization threshold near 68 eV to the 3*d* threshold near 680 eV. An interesting effect occurs in that the outer-shell 5*p* and 5*s* photoemission processes are substantially influenced by electron-correlation interactions with the 4*d* subshell, both above and below the 4*d* ionization threshold. This pronounced interchannel-coupling effect, predicted theoretically (Amusia and Ivanov, 1976; Johnson and Cheng, 1979), is confirmed by such data as shown in Fig. 11 where the small 5*s* photoionization cross section is seen to be significantly enhanced by intershell correlations over the region of the shape resonance in the 4*d* cross section (West *et al.*, 1976; Adam *et al.*, 1978).

A striking example of the power of tunable synchrotron radiation to elucidate many-body effects is found in recent measurements of the asymmetric angular distribution of Xe 5*s* photoelectrons, conducted at the DORIS storage ring in Hamburg and at the Tantalus I ring in Wisconsin (Derenbach and Schmidt, 1983; Fahlman *et al.*, 1983). The scarce data on the $5s \rightarrow \varepsilon p$ asymmetry that were previously available seemed to indicate that the relativistic random-phase approximation (RRPA) adequately predicts the angular distribution if the interchannel correlation between the 4*d*-, 5*s*-, and 5*p*-subshell channels is included (Fig. 12). But more detailed data obtained subsequently show that this conclusion was fortuitous, and

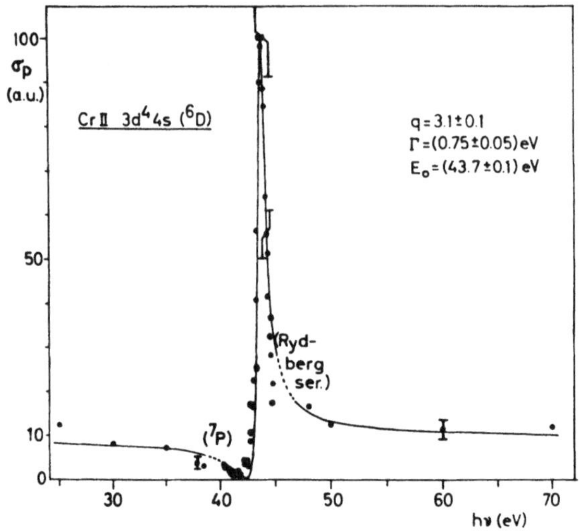

FIGURE 10. Intensity of the Cr $3p^63d^44s$ ^6D photoemission line, as a function of photon energy, showing the "Fano shape" effect of autoionization (from Bruhn *et al.*, 1982).

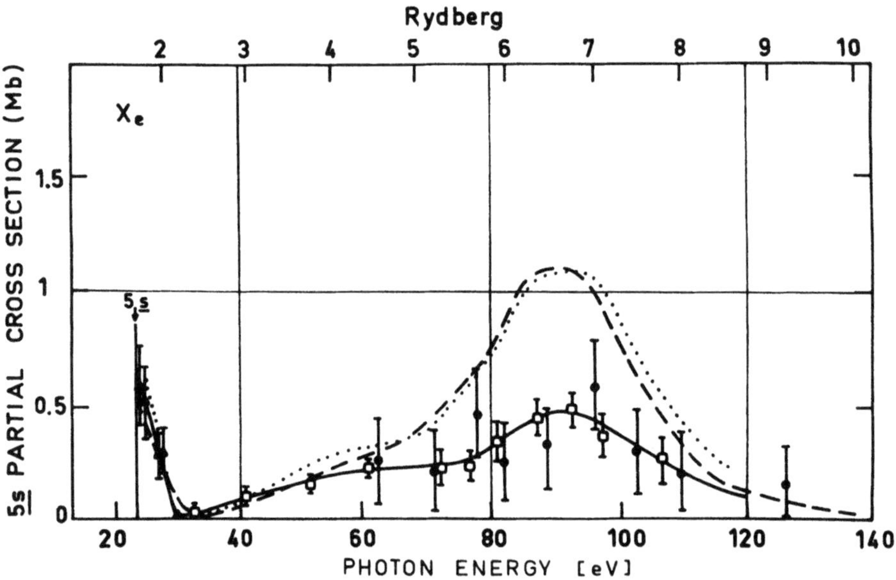

FIGURE 11. The Xe 5*s* photoionization cross section, as a function of photon energy. Curves indicate theoretical results from Amusia and Ivanov (1976) and Johnson and Cheng (1979); experimental data are from West *et al.* (1976) (solid dots) and Adam *et al.* (1978) (open squares) (from Adam *et al.*, 1978).

that more extensive many-body effects must be included in the calculation. It seems quite clear that two-electron promotion channels must be taken into account (Fahlman *et al.*, 1983), which goes beyond the capabilities of RRPA calculations in their present form.

Correlation satellites in photoelectron spectra arise when an incident photon causes the promotion of two atomic electrons, at least one of them into the continuum or to an autoionizing state imbedded in the continuum. We have already mentioned the discovery of such autoionizing doubly-excited states in the absorption spectrum of He (Fig. 4) (Madden and Codling, 1963). For atoms of higher atomic numbers, core-level photoelectron satellite spectra can yield unique information about ground-state correlation. Progress has been made in calculating some double-excitation electron peaks in the sudden approximation; both initial-state and final-state configuration interaction must be included (Wuilleumier, 1976; Martin and Shirley, 1976). Even in the photoelectron spectra of atomic inner shells, for

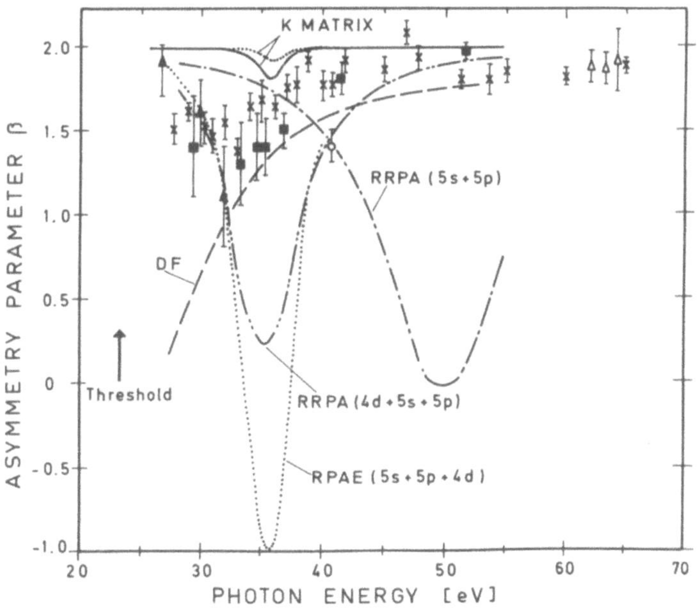

FIGURE 12. Photoelectron angular asymmetry parameter β for the 5*s* subshell of Xe. Earlier data, represented by the open circle, seemed to support the relativistic random phase (RRPA) calculation, but more complete data obtained with synchrotron radiation (Derenbach and Schmidt, 1983; Fahlman *et al.*, 1983) show the inadequacy of that approach and the need for including additional many-body features (from Derenbach and Schmidt, 1983).

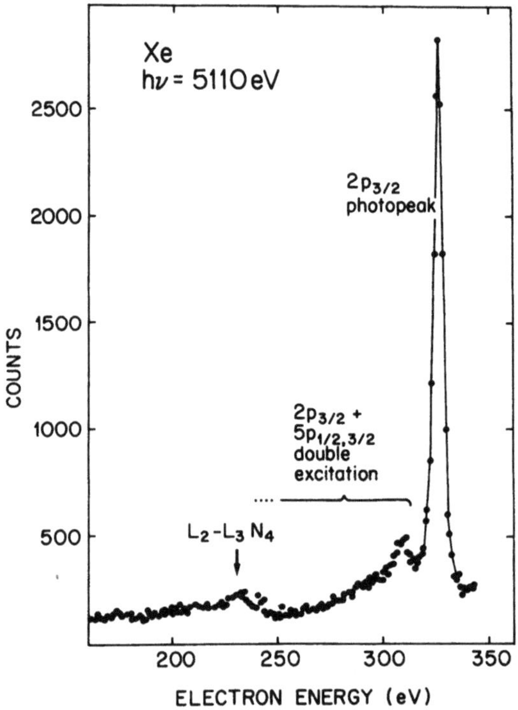

FIGURE 13. Photoelectron spectrum of Xe atoms irradiated with 5.110-keV X rays at SSRL. Below the L_3 photopeak, a satellite distribution arises from the simultaneous excitation or ionization of two electrons (from Armen *et al.*, 1983).

which an independent-particle description tends to be relatively successful, we find signs of double ionization (Fig. 13) (Armen *et al.*, 1983, 1985; Deslattes *et al.*, 1983). Recent measurements of the asymmetry of satellite photoelectron angular distributions add information in a field where experiment is well ahead of theory (Larkins *et al.*, 1982; Shirley *et al.*, 1982).

5. EXCITATION/DEEXCITATION DYNAMICS

The production of excited states of atoms and their deexcitation under emission of photons or Auger electrons is usually considered a two-step process: the decay is separated from the excitation phase through intervening relaxation of the electron cortège, and the photon and electron emissions are treated as independent processes by means of first-order time-dependent perturbation theory. This two-step model of atomic excitation/deexcitation has the advantages of simplicity and tractability, but it breaks down in particular near the thresholds of inner-shell excitation and when interference arises between competing deexcitation channels. An important advance in the fundamental understanding of atomic transitions will be brought about by a description that unifies the formation and decay of autoionizing and inner-shell vacancy states.

Substantial inroads have already been made in the difficult problem of creating a unified theory of atomic excitation and deexcitation. Åberg and collaborators have drawn upon the fact that any excitation process which gives rise to emission of particles with characteristic energies, including photons, is an example of a resonance scattering process (Åberg, 1980; Åberg and Howat, 1982; Åberg and Tulkki, this volume, Chap. 10). These authors have hence worked toward an understanding of the decay of autoionizing and inner-shell vacancy states by electron and photon emission from a unified point of view by using time-independent multichannel scattering theory. Armstrong and collaborators have computed interference effects between radiative and radiationless decay modes of atomic excited states; such interference arises because the decay channels are not in fact independent (Armstrong *et al.*, 1978). The final-state coupling between the electron and photon continua has been further examined and interpreted by Haan and Cooper (1983). In an attempt to arrive at a theory that can encompass relaxation and interference effects in a consistent and systematic way, Wendin and collaborators have used a diagrammatic kind of random-phase approximation with exchange (Ohno and Wendin, 1979; Wendin, 1982). A rather complete one-step model of X-ray emission and Auger processes has been developed by Gunnarsson and Schönhammer (1982), who have derived a formalism which describes these processes as a single quantum-mechanical event.

These difficult theoretical efforts are mostly still at a conceptual stage where quantitative predictions are not easily extracted, and where the guidance of experimental data is of the essence for further elaboration. Synchrotron radiation has provided the means for measuring the threshold-excitation and interference effects that signal the breakdown of the two-step model.

The resonant Raman effect represents the epitome of the failure of the two-step approximation: here excitation and deexcitation are described by a single matrix element. Resonant Raman scattering can be thought of as the inelastic analog of resonance fluorescence; both the radiative or X-ray (Eisenberger *et al.*, 1976a, b; Briand *et al.*, 1981) and radiationless or Auger (Brown *et al.*, 1980) versions of the process have now been seen with synchrotron light. In the resonant Raman effect, a photon of exactly the right energy (within the lifetime width) promotes an atomic electron to an excited state which decays through an X-ray or Auger transition, all of this occurring in a single step. The emission line exhibits linear dispersion and subnatural linewidth at resonance.

In the other extreme, a photon with large excess energy can remove an atomic electron rapidly into the continuum, the atomic wave functions adjust themselves to the altered potential, and subsequent deexcitation occurs in a quite separate step. The two opposite regimes, resonant Raman vs. wholly distinct excitation and deexcitation, are not separated by a drastic discontinuity. As one might expect, there is a phenomenon that links the two and leads gradually from one into the other.

Postcollision interaction is the mechanism that bridges excitation and deexcitation to varying extent, feeding energy across and smoothly linking resonant Raman scattering with the opposite two-step extreme. Postcollision interaction arises in the near-threshold region when the slowly receding

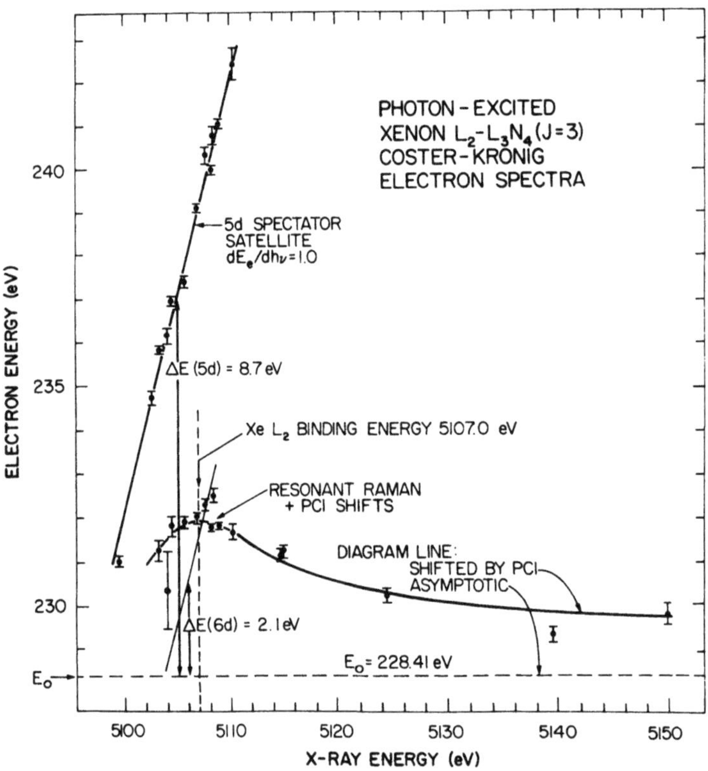

FIGURE 14. Threshold behavior of a L_2–L_3N_4 Coster–Kronig line from photoionized Xe. Incident X rays near 5107 eV (the Xe L_2 binding energy) produce Auger resonant Raman scattering, with linear dispersion of the Coster–Kronig electron energy. Above threshold, the Coster–Kronig electrons have gained an energy increment from postcollision interaction; only in the high-X-ray-energy limit does the Coster–Kronig energy approach its "diagram" value of 228.4 eV, and the deexcitation becomes independent of the excitation process (from Crasemann *et al.*, 1983).

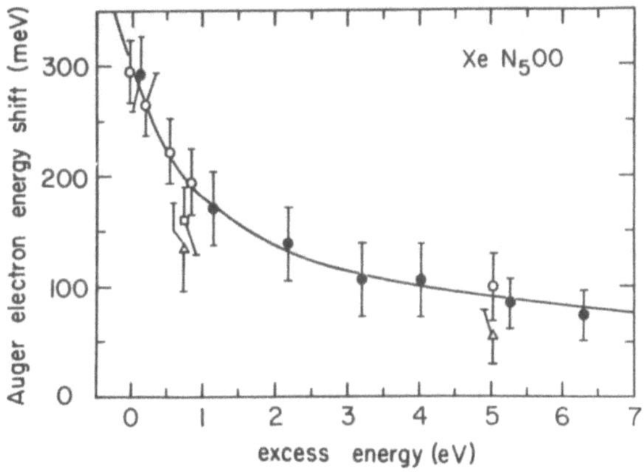

FIGURE 15. Postcollision interaction shift of Xe N_5OO Auger electrons, as function of photon energy above the $4d_{5/2}$ ionization threshold (from Southworth *et al.*, 1983).

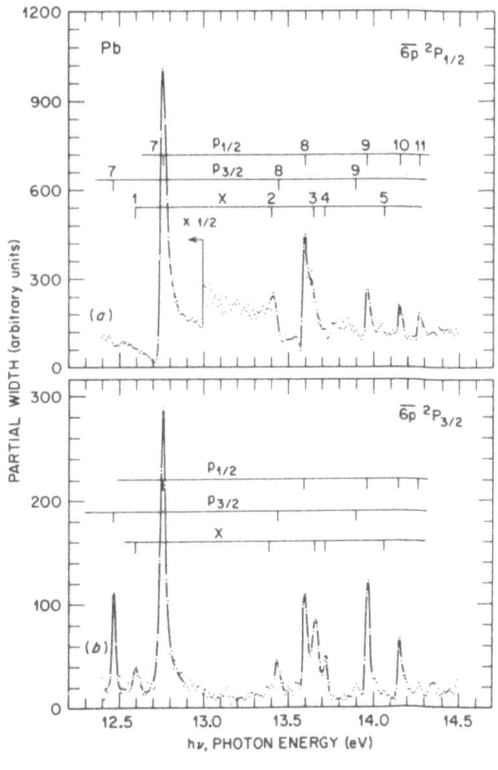

FIGURE 16. Partial widths of the two exit channels in the $6s^26p^2 \rightarrow 6s6p^2(^4P_{1/2})$, $np(1/2; 3/2)$ series of Pb. The photon bandpass was approximately 22 meV (from Krause *et al.*, 1983).

photoelectron is still within, or near, the residual singly charged ion as the subsequent decay occurs. If this decay consists of Auger-electron emission, the ion suddenly becomes doubly charged. The photoelectron instantaneously (in the sudden approximation) finds itself in a deeper potential well; since the photoelectron's kinetic energy cannot increase instantaneously, the energy difference is transferred (through a virtual photon) to the Auger electron. The Auger electron thus gains energy and the photoelectron, already slow to start out with, may end up bound in a Rydberg orbit (Niehaus, 1977; 1985; Schmidt, 1982; Mehlhorn, 1985).

By slowly tuning highly monochromatized X rays through an atomic inner-shell threshold and observing the energy of a suitable Auger peak, the resonant Raman effect and its continuation in the form of postcollision interaction can be traced (Fig. 14) (Crasemann *et al.*, 1983). In Fig. 15, we illustrate the effect of postcollision interaction in the Auger decay of a 4*d*

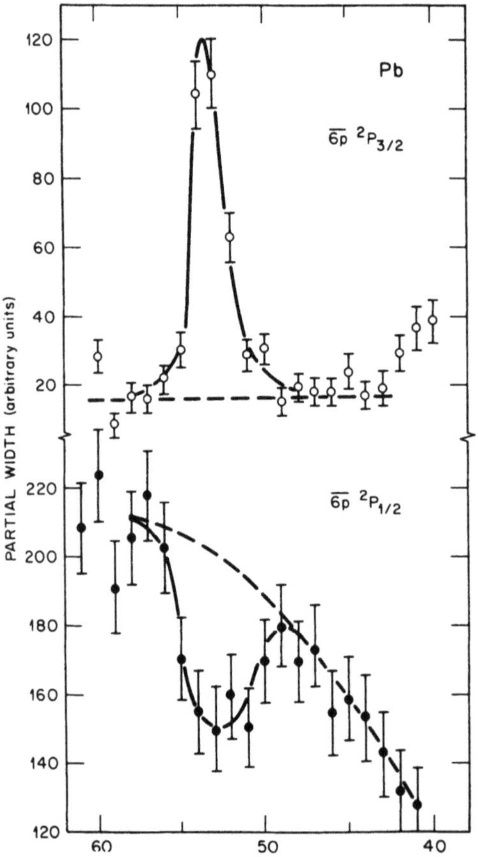

FIGURE 17. Decay of Pb $[6s]np$ $(n = 7)$ states into $[6p]$ $^2P_{1/2}$ and $^2P_{3/2}$ final states. Constructive interference between autoionization and direct photoemission channels enhances $^2P_{3/2}$ (top), while $^2P_{1/2}$ exhibits the effect of destructive interference (bottom) (from Krause *et al.*, 1983).

vacancy in Xe atoms (Schmidt *et al.*, 1977, 1981; Southworth *et al.*, 1983); only far above threshold does the Auger energy drop to its "diagram" value.

Not only are atomic excitation and deexcitation tied together through such mechanisms as resonant Raman scattering and postcollision interaction, but the various channels are not truly independent and can, in suitable circumstances, be observed to interact and interfere with one another. A demonstration of this effect has been obtained by M. O. Krause and collaborators (1983) at the Wisconsin Synchrotron Radiation Center in the excitation of a Pb $6s$ electron to a Rydberg-like np state. The excited state undergoes radiationless decay into either of two single-electron channels: $[6p]\ ^2P_{1/2}$ or $[6p]\ ^2P_{3/2}$ (where the square brackets denote hole states). These same final states can, however, also be reached through direct photoionization of a $6p$ electron. The analysis of the electrons emitted in the decay of the $6s^2 6p^2 + h\nu \rightarrow 6s6p^2(^4P_{1/2})\, np_{3/2}$ resonance series of Pb shows surprisingly strong constructive interference of the autoionization resonance with the $[6p]\ ^2P_{3/2}$ direct ionization channel and destructive interference with the $[6p]\ ^2P_{1/2}$ channel, in a most striking demonstration of interchannel interaction (Figs. 16, 17).

6. PHOTOIONIZATION OF ATOMS IN EXCITED STATES

Until now, the vast majority of photoionization measurements has been performed on atoms and molecules in their ground states. Very valuable information has been gained, as noted in the preceding sections of this chapter, yet the field has been limited: dipole selection rules confine the class of final states that can be reached, and the initial states often are ensembles of near-degenerate levels (particularly in open-shell atoms), which complicates the interpretation. If atoms can be prepared in an excited initial state for photoionization, these limitations can be circumvented, and furthermore, knowledge can be gained on the behavior of excited atoms and molecules which is of great potential importance in such fields as plasma physics, astrophysics, and photochemistry.

In a few cases it has been possible to prepare atoms in specific initial states with the use of high-power, frequency-tunable lasers, and to photoionize these laser-excited states (Wuilleumier, 1982). Available laser energies and tunability ranges have restricted these experiments to outer electrons of alkali and alkaline-earth atoms. In some atoms placed into excited states by means of a pulsed laser, it has been possible to measure inner-shell photoabsorption over an energy range from 30 to 100 eV by using ultraviolet continuum radiation from a "BRV" (Ballofet-Romand-Vodar) source (Sugar *et al.*, 1979).

The high intensity and broad spectrum of synchrotron radiation invites an extension of these scarce excited-state photoionization measurements. Such work was recently initiated at Orsay (Bizau *et al.*, 1982); we describe two examples of early results. Light in the 20– to 200-eV photon-energy range from the ACO ring was monochromatized with a toroidal grating monochromator and made to impinge upon a weakly collimated metal-vapor beam, in the source volume of a cylindrical-mirror electron-energy analyzer (Fig. 18). A tunable ring dye laser was placed so as to deliver up to one watt of power (approximately 10 W/cm^2) in the 5000– to 6000-Å region to the source volume. For measurements on Na atoms, the laser was locked to the $3\,^2S_{1/2}\,(F=2) \rightarrow 3\,^2P_{3/2}\,(F=3)$ transition, and for studies of Ba atoms, to the $6s^2\,^1S_0 \rightarrow 6s6p\,^1P_1$ transition.

Typical results of the Na experiment are shown in Fig. 19 (Bizau *et al.*, 1983a). In the absence of laser radiation (top frame), the spectrum arises from ionization of the $1s^22s^22p^63s\,^2S$ ground state of Na. The 31.78-eV fundamental transmitted by the monochromator ionizes $3s$ electrons (binding

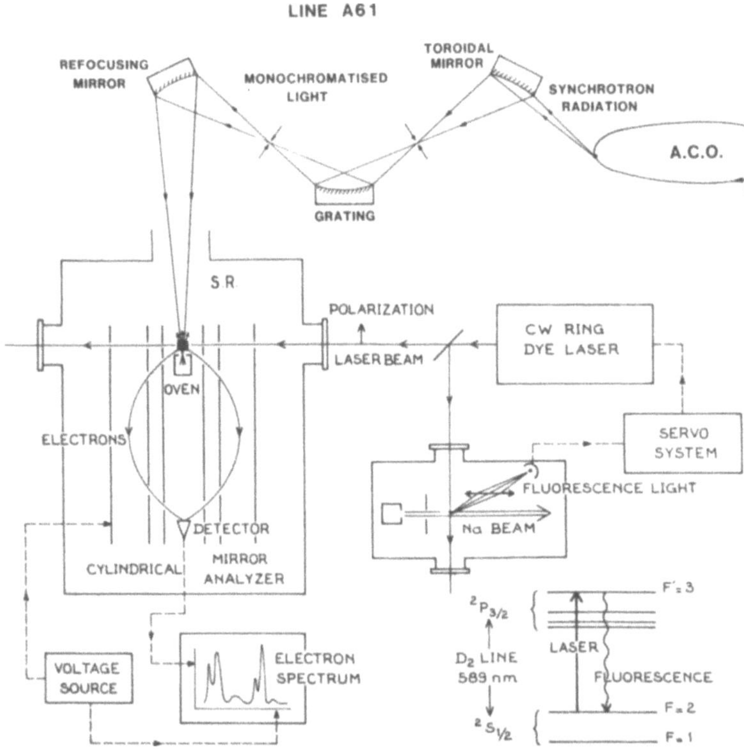

FIGURE 18. Experimental arrangement for studying the photoionization of excited atoms by combining the use of synchrotron and laser radiations (from Bizau *et al.*, 1983a).

energy 5.14 eV); the 63.56-eV first harmonic causes ionization of $2p$ electrons (binding energies 41–43 eV). The structure observed in Fig. 19 between 41 and 43 eV binding energy arises from cases in which the residual positive ions, ionized in the $2p$ subshell, are left in a $2p^5 3p\ ^2P$ excited state because of electron correlations.

Two new features stand out when the laser is turned on (Fig. 19, lower frame). The first is a photoelectron peak that appears at a binding energy of 40 eV, corresponding to direct ionization of a $2p$ electron in the laser-excited

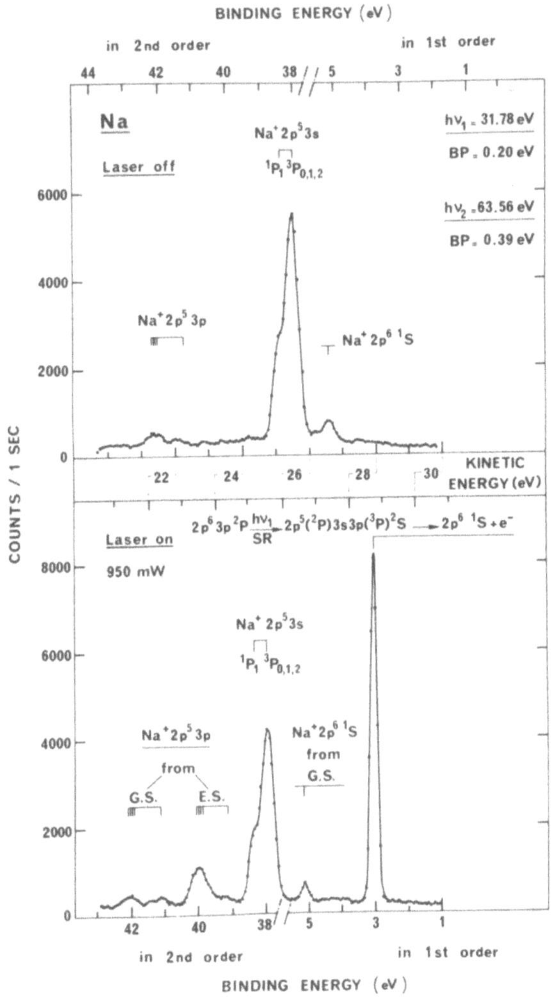

FIGURE 19. Spectra of photoelectrons ejected from Na atoms in the ground state (top, laser off) and in the excited $2p^6 3p\ ^2P_{3/2}$ state (bottom, laser on) (from Bizau *et al.*, 1983a).

$2p^63p\ ^2P$ Na atoms. Because screening of the $2p$ electrons by the outer electron is reduced in this excited state, the $2p$ binding energy is greater than in ground-state atoms. The intensity ratio of the peaks at 40 and 38 eV binding energy indicates the relative population of atoms in the excited state (Carré *et al.*, 1983).

The second new feature in the lower frame of Fig. 19 is the intense electron peak at 3.03 eV binding energy. This line arises from the decay of an even-parity autoionizing state excited by stepwise laser–vuv two-photon absorption; the transition scheme is noted in the figure. When the

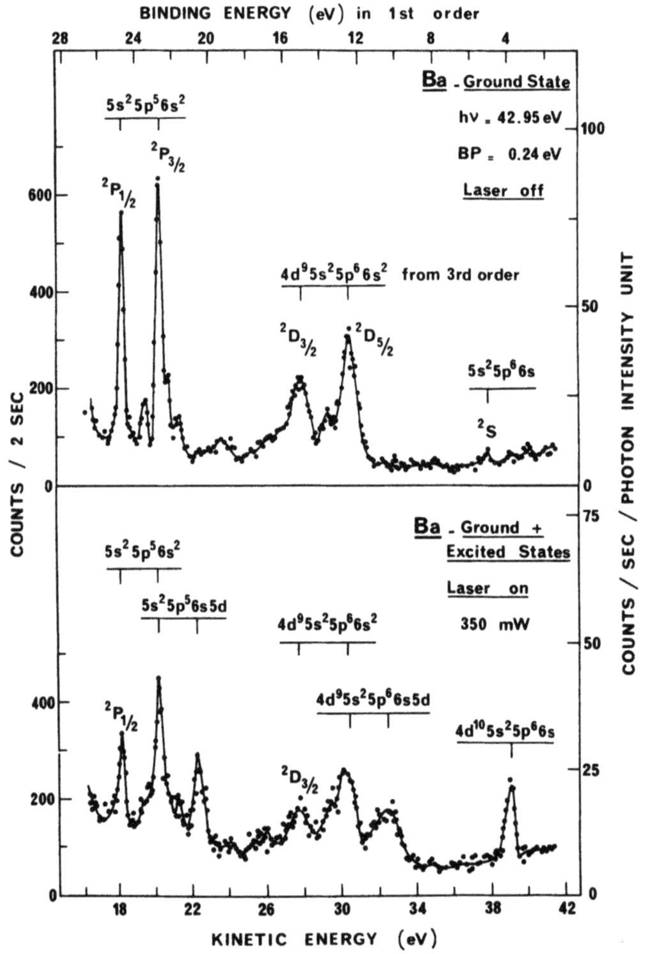

FIGURE 20. Spectra of photoelectrons ejected from Ba in the ground state (top, laser off) and in the excited $6s5d\ ^2D_1$ state (bottom, laser on) (from Bizau *et al.*, 1983b).

monochromator was tuned slightly off the 31.78-eV energy, this peak disappeared because the nonresonant photoemission of $3p$ electrons is very weak at this photon energy. To the extent that the radiative decay of the doubly excited state is negligible, the oscillator strength for the excitation of such even-parity resonances via inner-shell photoabsorption can be determined by comparing the relative intensities of the photoelectron peak (with known photoionization cross section) and of the autoionization peak. Once such a determination has been made, the intensity of the autoionization peak can be used as a very sensitive indicator of the density of excited states in the medium.

It can be noted in Fig. 19 that the kinetic energy of all electrons is reduced when the laser is on. This effect is due to a plasma potential drop that develops in the laser-excited medium through purely collisional processes—associative ionization and Penning ionization—that produce molecular and atomic ions. The use of synchrotron radiation for calibration purposes in collision experiments appears to be very useful in deriving quantitative information on such processes (Bizau *et al.*, 1983b).

As a second example of the combined use of synchrotron and laser light we show in Fig. 20 spectra of photoelectrons ejected from Ba atoms in ground and excited states. With the monochromator set at 42.95 eV in first order, but transmitting higher orders as well, the $6s$, $5d$, $5p$, and $4d$ subshells are ionized. Most of the excited atoms are in the $6s5d\,^2D_1$ metastable state, hence the binding energies of the inner electrons in the excited atoms are shifted to lower values: the mean radius of a $5d$ orbital is smaller than that of a $6s$ orbital. It is worthwhile noting that the spectra in Fig. 20 show the direct photoionization of excited electrons to the continuum. The photoionization cross section of the $5d$ excited electron is greatly enhanced over that of $6s$ electrons. From spectra such as these, it is possible to extract the photoionization cross sections, as functions of energy over a considerable range, of $5d$ and $4d$ electrons in excited atoms.

7. MOLECULAR PHYSICS

Applications of synchrotron radiation to the study of molecules is an active field so closely interlaced with the main subject of this chapter that it needs to be surveyed briefly here. To be sure, molecular absorption spectroscopy has been pursued long before the advent of synchrotron radiation facilities, and even with conventional light sources has led to significant observations, e.g., of the effect of autoionization on vibrational branching ratios and angular distributions. The advent of high-intensity, high-resolution monochromatic beams of vuv photons from synchrotron radiation has, however, made it possible to explore a large variety of dynamical processes in molecular photoionization over wide energy ranges. The growing expansion of these investigations and their importance for such related disciplines as atmospheric physics, photochemistry, and the study of reactive collisions would justify a thorough review. We limit ourselves here, however, to pointing out some special features that were introduced into photoionization studies through the molecular field, and we give a few examples that illustrate the present status and possible future developments of research in this area.

Most of the progress in molecular spectroscopy with synchrotron radiation mirrors work in atomic photoionization. The aim is to gain insight into the processes that occur during photoexcitation and escape of the photoelectron through the anisotropic molecular field, in terms of such physical observables as photoionization cross sections and branching ratios, photoelectron angular distributions, and spin polarization of the photoelectrons (Dehmer *et al.*, 1983). Additional concepts and techniques are required to deal with the strictly molecular aspects of the problem, namely, the anisotropy of the multicenter molecular field, the interaction between rovibronic modes, and the new channels opened in the final states of the processes by the possible dissociation of the molecular ions. In particular, large progress in the study of the various fragmentation channels was made possible by the development of the specific-threshold photoelectron–photoion coincidence technique (Nenner *et al.*, 1980).

Shape resonances in the photoionization spectra of molecules are one of the concepts that have attracted much interest during recent years (Dehmer and Dill, 1976). Shape resonances result from the admixture of high-angular-momentum components in the outgoing electron wave. The effective potential seen by the photoelectron can then be rendered repulsive by centrifugal effects on these high-l components, over some range of the electron coordinates. This barrier potential can temporarily trap the photoelectron in a region of the molecular core and delay penetration of the continuum wave until sufficiently high kinetic energy has been gained by the photoelectron. First discovered in atoms, this effect shows up most clearly in the absorption spectra of certain molecules. As an example, the sulfur L-shell absorption spectrum of SF_6 is shown in Fig. 21; the H_2S sulfur L spectrum is also shown for reference, because hydrogen atoms do not contribute appreciably to shape resonance effects (Zimkina and Vinigradov, 1971). The latter photoabsorption spectrum exhibits the qualitative behavior that one might expect for a core-electron spectrum, showing intense discrete excitation of a $2p$ electron to an empty valence orbital, followed by partially resolved Rydberg structure which converges to a smooth continuum. Even the "delayed onset" of the $2p \rightarrow \varepsilon d$ transition to the continuum can be noted. In sharp contrast, the photoabsorption spectrum of the same sulfur $2p$ subshell in SF_6 (Blechschmidt *et al.*, 1972) shows three

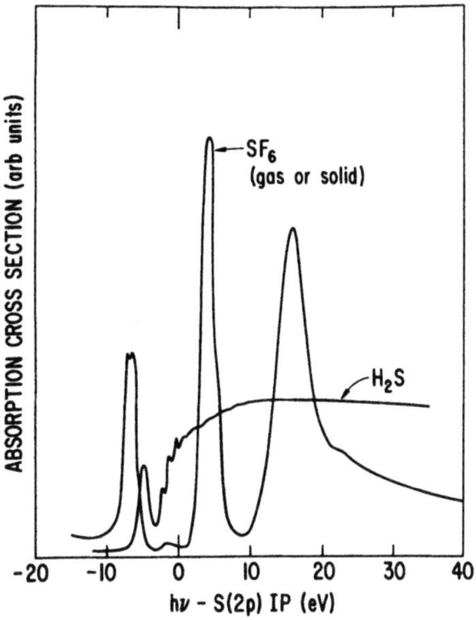

FIGURE 21. Photoabsorption spectra of H_2S (Zimkina and Vinigradov, 1971) and SF_6 (Blechschmidt *et al.*, 1972) near the sulfur $L_{2,3}$ edge (from Dehmer *et al.*, 1983).

intense, broad peaks, one below the ionization threshold and two above, without any apparent Rydberg structures and accompanied by a reduction of the continuum cross section elsewhere. This dramatic difference in the photoabsorption spectrum can be attributed to potential-barrier effects that produce three shape-resonantly enhanced final-state features of different symmetry.

Shape resonances show up clearly in photoelectron branching ratios, partial photoionization cross sections, and photoelectron angular distributions, including those that are vibrationally resolved. Triply differential measurements are providing an impressive amount of data of this type for a large number of molecules. The trapping of the excited electron can result in coupling of the electronic and nuclear motions, leading to a breakdown of the Born–Oppenheimer approximation. Consequently the electronic transition matrix element is no longer independent of the vibrational quantum numbers, affecting the relative strengths of the various vibrational transitions and the angular distribution of the photoelectrons. With sufficiently high energy resolution of both the monochromator and the electron spectrometer, it is possible to investigate such vibrationally resolved effects. In Fig. 22, the relative branching ratio for the first two v levels of $N_2^+ X^2\Sigma_g^+$ is shown as a function of photon energy (West *et al.*, 1980). As predicted by theory, the vibrational branching ratio oscillates about the Franck–Condon factor over a range of some 30 eV. One observes here the breakdown of the

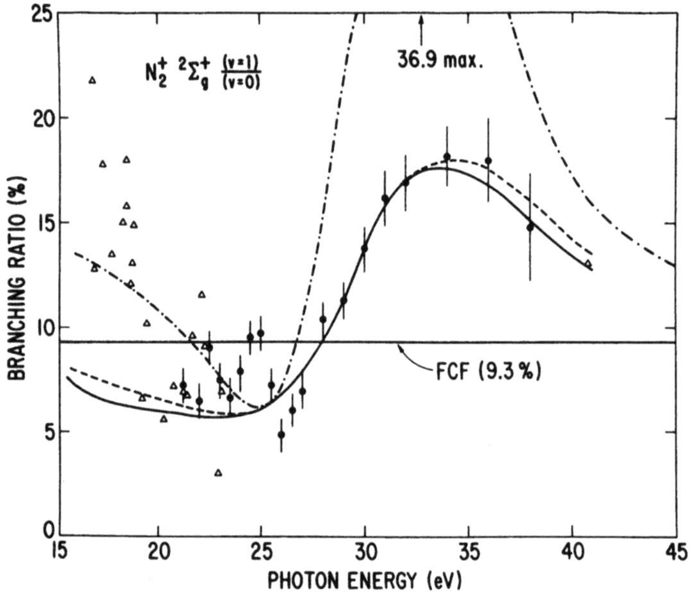

FIGURE 22. Branching ratios for the production of the $v = 0, 1$ levels of $N_2^+X\Sigma_g^+$ by photoionization of N_2 (from West *et al.*, 1980).

Born–Oppenheimer approximation caused by the shape resonance in the $3\sigma_g$ channel.

Angle-resolved photoelectron spectrometry with synchrotron radiation is also used to study such effects in molecules as manifestations of Cooper minima that can arise in atomic orbitals which possesses a radial node. In particular, the asymmetry parameters are strongly affected and undergo a fast change near a Cooper minimum, as in the atomic case. As an example, we show in Fig. 23 the asymmetry parameter β for photoionization of the first three orbitals of chlorine, $2\pi_g$, $2\pi_u$, and $5\sigma_g$ (Carlson *et al.*, 1983). Deep dips in β are observed in the vicinity of the Cooper minimum. Calculations based on the MSXα method have led to good qualitative agreement with these measurements.

Autoionization in molecular photoionization has not yet been studied as extensively as the properties of the continuum wave functions in the single-particle model. As is well known, autoionization is a multichannel process in which a resonantly excited discrete state from one channel couples to the underlying electronic continua of one or more channels to produce ionization. It was originally discovered in atoms (cf. Figs. 3, 10), for which the famous Fano–Beutler profiles in the photoionization cross sections have been known for many years. In molecules, manifestations of autoionization

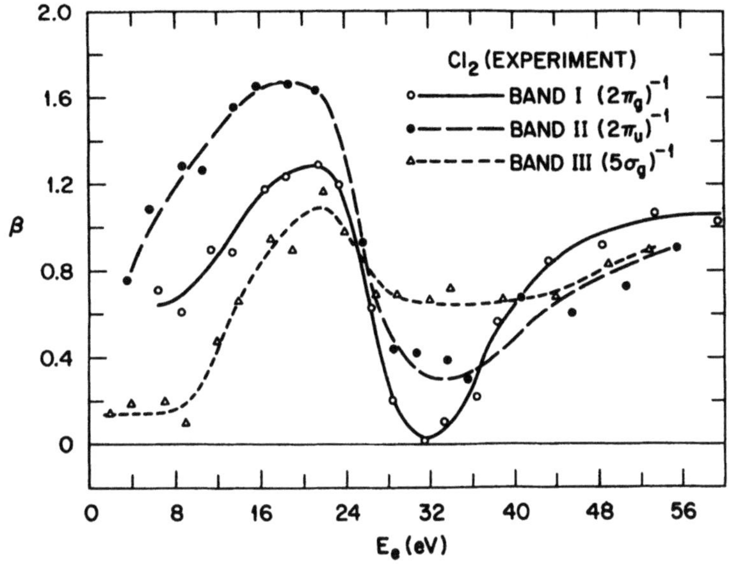

FIGURE 23. Angular-distribution asymmetry parameters *β*, plotted as functions of photoelectron energy for the first three bands in the photoelectron spectrum of Cl_2 (from Carlson *et al.*, 1983).

in photoionization branching ratios and photoelectron angular distributions have only recently come into focus. Very high resolution is necessary for such measurements, and the best experimental case to date has been studied with laboratory light sources (Dehmer and Chupka, 1976). The requisite resolution is, however, now being approached in some synchrotron radiation laboratories, and some data are becoming available from triply differential studies carried out on complex autoionization structures in O_2 (Morin *et al.*, 1982), CO (Ederer *et al.*, 1981), and CO_2 (Parr *et al.*, 1982). In some of these experiments, it was possible to resolve the effects arising from the competition between overlapping and autoionizing resonances.

Even in as short a review of molecular dynamics as this one, we must point out the vast possibilities opened by synchrotron radiation for the study of dissociation processes, occurring either in competition with or subsequent to molecular photoionization. Two unique features of synchrotron radiation come into play: tunability that permits preparation of a molecular ion with a well-defined internal energy, and the sharp time structure of the light that permits measurements of threshold photoelectron-photoion coincidences by time-of-flight spectrometry (Nenner *et al.*, 1980). The efficiency of the technique is illustrated in Fig. 24 for the case of N_2O. The threshold photoelectron spectrum (TPES) is shown, as well as the total ion yield curve; some coincidence spectra recorded at selected photon energies are included to

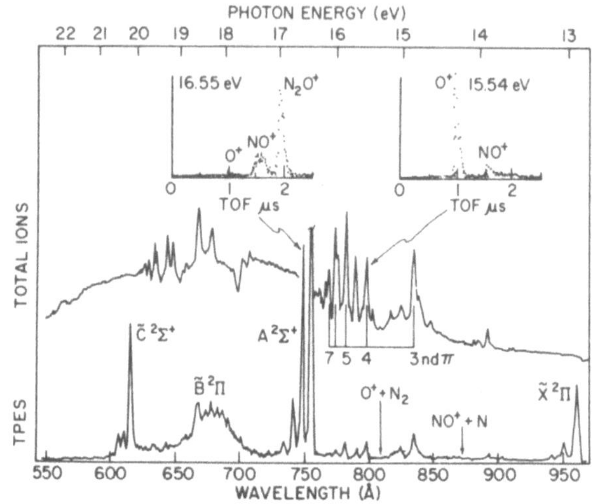

FIGURE 24. Threshold photoelectron (TPES) and total-ion spectra of N_2O. Insets are the coincidence mass spectra at the indicated energies (from Govers *et al.*, 1983).

illustrate how the branching ratios between various dissociation channels vary dramatically with wavelength of the incident radiation.

A very interesting result has been obtained by Eberhardt and collaborators (1983) in studies of the fragmentation of small molecules under soft-X-ray excitation. These workers found that in the region of the carbon 1*s* absorption threshold organic molecules decompose into ionic fragments not only after direct core ionization but also following core-to-bound-state (π^* or Rydberg) transitions. The fragmentation pattern changes completely between these types of excitation or ionization. Even more importantly, it was found that in larger molecules like acetone [$(CH_3)_2CO$] site-specific core-to-π^* excitations cause a break in the molecular structure at the site where the optical excitation occurred. This site-specific behavior led the authors to "foresee the use of tunable soft X rays as scalpel-like tools to break large organic molecules around certain selectable atoms (C, N, or O) within the molecule," opening a new area of X-ray-induced chemical reactions (Eberhardt *et al.*, 1983).

Finally, we show in Fig. 25 the result of a preliminary experiment that opens a new class of studies with wide perspectives (Govers *et al.*, 1983). A monochromatic beam of synchrotron radiation from the ACO storage ring was used to prepare N_2^+ ions in a selected state with well-defined internal energy, by means of the threshold photoelectron–photoion coincidence technique. The N_2^+ ions can thus be prepared in various vibrational states ($v=0$ to 4 for the *X* state, $v=0$ to 6 for the *A* state). The beam of

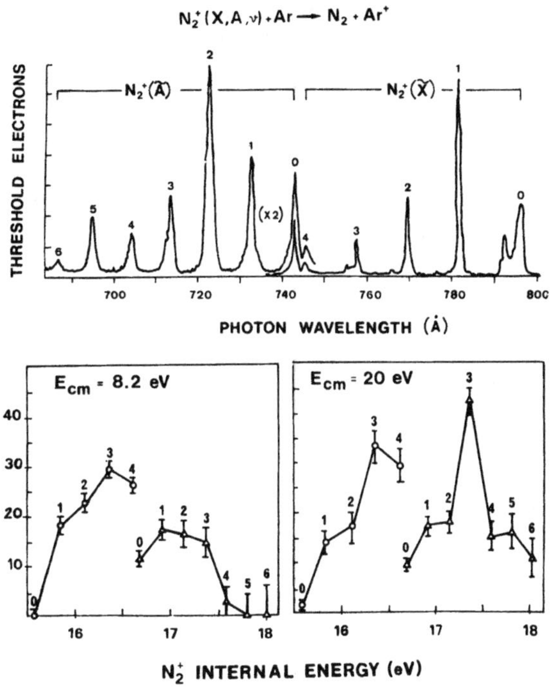

FIGURE 25. Cross section for charge transfer between $N_2^+(X, A, v)$ ions in collisions with Ar atoms, as function of the internal energy of the ion (from Govers *et al.*, 1983).

$N_2^+(X, A, v)$ ions was then allowed to collide with a beam of Ar atoms, and the cross section for charge transfer between N_2^+ ions and Ar atoms was measured as a function of translational energy. Figure 25 shows the threshold electron spectrum used to prepare N_2^+ ions in a selected state, and the variation of the cross section for charge transfer for the various excited states of N_2^+, demonstrating a strong v dependence. As in the case of collision studies with excited Na atoms (Carré *et al.*, 1983), in which the synchrotron radiation was used to characterize the partners in the collision, synchrotron light is used here as a very efficient tool to prepare one of the colliding systems in a known selected state.

8. CONCLUSION

In the two decades since the power of synchrotron radiation in atomic spectroscopy was first demonstrated in the discovery of the autoionizing states of He, use of the radiation from storage rings has expanded and lent access to new areas of absorption and photoemission spectroscopy and scattering experiments. Measurements of Auger and photoelectron angular distributions and spin polarization are lending new insights into many-electron phenomena. Complete characterization of initial and final states has become increasingly possible for atoms as well as molecules. Inner shells of heavy atoms have become accessible, where relativistic and quantum electrodynamic effects play a major role. The dynamics of excitation and deexcitation of deep hole states has become amenable to exploration, as have processes that link these two stages. Central problems of atomic and molecular physics concerning elucidation of the structure and dynamics of many-electron systems are benefiting from the new techniques.

Despite this obvious progress, hard synchrotron light is not yet being used nearly as widely in atomic and molecular physics as in other areas such as surface science, solid state physics, and biology. The reason lies primarily in that gas-phase targets must be employed for the study of free atoms and molecules. The tenuity of these targets entails a need for very-high-flux photon beams; only with the use of undulators and wigglers is the signal-to-noise ratio for many experiments in the X-ray region becoming acceptable. Very elaborate differential-pumping arrangements are required to isolate gaseous targets from the ultrahigh vacuum of storage rings, unless the photon energy exceeds the $\sim$3.2-keV cutoff of Be windows. Source size and horizontal emission angle limit the resolution with which the radiation from bending magnets can be monochromatized.

These difficulties that make the use of synchrotron light for many research applications in atomic and molecular physics somewhat marginal at the present time will be resolved once insertion devices come into wider use in existing machines, and once the next generation of synchrotron radiation facilities is born, starting with Super-ACO in Orsay in 1986, and hopefully followed soon by other low-emittance, high-brilliance, all-insertion-device machines in the U. S. and elsewhere.

ACKNOWLEDGMENTS

We are grateful to many colleagues for their help in assembling material for this review and to M. O. Krause, R. P. Madden, and H. Winick for constructive comments. Some sections of this chapter have been included in an article by authors in *Physics Today*, June 1984. The work was supported in part by the U. S. Air Force Office of Scientific Research.

REFERENCES

Åberg, T., 1980, *Phys. Scr.* **21**, 495.

Åberg, T., and G. Howat, 1982, in *Handbuch der Physik*, Vol. 31, edited by W. Mehlhorn (Springer, Berlin).

Adam, M. Y., F. Wuilleumier, N. Sandner, S. Krummacher, V. Schmidt, and W. Mehlhorn, 1978, *Jpn. J. Appl. Phys.* **17-2**, 170.

Amusia, M. Ya., and V. K. Ivanov, 1976, *Phys. Lett.* **59A**, 194.

Anon., Electrical World, 1896,

Anon., 1896, *Sci. Am.* **75**, 142, August 8.

Armen, G. B., G. S. Brown, M. H. Chen, B. Crasemann, G. E. Ice, and J. C. Levin, 1983, private communication.

Armen, G. B., T. Åberg, K. R. Karim, J. C. Levin, B. Crasemann, G. S. Brown, M. H. Chen, and G. E. Ice, 1985, *Phys. Rev. Lett.* **54**, 182.

Armstrong, L., Jr., C. E. Theodosiou, and M. J. Wall, 1978, *Phys. Rev. A* **18**, 2538.

Basov, N. G., 1975, Ed., *Sinkhrotronnoe izluchenie*, Proceedings of the P. N. Lebedev Institute, Vol. 80 (Nauka Press, Moscow) [English transl.: *Synchrotron Radiation*, Consultants Bureau, New York, 1976].

Benoist d'Azy, O., R. Lopez Delgado, and A. Tramer, 1975, *J. Chem. Phys.* **9**, 327.

Bienenstock, A., and H. Winick, 1983, *Phys. Today*, June, p. 48.

Bienenstock, A. I., T. Elioff, and E. E. Haller, 1983, Eds., Report of the ALS/SSRL Users Workshop (Lawrence Berkeley Laboratory report No. PUB-5095).

Billardon, M., P. Elleaume, J. M. Ortega, C. Bazin, M. Bergher, M. Velghe, Y. Petroff, D. A. G. Deacon, K. E. Robinson, and J. M. Madey, 1983, *Phys. Rev. Lett.* **51**, 1652.

Bizau, J. M., F. Wuilleumier, P. Dhez, D. Ederer, J. L. Picqué, J. L. LeGouët, and P. Koch, 1982, in *Laser Techniques for Extreme Ultraviolet Spectroscopy—1982*, edited by T. J. McIlrath and R. Freeman, AIP Conference Proceedings No. 90 (American Institute of Physics, New York), p. 331.

Bizau, J. M., F. Wuilleumier, P. Dhez, P. Gerard, D. Ederer, J. C. Keller, J. L. LeGouët, J. L. Picqué, B. Carré, and P. Koch, 1983a, private communication.

Bizau, J. M., B. Carré, P. Dhez, D. Ederer, P. Gerard, J. C. Keller, P. Koch, J. L. LeGouët, J. L. Picqué, G. Wendin, and F. Wuilleumier, 1983b, in XIII ICPEAC, Abstracts of Contributed Papers, edited by J. Eichler, W. Fritsch, I. V. Hertel, N. Stolterfoht, and U. Wille, Berlin, p. 27.

Blechschmidt, D., R. Haensel, E. E. Koch, V. Nielsen, and T. Sagawa, 1972, *Chem. Phys. Lett.* **14**, 33.

Blokhin, M. A., 1957, *The Physics of X Rays* (Government Publishing House of Technical-Theoretical Literature, Moscow).

Breinig, M., M. H. Chen, G. E. Ice, F. Parente, B. Crasemann, and G. S. Brown, 1980, *Phys. Rev. A* **22**, 520.

Briand, J. P., D. Girard, V. O. Kostroun, P. Chevalier, K. Wohrer, and J. P. Mossé, *Phys. Rev. Lett.* **46**, 1625.

Brown, G., K. Halbach, J. Harris, and H. Winick, 1983, *Nucl. Instrum. Methods* **208**, 65.

Brown, G. S., M. H. Chen, B. Crasemann, and G. E. Ice, 1980, *Phys. Rev. Lett.* **45**, 1937.

Bruhn, R., E. Schmidt, H. Schröder, and B. Sonntag, 1982, *J. Phys. B* **15**, 2807.

Carlson, T. A., M. O. Krause, F. A. Grimm, and T. A. Whitley, 1983, *J. Chem. Phys.* **78**, 638.

Carré, B., J. M. Bizau, P. Dhez, D. Ederer, P. Gerard, J. C. Keller, P. M. Koch, J. L. LeGouët, J. L. Picqué, G. Spiess, and F. Wuilleumier, 1984, *Opt. Commun.* **52**, 29.

Chandesris, D., C. Guillot, G. Chauvin, J. Lecante, and Y. Petroff, 1981, *Phys. Rev. Lett.* **47**, 1273.

CODLING, K., 1979, in *Synchrotron Radiation, Techniques and Applications*, edited by C. Kunz (Springer, Berlin), Chap. 5.

COOPER, J. W., U. FANO, AND F. PRATS, 1963, *Phys. Rev. Lett.* **10**, 518.

CRASEMANN, B., M. H. CHEN, G. B. ARMEN, J. LEVIN, G. E. ICE, AND G. S. BROWN, *Stanford Synchrotron Radiation Laboratory Report No. SSRL 83/01*, p. VII-54.

CRASEMANN, B., AND G. E. ICE, 1985, *Synchrotron Radiation—A Tool for Research* (Academic, New York), in preparation.

DEHMER, J. L., AND D. DILL, 1976, *J. Chem. Phys.* **65**, 5327.

DEHMER, J. L., D. DILL, AND A. C. PARR, 1983, in *Photophysics and Photochemistry in the Vacuum Ultraviolet*, edited by S. McGlynn, G. Findley, and R. Huebner (Reidel, Dordrecht).

DEHMER, P. M., AND W. A. CHUPKA, 1976, *J. Chem. Phys.* **65**, 2243.

DERENBACH, H., AND V. SCHMIDT, 1983, *J. Phys. B* **16**, L337.

DESLATTES, R. D., R. E. LAVILLA, P. L. COWAN, AND A. HENINS, 1983, *Phys. Rev. A* **27**, 923.

EBERHARDT, W., T. K. SHAM, R. CARR, S. KRUMMACHER, M. STRONGIN, S. L. WENG, AND D. WESNER, 1983, *Phys. Rev. Lett.* **50**, 1038.

EDERER, D. L., A. C. PARR, B. E. COLE, R. STOCKBAUER, J. L. DEHMER, J. B. WEST, AND K. CODLING, 1981, *Proc. R. Soc. London Ser. A* **378**, 423.

EISENBERGER, P., P. M. PLATZMAN, AND H. WINICK, 1976a, *Phys. Rev. Lett.* **36**, 623.

EISENBERGER, P., P. M. PLATZMAN, AND H. WINICK, 1976b, *Phys. Rev. B* **13**, 2377.

FAHLMAN, A., T. A. CARLSON, AND M. O. KRAUSE, 1983, *Phys. Rev. Lett.* **50**, 1114.

FARGE, Y., 1980, *Appl. Opt.* **19**, 4021.

GOVERS, T. R., P. M. GUYON, T. BAER, K. COLE, H. FRÖHLICH, AND M. LAVOLLEE, 1984, *Chem. Phys.* **87**, 373.

GUNNARSSON, O., AND K. SCHÖNHAMMER, 1982, in *X-Ray and Atomic Inner-Shell Physics—1982*, edited by B. Crasemann, AIP Conference Proceedings No. 94 (American Institute of Physics, New York), p. 517.

HAAN, S. L., AND J. COOPER, 1983, *Phys. Rev. A* **28**, 3349.

HEINZMANN, U., AND F. SCHÄFERS, 1980, *J. Phys. B* **13**, L415.

ICE, G. E., M. H. CHEN, AND B. CRASEMANN, 1978, *Phys. Rev. A* **17**, 650.

JOHNSON, W. R., AND K. T. CHENG, 1979, *J. Phys. A* **20**, 978.

KRAUSE, M. O., 1981, *IEEE Trans. Nucl. Sci.* **NS-28**, 1215.

KRAUSE, M. O., F. CERRINA, A. FAHLMAN, AND T. A. CARLSON, 1983, *Phys. Rev. A* **51**, 2093.

KUNZ, C., 1979, Ed., *Synchrotron Radiation, Techniques and Applications* (Springer, Berlin).

LARKINS, F. P., 1982, in *X-Ray and Atomic Inner-Shell Physics—1982*, edited by B. Crasemann, AIP Conference Proceedings No. 94 (American Institute of Physics, New York), p. 530.

MADDEN, R. P., AND K. CODLING, 1963, *Phys. Rev. Lett.* **10**, 516.

MARTIN, R. L., AND D. A. SHIRLEY, 1976, *Phys. Rev. A* **13**, 1475.

MEHLHORN, W., 1985, Chapter 4 of this volume.

MORIN, P., I. NENNER, M. Y. ADAM, M. J. HUBIN-FRANSKIN, J. DELWICHE, H. LEFEBVRE-BRION, AND A. GIUSTI-SUZOR, 1982, *Chem. Phys. Lett.* **92**, 609.

NENNER, I., P. M. GUYON, T. BAER, AND T. R. GOVERS, 1980, *J. Chem. Phys.* **72**, 6587.

NIEHAUS, A., 1985, Chapter 9 of this volume.

NIEHAUS, A., 1977, *J. Phys. B* **10**, 1845.

OHNO, M., AND G. WENDIN, 1979, *J. Phys. B* **12**, 1305.

PARR, A. C., D. L. EDERER, J. L. DEHMER, AND D. M. P. HOLLAND, 1982, *J. Chem. Phys.* **77**, 111.

POLIAKOFF, E. D., J. L. DEHMER, D. DILL, A. C. PARR, K. H. JACKSON, AND R. N. ZARE, 1981, *Phys. Rev. Lett.* **46**, 907.

SAH, R. C., AND THE ALS DESIGN GROUP, 1983, *IEEE Trans. Nucl. Sci.* **NS-30**, 3100.

SCHMIDT, V., 1982, in *X-Ray and Atomic Inner-Shell Physics—1982*, edited by B. Crasemann, AIP Conference Proceedings No. 94 (American Institute of Physics, New York), p. 544.

SCHMIDT, V., N. SANDNER, W. MEHLHORN, M. Y. ADAM, AND F. WUILLEUMIER, 1977, *Phys. Rev. Lett.* **37**, 63.

SCHMIDT, V., S. KRUMMACHER, F. WUILLEUMIER, AND P. DHEZ, 1981, *Phys. Rev. A* **24**, 1803.

SHIRLEY, D. A., P. H. KOBRIN, D. W. LINDLE, C. M. TRUESDALE, S. H. SOUTHWORTH, U. BECKER, AND H. G. KERKHOFF, 1982, in *X-Ray and Atomic Inner-Shell Physics—1982*, edited by B. Crasemann, AIP Conference Proceedings No. 94 (American Institute of Physics, New York), p. 569.

SOKOLOV, A. A., AND I. M. TERNOV, 1968, *Synchrotron Radiation* (Akademie-Verlag, Berlin).

SONNTAG, B., AND F. WUILLEUMIER, 1983, *Nucl. Instrum. Methods* **208**, 735.

SOUTHWORTH, S., U. BECKER, C. M. TRUESDALE, P. H. KOBRIN, D. W. LINDLE, S. OWAKI, AND D. A. SHIRLEY, 1983, *Phys. Rev. A* **28**, 261.

SPENCER, J. E., AND H. WINICK, 1980, in *Synchrotron Radiation Research*, edited by H. Winick and S. Doniach (Plenum Press, New York), Chap. 21.

SUGAR, J., T. B. LUCATORTO, T. J. MCILRATH, AND A. WEISS, 1979, *Opt. Lett.* **4**, 109.

SUPER-ACO STORAGE RING, THE, 1983, LURE Report, Orsay.

THOMPSON, D., R. COÏSSON, J. LEDUFF, F. DUPONT, M. ERICKSSON, A. HOFMANN, D. HUSMANN, G. MULHAUPT, M. POOLE, M. RENARD, M. SOMMER, V. SULLER, S. TAZZARI, AND F. WANG, 1981, *IEEE Trans. Nucl. Sci.* **NS-28**, 3153.

WENDIN, G., 1982, in *X-Ray and Atomic Inner-Shell Physics—1982*, edited by B. Crasemann, AIP Conference Proceedings No. 94 (American Institute of Physics, New York), p. 495.

WEST, J. B., A. C. PARR, B. E. COLE, D. L. EDERER, R. STOCKBAUER, AND J. L. DEHMER, 1980, *J. Phys. B* **13**, L105.

WEST, J. B., P. R. WOODRUFF, K. CODLING, AND R. G. HOULGATE, 1976, *J. Phys. B* **9**, 407.

WINICK, H., AND S. DONIACH, 1980, Eds., *Synchrotron Radiation Research* (Plenum, New York).

WINICK, H., H. WIEDEMANN, I. LINDAU, K. HODGSON, K. HALBACH, J. CERINO, A. BIENENSTOCK, AND R. BACHRACH, 1983, *IEEE Trans. Nucl. Sci.* **NS-30**, 3097.

WOLFF, H. W., K. RADLER, B. SONNTAG, AND R. HAENSEL, 1972, *Z. Phys.* **257**, 353.

WUILLEUMIER, F. J., 1976, ed., *Photoionization and Other Probes of Many-Electron Interactions* (Plenum, New York).

WUILLEUMIER, F. J., 1981, in *Atomic Physics 7*, edited by D. Kleppner and F. M. Pipkin (Plenum, New York), p. 491.

WUILLEUMIER, F., 1982, in *X-Ray and Atomic Inner-Shell Physics—1982*, edited by B. Crasemann, AIP Conference Proceedings No. 94 (American Institute of Physics, New York), p. 615.

ZIMKINA, T. M., AND A. C. VINIGRADOV, 1971, *J. Phys. (Paris)* **C32**, 3.

CHAPTER 8

Investigations of Inner-Shell States by the Electron Energy-Loss Technique at High Resolution

George C. King and Frank H. Read

1. INTRODUCTION

1.1. *Early Inner-Shell Excitation Studies*

Inner-shell excited states of neutral atoms and molecules are formed when an inner-shell electron is promoted to an unfilled valence or Rydberg orbital. The promotion can be induced by photoabsorption or by charged-particle impact.

The first observation of a molecular inner-shell excitation is that reported by Prins (1934), who used a bremsstrahlung continuum source to investigate X-ray absorption in N_2 and other gases. He found that a strong absorption peak occurs at an energy of approximately 401 eV, 9 eV below that of the K-shell ionization energy, and he correctly attributed this to the excitation of an atomic K-shell electron to the first unfilled $2p\pi$ orbital of the molecule, giving the configuration $(1s\sigma_N)^{-1}(2p\pi)\ {}^1\Pi$. He suggested also that the outer electron moves in an effective field similar to that experienced by the valence electron of the NO molecule, since the removal of the K-shell electron from the vicinity of a nitrogen nucleus is equivalent to an increase by unity in the charge number of that nucleus. This causes the valence and Rydberg structure of the inner-shell excited $N_2{}^{K^*}$ molecule (where the notation K^* indicates excitation of a K-shell electron) to simulate that of the valence-excited NO molecule. This model is now known as the equivalent-core model or the $Z+1$ core analogy model.

George C. King and Frank H. Read • Department of Physics, Schuster Laboratory, University of Manchester, Manchester M13 9PL, England.

The next major advance was the use of synchrotron radiation for inner-shell absorption studies. Nakamura *et al.* (1969) investigated N_2 in this way and found a high absorption strength for the transition to the $(1s\sigma_N)^{-1}(2p\pi)\ ^1\Pi$ state. Soon afterward van der Wiel *et al.* (1970) used electron impact to excite inner-shell transitions. The process is

$$M + e_{\text{inc}} \rightarrow M^* + e_{\text{scat}}, \tag{1}$$

where M represents a free atom or molecule. The energy lost by the inelastically scattered electrons gives the excitation energy of the inner-shell state. By measuring the yield of inelastically scattered electrons as a function of their energy loss, van der Wiel *et al.* were able to obtain oscillator strengths for the transitions to the $(1s)^{-1}(2p\pi)$ states of $N_2{}^{K*}$, $C^{K*}O$, and CO^{K*}. Their spectra are shown in Fig. 1. This work was followed by a

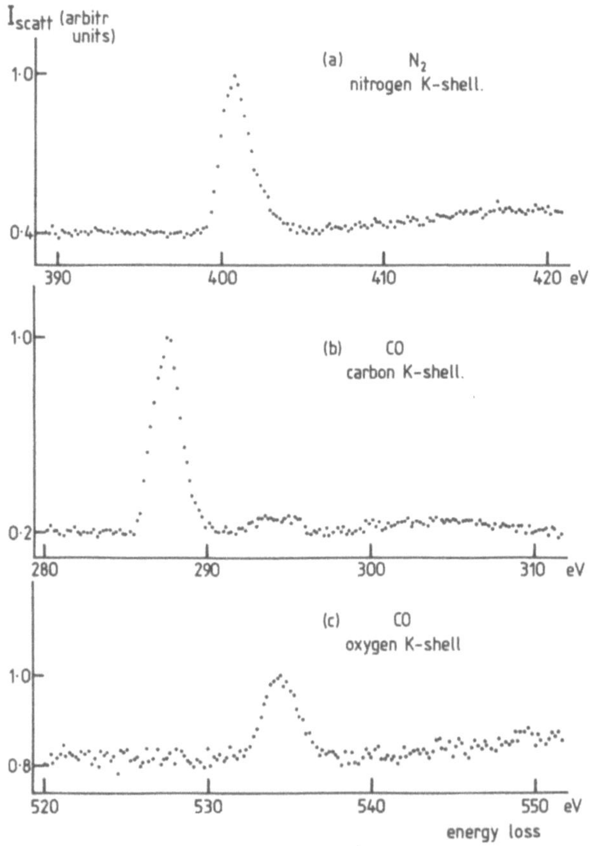

FIGURE 1. Electron energy-loss spectra showing the transitions to the $(1s)^{-1}(2p\pi)$ states of (a) N_2^{K*}, (b) $C^{K*}O$, and (c) CO^{K*}. (After van der Wiel *et al.*, 1970.)

systematic and comprehensive study of inner-shell excitation in molecules, by Brion and collaborators, using the electron energy-loss technique at an energy resolution of approximately 0.5 eV (for a recent review see Brion, 1982).

An early spectrum from Brion's group, obtained by Wight *et al.* (1972/3), for N_2, is shown in Fig. 2. The most noticeable features of the spectrum are (i) that the oscillator strength for the lowest transition, to the $(1s\sigma_N)^{-1}(2p\pi)\ {}^1\Pi$ state, is much larger (by a factor of approximately 10) than the sum of the oscillator strengths for the higher-energy discrete transitions; (ii) that the term values of the discrete structures correspond approximately to those of the NO molecule (as required by the equivalent-core model); and (iii) that structure exists between the first and second ionization potentials (region II) and also above the second ionization potential (region III). This distribution of intensity in the spectrum has been explained by Dehmer and Dill (1976) in terms of the shape and strength of the effective potential experienced by the promoted electron. A centrifugal barrier causes shape resonances in the high angular momentum components ($l \geqslant 2$) of the final-state wave function giving concentrations of oscillator strength in particular discrete and continuum regions. For example, the centrifugal barrier in the *d*-wave component of the $2p\pi$ orbital causes this orbital to be highly localized and the corresponding oscillator strength to be large. The calculated value of the oscillator strength is 0.118 (Rescigno and Orel, 1979) and the measured value is 0.195 ± 0.020 (Kay *et al.*, 1977). The centrifugal barrier that gives this high strength also causes an enhancement

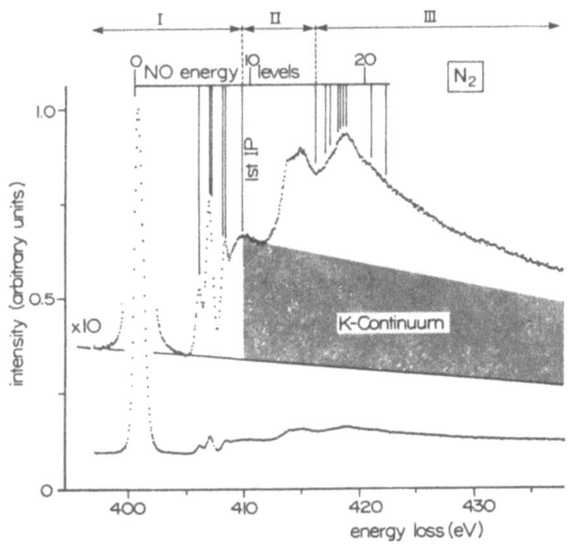

FIGURE 2. *K*-shell energy-loss spectrum of molecular nitrogen. (After Wight *et al.*, 1972/3.)

of the cross section in a different type of experiment, the scattering of low-energy electrons by N_2, giving rise to the well-known shape resonance at 2.3 eV (Schulz, 1973). Many unusual and interesting distributions of oscillator strength have been observed by Brion and co-workers in the inner-shell spectra of a wide range of molecules.

1.2. High-Resolution Studies

Up to this point studies of inner-shell transitions, both by photoabsorption and electron impact, had used fairly low energy resolutions because of practical considerations. Wight *et al.* (1972/3), for example, had worked with a resolution of approximately 0.5 eV giving a resolution, at an excitation energy of a few 100 eV, that was comparable to or better than that obtainable at that time in photoabsorption measurements using synchrotron radiation. This resolution was adequate for observing the more prominent energy-loss peaks and for studying the gross structures associated with centrifugal and other barriers.

Improving the energy resolution of the measurements by an order of magnitude allows the study of the more detailed structure in the spectra, such as the vibrational structure of inner-shell excited molecules and series of Rydberg states of inner-shell excited atoms. Improved resolution also allows the accurate determination of natural linewidths. This improvement in resolution was achieved by the Manchester group (for example, Tronc *et al.*, 1976; King *et al.*, 1977b), who obtained a resolution of 0.07 eV in electron energy-loss measurements in a number of atoms and molecules. The energy resolution now available in inner-shell electron energy-loss studies is as good as, and in many cases better than, that currently available in photoabsorption studies using synchrotron radiation sources (Wuilleumier, 1981). We will therefore concentrate in the present chapter on the studies that take advantage of the superior resolution. We will not attempt to give complete coverage of the subject but will instead illustrate each aspect with one or two specific examples. A comprehensive bibliography of all the experimental (electron energy-loss and photoabsorption) and theoretical studies of inner-shell excitations of free atoms and molecules has been given by Hitchcock (1982).

A further advantage that electron impact excitation has over photoabsorption studies is that it offers the ability to excite optically forbidden transitions. This aspect will also be treated in the present chapter. Yet another advantage is that energy-loss scales can be accurately calibrated by direct voltage measurement. This, for example, enables accurate ionization energies to be established, by observation and analysis of Rydberg series. Values of inner-shell ionization energies obtained from high-resolution electron energy-loss studies are summarized in Table I.

TABLE I
Inner-shell ionization potentials obtained from high-resolution electron energy-loss measurements.

Species	Configuration	Ionization energy (eV)[a]	Reference
Ar^+	$(2p_{3/2})^{-1}$	248.628(11)	King *et al.* (1977b)
	$(2p_{1/2})^{-1}$	250.776(10)	King *et al.* (1977b)
Kr^+	$(3d_{5/2})^{-1}$	93.788(12)	King *et al.* (1977b)
	$(3d_{3/2})^{-1}$	95.038(14)	King *et al.* (1977b)
Xe^+	$(4d_{3/2})^{-1}$	67.548(10)	King *et al.* (1977b).
	$(4d_{5/2})^{-1}$	69.537(10)	King *et al.* (1977b)
CO^+	$(1s\sigma_c)^{-1}$	296.07(4)	Shaw *et al.* (1984b)
CH_4^+	$(1s\sigma_c)^{-1}$	290.76(3)	Tronc *et al.* (1979)
CO_2^+	$(1s\sigma_c)^{-1}$	297.66(4)	Tronc *et al.* (1979)
$C_2H_2^+$	$(1s\sigma_c)^{-1}$	291.19(3)	Tronc *et al.* (1979)
$C_2H_4^+$	$(1s\sigma_c)^{-1}$	290.74(4)	Tronc *et al.* (1979)
N_2^+	$(1s\sigma_c)^{-1}$	409.95(5)	Tronc *et al.* (1980)
Cl_2^+	$(2p_{3/2})^{-1}$	207.80(2)	Shaw *et al.* (1980)
	$(2p_{1/2})^{-1}$	209.42(2)	Shaw *et al.* (1980)
HCl^+	$(2p_{3/2})^{-1}$	207.40(3)	Shaw *et al.* (1984a)
	$(2p_{1/2})^{-1}$	209.03(3)	Shaw *et al.* (1984a)
HBr^+	$(3d_{5/2})^{-1}$	77.12(3)	Shaw *et al.* (1984a)
	$(3d_{3/2})^{-1}$	78.23(3)	Shaw *et al.* (1984a)

[a] The numbers in parentheses show the errors in the last significant figures. These errors include statistical errors arising from the Rydberg series analysis and an estimated absolute calibration error.

In the high-resolution electron impact experiments an electron beam is produced and energy selected to provide a well-defined beam with an energy spread of typically 0.03–0.05 eV and with an adjustable energy. The ability to change the energy of the incident electron is very powerful and is a degree of freedom not available in photoabsorption measurements. The electron beam is passed through a gas cell containing the target atoms or molecules, and those electrons that are inelastically scattered through a small angular range in the forward direction are analyzed in energy, again with a resolution of typically 0.03 to 0.05 eV. A spectrum is thus obtained of the yield of inelastically scattered electrons versus the amount of energy they have lost. The peaks in the energy-loss spectrum correspond directly to the energy levels of the target gas. The peaks have an energy width that is typically 0.05 to 0.07 eV, obtained by combining, approximately in quadrature, the energy spread of the incident beam and the energy resolution of the analyzer.

The inner-shell states that can be studied most usefully with an energy resolution of approximately 0.05 eV are those having natural decay widths Γ that are in the range from approximately 0.05 to 0.2 eV. Many states having a vacancy in the subshell immediately below the valence subshell (for example, a vacancy in the 3*s* subshell of Ar) have values of Γ in this range, as do most states having vacancies in the next lower lying subshell or full shell (for example a 2*p* vacancy in Ar or a 3*d* vacancy in Kr). It is these types of states that will be covered in the present chapter. Their excitation energies are typically in the range from 50 to 500 eV. States with deeper lying vacancies tend to have natural widths greater than 0.2 eV and are therefore better studied by lower-resolution techniques.

1.3. Comparison of the Photoabsorption and Electron Energy-Loss Techniques

The absorption or emission of electromagnetic radiation has been the traditional method of making spectroscopic measurements in atoms and molecules. There is, however, relatively little information on transitions in the far uv and X-ray regions of the spectrum. This is a significant gap since these ranges cover many highly excited states formed by promoting valence electrons, as well as all inner-shell transitions. Although there has been a considerable amount of work on photoelectric studies using the resonance line sources He I (21.21 eV) and He II (40.8 eV), there have been fewer photoabsorption measurements at continuous photon energies. The reason for this has been the limited availability, until recently, of intense continuum light sources above about 20 eV. Conventional light sources using hydrogen or noble gas continua have provided useful though rather weak radiation up to about 20 eV and also a few weak bremsstrahlung continua have been employed in restricted regions. The situation changed considerably with the advent of intense synchrotron radiation sources. These provide a smooth continuum from the infrared to the X-ray regions. Even here, however, there still exist significant difficulties which limit the resolution available at photon energies above about 200 eV. These include the difficulty of dispersing short-wavelength radiation, the low reflectivity of mirrors and gratings at these wavelengths, and order overlapping and absorption by optical components (particularly near the carbon *K*-edge, due to surface contamination). Additional difficulties are associated with calibration of the photon energy scale and with line saturation effects (Inokuti, 1971). Both these latter difficulties are avoided in the electron energy-loss technique, the first because the energy scale can be established with accurate voltmeters, and the second because the electron excitation process is nonresonant.

The wavelength spread $\Delta\lambda$ in the photoabsorption process

$$M + h\nu \rightarrow M^* \tag{2}$$

corresponds to an energy width ΔE in the electron energy-loss process (1) given by

$$\Delta E = E\frac{\Delta\lambda}{\lambda} = E^2\frac{\Delta\lambda}{hc} \tag{3}$$

The energy widths obtainable in the two processes are compared in Fig. 3. The hatched portion of the figure shows the range of excitation energy E and natural decay width Γ possessed by inner-shell excited states. The boundaries of this region are not well defined, but the right-hand boundary corresponds to the measured or calculated widths of K-shell excited states (Keski-Rahkonen and Krause, 1974), the left-hand boundary represents the minimum energy for inner-shell excited states, and the upper boundary represents the position at which the states cease to exist as discrete states because the widths become greater than the typical energy spacings between

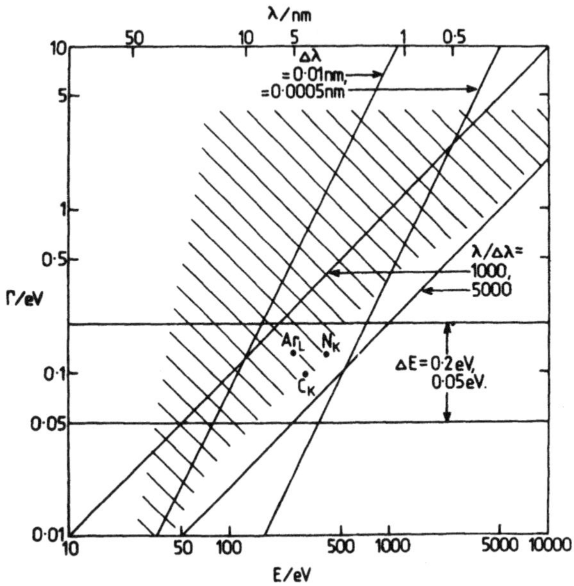

FIGURE 3. The hatched area shows the approximate ranges of excitation energy E and decay width Γ possessed by inner-shell excited levels. Three specific levels are shown within this area. The solid lines show some conditions in which Γ is equal to the experimental energy width ΔE in inelastic electron scattering or photabsorption experiments. The width of a specific level can be measured only if the representative point lies above the relevant line. (After Read, 1978.)

the states. The narrow region extending to small values of Γ contains states such as Ne $1s^2 2s 2p^6 nl$, which tend to be narrow for high values of n because the decay of these states depends predominantly on the overlap of the Rydberg orbitals with the core.

The solid lines in Fig. 3 show conditions in which the natural width Γ is equal to the experimental energy width in reactions (1) and (2). For example the line marked $\Delta\lambda = 0.01$ nm passes through the hatched region dividing it into the region in which the states have natural widths Γ greater than the apparatus width ΔE (namely, the region below and to the right) and in which $\Gamma < \Delta E$. This value of $\Delta\lambda$ would therefore not be small enough to determine the widths of states in the latter region, such as the three specific states marked in the figure. On the other hand the three conditions $\Delta\lambda = 0.0005$ nm, or $\lambda/\Delta\lambda = 5000$, or $\Delta E = 0.05$ eV are all conditions under which the widths of these three specific states can be determined. Figure 3 is also useful in showing the necessary conditions for resolving vibrational structure of inner-shell excited states of molecules. For example, if $\Delta\lambda = 0.01$ nm and the vibrational spacing is 0.2 eV (and assuming that the decay width is less than this spacing), we see that only those states having an excitation energy $\leqslant 150$ eV can be resolved.

An example of what can be observed with the electron energy-loss technique at high resolution (0.075 eV) is shown in Fig. 4 (King *et al.*, 1977c). The vibrational structure of the $(1s)^{-1}(2p\pi)\ ^1\Pi$ state of N_2 is clearly resolved. The energy resolution is less than the natural width Γ of the peaks, which was deduced to be 0.123 ± 0.010 eV (Shaw *et al.*, 1982). To obtain the same resolution in a conventional absorption spectrum would require a spectral resolution of 0.0006 nm at 3.1 nm, which appears to be unobtainable at the present time.

We have suggested above that an energy-loss spectrum is often analogous to a photoabsorption spectrum. This analogy was first established by Bethe (1930), who showed that for fast electrons there is a quantitative relationship between the differential electron scattering cross section and the generalized oscillator strength $f_n(K)$ (see also Inokuti, 1971; Hall and Read, 1984),

$$\frac{d\sigma}{d\Omega} = \frac{4k_n R f_n(K)}{E_n k_0 K^2} \tag{4}$$

Here $\hbar\mathbf{k}_0$ and $\hbar\mathbf{k}_n$ are the momenta of the incident and scattered electrons respectively, $\hbar\mathbf{K}$ is the momentum transferred ($\hbar\mathbf{k}_0 - \hbar\mathbf{k}_n$) to the target atom or molecule, R is the Rydberg energy, and E_n is the excitation energy. When K is small compared with a_0^{-1} (where a_0 is the Bohr radius) the generalized oscillator strength can be expanded in powers of K (see also Huo, 1979):

$$f(K) = f^{\text{opt}} + Kf^{(1)} + K^2 f^{(2)} + \cdots \tag{5}$$

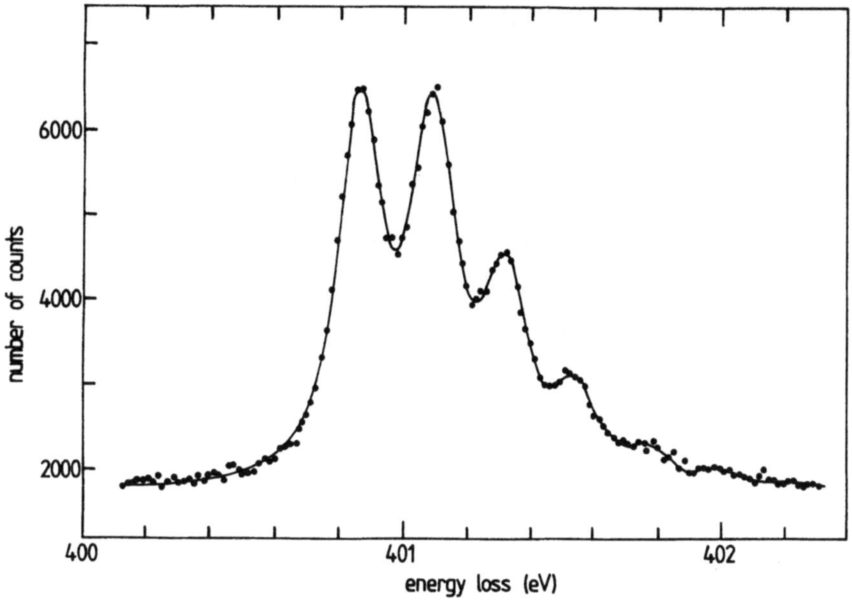

FIGURE 4. Vibrational structure of the electronic state of N_2 formed by the promotion of an atomic $1s$ electron to the first unoccupied valence orbital ($2p\pi$). The spectrum has been obtained by inelastic electron scattering in the forward direction, at an incident energy of 1.5 keV and with an energy resolution of 75 meV. The solid curve shows a least-squares computer fit to the data. (After King *et al.*, 1977c.)

where f^{opt} is the familiar optical oscillator strength. The second and higher-order terms of the expansion become negligible when E_n is much smaller than the incident electron energy and when the angle of scattering is small, and then the electron scattering cross section is directly related to the optical oscillator strength. An electron energy-loss spectrum then effectively simulates the photoabsorption spectrum.

In inner-shell excitation experiments K is sometimes large enough that the higher-order terms in Eq. (5) cannot be ignored, and then the electron energy-loss spectrum contains peaks due to non-electric-dipole transitions. This was first exploited by King *et al.* (1977b), who observed electric-quadrupole transitions in Ar, Kr, and Xe. The effect can be enhanced if the value of K in increased by a judicious choice of the incident electron energy and the angle of scattering, as will be discussed in more detail in Sec. 5. When the incident energy is near to the excitation energy, electron exchange becomes possible, and then spin-forbidden transitions can be excited. The observation of such transitions will also be discussed in Sec. 5.

2. EXPERIMENTAL DETAILS

In the electron energy-loss technique a beam of energetic electrons is used to excite inner-shell transitions in free atoms or molecules. The energy lost by the inelastically scattered electrons gives the excitation energy of the inner-shell state, and the yield of the detected electrons gives information about the relative cross section for excitation of the inner-shell state at that value of incident electron energy. To illustrate the experimental aspects of the technique an electron spectrometer that has recently been used at Manchester will be described in some detail (Shaw *et al.*, 1984b). The general problems involved in the design of electron energy selector systems have been extensively discussed by Read *et al.* (1974).

2.1. Electron Source and Energy Selector

The spectrometer is illustrated in Fig. 5. The source of electrons is a directly heated tungsten hairpin filament, used in a triode gun configuration. This source has the advantage of being more resistant to poisoning than oxide cathodes, which is an important factor in studies of corrosive gases.

The heart of the first half of the spectrometer is the energy dispersing element, or energy selector, which is a hemispherical electrostatic deflector of mean radius 2.54 cm. This type of selector has been described in detail

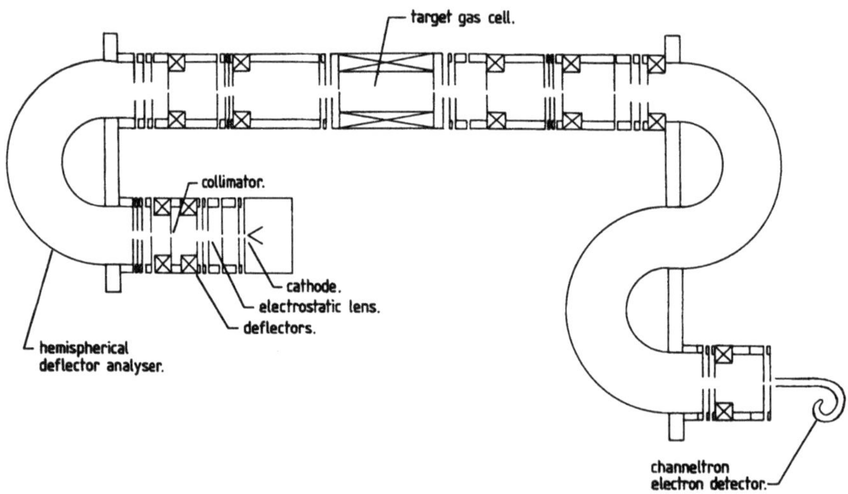

FIGURE 5. Schematic diagram of a high-resolution electron energy-loss spectrometer (Shaw *et al.*, 1984b).

(Purcell, 1938; Simpson, 1964; Kuyatt and Simpson, 1967; Imhof *et al.*, 1976). It has the advantages that it has first-order focusing and is suited to the cylindrical geometry of the electron beam and the lens optics. The electron beam radius at the entrance and exit planes of the energy selector is 0.5 mm, while the pencil half-angle of the beam is 2°. At an analyzing energy of 7 eV this provides an electron beam of 50 meV energy width, which is ideally suited to inner-shell studies. A better energy resolution ΔE can be obtained by using lower values of the analyzing energy, but at the expense of reduced transmitted beam current, since the current is approximately proportional to $(\Delta E)^{5/2}$ (Read *et al.*, 1974). At an energy width of 50 meV the current at the target is typically 10 nA. One point of note is the use of virtual apertures at the entrance and exit planes of the hemispherical deflector. These virtual apertures are images of real apertures or collimators placed in regions of higher potential in the electrostatic lens system, as explained below. The use of virtual apertures enables more accurate correction of fringe-field effects and, more importantly, it minimizes the problems (for example due to patch fields) associated with passing low-energy electrons through real defining apertures. The second half of the spectrometer, the energy analyzing section, contains two further energy dispersing elements placed in tandem (see below), and virtual entrance and exit apertures are again used.

Combinations of triple-aperture electrostatic lenses are used to control and transport the electron beam from the electron source to the energy selector and from the selector to the target region, and similarly through the analyzer section of the spectrometer. The focusing properties of these lenses are given by Harting and Read (1976). In the first section a combination of two lenses images the electron source onto the entrance slit of the hemispherical deflector. Two collimators between the lenses serve to define the radial and angular extent, respectively, of the electron beam, and are situated in regions of high electron energy (typically a few tens of eV). The image of one of these collimators provides the virtual aperture at the entrance plane of the deflector. A second set of lenses serves to image the exit of the hemispherical deflector onto the target interaction region. Again the image of a defining aperture provides the virtual aperture at the exit of the deflector. Three lenses are needed in this section to increase the electron energy from a few eV to the final bombarding energy, the value of which is chosen for optimum excitation of the state of interest and ranges from a few hundred eV (for the excitation of electric-diple forbidden transitions) to a few keV (for the excitation of optically allowed transitions).

The incident electron beam is passed through a cell containing the gas of interest. The cell is 25 mm long and contains entrance and exit apertures (1 mm diameter) that allow the passage of the electron beam but are small enough to maintain a high target-to-background pressure ratio, typically

300:1. This minimizes the possible deleterious effects of corrosive gases, and of scattering from background gas in the other regions of the spectrometer. A gas cell is preferred to a gas beam because is gives a longer interaction path and hence a higher signal. A disadvantage of a gas cell target geometry is that angular information about the scattering process is more difficult to obtain than with a crossed beam geometry. Incorporated into the gas cell are two pairs of electrostatic deflectors which may be used to deflect the unscattered primary beam so that it does not pass into the analyzer section. The inelastically scattered electrons are also deflected, so that only those electrons that have been scattered through a small angle are able to pass into the analyzer section, but the kinematics of inner-shell excitation are such that at the incident and excitation energies used in these experiments the yield of inelastically scattered electrons does not fall off rapidly as the scattering angle increases, in contrast to the behavior for valence-shell excitation. The deflection at the target therefore considerably reduces the yield of the unscattered primary beam but has a relatively small effect (a reduction of typically a factor of 2) on the yield of the inelastically scattered electrons.

2.2. Energy Analyzer and Detection System

The inelastically scattered electrons that pass through to the analyzer section are focused by a further combination of three triple-aperture lenses onto the entrance plane of the energy analyzer, which consists of two hemispherical deflectors placed in tandem. This arrangement helps to minimize the detection of the secondary and reflected electrons that arise when the wings of the primary and scattered beams encounter various surfaces in the apparatus. In this connection it is useful to note that the primary electron beam consists of approximately 10^{11} particles per second while the signal of interest might contain only about 10 particles per second. The tandem arrangement leads to high values of the signal-to-noise ratio in the spectra obtained, while having little effect on the transmission efficiency of the analyzer section. A slot in the back of the first hemispherical deflector, and a Faraday cup that is behind it, are further measures to avoid surface reflection of the unscattered primary beam.

The energy-analyzed electrons at the exit of the second hemispherical deflector are focused by the final set of electrostatic lenses onto the entrance of a channel electron multiplier (Mullard B310 BL/01) and detected. The pulses from the electron multiplier are amplified, shaped, and discriminated using conventional electronic techniques before being fed to a multichannel analyzer operating in the multiscaling mode. In order to obtain an energy-loss spectrum a linear voltage ramp, derived from the multichannel analyzer, is applied between the target region and the local earth of the electron energy analyzer. It is important that this counting system should be able to register, without saturation or distortion, the high count rates observed in valence shell spectra (up to a few MHz) and yet be able to handle the low count rates of the inner-shell energy-loss spectra without letting too much spurious noise through the system. In practice a pulse stretching circuit is interposed in the counting system when inner-shell spectra are being accumulated. The length of the stretched pulse, typically 3 ms, minimizes the effects of electrical interference, which consists mainly of short bursts of rf noise, but without leading to distortion of the energy-loss spectra. The overall resolution of the spectrometer, that is the width (FWHM) of the peaks in the energy-loss spectra, is typically 55–75 meV.

Molybdenum is used in the construction of all the electron optical components of the spectrometer, including the hemispherical deflectors, since this material is found to give stable operation of the spectrometer. Sources of drift that can seriously affect the quality of energy-loss spectra are short- and long-term variations in the voltage sources that are applied to the electron spectrometer. Of particular importance is the power supply that determines the value of the electron energy loss (see below). In the present apparatus highly stabilized power supplies are used, and where necessary these are maintained in a temperature-controlled environment. These precautions result in drifts that are less than about 5 mV over a typical data accumulation period of 48 h. The minimization of such drifts is particularly important when the natural width of a state is to be determined from the energy-loss spectra.

2.3. *Energy Calibration*

The accuracy of the energy-loss measurements depends essentially on the precision with which the energy-loss scale can be determined. This scale is calibrated against neutral valence excited states whose energies are accurately known from spectroscopic data and it is assumed that there are no experimental energy shifts between this region of energy loss (about 10 eV) and those of interest up to about 500 eV. Support for this assumption has come from the inner-shell studies of Tronc *et al.* (1976). These authors determined the position of the $(1s\sigma)^{-1}2p\pi\ ^1\Pi$ state of CO by using a mixture of CO and He as the target gas and calibrating the energy scale against the 2^1P and 3^1P states of He and also the $(2s2p)^1P^0$ autoionizing state of He (Madden and Codling, 1965). Since the latter state is nearer in energy to the region of interest and also lies above the first ionization potential of He, where the effects of positive ions may be present, it may be expected to give the more reliable calibration. In practice both calibrations agreed to within 10 meV. The energy differences between the two regions of energy loss are measured by means of a precision digital voltmeter. Most recently, at Manchester, a Datron Ltd. model 1059 voltmeter has been used having an accuracy of 0.005%, or 12.5 mV at an energy loss of 250 eV. Support for the accuracy of the technique comes also from the fact that some of the peak energies have been determined using independent pieces of apparatus and voltmeters and the values obtained are in substantial agreement (for example, the $(1s\sigma)^{-1}2p\pi\ {}^1\Pi$ state of CO, Tronc *et al.*, 1976; Brion *et al.*, 1982; Shaw *et al.*, 1984a; Shaw *et al.*, 1984b). It has thus been suggested that the present technique provides in principle the possibility of accurate and direct high-voltage (100–1000 V) calibration standards (Brion *et al.*, 1982). Comparable accuracy has recently been achieved in the measurement of core-electron binding energies by Pettersson *et al.* (1982) using soft X-ray emission spectroscopy and calibrating the wavelength scale against known lines excited in a sliding spark discharge source.

2.4. *Multidetection Techniques*

The main difficulty associated with inner-shell studies is that the yields of inelastically scattered electrons are very low. This arises because the cross section for excitation of a state of energy E is approximately proportional to E^{-3}, and for high-lying Rydberg states this can result in counting rates of 1 Hz or less. One obvious way to improve the situation is to use a multidetector system placed at the exit plane of the electron energy analyzer. Thus instead of looking at only one value of energy loss at any one time, all the inelastically scattered electrons, over the range of interest, are detected simultaneously. This would improve the detection efficiency by a factor of approximately 50 and would significantly extend the range of the present technique. Such multidetection systems have recently been described for valence-shell studies (Hicks *et al.*, 1980), and may be expected to be introduced soon for inner-shell studies.

3. HIGH-RESOLUTION STUDIES OF ATOMS

Several early studies of the absorption of soft X rays by atoms, in particular the rare gases (for example, Lukirskii *et al.*, 1965; Codling and Madden, 1964; Nakamura *et al.*, 1968), revealed the existence of structure near some of the inner-shell ionization potentials. The structure arises from the promotion of one of the inner-shell electrons to an unfilled orbital. These studies were, however, subject to the limitations, discussed in Sec. 1.3, inherent in photoabsorption measurements, notably lack of resolution. In this section we illustrate what can be achieved with high-resolution electron-impact excitation by discussing experimental studies of the energy regions near the $L_{2,3}$ edges in Ar, the $M_{4,5}$ edges in Kr, and $N_{4,5}$ edges in Xe, at an incident electron energy of 1.5 keV. These studies yield information on excitation energies, ionization and binding energies, and natural widths. The use of electron impact excitation to excite electric-dipole-forbiden transitions is discussed in Sec. 5.

3.1. Analysis of Energy-Loss Spectra

Examples of inner-shell energy-loss spectra, obtained in argon, krypton, and xenon (King *et al.*, 1977b; Shaw *et al.*, 1985) are shown in Fig. 6, 7, and 8, respectively. Two sets of peaks, each associated with one of the two ion cores states ($^2P_{1/2}$ and $^2P_{3/2}$), can be discerned in each spectrum. Each peak is assumed to have a Voigt line shape, being the convolution of a Lorentzian line shape with an apparatus function which is assumed to have a Gaussian shape. A least-squares procedure is applied to limited portions of a spectrum, containing six or fewer lines on a constant background, in order to determine the various unknown quantities, namely, (i) the mean position of each peak, (ii) the natural width Γ of the Lorentzian line shape, (iii) the area of each peak, and (iv) the height of the flat background. The width Γ is assumed to be the same for each of the peaks of the portion analyzed, although it may be different for other portions of the same spectrum. The normalized values of χ^2 (that is, the normalized sum of weighted squared

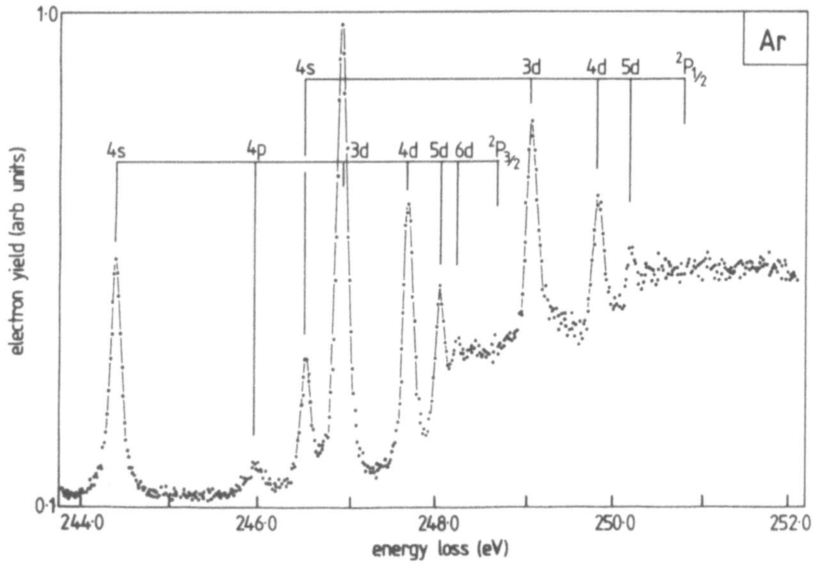

FIGURE 6. Energy-loss spectrum of Ar obtained at an incident electron energy of 1.5 keV and with a resolution of 65 meV. (From Shaw *et al.*, 1985.)

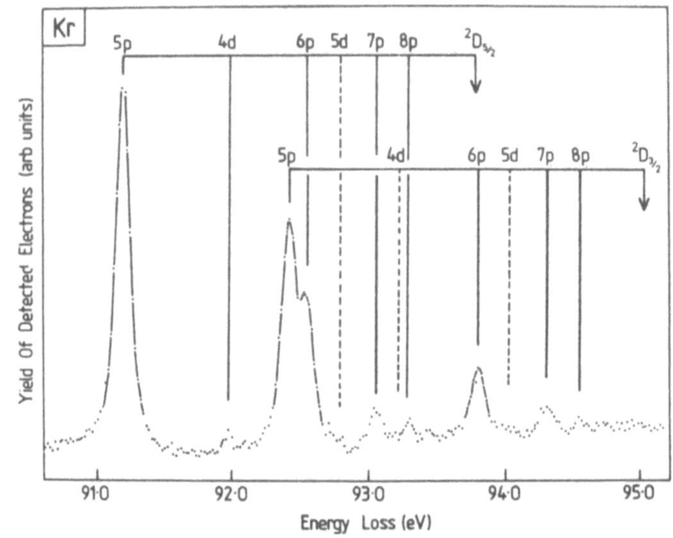

FIGURE 7. Energy-loss spectrum of Kr obtained at an incident electron energy of 1.5 keV and with a resolution of 70 meV. (From King *et al.*, 1977b.)

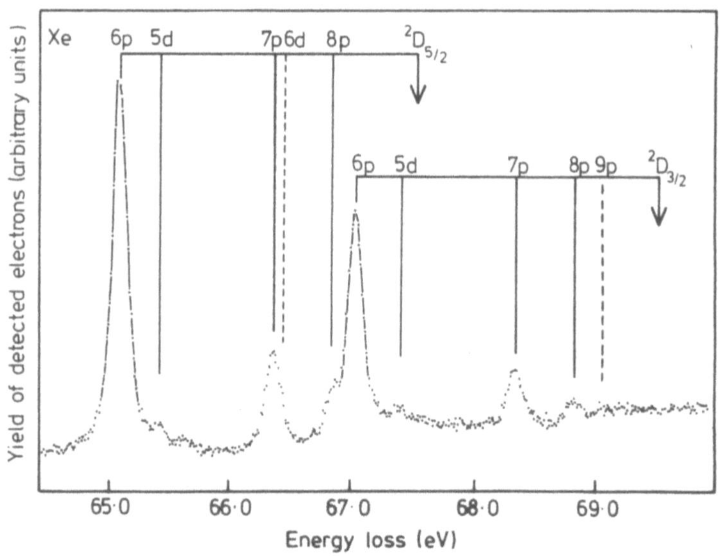

FIGURE 8. Energy-loss spectrum of Xe obtained at an incident-electron energy of 1.5 keV and with a resolution of 70 meV. (From King *et al.*, 1977b.)

residuals) is found to be near to unity for all the analyses (the mean value being 1.11), which is an indication of the good quality of the computed fits.

The energies of the ionization limits are important quantities to be derived from the spectra. First the equivalent-core ($Z+1$ core-analogy) model (Nakamura *et al.*, 1968; Schwarz, 1974, 1976) is assumed to be valid, so that an inner-shell electron in a neutral atom of nuclear charge Z, which has been promoted to an outer orbital nl, would have an effective quantum number n^* which is expected to be similar to the effective quantum number of an excited nl electron in the neutral atom of nuclear charge $Z+1$. In the case of Xe, for example, it is assumed that

$$n^*[\mathrm{Xe}\ 4d^9(^2D_{3/2,5/2})\ 5s^25p^6nl] = n^*(\mathrm{Cs}\ 4d^{10}5s^25p^6nl) + \Delta_l \tag{6}$$

with the further assumption that the small correction Δ_l depends on l but is independent of n or of the value of j(3/2 or 5/2) of the xenon ion core. The same assumptions have also been made for the inner-shell excited states of Ar and Kr. Further reasonable assumptions are also made concerning fine-structure splitting (King *et al.*, 1977b). A least-squares program is used to find the values of Δ_l and the ionization limit for each of the observed Rydberg series. It will be seen below that this procedure is justified by the consistency of the values of Δ_l obtained for the three atoms Ar, Kr, and Xe.

3.2. *Argon*

The energy-loss spectrum of Ar in Fig. 6 was obtained at an incident electron energy of 1.5 keV and with a resolution of 65 meV (Shaw *et al.*, 1985). It shows structure near the $L_{2,3}$ edges of Ar whose assignment is indicated on the figure. The ground-state configuration of Ar is $1s^22s^22p^63s^23p^6$ and optically allowed transitions of a $2p$ electron to one of the orbitals ns ($n \geqslant 4$) or nd ($n \geqslant 3$) are observed. A weaker, parity-forbidden transition to the $4p$ orbital is also observed. This forbidden transition occurs because the selection rules are sufficiently relaxed at the value of incident electron energy used here, which is only six times the excitation energy. The $2p$ hole can have $j = 3/2$ or $1/2$ and consequently two Rydberg series are observed, each converging to one of these two states of the ion. The ratio of the intensities of corresponding peaks in the two series is observed to be approximately proportional to the ratio of the statistical weights $2j+1$.

The features observed in this energy-loss spectrum have been discussed and analyzed by King *et al.* (1977b), and their results, which are in good agreement with the later, more accurate, results of Shaw *et al.* (1983), are summarized in Table II. The first and second columns of the table show the assignments and energies of the observed transitions. The energy of the first peak, ($2p_{3/2}^{-1}4s$), was calibrated against the $4s\,^3P_1$ and $4s\,^1P_1$ states of

TABLE II

Measured energies E and widths Γ, and deduced effective quantum numbers n^* and relative spectral densities of generalized oscillator strength df/dE, for the observed transitions in Ar, Kr, and Xe. The values of n^* and Δ_l and its error have been deduced for the members of the principal series of each atom by assuming that the generalization of Eq. (6) is correct. The numbers in brackets show the errors in the last significant figures. After King *et al.* (1977b), Shaw *et al.* (1982).

State		E (eV)	Γ (meV)	n^*	Δ_l	$E_{calc} - E_{obs}$ (eV)	Relative df/dE
Ar $2p_{3/2}^{-1}$	$4s$	244.390(10)[a]	116(3)	1.792	0.022(1)		0.11(1)
	$4p$	245.956(9)		2.259	0.022(4)		0.04(1)
	$3d(+5s)$	246.927(1)	118(4)	2.828	−0.026(4)	0.000	1.0
	$4d(+6s)$	247.669(4)	142(14)	3.771	−0.026(4)	0.002	1.43(9)
	$5d(+7s)$	248.026(5)		4.743	−0.026(4)	−0.003	1.72(18)
	edge	248.628(4)					1.53(9)
$2p_{1/2}^{-1}$	$4s$	246.514(4)		1.787	0.017(1)		0.06(1)
	$3d(+5s)$	249.074(3)	132(10)	2.828	−0.026(4)	0.001	0.49(3)
	$4d(+6s)$	249.819(4)	132(10)	3.771	−0.026(4)	0.000	0.67(6)
	$5d(+7s)$	250.171(12)		4.743	−0.026(4)	0.000	0.62(13)
	edge	250.776(1)					0.91(7)
Kr $3d_{5/2}^{-1}$	$5p$	91.200(10)[a]	83(4)	2.293	0.000(4)	0.000	1.0
	$6p$	92.560(2)		3.330	0.000(4)	0.001	1.16(4)
	$7p$	93.063(4)		4.342	0.000(4)	0.003	0.84(6)
	$8p$	93.301(20)		5.348	0.000(4)	0.011	1.06(45)
	$4d$	91.987(16)		2.749	−0.018(13)		0.06(1)
	edge	93.788(7)					0.85(19)
$3d_{3/2}^{-1}$	$5p$	92.425(2)		2.282	0.000(4)	0.000	0.62(1)
	$6p$	93.809(4)	98(12)	3.319	0.000(4)	−0.006	0.71(4)
	$7p$	94.319(8)		4.331	0.000(4)	−0.006	0.61(7)
	$8p$	94.567(19)		5.337	0.000(4)	−0.007	0.45(18)
	$5d$	94.037(30)		3.687	−0.019(56)		0.16(5)
	edge	95.038(10)					0.52(10)
Xe $4d_{5/2}^{-1}$	$6p$	65.110(10)[a]	111(4)	2.362	0.000(1)	0.000	1.0
	$7p$	66.375(2)	128(9)	3.406	0.000(1)	0.000	1.03(2)
	$8p$	66.854(4)		4.421	0.000(1)	−0.002	0.77(5)
	$5d$	65.446(6)		2.544	−0.011(4)		0.04(1)
	edge	67.548(1)					0.81(13)
$4d_{3/2}^{-1}$	$6p$	67.039(2)	119(8)	2.334	0.000(1)	0.000	0.71(2)
	$7p$	68.345(3)	133(15)	3.379	0.000(1)	0.000	0.68(2)
	$8p$	68.838(12)		4.394	0.000(1)	−0.006	0.62(8)
	$5d$	67.411(9)		2.530	−0.018(5)		0.04(1)
	edge	69.537(2)					0.64(4)

[a] All errors shown in the table are standard deviations, except these three which are the estimated systematic errors of the energy calibrations. The other energy errors in the table are relative to the energy of the first peak in each atom, and do not contain the calibration errors.

neutral Ar, with an estimated uncertainty of 0.01 eV. The energies of the remaining peaks were then obtained relative to that of the first peak by means of the least-squares analysis described above. The *nd* peaks all contain an admixture of $(n+2)s$ peaks having similar energies, but it is clear from the general pattern of peak heights and from the values of the densities of oscillator strength df/dE (see below) that this admixture is probably small. This is substantiated by the calculations of Hahn (1976), showing that the ratio of cross sections for exciting the $(n+2)s$ and *nd* states is approximately 0.045.

The third column of Table II shows natural widths Γ deduced from the widths of those peaks that are prominent and that are sufficiently well separated from neighboring peaks. The tabulated values for the $2p_{3/2}^{-1}nd$ states contain unknown (although probably small) contributions from the fine-structure splitting of the $J=1$ states. The small admixtures of $(n+2)s$ peaks are not expected to increase the apparent values of Γ appreciably [since the separation in energy of the *nd* and $(n+2)s$ states is much smaller than Γ, see below]. This value of Γ agrees reasonably well with the approximate value calculated by Keski-Rahkonen and Krause (1974), namely, 160 meV. According to the same authors the width of the $2s^{-1}(L_1)$ state is considerably broader (360 meV), which would explain the inability to see discrete structure near the L_1 edge.

The values of n^* and Δ_d and the ionization limits are obtained by the procedure described above. Note that the value of Δ_d is small and that the same value is obtained for both the $2p_{3/2}^{-1}$ and $2p_{1/2}^{-1}$ sets of states. The differences between the observed energies of the *nd* states and the energies calculated from the tabulated values of n^* and the ionization limits are shown in the sixth column of Table II . These differences are, by accident, considerably smaller than the uncertainties in the measured energies themselves.

The ionization limits together with the measured energies of the $2p^{-1}4s$ and $2p^{-1}4p$ states enable the values of n^* and $\Delta_{s,p}$ of these states to be deduced, as shown in the table. The values of $\Delta_{s,p}$ are again small. The small difference in the values of Δ_s for the $2p_{3/2}^{-1}\,4s$ and $2p_{1/2}^{-1}4s$ states presumably show the existence of magnetic interactions. To estimate the importance of these interactions Shaw *et al.* (1985) have used the nonrelativistic Hartree–Fock program of Froese-Fischer (1978) to calculate the energy separation of the two levels (having $J=1$ and 2) of the $(2p_{3/2})^{-1}4s$ configuration. Their result is 30 meV, which is significantly smaller than the natural width of the levels. The separation of the levels of the $(2p_{1/2})^{-1}4s$ configuration is of the same order. The $(2p)^{-1}3d$ configurations also have fine structures: for example, Shaw *et al.* (1985) calculate that the eight levels of the $(2p_{3/2})^{-1}3d$ configuration are spaced over 40 meV, and that the separation of the two levels having $J=1$, which are reached by electric-dipole transitions, is only

30 meV. The fine structure of these inner-shell excited states would therefore not be visible, although it could give rise to small variations in n^* and Δ. The fine structures of the $(2p)^{-1}4p$ levels will be discussed in Sec. 5.

The final column of the table shows relative values of the spectral density of generalized oscillator strength, df/dE, for the discrete transitions observed and also for the two ionization continua. Fano and Cooper (1968) have suggested that for discrete Rydberg states of principal quantum number n and energy E_n the spectral density of oscillator strength should be defined as

$$\frac{df}{dE} = f_n \left(\frac{dE_n}{dn} \right)^{-1}. \tag{7}$$

In terms of the effective quantum number n^* this gives

$$\frac{df}{dE} = \frac{f_n n^{*3}}{2R(dn^*/dn)} \tag{8}$$

where R is the Rydberg energy. The relative values of the generalized oscillator strengths f_n have been obtained by multiplying the observed peak areas by E_n^3 (Inokuti, 1971). The generalized oscillator strength density df/dE is expected to vary in a continuous manner as the high Rydberg states merge towards the ionization limit, its value at the ionization limit going smoothly into the continuum oscillator strength per unit energy interval of that part of the continuum possessing the same angular momentum quantum numbers as the Rydberg series in question (Fano and Cooper, 1968). It can be seen from the table that this is the observed behavior for both series of nd states, within the experimental uncertainties, except that the strengths of the $5d$ lines are slightly abnormal.

In the absence of configurational mixing in the ion core states, and with the neglect of any variations with energy of transition strengths, the observed oscillator strengths for equivalent transitions in the two series corresponding to the ion core states $j = 3/2$ and $j' = 1/2$, respectively, should be in the ratio $(2j + 1)/(2j' + 1)$. This should be true also for the ratio of continuum oscillator strengths. The observed ratios for argon are shown in Table III. The continuum ratio is different from 2.0, which appears to indicate that the ion core states do not have pure $2p^{-1}$ configurations and the $5d$ states appear to be anomalous, as noted above.

The values of the two series limits are given in Table IV and compared with the values obtained from photoabsorption (Nakamura *et al.*, 1968), ESCA (Siegbahn *et al.*, 1969; Johansson *et al.*, 1973) and X-ray emission (Pettersson *et al.*, 1982) measurements. The agreement with the recent ESCA and X-ray emission work is particularly good.

TABLE III

Observed ratios of oscillator strengths of equivalent transitions based on the ion core states of angular momenta j and j'.[a]

Atom	State	$\frac{2j+1}{2j'+1}$	Observed ratio
Ar	$2p^{-1}$ $4s$	2.0	1.8(1)
	$3d(+5s)$		2.0(1)
	$4d(+6s)$		2.1(2)
	$5d(+7s)$		2.7(7)
	Continuum		1.7(2)
Kr	$3d^{-1}$ $5p$	1.5	1.6(1)
	$6p$		1.6(1)
	$7p$		1.4(2)
	$8p$		2.4(14)
	Continuum		1.6(5)
Xe	$4d^{-1}$ $6p$	1.5	1.4(1)
	$7p$		1.5(1)
	$8p$		1.2(2)
	$5d$		1.0(2)
	Continuum		1.3(2)

[a] From King *et al.* (1977b).

The small peak at 245.96 eV in Fig. 6 has been assigned to the optically forbidden transition to the $(2p_{3/2})^{-1}4p$ state. The resulting value of Δ_p is small, thus justifying this assignment. The corresponding transition to the $(2p_{1/2})^{-1}4p$ state would lie at approximately 248.10 eV, and would thus be obscured by the larger $(2p_{3/2})^{-1}5d$ peak. The optically forbidden transitions will be discussed in Sec. 5.

3.3. *Krypton*

The ground-state configuration of Kr is $KL\ 3s^2 3p^6 3d^{10} 4s^2 4p^6$, and transitions of the $3d_{3/2}$ and $3d_{5/2}$ electrons to unfilled p orbitals are observed. Transitions to f orbitals are also optically allowed, but the high centrifugal barrier in the effective mean potential for f electrons (Rau and Fano, 1968; Fano and Cooper, 1968) gives these transitions a very small oscillator strength, and they are not observed in the present work. Optically forbidden transitions to outer d orbitals are also observed. An energy-loss spectrum of Kr in the region of the $3d_{5/2}^{-1}$ and $3d_{3/2}^{-1}$ edges is shown in Fig. 7 (King *et al.*, 1977b). The positions of the features observed in Fig. 7 are given in

TABLE IV

Comparison of measured inner-shell ionization energies in Ar, Kr, and Xe.

		Electron impact[a]	Photo-absorption[b]	ESCA[c]	ESCA[d]	X-ray[e] emission
Argon	L_3	248.628(11)	248.52(5)	248.45(7)	248.62(8)	248.629(10)
	L_2	250.776(10)	250.55(5)	250.56(7)		250.777(10)
Krypton	M_5	93.788(12)	93.83(7)	93.7(2)	93.80(1)	
	M_4	95.038(14)	95.04(7)	94.9(2)		
Xenon	N_5	67.548(10)	67.55(2)	67.5(1)		
	N_4	69.537(10)	69.53(2)	69.5(1)		

[a] King *et al.* (1977b).
[b] Ar, Nakamura *et al.* (1968); Kr, Xe, Codling *et al.* (1964).
[c] Siegbahn *et al.* (1969).
[d] Johansson *et al.* (1973).
[e] Pettersson *et al.* (1982).

Table II, together with the derived quantities n^* and Δ_p. The energies of the optically allowed transitions have been measured previously by Codling and Madden (1964) by optical spectroscopy, with a resolution of approximately 35 meV, and the relative and absolute values agree within the experimental errors.

Only two peaks are sufficiently well separated from their neighbors to allow their natural widths Γ to be derived. These widths are approximately equal to the value calculated for the ion state by Keski-Rahkonen and Krause (1974). The widths of the $3p_{3/2}^{-1}$ (M_3) and $3p_{1/2}^{-1}$ (M_2) hole states are approximately 2.2 eV (Keski-Rahkonen and Krause, 1974; Yin *et al.*, 1974), and for this reason structure near the $M_{2,3}$ edges was not observed.

The methods described above have been used to deduce the values of n^* and Δ_p shown in Table II. The value of Δ_p was constrained to be the same for both series of *np* states, and was found to be nearly zero. The separate values of Δ_p for the two Rydberg series are still small without this constraint, having the values -0.002 and 0.004 for the $d_{5/2}^{-1}$ and $d_{3/2}^{-1}$ series, respectively. Codling and Madden (1964) obtained values of n^* that are almost the same as those obtained by King *et al.* (1977b), while for the $3d_{3/2}^{-1}$ series their values are all slightly larger (by 0.02 to 0.06).

The values of df/dE given in Table II for Kr show the expected smooth changes in values through the *np* series and into the continua. The ratios of generalized oscillator strengths shown in Table III also change smoothly, within the experimental uncertainties, and are approximately equal to the ratio of the statistical weights of the ion core states.

The deduced values of the energies of the ionization limits are listed in Table IV, where they are compared with those of Codling and Madden (1964), Siegbahn *et al.* (1969), and Johansson *et al.* (1973), and the agreement can be seen to be satisfactory.

The small peak at 91.99 eV in Fig. 7 has been assigned to the parity-forbidden transition to the $3d_{5/2}^{-1}4d$ state. The resulting value of Δ_d is small (see Table II), thus justifying this assignment. Transitions to $3d^{-1}ns$ states are also optically forbidden. The $3d_{5/2}^{-1}5s$ state, for example, might be expected to have an energy of approximately 89.7 eV if the value of Δ_s is approximately 0.02 (as in Ar). In fact, this state has been observed close to this energy in the more recent work of Shaw *et al.* (1985).

3.4. Xenon

The ground-state configuration of Xe is $KLM4s^24p^64d^{10}5s^25p^6$ and optically allowed transitions of $4d_{5/2,3/2}(\mathcal{N}_{4,5})$ electrons to unfilled p orbitals are observed, see Fig. 8 (King *et al.*, 1977b). Optically forbidden transitions to the $5d$ orbital are also observed. As in the case of Kr, optically allowed transitions to the $4d^{-1}nf$ states of Xe are expected to be very weak (Fano and Cooper, 1968), and they are not observed. The measured energies of the features observed in Fig. 8 are listed in Table II together with the quantities Γ, n^*, Δ_l, and df/dE. The energies of the optically allowed transitions have been measured by Codling and Madden (1964) by optical spectroscopy, with a resolution of approximately 25 meV, and their values agree with those of King *et al.* within the experimental errors.

It has been possible to deduce the natural width Γ of four of the states, as shown in Table II. The widths of these same states have also been measured by Ederer *et al.* (1976) from widths of photoabsorption lines, and they find the values 114 ± 8, 90 ± 20, 130 ± 30, and 110 ± 30 meV, respectively. The widths of the $4d_{5/2,3/2}$ hole states are presumably both approximately equal to the value of Γ of the states with the highest n values, namely, the $4d^{-1}7p$ states, having a mean width of 129 ± 8 meV. This is consistent with the approximate value of 100 meV calculated by Keski-Rahkonen and Krause (1974). The widths of the higher-lying $4p_{3/2,1/2}(\mathcal{N}_{2,3})$ states are calculated to be much larger. Keski-Rahkonen and Krause (1974) give approximately 2.5 eV and Chen *et al.* (1976) give 2.38 eV, thus explaining the failure to detect structure near the $\mathcal{N}_{2,3}$ edges.

As in the case of Kr, the spectral densities of oscillator strengths (Table II) and their ratios (Table III) show the expected values and behavior. The absolute values of the optical oscillator strengths of the lines have been determined by Ederer *et al.* (1976). Their relative values of df/dE [after inclusion of the term dn^*/dn, see Eq. (8)] have greater uncertainties than those of King *et al.* (1977b), but are on the whole consistent with them.

Shannon *et al.* (1977) have measured the continuum $d_{5/2}^{-1}/d_{3/2}^{-1}$ branching ratio by means of photoelectron spectroscopy, and have found that this ratio has the expected statistical value 1.5 at high photon energies, but that it varies between 1.3 and 1.6 at a few tens of eV above threshold, having the value 1.4 ± 0.1 at 6.5 eV above threshold (using the appropriate threshold for each exit channel). The value of 1.3 ± 0.2 at the threshold itself is consistent with the behavior observed by Shannon *et al.*

Values of the ionization limits are listed in Table IV, and it can be seen that there is good agreement between the electron energy-loss values and those of Codling and Madden (1964).

The small peaks at 65.45 and 67.41 eV in Fig. 8 have been assigned to the optically forbidden transitions to the $4d_{5/2,3/2}^{-1}5d$ states. The deduced values of Δ_d are -0.011 and -0.018, respectively. The optically forbidden transition to the $4d_{5/2}^{-1}6s$ state is expected at about 63.7 eV and has been observed in the recent work of Shaw *et al.* (1985).

3.5. Equivalent-Core Model

One of the important results of the studies by King *et al.* (1977b) is that the accurate measurement of energies of optically allowed and forbidden transitions of three related atoms allows the equivalent-core model to be tested much more thoroughly than was previously possible. The result of this test is that the model provides, for these three atoms at least, an accurate and reliable means of estimating the energies of inner-shell excited states. The differences Δ_l [Eq. (6)] between the effective quantum numbers n^* of the inner-shell-excited atom and the equivalent-core atom are quite small, which indicates that the effective field experienced by the outer excited electron is indeed very similar in these two types of atom. A pattern which emerges from the values obtained is that Δ_d tends to be small and negative (~ -0.02), that Δ_p tends to be nearly zero, and that Δ_s tends to be small and positive (~ 0.02).

This pattern of behavior of the values of Δ_l might be explained in terms of two separate regions of the inner-shell-excited atoms, namely, the regions lying inside and outside the average radius r_h of the inner-shell hole. The inner region would have an effective mean potential less deep than that of the equivalent-core atom (because the nuclear charge is less), and since electrons of low angular momentum penetrate this region more effectively than those of higher angular momentum one would expect Δ_s to be increased by this penetration but Δ_p to be less affected and Δ_d to be relatively unaffected. This might then explain the observation that $\Delta_s > \Delta_p > \Delta_d$. On the other hand the region outside r_h, which is the region of the outer full shell, is likely to be penetrated to approximately the same extent by the *s*, *p*, and *d* outer electrons. Another factor that also ought to be taken into account is the effect of the exchange interactions between the inner and outer electrons (e.g., Schwartz, 1974, 1975). Although the magnitudes of these interactions are not known the observations on electric-dipole-forbidden transitions that are described in Sec. 5.3 indicate that the fine-structure splittings to which these interactions contribute are probably of the same order as the natural widths of the levels, in which case the exchange interactions might contribute significantly to the observed values of Δ_s, Δ_p, and Δ_d.

Another result of the work of King *et al.* (1977b) and Shaw *et al.* (1985) is the accurate measurement of the widths and lifetimes of the inner-shell-excited states; that for Ar represent the first accurate measurement of any atomic lifetime in this range of excitation energy. In fact very little experimental data of any kind exist on the lifetime of X-ray levels in this range of excitation energy. These lifetimes are relevant to the subject of postcollision interactions (for a review see Read, 1975).

4. HIGH-RESOLUTION STUDIES OF MOLECULES

High-resolution studies of inner-shell excitation in molecules provide information similar to that obtained in atoms plus further information that is characteristic of molecules only, such as vibrational spacings and intensities. Thus as well as being able to obtain excitation energies, relative intensities, and natural decay widths, potential curves can sometimes also be deduced. Another difference concerns the equivalent-core model: for inner-shell excitation of atoms the relevant properties of the equivalent-core atom are invariably already known, whereas for inner-shell excitation of molecules the equivalent-core molecule usually has an odd number of electrons and does not exist as a free molecule. Simulating this molecule by inner-shell excitation can therefore provide useful information that is difficult or impossible to obtain by other means. These topics will be dealt with in the present section. Discussion of the excitation of spin-forbidden transitions will be deferred to Sec. 5.

Perhaps the best-studied molecular inner-shell spectrum is that of N_2, and so it is appropriate to start with a discussion of this. Inner-shell excitation of the diatomic molecules CO and Cl_2 will then be considered. The diatomic molecules NO, HCl, HBr and Br_2 have also been studied at high resolution (Tronc *et al.*, 1980; Shaw *et al.*, 1984). The section will finish with a discussion of inner-shell excitation of polyatomic molecules containing N or C atoms.

4.1. *Nitrogen*

The experimental work of Brion and co-workers (see, for example, Wight *et al.*, 1972/3) shows clearly that much can be learned about the fields that exist within molecules from studies of the promotion of atomic inner-shell electrons to unfilled outer molecular orbitals or to continuum states close to the ionization threshold. The relative probabilities with which the available bound or continuum states can be reached, either by photoabsorption or by electron impact excitation, depend partly on the probability with which an electron can be ejected with the necessary energy from its atomic inner shell, but more importantly on the "focusing" effect of the molecular field. Dehmer and Dill (1976) have discussed these points in full, and have given references to the earlier experimental and theoretical work, in connection with their calculations of the cross section for photoabsorption by the atomic K shell electrons of N_2.

The inner-shell photoabsorption spectrum of N_2 (Nakamura *et al.*, 1969; van der Wiel *et al.*, 1970; Wight *et al.*, 1972/3; van der Wiel *et al.*, 1972) provides an excellent illustration of the influence of the effective molecular field on the photoabsorption process. The most unusual aspect of the spectrum (see Fig. 2) is that the lowest energy peak, corresponding to the promotion of a $1s$ electron to the first unfilled $(2p\pi)$ orbital, has an area that is much greater (by a factor of the order of 10) than the sum of the areas of all the other discrete peaks below the ionization threshold. As mentioned in the Introduction, this is attributable (Dehmer and Dill, 1976) to the fact that the $2p\pi$ orbital is highly localized, being confined to the vicinity of the molecular core by a centrifugal barrier in its d-wave component. The existence of this barrier is also well known from the work on shape resonances in electron scattering on N_2 (Schulz, 1973), in which it was found that electrons of energy approximately 2.3 eV could occupy the unbound $2p\pi$ orbital to form the temporary negative ion N_2^-.

The gross features of the $(1s)^{-1}(2p\pi)\ {}^1\Pi$ state of N_2 have also been studied in X-ray emission and Auger electron experiments (see King *et al.*, 1977c, and references therein). The vibrational structure of the state was first revealed by the high-resolution (75-meV) electron energy-loss studies of King *et al.* (1977c). This study was repeated later by Hitchcock and Brion (1980) with a slightly poorer resolution and with an energy calibration that differs by 0.16 eV from that of King *et al.* The spectrum of King *et al.* is shown in Fig. 4. To obtain the same resolution in a conventional photoabsorption spectrum would require a spectral resolution of 0.0006 nm at 3.1 nm. Table V shows the peak energies and relative areas deduced from the electron energy-loss spectrum, by means of a least-squares fitting program in which the peak energies are treated as variable parameters, but with the restriction that they be related by

$$E(v) = E_0 + \omega(v + 1/2) - \omega x(v + 1/2)^2. \tag{9}$$

The deduced values of ω and ωx are also shown in the table. The deduced

TABLE V

Measured and deduced data for the vibrational levels of the inner-shell state $N_2^{K*}(1s\sigma)^{-1}(2p\pi)\ ^1\Pi$.[a]

State	v	E (eV)	Normalized area	ω (meV)	ωx (meV)	R_e (nm)	Γ (meV)
$N_2^{K*}(1s\sigma)^{-1}(2p\pi)$	0	400.86(3)[b]	0.358(4)	236(2)	1.8(4)	0.1162(3)	123(10)
	1	401.09(1)	0.337(4)				
	2	401.31(1)	0.185(3)				
	3	401.54(1)	0.080(3)				
	4	401.76(1)	0.029(3)				
	5	401.98(2)	0.010(3)				
	6	402.19(2)	0.001(3)				
NO $gs(2p\pi)$				236.05	1.73	0.11508	

[a] After King *et al.* (1977c), Shaw *et al.* (1982).

[b] This is the estimated uncertainty in the absolute calibration of the energy scale. The uncertainties in the energies of the remaining vibrational peaks are relative to the energy of the first peak. The numbers in parentheses indicate the standard error in the last significant figure.

peak areas have been normalized to a total area of unity. The center of gravity of the vibrational peaks is at an energy of 401.12 ± 0.02 eV, which should be approximately equal to the midplateau position of the truncated peak observed by Nakamura *et al.* (1969), namely, 400.84 eV. The peak shapes observed are assumed to be Voigt profiles, being the convolution of a Lorentzian shape whose width is the natural decay width Γ of the state with a Gaussian shape representing the shape of the apparatus energy profile. The width Γ is treated as an additional variable in the peak fitting program. Its deduced value is shown in Table V. The least-squares fit for the spectrum shown in Fig. 4 has a normalized chi-squared of 1.05, and is shown by the solid line in the figure. Only a single vibrational sequence appears to be present, indicating that contributions from transitions to other unfilled orbitals (Wuilleumier and Krause, 1972) are negligible, and indicating also that separate transitions from the inner $1s\sigma_g$ and $1s\sigma_u$ orbitals are probably not both present (see the discussion below). Recently Barth and Schirmer (1985) have calculated the energies and oscillator strengths for inner-shell transitions in N_2 and have obtained good agreement with the results of King *et al.* (1977c).

Information about the mean internuclear separation R_e in the inner-shell excited molecule is obtained from the relative areas of the vibrational peaks by assuming that these areas are proportional to the Franck–Condon overlap integrals $q_{v'v''}$ between the vibrational wave function of the ground vibrational state ($v'' = 0$) of the target molecule N_2 and the vibrational wave functions of the inner-shell excited states (v'). For ease of calculation the initial and final states are both assumed to have Morse potential functions. The value of R_e that gives the best fit between the calculated and observed relative intensities is shown in Table V.

The changes in ω and R_e in going from N_2 to N_2^{K*} can be understood in terms of the antibonding character of the outer π orbital and the relaxation of the inner orbitals. For example in going from N_2 to N_2^- the addition of the antibonding orbital causes R_e to increase (from 0.1094 to 0.1189 nm), and ω to decrease (from 292 to 244 meV) (Schulz, 1973). In going from N_2 to N_2^{K+}, on the other hand, only the relaxation effects are present, and R_e decreases (from 0.1094 to 0.1059 nm) while ω increases (from 292 to 330 meV). These two opposing effects are both present in going from N_2 to N_2^{K*}, although the values of R_e and ω (0.1162 nm and 236 meV, respectively) seem to indicate that the increased antibonding character is the more important.

The fact that the lifetime $\tau_{K*}(=\hbar/\Gamma=5.1\times10^{-15}\,\mathrm{s})$ of the state is approximately one-third as long as the classical vibrational period $\tau_{\mathrm{vib}}(=h/hc\omega=1.8\times10^{-14}\,\mathrm{s})$ of the molecule means that the vibrational wave functions are not those of stationary states, as is assumed in the Franck–Condon analysis discussed above. Similar situations are encountered in the analysis and interpretation of the vibrational structure of short-lived negative-ion resonances (Schulz, 1973). The N_2^- ($2p\pi$) resonance, for example, has a vibrational spacing of approximately 0.244 eV and a decay width of 0.57 eV at the equilibrium internuclear separation (Schulz, 1973), which means that the molecule is usually able to undergo only a small part of one complete vibration during its lifetime. A time-dependent analysis must then be used to analyze the observed vibrational structure. Furthermore the decay width Γ for this resonance is a strong function of the internuclear distance R, being greatest at small values of R, so that the observed peak widths represent only an average value of Γ. In the case of the present inner-shell excited state, however, we expect Γ to be approximately independent of R, since the largest contribution to Γ comes from Auger decays involving primarily the $2p\pi_u$ and $2s\sigma_g$ orbitals (Wuilleumier and Krause, 1972), and the relevant properties of these orbitals (such as the degree of correlation they possess) are probably not strong functions of R. Andrić and Čadež (1978) have carried out the appropriate time-dependent analysis and have shown that the decay process does not change the relative areas of the vibrational peaks.

As discussed in Sec. 3.5, the high-resolution studies by King *et al.* (1977b) of inner-shell excited states of Ar, Kr, and Xe have demonstrated the usefulness of the equivalent-core model for these states. In the case of inner-shell excited states of homonuclear diatomic molecules, however, an ambiguity exists in the definition of the equivalent core molecule, depending on whether the inner-shell hole is regarded as being localized or delocalized (see, for example, Cederbaum and Domcke, 1977; Lozes *et al.*, 1979; Dutta and Huang, 1980). If the inner-shell hole of N_2^{K*} is delocalized (having the configuration $1s\sigma_g^{-1}$ or $1s\sigma_u^{-1}$) the "core" can be taken to be the two nitrogen nuclei and the three accompanying $1s$ electrons, which is thus approximately equivalent to two separated charges of magnitude 5.5 each, but if the inner-shell hole is localized in the vicinity of one of the nuclei [having the configuration $(\mathcal{N}1s)^{-1}$] then this nucleus and its single $1s$ electron has a total charge of 6 while the core charge at the other center is 5. Thus for a localized hole the outer electrons of the molecule move in a mean electrostatic potential similar to that of the outer electrons of the neutral NO molecule, which is thus the "equivalent core molecule." Although it has usually been assumed in analyses of inner-shell excited states of homonuclear diatomic molecules that the inner-shell holes are localized (see for example Wight *et al.*, 1972/3), the only other information on this localization appears to be that obtained from calculations of the ionization potentials of these states. Thus Bagus and Schaefer (1972) have found it necessary to assume that the $1s$ hole state of O_2^+ is localized in order to obtain agreement with the experimentally measured ionization potential, and Snyder (1971) has argued that the fact that the electronic relaxation energies of N_2 and NH_3 are similar implies that the $(\mathcal{N}1s)^{-1}$ hole state is localized in N_2 as it is in NH_3.

Further information about the localization can be obtained by considering the ratio τ_{K*}/τ_h, where $\tau_h(=\hbar/\Delta E)$ is the characteristic "hopping time" between the orbitals $1s\sigma_g$ and $1s\sigma_u$, differing in energy by ΔE. Rescigno and Orel (1979) have calculated ΔE to be 0.06 eV, but the experimental results at low incident energies (see Sec. 5) suggest that it may be smaller than this. Thus $\tau_h \geqslant 2\tau_{K*}$, which implies that if the initial vacancy is localized at one of the nuclei then it will remain approximately localized during the lifetime of the state. In fact the initial vacancy is almost certainly localized because, as in the process of photoionization above an absorption edge, the electrons most likely to be ejected are those that are nearest to a nucleus since this facilitates conservation of momentum (Read, 1980).

The validity of the equivalent-core model is also affected by the existence of exchange interactions between the inner and outer electrons (e.g., Schwartz, 1976), as mentioned above. In the case of N_2^{K*} these interactions are responsible for the splitting between the singlet and triplet states of the $(1s)^{-1}(2p\pi)$ configuration, and this splitting is known (see Sec. 5.3) to be 0.82 eV. The $1s$–$2p\pi$ exchange interaction presumably affects the term value of the state more than it affects the form of the potential function, and it will in any case be much smaller for the higher configurations $(1s)^{-1}(nl)$. We will return to this topic in Sec. 5.

Assuming complete localization and ignoring the effects of exchange interaction, the values of ω, ωx, and R_e deduced for the $N_2(1s)^{-1}(2p\pi)\ {}^1\Pi$ state should be similar to those of the ground electronic state of NO. It can be seen from Table V that there is indeed a close similarity between the two sets of parameters. It is interesting to note that although the lifetime τ_{K^*} apears to be short enough for the vacancy to remain localized and for the NO equivalent-core model to have some validity, it is nevertheless long enough to support a discrete vibrational structure for the $N_2^{K^*}$ molecule.

Nitrogen, Higher Rydberg States. The high-resolution measurements of King *et al.* (1977c) have been extended by Tronc *et al.* (1980) to cover the higher-lying Rydberg states. The results of these authors are summarized in Table VI, which also gives the relevant parameters of the analogous states of CO^{K^*} (see below) and of the equivalent-core molecule of $N_2^{K^*}$ and $C^{K^*}O$, which in both cases is NO. The similarity between the parameters of the three molecules indicates the degree of validity of the equivalent-core model. The results for the $(1s\sigma_N)^{-1}(3p\pi)$ state of $N_2^{K^*}$ can be compared with the results of the *ab initio* calculations of Clark and Muller (1976): they find essentially the same Frank–Condon factors (0.86, 0.125 and 0.01 for $v=0$, 1, and 2, respectively), but somewhat different values of ω and R_e (330 meV and 0.1057 nm, respectively). Also shown in Table VI are the term values and quantum defects δ for the three molecules, calculated using the value 409.93 eV (Johansson *et al.*, 1973) for the $1s$ ionization energy of N_2. The quantum defects of the states of $N_2^{K^*}$ are similar to those of the analogous states of $C^{K^*}O$, and are consistently slightly lower than those of the equivalent core states. The difference

$$\Delta = \delta(\mathrm{NO}) - \delta(N_2^{K^*}) \tag{10}$$

between the quantum defects of the equivalent core state and the analogous inner-shell excited state is largest for the $3s\sigma$ orbital (if we ignore the results of low accuracy): this behavior is similar to that observed (King *et al.*, 1977b) for the inner-shell excited states of Ar, Kr, and Xe, for which the values of Δ are greatest for s orbitals. By requiring that the $2p\pi$, $3p\pi$, and $4p\pi$ orbitals should have the same quantum defect, within the experimental errors, we find this quantum defect to be 0.777 ± 0.003, giving a $1s$ ionization energy of 409.95 ± 0.05 eV, which is consistent with that found by Johansson *et al.* (1973).

The peak areas (summed over the vibrational transitions) are proportional to the generalized oscillator strengths of the transitions. The deduced relative values of the generalized oscillator strengths are shown in Table VII, together with the values calculated by Dehmer and Dill (1976) and Rescigno and Langhoff (1977) for the relative values of the optical oscillator strengths. The extent of the agreement is essentially within the experimental and calculational uncertainties.

TABLE VI

Parameters of inner-shell Rydberg excited states of N_2 and CO.

Assignment	Energies N_2	Energies CO	ω (meV) and R_e (nm) N_2^{K*}	ω (meV) and R_e (nm) $C^{K*}O$	ω (meV) and R_e (nm) NO	Term values (eV) N_2^{K*}	Term values (eV) $C^{K*}O$	Term values (eV) NO	Quantum defects N_2^{K*}	Quantum defects $C^{K*}O$	Quantum defects NO
$(1s\sigma)^{-1}(2p\pi)\ v=0$	400.86	287.40	$\omega=235(2)$	258(4)	236	9.07(3)	8.67(2)	9.26	0.776(2)	0.747(2)	0.79
1	401.09	287.65	$R_e=0.1162(3)$	0.1104(2)	0.1151						
2	401.31	287.91	$\Gamma=128(6)$	85(3)							
$(1s\sigma)^{-1}(3s\sigma)\ v=0$	406.29	292.37	$\omega=-$	306(6)	294	3.64(3)	3.70(2)	3.79	1.07(1)	1.08(1)	1.11
1		292.66	$R_e=0.1083(4)$	0.1079(2)	0.1064						
2		292.97	$\Gamma=200(40)$	87(4)							
$(1s\sigma)^{-1}(3p\pi)\ v=0$	407.18	293.33	$\omega=249(30)$	295(12)	295	2.75(3)	2.74(2)	2.78	0.78(1)	0.77(1)	0.79
1	407.43	293.63	$R_e=0.1068(4)$	0.1075(3)	0.1062						
2		293.91	$\Gamma=200(40)$	102(2)							
3		294.19									
$(1s\sigma)^{-1}(3d\pi)\ v=0$	408.44	294.59	$\Gamma=-$	122(6)		1.49(3)	1.48(3)	1.50	−0.02(3)	−0.04(2)	−0.01
$(1s\sigma)^{-1}(4p\pi)\ v=0$	408.67	294.78				1.26(4)	1.29(3)	1.29	0.71(5)	0.75(3)	0.76
		295.07									
$(1s\sigma)^{-1}(5p\pi)\ v=0$	409.12	295.30				0.81(4)	0.77(3)	0.76	0.90(10)	0.80(6)	0.78
$(1s\sigma)^{-1}(6p\pi)\ v=0$	409.49	295.55				0.44(4)	0.52(3)	0.49	0.44(24)	0.89(10)	0.75

[a] From Tronc *et al.* (1980); Shaw *et al.* (1984).

TABLE VII

Relative oscillator strengths of electronic transitions for N_2^{K*}.[a]

Outer orbital	$2p\pi$	$3s\sigma$	$3p\pi$	$3d\pi$	$4p\pi$	$5p\pi$	$6p\pi$
Tronc *et al.* (1980)	1.00	0.012(1)	0.029(3)	0.011(1)	0.006(1)	0.006(1)	0.004(1)
Dehmer and Dill (1976)[b]	1.00	0.010	0.030	0.009	0.010	0.005	0.003
Rescigno and Langhoff (1977)[c]	1.00	0.009	0.057	0.006	0.016	—	—

[a] After Tronc *et al.* (1980).
[b] Relative generalized oscillator strengths at momentum transfers of 1.51–1.55 a.u.
[c] Relative optical oscillator stengths at zero momentum transfer.

The absolute values of the oscillator strengths can be estimated by comparing that of the $(1s\sigma_N)^{-1}(2p\pi)\ ^1\Pi$ transition with the measured (Wuilleumier and Krause, 1972; Kay *et al.*, 1977) or calculated (Dehmer and Dill, 1976; Rescigno and Langhoff, 1977; Rescigno and Orel, 1979) values. Table VIII shows these oscillator strengths, together with the momentum transfer to which they refer. In the experiment of Tronc *et al.* (1980) the momentum transfer was 1.51 a.u.

4.2. *Carbon Monoxide*

Excitation of a carbon $1s\sigma$ electron in CO has been observed by van der Wiel *et al.* (1970) and by Wight *et al.* (1972/3) in energy-loss spectra at electron impact incident energies of 10 and 2.5 keV, respectively. High-

TABLE VIII

Absolute generalized oscillator strength for the transition to the $(1s\sigma)^{-1}(2p\pi)\ ^1\Pi$ state of N_2^{K*}.[a]

Momentum transfer (atomic units)	0	0.61	1.51
Optical absorption (Wuilleumer and Krause, 1972)	0.12 ± 0.05		
Inelastic electron scattering (Kay *et al.*, 1977)		0.195 ± 0.02	
Inelastic electron scattering (Johnson and Bonham, 1979)	0.15 ± 0.03[b]		
Calculated (Dehmer and Dill, 1976)	0.23		
Calculated (Rescigno and Langhoff, 1977)	0.257		
Calculated (Rescigno and Orel, 1979)	0.118	0.114	0.095

[a] After Tronc *et al.* (1980).
[b] Extrapolated to $K = 0$.

resolution energy-loss spectra have been reported in CO by Tronc *et al.* (1976), who were the first authors to resolve the vibrational structure of the $(1s\sigma_C)^{-1}(2p\pi)\ ^1\Pi$ state and more recently by Tronc *et al.* (1979), Hitchcock and Brion (1980), and Shaw *et al.* (1984). Van der Wiel and El Sherbini (1972) and Kay *et al.* (1977) have studied the production of the ions CO^+, CO^{2+}, C^+, C^{2+}, and O^+ following carbon K excitation and ionization of CO, and have derived the corresponding oscillator strengths from the small-angle inelastic scattering of high-energy electrons observed in coincidence with these ions. No photoabsorption spectrum exists in the energy range of carbon K-shell excitation to discrete states, presumably because of problems arising from carbon contamination of optical components.

Energy-loss spectra of CO, corresponding to the excitation of a $1s\sigma_C$ electron, are shown in Figs. 9 and 10 (Shaw *et al.*, 1984). Figure 9 shows the vibrational structure observed when the $1s$ carbon electron is promoted to the first unfilled valence orbital $2p\pi$, while Fig. 10 shows transitions of the $1s$ electron to Rydberg orbitals. Molecular parameters deduced from these data are given in Table VI.

An important point to note is that the $(1s\sigma_C)^{-1}np\pi$ configurations are excited more strongly than other configurations, and that the oscillator strength for the lowest transition of this type is exceptionally high, due to the fact that the $2p\pi_g$ orbital is highly localized since it is confined to the vicinity of the molecular core by a centrifugal barrier in its d-wave com-

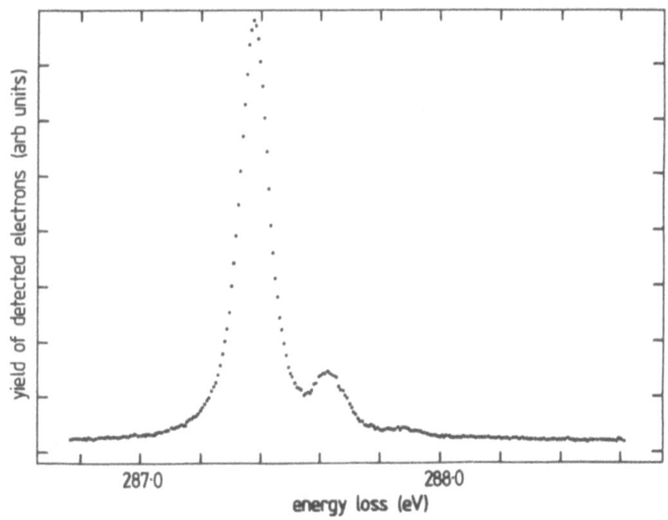

FIGURE 9. Electron energy-loss spectrum of CO obtained at an incident energy of 1.5 keV and with an energy resolution of 55 meV. The state observed is the $(1s\sigma_C)^{-1}(2p\pi)\ ^1\Pi$ carbon K-shell excited state of CO. (From Shaw *et al.*, 1984.)

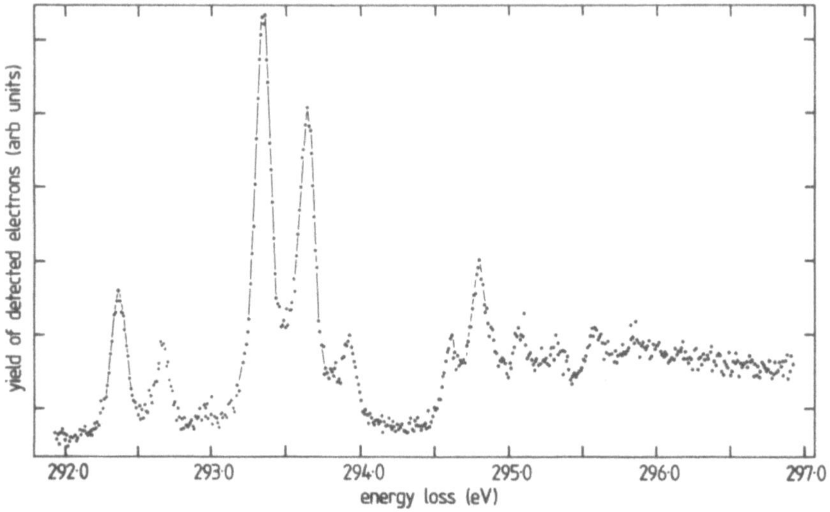

FIGURE 10. Electron energy-loss spectrum of CO obtained at an incident energy of 1.5 keV and with an energy resolution of 65 meV. The peaks correspond to transitions of a $1s\sigma_C$ electron to Rydberg orbitals. (From Shaw *et al.*, 1984.)

ponent (Dehmer and Dill, 1976). A second feature is that the derived values of ω, ωx, and R_e are similar to those of the equivalent-core molecule NO. A third feature of the results is that the natural width Γ of the states appears to increase as the binding energy decreases. Similar trends can be seen in the natural widths of N_2 and of the atoms Ar, Kr, and Xe (Table II). However, Shaw *et al.* (1984) have pointed out that this effect in the earlier work may have been partly instrumental since the states of lower binding energy have lower intensity and therefore usually require longer collection times and hence are more prone to drifts in the energy scale. The final feature of note is that the quantum defects of the inner-shell-excited states are systematically slightly lower than those of the equivalent-core states, the differences all being generally within the range 0.03 ± 0.02. This result is similar to that found (King *et al.*, 1977b) for inner-shell excited states of Ar, Kr, and Xe, for which the differences are also small, in the range 0.00 ± 0.03. This is further evidence that the equivalent-core model gives a reasonably good representation of these inner-shell-excited states.

4.3. *The Application of the Equivalent-Core Model to Chlorine*

Having shown the degree of validity of the equivalent-core model it is appropriate to see if the model can be usefully applied to the study of molecules that are difficult to study directly. An example of such a molecule is ArCl, which is the equivalent-core molecule of Cl_2^{L*}. ArCl is one of a class of molecules used in ultraviolet lasers (Waynant, 1977) but about which little is known (Golde and Thrush, 1974; Golde, 1975; Gundel *et al.*, 1976; King *et al.*, 1977a; Olsen and Liu, 1978; Castex *et al.*, 1980), due to the repulsive nature of the ground state of the molecule. However, by studying inner-shell excitation of Cl_2 it is possible, using the equivalent core model, to obtain information about ArCl. Such a study has been reported by Shaw *et al.* (1980) and a similar study on inner-shell excitation of F_2 has been undertaken by Hitchcock and Brion (1981) at an energy resolution of 0.5 eV, to obtain information on the equivalent-core molecule NeF.

A spectrum covering energy losses in Cl_2 from 197 to 210 eV is shown in Fig. 11 (Shaw *et al.*, 1980). The observed features correspond to the excitation of an $L_{2,3}$ electron to an unoccupied valence or Rydberg orbital of Cl_2. The energies of the observed peaks are given in Table IX.

A notable property of the spectrum is that it consists of two essentially identical series that are separated in energy and have different relative heights. The solid and broken curves in Fig. 11 show this decomposition. A detailed analysis shows that the energy separation of the two series is 1.62 ± 0.02 eV and that the relative heights are $1:0.60 \pm 0.05$. The energy separation is consistent with the atomic L_2–L_3 energy difference found

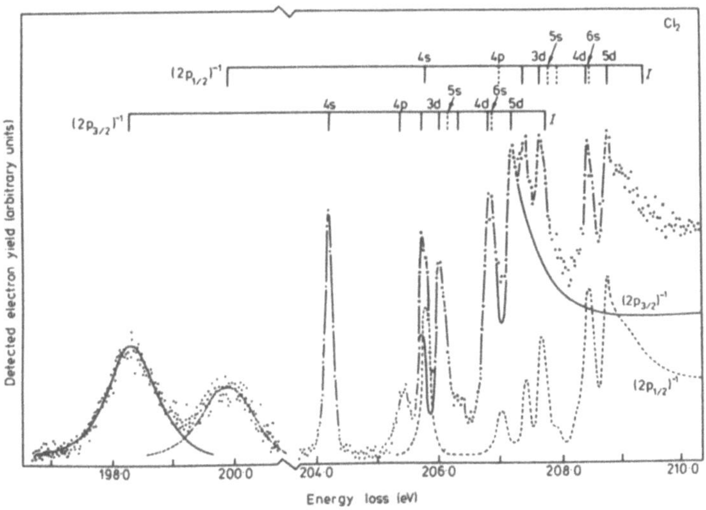

FIGURE 11. Electron energy-loss spectrum of Cl_2 at an incident energy of 1.5 keV and a resolution of 65 meV. The peaks correspond to the excitation of an $L_{2,3}$ electron to an unoccupied valence or Rydberg orbital. The full and broken curves show the decomposition of the observed spectrum into two identical series corresponding to the $(2p_{3/2})^{-1}$ and $(2p_{1/2})^{-1}$ vacancies. (From Shaw *et al.*, 1980.)

TABLE IX

Measured and derived information for inner-shell excited states in molecular chlorine. The number in parentheses show the errors in the last significant figures. The ionization limits have been deduced from the $s+d$ Rydberg series, assuming a constant value of the quantum defect.

State	E(eV) $(2p_{3/2})^{-1}$	E(eV) $(2p_{1/2})^{-1}$
$(2p)^{-1}$: $(3p\sigma)\ ^1\Sigma$	198.305 (20)	199.929 (30)
$4s$	204.240 (20)[a]	205.868 (10)
$4p$	205.435 (5)	—
	205.789 (10)	207.429 (5)
$3d$	206.076 (5)	207.709 (7)
	206.377 (16)	—
$4d$	206.857 (8)	208.480 (5)
$5d$	207.225 (10)	208.816 (10)
Ionization edge	207.800 (10)	209.420 (10)

$\delta(d)=0.19(1)$, $\delta(4s)=2.04(1)$
$I(2p_{1/2})-I(2p_{3/2})=1.62(2)$ eV
$\Gamma=0.130(10)$ eV.

[a] All the errors shown in the table are standard deviations except this one, which is the estimated maximum systematic error of the energy calibration. The other energy errors shown in the table are relative to the energy of the $(2p_{3/2})^{-1}4s$ peak, and do not contain the possible calibration error.

previously in lower-resolution X-ray emission and photoelectron studies on chlorine in the solid-state form (for example, NaCl vacuum-evaporated on Al, see Nordberg *et al.*, 1966): Bearden (1967) reports 1.6 ± 0.1 eV, Bearden and Burr (1967), 1.6 ± 0.3 eV, and Shirley *et al.* (1977), 2.0 eV. The relative height corresponds approximately to the relative statistical weight of the atomic $2p_{3/2}$ and $2p_{1/2}$ vacancies (namely, 1:0.5). The degeneracies of the $2p_{3/2}$ and $2p_{1/2}$ vacancies would be split if the vacancies were to have an appreciable molecular character, but the spectrum shows that the degeneracy splitting is less than the width (130 meV) of the observed peaks. Since this width corresponds to the expected natural with of a $2p$ vacancy in chlorine (the calculated width is approximately 100 meV, Krause, 1979; and Shaw *et al.*, 1982, find a natural width of 117 ± 5 meV for the $2p$ vacancy of Ar) it can be inferred that the degeneracy splitting is significantly smaller than 130 meV and that the vacancy is essentially atomic in character. The smallness of the splitting also implies that the "hopping-time" for the redistribution of the vacancy site (see discussion on Sect. 4.1) is of the order of or longer than the natural lifetime of the state, so that a vacancy which is initially created as an atomic vacancy will remain so during the existence of the inner-shell excited state. From these arguments it can be seen that the equivalent core molecule for Cl_2^{I*} is that in which one of the chlorine atoms is effectively increased in charge by one unit, namely, the inert-gas halide ArCl.

To assist in classifying the discrete transitions it is useful to consider the available information on the potential functions of the ArCl molecule, given in Fig. 12. The potential functions shown for the repulsive $X^2\Sigma$ and $A^2\Pi$ states are those calculated by Olsen and Liu (1978), neglecting the small spin–orbit splitting of the Cl ground state (0.109 eV). The two broad low-energy peaks in Fig. 11 clearly correspond to transitions to a repulsive state. The measured width of the peaks, 0.94 ± 0.03 eV, implies (Shaw *et al.*, 1982; Ewing and Brau, 1975) that the slope of the repulsive curve is 99 ± 3 eV nm^{-1} at $R = 0.1988$ nm (the equilibrium separation for Cl_2). The calculations of Olsen and Liu (1978) give slopes of approximately 110 and 190 eV nm^{-1} for the $X^2\Sigma$ and $A^2\Pi$ curves, respectively. The lower of these values is the closer to the slope deduced from the inner-shell energy loss spectrum, and so it is assumed that the $X^2\Sigma$ state is the appropriate upper state for the observed transition. Better agreement is obtained if correlation and relativistic effects are taken into account (Olsen, 1980) since this is expected to shift the repulsive curves by about 0.005 nm to the left. This has been taken into account in plotting the $X^2\Sigma$ and $A^2\Pi$ curves in Fig. 11. Having established the point corresponding to the lowest energy transition it is then possible to position the energies of the upper states of the lowest energy transitions as shown in the figure. The potential function of the

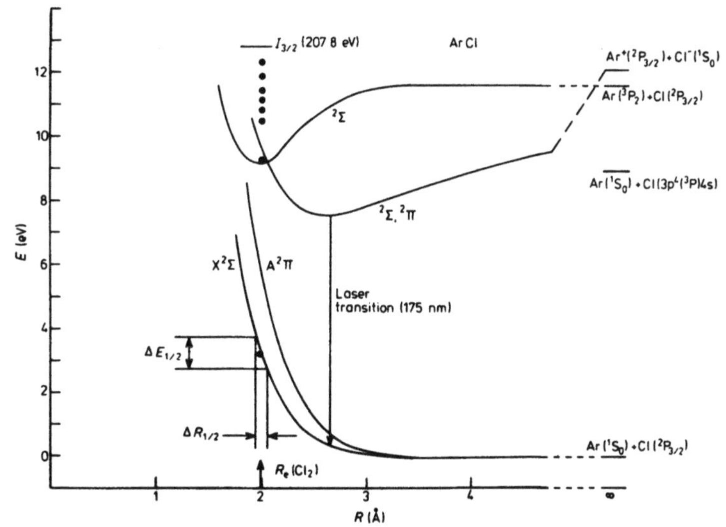

FIGURE 12. Observed inner-shell excitation energies of Cl_2 and constructed potential energy curves of ArCl. (From Shaw *et al.*, 1980.)

lowest attractive state has been constructed by assuming (see the discussions of Ewing and Brau, 1975; Gundel *et al.*, 1976) that it is approximately the same as that of the ground state of KCl, that its height above the $X^2\Sigma$ state corresponds to the measured laser transition energy (Waynant, 1977), and that at large internuclear separation R it has a Coulombic form tending to the energy of the dissociation products $Ar^+(^2P_{3/2}) + Cl^-(^1S_0)$. The three possible symmetries of this ionic state are $^2\Sigma_{1/2}$, $^2\Pi_{1/2}$, and $^2\Pi_{3/2}$, but the small energy differences between these have been ignored and a single curve only has been drawn. The remaining potential curve shown in Fig. 12, that of the $^2\Sigma$ state with the dissociation products $Ar(^3P_2) + Cl(^2P_{3/2})$, is the curve estimated by Gundel *et al.* (1976).

The peaks labeled 3*d*, 4*d*, and 5*d* in Fig. 11 appear to represent Rydberg series converging to the $2p_{3/2}$ and $2p_{1/2}$ ionization limits. By assuming that the quantum defect δ is the same for all these upper states a value of 0.19 ± 0.01 is obtained for it while the ionization energies are as given in Table IX. The sharp peaks labeled 4*s* are identified by their quantum defects (Shaw *et al.*, 1982) and the peak at 205.435 eV is tentatively identified as the electric-dipole-forbidden transition to the $(2p_{3/2})^{-1}4p$ state.

4.4. *Polyatomic Molecules*

There has been much work on the study of energy-loss spectra in polyatomic molecules, particularly by Brion and co-workers. Most of this work, however, has been performed with limited resolution. High-resolution energy-loss measurements have been made of CH_4, CF_4, CO_2, COS, C_2H_2, and C_2H_4 by Tronc *et al.* (1979), and also of N_2O (Tronc *et al.*, 1980). These studies are concerned with the excitation of a carbon or nitrogen *K*-shell electron and so the observed states have excitation energies just below their respective 1*s* ionization potentials. To illustrate what can be learned from high-resolution studies we now present one example, that of methane (Tronc *et al.*, 1979).

Methane. Figure 13 shows the energy-loss spectrum of methane in the $1sa_1$ electron excitation energy region (Tronc *et al.*, 1979). The peak positions are given in Table X, where they are compared with electron impact measurements at higher incident energy (Wight and Brion, 1974b), with optical data (Chun, 1969; Eberhardt *et al.*, 1976), and with theoretical calculations (Bagus *et al.*, 1973; Deutsch and Kunz, 1973, 1975).

The calculations of Bagus *et al.* (1973) show that the low-intensity peak at 287.05 eV can be attributed to the $a_1 2s \rightarrow a_1 3s$ electric-dipole-forbidden transition. This transition is observed both in electron impact spectra and in photoabsorption spectra because of the vibrational electronic coupling between the excited Rydberg state and the ground state by the ν_3 and ν_4 vibrational modes of the CH_4 molecule. Measurements of isotope effects on

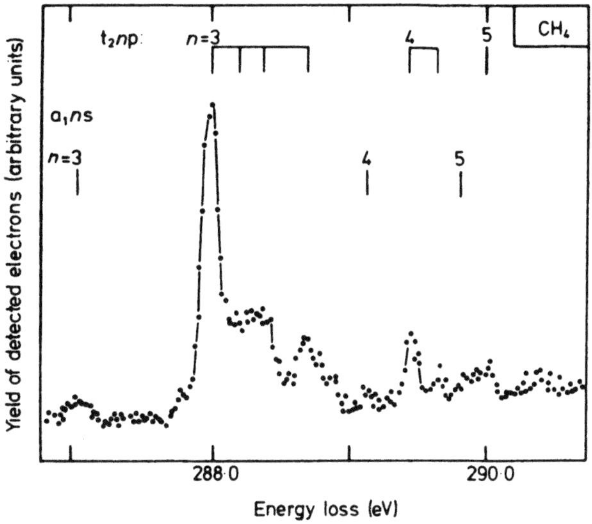

FIGURE 13. Electron energy-loss spectrum of CH_4 obtained at an incident energy of 1.5 keV and with an energy resolution of 70 meV. (From Tronc *et al.*, 1979.)

intensities in the *K*-shell excitation spectra of CH_4 and CD_4 (Hitchcock *et al.*, 1977) are compatible with a Herzberg–Teller vibronic-coupling mechanism for this transition. The energy of this lowest a_1 1*s* hole state is in good agreement with an unrestricted Hartree–Fock energy calculation by Deutsch and Kunz (1975).

The most intense peak, at 288.00 eV, corresponds to the excitation of a a_1 1*s* carbon electron to the first *p* Rydberg orbital ($t_2 3p$) (Wight and Brion, 1974b; Chun, 1969; Bagus *et al.*, 1973). The natural width of this peak, estimated from a least-squares fitting, is 94 ± 6 meV, giving a lifetime of $(7.0 \pm 0.5) \times 10^{-15}$ s.

The structures at 288.24 and 288.37 eV are understood as vibrational levels of the $(a_1 1s)^{-1} (t_2 3p)\, {}^1T_2$ excited state. In the excitation of the non-bonding a_1 1*s* electron to the non-bonding $t_2 3p$ Rydberg orbital, the 3*p* orbital can mix with antibonding valence orbitals, leading to excitation of those modes (ν_2, ν_3, and ν_4) which distort the CH_4 molecule (see also Gelius, 1974).

The peak observed at 288.70 eV could result from the splitting of the degenerate ${}^1T_2$ excited state. This splitting of 0.70 eV is in good agreement with the Jahn–Teller splitting observed by Harshbarger and Lassettre (1973) in valence-shell excitation, and with the splitting observed in photoelectron spectroscopy for the ground state of the CH_4^+ ion by Brundle *et al.* (1970). The higher energy structures in Fig. 13 correspond to transitions to higher Rydberg *s* and *p* orbitals and their assignments are given in Table X.

TABLE X

Data for carbon K-shell-excited states of CH_4. After Tronc *et al.* (1979).

Electron impact		Optical data			Theory		
Tronc *et al.* (1979)	Wight and Brion (1974b)	Chun (1969)	Eberhardt *et al.* (1976)	Brown *et al.* (1978)	Deutsch and Kunz (1975)	Bagus *et al.* (1973)	Assignment
287.05(3)	287.0	287.2	286.9	287.0	287.07	287.3	$(a_1 1s)^{-1}(a_1 3s)$
288.00(2)	288.0		288.0	288.1			$(a_1 1s)^{-1}(t_2 3p)$
288.24(2)							$(a_1 1s)^{-1}(t_2 3p)+\nu$
288.37(2)		288.3			288.5	288.4	$(a_1 1s)^{-1}(t_2 3p)+\nu$
288.70(2)							
289.14(3)							$(a_1 1s)^{-1}(a_1 4s)$
289.44(3)	289.4		289.4				$(a_1 1s)^{-1}(t_2 4p)$
289.65(3)	289.8						$(a_1 1s)^{-1}(t_2 4p)+\nu$
289.83(5)							$(a_1 1s)^{-1}(a_1 5s)$
290.01(4)							$(a_1 1s)^{-1}(t_2 5p)$

5. THE OBSERVATION OF ELECTRIC-DIPOLE-FORBIDDEN INNER-SHELL TRANSITIONS

At sufficiently high values of incident energy, electron-impact excitation is analogous to photoabsorption and so optical selection rules apply (Sec. 1.3). However, at values of the incident electron energy comparable to the excitation energy of the target state, these selection rules are considerably relaxed, and then electron-impact excitation has the great advantage than it is able to induce electric-dipole-forbidden transitions. This advantage has only recently been exploited, when inner-shell electric-dipole-forbidden transitions were observed in the rare gases by King *et al.* (1977b). Shaw *et al.* (1982) have made a systematic study of the parity-forbidden $2p \rightarrow 4p$ transition in Ar and of an inner-shell spin-forbidden transition in N_2, and it is this work that forms the basis of the present discussion. More recently Shaw *et al.* (1985) have studied inner-shell electric-dipole-forbidden transitions on Kr, Xe, CO_2, and N_2O, while Ungier and Thomas (1983) have observed such transitions in N_2 and CO in their low-resolution Auger spectra.

5.1. *Experimental Details*

The apparatus used for these studies is essentially the same as that shown schematically in Fig. 5. There are two points of note. The spectrometer is operated at much lower values of incident electron energy, typically less than twice the excitation energy of the state of interest. Also the yields of scattered electrons tend to be extremely small at these low incident energies.

5.2. *Electric-Dipole-Forbidden Inner-Shell Transitions in Atoms*

The inner-shell studies of the rare gases by King *et al.* (1977b) provided examples of electric-dipole-forbidden transitions in argon, krypton, and xenon. Such transitions had not been seen previously in photoabsorption measurements. For example, in Ar (Fig. 6) the parity-forbidden transition, $2p \rightarrow 4p$, can be seen. In this measurement the incident electron energy was approximately six times the excitation energy of the transition. More recently Shaw *et al.* (1982) have studied this region of argon in more detail and have measured the energy-loss spectrum as a function of incident electron energy. Their results are shown in Fig. 14. The peaks in the spectra correspond to the promotion of a $2p_{3/2}$ or $2p_{1/2}$ electron to an unoccupied Rydberg orbital. The features observed in the 1500-eV incident energy spectrum are similar to those described by King *et al.* (1977b). The major point of note is the dramatic rise in the intensity of the parity-forbidden $2p \rightarrow 4p$

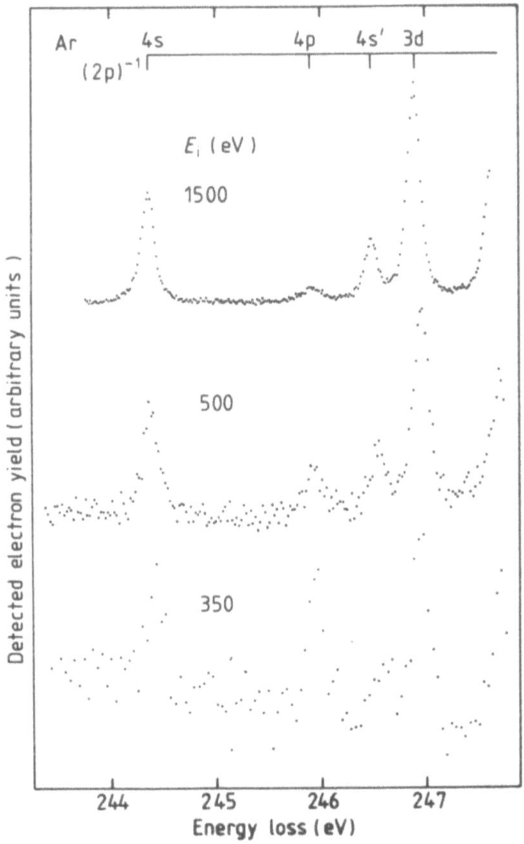

FIGURE 14. Energy-loss spectra of Ar obtained at the indicated values of incident electron energy, and with a resolution of 65 meV. (From Shaw *et al.*, 1982.)

transition relative to the optically allowed transitions as the incident electron energy is reduced.

Although only one peak is observed, corresponding to the transition to the $(2p_{3/2})^{-1}4p$ configuration, this configuration must contain six levels, having J values of 0, 1 (twice), 2 (twice), and 3. The energy separations of these levels can be expressed in terms of quantities F_2, G_0, and G_2 representing contributions to the electrostatic energy, and a quantity ζ_{4p} representing the spin–orbit energy (Condon and Shortley, 1951; Cowan, 1981; Edlén, 1964). The values of F_2, G_0, G_2, and ζ_{4p} were calculated by Shaw *et al.* (1982) using the nonrelativistic Hartree–Fock program of Froese-Fischer (1978). Then using these values they found that five of the six levels are approximately uniformly spaced over 50 meV whilst the sixth level, $J=0$, lies about 75 meV above the mean energy of the other five levels.

Shaw *et al.* (1982) looked for evidence in their spectra of the existence of this fine structure in the $(2p_{3/2})^{-1}4p$ configuration, and although there is a slight indication of an asymmetry in the peak shape (with an increased yield on the high-energy side) the limited statistical accuracy does not permit a more definite conclusion. The estimated splitting is in any case smaller than the natural width of the levels, which increases the difficulty of observing the fine structure. Shaw *et al.* also found that the energy at which the peak maximum occurs does not depend on the incident electron energy. Shaw *et al.* looked for evidence of the fine structure of the $(2p_{3/2})^{-1}4s$ and $(2p_{3/2})^{-1}3d$ configurations, but did not find any, presumably because again the splittings are less than the natural linewidths. More recently however, Shaw *et al.* (1985) have observed the analogous fine structure in Kr and Xe.

The ratio of the observed intensities of the $2p_{3/2} \rightarrow 4p$ and $2p_{3/2} \rightarrow 4s$ transitions is displayed in Fig. 15 as a function of the dimensionless quantity Ka_0, where $\hbar K$ is the momentum transferred to the atom by an electron scattered in the forward direction and a_0 is the Bohr radius. From the figure it may be seen that the intensity ratio is approximately proportional to K^2. The ratios have not been corrected for the variation with scattered electron energy of the analyzer acceptance angle, since this angle is difficult to determine precisely and also because the angular dependences of the forbidden and allowed transitions are not known.

Three different ways of viewing the selection rules and relative intensities for the transitions to the various levels of the $(2p_{3/2})^{-1}4s$ and $(2p_{3/2})^{-1}4p$ configurations are summarized in Table XI. In the first approach the selection rules for electromagnetic transitions (e.g., Read,

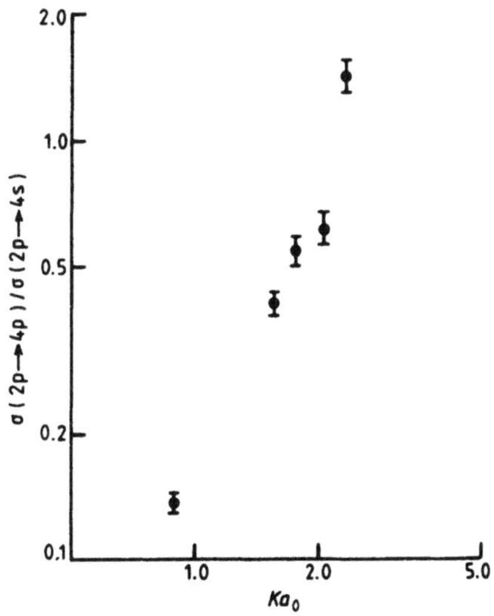

FIGURE 15. The dependence of the ratio of intensities of the observed peaks corresponding to the transitions $2p$–$4p$ (parity forbidden) and $2p$–$4s$ (electric-dipole allowed) on the dimensionless quantity Ka_0, where $\hbar K$ is the momentum transfer and a_0 is the Bohr radius. (From Shaw *et al.*, 1982.)

TABLE XI

Classifications of photon and electron impact transitions to the levels of the $(2p_{3/2})^{-1}4s$ and $(2p_{3/2})^{-1}4p$ configurations of Ar. The third column shows the classification of the optical transitions, the fourth column shows the lowest power of K in the electron impact cross section when the Born approximation is valid, while in the last column D and X denote the direct and exchange contributions respectively, and the numbers refer to the values of $(l_i + l_f)_{\min}$ (see text).[a]

Configuration	J_f	Optical transitions	Lowest power of K at high E_i	Electron impact transitions at low E_i
$(2p^5\ {}^2P_{3/2})\,4s$	1	$E1$	K^{-2}	$D1$, $X1$
	2	$M2$	—	$D3$, $X1$
$(2p^5\ {}^2P_{3/2})\,4p$	0	—	K^0	$D0$, $X2$
	1	$M1$	—	$D2$, $X0$
	2	$E2$	K^0	$D2$, $X2$
	3	$M3$	—	$D4$, $X2$

[a] From Shaw *et al.* (1982).

1980) are considered, and in the table are given the relevant type of transition, EL or ML (where E and M denote electric and magnetic transitions respectively, and L denotes the multipolarity). Only the transitions to the $[(2p_{3/2})^{-1}4s]_1$ level are of the electric dipole ($E1$) type, all the others being forbidden, to varying degrees. In the second approach use is made of the Born approximation, which is valid only at high incident energies, E_i, to give the leading power of K in an expansion of the differential cross section in powers of K^2 (Inokuti, 1971). It may be seen that the $J = 0$ and $J = 2$ levels of the $(2p_{3/2})^{-1}4p$ configuration have differential cross section proportional to K^0, corresponding to electric quadrupole transitions, and that the ratio of this cross section to that for the $[(2p_{3/2})^{-1}4s]_1$ level should be proportional of K^2 at sufficiently high energies. This is in fact approximately the observed dependence, although the Born approximation may apply only poorly at the incident energies used in the work of Shaw *et al.* In the third approach the range of values of l_i and l_f, the orbital angular momenta of the incident and outgoing electrons, respectively, are considered, since it is these angular momenta that determine the degree of penetration into the atomic core when the incident and scattered electron energies are low. For example, the lower the value of l_i the more likely is the incident electron to overlap, and interact, with the target $2p$ electron. The possible combinations of l_i and l_f that can occur depend (Shaw *et al.*, 1982) upon (i) the value of ΔJ, (ii) the parities of the initial and final states, and (iii) whether there is direct or exchange scattering (indicated by D and X, respectively, in

Table XI). A single parameter, $(l_i + l_f)_{\min}$, is used in Table XI to characterize the excitation process. The transitions of greatest intensity are expected to be those that can proceed through direct scattering and that have the lowest values of $(l_i + l_f)_{\min}$, except that at near-threshold energies the direct and exchange contributions will be comparable.

From the information in Table XI, the transition to the $[(2p_{3/2})^{-1}4p]_3$ level is probably relatively weak, but the transitions to the other levels of the $(2p_{3/2})^{-1}4p$ configuration are all somewhat stronger. Approximately this same strength is also expected for the transition to the $[(2p_{3/2})^{-1}4s]_1$ level, while the transition to the $[(2p_{3/2})^{-1}4s]_2$ level is expected to be significantly weaker. At the lower incident energy of 350 eV (lowest spectrum of Fig. 14) it would be expected that the direct and exchange contributions have a similar strength, causing all the transitions to the $(2p_{3/2})^{-1}4s$ and $(2p_{3/2})^{-1}4p$ levels to have similar strengths.

5.3. *Electric-Dipole-Forbidden Inner-Shell Transitions in N_2*

Electron-energy-loss spectra obtained in N_2 (Shaw *et al.*, 1982) for a range of incident electron energies (indicated alongside each spectrum) are shown in Fig. 16. The high-energy feature corresponds to excitation of the $(1s)^{-1}(2p\pi)\ ^1\Pi$ state of N_2 (Sec. 3), and at incident energies above about 1000 eV it is the only structure observed in this region. As the incident energy is reduced, however, a low-energy structure becomes evident and at the lowest values of incident energy used, becomes dominant. This structure corresponds to the promotion of a $1s$ nitrogen electron to the $2p\pi$ orbital to give the $(1s)^{-1}(2p\pi)\ ^3\Pi$ state of N_2. The spectra represent the first observation of an inner-shell singlet-to-triplet transition.

The spectra were analyzed to obtain the molecular constants of the singlet and triplet states which are given in Table XII. The statistical accuracy of the data was insufficient to be able to detect any difference between the natural widths of the singlet and triplet states or any dependence on incident energy of the apparent natural width of either of these states. In fact Krause (1979) has shown that the radiative decay width for the singlet K-shell state of N_2 is two orders of magnitude smaller than the Auger decay width. Therefore, although the radiative transition from the singlet state is allowed while that of the triplet state is forbidden, the two states are expected to have essentially the same total widths.

The values obtained for the parameters ω and Γ of the $^1\Pi$ state (Table XII) are in good agreement with the earlier work of King *et al.* (1977c). As may be seen, the value of R_e is slightly smaller for the $^3\Pi$ state than for the $^1\Pi$ state. The difference in energy between the $v = 0$ levels of the two states is measured to be 0.82 ± 0.01 eV. This is close to the value 0.9 eV calculated by Rescigno (1984) and agrees with the value 0.8 eV

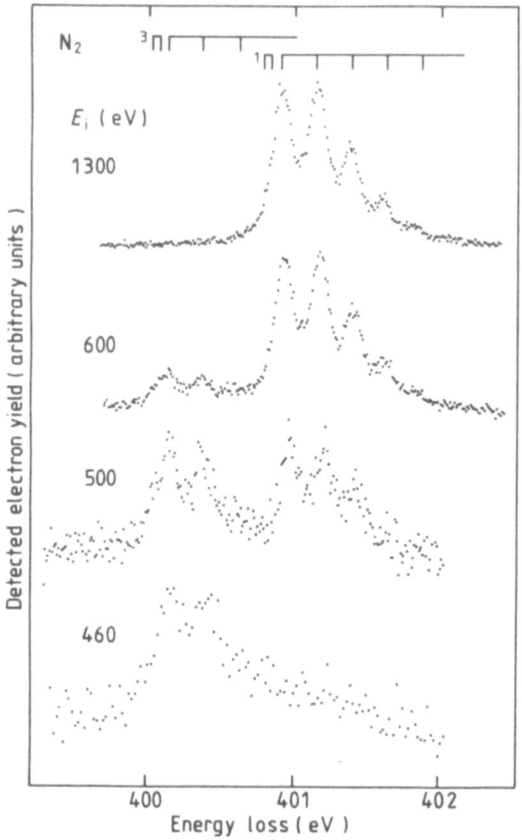

FIGURE 16. Energy-loss spectra of N_2 obtained at the indicated values of incident electron energy, and with an energy resolution of 65 meV. The lower- and higher-energy vibrational structures correspond to excitation of the $(1s)^{-1}(2p\pi)\,^3\Pi$ and $(1s)^{-1}(2p\pi)\,^1\Pi$ states of N_2, respectively. (From Shaw *et al.*, 1982.)

TABLE XII

Derived properties of two inner-shell excited states of N_2. The numbers in brackets indicate the standard errors in the last significant figures.[a]

State	ω (meV)	ωx (meV)	R_e (nm)	Γ (meV)
$(1s)^{-1}(2p\pi)\,^3\Pi$	242(4)	1.8	0.1155(6)	123(10)
$(1s)^{-1}(2p\pi)\,^1\Pi$	236(2)	1.8	0.1162(3)	123(10)

[a] After Shaw *et al.* (1982).

obtained from the low-resolution Auger electron measurements of Ungier and Thomas (1983). Recently Barth and Schirmer (1985) have calculated the singlet–triplet splitting to the 0.41 eV.

The ratio of triplet-to-singlet state yields was also determined. The yield of each state is proportional to the sum of the areas of all the vibrational peaks corresponding to that state. The triplet-to-singlet ratio was found to be 0.030(1), 0.180(3), and 0.96(9) at incident energies of 800, 600, and 500 eV, respectively. It was not possible to determine the relative normalization of the spectra taken at different incident energies, since, in order to optimize the yields of inelastically scattered electrons, significant changes in the apparatus tuning were required, particularly at the lower values of the incident electron energy.

There has been much discussion in the literature (Sec. 4.1) about whether the $1s$ hole is localized at the site of one of the nitrogen nuclei. If the hole is delocalized then u–g symmetry is appropriate and a u–g splitting in energy would be expected. Rescigno and Orel (1979), for example, have calculated this splitting to be 0.06 eV. The $1s\sigma_u \rightarrow 2p\pi_g$ transition would be electric-dipole-allowed while the $1s\sigma_g \rightarrow 2p\pi_g$ transition would be electric-dipole-forbidden. In the earlier work on the $(1s)^{-1}(2p\pi)\ {}^1\Pi$ state of N_2 King *et al.* (1977c) did not see any evidence of the occurrence of two close-lying states, although the predicted splitting is smaller than the natural widths of the states (123 meV) which would have made such a splitting difficult to discern. In the work of Shaw *et al.* (1982), however, an enhancement of electric-dipole-forbidden transitions (as predicted by Rescigno and Orel) might be expected since it was possible to achieve a large value of momentum transfer, increasing from 1.65 a.u. at an incident energy of 1300 eV to 3.73 a.u. at 460 eV. This enhancement might lead, for example, to a change in the relative areas of the vibrational levels, or to changes in the peak energies or widths. However, within the limited statistical accuracy of the measurements it was not possible to detect any changes of this type.

Other experimental studies of this region of N_2 include the low-resolution election impact work of Wight *et al.* (1972/3), Fantoni and Stefani (1978), and Ungier and Thomas (1983), and the photoabsorption measurements of Nakamura *et al.* (1969). Wight *et al.* (1972/3) used an incident energy of 2.5 keV, which is too high for the excitation of the triplet state. Similarly, in the photoabsorption work the optically forbidden singlet–triplet transition was not seen. Fantoni and Stefani (1978) have made electron energy-loss measurements in this region of N_2, at high incident energy (1600 eV) and large scattering angles, but with an unselected electron beam, giving a much lower resolution in the measurements. By using large angles they obtained large values of the momentum transfer and hence were able to measure the generalized oscillator strength of the $(1s)^{-1}(2p\pi)\ {}^1\Pi$ state of N_2. Under these conditions parity-forbidden transitions may have been observed but again their incident energy was too high for the triplet state to be excited. Ungier and Thomas (1983) have obtained evidence for the excitation of the triplet state in Auger election spectra.

Singlet–triplet splittings of inner-shell states are observed in ESCA measurements. For example when a nitrogen 1*s* electron is photoejected from the molecule the spins of the remaining nitrogen 1*s* electron and the unpaired valence 2*pπ* electron combine to form either a singlet or triplet state of the NO^+ ion. Due to the exchange interaction between the core electron and the unpaired 2*pπ* electron these two states have energies that differ by 1.5 eV (Siegbahn *et al.*, 1969). Wight and Brion (1974a) have calculated the energy difference between the $^4\Sigma^-$ and $^2\Sigma^-$ states of NO, arising from the excitation of a nitrogen 1*s* electron to the 2*pπ* orbital, to be 1.4 eV. These splittings are similar in magnitude to the singlet–triplet splitting observed in the energy-loss memasurements of Shaw *et al.* (1982).

5.4. *Electric-Dipole-Forbidden Transitions in Other Molecules*

Shaw *et al.* (1984, 1985) have observed electric-dipole-forbidden transitions in the molecules CO, CO_2, and N_2O. Their results for CO are shown in Fig. 17, which shows energy-loss spectra obtained at two different values

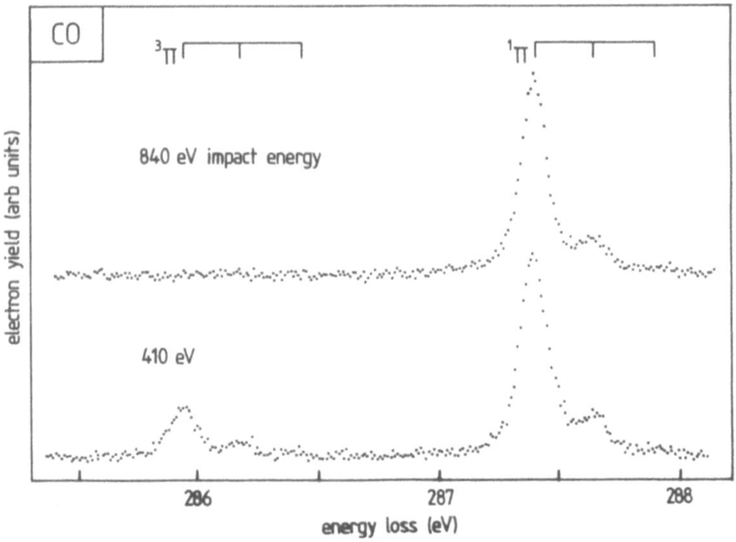

FIGURE 17. Energy-loss spectra of CO, obtained at the indicated values of incident electron energy, and with an energy resolution of 65 meV. The lower- and higher-energy vibrational structures correspond to excitation of the $(1s\sigma_C)^{-1}(2p\pi)\ ^3\Pi$ and $(1s\sigma_C)^{-1}(2p\pi)\ ^1\Pi$ states of CO, respectively.

of incident electron energy. Again, at the lower value the triplet state $(1s\sigma_C)^{-1}(2p\pi)\ ^3\Pi$ is observed.

Of note is the singlet–triplet splitting, which has a value of 1.46 ± 0.01 eV (Shaw *et al.*, 1984). Firstly this is quite different from the singlet–triplet splitting of the analogous inner-shell states of N_2, which is isoelectronic with CO and which has the same equivalent core molecule. Presumably this occurs because the 1*s* hole has a smaller mean radius in the nitrogen atom than it has in the carbon atom. Consequently the overlap between the hole and the excited $2p\pi$ orbital, the latter being the same in both cases, will differ, resulting in a smaller exchange interaction in N_2. Secondly the splitting also gives a direct measure of K, the exchange integral for the inner-shell 1*s* hole orbital and the $2p\pi$ orbital since K is equal to half the singlet–triplet splitting. The measured value of 1.46 eV compares with the theoretical values of 2.20 eV (Kondratenko, 1977), 1.15 eV (Schwartz, 1976) and 1.29 eV (Barth and Schirmer, 1985).

The energies of the singlet and triplet $(1s)^{-1}\ (2p\pi)$ states of N_2 and CO can be used to calculate the weighted mean energy of the $(1s)^{-1}2p\pi$ configuration of each molecule, and hence to calculated the mean binding energy of this configuration with respect to the ion core configuration $(1s)^{-1}$. This mean binding energy is found to be 9.70 and 9.76 eV, for N_2 and CO respectively. This appears to indicate that the ion cores N_2^{K+} and $C^{K+}O$ provide approximately the same mean field for the $2p\pi$ electron as they would in the equivalent-core model, since they are then both equivalent to NO^+.

6. INNER-SHELL RESONANCES

An unusual development in inner-shell studies has been the observation of negative ion resonances associated with inner-shell vacancies (King *et al.*, 1977d; Ziesel *et al.*, 1979; Teillet-Billy and Ziesel, 1979; King *et al.*, 1980). These resonances appear in electron-impact cross sections, and indicate the existence of short-lived excited states of negative ions, formed as intermediate complexes. The negative ion resonance quickly dissociates into any of the energetically accessible decay channels. For example,

$$\begin{aligned} e + N_2 \rightarrow (N_2^{K*-})_{res} &\rightarrow N_2^* + e \\ &\quad \text{or } N_2^+ + e + e \\ &\quad \text{or } N + N + e \\ &\quad \text{or } N + N^+ + e + e, \text{ etc.} \end{aligned} \qquad (11)$$

Previously all negative ion resonances that had been studied (see, for example Schulz, 1973) had been associated with either (a) ground electronic states, (b) valence excited states, (c) autoionizing doubly excited valence states, or (d) autoionizing states in which an electron has been excited from the shell or subshell immediately below the valence subshell. This new class of inner-shell negative ion resonances is associated with states in which an electron has been excited from deeper-lying inner shells.

The inner-shell resonance in N_2 (King *et al.*, 1977d; Teillet-Billy and Ziesel, 1979) will serve as an example. As seen in Sec. 4.1, the transition in N_2 to the lowest-lying K-shell excited state $(1s)^{-1}(2p\pi)\ ^1\Pi$ has an oscillator strength much greater than that of the higher states due to the existence of a centrifugal barrier in the d-wave component of the $2p\pi$ orbital, making it highly localized. This consideration led King *et al.* (1977d) to conjecture that two electrons could be contained in the $2p\pi$ orbital, one being the $1s$ electron that is promoted from the K shell while the other is the incident electron that causes the excitation. An inner-shell resonance having the configuration $N_2^-(1s)^{-1}(2p\pi)^2$ would then be formed. This resonance would quickly decay into various channels, and in principle it could be observed in any one of them. King *et al.* initially observed it in the yield of positive ions, and this work is described below.

6.1. Experimental Details

The apparatus used in the studies of King *et al.* (1977d) and King *et al.* (1980) is a modified form of that shown schematically in Fig. 5. An electron beam having a diameter of 1 mm, an adjustable energy, and an energy spread of typically 0.05 eV is crossed at right angles with a target gas beam that emanates from a hypodermic needle. The target region is surrounded by a cylindrical copper mesh (90% transparency) that shields the target

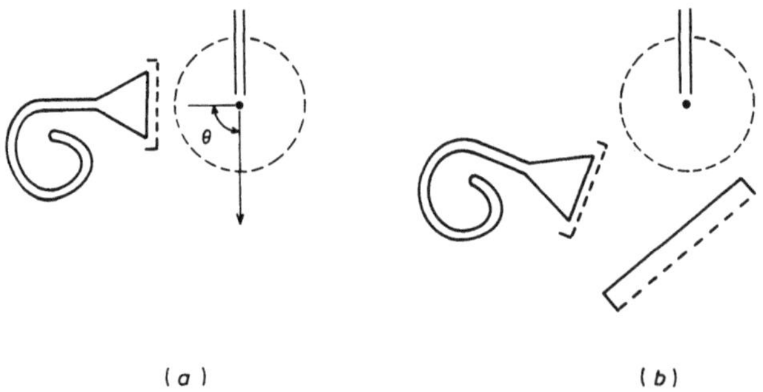

FIGURE 18. Target and detector geometries used in the inner-shell resonance studies of King *et al.* (1977d). The electron beam direction is normal to the plane of the figure and the black dot represents the point of intersection of the electron and molecular beams. (a) Geometry used for detecting positive ions, (b) geometry allowing crude energy analysis of the scattered and ejected electrons.

from field penetration due to extraneous potentials. A channel electron multiplier having a conical input of diameter 10 mm is used to detect the reaction products of interest. The target and detector geometry that was used for detecting positive ions is shown in Fig. 18a. A negative potential is applied to the mesh in front of the detector to prevent detection of scattered electrons, and the potential of the input end of the detector is varied to maximize the detection efficiency for the positive ions. Various values of θ, the angle between the detector and molecular beam, were tried in order to maximize the resonance signal-to-background ratio and a value of 90° was found to be the most suitable. This implies that the resonance is observed in a channel corresponding to the production of energetic atomic ions by molecular dissociation. No attempt was made with this configuration to analyze the energy, charge, or mass-to-charge ratio of the detected ions. A detection geometry was also tried that did allow crude energy selection of the reaction products, and this is shown in Fig. 18b. It employs an energy-dependent electrostatic mirror that is similar in construction to the one used by Cvejanović and Read (1974) for threshold spectroscopy studies. The mirror can be biased so that electrons or positive ions can be directed towards the detector.

Calibration of the observed features was achieved by using a mixture of the gas under investigation and either Ar or N_2, both of which have sharp structures at known energies in their cross section for metastable excitation (Brunt *et al.*, 1976, 1978). Such structure exists in these measurements because the detector intercepts a small fraction of the metastable atoms or molecules produced in the target region.

In the work of Ziesel *et al.* (1979) the authors also used electron impact excitation and extracted ions from the target ragion at an angle of 90° to the electron beam. These authors were able to mass analyze the ions produced by a 60° magnetic field analyzer and thus isolate the possible ionization routes of the resonances.

6.2. *Inner-Shell Resonance in* N_2

The inner-shell resonance structure observed by King *et al.* (1977d), including that in N_2 is shown in Fig. 19. The spectra represent measured yields of positive ions as a function of incident electron energy. The energy calibrations of these spectra have been slightly changed by the more recent work of King *et al.* (1980) who give the onset of the resonance at 400.4 ± 0.2 eV, the maximum at 401.7 eV and the width (FWHM) of the peak as 1.2 ± 0.1 eV. King *et al.* (1980) also measured the yield of low-energy ($\leqslant 40$ eV) scattered and ejected electrons using the target and detector geometry shown in Fig. 18b. Their spectrum is shown in Fig. 20. The major contribution to this yield comes from the direct ionization process. At the resonance energy a dip in the yield was observed presumably showing interference between the resonant and direct ionization processes. The inner-shell resonance in N_2 has also been observed by Teillet-Billy and Ziesel (1979). These authors measured the yield of N^+ and N_2^{2+} ions following electrons impact on N_2. The energy positions of the resonance obtained by Teillet-Billy and Ziesel (1979) and King *et al.* (1980) are essentially in agreement. All the measurements indicate that the peak intensity of the resonance is typically 0.1% of the total ion yield at the resonance energy.

The equivalent-core model suggests that the inner-shell resonance structure should resemble the shape resonance structure of the equivalent-core molecule. For example, the inner-shell resonance structure in N_2 could be expected to resemble the low energy shape resonance associated with the ground state of NO. The lowest-energy electron configuration of NO^- is π^2 and so three electronic states are expected: $^3\Sigma^-$, $^1\Delta$, and $^1\Sigma^+$. Tronc *et al.* (1975) have measured vibrational excitation of the first three vibrational levels of the ground state of NO and have observed a regularly spaced series of peaks which they attributed to the NO^- $^3\Sigma^-$ ($v = 1$, 2, 3, and 4) resonance. At higher energies they also observed the NO^- $^1\Delta$ resonance. The inner-shell resonance of N_2 should therefore have a structure consisting

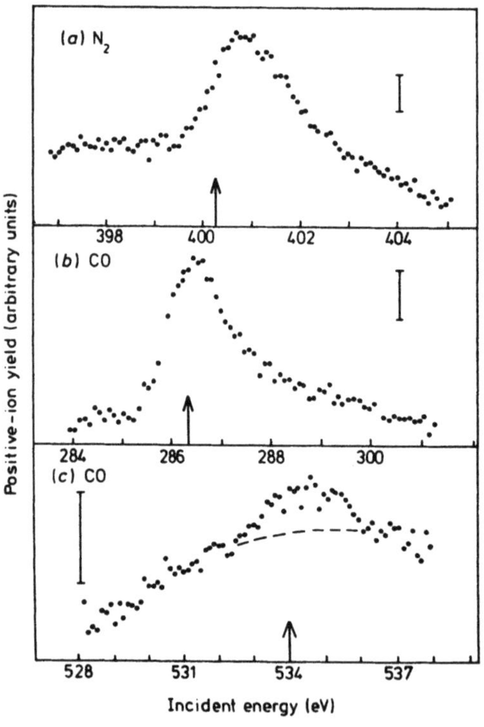

FIGURE 19. The yield of positive ions from electron impact of N_2 (a) and CO (b) and (c) as a function of incident energy (King *et al.*, 1977d). The vertical bars represent 0.1% of the total number of positive ions detected. The vertical arrows show the positions of the center of gravity of the lowest vibrational levels of the inner-shell singlet and triplet excited states $(1s)^{-1}(2p\pi)$ of the neutral $\mathcal{N}_2^{K*}$ and $C^{K*}O$ molecules, and the position of the peak maximum of the corresponding state of CO^{K*}.

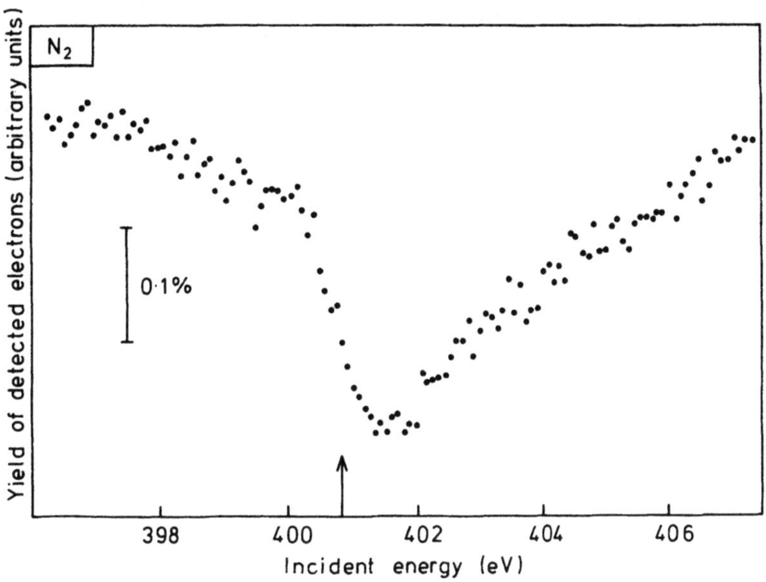

FIGURE 20. The yield of ejected and scattered electrons from electron impact excitation of N_2 (King *et al.*, 1980). The vertical bar indicates the percentage of the total number of particles detected. The vertical arrow is at the energy of the center-of-gravity of the lowest vibrational levels of the inner-shell singlet and triplet states $(1s)^{-1}(2p\pi)\ ^{1,3}\Pi$ of the N_2^{K*} molecule.

of three overlapping vibrational series with the first members spanning approximately 1.1 eV, in reasonable agreement with the observed structure. This overlapping may also explain the failure to see vibrational structure in the resonance.

The inner-shell states of N_2 that may be expected to be the parent states of the inner-shell resonance are the $(1s)^{-1}(2p\pi)\ ^{1,3}\Pi$ states (Sec. 5). The energy-loss spectra taken at different values of incident electron energy show that the intensity of the triplet states relative to the singlet state increases dramatically as the incident energy approaches threshold, indicating that the near-threshold excitation cross section for the triplet state is much larger than that of the singlet state. If the resonance can be considered to consist of an extra electron attached to the parent state, then it might reasonably be expected that the resonance is associated with the triplet state. On the other hand, the resonance can be interpreted in terms of the grandparent model (see e.g., Brunt *et al.*, 1978), so that the two outer electrons couple together to form a correlated pair that couples to the N_2^{K+} core. The energy with which the $(2p\pi)^2$ pair are bound to the ion core is then 9.55 eV (using the resonance onset energy), which is close to the mean binding energy of a single $(2p\pi)$ election to the core, namely, 9.70 eV (Sec. 5.3).

6.3. Inner-Shell Resonances in Other Molecules

Resonances associated with inner-shell vacancies have been observed in a number of other molecules including CO (King *et al.*, 1977d, 1980; Ziesel *et al.*, 1979), NO, N_2O, and CO_2 (King *et al.*, 1980). All of the molecules in which inner-shell resonance structure has been observed are characterized by an inner-shell state that has a large oscillator strength.

The spectra obtained by Ziesel *et al.* (1979) in their study of CO are shown in Fig. 21. These are the ionization-efficiency curves for the formation of C^+, O^+, CO^{2+}, C^{2+}, and O^{2+} ions by electron impact. A resonance structure is observed in all five decay channels. Its energy position and shape are similar to that of the structure observed by King *et al.* (1977d). The relative intensity of the structure is small in every channel and always less than 1%. By looking at various decay routes, Ziesel *et al.* were able to deduce branching ratios for the resonance. The relative abundances for the

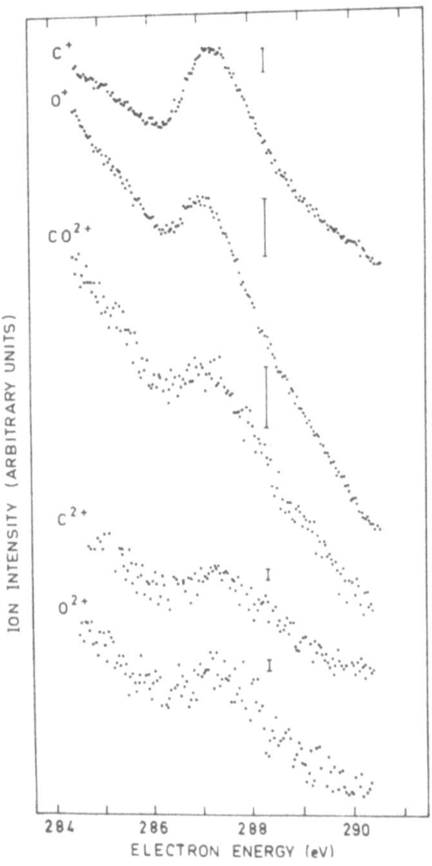

FIGURE 21. Ionization-efficiency curves for the formation of C^+, O^+, CO^{2+}, C^{2+}, and O^{2+} ions. The vertical bar for each curve represents 0.1% of the total signal for the ion. The energy step is 38 meV. (From Ziesel *et al.*, 1979.)

resonance decay indicate that the decay occurs mainly to dissociative states of CO^+ by double Auger transitions in which one electron is ejected.

Resonance structure associated with an inner-shell state of oxygen in the CO molecule has also been observed (King *et al.*, 1977d). This is shown in Fig. 19c. The structure occurs at an energy close to the inner-shell state $(01s)^{-1}(2p\pi)\ {}^1\Pi$ of CO (Wight *et al.*, 1972/3) and therefore it is considered to have the configuration $(01s)^{-1}(2p\pi)^2$. In this case the resonance would be expected to be similar to the shape resonance of CF, which illustrates the point that these studies should allow the investigation of resonances that are unobtainable otherwise due to the unstable nature of the parent species.

The CO and N_2 inner-shell resonances both have NO^+ as their equivalent-core grandparent state and so it is interesting to compare the inner-shell resonance structure in the two molecules. In fact the binding energies to the ion core of the two resonances are approximately 9.97 and 9.55 eV, respectively (using the resonance onset energies). Part of this small difference may be due to the effect of Frank–Condon overlap factors. The internuclear separations of the ground states of CO and N_2 are 0.1128 nm and 0.1098 nm, respectively, while that of the resonances is presumably similar to that of NO^+, namely, 0.1063 nm. The shorter internuclear separation of N_2 could account for the higher, relative onset of the N_2 inner-shell resonance. On the other hand interpreting the resonances in terms of

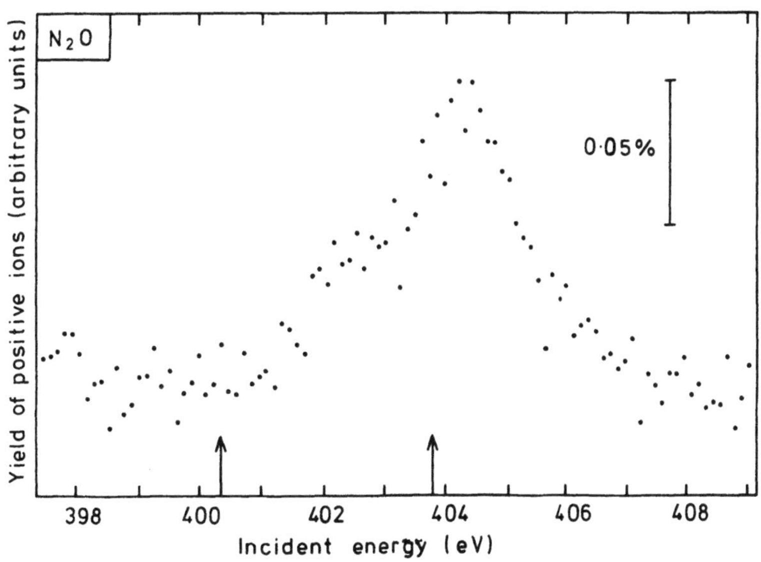

FIGURE 22. The yield of positive ions from electron impact excitation of N_2O (King *et al.*, 1980). The vertical arrows are at the energies of the lowest vibrational levels of the $(1s_{\mathrm{N\ terminal}})^{-1}(3\pi)$ and the $(1s_{\mathrm{N\ central}})^{-1}(3\pi)$ inner-shell excited states of the $N_2^{\ast}O$ molecule.

the parent-state model, Teillet-Billy and Ziesel (1979) point out that the binding energy of the resonance to its parent state is affected by the magnitude of the exchange integral involving the singly occupied $1s$ atomic orbital and the valence $2p$ orbital. Using the independent-electron model and the frozen-orbitals approximation they find that the binding energy of $[N_2^{K*}]^-$ is approximately 0.35 eV lower than that of $[C^{K*}O]^-$, which is approximately in agreement with experimental results (King *et al.*, 1980).

An interesting example of inner-shell resonance structure is provided by N_2O. This molecule is linear in its ground state and has the form N–N–O making the two nitrogen nuclei nonequivalent. The inner-shell energy-loss spectrum of N_2O has been reported by King *et al.* (1980) and indeed shows two prominent peaks, separated by 3.64 eV, associated with K-shell excitation of the two nonequivalent nuclei. The observed yield of positive ions as a function of incident electron energy, in N_2O, also shows evidence of two separate resonance features. This spectrum (Tronc *et al.*, 1980) is shown in Fig. 22. The expected configuration of the resonance is $(1sN)^{-1}(3\pi)^2$. Although there is no information available on the geometry of the equivalent-core molecule ($O\text{–}N\text{–}O^-$, or the peroxy isomer $N\text{–}O\text{–}O^-$), it may be expected that the double occupancy of the 3π orbital causes the molecule to be bent. This would cause the Franck–Condon factors for the transitions to the lower resonance levels to be small, which would delay the apparent onset of the resonance and lead to the observed low values of the resonance binding energy.

REFERENCES

ANDRIĆ, L., AND I. ČADEŽ, 1978, *SPIG-78 Dubrovnik Contributed Papers*, edited by R. K. Janev (Institute of Physics, Belgrade), p. 39.

BAGUS, P. S., AND H. F. SCHAEFER, 1972, *J. Chem. Phys.* **56**, 224.

BAGUS, P. S., M. KRAUSS, AND R. E. LAVILLA, 1973, *Chem. Phys. Lett.* **23**, 13.

BARTH, A., AND J. SCHIRMER, 1985, *J. Phys. B* **18**, 867.

BEARDEN, J. A., 1967, *Rev. Mod. Phys.* **39**, 78.

BEARDEN, J. A., AND A. F. BURR, 1967, *Rev. Mod. Phys.* **39**, 125.

BETHE, H., 1930, *Ann. Phys. (Leipzig)* **5**, (5), 325.

BRION, C. E., 1982, in *Physics of Electronic and Atomic Collisions*, edited by S. Datz (North-Holland, Amsterdam), p. 579.

BRION, C. E., S. DAVIEL, R. SODHI, AND A. P. HITCHCOCK, 1982, in *X-Ray and Atomic Inner-Shell Physics, 1982*, AIP Conference Proceedings Series No. 94, edited by B. Crasemann (American Institute of Physics, New York), p. 429.

BROWN, F. C., R. Z. BACHRACH, AND A. BIANCONI, 1978, *Chem. Phys. Lett.* **54**, 425.

BRUNDLE, C. R., M. B. ROBIN, AND H. BASCH, 1970, *J. Chem. Phys.* **53**, 2196.

BRUNT, J. N. H., G. C. KING, AND F. H. READ, 1976, *J. Phys. B* **9**, 2195.

BRUNT, J. N. H., G. C. KING, AND F. H. READ, 1978, *J. Phys. B* **11**, 173.

CASTEX, M. C., J. LE CALVE, D. HAAKS, B. JORDAN, AND G. ZIMMERER, 1980, *Chem. Phys. Lett.* **70**, 106.

CEDERBAUM, L. S., AND W. DOMCKE, 1977, *J. Chem. Phys.* **66**, 5084.

CHEN, M. H., B. CRASEMANN, L. I YIN, T. TSANG, AND I. ADLER, 1976, *Phys. Rev. A* **13**, 1435.

CHUN, H. U., 1969, *Phys. Lett.* **A30**, 445.

CLARK, D. T., AND J. MULLER, 1976, *Theor. Chim. Acta* **41**, 193.

CODLING, K., AND R. P. MADDEN, 1964, *Phys. Rev. Lett.* **12**, 106.

CONDON, E. U., AND G. H. SHORTLEY, 1951, *The Theory of Atomic Spectra* (Cambridge University Press, London).

COWAN, R. D., 1981, *The Theory of Atomic Structure and Spectra* (Univ. of California, Berkeley).

CVEJANOVIĆ, S., AND F. H. READ, 1974, *J. Phys. B* **10**, 1180.

DEHMER, J. L., AND D. DILL, 1976, *J. Chem. Phys.* **70**, 3390.

DEUTSCH, P. W., AND A. B. KUNZ, 1973, *J. Chem. Phys.* **59**, 1155.

DEUTSCH, P. W., AND A. B. KUNZ, 1975, *J. Chem. Phys.* **62**, 4069.

DUTTA, C. M., AND H. W. HUANG, 1980, *Phys. Rev. Lett.* **44**, 643.

EBERHARDT, W., R. P. HAELBICH, M. IWAN, E. E. KOCH, AND C. KUNZ, 1976, *Chem. Phys. Lett.* **40**, 180.

EDERER, D. L., T. B. LUCATORTO, E. B. SALOMAN, R. P. MADDEN, M. MANALIS, AND J. SUGAR, 1976, in *Electron and Photon Interactions with Atoms*, edited by H. Kleinpoppen and M. R. C. McDowell (Plenum Press, New York), p. 69.

EDLÉN, B., 1964, *Handbuch der Physik.* (Springer, Berlin), Vol. 27, p. 80.

EWING, J. J., AND C. A. BRAU, 1975, *Phys. Rev. A* **12**, 129.

FANO, U., AND J. W. COOPER, 1968, *Rev. Mod. Phys.* **40**, 441.

FANTONI, R., AND G. STEFANI, 1978, 4th General Conf. of the European Physical Society, Trends in Physics, York, England.

FROESE-FISCHER, C., 1978, *Comput. Phys. Commun.* **14**, 145.

GELIUS, U., 1974, *J. Electron. Spectrosc.* **5**, 985.

GOLDE, M. F., 1975, *J. Mol. Spectrosc.* **58**, 261.

GOLDE, M. F., AND B. A. THRUSH, 1974, *Chem. Phys. Lett.* **29**, 486.

GUNDEL, L. A., D. W. SETSER, M. A. A. CLYNE, J. A. COXON, AND W. NIP, 1976, *J. Chem. Phys.* **64**, 4390.

HAHN, Y., 1976, *Phys. Rev. A* **13**, 1326.

HALL, R. I., AND F. H. READ, 1984, in *Electron–Molecule Collisions*, edited by Shimamura and Takayanagi (Plenum Press, New York).

HARSHBARGER, W. R., AND E. N. LASSETTRE, 1973, *J. Chem. Phys.* **58**, 1505.

HARTING, E., AND F. H. READ, 1976, *Electrostatic Lenses* (Elsevier, Amsterdam).

HICKS, P. J., S. DAVIEL, B. WALLBANK, AND J. COMER, 1980, *J. Phys. E* 713.

HITCHCOCK, A. P., 1982, *J. Electron. Spectrosc.* **25**, 245.

HITCHCOCK, A. P., AND C. E. BRION, 1980, *J. Electron. Spectrosc.* **18**, 1.

HITCHCOCK, A. P., AND C. E. BRION, 1981, *J. Phys. B* **14**, 4399.

HITCHCOCK, A. P., M. POCOCK, AND C. E. BRION, 1977, *Chem. Phys. Lett.* **49**, 125.

HUO, W. M., 1979, *J. Chem. Phys.* **71**, 1593.

IMHOF, R. E., A. ADAMS, AND G. C. KING, 1976, *J. Phys. E* **9**, 138.

INOKUTI, M., 1971, *Rev. Mod. Phys.* **43**, 297.

JOHANSSON, G., J. HEDMAN, A. BERNDTSSON, M. KLASSON, AND R. NILSSON, 1973, *J. Electron. Spectrosc.* **2**, 295.

JOHNSON, J. B., AND R. A. BONHAM, 1979, private communication.

KAY, R. B., P. E. VAN DER LEEUW, AND M. J. VAN DER WIEL, 1977, *J. Phys. B* **10**, 2513; 2521.

KESKI-RAHKONEN, O., AND M. O. KRAUSE, 1974, *At. Data Nucl. Data Tables* **14**, 139.

KING, D. L., L. G. PIPER, AND D. W. SETSER, 1977a, *J. Chem. Soc. Faraday Trans. II*, **73**, 177.

KING, G. C., M. TRONC, F. H. READ, AND R. C. BRADFORD, 1977b, *J. Phys. B* **10**, 2479.

KING, G. C., F. H. READ, AND M. TRONC, 1977c, *Chem. Phys. Lett.* **52**, 50.

KING, G. C., J. W. MCCONKEY, AND F. H. READ, 1977d, *J. Phys. B* **10**, L541.

KING, G. C., J. W. MCCONKEY, F. H. READ, AND B. DOBSON, 1980, *J. Phys. B* **13**, 4315.
KONDRATENKO, A. V., L. N. MAZALOV, F. KH. GEL'MUKHANOV, V. I. AVDEEV, AND E. A. SAPRYKHINA, 1977, *J. Struct. Chem.* **18**, 437 (*Zh. Strukt. Khim.* **18**, 546).
KRAUSE, M. O., 1979, *J. Phys. Chem. Ref. Data* **8**, 307.
KUYATT, C. E., AND J. A. SIMPSON, 1967, *Rev. Sci. Instrum.* **38**, 103.
LOZES, R. L., O. GOSCINSKI, AND U. I. WAHLGREN, 1979, *Chem. Phys. Lett.* **63**, 77.
LUKIRSKII, A. P., T. M. ZIMKINA, AND I. A. BRYTOV, 1965, *Bull. Acad. Sci. USSR Phys.* **28**, 681.
MADDEN, R. B., AND K. CODLING, 1965, *Astrophys. J.* **141**, 364.
NAKAMURA, M., *et al.*, 1968, *Phys. Rev. Lett.* **21**, 1303.
NAKAMURA, M., S. SASANUMA, S. SATO, M. WATANABE, H. YAMASHITA, Y. IGUCHI, A. EJIRI, S. NAKAI, S. YAMAGUCHI, T. SAGAWA, Y. NAKAI, AND T. OSHIO, 1969, *Phys. Rev.* **178**, 80.
M. NAKAMURA, Y. MORIOKA, E. HAYAISHI, E. ISHIGURA, AND M. SASANUMA, 1971, in *Proc. 3rd Int. Vac. UV Rad. Phys. Conf., Tokyo.*
NORDBERG, R., K. HAMRIN, A. FAHLMAN, C. NORDLING, AND K. SIEGBAHN, 1966, *Z. Phys.* **192**, 462.
OLSEN, R. E., 1980, private communication.
OLSEN, R. E., AND B. LUI, 1978, *Phys. Rev. A* **17**, 1568.
PETTERSSON, L., J. NORDGREN, L. SALANDER, C. NORDLING, AND K. SIEGBAHN, 1982, *J. Electron. Spectrosc.* **27**, 29.
PRINS, M. J.-A., 1934, *Physica* **1**, 1174.
PURCELL, E. M., 1938, *Phys. Rev.* **54**, 818.
RAU, A. R. P., AND U. FANO, 1968, *Phys. Rev.* **167**, 7.
READ, F. H., 1975, *Radiat. Phys.* **64**, 23.
READ, F. H., 1978, *J. Phys. (Paris)* **39**, Suppl. Cl, 82.
READ, F. H., 1980, *Electromagnetic Radiation* (Wiley, New York).
READ, F. H., J. COMER, R. E. IMHOF, J. N. H. BRUNT, AND E. HARTING, 1974, *J. Electron. Spectrosc.* **4**, 293.
RESCIGNO, T. N., AND P. W. LANGHOFF, 1977, *Chem. Phys. Lett.* **51**, 65.
RESCIGNO, T. N., AND OREL, A. E., 1979, *J. Chem. Phys.* **70**, 3390.
RESCIGNO, T. N., AND A. E. OREL, 1984, in *Lecture Notes in Chemistry*, edited by G. Berthier *et al.*, vol. 35 (edited by F. A. Gianturco and G. Stefani), p. 215 (Springer-Verlag, Berlin).
SCHULZ, G. J., 1973, *Rev. Mod. Phys.* **45**, 378.
SCHWARZ, W. H. E., 1974, *Agnew Chem. Int.* **13**, 454.
SCHWARZ, W. H. E., 1975, *Chem. Phys.* **11**, 217.
SCHWARZ, W. H. E., 1976, *Chem. Phys.* **13**, 153.
SHANNON, S. P., K. CODLING, AND J. B. WEST, 1977, *J. Phys. B* **10**, 825.
SHAW, D. A., G. C. KING, AND F. H. READ, 1980, *J. Phys. B* **13**, L723.
SHAW, D. A., G. C. KING, F. H. READ, AND D. CVEJANOVIĆ, 1982, *J. Phys. B* **15**, 1785.
SHAW, D. A., D. CVEJANOVIĆ, G. C. KING, AND F. H. READ, 1984a, *J. Phys. B* **17**, 1173.
SHAW, D. A., G. C. KING, D. CVEJANOVIĆ, AND F. H. READ, 1984b, *J. Phys. B* **17**, 2091.
SHAW, D. A., G. C. KING, AND. F. H. READ, 1985, private communication.
SHIRLEY, D. A., R. L. MARTIN, S. P. KOWALCZYK, F. R. MCFEELY, AND L. LEY, 1977, *Phys. Rev. B* **15**, 544.
SIEGBAHN, K., *et al.*, 1969, *ESCA Applied to Free Molecules* (North-Holland, Amsterdam).
SIMPSON, J. A., 1964, *Rev. Sci. Instrum.* **35**, 1698.
SNYDER, L. C., 1971, *J. Chem. Phys.* **55**, 95.
TEILLET-BILLY, D., AND J. P. ZIESEL, 1979, in *Electronic and Atomic Collisions, Abstracts of Contributed Papers*, edited by K. Takayanagi and N. Oda (Society for Atomic Collision Research, Japan), p. 404.
TRONC, M., A. HUETZ, M. LANDAU, F. PICHOU, AND J. REINHARDT, 1975, *J. Phys. B* **8**, 1160.
TRONC, M., G. C. KING, R. C. BRADFORD, AND F. H. READ, 1976, *J. Phys. B* **9**, L555.

TRONC, M., G. C. KING, AND F. H. READ, 1979, *J. Phys. B* **12**, 137.
TRONC, M., G. C., KING, AND F. H. READ, 1980, *J. Phys. B* **13**, 999.
UNGIER, L., AND T. D. THOMAS, 1983, *Chem. Phys. Let.* **96**, 247.
VAN DER WIEL, M. J., AND T. M. EL-SHERBINI, 1972, *Physica* **59**, 453.
VAN DER WIEL, M. J., T. M. EL-SHERBINI, AND C. E. BRION, 1970, *Chem. Phys. Lett.* **7**, 161.
WAYNANT, R. W., 1977, *Appl. Phys. Lett.* **30**, 234.
WIGHT, G. R., AND C. E. BRION, 1974a, *J. Electron. Spectrosc.* **4**, 313.
WIGHT, G. R., AND C. E. BRION, 1974b, *J. Electron. Spectrosc.* **4**, 25.
WIGHT, G. R., C. E. BRION, AND M. J. VAN DER WIEL, 1972/3, *J. Electron. Spectrosc.* **1**, 457.
WUILLEUMIER, F., 1981, in *Inner-shell and X-ray Physics of Atoms and Solids*, edited by D. J. Fabian, H. Kleinpoppen, and L. M. Watson (Plenum Press, New York), p. 395.
WUILLUMIER, F., AND M. O. KRAUSE, 1972, in Proceedings of the International Conference on Inner-Shell Ionization Phenomena and Future Applications, ed. by R. W. Fink, S. T. Manson, J. M. Palms, and P. Venugopala Rao, U.S. Atomic Energy Commission report No. CONF-720404, p. 773.
YIN, L. I, I. ALDER, T. TSUNG, M. H. CHEN, D. A. RINGERS, AND B. CRASEMANN, 1974, *Phys. Rev. A* **9**, 1070.
ZIESEL, J. P., D. TEILLET-BILLY, L. BOUBY, AND R. PAINEAU, 1979, *Chem. Phys. Lett.* **63**, 47.

CHAPTER 9

Coherence Effects in Electron Emission by Atoms

A. Niehaus

1. GENERAL INTRODUCTION

In experimental collision physics one obtains information by observing the changes of a system from some experimentally prepared initial state to some detected final state. This information is usually formulated in terms of transition probabilities. In order to obtain well-defined transition probabilities experimentally, one tries to define initial and final states as well as possible by the experimental setup. In an ideal case, initial and final states of the collision system are completely determined in the quantum-mechanical sense. The corresponding transition probability we denote by the symbol I_{if}; it contains all the information that can possibly be obtained on the transition. Usually, a transition $i \rightarrow f$ can proceed in "different ways," or in the quantum-mechanical language, via different intermediate substates. Since the system evolves in such a case from a completely defined initial state, the state vector representing the intermediate state is a coherent sum of the substate vectors, and consequently the transition probability I_{if} is a coherent sum, squared, of amplitudes belonging to the different intermediate substates. Denoting the set of quantum numbers that define the intermediate substates by ρ, we can therefore write

$$I_{if} = \left| \sum_{\rho} A_{f\rho i} e^{i\delta_{f\rho i}} \right|^2 \tag{1}$$

In case of a collision process, some of the quantum numbers defining initial and final states are continuously variable, such as, for example, value and

A. Niehaus • Fysisch Laboratorium, Rijksuniversiteit Utrecht, 3508 Utrecht, The Netherlands.

direction of initial or final relative momenta of the collision partners. The functional dependence of the modulus A and of the phase δ on such a continuous variable then leads to characteristic interference patterns in I_{if}, from which indirect information on the collision system in the intermediate state can be obtained. The described interference patterns are, in principle, of the same type as the interference patterns observed for the intensity of monochromatic photons scattered by two slits, the photon within the two slits representing the intermediate state. Because of this similarity we will call interferences in the transition probability I_{if} "two-slit" interferences.

The difficulty of experimentally defining initial and final states "completely" increases of course with the complexity of the collision process. Usually, in a real experiment, the system is prepared with certain probabilities ω_i in each of different initial states, and the detectors do not distinguish between all final states. Since events belonging to different transitions $i \rightarrow f$ do not have defined phase relations, the transition probability belonging to such a real experiment is the sum of the intensities belonging to the different events:

$$\bar{I} = \sum\!\!\!\!\!\!\int_i \omega_i \sum\!\!\!\!\!\!\int_f I_{if} = \sum\!\!\!\!\!\!\int_i \omega_i \sum\!\!\!\!\!\!\int_f \left| \sum_p A_{fpi} e^{i\delta_{fpi}} \right|^2 \tag{2}$$

The symbol $\sum\!\!\!\!\!\!\int$ denotes the summation and integration over a certain range of discrete or continuous quantum numbers. Depending on how strongly the moduli, and especially the phases, of the individual contributions in Eq. (2) vary with the quantum numbers over which the averaging has to be performed, and depending on the ranges of these quantum numbers, the interferences will be more or less strongly masked, leading to a more or less significant loss of information obtainable from the experiment. This is the reason why "two-slit" interferences are well known for simple processes, such as for instance elastic scattering, but hardly at all known for more complicated processes in which ionization or dissociation occurs.

Recently, however, due to improved experimental techniques, especially due to the introduction of coincidence techniques, it has become more and more possible to observe "two slit" interferences also for rather complex processes. In this chapter we discuss such recently observed interferences for collision processes leading to electron emission. The main goal in this discussion is to point out the physical origin and the information content of the different effects. This will be done by deriving, for each of five different cases, an approximate theoretical formula for the measured transition probability, and by comparing the experimental results with predictions of these formulas. Since the derivation of the approximate theoretical formulas implies the assumption of certain "mechanisms," the agreement found for each of the five cases between experiment and theory may be taken as a proof of the proposed "mechanisms." The five different cases we distinguish by the type of intermediate states that contribute to the final state.

2. INTERFERENCE OF CONTRIBUTIONS FROM DIFFERENT MAGNETIC SUBSTATES—ANGULAR ELECTRON INTENSITY DISTRIBUTION

2.1. *Introduction*

A He atom, which has been excited into one of the lower-lying doubly excited states by some collision process, decays by electron emission with lifetimes of the order of $\tau = 100\text{–}1000$ a.u. (Hicks *et al.*, 1974) (a.u. = atomic units). If such an autoionizing excited state has orbital angular momentum $L \neq 0$, the electron intensity observed at some angle will contain contributions from different substates M_L. In the absence of interaction of the autoionizing atom with some external field or with some other particle the substates are degenerate, hence the energy of the electrons belonging to the decay of the different substates is the same, and interference of the "two-slit" type is possible in principle. We will discuss the case of excitation of He in collisions with He^+ in the keV energy range, where double excitation predominantly occurs into the $He(2p^2)\ {}^1D$ state. The "two-slit" situation for this case may be indicated by the scheme

$$\underset{\text{(initial)}}{He^+ + He} \rightarrow \underset{\text{(intermediate)}}{\begin{Bmatrix} He^{**}(2p^2)(L=2;\ M_L) + He^+ \\ He^{**}(2p^2)(L=2;\ M'_L) + He^+ \end{Bmatrix}} \rightarrow \underset{\text{(final)}}{He^+ + He^+ + e^-}. \tag{3}$$

Measurements of the transition probability I_{if} have been performed as a function of the observation angle of the ejected electron (Kessel *et al.*, 1979). The interference pattern observed in such an experiment is the angular distribution belonging to a well-defined event. The theory that relates the angular distribution belonging to a well-defined event to the population amplitudes of the magnetic substates is rigorous. For the general case the problem has been formulated by Eichler and Fritsch (1976) and an application of the formulation to the rather complicated case of the autoionization $Ne^{**}(2p^4 3s^2)\ {}^1D \rightarrow Ne^+(2p^5)\ {}^2P_{1/2,3/2} + e^-$ has been presented recently by Boskamp *et al.* 1982a. In the present case of process (3) we can use a simpler formulation, because the state of the ion formed in the autoionization process has zero angular momentum. On the other hand, we have to account for the fact that the autoionizing He atom from process (3) is not at rest but has a finite velocity $\mathbf{V}_p$ with respect to the electron detector, leading to an angle-dependent shift of the energy ε_e of the ejected electron in the emitter frame with respect to the energy ε measured at the detector. In the following we will call this shift the "Doppler shift." In principle, one also must account for the fact that in process (3) the autoionization arises when the collision is not yet completely over: at a relative collision velocity of ~ 0.1 a.u.—which corresponds to a collision energy of ~ 1 keV in case (3)—the average distance between the collision partners at the instant of electron ejection is only ~ 38 a.u. at the lifetime $\tau = 380$ a.u. of the ^1D state. The Coulomb interaction with the receding He^+ will thus influence the autoionization to a certain extent. However, since the main effect of this "postcollision interaction" (PCI) is a modification of the autoionization line shape, we will neglect this complication in the present discussion of the angular distribution. The effects of PCI are the subject of Secs. 3–5 of this chapter.

2.2. *Theory*

The interactions leading to excitation and ionization in process (3) are purely Coulombic and therefore independent of spin. Furthermore, the excited atomic state is a singlet and the spin projections of the He^+ ions and of the electron in the final state remain unobserved. We therefore may ignore the spin variables altogether in our derivation of I_{if}. The wave function of the pure intermediate state—formed via the time-dependent perturbation during the collision from the completely defined initial state—can be described by the product of the electronic wave functions of the two collision partners and the wave function of their translational motion. Since we want to describe a process in which the He^+ projectile remains in the ground state and does not change its translational motion in the intermediate state, we can omit the projectile part of the wave function completely and write

$$|\psi_\rho\rangle = \sum_M |LM; \mathbf{V}_\rho\rangle\, a_M, \tag{4}$$

with L, M indicating the electronic state of the target, and $\mathbf{V}_\rho$ indicating its translational motion. The population amplitudes a_M depend on the collision energy E and on the scattering angle Θ. These amplitudes contain all the information on the collision. We write the wave function of the completely defined final state, reached by emission of an electron of wave vector $\mathbf{k}$ in the laboratory system, similarly as

$$|\psi_f\rangle = |\mathbf{k}; L_f M_f; \mathbf{V}_f\rangle, \tag{5}$$

with $L_f M_f$ indicating an electronic state with angular momentum L_f and projection quantum number M_f. The final velocity $\mathbf{V}_f$ differs slightly from the velocity of the emitter $\mathbf{V}_\rho$. To calculate the transition probability I_{if} we need the matrix element

$$V_{\rho f} = \langle \psi_f | V | \psi_\rho \rangle, \tag{6}$$

with V being the perturbation leading to autoionization. In the calculation of Eq. (6) there arises a problem because the free-electron wave function in the final state is defined in the laboratory system. In order to express the matrix element in terms of the transition amplitude defined in the frame of the moving atom, one must transform the wave function of wave vector $\mathbf{k}$ in the laboratory into the moving frame. This Galiliean transformation yields, in the moving frame, a wave of wave vector $(\mathbf{k} - \mathbf{V}_\rho)$. The matrix element (6) can be separated into an electronic part and a part belonging to the translational motion. The latter reduces to an overlap integral of plane waves with wave numbers K_ρ and K_f, because V does not act on the coordinates of the decaying atom. The wave numbers are given by energy conservation. When we introduce the complex electronic energy $(\varepsilon_0 - i\Gamma/2)$ of the excited atom with lifetime $\tau = \Gamma^{-1}$, this leads to

$$\begin{aligned} E_{\text{tot}} &= \varepsilon_0 + \frac{K_\rho^2}{2m} - \frac{i\Gamma}{2} = \frac{(\mathbf{k} - \mathbf{V}_\rho)^2}{2} + \frac{K_f^2}{2m} \\ &= \varepsilon - D + \frac{K_f^2}{2m}, \end{aligned} \tag{7}$$

where E_{tot} is the sum of electronic and translational energy, ε equals $k^2/2$, and D is the Doppler shift between the electron energy in the emitter frame and the electron energy ε in the laboratory. Atomic units will be used throughout this chapter. Due to the imaginary part in the electronic energy of the intermediate state, the value of K_ρ is complex. This leads to an exponential decay of the intermediate state along the trajectory. To calculate the overlap integral of the plane waves with wave numbers K_ρ and K_f given by Eq. (7), we make use of the fact that $E_{\text{tot}} \gg \varepsilon_0$, ε, Γ, D, and approximate

$$K_\rho = \left[2m \left(E_{\text{tot}} - \varepsilon_0 + \frac{i\Gamma}{2} \right) \right]^{1/2} \approx (2mE_{\text{tot}})^{1/2} - \frac{(\varepsilon_0 - i\Gamma/2)}{V_0}, \tag{8}$$

$$K_f = [2m(E_{\text{tot}} - \varepsilon + D)]^{1/2} \approx (2mE_{\text{tot}})^{1/2} - \frac{(\varepsilon - D)}{V_0}, \tag{9}$$

with $V_0 = (2mE_{\text{tot}})^{1/2}$. With the foregoing arguments we can write for the transition probability

$$I_{if} = \mathcal{N} \left| \sum_M a_M \langle \mathbf{k} - \mathbf{V}_\rho ; L_f M_f | V | LM \rangle \right|^2 |\langle \mathbf{K}_f | \mathbf{K}_\rho \rangle|^2, \tag{10}$$

where N is a normalization constant which will be determined below. The first term—the atomic transition probability in the emitter frame—can be separated into an M-independent transition strength and an angular distribution. By expanding the final state with wave vector $(\mathbf{k}-\mathbf{V}_\rho)$ into partial waves, as

$$|\mathbf{k}-\mathbf{V}_\rho; L_f M_f\rangle \sum_{l,m} |\varepsilon - D; l, m, L_f M_f\rangle \, Y^*_{lm}(\theta_e, \phi_e), \tag{11}$$

and making use of the fact that $L_f = M_f = 0$ (so that $l = L$, and $m = M$), the angular part becomes

$$|f(\omega_e)|^2 = \left| \sum_{M=-2}^{+2} a_M Y_{lM}(\theta_e, \phi_e) \right|^2, \tag{12}$$

where $\omega_e = (\theta_e, \phi_e)$ are the angles between $(\mathbf{k}-\mathbf{V}_\rho)$ and the quantization axis. The angle-independent internal transition strength is given by the "Golden Rule" (Messiah, 1964) as $\Gamma/2\pi$. The second term in Eq. (10)—the overlap integral with K_ρ and K_f given by Eqs. (8) and (9)—can be integrated directly and yields a Lorentzian line intensity around $\varepsilon_0 + D$. By requiring that I_{if} integrated over ε and over the angles $\omega = (\theta, \phi)$ becomes unity, we determine the normalization. The final result is

$$I_{if} = \frac{\Gamma/2\pi}{(\varepsilon - D - \varepsilon_0)^2 + (\Gamma/2)^2} \left| \frac{d(\varepsilon - D)}{d\varepsilon} \right| \left| f(\omega_e) \right|^2 \left| \frac{d\omega_e}{d\omega} \right|. \tag{13}$$

The geometry of a collision event for which Eq. (13) has been derived is shown in Fig. 1. Interference due to contributions from the different sublevels of the intermediate state arises only in the internal angular distribution given by relation (12). To simulate a measurement in which this angular distribution is determined by integrating over the Lorentzian lines at the angle-dependent positions $(\varepsilon_0 + D)$, we must integrate Eq. (13) over ε. This yields

$$\int I_{if}\, d\varepsilon = |f(\omega_\varepsilon)|^2 \left| \frac{d\omega_e}{d\omega} \right|. \tag{14}$$

The Jacobian factor $|d\omega_e/d\omega|$, caused by the transformation from the laboratory to the moving atom, is close to unity since $|k| \gg V_\rho$. It is seen that the measured angular distribution belonging to a well-selected event contains very direct information on the population amplitudes a_M. These complex numbers depend on collision energy and scattering angle Θ, and contain all the information on the excitation event.

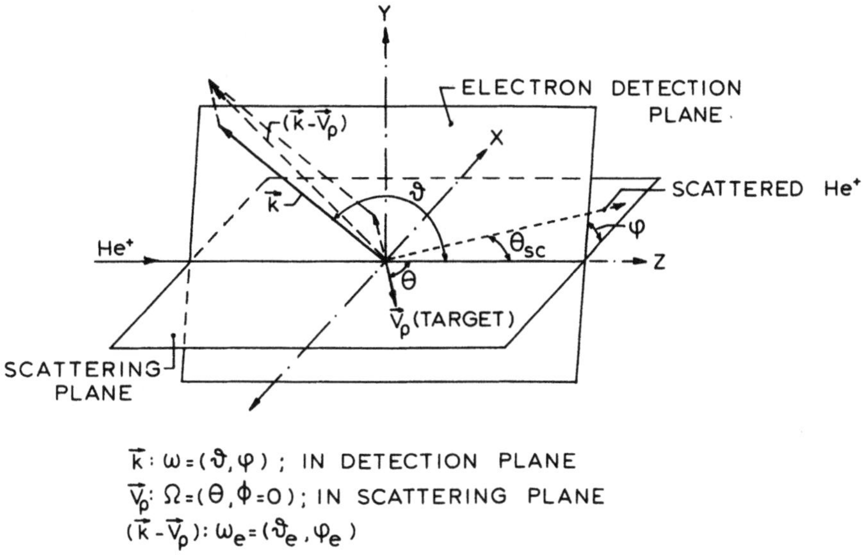

FIGURE 1. Geometry of a collision event, showing the relations of the various quantities defined in the text.

*2.3. Comparison with Experimental Results for Autoionization of He**($2p^2$) ^1D*

In order to measure the angular distribution given by Eq. (14), it is necessary to perform an experiment in which both the scattering plane and the electron detection plane are defined by measuring the electron in coincidence with one of the scattered heavy particles. In a noncoincident experiment, where only the ejected electron angular distribution is measured, neither Θ nor the angle between scattering plane and electron detection plane are defined. The corresponding theoretical angular intensity distribution is obtained from Eq. (14) by carrying out the appropriate integrations over the undistinguished events. Since the phase of the spherical harmonics in Eq. (12) varies by 2π if ϕ_e is varied over the integration range of 2π, it is obvious that this leads to a complete loss of coherence between the contributions from the different sublevels. Phase relations between the amplitudes cannot be determined in such an experiment.

Coincident angular distributions that can directly be interpreted in terms of relation (14) have been measured by Kessel *et al.* (1979). In practice there arises a difficulty because, in case of reaction (3), either the target atom or the projectile atom can be doubly excited after the collision. For each of the two possibilities we have an angle-dependent energy distribution given by Eq. (13). Due to the different velocity vectors of target and projectile the lines will in general be Doppler-shifted by different amounts;

however, due to the finite widths of the lines, there are angular ranges where the two contributions overlap. In these ranges the theoretical expression (13) must be modified. This is a subject of Sec. 6. For the present example we restrict measurements and theoretical interpretation to angular regions in which overlap of target and projectile contributions can be neglected.

The experimental arrangement used by Kessel *et al.* (1979) is sketched in Fig. 2. The center-of-mass scattering angle Θ is fixed by choosing a fixed scattering angle Θ_{sc} of the fast projectile with the help of an annular slit. The angle ϕ between the fixed electron-detection plane and the scattering plane is defined by an opening which admits ions within an angular range of $\Delta\phi = 15°$. The azimuthal position of this opening can be rotated by 2π. The electron detector is a hemispherical electrostatic energy analyzer, which can be rotated—within the plane defined by $\phi = 0°$, $180°$—to different polar angles θ between $60°$ and $135°$. Measurements have been performed for several collision energies E, and for several scattering angles Θ_{sc}. For a fixed combination (E, Θ_{sc})—i.e., for a well-selected heavy-particle trajectory—an angular distribution is measured by varying θ and ϕ. This can be done in different ways. In Fig. 3 we show as an example the interference pattern observed if θ is fixed at $135°$ and ϕ is varied from 0 to 2π, at a collision energy $E = 2$ keV and a scattering angle $\Theta_{sc} = 6°$. The measured points are compared with a calculation using relations (14) and (12) with assumed complex numbers a_M. By performing such comparisons for different "cuts" through the angular distribution belonging to a selected trajectory, the amplitudes can be determined.

We notice in Fig. 3 that the angular distribution is symmetric with

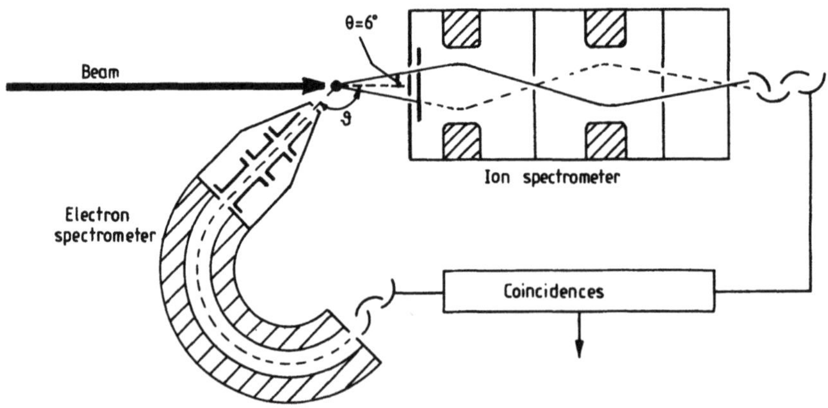

FIGURE 2. Sketch of the apparatus of Kessel *et al.* (1979), used for measurements of scattered ions and ejected electrons in coincidence.

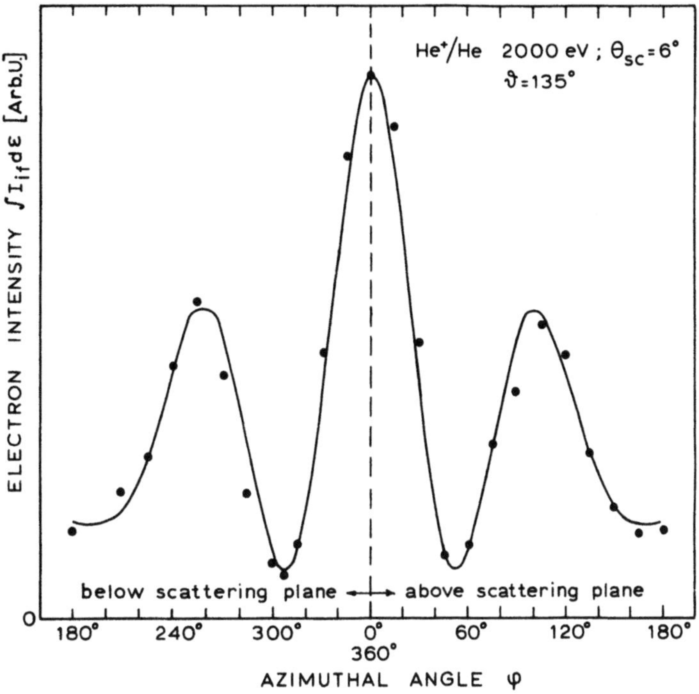

FIGURE 3. Interference pattern in the ejected-electron intensity as a function of azimuthal angle ϕ with respect to the scattering plane, at a polar ejection angle $\theta = 135°$. Measured data are indicated by dots; the curve is calculated using the complex magnetic sublevel population amplitudes as parameters for the fit.

respect to the collision plane ($\phi = 0°$, $180°$). In Eq. (12) this symmetry is achieved by setting

$$a_M = (-)^M a_{-M}. \tag{15}$$

The observed symmetry is a consequence of the fact that the perturbation leading to excitation of the intermediate state commutes with the operator of reflection at the collision plane. Once the amplitudes a_M are determined, the complete angular distribution can be calculated. A perspective representation of such a distribution for the present example is shown in Fig. 4.

In this example of the autoionization $\mathrm{He}(2p^2)\,{}^1D \rightarrow \mathrm{He}^+({}^2S) + e^-$, the final atomic state has zero angular momentum, hence the angular intensity distribution of the ejected electron has the same structure and orientation as the electron density of the excited atoms; it can therefore be regarded as a "picture" of the atom excited in a well-defined event.

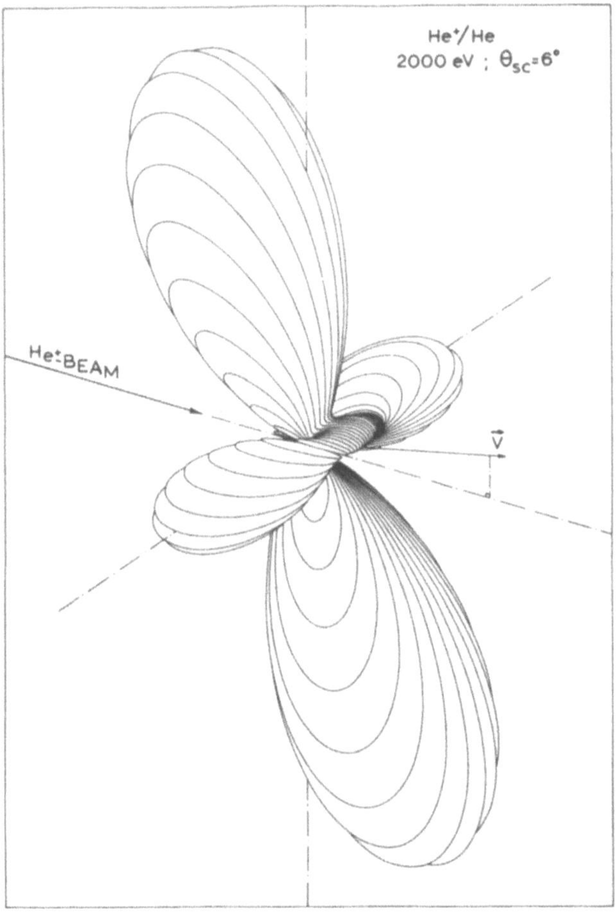

FIGURE 4. Perspective representation of the $He(2p^2)\ ^1D$ electron-intensity distribution for a well-defined collision event. The scattering plane is defined by the He^+ beam and by the final relative velocity vector **V**.

3. INTERFERENCE OF CONTRIBUTIONS FROM DIFFERENT STATES IN ONE ATOM

3.1. Introduction

Let us consider the situation when a He atom is first excited by some collision process into intermediate states of different energies, and subsequently decays by autoionization into a certain final state. The ejected electron has different energies for different intermediate states if the individual lines do not overlap. If the collision leading to excitation is such that the decay of the excited atom occurs "isolated in space"—e.g., after electron-impact excitation well above the excitation threshold—then the individual lines are Lorentzian with a width $\Gamma_n = \tau_n^{-1}$, and overlap does only occur if the energy separations between the intermediate states are of the same order as the line widths. To be definite, we consider the excited states $2s^2(^1S)$ and $2p^2(^1D)$ of He, which have lifetimes of 150 and 380 a.u. and excitation energies of 57.80 and 59.90 eV, respectively. The energy separation of 2.1 eV is an order of magnitude larger than the widths of $\Gamma(^1S) = 0.18$ eV and $\Gamma(^1D) = 0.072$ eV so that there is essentially no overlap, and consequently, no possibility of interference among contributions from the two states. If, however, the He is excited in a collision process in which after the excitation a charged collision partner recedes slowly from the excited atom, the autoionization process is influenced by the Coulomb interaction with this collision partner, leading to a broadening and shifting of the individual lines, and thus to possible overlap and interference. The phenomena caused by this Coulomb interaction on an autoionizing atom are usually called "postcollision interaction" (PCI) phenomena (Hicks *et al.*, 1974; Niehaus, 1978). We discuss here the "two-slit" type interferences caused by PCI in the case of the collision system $Li^+ + He$. In the keV collision-energy range, the main double excitation leads into the states $2s^2(^1S)$ and $2p^2(^1D)$ of He. The PCI effect of Li^+ broadens the corresponding lines sufficiently so that overlap occurs. The corresponding "two-slit" situation can be indicated as

$$\mathrm{Li}^+ + \mathrm{He} \rightarrow \left\{ \begin{matrix} \mathrm{Li}^+ + \mathrm{He}^{**}2s^2(^1S) \\ \mathrm{Li}^+ + \mathrm{He}^{**}2p^2(^1D) \end{matrix} \right\} \xrightarrow{\mathrm{PCI}} \mathrm{Li}^+ + \mathrm{He}^+ + e^-(\varepsilon). \qquad (16)$$

Before we derive an expression for the transition probability we make some simplifying assumptions. First of all, the velocity of relative motion of the heavy particles is small compared to the velocity of the electrons. We therefore assume that the relative motion can be separated and described in terms of potential curves, with the potentials given by the total electronic energy at fixed distance R. Secondly, the relevant distances R at which transitions from the intermediate states to the final state occur are large compared with the impact parameter of the selected collisions. We therefore omit the centrifugal term in the effective-potential curves, which in this way become independent of the impact parameter. Thirdly, the relevant distances are considered large enough to omit all interactions except for the long-range Coulomb repulsion between Li^+ and He^+ in the final state. We thus arrive at the following simple potential curves for the system in the intermediate states and in the final state, respectively:

$$V_n(R) = \varepsilon_0^n - i\Gamma_n/2, \qquad (17)$$
$$V_f(R) = \varepsilon + 1/R.$$

Here, we have again introduced the imaginary parts Γ_n into the electronic energy of the intermediate states in order to account for their decay. The zero of the electronic energy is chosen to be the energy of He^+ and Li^+ at infinite separation, with an electron of zero energy also at infinity. In Fig. 5 we show a potential-curve diagram. The total time-independent energy of the system is

$$E_{tot} = V_n(R) + K_n^2/2\mu = V_f(R) + K_f^2/2\mu = \varepsilon_0^n + E_n. \qquad (18)$$

Here, E_n is the asymptotic relative kinetic energy of the heavy particles in the intermediate states, and K_n and K_f are the R-dependent wave numbers of relative motion in the intermediate and final states; K_f is a real number while K_n is complex. From Eqs. (17) and (18) we find

$$K_f = [2\mu(E_{tot} - \varepsilon - 1/R)]^{1/2}, \qquad (19)$$
$$K_n = [2\mu(E_{tot} - \varepsilon_0^n + i\Gamma_n/2)]^{1/2}.$$

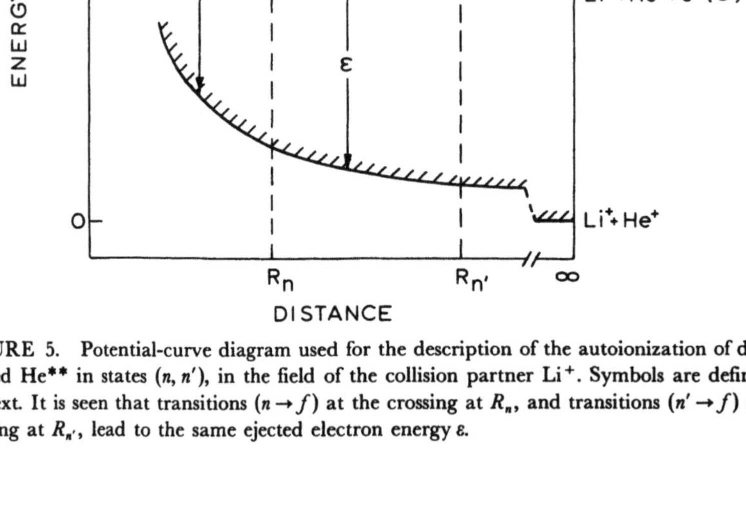

FIGURE 5. Potential-curve diagram used for the description of the autoionization of doubly excited He** in states (n, n'), in the field of the collision partner Li^+. Symbols are defined in the text. It is seen that transitions $(n \to f)$ at the crossing at R_n, and transitions $(n' \to f)$ at the crossing at $R_{n'}$, lead to the same ejected electron energy ε.

In our simplified description the autoionization step is the transition $V_n \rightarrow V_f$. If this transition occurs at sufficiently large R, the instantaneous relative motion will not change. We assume this to be true as well for the whole region of distances relevant for our problem. Classically, this means that transitions can only occur at the crossings of $V_n(R)$ and $V_f(R)$. The corresponding crossing points R_n and $R_{n'}$ are indicated in Fig. 5 for an arbitrarily chosen energy ε of the ejected electron. Obviously, by choosing ε one chooses the distances at which transitions from the different intermediate states can contribute to the observed intensity. Alternatively, since the distance is given by the product of relative velocity and time, one can say that, at a certain ε, contributions due to the decay of the different intermediate states at different defined time points are sampled. This leads to interferences as a function of ε because the phase factors evolve differently in time for the different intermediate states. These interferences can be considered a sort of "quantum beats," with characteristic times $\Delta t \sim (\Delta E)^{-1} \sim 10$ a.u. ($\approx 2.4 \times 10^{-16}$ s).

3.2. *Theory*

In order to derive an approximate expression for I_{if} we start from the simplified potential-curve diagram shown in Fig. 5. The wave functions for the intermediate and final states can then be written

$$\psi_\rho = \sum_n |n\rangle\, |K_n\rangle, \qquad \psi_f = |\mathbf{k}\rangle\, |K_f\rangle, \tag{20}$$

with $|n\rangle$ and $|\mathbf{k}\rangle$ indicating the electronic parts, and $|K_n\rangle$ and $|K_f\rangle$ the parts belonging to the relative motion. For I_{if} we thus have

$$I_{if} \propto \left| \sum_n a_n \langle \mathbf{k}|\, v\, |n\rangle \langle K_f \mid K_n\rangle \right|^2. \tag{21}$$

As in Sec. 2 we express the atomic transition matrix element—the first bracket—by an angular distribution times the transition strength:

$$\langle \mathbf{k}|\, v\, |n\rangle \propto f_n(\omega)(\Gamma_n/2\pi)^{1/2}. \tag{22}$$

The overlap integral—the second bracket—which in the preceding section yielded the Lorentzian line shape, now yields the characteristic PCI line shape. The integral cannot be evaluated without further approximation. We carry out the evaluation as follows. We use WKB wave functions with wave numbers obtained from Eqs. (19) by approximating for large E_{tot}

$$K_f = (2\mu E_{\text{tot}})^{1/2} - (\varepsilon - R^{-1})[(2/\mu)\, E_{\text{tot}}]^{-1/2}, \tag{23}$$

$$K_\rho = (2\mu E_{\text{tot}})^{1/2} - (\varepsilon_0^n + i\Gamma_n/2)[(2/\mu)\, E_{\text{tot}}]^{-1/2}. \tag{24}$$

The resulting integral is then evaluated by the standard stationary-phase method (Child, 1976), the stationary phase points being the crossing points of the potentials (17), namely,

$$R_n = (\varepsilon_0^n - \varepsilon)^{-1}. \tag{25}$$

Finally we account for the fact that the emitter moves with respect to the laboratory frame by replacing ε with $(\varepsilon - D)$, and $f_n(\omega)$ with $f_n(\omega_e)$—as outlined in Sec. 2—and normalize by requiring $\int I_{if}\, d\omega\, d\varepsilon = 1$. The final result is

$$\begin{aligned} I_{if} = \Bigg| \sum_n a_n f_n(\omega_e) \left|\frac{d\omega_e}{d\omega}\right|^{1/2} \\ &\times (\Gamma_n/V_0)^{1/2} \frac{\exp\{-(\Gamma_n/2V_0)\{[\varepsilon_0^n - (\varepsilon - D)]^{-1} - \varepsilon_0^{n-1}\}\}}{\varepsilon_0^n - (\varepsilon - D)} \\ &\times |d(\varepsilon - D)/d\varepsilon|^{1/2} \exp\left\{-\frac{i}{V_0} \ln[\varepsilon_0^n - (\varepsilon - D)]\right\}\Bigg|^2, \end{aligned} \tag{26}$$

where we have $V_0 = (2E_{\text{tot}}/\mu)^{1/2}$, and $f_n(\omega_e)$ is

$$f_n(\omega_e) = \sum_{M=-L_n}^{L_n} b_M^n Y_{L_n M}(\theta_e, \phi_e), \qquad \int f_n(\omega_e)\, d\omega_e = 1. \tag{27}$$

The angles $\omega_e = (\theta_e, \phi_e)$ and $\omega = (\theta, \phi)$ are defined as in Sec. 2 (see Fig. 1). The new feature of Eq. (26) as compared with the corresponding expression for the case without PCI is that, at a given ε and direction ω, in general I_{if} is made up of a coherent sum of contributions from different states. How significantly the contributions overlap depends on their widths and energy positions, which are given by the expression given on the second line in Eq. (26). Within a region of overlap Eq. (26) predicts interference structures because the relative phase of the contributions varies with ε. These interference structures are the "quantum beats" whose existence was already suspected from the qualitative discussion in the Introduction.

3.3. Comparison with Experimental Results for Collisions of Li^+ with He

Formula (25) could be applied directly to the description of electron spectra from the decay of doubly excited He formed in collisions with Li^+, provided the electrons are measured in a well-defined direction **k** with respect to the collision plane and for a well-defined trajectory of the heavy-particle motion in the collision plane. In such a case no averaging of Eq. (26) would be necessary to simulate the experimental data, and the predicted interference structure should be observed. This fact has indeed been verified for electron spectra of the $He^+ \rightarrow He$ collision system. But we want to discuss here a nonideal case in order to demonstrate that sometimes coherence, and the interference structures caused by the coherence of contributions, survive rather drastic averaging. Such nonideal experimental data are the noncoincident electron spectra of $Li^+ \rightarrow He$ in a certain direction **k**. In order to simulate such spectra theoretically, Eq. (26) must be averaged over all possible relative orientations of the electron-detection plane with respect to the scattering plane—i.e., over the angle ϕ—and over all trajectories leading to population of the doubly excited states—i.e., over the angle Θ.

The integration over ϕ makes the electron intensity cylindrically symmetric, but of course there can still be interference as a function of ε, if only the dependence of the other quantities in Eq. (25) on Θ and ϕ is weak enough. The crucial quantity is the Doppler shift D, because it determines the energy separation of the different contributions and their relative phase. The dependence on D in Eq. (26) is such that one gets nearly identical interference structures for different D values, the structures being only shifted by an amount D. Thus the integration over Θ, ϕ leads to a masking of structures which have widths smaller than the range of D values that occurs. This range, on the other hand, depends on the observation angle θ chosen for the electrons. In particular, at angles $\theta = 0°$ and $180°$ the dependence of D on ϕ vanishes completely, i.e., the Doppler shift is the same for all orientations of the scattering plane in space. At the same time, the dependence of $f(\omega)$ on ϕ vanishes for these angles, so that integration over ϕ has no effect at all on the structures predicted by Eq. (26) for $\theta = 0°$ and $180°$. Averaging over the indistinguished final states of the heavy-particle motion therefore affects the coherence only via the different possible scattering angles. These angles, on the other hand, are restricted to a rather narrow range by the process itself (Zwakhals *et al.*, 1982). We therefore expect that interferences will be observed in noncoincidence measurements at $\theta = 0°$ or $180°$.

Such spectra—due to $Li^+ \rightarrow He$ collisions—have been measured at several collision energies (Zwakhals *et al.*, 1982). At not-too-low collision energies the spectra are dominated by process (16). An example is shown in Fig. 6. The peaks belong to the atomic autoionization of the He, and the broad background is due to molecular autoionization of the collision system (Yagishita *et al.*, 1980; Zwakhals *et al.*, 1982). In the absence of PCI and Doppler shift the atomic autoionization part of the spectrum would consist of three peaks in the range of electron energies between 30 and 36 eV. These

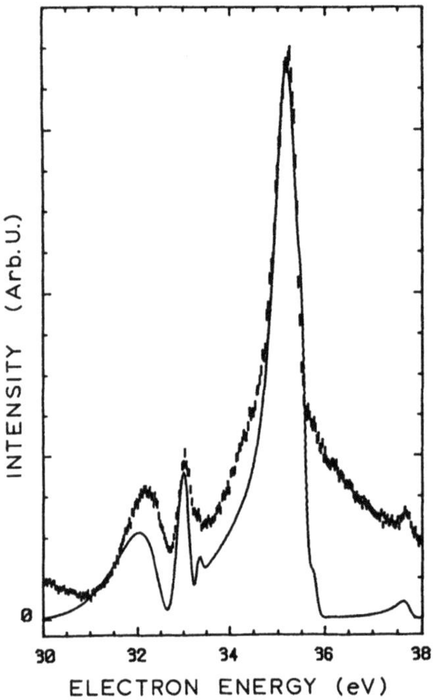

FIGURE 6. Electron-energy distribution for Li^{+}–He collisions at 3.3 keV, at a detection angle $\theta = 0°$. Experimental data points with error bars are shown; the solid curve represents a calculation of the atomic autoionization part of the spectrum assuming coherence of the contributions from different intermediate states.

peaks would have Lorentzian shapes and would be positioned at $\varepsilon_0^1 = 33.2$ eV, $\varepsilon_0^2 = 35.3$ eV, and $\varepsilon_0^3 = 35.54$ eV, corresponding to decay of the doubly excited He states $2s^2(^1S)$, $2p^2(^1D)$, and $2s2p(^1P)$, respectively. With the influence of PCI and Doppler shift, but with random phase relations between the three distributions, the spectrum would look like the spectrum shown in Fig. 7. Clearly, there appears structure in the measured spectrum which cannot be explained by the incoherently composed spectrum. The solid line in Fig. 6 represents a calculation based on Eq. (26), assuming full coherence and regarding the amplitudes a_n as free parameters. The agreement with the measured spectrum is good enough to prove that the additional measured features are due to interference of the contributions from the different intermediate states. The prominent oscillation below ~ 33 eV is caused by interference of the PCI-broadened contributions of the states $2p^2(^1D)$ and $2s^2(^1S)$.

As long as a measured spectrum can be explained completely on the basis of Eq. (26), the information obtainable is the following: modulus and phase of the population amplitudes a_n. However, if there arise significant deviations which have to be attributed to approximations made in the derivation of Eq. (26), an analysis with an improved theory will yield additional information. Especially, it is to be expected that for electron

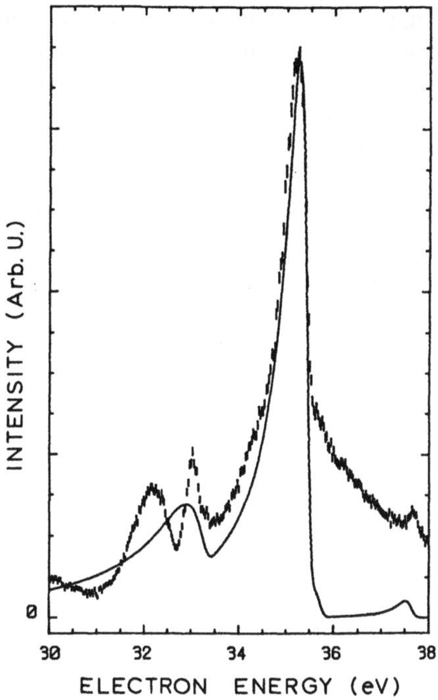

FIGURE 7. Comparison of the experimental Li^+–He electron spectrum (same as Fig. 6) with a calculation using incoherent summation of PCI-broadened lines belonging to different intermediate states.

energies far away from the nominal energies ε_0^n, i.e., for transitions occuring at small separations R, the potential curves (17) have to be modified, and also, a dependence of the width Γ_n on R has to be introduced. Analysis of structures using such modifications, therefore, will yield information on these quantities. Inspection of the agreement between the measurement and the simple theory, shown in Fig. 6, indeed suggests that deviations which could be analyzed in the described way are present.

4. INTERFERENCE OF CONTRIBUTIONS FROM DIRECT AND INDIRECT PROCESSES

4.1. *Introduction*

The kind of interference considered here is well known for all kinds of elastic scattering processes. It arises in the scattered intensity at a certain scattering angle as a function of primary energy, in an energy range where a scattering resonance can be formed. A prominent example is the elastic scattering of electrons off He in the energy of the $1s2s^2$ He resonance at 19.3 eV (Schulz, 1963). The electron intensity in such a case is described by a

Wigner shape function, caused by the coherent superposition of the contributions from direct scattering and scattering via the intermediate compound state. We want to discuss here a much more complicated case which is not so well known, namely, the electron impact excitation of singly excited states of an atom in the region of impact energies where double excitation followed by autoionization is possible. As an example we take again the He atom. The indirect process leading to single excitation is constituted by double excitation followed by autoionization and capture of the slow inelastically scattered electron (Heideman *et al.*, 1974; Smith *et al.*, 1974). The two-slit situation can be indicated by the scheme

$$e^- + \mathrm{He} \rightarrow \begin{cases} \mathrm{He}^{**}(2p^2)\ {}^1D + e^-(s) \xrightarrow{\mathrm{PCI}} \mathrm{He}^*(1snl) + e^-(f) \\ \mathrm{He}^*(1snl) + e^-(f) \nearrow \end{cases} \tag{28}$$

The lifetime of the $2p^2\ {}^1D$ state is 380 a.u., long enough so that the formation of this state can be considered an intermediate step. In order to populate the final state via the indirect process, the autoionization $\mathrm{He}^{**}(2p^2)\ {}^2D \rightarrow \mathrm{He}^*(1s) + e^-(f)$ has to take place in such a way that the slow electron—$e^-(s)$ of the scheme (28)—is captured by the newly formed $\mathrm{He}^+(1s)$ to yield the finally observed single-excited $\mathrm{He}^*(1snl)$. This second step in the indirect process can again be considered as PCI-influenced autoionization, the PCI occurring due to Coulomb interaction with the slow electron.

In order to derive an approximate expression for I_{if} in the case of reaction (28), we follow a procedure similar to that outlined in Sec. 3 for the case of PCI caused by a slowly receding heavy particle. This approach may be questionable, but the result agrees well with experimental data (Niehaus, 1978).

The essential part of the derivation is the assumption that the motion of the slow electron can be separated from the motion of the other electrons, so that we can define Born–Oppenheimer potential curves for the motion of the slow electron in the intermediate and final states. These potentials are the electronic energies of the system without the slow electron, at certain distances of the slow electron from the atom or ion, respectively. Restricting ourselves again to the Coulomb interactions, and omitting the angular-momentum containing parts in the one-dimensional potentials, arguments presented in Sec. 3 lead to expressions for these potentials for the intermediate and final states, respectively,

$$\begin{aligned} V_\rho(R) &= \varepsilon_0 - i\Gamma/2, \\ V_f(R) &= -1/R + \varepsilon, \end{aligned} \tag{29}$$

with an energy normalization which makes the electronic energy of

$He^+ + e^-$ zero at infinite separation. A potential-curve diagram showing the relation of the relevant quantities is shown in Fig. 8. For the indirect channel to be open, the total energy of the whole system, E_{tot}, must be larger than the excitation energy ε_0 of the doubly excited state, which in our normalization is equal to the "nominal" energy of the autoionization electron, i.e., the energy that the autoionization electron would have in the absence of PCI. The total energy is

$$E_{tot} = V_f(R, \varepsilon) + K_f^2/2 = V_\rho(R) + K_\rho^2/2 = \varepsilon_0 + E, \tag{30}$$

where E is the excess energy, i.e., the asymptotic kinetic energy of the slow electron before autoionization. The quantity B in Fig. 8 is the binding energy of the slow electron in the final state,

$$B = \varepsilon - E_{tot} = \varepsilon - (\varepsilon_0 + E). \tag{31}$$

From Fig. 8 it is clear that, in a collision system prepared at an energy E_{tot}, the population of a Rydberg state with binding energy B implies that the free electron—the fast electron $e^-(f)$ of scheme (28)—has energy $\varepsilon =$

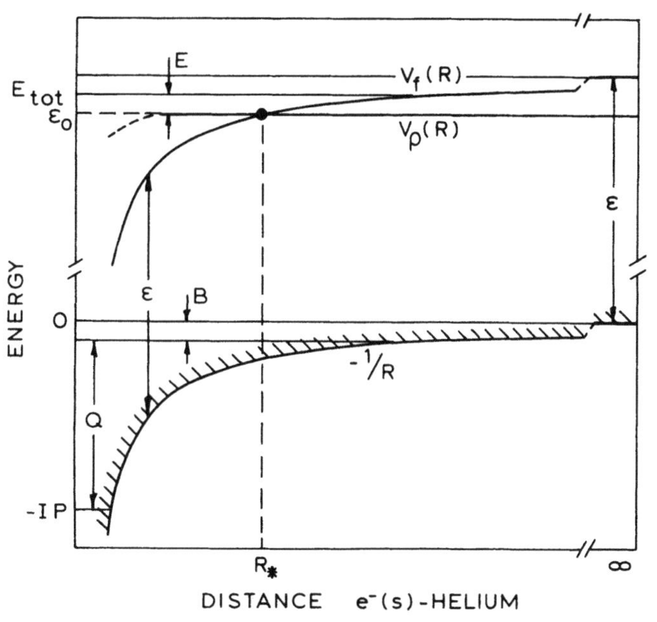

FIGURE 8. Potential-curve diagram used for the description of autoionization of an atom in the Coulomb field of a slowly receding electron. Potentials are electronic energies of the system minus the kinetic energy of the slow electron as a function of distance of the slow electron from the atom. Transitions $\rho \to f$ occur at the crossing R_*. The notation is defined in the text.

$E_{tot}+B$. A population of this final state via the intermediate state is classically possible only at the crossing of $V_\rho(R)$ with $V_f(R, \varepsilon)$. This crossing distance $R_* = (\varepsilon - \varepsilon_0)^{-1}$ changes with E_{tot}, so that the semiclassically calculated contribution corresponding to transitions around R_* will have a phase that varies with E_{tot}. On the other hand, the phase of contribution from the direct population of the final state will be nearly constant within a range of $\Delta E_{tot} \ll \varepsilon_0$ where population via the indirect process is significant. The result is interference in I_{if} as a function of E_{tot}. Experimentally, the final state is determined by measuring the fast electron at energy $\varepsilon = B + E_{tot}$, with E_{tot} given by the energy of the impacting electron $E_{imp} = E_{tot} + IP$ (IP = ionization energy of He). One may consider this as a measurement of electrons which have suffered a certain energy loss

$$Q = E_{imp} - \varepsilon = E_{tot} + IP - B - E_{tot} = IP - B. \qquad (32)$$

The interference is expected to appear in the intensity of electrons having suffered an energy loss Q as a function of impact energy E_{imp}.

4.2. *Theory*

We now derive an expression applicable to the analysis of the constant-energy-loss measurements described above. The intermediate-state wave function of the system—see scheme (28)—is

$$|\psi_\rho\rangle = a\,|n\rangle\,|K_\rho\rangle + b\,|\mathbf{k}\rangle\,|K_f\rangle, \qquad (33)$$

where $|n\rangle$ is the wave function of $He^{**}(2p^2)\ ^1D$, and $|K_\rho\rangle$ is the wave function of the slow electron. The wave number K_ρ is given by Eq. (30). The second term in Eq. (33) accounts for the fact that the final state, with wave function

$$|\psi_f\rangle = |\mathbf{k}\rangle\,|K_f\rangle, \qquad (34)$$

is populated directly with amplitude b. Here $|\mathbf{k}\rangle$ is the wave function of $He^*(1s) + e^-(f)$, and $|K_f\rangle$ is the wave function of the slow electron in the final state, K_f being given by Eq. (30). Since the autoionization operator does not act on the final-state wave function, we simply have for the transition probability

$$I_{if} = |a\langle\mathbf{k}|\,V\,|n\rangle\langle K_f\,|\,K_\rho\rangle + b|^2. \qquad (35)$$

The first bracket can again be replaced by $f(\omega)(\Gamma/2\pi)^{1/2}$. Since no direction of the slow electron is specified in case of the constant-energy-loss measurements, however, the angular-dependent amplitude must be averaged over ϕ, resulting in a cylindrically symmetric function $f(\theta)$ to describe the experimental transition probability. In principle we must distinguish in Eq. (5) different angular momenta of the slow electron. In the final Rydberg state, the corresponding states have different energies. In measurements, on the other hand, these states are not distinguished, due to lack of energy resolution. In order to obtain from Eq. (35) an expression comparable with the experimental data, incoherent summation over contributions from different angular momenta must be carried out. In our description of the motion of the slow electron—using the potentials of Eq. (29)—we have neglected the influence of the angular momentum altogether. By calculating the overlap integral in Eq. (35) with WKB wave functions with K_f and K_ρ given by Eqs. (29) and (30), therefore, we can expect to obtain immediately an expression comparable with the experimental averaged transition probability. We approximate as the complex wave number K_ρ

$$K_\rho = [2(E + i\Gamma/2]^{1/2} \approx (2E)^{1/2} + (i\Gamma/2)(2E)^{-1/2}. \tag{36}$$

We evaluate the overlap integral $\langle K_f | K_\rho \rangle$ again by the method of stationary-phase approximation. For given ε one has one point of stationary phase of the integrand as a function of R, namely, the real crossing point of the potentials of Eq. (29):

$$R_* = (\varepsilon - \varepsilon_0)^{-1} = (E + B)^{-1}. \tag{37}$$

The value of the integral, together with the factor $f(\theta)(\Gamma/2\pi)^{1/2}$, stemming from the atomic transition matrix element, then yields the averaged transition probability

$$\begin{aligned} \bar{I} = |af(\theta)(E+B)^{-1}\, \Gamma^{1/2}(2E)^{-1/4} \exp\{-(\Gamma/2)(2E)^{-1/2}(E+B)^{-1}\} \\ \times \exp\{i(2/B)^{1/2} \tan^{-1}(E/B)^{1/2})\} + b|^2, \end{aligned} \tag{38}$$

which can be used to describe the structure in the excitation functions above the threshold ε_0 for double excitation, measured by the described constant-energy-loss mode. The population amplitudes a and b are, in principle, dependent on the excess energy E. In the case of the direct excitation amplitude b, this dependence will be very weak, so that b can be approximated by a constant for a given observation angle θ. Although it is difficult to predict the variation of the indirect amplitude above the threshold, it is to be assumed that the main structure in the interference pattern is given by the explicit energy dependence of the amplitude and phase functions in Eq. (38).

4.3. Comparison with Experimental Results for the Process $e^- + He \rightarrow He^*(1snl) + e^-$

In Fig. 9 we compare experimental data of King *et al.* (1975) with calculations using our formula (38). In the calculations only the amplitudes a and b are unknown. They are taken to be independent of energy, so that only two parameters were varied to fit the experimental data, namely, the ratio $|a|/|b|$, and the relative phase $\arg(a)-\arg(b)$. It turned out that these parameters needed not be varied to describe the excitation functions for the different Rydberg states shown in Fig. 9. In the constant-energy-loss mode of

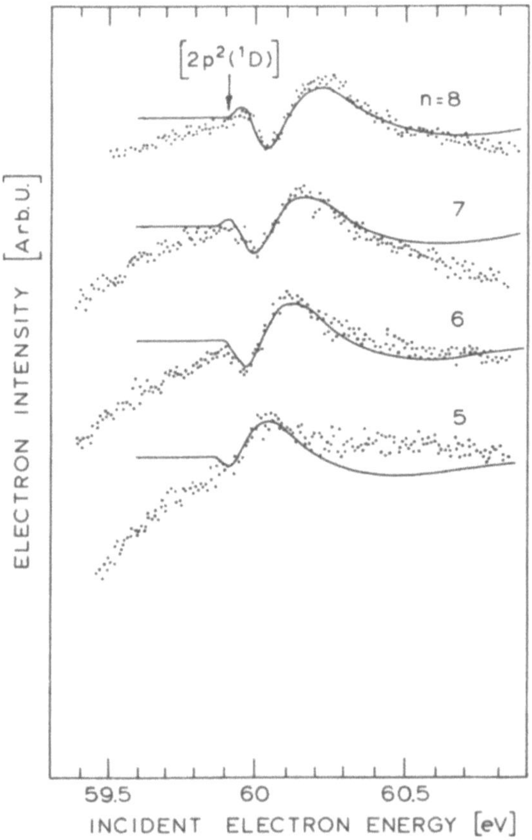

FIGURE 9. Excitation functions for $He^*(1snl)$ in e^-–He collisions. Experimental data are from King *et al.* (1975). The solid curve indicates results of a calculation assuming coherent superposition of the direct excitation amplitude and the indirect amplitude of excitation of $He^{**}(2p^2)\,{}^1D$ followed by autoionization with capture of the slow scattered electron. The threshold for excitation of $He^{**}(2p^2)\,{}^1D$ is indicated.

the measurement of excitation functions of Rydberg states, the principal quantum numbers n, indicated in Fig. 9, are related to the experimentally fixed energy loss Q by

$$(2n^2)^{-1} = B = IP - Q. \tag{39}$$

The essential features of the measured structures are well reproduced. The main deviations are due to a slow variation of the direct amplitude b with excess energy, although some variation of a cannot be excluded. In any case, the agreement is good enough to prove that the observed structures are due to the interference of contributions from direct and indirect excitation of the type indicated in scheme (28). The information one can extract from an analysis of such structures is contained in the complex amplitudes a and b, if the simple formula (38) is used. More precise measurements may necessitate the development of an improved theoretical description, which includes the possibility of angular-momentum exchange. Analysis of the interference structures may then also yield information on this process.

Constant-energy-loss measurements can also be carried out for values of the energy loss Q such that $Q > IP$. The electron intensity of an energy $\varepsilon = E_{\text{imp}} - Q$ then corresponds to the probability to create a slow ionization electron of energy $(-B)$. For this process as well, one can have direct and indirect contributions and the corresponding interferences. The theoretical expression which is analogous to Eq. (38) is simply obtained by carrying out the stationary-phase evaluation for the case $B < 0$; this gives, instead of the phase $\phi_* = (2/B)^{1/2} \tan^{-1}(E/B)^{1/2}$ in Eq. (38),

$$\phi_* = (-2B)^{-1/2} \ln\{[E^{1/2} - (-B)^{1/2}]/[E^{1/2} + (-B)^{1/2}]\}. \tag{40}$$

Comparison with experimental data of Hicks *et al.* (1974) shows that also for these cases the measured structures are well reproduced (Niehaus, 1978). In conclusion we note that extension of Eq. (38) to the case of more than one intermediate electronic state is possible and leads to a result completely analogous to Eq. (26).

5. INTERFERENCE OF CONTRIBUTIONS FROM DIFFERENT DISTANCES ON ONE POTENTIAL CURVE

5.1. *Introduction*

In the two preceding sections we have seen that the interaction of the autoionizing atom with a slowly receding charged particle can be described in terms of potential curves. The interferences in the electron spectra from the Li^+–He system could be interpreted as occurring as a result of coherent contributions of transitions from different potential curves to a common final potential curve. Classically, the contribution from one potential curve comes from transitions at a well-defined distance, namely, the distance at which the potential curve in question crosses the final-state potential curve. In the cases discussed so far there was only one such crossing of an intermediate potential with a final potential; we will now treat the case in which there are two crossings. Then we have two contributions to the observed intensity from two different distances on one potential curve. In general, i.e., for a well-selected event, these contributions will have well-defined relative phases leading to interferences of the "two-slit type," the "two slits" being the two crossing distances R_1 and R_2. An example in which these interferences have been observed is the system He^+–Ca (Hultzsch *et al.*, 1979). Since the recombination energy of $He^+(1s)$ of 24.6 eV is larger than the energy needed to remove the two outer s electrons of Ca, the system can spontaneously eject an electron as soon as the electron clouds overlap in the course of a collision. The final state of the system after this spontaneous ionization process is $He^+ + Ca^{++} + e^-(\varepsilon)$. The two-slit scheme for this example can be indicated as

$$He^+ + Ca \rightarrow \begin{Bmatrix} (He - Ca^+)^*(R_1) \\ (He - Ca^+)^*(R_2) \end{Bmatrix} \rightarrow He + Ca^{++} + e^-(\varepsilon). \qquad (41)$$

The potential-curve diagram for process (41) is shown in Fig. 10 along with a measured electron spectrum. The potential for the intermediate state is obtained from the measured spectrum via a model calculation. The result of the calculation—into a which a model potential and a model transition probability enter—is shown as the solid line. One notices that the electron spectrum exhibits a "rainbowlike" structure around the energy ε_* for which the two crossing points of the intermediate and final-state curves coalesce. For $\varepsilon < \varepsilon_*$—the classically forbidden region without crossing point—the intensity decreases exponentially yielding the "dark side" of the rainbow, and for $\varepsilon > \varepsilon_*$, there are two crossing points yielding the first and supernumerary rainbows by interference of the two contributions. In the measured spectrum the interference is partially masked because no selection of the impact parameters by a coincidence method was applied. However, since at the low collision energy of 2.3 eV the radial velocity of the heavy particles in the critical region of distances is to a large extent given by the rather strong attractive force, some of the interference structure predicted by the theory survives the averaging over impact parameters. The damping of the oscillations in the theoretical spectrum is due to a rather strong decrease of the transition-probability function $\Gamma(R)$ with distance. Because of this strong R dependence the derivation of an approximate theoretical formula for the transition probability I_{if} is somewhat complicated and will not be

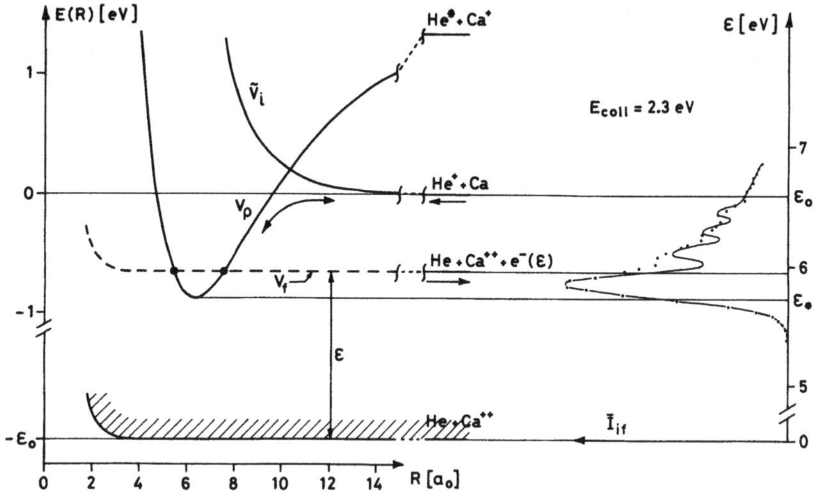

FIGURE 10. Potential-curve diagram for the system He^+–Ca, with an electron spectrum for the collision energy of 2.3 eV. Experimental points are from Hultzsch *et al.* (1979); the solid curve represents a calculation that shows the "rainbow" interference structure caused by the minimum in the potential curve of the intermediate state.

given here. A derivation of the expression leading to the solid curve in Fig. 10 is given by Hultzsch *et al.* (1979).

We will treat here, instead, another example of a rainbow structure in an electron spectrum, which has very recently been discovered (Niehaus and Zwackhals, 1983), and which is much simpler to describe theoretically: Inner-shell photoionization followed by PCI-influenced Auger decay. The quite different nature of this process as compared to our example of Eq. (41) may look surprising, but we will show that regarding the rainbow effect the physical situation is quite analogous. We first indicate the two-slit situation for the case of L-shell photoionization of Ar by the scheme

$$h\nu + \mathrm{Ar} \rightarrow \left\{ \begin{matrix} [\mathrm{Ar}^+(L^{-1}) - e^-(s)(R_1)] \\ [\mathrm{Ar}^+(L^{-1}) - e^-(s)(R_2)] \end{matrix} \right\} \rightarrow \mathrm{Ar}^{++}(M^{-2}) + e^-(s) + e^-(f). \tag{42}$$

The scheme (42) is based partly on the same argument as that of PCI-influenced autoionization, namely, (i) the lifetime of the Ar ion is long enough to allow an intermediate state of the system $\mathrm{Ar}^+ - e^-(s)$ to be defined; (ii) the velocity of the slow electron is low enough, and its distances from the decaying ion large enough, to allow definition of Born–Oppenheimer potential curves for its motion. The special feature of two crossing points, however, arises if the angular-momentum terms in the approximate one-dimensional potential curves are included. The potential curves then are

$$V_\rho(R) = \varepsilon_0 - i\Gamma/2 - 1/R + l_\rho(l_\rho + 1)/2R^2,$$
$$V_f(R) = \varepsilon - 2/R + l_f(l_f + 1)/2R^2. \tag{43}$$

With these potentials we have two real crossing points, R_1 and R_2, if l_f is

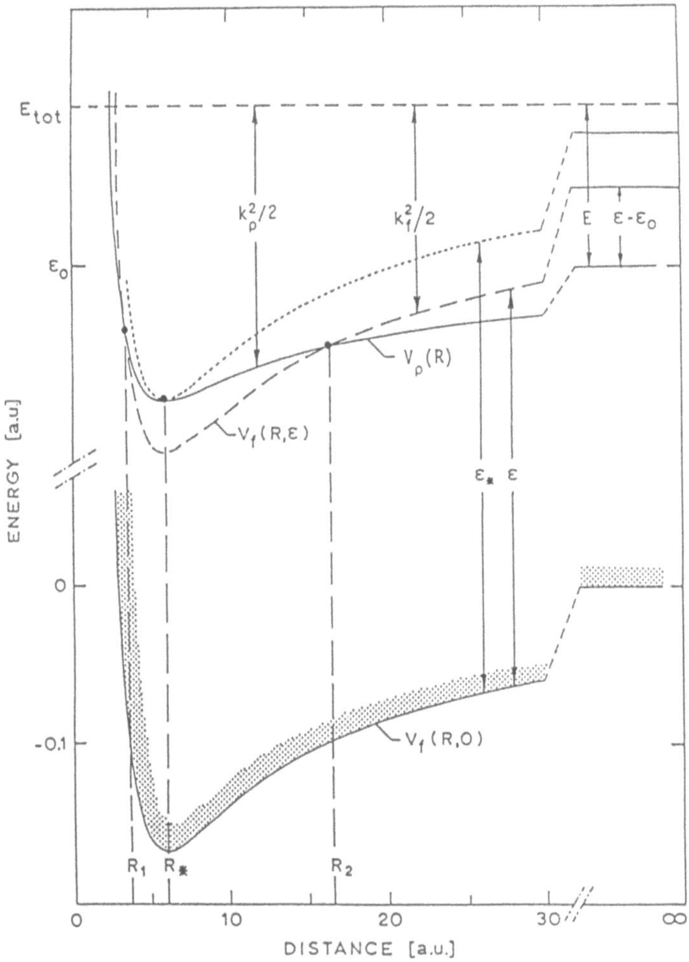

FIGURE 11. Effective potentials for the slow photoelectron after inner-shell photoionization of an atom, as a function of distance from the atom. The potential before Auger decay is $V_\rho(R)$; $V_f(R)$ is the potential after the Auger decay in which an electron of energy ε has been ejected. The two crossing points of $V_\rho(R)$ with $V_f(R)$ are indicated, also the "rainbow" energy ε_* for which the two crossing points coalesce at R_*.

larger than l_p and if the ejected-electron energy ε is in the range $\varepsilon_0 < \varepsilon < \varepsilon_*$, with ε_* defined as the energy for which the two crossing points coalesce in R_*. The situation is illustrated in Fig. 11 for the case $l_p = 2$; $l_f = 3$. For $\varepsilon > \varepsilon_*$, there is no real crossing. The position of the rainbow is defined by ε_*, which in the present case has the dark side at high electron energies, in contrast to example (41) (see Fig. 10). The rainbow position can be calculated from the relations

$$d[V_p(R) - V_f(R)]/dR\,|_{R=R_*} = 0, \qquad V_p(R_*) - V_f(R_*) = 0, \tag{44}$$

which lead to

$$R_* = l_f(l_f + 1) - l_p(l_p + 1), \tag{45}$$

$$\varepsilon_* = \varepsilon_0 + [2(l_f + 1)\,l_f - 2(l_p + 1)\,l_p]^{-1}. \tag{46}$$

When we consider potentials of the form (43) with $l_p \neq l_f$, we assume angular-momentum exchange to occur between the slow photoelectron and the fast Auger electron. Whether this assumption is realistic is not known as yet from theoretical calculations. Recently observed structures in the PCI-influenced line shapes due to Auger decay of $Ar(L^{-1})$ following near-threshold electron impact ionization (Huster and Mehlhorn, 1982; Hedman *et al.*, 1982) have, however, been convincingly interpreted as rainbows of this type (Niehaus and Zwakhals, 1983), so that we may infer that angular-momentum exchange does indeed occur. An expression for the transition probability I_{if} derived under the assumption of angular-momentum exchange can thus be used to obtain semiempirical information on the process. We will derive such an expression for the case of defined l_p and l_f. Modifications necessary to apply the expression to a real process with, in general, more than one value of l_p and many values of l_f, will be discussed thereafter.

5.2. *Theory*

Following exactly the same procedure as in the preceding sections we arrive at the general expression for the transition probability:

$$I_{if} = \left| \left(\frac{\Gamma}{2\pi}\right)^{1/2} f(\omega) \int_0^\infty |K_p K_f|^{-1/2} \exp i \left\{ \int_{R_{T_p}}^R K_p\, dR' - \int_{R_{T_f}}^R K_f\, dR' \right\} dR \right|^2 \tag{47}$$

The integral over distances is the overlap integral $\langle K_f \mid K_p \rangle$, which also appeared in our other derivations with WKB wave functions for the motion of the slow electron in intermediate and final state. The R-dependent wave numbers follow directly from the potentials (43) and from energy conservation (Fig. 11):

$$K_\rho=\left\{2\left[E+\frac{i\Gamma}{2}-\frac{l_\rho(l_\rho+1)}{2R^2}+\frac{1}{R}\right]\right\}^{1/2},$$
$$K_f=\left[2\left(E-(\varepsilon-\varepsilon_0)-\frac{l_f(l_f+1)}{2R^2}+\frac{2}{R}\right)\right]^{1/2}. \quad (48)$$

In Eq. (47), R_{T_ρ} and R_{T_f} are the turning points on the potentials V_ρ and V_f. In the preceding sections we separated real and imaginary parts of K_ρ and evaluated the integral by the stationary-phase method, with the real points of stationary phase given by the real crossings of the potential curves. This procedure relies on the approximation $E \gg \Gamma$, and has the disadvantage that the resulting expression does not have the correct behavior for large E. For $E=\infty$ it yields a δ function instead of the correct Lorentzian function of width Γ. We found (Niehaus and Zwakhals, 1983) that the procedure can be improved by allowing K_ρ, K_f, and R to have complex values. K_ρ and K_f then need not be approximated, and the complex points of stationary phase R_1 and R_2 are simply obtained from the relation

$$K_\rho(R)=K_f(R). \quad (49)$$

The integral becomes the sum of the two contributions from the points of stationary phase:

$$I_{if}=|\Gamma^{1/2}f(\omega)[e^{i\phi_1}K_1^{-1}\phi_1''^{-1/2}+e^{i\phi_2}K_2^{-1}(-\phi_2'')^{-1/2}]|^2, \quad (50)$$

with

$$\phi_1=\int_{R_{T_\rho}}^{R_1}K_\rho\,dR'-\int_{R_{T_f}}^{R_1}K_f\,dR'; \qquad \phi_2=\int_{R_{T_\rho}}^{R_2}K_\rho\,dR'-\int_{R_{T_f}}^{R_2}K_f\,dR'; \quad (51)$$

$$K_1=K_\rho(R_1)=K_f(R_1), \qquad K_2=K_\rho(R_2)=K_f(R_2); \quad (52)$$

$$\phi_1''=\left.\frac{d^2\phi}{dR^2}\right|_{R=R_1}, \qquad \phi_2''=\left.\frac{d^2\phi}{dR^2}\right|_{R=R_2} \quad (53)$$

The turning points R_{T_ρ}, R_{T_f} are obtained as the smallest values for which $K_\rho=0$ and $K_f=0$, respectively. The operations (51) and (53) are easily carried out for the case of the wave numbers (48). The resulting expressions are somewhat lengthy and are not given here. They are reported in Niehaus and Zwakhals (1983). In Fig. 12 we show calculations of a PCI-broadened line for the case $l_\rho=3$, $l_f=4$, and for various excess energies E. The line

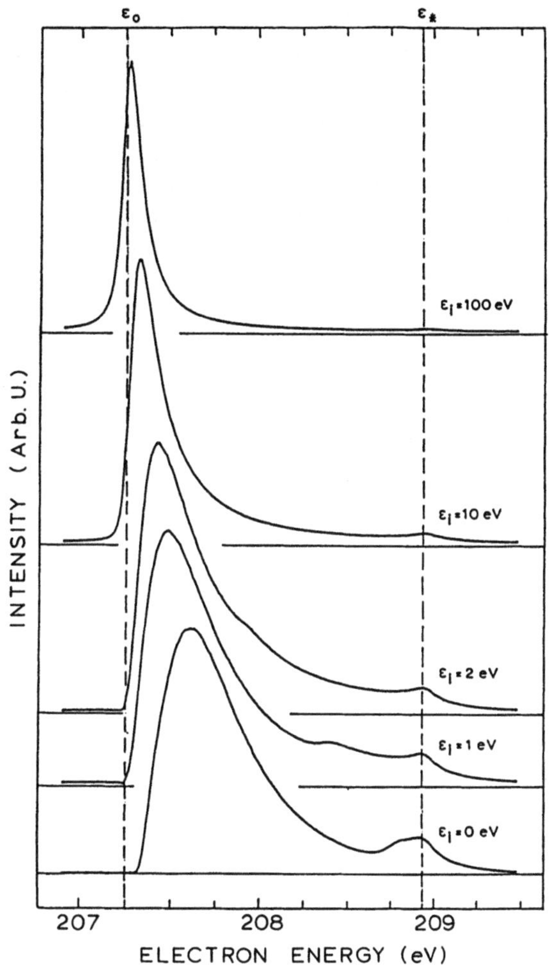

FIGURE 12. Calculated PCI-broadened Auger line shapes for $l_p = 3$ and $l_f = 4$, and for various excess energies E. The line position $\varepsilon_0 = 207.25$ eV and the lifetime $\tau = \Gamma^{-1} = 227$ a.u. correspond to the Ar L_2–$M_{23}M_{23}(^3P_{0,1,2})$ transition. Also indicated is the rainbow position ε_*.

position $\varepsilon_0 = 207.25$ eV, and the width $\Gamma = 0.12$ eV correspond to the Ar L_2–$M_{23}M_{23}(^3P_{0,1,2})$ transition (Hedman *et al.*, 1982). The rainbow caused by the maximum of the difference potential $(V_p - V_f)$ at the energy ε_* given by Eq. (46) is clearly seen, especially at low excess energies, where the velocity of the slow electron is low enough so that a significant percentage of the systems formed in the intermediate state decay in the neighborhood of R_*. The modulation caused by constructive and destructive interference of the contributions from the two crossing points $R_1 < R_*$

and $R_2 > R_*$ is not very strong. The second rainbow is best seen at $E = 1$ eV and lies about 0.5 eV below the first one. The reason for the rather weak modulation is the fact that, with increasing energy separation from ε_*, the ratio of the amplitudes from R_1 and R_2 decreases rapidly. The reason for this strong decrease can immediately be recognized by inspection of the potential curves: the ratio of the slopes of the difference potential at R_1 and R_2 increases rapidly as the crossing points separate from R_*.

The effect of the maximum of the difference potential on the shape and the position of the main Auger peak is obviously very weak. We have found in our calculations for $l_p = 0$, 1, 2, 3 and $\Delta l = l_p - l_f = 0$, 1, that the peak position ε_p is to a very good approximation independent of l_p and Δl, for excess energies varying from zero to infinity, and is still given by the relation obtained in a derivation in which the angular-momentum terms in the potentials are omitted, namely, by (Niehaus, 1977)

$$[2(\xi_p + \delta)]^{1/2} - \xi_p(5\xi_p + 4\delta) = 0, \tag{54}$$

with the scaled quantities $\xi_p = (\varepsilon_p - \varepsilon_0)/\Gamma^{2/3}$ and $\delta = E/\Gamma^{2/3}$. The very weak dependence of the shape of the main Auger peak on the angular momentum of the slow electron, and on the angular-momentum exchange with the Auger electron, explains why experimental data proved to be very well explainable in terms of the line-shape formula obtained by omitting the angular momentum.

5.3. *Comparison with Experimental Data for Inner-Shell Electron-Impact Ionization of Ar L Followed by Auger Decay*

In Auger spectra of Ar L vacancy states excited by electron impact recently "shoulders" of the PCI-broadened lines were observed (Huster and Mehlhorn, 1982; Hedman *et al.*, 1982). As an example we show in Fig. 13 the distribution belonging to the Ar L_2–$M_{23}M_{23}(^3P_{0,1,2})$ transitions following electron-impact ionization at an excess energy of 30 eV. The distribution is reproduced from a figure published by Hedman *et al.* (1982). It has been shown that the PCI influence of the two electrons, receding from the ion in the case of electron-impact ionization, can be very well approximated by the influence of one electron (Huster and Mehlhorn, 1982), i.e., the PCI-broadened lines in the case of electron impact can be described by the PCI line shapes derived for photoionization, only an equivalent excess energy of the photoelectron has to be used. From a comparison of the experimental main Auger line belonging to the Ar L_3–$M_{23}M_{23}(^1S_0)$ transition following electron impact at an excess energy of 30 eV, with the theoretical Auger line for photon impact at various excess energies, Hedman *et al.* (1982) found that an excess energy of 3 eV of the photoelectron should be included in the

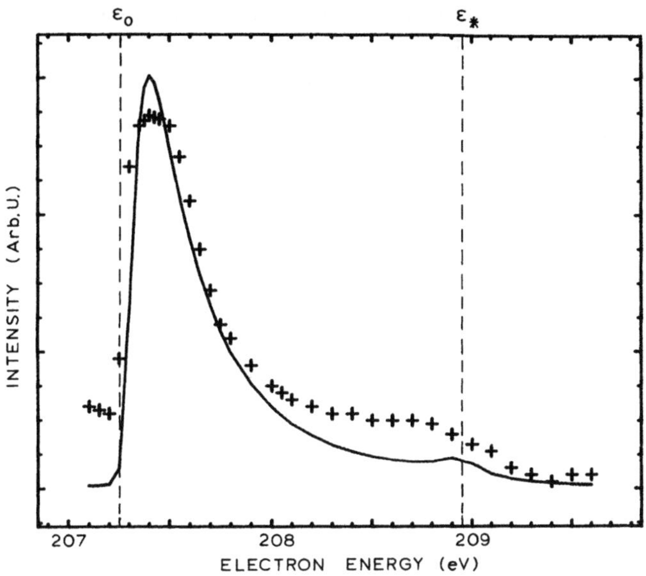

FIGURE 13. Ar L_2–$M_{23}M_{23}(^3P_{0,1,2})$ electron distribution for L_2 ionization by electrons of 30 eV excess energy. Experimental points are reproduced from a paper by Hedman *et al.* (1982). The solid line represents a calculation using Eq. (54) with $E=3$ eV, $l_\rho=3$, $l_f=4$.

calculations to simulate the electron-impact data at 30 eV excess energy. The Ar L_3–$M_{23}M_{23}(^1S_0)$ line does not have a shoulder and therefore could be described by the theory which does not take into account angular-momentum exchange. In order to find out whether the observed shoulder in the Ar L_2–$M_{23}M_{23}(^3P)$ distribution shown in Fig. 13 can be identified as a rainbow caused by angular-momentum exchange, a comparison should be made with a calculation, using our Eq. (50), with the known values of ε_0 and Γ for the Ar L_2–$M_{23}M_{23}(^3P)$ transition, at an excess energy of $E=3$ eV; the values of l_ρ and Δl should be considered as free parameters, although the energy of the observed shoulder suggest that the main contribution comes from a transition with $l_\rho=3$ and $\Delta l=1$ [Eq. (46)]. The result of the calculation of the line shape with these values yields the solid curve in Fig. 13. There is good qualitative agreement. While it is quite natural that $\Delta l=1$ should be the most probably case, except for $\Delta l=0$, which does not give a rainbow, it may look surprising that the angular momentum of the slow electron should be $l_\rho=3$ and not, for instance, $l_\rho=0$, 1, or 2. In fact, this conclusion may not follow from the comparison. For $\Delta l=1$ and $l_\rho=0$, 1, 2, the energy separation of the rainbow from the nominal position is $(\varepsilon_*-\varepsilon_0)=6.8$, 3.4, and 2.27 eV, respectively [cf. Eq. (46)]; in other words, the rainbows for these cases are at energies where there is almost no intensity left over of the PCI-broadened peak. The corresponding weak features therefore cannot be observed. The rainbow position for $\Delta l = 1$ and $l_\rho = 4$ lies at $\varepsilon_* = \varepsilon_0 + 1.36$ eV, and a contribution to the observed distribution is possible regarding the deviation between the experimental data and the solid line in Fig. 13.

This discussion was mainly intended to explain the physical origin of the rainbow interference feature and to show that its analysis can, in principle, yield information on the angular-momentum exchange by PCI. An analysis requires, however, a modification of Eq. (50). In photoionization, and even more so in electron-impact ionization, the slow electron is in a superposition of several angular-momentum states l_ρ. Over the contributions of these different intermediate states we must sum coherently. Furthermore, if the final state of the slow electron lies in the continuum, only the wave vector $\mathbf{K}_f$ is defined, but not the angular momentum l_f. The wave in the direction $\mathbf{K}_f$ can then be expanded in partial waves with angular momenta l_f. The contributions to states of different l_f must be summed coherently. Contributions from intermediate states of different energy (n) can overlap—as in the Li^+–He case treated in Sec. 3—so that we must also sum over these contributions coherently. We now define an amplitude $a^n_{l_\rho l_f}$ which contains the matrix element for excitation of a certain intermediate state (l_ρ, n), and the matrix element for decay into a certain final state with the slow electron in the partial wave of angular momentum l_f. With these amplitudes the transition probability is

$$I_{if} \propto \left| \sum_n \Gamma_n^{1/2} \sum_{l_\rho l_f} a^n_{l_\rho l_f} f^n_{l_\rho l_f}(\omega)\, F(n, l_\rho, l_f) \right|^2, \tag{55}$$

where $F(n, l_\rho, l_f)$ is the expression in brackets in Eq. (50), and $f^n_{l_\rho l_f}(\omega)$ indicates the angular dependence of the emission amplitude of the Auger electron. Equation (55) corresponds to a coincidence measurement of the slow electron in a direction $\hat{K}_f$ and of the Auger electron in a direction $(\omega) = (\theta, \phi)$, where ϕ is the angle between the detection planes of the two electrons. Such measurements have not been performed as yet. The noncoincident spectra which have been measured correspond to the integral of Eq. (55) over the directions $\hat{K}_f$. Finer details, especially of the angular structure predicted by Eq. (55), will be lost due to this integration. But also, it may be speculated that the quantum beat structure in the energy spectrum predicted by (55)—and, indeed, also predicted by the earlier formulation (Niehaus, 1977)—is probably completely masked in the noncoincident spectra. For the analysis of one line, corresponding to one state (n), which has been measured in a noncoincident experiment, it is probably a good approximation to consider the line as an incoherent sum of individual spectra belonging to definite l_ρ and l_f.

6. INTERFERENCE OF CONTRIBUTIONS FROM DIFFERENT ATOMS

6.1. *Introduction*

We again consider the collision of He^+ with He in the keV region. After such a collision the neutral He may be doubly excited and consequently may autoionize. Because of the possibility of electron exchange in the collision, the autoionizing atom can be either the slowly recoiling target particle or the fast scattered projectile. Due to the different velocities of these two particles, the energy of an autoionization electron corresponding to a certain transition will suffer different Doppler shifts. An electron spectrum corresponding to one transition of energy ε_0 in the emitter frame will therefore consist of two autoionization lines centered around the energies $(\varepsilon_0 + D_T)$ and $(\varepsilon_0 + D_P)$, respectively, with D_T and D_P the Doppler shifts corresponding to emission of the electron by the target and by the projectile. Since the Doppler shifts change with the electron detection angles (θ, ϕ), and since the magnitude of D_P is large compared with the width of the autoionization lines of doubly excited He, the corresponding target and projectile lines are well separated over most of the total (θ, ϕ) range. For this reason it was possible to determine the angular electron distribution for the target without interference from the projectile contribution, as described in Sec. 2. There are, however, also angular regions (θ, ϕ) for which the two lines overlap. In particular, for each azimuth ϕ there is a critical angle θ_c for which the Doppler shifts are identical. At an electron energy for which the two lines overlap, the experiment can in principle not distinguish between the two possibilities, namely, that the electron comes from the target or from the projectile. We thus have again a two-slit situation, which can be indicated as

$$He^+ + He \rightarrow \begin{Bmatrix} He^+ + He^{**} \\ He^{**} + He^+ \end{Bmatrix} \rightarrow He^+ + He + e^-(\varepsilon), \tag{56}$$

and which makes it possible that interference arises between the contributions to the measured electron intensity from target and projectile.

6.2. *Theory*

For convenience we assume that only one doubly excited state of He plays a role. The individual line, belonging to decay of the target or of the projectile, is broadened and shifted by PCI of the decaying particle with the receding He^+ ion, and in addition, it is shifted and deformed by the Doppler effect. The transition probabilities I_{if} for this situation have already been derived in Sec. 3 [Eq. (26)]. For the present problem of two overlapping contributions, we only need to modify this result appropriately. Especially, it is necessary to find out in which way the contributions to a well-defined final state of the system must be combined. The key in this connection is that the He^+–He collision system must be described by a superposition of "gerade" (g) and "ungerade" (u) substates, where (g) denotes an electronic state that is invariant under reflection of all electron coordinates at the center of the homonuclear collision system, and (u) denotes a state that changes sign under this symmetry operation. In the initial state of the collision system $He^+ + He$, the amplitudes of the (g) and (u) substates are equal. The population of the two substates in the intermediate state of the system $He^+ + He^{**}$ is determined by the different electronic interactions during the collision.

In the present case of collisions in the low-energy regime in which the interactions can be expressed in terms of potential curves and by coupling between them, we can qualitatively predict which interaction—the one in the (g) channel or the one in the (u) channel—will be mainly responsible for double excitation of He. This we do by considering the one-electron correlation diagram for He^+–He, which correlates the separated-atom states with the united-atom state (Lichten, 1967). The (g) substate in the initial state of the system $He^+(1s) + He(1s^2)$ is given by the product $1s\sigma_g 2p\sigma_u^2(^2\Sigma_g)$ of one-electron molecular orbitals of defined symmetry. This $^2\Sigma_g$ substate correlates to the united-atom state $1s2p^2$. The (u) substate is given by $1s\sigma_g^2 2p\sigma_u(^2\Sigma_u)$ and correlates to $1s^2 2p$. Without going into details of the excitation mechanism it is obvious that double excitation is very improbable via the (u) substate, because only one electron is promoted, while excitation via the (g) substates seems probable because the two promoted electrons can become excited electrons after separation of the collision partners. Based on these arguments one can predict that predominantly the (g) substate is populated in the intermediate state of the collision system. This prediction is supported by experimental data, namely, by the absence of charge-exchange oscillations in the differential cross section for double excitation in $He^+ + He$ collisions (Barat *et al.*, 1972). We therefore can assume that the intermediate state of our collision system has pure (g) symmetry. In such a state the amplitudes belonging to the two situations—excited target or excited projectile, respectively—have equal absolute value, and consequently, also the contributions to a well-defined final state of given free-electron wave number have equal absolute values for target and projectile in the frame moving with the center of the homonuclear system. It is important to account for the fact that the molecular parity is a good quantum number also for the final state. The (g) symmetry can be realized in two ways: (i) the electronic wave function of the $He^+ + He^+$ subsystem is "gerade" and the free-electron wave function is also "gerade," or (ii) both these wave functions are "ungerade" leading to a (g) state of the whole system. As is well known, in a homonuclear system with two electrons—like $He^+ + He^+$—the molecular parity is connected with the total spin by

$$\begin{aligned} &\text{"Gerade" } (g) \leftrightarrow \text{singlet} \\ &\text{"Ungerade" } (u) \leftrightarrow \text{triplet.} \end{aligned} \tag{57}$$

Since there is (essentially) no interaction at the instant of decay from the intermediate to the final state, the population of the (g) and of the (u) state of the subsystem He^+–He^+ is given by the statistical weights, namely, 3:1 in favor of the "ungerade" state. Summarizing, we may immediately write down an expression for the transition probability pertaining to reaction (56) for the case that spin states are not distinguished:

$$\bar{I}_{if} = \tfrac{1}{4}|A_P + A_T|^2 + \tfrac{3}{4}|A_P - A_T|^2. \tag{58}$$

Here, A_P and A_T are the decay amplitudes belonging to projectile and target. Except for a relative phase factor these expressions are identical—in principle—with the appropriately modified decay amplitude given by Eq. (26). The first term in Eq. (58) belongs to the "gerade" free-electron wave function, and the second term belongs to the "ungerade" free-electron wave function. The signs used to combine projectile and target amplitudes are determined by the symmetry—(g) or (u)—of the respective free-electron wave functions, and by the condition that in the center-of-mass system A_P transforms to A_T under the molecular parity operation.

When Eq. (58) is used with the expressions A_P and A_T as resulting from

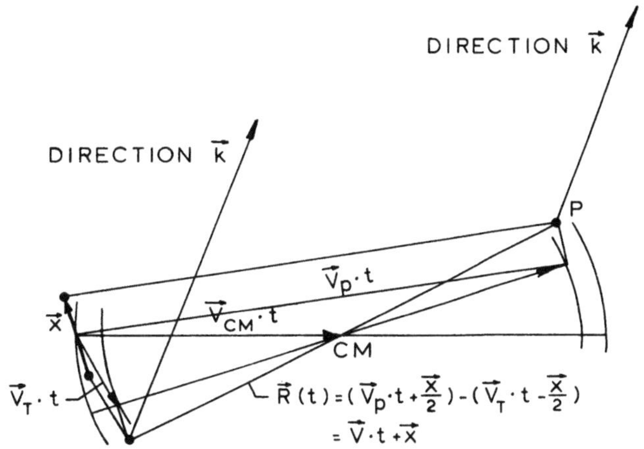

FIGURE 14. Schematic drawing showing the relation between the vector **R** at some time t, for given velocity vectors $\mathbf{V}_P$ and $\mathbf{V}_T$.

Eq. (26) by using the appropriate velocities $\mathbf{V}_P$ and $\mathbf{V}_T$ to calculate the Doppler shifts, and by replacing V_0 by $V = |\mathbf{V}_P - \mathbf{V}_T|$, one arrives at an approximation which does not yet account for the fact that the distance vector $\mathbf{R}$ between target and projectile is to a certain extent also determined by the vector of closest approach $\mathbf{X}$ belonging to the observed event. The influence of $\mathbf{X}$ on $\mathbf{R}$ at a certain time after the collision is demonstrated in Fig. 14.

The approximation mentioned above corresponds to $|\mathbf{X}| = 0$. If the main influence of the finite size of $\mathbf{X}$ is taken into account one obtains the following expressions for A_P and A_T to be used in Eq. (58) (Boskamp *et al.*, 1982b):

$$A_P = 2^{-1/2} \left(\frac{\Gamma}{V}\right)^{1/2} f(\omega_P) \left|\frac{d\omega_P}{d\omega}\right|^{1/2} \exp\left[-\left(\frac{\Gamma}{2}\right)\frac{R_P - R_c}{V}\right] \tag{59a}$$

$$\times\ (\varepsilon_0 - \varepsilon + D_P)^{-1} \left|\frac{d(\varepsilon - D_P)}{d\varepsilon}\right|^{1/2} \exp i[V^{-1} \ln R_P - \tfrac{1}{2}(\mathbf{k} - \mathbf{V}_P)\cdot\mathbf{X}],$$

$$A_T = 2^{-1/2} (-)^{\Sigma l_i} \left(\frac{\Gamma}{V}\right)^{1/2} f(\omega_T) \left|\frac{d\omega_T}{d\omega}\right|^{1/2} \exp\left[-\left(\frac{\Gamma}{2}\right)\frac{R_T - R_c}{V}\right]$$

$$\times\ (\varepsilon_0 - \varepsilon + D_T) \left|\frac{d(\varepsilon - D_T)}{d\varepsilon}\right|^{1/2} \exp i[V^{-1} \ln R_T + \tfrac{1}{2}(\mathbf{k} - \mathbf{V}_T)\cdot\mathbf{X}]. \tag{59b}$$

Here, we have $R_P = (\varepsilon_0 - \varepsilon + D_P)^{-1}$, $R_T = (\varepsilon_0 - \varepsilon + D_T)^{-1}$; D_P and D_T are the Doppler shifts for velocities $\mathbf{V}_P$ and $\mathbf{V}_T$, respectively [Eq. (7)], and $V = |\mathbf{V}_P - \mathbf{V}_T|$. The internal emission angles ω_P and ω_T pertain to the laboratory detection angles ω [Eq. (12)]. The symbol $(-)^{\Sigma l_i}$ accounts for the atomic parity of the excited He** and guarantees the correct behavior of the amplitudes under the molecular parity operation. Equations (58) and (59) can be generalized to more than one intermediate state by simply replacing A_P and A_T by the corresponding sums $\sum_n A_P^n$ and $\sum_n A_T^n$, respectively.

6.3. *Comparison with Experimental Data*

As already discussed in Sec. 2 in He^+–He collisions in the keV region mainly the autoionizing state $He(2p^2)\,{}^1D$ is excited. For this state the angular distribution was determined from coincidence measurements in an angular range of electron detection where the PCI-influenced, Doppler-shifted lines belonging to target and projectile do not overlap. From these measurements the functions $f(\omega_P)$ and $f(\omega_T)$ appearing in Eq. (59) can be derived. For a calculation of I_{if} in the overlap region via Eqs. (59) and (58) there remains, therefore, only one unknown quantity, viz., the distance of closest approach X; all other quantities are determined by the chosen experimental conditions. Even X can be estimated reasonably well for a defined scattering event from the approximately known scattering potential. Using the He^+–He potential given by Marchi and Smith (1965) for the incoming part of the trajectory and a shielded Coulomb potential $V = 2.2 + 4 \times [\exp(-R/0.055)]/R$ for the outgoing trajectory (Boskamp *et al.*, 1982), one obtains $X = 0.39$ a.u. for a laboratory scattering angle $\Theta_{sc} = 6°$ at a laboratory collision energy of 2 keV. These are the conditions of the experiment we want to use for comparison. The expressions (59) and (58) allow one to calculate an electron spectrum at any electron detection angle $\omega = (\theta, \phi)$. Such a spectrum contains the target line and the projectile line with possible interference structure in the overlap region. To demonstrate the situation we show, in Fig. 15, intensity-normalized spectra calculated for the case of random phase relation between target and projectile contributions. Spectra are shown for various angles θ, with the detector in the collision plane ($\phi = 0°$). One notices that the PCI-broadened lines overlap considerably in a certain angular range around the critical angle at which the lines coincide exactly. This angle is determined by the condition $D_P = D_T$, and for the mentioned experimental conditions turns out to be $\theta_c = 99.8°$ ($\phi = 0°$). For $\phi = 180°$ one obtains $\theta_c = 75.5°$. Inspection of Eq. (59) shows that for $D_P = D_T$ the relative phase between A_P and A_T is small and practically constant across the whole line, given by $(\mathbf{k} - \mathbf{V}) \cdot \mathbf{X}$. Consequently, the intensity belonging to the "ungerade" free-electron state, as calculated from Eq. (58), almost vanishes at θ_c, while the "gerade" intensity is approximately twice as large as in regions where there is no overlap.

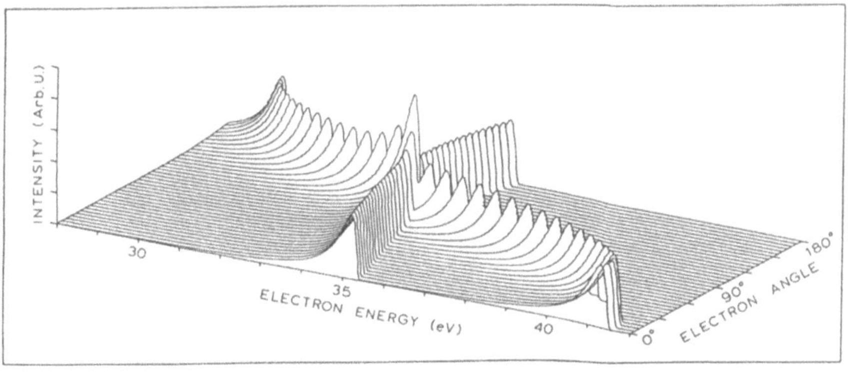

FIGURE 15. Doppler shift of PCI-broadened target and projectile lines from autoionization of $He(2p^2)\ ^1D$ excited in 2-keV He^+–He collisions in which the projectile is scattered through $\Theta = 6°$. The spectra were calculated for various electron observation angles θ in the scattering plane. The individual lines are normalized and incoherently superimposed in the overlap regions.

For the total intensity at θ_c, integrated over electron energies, we find approximately

$$\int \bar{I}_{if}(\theta_c)\, d\varepsilon \approx \tfrac{1}{4}|A_P + A_P|^2 = A_P^2, \tag{60}$$

and in region of no overlap $(\theta \neq \theta_c)$ we have

$$\int \bar{I}_{if}(\theta \neq \theta_c)\, d\varepsilon \approx \tfrac{1}{4}(2A_P^2) + \tfrac{3}{4}(2A_P^2) = 2A_P^2. \tag{61}$$

It is therefore to be expected that the coherence of target and projectile contribution causes a pronounced dip in the angular distribution of the total electron intensity at θ_c, integrated over energy.

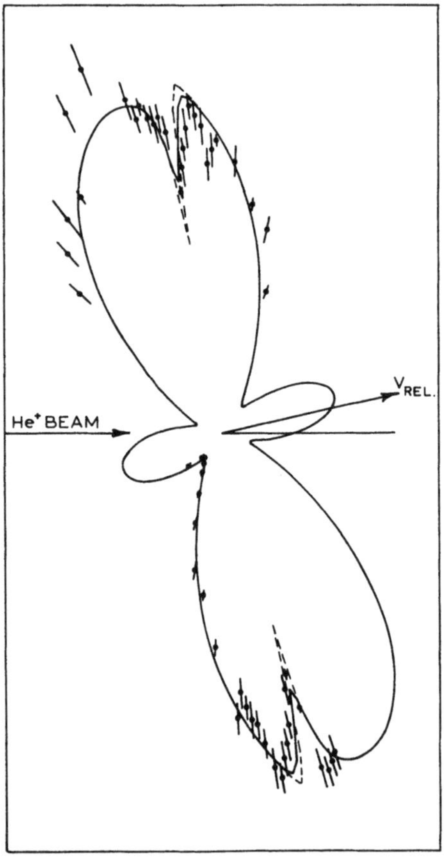

FIGURE 16. Experimental and theoretical angular electron distribution in the scattering plane for He^+–He collision systems at 2 keV and 6° projectile scattering angle. The interference dip is clearly seen around the critical angle where projectile and target contributions overlap completely.

Measurements of such angular distributions have been carried out for several collision energies, several scattering angles Θ, and several angles ϕ (Boskamp *et al.*, 1982, 1984). The predicted interference structure around θ_c was indeed observed. An example for 2-keV collision energy, $\Theta = 6°$, and $\phi = 0°$ and 180° is shown in Fig. 16. The pronounced dips at θ_c are clearly seen in the experimental data. The solid curve is calculated from Eqs. (58) and (59) with the known internal angular distribution of the $(2p^2)\ ^1D$ state and with an estimated value of $X = 0.39$ a.u. To account for the experimental angular resolution of approximately 1°, the theoretical result was smoothed with a Gaussian. The dashed line shows the unsmoothed result. We notice agreement between theory and experiment within experimental error. The agreement in the overlap regions around θ_c proves the physical interpretation of the interference dip as being due to interference of contributions from target and projectile.

On the other hand, the full coherence of the amplitudes belonging to decay of the noninteracting particles is a consequence of the quantum-mechanical prescription to assign a definite molecular parity to the collisionally excited system, and therefore may not appear surprising. On the other hand, however, it is worth mentioning that this coherence implies that, although the particles do not interact anymore, the excitation caused by the collision event is not localized on one of the particles, and the ejection of the electron cannot be ascribed to one of the particles. This is in clear contradiction to any classical interpretation, and to an interpretation based on local hidden-variable theories. (For a discussion of related problems, see, e.g., Pipkin, 1978.)

REFERENCES

BARAT, M., D. DHUICQ, R. FRANÇOIS, R. MCCARROLL, D. PIACENTINI, AND A. SALIN, 1972, *J. Phys. B* **5**, 1343.

BOSKAMP, E., O. GRIEBLING, R. MORGENSTERN, AND G. NIENHUIS, 1982a, *J. Phys. B* **15**, 3745.

BOSKAMP, E., R. MORGENSTERN, AND A. NIEHAUS, 1982b, *J. Phys. B* **15**, 4577.

BOSKAMP, E., R. MORGENSTERN, P. V. D. STRATEN, AND A. NIEHAUS, 1984, *J. Phys. B* **17**, 2823.

CHILD, M. S., 1976, Semiclassical Methods in Molecular Collisions, in *Dynamics of Molecular Collisions, Part B*, edited by H. Miller (Plenum Press, New York), p. 171.

EICHLER, J., AND W. FRITSCH, 1976, *J. Phys. B* **9**, 1477.

HEDMAN, S., K. HELENELUND, L. ASPLUND, K. GELIUS, AND K. SIEGBAHN, 1982, *J. Phys. B* **15**, L799.

HEIDMAN, H. G. M., G. NIENHUIS, AND T. VAN ITTERSUM, 1974, *J. Phys. B* **7**, L493.

HICKS, P. J., S. CVEJANOVIC, J. COMER, R. H. READ, AND J. M. SHARP, 1974, *Vacuum* **24**, 573.

HULTZSCH, W., W. KRONAST, A. NIEHAUS, AND M. W. RUF, 1979, *J. Phys. B* **12**, 1821.

HUSTER, R., AND W. MEHLHORN, 1982, *Z. Phys. A* **307**, 67.

KESSEL, Q. C., R. MORGENSTERN, B. MÜLLER, AND A. NIEHAUS, 1979, *Phys. Rev. A* **20**, 804.

KING, G. C., F. H. READ, AND R. C. BRADFORD, 1975, *J. Phys. B* **8**, 2210.

MARCHI, R. P., AND F. T. SMITH, 1965, *Phys. Rev.* **139A**, 1025.
NIEHAUS, A., 1977, *J. Phys. B* **10**, 1845.
NIEHAUS, A., 1978, Post-Collision Interaction in Atomic Excitation Processes, in *Electronic and Atomic Collisions*, edited by G. Watel (North-Holland, Amsterdam), p. 185.
NIEHAUS, A., AND C. J. ZWAKHALS, 1983, *J. Phys. B* **16**, L135.
PIPKIN, F. M., 1978, Atomic Physics Tests of the Basic Concepts in Quantum Mechanics, in *Advances in Atomic and Molecular Physics*, Vol. 14, edited by D. R. Bates and B. Bederson (Academic, New York), p. 281.
SCHULZ, G. J., 1973, *Rev. Mod. Phys.* **45**, 378.
ZWAKHALS, C. J., R. MORGENSTERN, AND A. NIEHAUS, 1982, *Z. Phys. A* **307**, 41.

PART B

Scattering and Collision-Induced Processes

CHAPTER 10

Inelastic X-Ray Scattering Including Resonance Phenomena

T. Åberg and J. Tulkki

1. INTRODUCTION

The development of intense tunable synchrotron radiation sources will undoubtedly stimulate the study of inelastic X-ray scattering by bound electrons in much the same way as the progress in laser technology has revived Raman scattering investigations in the optical region. Using conventional X-ray tubes the measurement of inelastically scattered photons has been tedious under off-resonance conditions. The fundamental reason is the low electronic cross section, characterized by the square of the classical electron radius r_0 ($=2.82\times10^{-15}$ m). Tunability changes the situation drastically since under resonance conditions differential cross sections can typically increase 10^6 fold. Wiggler and undulator magnets in dedicated storage rings should make it feasible to study off-resonance scattering both in the X-ray (Eisenberger, 1980) and the ultraviolet region.

With this in mind we shall address ourselves to resonance features of inelastic X-ray cross sections and particularly to their behavior *below* and *at* the K absorption edge of the target atoms. This excludes the impulse-approximation treatment of the Compton effect with all its consequences for determining electron momentum distributions in molecules and solids using the inelastic scattering of hard X rays or γ rays. These aspects of inelastic photon scattering have recently been reviewed by Williams and others (1977) and by Bushuev and Kuz'min (1977). The latter authors also discuss

T. Åberg and J. Tulkki • Laboratory of Physics, Helsinki University of Technology, 02150 Espoo 15, Finland.

Raman, plasmon, and parametric scattering in crystals including an ample list of references. In our work we will, following some recent developments (Tulkki and Åberg, 1980a, b, 1982; Tulkki, 1983), emphasize the general theory including resonance phenomena and the angular distribution and polarization of the inelastically scattered radiation. Hence, we shall explicitly demonstrate the evolution of the inelastic scattering into fluorescence across an X-ray absorption edge. Our treatment generalizes and illuminates a number of analytical results (Gavrila, 1972a, b; Costescu and Gavrila, 1973; Gavrila and Tugulea, 1975) concerning the inelastic scattering of photons accompanied by the ionization of K and L electrons in hydrogenlike ions.

We shall limit our discussion to X rays with energies in the keV range which are typical for synchrotron radiation experiments. Relativistic effects, although important in many cases, are only briefly considered. Radiative corrections to the inelastic X-ray scattering cross section are likely to be small in this region as judged from recent work on these effects in photoelectron and bremsstrahlung spectra (McEnnan and Gavrila, 1977a, b; Botto and Gavrila, 1982). Hence they are not discussed except for the infrared divergence. This phenomenon has recently been related to off-resonance features in inelastic X-ray scattering spectra (Bannett *et al.*, 1976, 1977).

The terminology used in the literature for the inelastically scattered X rays under different experimental conditions is diverse. Especially the difference between Compton and Raman scattering has been defined in various ways (e.g., Bushuev and Kuz'min, 1977). Hence we shall in the following use inelastic X-ray scattering (abbreviated as IXS) to denote *any* scattering in which there is a change in the internal energy of the target *electrons* and refer to the common terminology at appropriate places. Although emphasis will be on IXS by atoms and molecules, most of the results are also relevant to solid-state systems.

2. THEORY

Inelastic X-ray scattering is considered in the weak-photon field limit by attributing the scattering to single collisions between photons and electrons. The IXS amplitude is derived using a time-independent multichannel scattering approach (Åberg, 1980; Åberg and Howat, 1982) to avoid any ambiguities with regard to the lifetimes of the intermediate states. This leads to an amplitude which is valid for an arbitrary energy separation of the intermediate states. In this section, only the nonrelativistic amplitude in the Coulomb gauge is considered. The relativistic amplitude and the choice of gauge are briefly discussed in Sec. 5.

2.1. *Resonant Scattering Involving Photons and Electrons*

A complete analysis of inelastic scattering of photons by bound electrons requires a very general formalism of scattering theory since the photon and electron fields should in principle be treated on the same footing. The final scattering states may consist of several ejected photons and electrons. The initial and final states may also overlap in total energy with several other states in various particle spaces. Since neither the polarization properties of the incoming photon beam nor the target orientation is always fully specified, the density-matrix formalism is required (Tulkki and Åberg, 1980a).

We shall restrict ourselves to final scattering states involving at most two ejected particles. Consequently, inelastic scattering is considered to result in one emitted photon with an electron transferred either into a bound state or into the continuum. In order to describe this process in the lowest-order approximation with respect to the photon–electron interaction, continua with $0 \leqslant n_e \leqslant 2$ and $0 \leqslant n_{\text{ph}} \leqslant 2$ have to be considered. Here n_e is the number of electrons and n_{ph}, the number of photons. The corresponding manifolds of continuum states are denoted by $\{n_e, n_{\text{ph}}\}$. The electron states in an atom are classified as usual by the energy τ of the ejected electron and its quantum numbers, the total angular-momentum quantum numbers (J, M), and by specifying a complete set of quantum numbers for the core. The photon states are classified either by the unit wave vector $\mathbf{k}^0$ and polarization vector $\mathbf{e}$ or by the multipolarity $(\mathcal{J}, M)$ and parity ξ (see Appendix) in addition to their energy $\hbar\omega$.

The photon and electron systems are described by the Hamilton operators H_1 and H_2 respectively, when they do not interact. Consequently H_1 refers to the free-photon field and H_2 to a central-field-type Hamiltonian which describes the full electron–electron interaction $(e^2/4\pi\varepsilon_0)\sum_{\mu>\nu} r_{\mu\nu}^{-1}$ in an average fashion. The composite Hamiltonian $H_0 = H_1 + H_2$ has an eigenfunction spectrum $|\mathcal{N}\rangle = |\psi_1\rangle\,|\psi_2\rangle$ which can be divided into the different manifolds $\{n_{\text{el}}, n_{\text{ph}}\}$. Each free-photon field solution $|\psi_1\rangle = |0,\dots, n_{\mathbf{k}},\dots\rangle$ is characterized by a set of occupation numbers $n_{\mathbf{k}}$ of the various photon states. Asymptotically, a continuum solution $|\psi_2\rangle = |\mathcal{N}_\tau\rangle$ approaches a many-electron standing wave solution when any one of the radial distances approaches infinity whereas the discrete solutions are square integrable. The eigenenergy $E_{\mathcal{N}} + \epsilon$ of the state $|\mathcal{N}\rangle$ can be divided into a discrete energy $E_{\mathcal{N}}$ and a continuous part ϵ which represents both the ejected electron and the photons.

The interaction operator V_{ph} between the photons and electrons will be defined in Sec. 2.3 and discussed in Sec. 5. It is needed for the description of the inelastic scattering process

$$\hbar\omega_1 + A_I \rightarrow \hbar\omega_2 + A_F, \tag{1}$$

where I refers to the initial state of the electronic system and F to the final state. Since we are dealing with scattering of X rays, inner-shell vacancy states that decay both radiatively and nonradiatively are involved. Hence the interacting photon–electron system also requires considerations of the electron–electron interaction operator $V_{el} = H_{el} - H_2$, where H_{el} is the complete Hamiltonian of the many-electron target. The total Hamiltonian $H = H_0 + V = H_0 + V_{ph} + V_{el}$ of the interacting system has scattering solutions $\psi_\epsilon^\pm$ such that

$$(H - E)\,\psi_\epsilon^\pm = 0, \tag{2}$$

where $\epsilon = E - E_{\mathcal{N}}$. For process (1), $E = \hbar\omega_1 + E_I = \hbar\omega_2 + E_F + \epsilon_e$, where ϵ_e represents the kinetic energy of any continuum electrons present in the final state. In order to solve Eq. (2) we must diagonalize the matrix H with respect to the states $|\mathcal{N}\rangle$ under both "outgoing wave" and "ingoing wave" boundary conditions, corresponding in Eq. (2) to the plus and minus sign, respectively. This can be done in two steps (Åberg, 1980; Åberg and Howat, 1982). First, the Lippmann–Schwinger equations

$$|\mathcal{N}_\varepsilon^\pm\rangle = |\mathcal{N}_\varepsilon\rangle + \lim_{\nu \to 0} \sum_{\mathcal{N}'} \int \frac{|\mathcal{N}'_\tau\rangle\langle\mathcal{N}'_\tau|\, V\,|\mathcal{N}_\varepsilon^\pm\rangle\, d\tau}{E_{\mathcal{N}} + \varepsilon - E_{\mathcal{N}'} - \tau \pm i\nu} \tag{3}$$

are solved for *each* set of manifolds $\{n_{el}, n_{ph}\}$. In Eq. (3), $E_{\mathcal{N}}$ refers to the bound-electron system and ε, to the kinetic energy which may be shared by a photon and an electron or may represent either particle. The matrix element that corresponds to transitions from $|\mathcal{N}_1\rangle = |A\rangle$ to $|\mathcal{N}_2\rangle = |B\rangle$, with the interaction within each manifold included according to Eq. (3), is given for $E = \epsilon' + E_F = \epsilon + E_I$ by

$$T_{BA}^- = T_{BA}^+ = \langle B_{\epsilon'}^-|\, V\,|A\rangle = \langle B|\, V\,|A_\epsilon^+\rangle. \tag{4}$$

The solutions (4) are not complete since the interaction between the various particle manifolds $\{n_e, n_{ph}\}$ has been neglected. Hence the second step consists of diagonalizing H with respect to the channel wave functions $|\mathcal{N}_\varepsilon^\pm\rangle$. This step involves nondiagonal matrix elements that reduce to elements which describe interactions between discrete and continuous states. The resulting transition amplitude between a given initial and final particle channel is

$$T_{FI}^+(E) = \langle F_{\epsilon'}^-|\, V\,|I_\epsilon\rangle + \sum_{\mathcal{N}} \frac{\langle F_{\epsilon'}^-|\, V\,|\mathcal{N}\rangle\langle\mathcal{N}|\, V\,|I_\epsilon^+\rangle}{E - \mathscr{E}_{\mathcal{N}}}, \tag{5}$$

where $F_{\epsilon'}^-$ and I_ϵ^+ are the energy-scale normalized solutions of Eq. (3) for

$\epsilon = E - E_I$ and $\epsilon' = E - E_F$, respectively. The quantities $\mathscr{E}_\mathcal{N}$ are complex energy-dependent solutions of the secular equation

$$\det[(E_{\mathcal{N}'} - \mathscr{E})\,\delta_{\mathcal{N}'\mathcal{N}} + \Delta_{\mathcal{N}'\mathcal{N}} - \tfrac{1}{2}i\Gamma_{\mathcal{N}'\mathcal{N}}] = 0. \tag{6}$$

The stationary-state energies $E_{\mathcal{N}'} = \langle \mathcal{N}' |\, H_0 \,| \mathcal{N}' \rangle$ may include contributions both from the photon and electron field. The level-shift matrix elements $\Delta_{\mathcal{N}'\mathcal{N}}$ are given by

$$\Delta_{AB}(E) = P \sum_{\mathcal{N}} \int \frac{\langle A |\, V \,| \mathcal{N}_\tau \rangle \langle \mathcal{N}_\tau |\, V \,| B \rangle \, d\tau}{E - \tau - E_{\mathcal{N}}} \tag{7}$$

for each pair of states $|\mathcal{N}\rangle$ appearing in Eq. (5). In Eq. (7), P indicates the principal Cauchy value of the integral. The widths $\Gamma_{\mathcal{N}'\mathcal{N}}$ are

$$\Gamma_{AB}(E) = 2\pi \sum_{\mathcal{N}} \langle A |\, V \,| \mathcal{N}_\epsilon \rangle \langle \mathcal{N}_\epsilon |\, V \,| B \rangle \tag{8}$$

for $\epsilon = E - E_{\mathcal{N}}$. Resonance behavior is to be expected whenever E is close to $\mathrm{Re}\,\mathscr{E}_\mathcal{N}$ of some states of particle spaces that do not include F_ϵ^- and I_ϵ^+. If $\Delta_{\mathcal{N}'\mathcal{N}} \simeq 0$ and $\Gamma_{\mathcal{N}'\mathcal{N}} \simeq \Gamma_{\mathcal{N}\mathcal{N}}\,\delta_{\mathcal{N}'\mathcal{N}}$, then $\mathscr{E}_\mathcal{N} \simeq E_\mathcal{N} - \tfrac{1}{2}i\Gamma_\mathcal{N}$, which is the usual approximation for nonoverlapping resonances.

Equations (4) and (5) represent the basic results of the time-independent multichannel scattering theory. They are given in a form which is convenient for a unified description of various inner-shell ionization phenomena involving both photons and electrons (Åberg, 1981). Equation (5) is a result of using "outgoing wave" boundary conditions for the corresponding resonant scattering matrix $\mathsf{S}(E) = [\mathsf{1} - \mathsf{T}(E)]/2\pi i$. Hence the transition matrix element (5) describes a resonant process in which a particle enters into a definite channel I_ϵ^+ and scatters with a certain probability into a given channel F_ϵ^-. Note that "ingoing wave" boundary conditions must be used to describe the escape of the particle into the final channel.

Equations (4) and (5) can be used for a description of both photon and electron scattering processes. However, there are two fundamental limitations associated with the use of these equations. First, with $V = V_{\mathrm{ph}}$ the iterative solutions of Eq. (3) and of the corresponding T (or S) matrix equations diverge after the lowest-order solution. In addition, the quantities $\Delta_{\mathcal{N}'\mathcal{N}}$ given by Eq. (7) also become divergent. The proper treatment of this problem requires the whole renormalization machinery of quantum electrodynamics (e.g., Jauch and Rohrlich, 1959). This eventually limits the nonphoton states to the one-electron and -positron solutions of the Dirac equation. Hence the radiative level shifts $\Delta_{\mathcal{N}\mathcal{N}}$ which are by no means negligible for inner-shell electrons can only be evaluated approximatively as one-electron quantities (Huang *et al.*, 1976). Second, the definitions of a fixed operator H_2 (and V_{el}) imply an orthogonal basis set for the many-electron states. This excludes an effective description of relaxation and postcollision phenomena. Appropriate generalizations of Eq. (5) for relaxation in nonradiative decay (Åberg and Howat, 1982) and for postcollision effects in photoionization (Åberg, 1980, 1981) have been considered.

2.2. *Statistical Formulation of Inelastic Scattering*

The target in process (1) usually consists of atoms in a molecular or solid environment. Hence it is not always possible to assume random orientation with respect to the polarization and direction of the incoming photon beam. However, orientational features may require a statistical description. Suppose that the energy eigenstates $|\alpha\rangle = |\gamma JM\rangle$ of the target atoms are given with respect to the body frame of a molecule (or crystal). Any molecular (or solid state) eigenfunction can be expressed as a superposition of these eigenstates. Consequently one may after introducing the matrix elements of finite rotations (see Apendix) also express the molecular eigenfunctions with respect to the laboratory frame, given by

$$|\phi_\mu(\hat{R})\rangle = \sum_\alpha |\alpha\rangle\, C_{\alpha\mu}(\hat{R}). \tag{9}$$

The coefficients $C_{\alpha\mu}(\hat{R})$ account both for the intrinsic molecular symmetry and for any fixed orientation $\hat{R}$ which the molecule may have with respect to the laboratory coordinate axes. Hence the orientation of the target molecules is described statistically by a density operator

$$\hat{\rho}_{\text{or}} = \sum_\mu \int w_\mu(\hat{R})\, |\phi_\mu(\hat{R})\rangle\langle\phi_\mu(\hat{R})|\, d\hat{R}. \tag{10}$$

This operator defines the density matrix of orientation $\boldsymbol{\rho}_{\text{or}}$ fulfilling the normalization condition $\text{Tr}\,\boldsymbol{\rho}_{\text{or}} = 1$. It has the elements

$$\rho_{\alpha\beta} = \sum_\mu \int w_\mu(\hat{R})\, C_{\alpha\mu}(\hat{R})\, C^*_{\beta\mu}(\hat{R})\, d\hat{R} \tag{11}$$

with respect to the atomic basis. The coefficients $w_\mu(\hat{R})$ are in principle determined by a complete set of linearly independent expectation values

$$\langle \hat{Q}_\nu \rangle = \text{Tr}[\boldsymbol{\rho}_{\text{or}} \mathbf{Q}_\nu] = \sum_\mu \int w_\mu(\hat{R})\, \langle\mu|\, Q_\nu\, |\mu\rangle\, d\hat{R}, \tag{12}$$

where $\hat{Q}_\nu$ represents for instance the dipole moment of the molecule.

The incoming photon beam may also be in a mixed polarization state which is described by the density operator

$$\hat{\rho}_{\text{ph}} = I(\omega_1)\,|\mathbf{e}\rangle\langle\mathbf{e}| + I'(\omega_1)\,|\mathbf{e}'\rangle\langle\mathbf{e}'|, \tag{13}$$

where the states $|\mathbf{e}\rangle$ and $|\mathbf{e}'\rangle$ represent the polarization of two independent beams. Expanding these states in the helicity states, represented by the spherical unit vectors (see Appendix), yields the two-by-two density matrix $\boldsymbol{\rho}_{\text{ph}}$ which is normalized according to

$$\text{Tr}\,\boldsymbol{\rho}_{\text{ph}} = I(\omega_1) + I'(\omega_1) = \frac{dN_b(\omega_1)}{d\omega_1}, \tag{14}$$

where $dN_b(\omega_1)/d\omega_1$ gives the frequency distribution of the incoming beam. The four elements $\rho_{q_1 q_1'}$ of $\boldsymbol{\rho}_{\text{ph}}$ can be expressed in terms of the observable Stokes parameters (e.g., Blum, 1981) for each interval $(\omega_1, \omega_1 + d\omega_1)$.

The initial density matrix is given by the direct product

$$\boldsymbol{\rho}_0 = \boldsymbol{\rho}_{\text{ph}} \otimes \boldsymbol{\rho}_{\text{or}} \tag{15}$$

and the final density matrix by

$$\boldsymbol{\rho} = \mathbf{T}\boldsymbol{\rho}_0\mathbf{T}^{\dagger}, \tag{16}$$

provided the target orientation remains unperturbed during the photon scattering.

For a given detector arrangement, the probability per unit time that a photon with frequency in the interval $(\omega_1, \omega_1 + d\omega_1)$ is scattered and detected is

$$dw = \frac{2\pi}{\hbar}\,\text{Tr}[\boldsymbol{\rho}\mathbf{F}]\,d\omega_1, \tag{17}$$

where the density matrix $\mathbf{F}$ is constructed from the detector operator

$$\hat{F} = \sum_{\gamma} \varepsilon_{\gamma}\,|\gamma\rangle\langle\gamma|. \tag{18}$$

Here $|\gamma\rangle$ is in general a product of a photon and an electron state. The efficiency ε_{γ} describes the relative probability of response of the detector to the state $|\gamma\rangle$. The index γ may involve integrations over continuous quantum numbers. For the molecular case there is in principle an integration over $\hat{R}'$, but the assumption of molecular rigidity introduces a factor $\delta(\hat{R} - \hat{R}')$ which leaves only the summation over a given set of final molecular states accepted by the detector. The photon part includes the description of the response of the detector to polarization and to the direction of the scattered photons entering the detector.

2.3. Nonrelativistic Cross Section

In order to describe the scattering process (1) we need a specification of the photon–electron interaction operator V_{ph}. As indicated in Sec. 5 (e.g., Akhiezer and Berestetskii, 1965), the nonrelativistic limit of Dirac's equation with an external electromagnetic field leads to the perturbation

$$V_{\text{ph}} = +\frac{e}{2m}(\mathbf{p}\cdot\mathbf{A}+\mathbf{A}\cdot\mathbf{p})+\frac{e^2}{2m}(\mathbf{A}\cdot\mathbf{A}), \tag{19}$$

where $\mathbf{A}$ is the vector potential of the field and $\mathbf{p} = -i\hbar\nabla$ the momentum operator. In Eq. (19) we have $e = |e|$. In the plane-wave representation of the photon field in a large box of volume L^3, $\mathbf{A}$ is given in the Coulomb gauge ($\nabla\cdot\mathbf{A} = 0$) by

$$\mathbf{A} = \sum_{\mathbf{k},q,\nu}\left(\frac{\hbar}{2\varepsilon_0\omega L^3}\right)^{1/2}(\mathbf{e}^{(q)}\mathbf{e}^{i\mathbf{k}\cdot\mathbf{r}_\nu}a_{\mathbf{k},q}+\mathbf{e}^{(q)*}e^{-i\mathbf{k}\cdot\mathbf{r}_\nu}a^{\dagger}_{\mathbf{k},q}), \tag{20}$$

where $\mathbf{e}^{(q)}(q = \pm 1)$ are the spherical unit vectors, perpendicular to the wave vector $\mathbf{k}$ (see Appendix), and where $a^{\dagger}_{\mathbf{k},q}$ and $a_{\mathbf{k},q}$ are the transverse-photon creation and destruction operators, respectively. The substitution of Eq. (20) in Eq. (19) specifies V_{ph}, where the summation over the position vectors $\mathbf{r}_\nu$ of each target electron is not indicated explicitly in the following. Note that $[\mathbf{p}, \mathbf{A}] = 0$ in the Coulomb gauge.

In accordance with Sec. 2.1 we restrict the initial and final states to one-photon states and consider the scattering of a photon, characterized by an *arbitrary* polarization vector $\mathbf{e}_1$ and the wave vector $\mathbf{k}_1$. The electron system is assumed to be in an arbitrary initial state $|I\rangle$. After the scattering the electrons are assumed to be in state $|F\rangle \neq |I\rangle$ and the photon in a state characterized by an *arbitrary* $\mathbf{e}_2$ and $\mathbf{k}_2$. Consequently, the only contribution to the amplitude (5) from the electron–electron interaction V_{el} in $V = V_{\text{ph}} + V_{\text{el}}$ appears in the secular equation (6). According to Eq. (7) and (8), V_{el} accounts for the nonradiative widths and level shifts of the intermediate states. In lowest order the $\mathbf{A}\cdot\mathbf{A}$ term of V_{ph} contributes only to the direct amplitude (4) in Eq. (5), whereas the $\mathbf{p}\cdot\mathbf{A}$ term appears in the sum over the intermediate states. This restricts the intermediate states to zero- and two-photon states. The lowest-order result with respect to V_{ph} is

$$T_{\text{FI}}(E) = \frac{\hbar e^2}{2\varepsilon_0 mL^3}\frac{1}{(\omega_1\omega_2)^{1/2}}\mathcal{T}_{FI}(\omega_1, \omega_2), \tag{21}$$

where $\mathcal{T}$ is given by

$$\mathcal{T}_{FI}(\omega_1,\omega_2)=\langle F^-|\exp(i\mathbf{k}\cdot\mathbf{r})|I\rangle(\mathbf{e}_1\cdot\mathbf{e}_2^*)$$
$$-\frac{1}{m}\sum_{\mathcal{N}}\left(\frac{\langle F^-|\,e^{i\mathbf{k}_1\cdot\mathbf{r}}(\mathbf{e}_1\cdot\mathbf{p})\,|\mathcal{N}\rangle\langle\mathcal{N}|\,e^{-i\mathbf{k}_2\cdot\mathbf{r}}(\mathbf{e}_2^*\cdot\mathbf{p})\,|I\rangle}{\mathscr{E}_{\mathcal{N}}-E_I+\hbar\omega_2}\right.$$
$$\left.+\frac{\langle F^-|\,e^{-i\mathbf{k}_2\cdot\mathbf{r}}(\mathbf{e}_2^*\cdot\mathbf{p})\,|\mathcal{N}\rangle\langle\mathcal{N}|\,e^{i\mathbf{k}_1\cdot\mathbf{r}}(\mathbf{e}_1\cdot\mathbf{p})\,|I\rangle}{\mathscr{E}_{\mathcal{N}}-E_I-\hbar\omega_1}\right), \quad (22)$$

with $\mathbf{k}=\mathbf{k}_1-\mathbf{k}_2$. For final bound states $\mathcal{T}$ is a dimensionless quantity, but it has the dimension $(\text{energy})^{-1/2}$ for one-electron continuum states. The corresponding differential cross section is obtained from the usual formula

$$\frac{d^2\sigma}{d\omega_2\,d\Omega}=\frac{2\pi}{\hbar}\frac{\rho(\hbar\omega_2)}{I}|T_{FI}|^2\,\delta\left(\omega_1-\omega_2+\frac{E_I-E_F}{\hbar}\right), \quad (23)$$

where $I=c/L^3$ is the unit flux and $\rho(\hbar\omega_2)=L^3\hbar^2\omega_2^2/(2\pi\hbar c)^3$ the density per unit energy of the final photon states. The result is

$$\frac{d^2\sigma}{d\omega_2\,d\Omega}=r_0^2\frac{\omega_2}{\omega_1}|\mathcal{T}_{FI}|^2\,\delta\left(\omega_1-\omega_2+\frac{E_I-E_F}{\hbar}\right). \quad (24)$$

For final continuum states we replace $d^2\sigma/d\omega_2\,d\Omega$ by $d^4\sigma/d\omega_2\,d\Omega\,d\epsilon\,d\Omega_e$, corresponding to the ejection of an electron into the solid angle $d\Omega_e$ with the excitation energy in the interval $(\epsilon,\epsilon+d\epsilon)$. We have indicated the final state in $\mathcal{T}_{FI}$ by F^- to stress that the final-channel interaction among the electronic continuum states should be evaluated using the "ingoing wave" boundary conditions. Equation (24) can be used to obtain scattering probabilities (17) for a large variety of target and detector arrangements. Equation (24) is valid for a ground or an excited initial state.

3. NONRESONANT SCATTERING

Suggestions that inelastic losses exist in X-ray scattering due to excitation of bound electrons originate from Smekal (1923) and Raman (1928). In the optical region the inelastic loss or gain is predominantly a consequence of phonon excitation or deexcitation, corresponding to the thermal diffuse scattering in the X-ray region. In early experiments the electronic X-ray Raman effect was studied under the same experimental conditions as the Compton effect. However, the results concerning the inelastic scattering of hard X rays by the K electrons of light solid targets were somewhat ambiguous until Suzuki (1967) observed the intensity distribution of this scattering as a function of the scattering angle θ_2. His experiment was followed by other studies in the frequency range $\omega_1 \gg \Omega_K$ (Suzuki *et al.*, 1970; Cohen *et al.*, 1973), including the verification of plasmon losses in the scattered radiation at small θ_2 (Priftis *et al.*, 1968; Suzuki and Tanokura, 1970; Eisenberger *et al.*, 1973). On the whole these experiments verified the theoretical predictions by Mizuno and Ohmura (1967) and Ohmura and Matsudaira (1964), although discrepancies exist on a detailed level. For a review of plasmon scattering and for further references, see Bushuev and Kuz'min (1977) and Suzuki (1982).

We shall assume in the following that $\omega_1 \gg \Omega_K$ and consider a target which consists of randomly oriented atoms. Since the ratio of the $\mathbf{p}\cdot\mathbf{A}$ to the $\mathbf{A}\cdot\mathbf{A}$ term in Eq. (22) approaches zero roughly as $(\omega_{FI}/\omega_1)^2$ when $\omega_1 \to \infty$, we shall, following Mizuno and Ohmura (1967), consider the cross section

$$\frac{d^2\sigma}{d\omega_2\, d\Omega} = r_0^2 \frac{\omega_2}{\omega_1} |\langle F^- | e^{i\mathbf{k}\cdot\mathbf{r}} | I\rangle (\mathbf{e}_1 \cdot \mathbf{e}_2^*)|^2 \, \delta(\omega_1 - \omega_2 - \Omega_{FI}), \tag{25}$$

where $\epsilon = \hbar\Omega_{FI} = (E_F - E_I)$. Equation (25) corresponds to an ideal polarized beam $(\mathbf{k}_1, \mathbf{e}_1)$ and an ideal detector, responding to $(\mathbf{k}_2, \mathbf{e}_2)$. Random orientation and no detection of the final electron states of a given energy ϵ imply an average over all degenerate initial states γ_I and a summation over all degenerate final states γ_F. Since γ_I and γ_F include the magnetic substates, the generalized oscillator strength (GOS)

$$f_{FI}(k, \epsilon) = \frac{\epsilon/R}{(ka_0)^2} \frac{1}{w_I} \sum_{\gamma_I \gamma_F} |\langle \gamma_F | e^{i\mathbf{k}\cdot\mathbf{r}} | \gamma_I \rangle|^2 \tag{26}$$

becomes independent of the direction of $\mathbf{k}$. In Eq. (26), w_I is the number of initial states. Following Inokuti (1971), ϵ is given in rydbergs ($1\ R = 13.605$ eV) and k in atomic units a_0^{-1}, where a_0 denotes the Bohr radius ($a_0 = 0.529 \times 10^{-10}$ m). If F is a continuum state, normalized per unit energy, $f_{FI}(k, \epsilon)$ should be replaced by $df_{FI}(k, \epsilon)/d\epsilon$ in which the summation over γ_F includes the summation over all final channels, consistent with energy conservation. The cross section corresponding to Eq. (26) is

$$\frac{d^2\sigma}{d\omega_2\, d\Omega} = r_0^2 \frac{\omega_2}{\omega_1} |(\mathbf{e}_1 \cdot \mathbf{e}_2^*)|^2 \frac{(ka_0)^2}{\epsilon/R} f_{FI}(k, \epsilon)\, \delta(\omega_1 - \omega_2 - \Omega_{FI}), \tag{27}$$

where $f_{FI}(k, \epsilon)$ becomes independent of the angle between $\mathbf{k}_1$ and $\mathbf{k}_2$ in the "optical" limit

$$\lim_{k \to 0} f_{FI}(k, \epsilon) = f_{FI}(\epsilon). \tag{28}$$

Here $f_{FI}(\epsilon)$ is the total oscillator strength for the transition $I \to F$ in the dipole approximation. Hence, if $\omega_1 \gg \Omega_K$ and $ka_0 \ll 1$ the inelastic scattering cross section is proportional to the total photoabsorption cross section for randomly oriented targets (Mizuno and Ohmura, 1967). In this case the angular dependence is determined by the factor $F(k, \mathbf{e}_1, \mathbf{e}_2) = |\mathbf{e}_1 \cdot \mathbf{e}_2^*|^2 (ka_0)^2$. If the incoming radiation is propagating in the z direction with the linear polarization vector along the x axis in the laboratory frame (see Fig. A.1 of the Appendix), then we have

$$|\mathbf{e}_1 \cdot \mathbf{e}_2^*|^2 = \sin^2 \varphi_2 + \cos^2 \varphi_2 \cos^2 \theta_2 \tag{29}$$

for a detector which is insensitive to polarization. Note that for $\theta_2 = 90°$ all emitted photons are polarized perpendicular to the scattering plane. This result is valid for any degree of polarization of the incoming radiation. If the incoming radiation is unpolarized, then the angular factor F is given by

$$F(\theta_2) = \tfrac{1}{2}(ka_0)^2(1 + \cos^2 \theta_2), \tag{30}$$

where $k^2 = k_1^2 + k_2^2 - 2k_1 k_2 \cos \theta_2 \simeq 4k_1^2 \sin^2 \theta_2/2$.

According to Eq. (27), nonresonant scattering is determined by the GOS which also describes the excitation and ionization of the target electrons in inelastic scattering of fast charged particles (Inokuti, 1971; Inokuti *et al.*, 1978). In that case the GOS must be multiplied by the Rutherford cross section. This illustrates the weakness of the photon scattering in comparison with the charged-particle scattering, characterized by the ratio $(r_0/a_0)^2 \cong 3 \times 10^{-7}$.

For inelastic scattering accompanied by ionization, Eq. (27) after integration over the Dirac delta function takes the form

$$\frac{d^2\sigma}{\hbar d\omega_2\, d\Omega} = r_0^2 \left(\frac{\omega_2}{\omega_1}\right) \frac{F(k, \mathbf{e}_1, \mathbf{e}_2)}{\epsilon/R} \frac{df(k, \epsilon)}{d\epsilon}\bigg|_{\epsilon = \hbar(\omega_1 - \omega_2)} \tag{31}$$

This result can be used to describe the transformation from the low-momentum-transfer regime $(k_1 - k_2)\, a_0 \leqslant ka_0 \ll 1$ to the high-momentum-transfer regime $1 \ll ka_0 \leqslant (k_1 + k_2)\, a_0$. As shown by Inokuti (1971), the hydrogenic GOS peaks at $(ka_0)^2 = \epsilon/R$ for large ϵ/R. It attains as a function of k its characteristic Doppler-broadened shape, described in general by the impulse-approximation formula

$$\frac{df(k, \epsilon)}{d\epsilon} = \frac{\epsilon/R}{(ka_0)^2} \frac{1}{w_I} \sum_I \int |\psi_I(\mathbf{p})|^2\, \delta\left(\frac{\hbar^2 k^2}{2m} + \frac{\hbar \mathbf{k} \cdot \mathbf{p}}{m} - \epsilon\right) d\mathbf{p}, \tag{32}$$

where $\psi_I(\mathbf{p})$ is the Fourier transform of the initial-state wave function. The relation $(ka_0)^2 = \epsilon/R$ coincides with the familiar Compton relation $\lambda_2 - \lambda_1 = (h/mc)(1 - \cos \theta_2)$ in the limit $\hbar\omega_1 \gg \epsilon$. According to Eq. (32), measurements of the IXS cross section as a function of k around $(ka_0)^2 = \epsilon/R$ yield, for large $\hbar\omega_1$, information about the momentum distribution of the loosely bound electrons in their initial state. For a discussion of the utility of Compton scattering, see Williams (1977), and for the validity of the impulse approximation, see Eisenberger and Platzman (1970) and Mendelsohn and Smith (1977).

If binding effects are negligible it follows from Eq. (31) and (32) that for an unpolarized incoming beam the inelastic scattering cross section is given for one electron by

$$\frac{d\sigma}{d\Omega} = r_0^2 \frac{1 + \cos^2\theta_2}{2}. \tag{33}$$

This is the Thomson total scattering cross section. The results (31) to (33) are nonrelativistic, implying the assumption $\hbar\omega_1 \ll mc^2$. The relativistic generalization of Eq. (33) is the Klein–Nishina formula which describes the inelastic scattering of high-energy photons by free electrons initially at rest. This result shows that actually $d\sigma/d\Omega$ goes to zero as $\omega_1 \to \infty$.

Experimentally one usually observes a superposition of inelastic scattering by electrons in various atomic subshells. The outer-shell electrons may give rise to typical Compton profiles in a situation where inner-shell electrons give rise to less intense and overlapping asymmetric structures. These should be reasonably well predicted by Eq. (27) in the limit (28) although ka_0 may be quite large. The spectrum of Cu $K\alpha$ IXS by metallic lithium at $ka_0 \simeq 3.6$ (Cohen *et al.*, 1973) serves as a typical example. The IXS by the 1*s* electrons seems to be described by Eq. (28) whereas the shape of the IXS spectrum of the conduction electrons is predicted approximately by Eq. (32). This result illuminates the complementarity of position and momentum space. If the binding energy of the electron is large, then the optical limit (28) provides an adequate description of GOS already for $ka_0 \gtrsim 1$. On the other hand, since conduction electrons correspond to rather well-defined initial momenta **p**, Eq. (32) may also be applicable in this range of k. Since $f_n(k, \epsilon) \sim k^{-2(l_I + l_F + 2)}$ asymptotically (Rau and Fano, 1967), transitions to final discrete states are generally suppressed in the limit of large ka_0 in contrast to continuum IXS distributions which tend to peak at $ka_0 \simeq (\epsilon/R)^{1/2}$.

The analysis given above requires a comment on the terminology involved in IXS. Although the term "Raman scattering" including various modifications has apparently been used in the literature for IXS involving ionization (Suzuki *et al.*, 1970; also Bushuev and Kuz'min, 1977) it originally refers to discrete excitations. The use of "Raman scattering" for ionization losses instead of "Compton scattering" merely reflects the contrast which exists between IXS spectra of bound and loose target electrons under the same experimental conditions. As noted above, high-energy IXS by tightly bound electrons is predominantly associated with ionization rather than excitation although there are in principle always contributions from both modes of transitions. Hence, unless the incoming photon energy is very large, the division of the IXS spectrum into a "Raman" and a "Compton" part is rather arbitrary since the detailed shape of the spectrum depends on the incoming photon energy and the momentum transfer.

4. RESONANT SCATTERING

In 1974 Sparks reported an observation of inelastic scattering of copper $K\alpha(\hbar\omega_1 \simeq 8$ keV) X rays by L electrons in Ni, Cu, Zn, Ge targets and by M electrons in a Ta target. The cross sections ranged from 1 to $7r_0^2$ and showed no angular dependence. These findings cannot be understood on the basis of the theory of the nonresonant scattering but require considerations of the resonant $\mathbf{p} \cdot \mathbf{A}$ terms in the cross section (24). As shown by Bannett and Freund (1975) the Ni, Cu, Zn, and Ge data (Sparks, 1974) for which $\omega_1 \gtrless \Omega_K$ could be explained by including only intermediate states with a K-shell hole. They named this IXS resonant X-ray Raman scattering.

Further investigations of IXS subject to the condition $\omega_1 \gtrless \Omega_K$ have been made by Bannett *et al.* (1977), Suortti (1979), and Kodre and Shafroth (1979) using the characteristic radiation from conventional X-ray tubes. Eisenberger *et al.* (1976a, b) and Briand *et al.* (1981) have used the technique of tunable synchrotron radiation for the study of IXS close to the K absorption edge.

Instead of observing the inelastically scattered X rays evolving into fluorescence the corresponding development of the nonradiative decay channel can also be studied. This has been demonstrated by Brown *et al.* (1980), who studied the Auger decay near the L_2 photoionization threshold in xenon.

Internal inelastic scattering of characteristic $K\alpha$ X rays, produced in solid targets, has also been observed (Åberg and Utriainen, 1969; Hall *et al.*, 1979). For this type of scattering the condition $\omega_1 = \Omega_K - \Omega_L \lesssim \Omega_K$ is well fulfilled in light targets. The internal IXS by L electrons appears as a faint structure near the K–LL radiative Auger edge. Note that the final state of the K–LL radiative Auger transitions (e.g., Åberg, 1975) involves two vacancies in the L shell of the *same* atom in contrast to the $K\alpha$ IXS which produces the final L vacancies in two separate target atoms. Hence these processes lead to energetically different final states.

4.1. Evolution of Resonant Scattering into Fluorescence

According to Eq. (22), the $\mathbf{A}\cdot\mathbf{A}$ term vanishes in the dipole approximation, $\exp(i\mathbf{k}\cdot\mathbf{r})\simeq 1$, unless $|I\rangle = |F^-\rangle$. If we in addition have $\hbar\omega_1 \simeq E_{\mathcal{N}} - E_I$, the corresponding resonant $\mathbf{p}\cdot\mathbf{A}$ terms for $\mathcal{N}$ and the adjacent intermediate states dominate. The nonresonant $\mathbf{p}\cdot\mathbf{A}$ terms having the denominator $\mathscr{E}_{\mathcal{N}} - E_I + \hbar\omega_2 \simeq E_{\mathcal{N}} - E_I - i\Gamma_{\mathcal{N}}/2 + \hbar\omega_2$ are smaller than these terms by a factor of $\Gamma_{\mathcal{N}}/4\hbar\omega_1$. This simplifies the treatment of the resonant IXS considerably.

As shown in Sec. 6.2, the cross section (24) with both second order amplitudes included is given in the dipole approximation by

$$\frac{d^2\sigma}{d\omega_2\, d\Omega} = [a + bf(\theta_2\varphi_2)], \tag{34}$$

where

$$f(\theta_2\varphi_2) = \cos^2\theta_2 + \sin^2\theta_2[\rho_{1-1}\exp(+2i\varphi_2) + \rho_{-11}\exp(-2i\varphi_2)](\rho_{11} + \rho_{-1-1})^{-1}, \tag{35}$$

for randomly oriented target atoms and for a polarization-insensitive detector. The coefficients a and b which involve the transition matrix elements in a reduced form describe the dynamics of IXS. According to Sec. 2.2, the elements $\rho_{\mu\nu}$ of the density matrix $\boldsymbol{\rho}_{\text{ph}}$ specify the polarization state of the incoming beam propagating in the z-direction of the laboratory frame. The

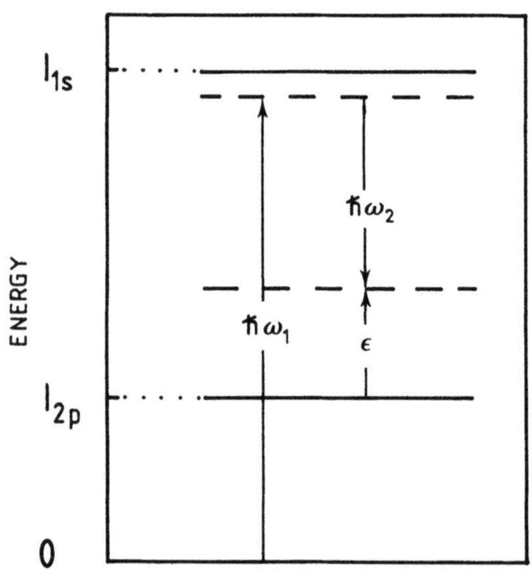

FIGURE 1. Schematic energy diagram of inelastic resonance scattering by $2p$ electrons.

angular dependence in Eq. (35) differs in general from that of the nonresonant scattering, given by Eq. (30) for an unpolarized incoming beam. If $b=0$, the cross section (34) is isotropic. This is approximately true for scattering close to the K absorption edge as indicated in Sec. 6.2.

We shall restrict ourselves to a study of the $\omega_1 \simeq \Omega_K$ region and inelastic scattering by $L_{2,3}(n=2)$ or $M_{2,3}(n=3)$ electrons. Figure 1 shows the corresponding energy diagram for $n=2$. The following assumptions are made:

1. Only the resonant part of the $\mathbf{p}\cdot\mathbf{A}$ term is included in Eq. (34) and the dipole approximation is used. Other choices of gauge than $\mathbf{p}\cdot\mathbf{A}$ are discussed in Sec. 5.3.
2. The target atoms are assumed to be in their ground states and any multiplet-splitting effects in the summation over final and intermediate states are neglected in the energy denominator of the resonant term.
3. The one-electron central-field approximation is applied to the description of the initial, intermediate, and final states. This implies orthogonality between the wave functions of the virtually excited electron and the excited or ionized electron in the final state. This excludes any postcollision effects (e.g., Åberg, 1981) from our analysis.
4. The ejected electron in the final state is not detected.

These assumptions restrict the final states to $(np_j)^{-1}\epsilon p$ $(n=2,3;$ $j=1/2, 3/2)$ and the intermediate states to $(ms)^{-1}\epsilon p$ and $(np_j)^{-1}\epsilon' s$ states. The excitation energies ϵ and ϵ' may take discrete or continuous values in contrast to m which includes the initially occupied s states. In the case of an initially incomplete $2p$ shell $(np_j)^{-1}$ denotes the additional vacancy. In the following only the $m=1$ term is considered. Corrections to this approximation are discussed in Sec. 4.2 and 5.3.

Exclusion of any of the assumptions listed above can introduce anisotropy effects. For example, if the nonresonant $\mathbf{p}\cdot\mathbf{A}$ term is included the interference term between it and the resonant part makes $b\neq 0$ in Eq. (34) and results in the cross section

$$\frac{d^2\sigma}{d\omega_2\, d\Omega}=\frac{1}{4\pi}\left(\frac{d\sigma_r}{d\omega_2}\right)\{(1+C)^2+C[f(\theta_2\varphi_2)-1]\} \tag{36}$$

for IXS by electrons in a np_j $(j=1/2$ or $3/2)$ subshell (Tulkki and Åberg, 1978). In Eq. (36) $d\sigma_r/d\omega_2$ is the isotropic K resonance cross section discussed below. The factor $C=(\Omega_{1s}-\Omega_{np_j}-\omega_2)/(\Omega_{1s}-\Omega_{np_j}+\omega_1)$ determines the ratio of the nonresonant to the resonant amplitude apart from the level width Γ_{1s}. The angular factor $f(\theta_2\varphi_2)$ is given by Eq. (35).

As a result of the four assumptions made above, the differential cross section per unit frequency for the K resonance scattering by np_j electrons can be obtained from Eq. (36). It is

$$\frac{d\sigma_r(\omega_1)}{d\omega_2} = 2\pi r_0^2 \int_0^\infty \left(\frac{\omega_2}{\omega_1}\right) \frac{(\Omega_{1s}-\Omega_{np_j})\, g_{np_j 1s}\, (\Omega_{1s}+\omega)\, dg_{1s}/d\omega}{(\Omega_{1s}+\omega-\omega_1)^2 + \Gamma_{1s}^2/4\hbar^2}$$
$$\times\, \delta(\omega_1 - \Omega_{np_j} - \omega - \omega_2)\, d\omega, \tag{37}$$

where $g_{np_j 1s}$ is the oscillator strength of the transition between the $1s^{-1}$ and np_j^{-1} hole states and $\hbar\omega$ the kinetic energy of the ejected electron. The oscillator density $dg_{1s}/d\omega$ is in general a superposition of discrete and continuous features. If the discrete part consists of transitions into Rydberg-like states $n'p$ $(n' > n)$, then the substitution

$$\frac{dg_{1s}}{d\omega} = g_{1sn'p}\, \delta(\Omega_{n'p} + \omega) \tag{38}$$

into Eq. (37) yields the corresponding partial cross section. In the following calculations known values of $g_{np_j 1s}$ (Scofield, 1974) have been used. The behavior of $dg_{1s}/d\omega$ as a function of ω is also known beyond the one-electron approximation in many cases. Hence the resonant cross section (37) is known more accurately than is anticipated by the one-electon assumption 3.

The theory outlined in Sec. 2 does not account for initial- and final-state lifetimes. However, following Eisenberger *et al.* (1976b), the latter contribution can be taken into account to lowest order in Eq. (37) by replacing the Dirac delta function $\delta(\omega_1 - \Omega_{np_j} - \omega - \omega_2)$ by the normalized density function

$$\frac{dN_f}{d\omega_2} = \frac{\Gamma_{np_j}/2\pi\hbar}{(\omega_1 - \Omega_{np_j} - \omega - \omega_2)^2 + \Gamma_{np_j}^2/4\hbar^2}. \tag{39}$$

The result (37) does not depend on the polarization state of the incoming beam. However, according to Eq. (14), (15), and (17), the final cross section is influenced by the frequency distribution $dN_b(\omega_1)/d\omega_1$ of the incoming photons. Hence, the observed cross sections correspond to the convolution

$$\frac{d\sigma(\omega_1^0)}{d\omega_2} = \int_0^\infty \left(\frac{d\sigma_r(\omega_1)}{d\omega_2}\right)\left(\frac{dN_b}{d\omega_1}\right) d\omega_1, \tag{40}$$

where $d\sigma_r(\omega_1)/d\omega_2$ is given by Eq. (37) after substituting the density (39). In Eq. (40), ω_1^0 is the frequency corresponding to the maximum of $dN_b/d\omega_1$.

Before discussing in detail (Tulkki and Åberg, 1982; Tulkki, 1983) the behavior of the cross section (40) as a function of ω_1^0 and ω_2 near the K edge, we shall illustrate the basic features of IXS over a broader energy range. The differential IXS cross section for neon $L_{2,3}$ electrons which in principle extends from $\omega_2=0$ to $\omega_2=\omega_1-\Omega_{2p}$ is used as an example. Figure 2 shows the calculated IXS spectrum as a function of ω_2 for three different ω_1^0 values below, at, and above the K edge. For simplicity, multiple excitation and spin–orbit interaction effects are excluded. Only the $n'=3$ Rydberg peak is taken into account in Eq. (38), although the observed Ne K absorption spectrum (Wuilleumier, 1971) is somewhat richer in details. The width of the elastic peak E at $\omega_2=\omega_1$ characterizes the chosen spectral bandwidth, $\Gamma_b=0.5$ eV, of the Lorentzian incoming-beam distribution function. The scattering angle is 90° and the incoming beam is assumed to be unpolarized. The dispersion correction corresponding to $|F\rangle=|I\rangle$ in the $\mathbf{p}\cdot\mathbf{A}$ term of Eq. (22) has been included in the calculation of the intensity of E which characterizes the elastic scattering of all ten electrons of Ne. At $\hbar\omega_1=852$ eV ($\omega_1<\Omega_K$) the IXS spectrum is given to good accuracy by Eq. (37), although the more accurate expression (36) has been used in Eq. (40). The contribution from the $\mathbf{A}\cdot\mathbf{A}$ term is only one percent of the $\mathbf{p}\cdot\mathbf{A}$ contribution at this energy. The IXS spectrum resembles an inverted K absorption spectrum although it approaches zero much faster. At $\hbar\omega_1=868$ eV ($\omega_1\simeq\Omega_K$), the continuum part of IXS looks more like an asymmetric peak and is nearly fluorescent. The $n'=3$ peak, corresponding to $2p^53p$ final states, has passed its resonance maximum since $\hbar\omega_1$ is somewhat larger than $\Omega_{1s}-\Omega_{3p}$. Since $\Gamma_{2p}=0$, the behavior of the $n'=3$ peak as a function of ω_1 is analogous to optical resonance fluorescence in the weak-field limit (Heitler, 1954; Wu *et al.*, 1975). It is important to note the enormous enhancement in the IXS intensity as $\omega_1\rightarrow\Omega_K$. At $\hbar\omega_1=910$ eV ($\omega_1>\Omega_K$), the fluorescent $K\alpha$ line is fully developed at $\omega_2=\Omega_{1s}-\Omega_{2p}$. It represents a resonance with halfwidth Γ_{1s} in the IXS cross section. In fact, it can be seen from Eq. (37) that it always peaks at this ω_2 value as long as $\omega_1>\Omega_{1s}$. From this resonance the cross section extends continuously up to $\hbar(\omega_1-\Omega_{2p})=892$ eV. The structure near this upper edge is almost entirely due to the $\mathbf{A}\cdot\mathbf{A}$ contribution (27). According to Eq. (28) and (40), its shape resembles the inverted $L_{2,3}$ absorption spectrum modified by the beam distribution function. Since the wavelength is $\lambda\simeq 26a_0$, the condition $ka_0\ll 1$ is well satisfied. For all ω_1 values used in Fig. 2 there is also another isolated resonance at $\omega_2=\omega_{2s}-\omega_{2p}$.

Figure 3 shows in detail the behavior of the $L_{2,3}$ differential IXS cross section across the Ar K edge for the energy bandwidth $\Gamma_b=0.1$, 0.5, and 1.0 eV, respectively (Tulkki, 1983). For simplicity only the $j=3/2$ component of the inelastic spectrum is shown. In this calculation the intermediate states of the scattering to discrete final states incorporate $1s^{-1}n'p$ Rydberg states, corresponding to $4\leqslant n'\leqslant 18$. Figure 3 clearly demonstrates

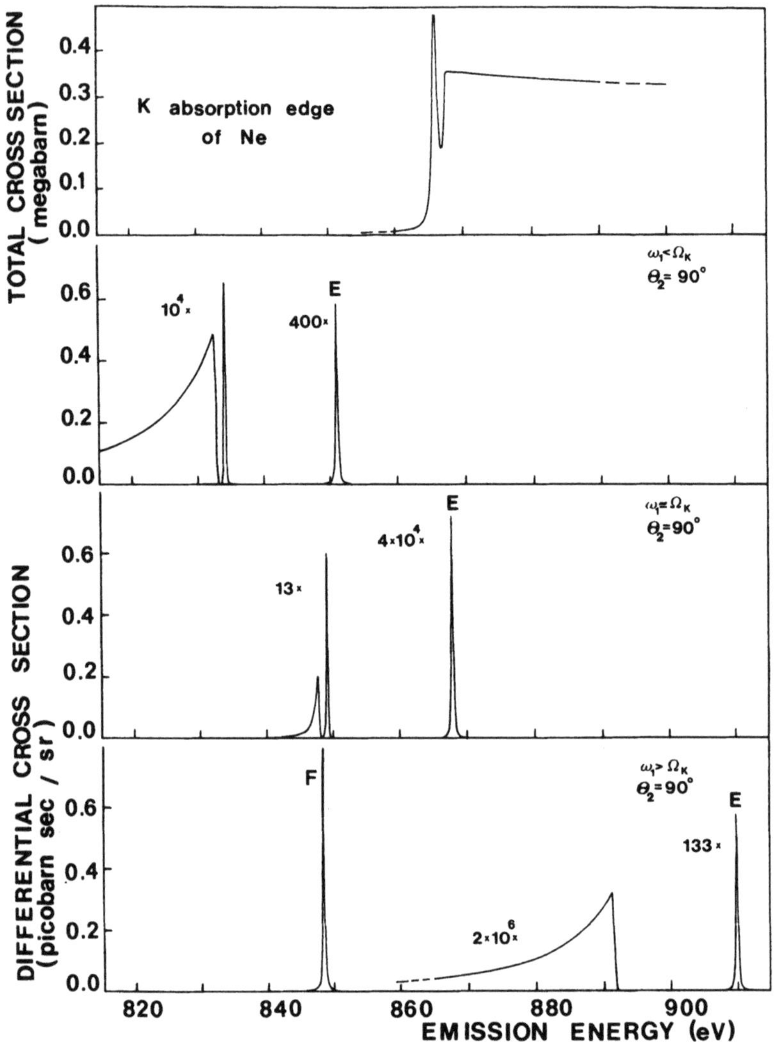

FIGURE 2. The inelastic X-ray scattering spectrum of Ne in the vicinity of the K ionization threshold. The schematic absorption spectrum, including instrumental broadening, has been obtained from the measurement of Wuilleumier (1971). Our inelastic X-ray scattering curves have been plotted using a spectral band width of $\Gamma_b = 0.5$ eV. The peak denoted by E refers to the elastically scattered photons, and F, to the fluorescence emission at $\hbar\omega_2 = 848.6$ eV. For clarity the peaks have been magnified to approximately the same size.

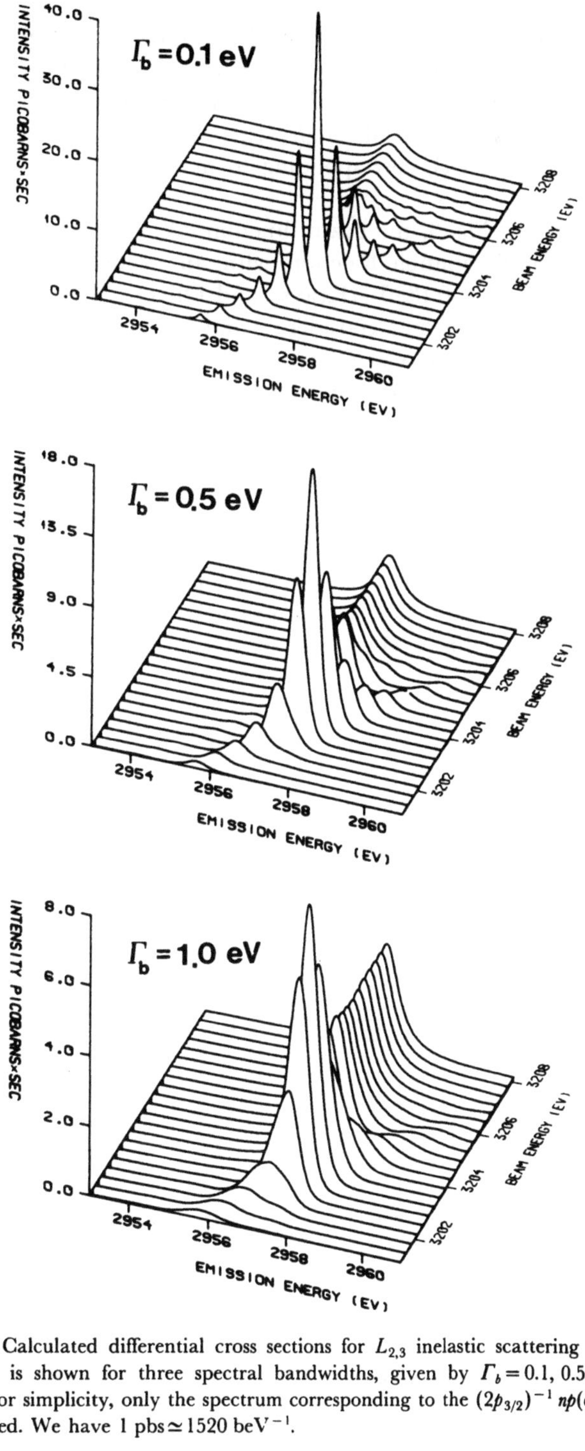

FIGURE 3. Calculated differential cross sections for $L_{2,3}$ inelastic scattering by Ar atoms. The spectrum is shown for three spectral bandwidths, given by $\Gamma_b = 0.1$, 0.5, and 1.0 eV, respectively. For simplicity, only the spectrum corresponding to the $(2p_{3/2})^{-1}\, np(\epsilon p)$ final states has been plotted. We have 1 pbs $\simeq$ 1520 beV^{-1}.

the influence of Γ_b on the discrete part of the spectrum. The behavior of any of the discrete peaks as a function of ω_2 is characterized by the enhancement of the peak-to-background ratio as Γ_b narrows and by the sharp rise and drop in intensity as ω_1 goes through $\Omega_{1s}-\Omega_{n'p}$. The linear dependence between ω_1 and ω_2(max), corresponding to maximum emission should be noted.

In constructing Fig. 3, the beam function $dN_b/d\omega_1$ has been simulated by a Lorentzian distribution function with an exponential cutoff. Since $\Gamma_{1s}=0.68$ eV and $\Gamma_{2p}=0.13$ eV (Krause and Oliver, 1979), the change in Γ_b from 1 to 0.1 eV results in a considerable increase in resolution. As shown by Fig. 3, the use of $\Gamma_b=0.1$ eV resolves the $n=4$, 5, and 6 peaks in contrast to $\Gamma_b=1$ eV which only results in a single broad structure with a maximum at $\omega_2=\omega_1-\Omega_{1s}+\Omega_{4p}$.

The structure developing at the $1s$ ionization threshold, $\hbar\omega_1=3206.0$ eV, is the $K\alpha_1$ fluorescence peak at $\hbar\omega_2=2957.7$ eV. In Fig. 4 the discrete structures have been depressed in order to demonstrate more clearly the evolution of the continuous L_3 inelastic scattering into the fluorescent $K\alpha_1$ peak with its characteristic half-width, equal to $\Gamma_{1s}+\Gamma_{2p}$. Each curve has been cut at the half-maximum value demonstrating the halfwidth-narrowing effect (Eisenberger *et al.*, 1976) below threshold.

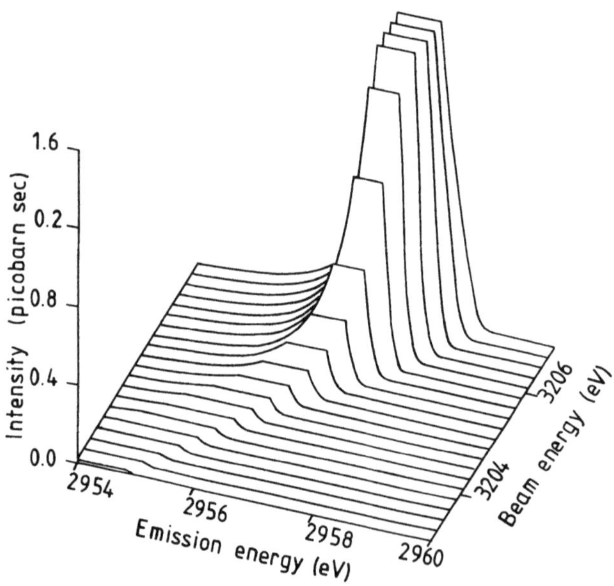

FIGURE 4. The half-width-narrowing effect in the evolution of the inelastic L_3 continuum scattering into $K\alpha_1$ fluorescence. Since each curve has been cut at the half-maximum value, the width of each cut is equal to the half-width.

According to Eqs. (39) and (40), the form of the threshold at $\omega_2 = \omega_1 - \Omega_{2p}$ is determined by Γ_{2p} and Γ_b rather than Γ_{1s} until $\omega_1 = \Omega_{1s}$ is reached. This effect is a consequence of energy conservation which prohibits the shape of the high-energy emission flank to be fully broadened by Γ_{1s} until ω_1 is well above threshold. Simultaneously the halfwidth of the fluorescent peak becomes independent of Γ_b. The distribution function $dN_b/d\omega_1$ continues to affect the shape of the IXS cross section at the onset $\omega_2 = \omega_1 - \Omega_{2p}$ rather than at $\omega_2 = \Omega_{1s} - \Omega_{2p}$. According to Eq. (40), the fluorescent peak is very accurately a convolution of the 1*s* and 2*p* Lorentz functions for $\omega_1^0 > \Omega_{1s}$.

The results of this section demonstrate an important advantage of IXS spectroscopy in comparison with ordinary X-ray absorption spectroscopy (Parratt, 1959). Depending on the chemical environment of the absorbing atom, the X-ray absorption near-edge structure (XANES) may exhibit boundlike features which are often classified as shape resonances (e.g., Dehmer *et al.*, 1983). This resonance is considered to be a consequence of localization due to the formation of a barrier in the effective potential of the excited electron. Although not a genuine bound state, the shape resonance is also influenced by the 1*s* lifetime which tends to reduce the effect of localization. As a function of ω_2, a shape resonance should therefore also exhibit to some extent similar behavior as the discrete structures in Fig. 3, as ω_1 is scanned through the corresponding frequency region in the *K* absorption spectrum. This effect would diminish as ω_1 approaches the extended X-ray absorption fine structure (EXAFS) region.

The recent synchrotron-radiation study of IXS in the vicinity of the Mn *K* absorption edge in $KMnO_4$ (Briand *et al.*, 1981) illustrates some of the features discussed above (Tulkki and Åberg, 1982). Here the strong resonance situated 15 eV below the *K* edge is almost an atomiclike 1*s*-to-4*p* excitation. Thus it exhibits as a function of ω_2 similar behavior as the Ar 1*s*-to-4*p* inelastic peak.

4.2. *Total Cross Section*

Integration of Eq. (40) over ω_2 yields, as a function of ω_1^0, an expression for the total IXS cross section which is valid close to the *K* absorption edge. It is equal to the absorption cross section multiplied by the ratio of the radiative $(1s)^{-1} \rightarrow (2p_j)^{-1}$ transition probability and the total decay rate for the $(1s)^{-1}$ hole state. The absorption cross section is a convolution of the beam distribution function and a Lorentz function having the halfwidth Γ_{1s}.

Figure 5 presents a calculation of the total Ar L_3 cross section as a function of the beam energy $\hbar\omega_1^0$ across the *K* absorption edge (Tulkki, 1983). The calculations were made for both $\Gamma_b = 0.1$ eV (solid line) and $\Gamma_b = 0.5$ eV (dashed line). The energy $\hbar\omega_1^0$ is given relative to the difference

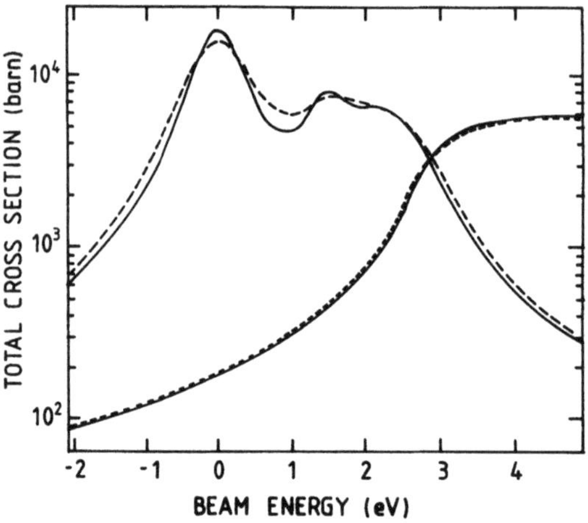

FIGURE 5. Total inelastic X-ray scattering cross sections for $2p$ electrons in Ar. The solid line corresponds to the bandwidth $\Gamma_b = 0.1$ eV, and the dashed line, to $\Gamma_b = 0.5$ eV. The cross section is given as a function of the difference between the beam energy and the $1s$–$4p$ excitation energy.

$\hbar(\Omega_{1s} - \Omega_{4p})$. Hence the contribution from the discrete states peaks at zero. The cross section shows some structure due to higher Rydberg states but is almost unaffected by the change of Γ_b from 0.1 to 0.5 eV. This reflects the ultimate resolution in the photoabsorption spectrum, governed by $\Gamma_{1s} = 0.68$ eV. With $\Gamma_b = 0.5$ eV the theoretical resolution limit for the absorption spectrum has already been reached, in contrast to the IXS spectrum shown in Fig. 3. The continuum part of the cross section has the typical arctangent shape characterizing the influence of Γ_{1s} at the onset of the continuous absorption.

TABLE I

Near-K-edge $L_{2,3}$ IXS cross sections, in units of r_0^2 ($r_0^2 \simeq 80$ mb), for Cu $K\alpha$, (Ni, Cu, Zn, Ge); Si $K\alpha$ (Si); and Ar $K\alpha$ (Ar) radiation.

Target	Calculations		Experiments		
atom	$\theta_2 = 0°$	$\theta_2 = 70°$	Sparks (1974)[a]	Bannett *et al.* (1977)[b]	Hall *et al.* (1979)[c]; Suortti (1979)[d]
Si	5.02	4.77			4.6 ± 1.6[c]
Ar	3.44	3.20			
Ni	6.81	6.51	7.0 ± 0.2	7.3 ± 0.2	6.8 ± 0.3[d]
Cu	2.05	1.90	2.5 ± 0.4	2.3 ± 0.2	2.1 ± 0.2[d]
Zn	1.14	1.01	1.3 ± 0.5	1.3 ± 0.1	
Ge	0.69	0.59	1.1 ± 0.4	0.62 ± 0.05	

[a] $\theta_2 \simeq 70°$.
[b] $\theta_2 \simeq 90°$.
[c] Internal IXS, total measured cross section (58 ± 20) r_0^2.
[d] $\theta_2 \simeq 20°$.

In Table I we compare total $L_{2,3}$ IXS cross sections with experimental values for some atoms. The calculations are based on Eq. (36) and employ relativistic Dirac–Fock $K\alpha$ oscillator strengths (Scofield, 1974) and tabulated semiempirical K oscillator densities. The calculations are given for two scattering angles in order to illustrate the anisotropy due to the interference term in Eq. (36). Our $\theta_2 = 0°$ values agree within 10% with those of Bannett and Freund (1975). According to Table I, the $L_{2,3}$ IXS cross section for Ar is $\sim 3 \times 10^{-3}$b at $\Omega_{1s} - \omega_1^0 \simeq 250$ eV, but it is $\sim 6 \times 10^3$b at the $1s$ threshold according to Fig. 5. This illustrates, as does Fig. 2 for Ne $L_{2,3}$ IXS, the enormous increase of the IXS cross section as ω_1^0 approaches the resonance frequency Ω_{1s}.

The calculations presented in Table I do not include the $\mathbf{A}\cdot\mathbf{A}$ term. Figure 6 illustrates its effect on the doubly differential Ar $L_{2,3}$ cross section for incident Ar $K\alpha$ X rays. The $\mathbf{A}\cdot\mathbf{A}$ cross section has been calculated from Eq. (27) in the limit (28) using Veigele's (1973) $L_{2,3}$ photoabsorption cross section and assuming an unpolarized incoming beam. Since the final states in the $\mathbf{A}\cdot\mathbf{A}$ amplitude are $2p^{-1}\epsilon l(^1P)\,(l=0, 2)$, in contrast to the final $2p^{-1}\epsilon p\,(^1S, {}^1D)$ state of the $\mathbf{p}\cdot\mathbf{A}$ term, there is no interference term. In general there is no interference between the $\mathbf{p}\cdot\mathbf{A}$ and $\mathbf{A}\cdot\mathbf{A}$ amplitudes due to parity reasons as long as the assumptions 1–4 of Sec. 4.1 are valid.

As shown by Fig. 6, the $\mathbf{A}\cdot\mathbf{A}$ contribution can be quite large at scattering angles close to 90°. Using the approximation $k_1 \simeq k_2$ it can be seen from Eq. (30) that the $\mathbf{A}\cdot\mathbf{A}$ cross section increases from zero to twice its value at $\theta_2 = 90°$ as θ_2 goes from 0° to 180°. As $\omega_1^0 \to 0$ the $\mathbf{p}\cdot\mathbf{A}$ cross section diverges (see Sec. 4.3), whereas the $\mathbf{A}\cdot\mathbf{A}$ term goes to zero. Hence, there is an intermediate off-resonance range below the K edge, where the $\mathbf{A}\cdot\mathbf{A}$ and $\mathbf{p}\cdot\mathbf{A}$ cross sections may both be of the order of r_0^2 at large scattering angles.

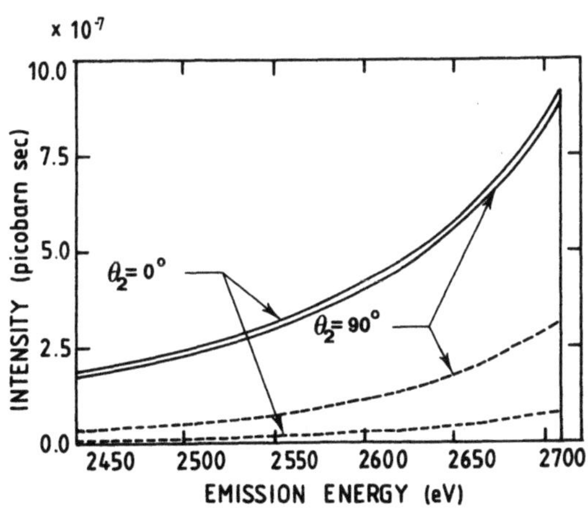

FIGURE 6. The angular dependence of the $\mathbf{p}\cdot\mathbf{A}$ (solid line) and $\mathbf{A}\cdot\mathbf{A}$ (dashed line) differential cross sections for $2p$ inelastic scattering by Ar atoms. The energy of the incident photons is equal to the $K\alpha$ energy of Ar.

As ω_1^0 approaches Ω_{2s} the contribution from the $2s^{-1}$ intermediate states becomes gradually significant in both the resonant and nonresonant $\mathbf{p}\cdot\mathbf{A}$ terms. However, in the Ar $K\alpha$-on-Ar case shown in Fig. 6, the ratio between the $2s$ and $1s$ amplitudes is only $\sim 10^{-4}$ for the cross-section maximum at $\omega_2 = \omega_1 - \Omega_L$. At $\omega_1 = \Omega_K$ this ratio has been reduced to 2×10^{-6}. This demonstrates the negligible contribution from other than the $1s^{-1}$ hole intermediate states in the resonance region. In this context it is worthwhile to stress that so far our conclusions have been based on the use of the velocity form $\mathbf{p}\cdot\mathbf{A}$. In Sec. 5.2 and 5.3 other choices of gauge are discussed.

Gavrila and Tugulea (1975) have calculated the doubly differential cross sections for IXS by $2s$ and $2p$ electrons in a hydrogenlike ion. They consider the case in which the electron is ejected into the continuum by the photon. A complete set of intermediate states is included in their dipole-approximation calculations. Gavrila and Tugulea predict a resonance at $\omega_2 = \omega(K\alpha)$ and an angular dependence which agrees with Eq. (35) for unpolarized incoming radiation. However, for many-electron atoms the hydrogenic approximation seems not to give very accurate values of IXS cross sections. Bannett *et al.* (1977) have made a comparison for Cu $K\alpha$ IXS by $L_{2,3}$ electrons in a series of elements Ni–Ge. According to their results, the hydrogenic cross section amounts to approximately 50% of the experimental value.

4.3. *Infrared Divergence*

We consider the behavior of the amplitude $\mathcal{T}_{FI}(\omega_2, \omega_1)$ in Eq. (22) for a fixed ω_1 as ω_2 approaches zero. The first and second terms go to zero, whereas the third term which describes the resonance behavior diverges. In order to see this we employ the energy conservation relation $\hbar\omega_1 + E_I = \hbar\omega_2 + E_F$ in Eq. (22) and, following Bannett *et al.* (1977), use the Born approximation for the ejected electron. For any atomic field (Gavrila, 1972a) the result is

$$\lim_{\omega_2 \to 0} \mathcal{T}_{FI}(\omega_1, \omega_2) = \frac{1}{m\omega_2} (\mathbf{e}_2^* \cdot \mathbf{k}_e) \langle F_{\mathbf{k}_e} | \, e^{i\mathbf{k}_1 \cdot \mathbf{r}} (\mathbf{e}_1 \cdot \mathbf{p}) \, | I \rangle, \tag{41}$$

where $\mathbf{k}_e$ denotes the wave vector of the ejected electron. This result shows that for $\omega_2 \to 0$ the corresponding differential cross section (24) approaches infinity as ω_2^{-1} so that the total cross section would diverge logarithmically. It demonstrates the inapplicability of the lowest-order single-scattering approach to inelastic soft-photon scattering. At this level of approximation any cross section involving scattering of photons and charged particles would show similar behavior, also known as the "infrared catastrophe" (e.g., Akhiezer and Berestetskii, 1965).

Using Eq. (41) in the cross section (24) under the assumption of random orientation and a polarization-insensitive detector yields the factorization

$$\lim_{\omega_2 \to 0} \frac{d^3\sigma}{d\omega_2\, d\Omega\, d\Omega_e} = \left[\frac{d\sigma_{\text{ph}}(\mathbf{k}_e)}{d\Omega_e}\right]\left(\frac{d^2P}{d\omega_2\, d\Omega}\right), \tag{42}$$

where $d\sigma_{\text{ph}}(\mathbf{k}_e)/d\Omega_e$ is the photoionization cross section for the ejection of an electron into the solid angle $d\Omega_e$ with the momentum $\mathbf{p}_e = \hbar\mathbf{k}_e$. The second factor describes the "bremsstrahlung" probability

$$\frac{d^2P}{d\omega_2\, d\Omega} = \frac{\alpha}{4\pi^2\omega_2}\left|\hat{k}_2 \times \frac{\mathbf{p}_e}{mc}\right|^2 \tag{43}$$

that the acceleration of the electron from zero velocity to the velocity $\mathbf{p}_e/m$ results in the emission of a photon with energy $\hbar\omega_2$ in the direction $\hat{k}_2$. Equation (41) implies integration over the Dirac delta function in Eq. (24). This leads to $p_e^2 = 2m\hbar(\omega_1 - \omega_2 - \Omega_t)$, where Ω_t is the threshold frequency of the ejected electron.

The integration over $d\Omega_e$ can be carried out in general using the differential cross section for photoionization of randomly oriented atoms by unpolarized radiation, which is

$$\frac{d\sigma_{\text{ph}}(\mathbf{k}_e)}{d\Omega_e} = \frac{\sigma}{4\pi}[1 + \beta P_2(\hat{k}_1 \cdot \hat{k}_e)] \tag{44}$$

in the dipole approximation. Here σ is the total cross section and β the asymmetry parameter (e.g., Manson and Starace, 1982). Using Eq. (44) in the integration of Eq. (42) over $d\Omega_e$ yields

$$\lim_{\omega_2 \to 0} \frac{d^2\sigma}{d\omega_2\, d\Omega} = \frac{1}{4\pi}\left(\frac{d\sigma}{d\omega_2}\right)\left[1 + \frac{\beta}{10} P_2(\hat{k}_1 \cdot \hat{k}_2)\right], \tag{45}$$

where

$$\frac{d\sigma}{d\omega_2} = \frac{4r_0\sigma}{3\pi c}\frac{\omega_1 - \omega_2 - \Omega}{\omega_2}. \tag{46}$$

Since we have $-1 \leqslant \beta \leqslant 2$, the anisotropy of the inelastic soft-photon scattering is small.

Evidently the infrared divergence can be avoided by a modification of the "classical" formula (43). This requires a consideration of the emission of any number of soft photons in the bremsstrahlung process. The scattering equation (2) must be solved up to all orders in the photon–electron interaction. This can be done if advantage is taken of the fact that the wavelength of a soft photon is much larger than the atomic dimensions. Consequently a classical treatment of the motion of the electrons is implied, which is equivalent to the assumption of photon emission with random phase. The emission of successive photons is statistically independent so that the probability of emission of n_k photons with frequency in the interval $(\omega_2, \omega_2 + d\omega_2)$ and with directions within a given $d\Omega_{\hat{k}}$ becomes

$$P(n_k) = e^{-\langle n_k \rangle} \frac{\langle n_k \rangle^{n_k}}{n_k!}. \tag{47}$$

In the Poisson distribution of Eq. (47), $\langle n_k \rangle$ is the average number of photons that are emitted with wave vectors $\mathbf{k}_2$. The probability of the simultaneous emission of any number of soft photons with total energy loss $\leqslant \Delta\epsilon$ is

$$P(\Delta\epsilon) = \sum_{n_k} \prod_k P(n_k), \tag{48}$$

subject to the constraint

$$\sum_k n_k(\hbar\omega_k) \leqslant \Delta\epsilon. \tag{49}$$

The proof of Eq. (48) is given in a number of textbooks (Akhiezer and Berestetskii, 1965; Jauch and Rohrlich, 1955) which also show that $\langle n_k \rangle \sim \omega_2^{-1}$. It follows that $\langle n \rangle = \sum_k \langle n_k \rangle$ is logarithmically divergent but that $P(\Delta\epsilon)$ becomes zero along with each individual $P(n_k)$ as $\Delta\epsilon$ approaches zero.

The modification of the result (42) by Eq. (48) shows that the ejection of a photoelectron is always accompanied by some radiative energy loss which in the limit $\Delta\epsilon \to 0$ is due to the emission of an arbitrary number of soft photons. If the energy resolution were better than $\Delta\epsilon$ this loss would in principle be observable as a distortion of the photoelectron peak (Botto and Gavrila, 1982). As $\Delta\epsilon$ is allowed to increase for a given energy resolution, this distortion eventually develop into an observable loss of the photoelectron energy due to the inelastic scattering of individual photons. For photon energies in the keV range $P(\Delta\epsilon)$ is very close to unity so that the cross section $d\sigma_{\text{ph}}(\mathbf{k}_e)/d\Omega_e$ itself is not very much affected by the infrared-divergence correction. This is also true for any other radiative corrections which do not become significant until $\hbar\omega_1$ reaches the MeV range (Botto and Gavrila, 1982).

Equation (42) implies that a correct description of photoionization involves consideration of simultaneous photon emission. This also concerns the observation of the inelastic scattering of individual photons by a detector arrangement which accepts photon energies larger than some limit $\Delta\epsilon$. In this case the efficiency for detecting a photon with an energy $\hbar\omega_2 > \Delta\epsilon$ is the same as for detecting this photon simultaneously with an arbitrary number of photons with total energy less than $\Delta\epsilon$. Hence the single-photon inelastic scattering cross section is also modified by the probability (48).

The considerations given above raise the question whether Eq. (45) has some validity outside the range $(0, \Delta\epsilon)$ of the undetected soft-photon emission. Recently Bannett *et al.* (1976, 1977) have studied the behavior of the Cu $K\alpha$ $L_{2,3}$ IXS cross section far below the threshold. The targets of the 8-keV Cu $K\alpha$ radiation were Ni, Cu, Zn, Ga, and Ge for which $\omega_1 \gtrsim \Omega_K$. Bannett *et al.* observed a plateaulike structure extending to the lower limit of their measurements at approximately 2.5 keV. This they took as evidence of the validity of Eq. (46) but other sources of photon emission are also possible. In the Ge spectrum the plateau is most prominent in comparison to the maximum of the cross section at $\omega_2 = \omega_1 - \Omega_L$. Following Sec. 4.2 we have therefore calculated, for $\theta_2 = 90°$, the ratio of the $\mathbf{A}\cdot\mathbf{A}$ and $\mathbf{p}\cdot\mathbf{A}$ cross sections at the maximum. The result is 0.83 which indicates that the $\mathbf{p}\cdot\mathbf{A}$ and $\mathbf{A}\cdot\mathbf{A}$ contributions are comparable also below the threshold $\omega_2 = \omega_1 - \Omega_L$, in analogy with the Ar case shown in Fig. 6. Hence the experimental findings of Bannett *et al.* (1976, 1977) may not be taken as final evidence for the validity of Eq. (45) without further investigations of the $\mathbf{A}\cdot\mathbf{A}$ contribution. This concerns also the total cross sections given in Table I. Note that the angular dependence (30) of the $\mathbf{A}\cdot\mathbf{A}$ cross section is different from that of Eq. (45), which would predict an almost isotropic behavior.

5. RELATIVISTIC AMPLITUDE FOR INELASTIC SCATTERING AND GAUGE INVARIANCE

In Sec. 2 the description of inelastic scattering was based on a nonrelativistic scattering approach in a restricted particle space. In particular, the states in the photon space were restricted to zero, one, or two photons. The positron states were not considered explicitly and the photon–electron interaction amplitude was only considered in the Coulomb gauge. In this section we shall briefly discuss the implications of removing these restrictions for inelastic photon scattering by bound many-electron systems.

5.1. *Relativistic Formulation*

In the quantum-electrodynamic (QED) formulation the photon–electron interaction operator

$$V_{\text{ph}} = -qc\gamma_\mu A^\mu(x) \tag{50}$$

is linear in the vector potential $A^\mu(x)$ and consequently also in the photon creation and destruction operators. Hence, if both the initial and final states of process (1) correspond to the lowest-order solution of Eq. (3) in the photon space, the intermediate states are automatically restricted to zero- and two-photon states in the transition amplitude (5). There is no direct amplitude which in the nonrelativistic approach is represented by the nonlinear $\mathbf{A}\cdot\mathbf{A}$ term.

In Eq. (50), γ_μ are the 4×4 matrices $\gamma_\mu = -i\beta\alpha_\mu$ ($\mu = 1, 2, 3$), $\gamma_4 = \beta$. The matrices α_μ and β are related to the 2×2 Pauli matrices $\sigma_\nu (\nu = x, y, z)$, and the unit matrix I, by

$$\alpha_\mu = \begin{pmatrix} 0 & \sigma_\mu \\ \sigma_\mu & 0 \end{pmatrix} (\mu = 1, 2, 3), \qquad \beta = \begin{pmatrix} I & 0 \\ 0 & -I \end{pmatrix}. \tag{51}$$

The γ_μ matrices operate in the transition amplitude on one-particle four-spinor solutions of the Dirac central-field equation including the positron states.

In the plane-wave representation, $A^\mu(x)$ is given by

$$A^\mu(x) = \sum_{\mathbf{k},\lambda} \left(\frac{\hbar}{2\varepsilon_0 \omega L^3}\right)^{1/2} e_\mu^{(\lambda)}(c_{k,\lambda} e^{ikx} + c_{k,\lambda}^\dagger e^{-ikx}), \tag{52}$$

where kx is the scalar product of the four-vector $x = (x, y, z, ict)$ and the four-dimensional wave vector $k = (\mathbf{k}, i\omega/c)$ for which $|\mathbf{k}|^2 = \omega/c$. The orthogonal four-dimensional polarization vectors $e^{(\lambda)}$ are related to k by the relations $e^{(\lambda)}k = 0$ ($\lambda = 1, 2$ transverse photons), $e^{(3)}k = \omega/c$ (longitudinal photons), and $e^{(4)}k = i\omega/c$ (scalar photons). The operators $c_{k,\lambda}$ and $c_{k,\lambda}^\dagger$ are the photon annihilation and creation operators, respectively. The specific form of the commutation rules for the scalar-photon operators depends on the choice of the metric (Akhiezer and Berestetskii, 1965). In the operator (50) with $A^\mu(x)$ given by Eq. (52) there is no specification of the gauge.

If the Coulomb gauge is chosen V_{ph} is given by $V_{\text{ph}} = -qc\boldsymbol{\alpha}\cdot\mathbf{A1}$, where the vector-potential operator (20) is denoted by **A**, and where **1** is the 4×4 unit matrix. In this gauge the nonrelativistic amplitude (21) is replaced by

$$\mathscr{T}_{FI}(\omega_1, \omega_2) = mc^2 \sum_{\mathscr{N}} \left(\frac{\langle F^- | \boldsymbol{\alpha} \cdot \mathbf{A}_1 | \mathscr{N} \rangle \langle \mathscr{N} | \boldsymbol{\alpha} \cdot \mathbf{A}_2^* | I \rangle}{\mathscr{E}_{\mathscr{N}} - E_I + \hbar\omega_2} + \frac{\langle F^- | \boldsymbol{\alpha} \cdot \mathbf{A}_2^* | \mathscr{N} \rangle \langle \mathscr{N} | \boldsymbol{\alpha} \cdot \mathbf{A}_1 | I \rangle}{\mathscr{E}_{\mathscr{N}} - E_I - \hbar\omega_1} \right), \tag{53}$$

where $\mathbf{A}_\lambda = \mathbf{A}_\lambda \mathbf{1}$ ($\lambda = 1, 2$) is for arbitrary polarizations a superposition of the transverse vector functions

$$\mathbf{A}_\lambda = \sum_{\nu=1}^{M} \mathbf{e}^{(q\lambda)} \exp(i\mathbf{k}_\lambda \cdot \mathbf{r}_\nu) \qquad (\lambda = 1, 2). \tag{54}$$

It is worth noting that in Eq. (53) the summation over $\mathscr{N}$ should also include the positron states and that the many-particle wave functions may be based on a Dirac central-field four-spinor basis. For $M = 1$ in Eq. (54) and for initial and final electron states the summation over intermediate positron states is simplified in the nonrelativistic limit, since $\mathscr{E}_{\mathscr{N}} - E_I \simeq 2mc^2 \gg \hbar\omega_{1(2)}$. This yields the $\mathbf{A} \cdot \mathbf{A}$ term in Eq. (22) after neglecting the small component in the four spinors (Akhiezer and Berestetskii, 1965). A similar approximation replaces $\boldsymbol{\alpha} \cdot \mathbf{A}_\lambda$ by $-(\mathbf{e}_\lambda \cdot \mathbf{p}) \exp(i\mathbf{k}_\lambda \cdot \mathbf{r})/mc$ and the four-spinors by the corresponding nonrelativistic wave functions in the summation over the intermediate electron states. This yields the $\mathbf{p} \cdot \mathbf{A}$ terms in Eq. (22). Obviously these approximations are also valid for $M > 1$, which shows the origin of the nonrelativistic approach, introduced in Sec. 2.2.

5.2. Gauge Invariance

In QED gauge invariance means that the fundamental equations including those describing the interaction between the electromagnetic and electron–positron fields remain invariant in the field-operator transformation

$$A^\mu = A^{\mu'} - \partial^\mu \eta, \tag{55a}$$

$$\psi^\mu = \exp(i\eta/\hbar)\, \psi^{\mu'}, \tag{55b}$$

$$\bar{\psi}^\mu = \exp(-i\eta/\hbar) \bar{\psi}^{\mu'}, \tag{55c}$$

where η is a scalar function satisfying

$$\Box \eta = \partial_\mu \partial^\mu \eta = 0 \tag{56}$$

but otherwise arbitrary. This assures that the Lorentz condition $\partial^\mu A_\mu = 0$ is valid for any choice of gauge. In general, we have

$$A^\mu = A_0^\mu + A_r^\mu, \tag{57}$$

where A_0^μ represents an external time-independent classical field $(\mathbf{A}, i\Phi_0/c)$ and where A_r^μ is the radiation field (52). The particle-field operators ψ^μ and $\bar{\psi}^\mu$ are related by $\bar{\psi} = \psi^* \beta$. They may be expanded in terms of eigenfunctions of a Dirac equation which includes A_0^μ or only a part of it with the free-particle Dirac equation as the limiting case.

Since there has been some confusion in the literature concerning the gauge invariance of transition amplitudes of one- and many-electron systems (e.g., Feuchtwang *et al.*, 1982) we shall briefly state the requirements that follow from this invariance. In accordance with Eq. (57) the Hamiltonian of a system interacting with photons can be divided into

$$H(A_0^\mu, A_r^\mu) = H_0(A_0^\mu) + V(A_0^\mu, A_r^\mu), \tag{58}$$

where H_0 includes the free fields and their interaction with the time-independent fields A_0^μ. Note that V may depend on A_0^μ if the photon–electron interaction operator V_{ph} is nonlinear in A_r^μ. Since the properties of the stationary electron (positron) states must be invariant in the gauge transformation of the time-independent field there is the first condition (I) that

$$i\hbar \frac{\partial \psi_n(A_0^\mu)}{\partial t} = H_0(A_0^\mu)\, \psi_n(A_0^\mu) = E_n \psi_n(A_0^\mu) \tag{59}$$

remains invariant in the transformation $A_0^\mu = A_0^{\mu'} - \partial^\mu \eta$, where $\partial\eta/\partial t = 0$. According to Eq. (55b), this also requires that the wave function transforms as $\psi_n(A_0^\mu) = \psi_n(A_0^{\mu'}) \exp(i\eta/\hbar)$. This is known as the form invariance of the wave equation (59) in a gauge transformation. This condition is fulfilled by the relativistic one-particle Dirac Hamiltonian

$$H_0^R(A_0^\mu) = c\boldsymbol{\alpha} \cdot (\mathbf{p} - q\mathbf{A}) + \boldsymbol{\beta} mc^2 + q\Phi_0 \mathbf{1} \tag{60a}$$

and also by the nonrelativistic one-electron Hamiltonian

$$H_0^{\text{NR}}(A_0^\mu) = \frac{1}{2m} (\mathbf{p} + e\mathbf{A})^2 - e\Phi_0, \tag{60b}$$

where we have put $q = -e$. Secondly it must be required that the transition amplitudes do not depend on the gauge used for A_0^μ. This condition (II), given by

$$\langle \psi_n(A_0^\mu)|\, V(A_0^\mu, A_r^\mu)\, |\psi_m(A_0^\mu)\rangle = \langle \psi_n(A_0^{\mu'})|\, V(A_0^{\mu'}, A_r^\mu)\, |\psi_m(A_0^{\mu'})\rangle, \tag{61}$$

is called the gauge invariance of transition amplitudes in a wave-function transformation. It should be noted that A_r^μ is the same in both amplitudes. Condition (61) is fulfilled by eigenfunctions of both Hamiltonians (60). Thirdly the transition amplitude must not change if A_0^μ remains fixed but A_r^μ is replaced by $A_r^{\mu'} - \partial^\mu \eta$. Hence condition III is

$$\langle \psi_n(A_0^\mu) | \, V(A_0^\mu, A_r^\mu) \, | \psi_m(A_0^\mu) \rangle = \langle \psi_n(A_0^\mu) | \, V(A_0^\mu, A_r^{\mu'}) \, | \psi_m(A_0^\mu) \rangle. \quad (62)$$

It is fulfilled by eigenfunctions of the relativistic Hamiltonian (60a) but *not* by eigenfunctions of its nonrelativistic counterpart (60b) if V includes the $\mathbf{A} \cdot \mathbf{A}$ term. In the latter case the most general form of V is

$$V(A_0^\mu, A_r^\mu) = \frac{e^2}{m} \mathbf{A} \cdot \mathbf{A}_r + \frac{e}{2m} (\mathbf{p} \cdot \mathbf{A}_r + \mathbf{A}_r \cdot \mathbf{p}) + \frac{e^2}{2m} \mathbf{A}_r \cdot \mathbf{A}_r - e\Phi_r, \quad (63)$$

which is gauge invariant only if the $\mathbf{A}_r \cdot \mathbf{A}_r$ term is excluded. Note that a static magnetic field represented by $\mathbf{A}$ is coupled to $\mathbf{A}_r$ in V and that there may be a scalar photon part Φ_r even in the nonrelativistic case.

The breakdown of condition III in the nonrelativistic description (22) of inelastic scattering in which the $\mathbf{A}_r \cdot \mathbf{A}_r$ is present can be traced back to the expansion of H_0^R in $(mc^2)^{-1}$. It leads to a nonrelativistic H_0^{NR} in which the linearity with respect to the vector part of the four-potential A^μ has been destroyed. In order to compensate for the lack of gauge invariance it is therefore necessary to include in Eq. (63) terms with coupling between $\mathbf{A}_r$ and $\mathbf{p}$ which are also nonlinear in $\mathbf{p}$ (Tulkki, unpublished).

5.3. *Length versus Velocity Forms*

It follows from Eq. (63) that in the absence of a static magnetic field the lowest-order gauge-invariant photon–electron interaction in powers of $(mc^2)^{-1}$ is

$$V(A_0^\mu, A_r^\mu) = \frac{e}{2m} (\mathbf{p} \cdot \mathbf{A}_r + \mathbf{A}_r \cdot \mathbf{p}) - e\Phi_r, \quad (64)$$

where for example we can choose $\Phi_r = 0$ so that $\mathbf{E}_r = -\partial \mathbf{A}_r / \partial t$. This implies that $\nabla \cdot \mathbf{A}_r = 0$ and $[\mathbf{p}, \mathbf{A}_r] = 0$, which is known as the Coulomb gauge. In the long-wavelength limit of Eq. (64) the gauge transformation $\mathbf{A}_r = \mathbf{A}_r' - \nabla \eta$, $\Phi_r = \Phi_r' + \partial \eta / \partial t$ with $\eta = \mathbf{r} \cdot \mathbf{A}_r(t)$ leads to

$$V(A_0^\mu, A_r^\mu) = e\mathbf{r} \cdot \mathbf{E}_r. \quad (65)$$

Here $\mathbf{E}_r$ is obtained in the Coulomb gauge from $\mathbf{E}_r = -\partial \mathbf{A}_r / \partial t$, where $\mathbf{A}_r(t)$ is the $\mathbf{r}$-independent vector potential.

The $\mathbf{p}\cdot\mathbf{A}_r$ representation is known as the velocity, and the $\mathbf{r}\cdot\mathbf{E}_r$ representation, as the length form in the dipole approximation of the photon–electron interaction. Since their relationship merely implies a gauge transformation their use should be consistent with condition III. Consequently, both forms yield identical transition amplitudes for eigenfunctions of H_0^{NR} with *local* potentials. As is well known this equivalence also follows from the commutator relation $[\mathbf{r}, H_0^{\mathrm{NR}}] = i\hbar\mathbf{v}$.

Following the considerations described above it can also be shown that for local potentials the velocity and length forms yield identical results for the sums over a *complete* set of intermediate states in Eq. (22), provided lifetime effects are neglected (Fried, 1973). However, for truncated sets of intermediate states the velocity and length forms do not yield identical results.

As a demonstration of the influence of truncation on the length and velocity forms we consider IXS by Ne $L_{2,3}$ electrons. The sum over intermediate states is restricted to $1s^{-1}$ and $2s^{-1}$ hole states. The ratio of the corresponding amplitudes is $r = R_{1s}F_{1s}/R_{2s}F_{2s}$, where

$$R_{ns} = \left[\left(\frac{dg_{ns}}{d\omega}\right) g_{2pns}\right]^{1/2}, \tag{66}$$

and where $F_{ns}(\mathbf{p}\cdot\mathbf{A})$ is

$$F_{ns} = [(\Omega_n + \omega_1 - \omega_2)\,\Omega_n]^{1/2}\left[\frac{1}{\Omega_n - \omega_2} + \frac{1}{\Omega_n + \omega_1}\right]. \tag{67}$$

In Eq. (67), $\Omega_n = \Omega_{ns} - \Omega_{2p}$. The length form $F_{ns}(\mathbf{r}\cdot\mathbf{E})$ is related to $F_{ns}(\mathbf{p}\cdot\mathbf{A})$ by

$$F_{ns}(\mathbf{r}\cdot\mathbf{E}) = [(\Omega_n + \omega_1 - \omega_2)\,\Omega_n]^{-1}\,F_{ns}(\mathbf{p}\cdot\mathbf{A}). \tag{68}$$

At $\omega_2 = \omega_1 - \Omega_L$, corresponding to the maximum of the IXS spectrum, we have $R_{1s}/R_{2s} \simeq 3.5$ according to known values of the oscillator densities and strengths. It follows that for $\omega_1 = 0.9\Omega_{1s}$ $r(\mathbf{p}\cdot\mathbf{A}) \simeq 7\times 10^3$ and $r(\mathbf{r}\cdot\mathbf{E}) \simeq 12$, respectively. This example illustrates that if only the resonant $1s^{-1}$ states are included in the sum over intermediate states, the velocity form gives more accurate results than the length form.

6. ANGULAR DISTRIBUTION AND POLARIZATION

The inelastically scattered radiation is in general polarized and varies in intensity with angle of emission. For randomly oriented target atoms, the anisotropic scattering pattern basically reflects the orientation of the incoming and outgoing radiation with respect to the electron orbitals in the intermediate states. For a spherically symmetric intermediate vacancy, the IXS would be essentially isotropic provided the interference between the resonant and nonresonant amplitudes is negligible. The classical analog is the absorption and reemission by a pair of randomly oriented dipoles with uncorrelated directions. The case of IXS by $L_{2,3}$ electrons close to the K absorption edge serves as an example of a nearly isotropic pattern of this type. Deviations from isotropy can be related to deviations from the one-electron central-field description of the relevant atomic states. In general, correlation effects in the intermediate states may influence the scattering pattern close to a resonance.

Orientation of the target atoms with respect to the incoming beam and polarization modifies the angular distribution and polarization of the inelastically scattered radiation. Since, as noted above, random orientation results in some cases in a nearly isotropic distribution, it puts angle-resolved IXS, especially in a combination with enhanced resolution, in a special position concerning studies of the target orientation. With the possibility of tuning the initial frequency, this technique may have distinct advantages over photoelectron-angular-distribution studies of oriented targets (Dill, 1976; Dill *et al.*, 1976). In contrast to IXS, differential photoelectron cross sections are never isotropic for random orientation.

In Sec. 6.1 we shall present the basic results of the theory (Tulkki and Åberg, 1980a) of inelastic photon scattering by oriented atoms, including an exposition of the $\mathbf{A}\cdot\mathbf{A}$ amplitude. In Sec. 6.2, the general form (35) of the doubly differential cross section for random orientation and for a polarization-insensitive detector is derived in the dipole approximation. The cross section for K-resonant IXS by $L_{2,3}$ electrons in an incomplete shell is generalized to include orientation. Resonant scattering by perfectly oriented molecules is also considered. We conclude with a remark concerning the relationship between our approach and the angular-momentum-transfer analysis (Starace, 1982) of fluorescent radiation subsequent to s-subshell photoionization.

6.1. General Formalism for Inelastic Photon Scattering

In the analysis of the scattering probability (17), maximum advantage should be taken of the internal symmetry of the scatterers. If the density operator (10) of orientation is given with respect to an atomic basis then the statistical tensor operators, defined by

$$\hat{T}_Q^{(K)}(\gamma J, \gamma' J') = \sum_{MM'} (-1)^{J'-M'} \langle JMJ'-M' \mid KQ \rangle \, |\gamma JM\rangle\langle\gamma' J' M'| \qquad (69)$$

(Blum, 1981) provide the framework. The operator (69) is the Qth component of an irreducible spherical tensor of rank K with the reduced matrix element $(2K+1)^{1/2}$. Using the properties of spherical tensors, one finds

$$\langle \hat{X}_Q^{(K)} \rangle = \mathrm{Tr}[\boldsymbol{\rho} \mathbf{X}_Q^{(K)}] = \sum_{\gamma \mathcal{J}, \gamma' \mathcal{J}'} (-1)^{\mathcal{J}-\mathcal{J}'+Q} \frac{\langle \gamma' \mathcal{J}' \| \hat{X}_Q^{(K)} \| \gamma \mathcal{J} \rangle}{(2K+1)^{1/2}} \times a_{-Q}^{(K)}(\gamma \mathcal{J}, \gamma' \mathcal{J}'), \tag{70}$$

where the state multipoles $a_Q^{(K)}(\gamma \mathcal{J}, \gamma' \mathcal{J}')$ are given by

$$a_Q^{(K)}(\gamma \mathcal{J}, \gamma' \mathcal{J}') = \mathrm{Tr}[\boldsymbol{\rho} \mathbf{T}_Q^{(K)}(\gamma \mathcal{J}, \gamma' \mathcal{J}')^\dagger]. \tag{71}$$

In Eq. (69), (70), and (71), γ represents the set of quantum numbers which is needed in addition to the total-angular-momentum quantum numbers $\mathcal{J}$ and M to specify the atomic state. Equations (70) and (71) relate the expectation value of any spherical tensor operator to the state multipoles describing the atomic orientation.

The expansion of the transition and density matrices in terms of the tensors (69) is particularly convenient for problems in which the orientation of the atoms is due to collisions or external fields. In principle, molecular orientation is best described using statistical multipoles which have been given with respect to the irreducible representations of the molecular point group. Nevertheless, the atomic expansion is also useful for an analysis of molecular IXS since an atomic basis often describes localized excitations in a molecule quite accurately and since the multipole expansion (A.1) of the transverse photon states can be used without further transformations.

Our general results are based in the following on the relativistic Coulomb-gauge amplitude (53), but with the summation over the positron states effectively excluded. By using the expansion (A.1) the photon–electron interaction operator $V_{\mathrm{ph}} = ec\boldsymbol{\alpha} \cdot \mathbf{A}$ is expressed in terms of the irreducible tensors (A.8), where the spherical photon states $\mathbf{A}_{\mathcal{J}M\omega}^{(\xi)}$ are defined by Eq. (A.4). The corresponding probability (17) is obtained by a recoupling procedure (Tulkki and Åberg, 1980a) which is based on the reasonable assumption that the initial-, intermediate-, and final-state energies are independent of the magnetic quantum numbers. For a given multipole interaction $(\mathcal{J}, \xi)$, the result is

$$d\left[\frac{d^2\sigma}{\hbar d\omega_2\, d\Omega}\right] = \left[\frac{\alpha c(2\mathcal{J}+1)}{8\pi}\right]^2 \left(\frac{\omega_2}{\omega_1}\right) \left[\sum_{H\eta} \sum_{\kappa\kappa'} X_\eta^{(H)}(\kappa\kappa')\, Y_\eta^{(H)}(\kappa\kappa')\right] d\omega_1. \tag{72}$$

This cross section describes the probability that a photon in the interval $(\omega_1, \omega_1 + d\omega_1)$ with the frequency distribution (14) is inelastically scattered into solid angle $d\Omega_2$ with the frequency in the interval $(\omega_2, \omega_2 + d\omega_2)$. The detector is assumed to respond to a definite circular polarization

$\mathbf{e}^{(q_2)}$ $(q_2 = \pm 1)$ and with the same efficiency to all final electronic states $|\gamma_F \mathcal{J}_F M_F\rangle$ which have the energy E_F. In this case the detector operator (18) is

$$\hat{F} = \sum_{\gamma_F \mathcal{J}_F M_F} |\mathbf{e}^{(q_2)} \exp(i\mathbf{k}_2 \cdot \mathbf{r})\rangle \, |\gamma_F \mathcal{J}_F M_F\rangle\langle\gamma_F \mathcal{J}_F M_F| \, \langle\mathbf{e}^{(q_2)} \exp(i\mathbf{k}_2 \cdot \mathbf{r})|, \quad (73)$$

where the summation must be consistent with conservation of energy. For this detector arrangement, the geometrical factor $X_\eta^{(H)}$ in Eq. (72) is

$$X_\eta^{(H)}(\kappa\kappa') = \sum_{\substack{\Lambda\lambda \\ \Lambda'\lambda'}} (-1)^{\kappa'} [(2\kappa+1)(2\kappa'+1)]^{1/2} \begin{Bmatrix} \mathcal{J} & \mathcal{J} & \Lambda' \\ \mathcal{J} & \mathcal{J} & \Lambda \\ \kappa & \kappa' & H \end{Bmatrix} X_\eta^H(\Lambda\lambda\Lambda'\lambda'), \quad (74)$$

where

$$\begin{aligned} X_\eta^{(H)}(\Lambda\lambda\Lambda'\lambda') = \sum_{q_1 q_1'} & [4\pi(2\Lambda'+1)]^{1/2} (-1)^{q_2+q_1'} \\ & \times \mathscr{Y}_{\Lambda\lambda}(\theta_2, \varphi_2)^* \rho_{q_1 q_1'} D^{(\Lambda')}_{\lambda' q_1 - q_1'}(\varphi_1, \theta_1, 0) \\ & \times \langle\Lambda'\lambda'\Lambda\lambda \mid H\eta\rangle\langle\mathcal{J} - q_1' \mathcal{J} q_1 \mid \Lambda' q_1 - q_1'\rangle\langle\mathcal{J} q_2 \mathcal{J} - q_2 \mid \Lambda 0\rangle. \end{aligned} \quad (75)$$

This result can be generalized to any state of polarization picked up by the detector.

The dynamical scattering factor $Y_\eta^{(H)}(\kappa\kappa')$ is given by Eq. (70), where the reduced matrix elements which are independent of orientation are evaluated by successive coupling from the multipole operators (A.8) including the projection over the intermediate states (Tulkki and Åberg, 1980a). The significance of the result (72) lies in the fact that it completely disentangles the dynamical effects from the geometrical angle- and polarization-dependent factors. The fluorescence pattern accompanying photoexcitation or -ionization follows as a special case.

As pointed out in Sec. 2.2, the density matrix $\boldsymbol{\rho}_{\text{or}}$ and consequently the state multipoles (71) are in principle only predictable through a complete set of linearly independent expectation values (12). In practice, the number of independent parameters is reduced by symmetry constraints. For example, for systems which are cylindrically symmetric around the laboratory z axis, we have $a_Q^{(K)} = \delta_{Q0} a_0^{(K)}$. Note also that statistical optimization methods involving a few parameters may often be necessary for the determination of the "best" $\boldsymbol{\rho}_{\text{or}}$.

From an experimental point of view, complete orientation measurements are hardly possible. For a fixed polarization of the incoming beam, the directions of the wave vectors $\mathbf{k}_1$ and $\mathbf{k}_2$ may be varied independently with respect to the laboratory frame. This provides values of the $(2J+1)^4$ independent geometrical functions $X_\eta^{(H)}$ in the cross section (72). Only if the number of these functions is larger than or equal to the number of independent parameters in $\boldsymbol{\rho}_{\rm or}$ can this density matrix be completely determined. Hence symmetry arguments combined with statistical models would usually be required for the interpretation of experimental data on the inelastic scattering by oriented systems.

According to Sec. 5.1, the summation over the positron states in the amplitude (53) leads to the $\mathbf{A}\cdot\mathbf{A}$ term in the nonrelativistic limit. Hence the angular distribution and polarization associated with this term could in principle be treated on the same footing as the inelastic multipole scattering discussed above. However, as noted in Sec. 3, the $\mathbf{A}\cdot\mathbf{A}$ term dominates for $\omega_1 \gg \Omega_K$ and can be treated, especially for low-momentum transfer values, in the dipole approximation. For random orientation, this leads to an angular dependence expressed by Eqs. (29) and (30).

The generalization to IXS by oriented target atoms is straightforward in the dipole approximation since the $\mathbf{A}\cdot\mathbf{A}$ and $\mathbf{p}\cdot\mathbf{A}$ amplitudes add incoherently in this case. We whall merely give the final result in terms of the spherical polarization tensor

$$\mathscr{E}_\eta^{(H)}(\mathbf{e}^{(q)}) = \sum_{MM'} \langle 1M1M' \mid H\eta \rangle \, D_{Mq}^{(1)}(\varphi, \theta, 0) \, D_{M'q}^{(1)}(\varphi, \theta, 0)^*, \tag{76}$$

the position tensor

$$r_\eta^{(H)} = \sum_{MM'} \langle 1M1M' \mid H\eta \rangle \, r_M P_F r_{M'}, \tag{77}$$

and the momentum-transfer tensor

$$G_\eta^{(H)} = k^2 \sum_{MM'} \langle 1M1M' \mid H\eta \rangle \, D_{M0}^{(1)}(\hat{k}) \, D_{M'0}^{(1)}(\hat{k})^*. \tag{78}$$

In Eq. (77), we have $P_F = \sum_{M_F} |FM_F\rangle\langle FM_F|$ and in Eq. (78), $\hat{k}$ is the direction of the momentum transfer with respect to the laboratory frame. The $\mathbf{A}\cdot\mathbf{A}$ cross section, corresponding to the multipole cross section (72), is

$$d\left[\frac{d^2\sigma}{\hbar d\omega_2 \, d\Omega}\right] = r_0^2 \frac{\omega_2}{\omega_1} \sum_{\substack{HH' \\ \eta\eta'}} \mathrm{Tr}[\boldsymbol{\rho}_{\rm ph} \mathscr{E}_\eta^{(H)}(\mathbf{e}^{(q_1)})] \, \mathrm{Tr}[\boldsymbol{\rho}_{\rm or} \mathbf{r}_{\eta'}^{(H')}] \times \mathscr{E}_\eta^{(H)}(\mathbf{e}^{(q_2)}) \, G_{\eta'}^{(H')}(\hat{k}) \, d\omega_1, \tag{79}$$

where the expectation values with respect to the tensors (76) and (77) are defined by Eq. (70). Symmetry constraints simplify Eq. (79) in analogy with Eq. (72). For example, if there is cylindrical and inversion symmetry with respect to the laboratory frame, the following relation holds independently of the polarization state of the incoming beam:

$$\frac{\left(\frac{d\sigma}{d\Omega}\right)_{0^\circ} - \left(\frac{d\sigma}{d\Omega}\right)_{54.7^\circ}}{\left(\frac{d\sigma}{d\Omega}\right)_{54.7^\circ}} = \frac{\mathrm{Tr}[\boldsymbol{\rho}_{\mathrm{or}} \mathbf{r}_0^{(2)}]}{r_0^{(0)}}, \tag{80}$$

where the angle θ of the cross section $(d\sigma/d\Omega)_\theta$ refers to the direction of the momentum transfer vector $\mathbf{k}$ with respect to the symmetry axis.

6.2. Scattering Patterns in Dipole Approximation

The assumption of random orientation implies $H = \eta = 0$ in cross section (72). Consequently, for $J = 1$ there are three geometrical coefficients $X_0^{(0)}(\kappa\kappa)$, corresponding to $\kappa = 0$, 1, and 2 in Eq. (74). In each coefficient the possible Λ values are $\Lambda = 0$, 1, and 2 out of which the assumption of a polarization-insensitive detector excludes the $\Lambda = 1$ value through the summation over q_2. Using Eq. (A.7), this yields, independently of the dynamical scattering factors $Y_0^{(0)}(\kappa\kappa)$, a cross section of the form (34), valid for an arbitrary polarization state of the incoming beam. For an unpolarized beam this cross section reduces to the familiar dipole form

$$\frac{d^2\sigma}{d\omega_2\, d\Omega} = a + b \cos^2 \theta_2. \tag{81}$$

Gavrila (1972b) has given, both analytically and numerically, in the interval $1.05 I_{1s} \leqslant \hbar\omega_1 \leqslant 20 I_{1s}$ the coefficients a and b for inelastic scattering by $1s$ electrons bound to hydrogenlike ions. These coefficients have also been tabulated for scattering by $2s$ and $2p$ electrons (Gavrila and Tugulea, 1975).

The influence of the target orientation on the radiation pattern is most easily detected using frequencies close to a resonance. Hence we shall follow earlier work (Tulkki and Åberg, 1980b) to give the cross section (72) in the nonrelativistic LS-coupling limit without the nonresonant term. The result is

$$d\left(\frac{d^2\sigma}{\hbar d\omega_2\, d\Omega}\right) = \left(\frac{r_0}{m\hbar}\right)^2 \left(\frac{\omega_2}{\omega_1}\right) \sum_{H\eta} \sum_{\substack{\Lambda\lambda \\ \Lambda'\lambda'}} X_\eta^H(\Lambda\lambda\Lambda'\lambda')\, Y_\eta^{(H)}(\Lambda\lambda')\, d\omega_1, \tag{82}$$

where the geometrical coefficient $X_\eta^H(\Lambda\lambda\Lambda\lambda')$ is obtained from Eq. (75) by putting $\mathscr{J}=1$, and where we have

$$\Upsilon_\eta^{(H)}(\Lambda\Lambda')=\sum_{\substack{\gamma_I L_I\\ \gamma'_I L'_I}}\sum_{\gamma_F L_F}\sum_{\substack{\gamma_N L_N\\ \gamma'_N L'_N}}(-1)^{L_I+L'_I+L_F+L'_N+\eta+1}$$
$$\times\begin{Bmatrix}\Lambda & L_N & L'_N\\ L_F & 1 & 1\end{Bmatrix}\begin{Bmatrix}L_I & 1 & L'_N\\ H & \Lambda' & \Lambda\\ L'_I & 1 & L_N\end{Bmatrix}B^*_{N'}(L'_I)\,B_N(L_I)\,a_{-\eta}^{(H)}(\gamma_I L_I,\gamma'_I L'_I). \tag{83}$$

The factors $B_N(L_I)$, given by

$$B_N(L_I)=\frac{\langle\gamma_F L_F\|\,\mathbf{p}\,\|\gamma_N L_N\rangle\langle\gamma_N L_N\|\,\mathbf{p}\,\|\gamma_I L_I\rangle}{\Omega_N-\Omega_I-\omega_1-i\Gamma_N/2\hbar}, \tag{84}$$

describe the effect of the dipole interactions and intermediate-state lifetimes on the cross section (82), in which $\Lambda(\Lambda')\leqslant 2$ and $H\leqslant 4$. It follows from Eq. (83) that, if $L_I=L'_I=0$, there are no orientational effects on the radiation pattern, since $H=0$.

In the case of IXS by np electrons in a complete or an incomplete $(np)^k$ shell, Eq. (83) leads to a simple result, provided γ_N is restricted to 1s-vacancy configurations. Using the one-electron central-field approximation, it follows from Eq. (83) that

$$\begin{aligned}\Upsilon_\eta^{(H)}(\Lambda\Lambda')&=\Upsilon_\eta^{(H)}(H0)\\ &=\sum_{L_I L'_I}(-1)^{\eta+1+L'_I+L_F}\left[\frac{(2L'_I+1)(2L_I+1)}{27(2H+1)}\right]^{1/2}\begin{Bmatrix}H & L_I & L'_I\\ L_F & 1 & 1\end{Bmatrix}\\ &\quad\times k(p^k S_I L'_I\{|p^{k-1}[S_F L_F]\,p)(p^k S_I L_I\{|p^{k-1}[S_F L_F]\,p)\\ &\quad\times|B_N|^2\,a_{-\eta}^{(H)}(L_I,L'_I),\end{aligned} \tag{85}$$

for a final $(np)^{k-1}S_F L_F$ term. The factor B_N is given by Eq. (84), but with the product $\langle 1s\|\,\mathbf{p}\,\|\epsilon p\rangle\langle 2p\|\,\mathbf{p}\,\|1s\rangle$ of *one-electron* reduced matrix elements in the denominator. Since we have $\Lambda'=0$ according to Eq. (85), $H=0$ leads to $\Lambda=\lambda=\lambda'=0$ in Eq. (75). Consequently, the one-electron central-field approximation predicts an isotropic K-resonant cross section (82) for IXS by $L_{2,3}$ electrons in randomly oriented atoms, in accordance with the result (36), used in our description of the origin of K fluorescence. Within this approximation there is obviously no anisotropy for any number of open shells. Hence any significant polarization or angular dependence of K-resonant or fluorescence radiation would indicate target orientation. It is also significant in this context that, according to Eq. (36), the anisotropic interference term between the resonant and nonresonant amplitude becomes very small close to the K resonance.

Since the statistical multipoles $a_{-\eta}^{(H)}(\gamma_I L_I, \gamma'_I L'_I)$ in Eq. (83) are in principle given with respect to a complete set of atomic basis functions $|\gamma SLM_S M_L\rangle$, our formalism can also be used for calculations of the radiation pattern of atoms in ligand fields (Tulkki and Åberg, 1980a) and in oriented molecules. Within the linear-combination-of-atomic-orbital (LCAO) approximation, the initial-state molecular wave function is expanded with respect to the scattering center and the corresponding multipoles $a_{-\eta}^{(H)}$ are derived. The same basis set is chosen to represent the final and intermediate states. This approximation is especially justified for K-resonant radiation or fluorescence since the $1s$ vacancy is localized at the scattering center and the corresponding intermediate-state energies are large in comparison with the molecular binding energies. In lowest order this procedure is related to the one-electron molecular-orbital method which has been used for the analysis of the intensity distribution of K X rays from randomly oriented molecules (Manne, 1971) and also from oriented molecules and crystalline solids (e.g., Wiech, 1982).

As an illustration we consider the scattering pattern of K-resonant IXS by valence electrons in perfectly oriented linear molecules. It is assumed that the detector is insensitive to polarization. It follows from Eqs. (82) to (85) that

$$\frac{d^2\sigma}{\hbar d\omega_2\, d\Omega} = \frac{1}{9}\left(\frac{r_0}{m\hbar}\right)^2 \frac{\omega_2}{\omega_1} \times \left[\mathcal{N}_{n\sigma}\, |B_{n\sigma}(\omega_2)|^2\, a_{n\sigma}^2 \sin^2\theta_2 + \mathcal{N}_{n\pi}\, |B_{n\pi}(\omega_2)|^2\, a_{n\pi}^2 \frac{1+\cos^2\theta_2}{2} \right], \tag{86}$$

where

$$|B_\nu|^2 = \frac{|\langle 1s\|\, \mathbf{p}\, \|2p\rangle\langle \epsilon p\|\, \mathbf{p}\, \|1s\rangle|^2}{(\Omega_K - \Omega_\nu - \omega_2)^2 + \Gamma_{1s}^2/4\hbar^2} \tag{87}$$

peaks at slightly different ω_2 values since $\Omega_{n\sigma}$ is in general larger than $\Omega_{n\pi}$ by, say, typically 10 eV. In Eq. (86), θ_2 is the scattering angle between the fixed molecular axis and $\mathbf{k}_2$. Both the σ and π electrons are assumed to have p symmetry with respect to the scattering center with the localized $1s$ hole. The corresponding coefficients in the one-electron LCAO expansion are denoted by $a_{n\sigma}$ and $a_{\pi\sigma}$, respectively. The number of the σ electrons is given by $\mathcal{N}_{n\sigma}$ ($\leqslant 2$), and the number of π electrons, by $\mathcal{N}_{n\pi}$ ($\leqslant 4$).

Equation (86) can be generalized to the scattering pattern of K-resonant IXS by valence electrons in perfectly oriented polyatomic molecules. In this case the molecular orbitals are identified according to the irreducible representations of the molecular point group. The corresponding angular functions are given in Table II as a function of the direction $\mathbf{k}_2$ with respect to the principal symmetry axes. Measurements of the anisotropic emission of K X rays (Wiech, 1982) seem to agree qualitatively with angular distributions given in Table II.

The cross section (82) applies to cases in which the orientational effects are stronger than spin–orbit splitting. If the orientation is caused by a field which is weaker than spin–orbit splitting, a $|\gamma SLJM\rangle$ basis is more appropriate. In this case, the energy-level structure may be resolved in the inelastic scattering, and Eq. (82) must be modified accordingly (Tulkki and Åberg, 1980a). We shall not reproduce these formulas here but point out an application which illustrates the generality of our approach.

The analysis of K-resonant IXS in Sec. 4.1, as well as the generalization (85) to systems with orientation, is essentially based on the one-electron approximation. In the case of random orientation the isotropy may be broken by correlation effects. Close to threshold, the polarization of the core by a slow continuum electron in the intermediate state can affect the angular distribution and polarization properties of the emitted radiation. This effect has recently been analyzed by Starace (1982) for fluorescence resulting from photoionization of an s subshell. Starace uses the Dill–Fano angular-momentum-transfer expansion which is appropriate for problems involving random target orientation. In our formalism, Starace's results should follow, provided only the resonant amplitude is included and the summation over the intermediate states is restricted to a *specific* energy level. This has been verified by a derivation (Tulkki, 1984) of the polarization factor

$$P_L(\theta_2) = \frac{I(\|) - I(\perp)}{I(\|) + I(\perp)}, \tag{88}$$

where θ_2 is the angle between $\mathbf{k}_2$ and the incident unpolarized beam propagating in the z direction of the laboratory frame in Fig. A.1. In

TABLE II

Angular functions for K-resonant scattering patterns from perfectly oriented polyatomic molecules.

Representation	Basis function	Angular dependence
A	x	$\sin^2 \varphi_2 + \cos^2 \varphi_2 \cos^2 \theta_2$
	y	$\cos^2 \varphi_2 + \sin^2 \varphi_2 \cos^2 \theta_2$
	z	$\sin^2 \theta_2$
E	x, y	$1 + \cos^2 \theta_2$
T	x, y, z	isotropic

Eq. (88), $I(\|)$ is the intensity measured by a detector which accepts linear polarization in the scattering plane, defined by $\mathbf{k}_1$ and $\mathbf{k}_2$. The intensity $I(\perp)$ corresponds to polarization perpendicular to the scattering plane. The derivation is based on formulas given by Tulkki and Åberg (1980a) and on the use of the detector operator

$$\hat{F}=\sum_{M_F}|S_F L_F J_F M_F\rangle\langle S_F L_F J_F M_F|\sum_{\alpha\alpha'}\rho^{(\nu)}_{\alpha\alpha'}|\mathbf{e}^{(\alpha)}\rangle\langle\mathbf{e}^{(\alpha')}|, \tag{89}$$

where the wave function $|S_F L_F J_F M_F\rangle$ refers to the final energy *level* in inelastic scattering by a $(np)^k$ subshell, and where the density matrices $\boldsymbol{\rho}^{(\|)}$ and $\boldsymbol{\rho}^{(\perp)}$ define the linear polarizations picked up by the detector. The result is

$$P_L(\theta_2)=\frac{3r\sin^2\theta_2}{2-r(3\cos^2\theta_2-1)}, \tag{90}$$

where

$$r=\frac{3}{2}(2L_c+1)(2J_c+1)(-1)^{J_f-S_c} \times\begin{Bmatrix}2 & 1 & 1\\ J_f & J_c & J_c\end{Bmatrix}\begin{Bmatrix}L_c & L_c & 2\\ 1 & 1 & 1\end{Bmatrix}\begin{Bmatrix}L_c & L_c & 2\\ J_c & J_c & S_c\end{Bmatrix}, \tag{91}$$

in complete agreement with Starace (1982), who also discusses the significance of Eq. (90). In Eq. (91) the indices f and c refer to the quantum numbers of the final and intermediate ionic core states, respectively. The result (90) implies that there is negligible overlap between adjacent resonant states.

ACKNOWLEDGMENTS

We thank A. Blomberg and M. Gavrila for reading the manuscript. This work was supported by the Finnish Academy of Sciences.

APPENDIX

The wave vectors $\mathbf{k}_1$ and $\mathbf{k}_2$ of the incoming and outgoing photons are defined with respect to the laboratory frame in Fig. A.1. In this frame, the transverse photon states can be expressed by the multipole expansion

$$\mathbf{e}^{(q)}e^{i\mathbf{k}\cdot\mathbf{r}}=\sum_{J=0}^{\infty}\sum_{M=-J}^{J}\left(\frac{2J+1}{8\pi}\right)^{1/2}D^{(J)}_{Mq}(\varphi,\theta,0)[(-q)\,\mathbf{A}^{(m)}_{JM\omega}+\mathbf{A}^{(e)}_{JM\omega}], \tag{A.1}$$

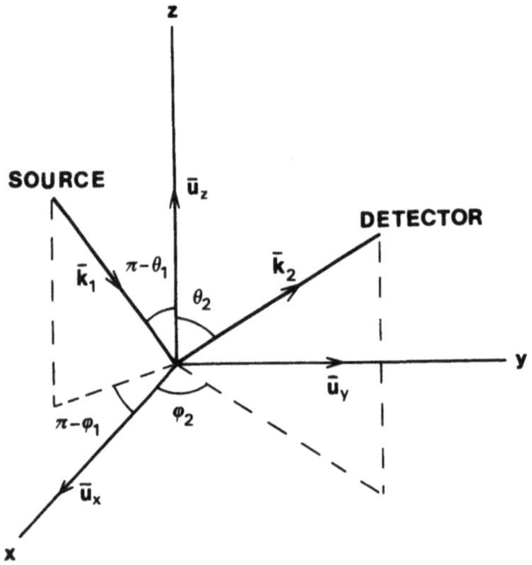

FIGURE A.1. Scattering geometry in the laboratory frame. The circular polarization or helicity states $\mathbf{e}^{(q_\nu)}$ ($q_\nu = \pm 1$) are the spherical unit vectors of the source ($\nu = 1$) and detector ($\nu = 2$) frames. These coordinate systems are obtained from the laboratory frame by a rotation corresponding to the Euler angles $\varphi_\nu, \theta_\nu, 0$ ($\nu = 1, 2$) as explained in the text.

where $\mathbf{e}^{(q)}$ are the spherical unit vectors ($q = \pm 1$) and where the matrix elements $D^{(J)}_{Mq}(\varphi, \theta, 0)$ of finite rotations refer to the rotation which carries the laboratory frame into the detector or source frame (Fig. A.1). In this Appendix, the multipolarity is denoted by J.

The spherical photon states $\mathbf{A}^{(\xi)}_{JM\omega}$ can be related to the vector spherical harmonics, defined by

$$\mathbf{Y}_{JLM}(\hat{k}) = \sum_{mq} \langle Lm1q \mid JM \rangle \, \mathcal{Y}_{LM}(\hat{k}) \, \mathbf{u}^{(q)}, \tag{A.2}$$

where $\mathcal{Y}_{LM}$ are the spherical harmonics and where $\mathbf{u}^{(q)}$ are the spherical unit vectors. The transverse and longitudinal combinations of $\mathbf{Y}_{JLM}$, given by

$$\begin{pmatrix} \mathbf{Y}^{(e)}_{JM} \\ \mathbf{Y}^{(m)}_{JM} \\ \mathbf{Y}^{(l)}_{JM} \end{pmatrix} = \begin{pmatrix} [J/(2J+1)]^{1/2} & 0 & [(J+1)/(2J+1)]^{1/2} \\ 0 & -1 & 0 \\ [(J+1)/(2J+1)]^{1/2} & 0 & -[J/(2J+1)]^{1/2} \end{pmatrix} \begin{pmatrix} \mathbf{Y}_{JJ+1\,M} \\ \mathbf{Y}_{JJM} \\ \mathbf{Y}_{JJ-1\,M} \end{pmatrix}, \tag{A.3}$$

reduce the definition of $\mathbf{A}^{(\xi)}_{JM\omega}$ ($\xi = e$, electric; $\xi = m$, magnetic; $\xi = l$, longitudinal) to

$$\mathbf{A}^{(\xi)}_{JM\omega}(\mathbf{r}) = \int d\Omega_{\hat{k}} \, \mathbf{Y}^{(\xi)}_{JM}(\hat{k}) \exp(i\mathbf{k}\cdot\mathbf{r}). \tag{A.4}$$

The parity ξ of the spherical photon states can be obtained from

$$-\mathbf{A}^{(\xi)}(-\mathbf{r}) = \pi_\xi \mathbf{A}^{(\xi)}(\mathbf{r}), \tag{A.5}$$

where we have $\pi_\xi = (-1)^J$ for $\xi = e, l$ and $\pi_m = (-1)^{J+1}$. It follows that

$$\mathbf{A}^{(\xi)*}_{JM\omega} = (-1)^{J+M+1}\,\mathbf{A}^{(\xi)}_{J-M\omega}. \tag{A.6}$$

In this work we have used the Condon–Shortley phase convention for the spherical harmonics and the following definition of Eulerian angles in the matrix elements $D^{(J)}_{MM'}(\alpha, \beta, \gamma)$: α indicates the rotation around the laboratory z axis, β the subsequent rotation around the new y axis, and γ the rotation around the z axis, resulting from the previous two rotations. Our convention is the same as used, for example, by Edmonds (1974) and by Brink and Satchler (1975). In the angular-distribution analysis we have repeatedly used the relations

$$\begin{aligned} D^{(J)}_{MM'}(\alpha, \beta, \gamma)^* &= (-1)^{M-M'}\,D^{(J)}_{-M-M'}(\alpha, \beta, \gamma), \\ D^{(J)}_{M0}(\alpha, \beta, \gamma) &= \left(\frac{4\pi}{2J+1}\right)\mathscr{Y}_{JM}(\beta, \alpha)^*. \end{aligned} \tag{A.7}$$

The recoupling procedures in Sec. 6 are essentially based on the fact that

$$\mathbf{Q}^{(J)}_M(\xi, \omega) = \boldsymbol{\alpha}\cdot\mathbf{A}^{(\xi)}_{JM\omega} \tag{A.8}$$

is a spherical tensor of rank J (Tulkki and Åberg, 1980a).

REFERENCES

ÅBERG, T., 1975, Two-Photon Emission, the Radiative Auger Effect, and the Double Auger Process, in *Atomic Inner-Shell Processes*, edited by B. Crasemann (Academic, New York), Vol. I, p. 353.

ÅBERG, T., 1980, *Phys. Scr.* **21**, 495.

ÅBERG, T., 1981, The Creation and Decay of Inner-Shell Vacancies as a Single Scattering Process, in *Inner-Shell and X-Ray Physics of Atoms and Solids*, edited by D. J. Fabian, H. Kleinpoppen, and L. M. Watson (Plenum, New York), p. 251.

ÅBERG, T., AND G. HOWAT, 1982, Theory of the Auger Effect, in *Handbuch der Physik*, Vol. XXXI, edited by W. Mehlhorn (Springer-Verlag, Berlin), p. 469.

ÅBERG, T., AND J. UTRIAINEN, 1969, *Phys. Rev. Lett.* **22**, 1346.

AKHIEZER, A. I., AND V. B. BERESTETSKII, 1965, *Quantum Electrodynamics* (Wiley, New York).

BANNETT, Y. B., AND I. FREUND, 1975, *Phys. Rev. Lett.* **34**, 372.

BANNETT, Y. B., J. I. GERSTEN, N. TZOAR, AND I. FREUND, 1976, *Phys. Rev. Lett.* **36**, 882.

BANNETT, Y. B., D. C. RAPAPORT, AND I. FREUND, 1977, *Phys. Rev. A* **16**, 2011.

BLUM, K., 1981, *Density Matrix Theory and Applications* (Plenum, New York).

BOTTO, D. J., AND M. GAVRILA, 1982, *Phys. Rev. A* **26**, 237.

BRIAND, J. P., D. GIRARD, V. O. KOSTROUN, P. CHEVALIER, K. WOHRER, AND J. P. MOSSÉ, 1981, *Phys. Rev. Lett.* **46**, 1625.

BRINK, D. M., AND G. R. SATCHLER, 1975, *Angular Momentum* (Clarendon, Oxford), 2nd Ed.

BROWN, G. S., M. H. CHEN, B. CRASEMANN, AND G. E. ICE, 1980, *Phys. Rev. Lett.* **45**, 1937.

BUSHUEV, V. A., AND R. N. KUZ'MIN, 1977, *Sov. Phys. Usp.* **20**, 406.

COHEN, G. G., N. G. ALEXANDROPOULOS, AND M. KURIYAMA, 1973, *Phys. Rev. B* **8**, 5427.
COSTESCU, A., AND M. GAVRILA, 1973, *Rev. Roum. Phys.* **18**, 493.
DEHMER, J. L., D. DILL, AND A. C. PARR, 1983, Photoionization Dynamics of Small Molecules, in *Photophysics and Photochemistry in the Vacuum Ultraviolet*, edited by S. McGlynn, G. Findley, and R. Huebner (Reidel, Dordrecht).
DILL, D., 1976, *J. Chem. Phys.* **65**, 1130.
DILL, D., J. SIEGEL, AND J. L. DEHMER, 1976, *J. Chem. Phys.* **65**, 3158.
EDMONDS, A. R., 1974, *Angular Momentum in Quantum Mechanics* (Princeton Univ. Press, Princeton, New Jersey).
EISENBERGER, P., 1980, Inelastic Scattering, in *Synchrotron Radiation Research*, edited by H. Winick and S. Doniach (Plenum, New York), p. 639.
EISENBERGER, P., AND P. M. PLATZMAN, 1970, *Phys. Rev. A* **2**, 415.
EISENBERGER, P., P. M. PLATZMAN, AND K. C. PANDY, 1973, *Phys. Rev. Lett.* **31**, 311.
EISENBERGER, P., P. M. PLATZMAN, AND H. WINICK, 1976a, *Phys. Rev. B* **31**, 2377.
EISENBERGER, P., P. M. PLATZMAN, AND H. WINICK, 1976b, *Phys. Rev. Lett.* **36**, 623.
FEUCHTWANG, T. E., E. KAZES, H. GROTCH, AND P. H. CUTLER, 1982, *Phys. Lett.* **93A**, 4.
FRIED, Z., 1973, *Phys. Rev. A* **8**, 2835.
GAVRILA, M., 1972a, *Phys. Rev. A* **6**, 1348.
GAVRILA, M., 1972b, *Phys. Rev. A* **6**, 1360.
GAVRILA, M., AND M. N. TUGULEA, 1975, *Rev. Roum. Phys.* **20**, 209.
HALL, J. M., K. H. JAMISON, O. L. WEAVER, P. RICHARD, AND T. ÅBERG, 1979, *Phys. Rev. A* **19**, 568.
HEITLER, W., 1954, *The Quantum Theory of Radiation* (Clarendon, Oxford).
HUANG, K. N., M. AOYAGI, M. H. CHEN, B. CRASEMANN, AND H. MARK, 1976, *At. Data Nucl. Data Tables* **18**, 243.
INOKUTI, M., 1971, *Rev. Mod. Phys.* **43**, 297.
INOKUTI, M., Y. ITIKAWA, AND J. E. TURNER, 1978, *Rev. Mod. Phys.* **50**, 23.
JAUCH, J. M., AND F. ROHRLICH, 1955, *The Theory of Photons and Electrons* (Addison-Wesley, Reading, Ma).
KODRE, A. F., AND S. M. SHAFROTH, 1979, *Phys. Rev. A* **19**, 675.
KRAUSE, M. O., AND J. H. OLIVER, 1979, *J. Phys. Chem. Ref. Data* **8**, 329.
MANNE, R., 1971, *J. Phys. (Paris)* **32 C4**, 151.
MANSON, S. T., AND A. F. STARACE, 1982, *Rev. Mod. Phys.* **54**, 389.
MCENNAN, J., AND M. GAVRILA, 1977a, *Phys. Rev. A* **15**, 1537.
MCENNAN, J., AND M. GAVRILA, 1977b, *Phys. Rev. A* **15**, 1557.
MENDELSOHN, L., AND V. H. SMITH, 1977, Atoms, in *Compton Scattering*, edited by B. H. Williams (McGraw-Hill, New York), p. 102.
MIZUNO, Y., AND Y. OHMURA, 1967, *J. Phys. Soc. Jpn* **22**, 445.
OHMURA, Y., AND N. MATSUDAIRA, 1964, *J. Phys. Soc. Jpn* **19**, 1355.
PARRATT, L. G., 1959, *Rev. Mod. Phys.* **31**, 626.
PRIFTIS, G., A. THEODOSSIOU, AND K. ALEXOPOULOS, 1968, *Phys. Lett.* **27A**, 577.
RAMAN, C. V., 1928, *Indian J. Phys.* **2**, 387.
RAU, A. R. P., AND U. FANO, 1967, *Phys. Rev.* **162**, 68.
SCOFIELD, J. H., 1974, *Phys. Rev. A* **9**, 1041.
SMEKAL, A., 1923, *Naturwissenschaften* **11**, 873.
SPARKS, C. J., JR., 1974, *Phys. Rev. Lett.* **33**, 262.
STARACE, A. F., 1982, *Phys. Rev. A* **25**, 842.
SUORTTI, P., 1979, *Phys. Stat. Sol.* (b) **91**, 657.
SUZUKI, T., 1967, *J. Phys. Soc. Jpn* **22**, 1139.
SUZUKI, T., 1982, X-Ray Raman Band and Plasmon Line in the Compton Spectrum, in *Advances in X-Ray Spectroscopy*, edited by C. Bonnelle and C. Mande (Pergamon, Oxford), p. 439.

SUZUKI, T., AND A. TANOKURA, 1970, *J. Phys. Soc. Jpn* **29**, 972.

SUZUKI, T., T. KISHIMOTO, T. KAJI, AND T. SUZUKI, 1970, *J. Phys. Soc. Jpn* **29**, 730.

TULKKI, J., 1983, *Phys. Rev. A* **27**, 3375.

TULKKI, J., AND T. ÅBERG, 1978, in Abstracts, 6th Int. Conf. on Atomic Physics, edited by E. Anderson, E. Kraulinya, and R. Peterkop, Riga, p. 393.

TULKKI, J., AND T. ÅBERG, 1980a, *J. Phys. B* **13**, 3341.

TULKKI, J., AND T. ÅBERG, 1980b, in VI Int. Conf. Vac. Ultrav. Rad. Phys., Extended Abstracts, II-69.

TULKKI, J., AND T. ÅBERG, 1982, *J. Phys. B* **15**, L435.

TULKKI, J., 1984, *J. Phys. B* **17**, 3493.

VEIGELE, W. M., 1973, *At. Data* **5**, 51.

WIECH, G., 1981, Anisotropic Emission of X Radiation, in *Inner-Shell and X-Ray Physics of Atoms and Solids*, edited by D. J. Fabian, H. Kleinpoppen, and L. M. Watson (Plenum, New York), p. 815.

WILLIAMS, B., ed., 1977, *Compton Scattering* (McGraw-Hill, New York).

WU, F. Y., R. E. GROVE, AND S. EZEKIEL, 1975, *Phys. Rev. Lett.* **35**, 1426.

WUILLEUMIER, F., 1971, *J. Phys. (Paris)* **32 C4**, 88.

CHAPTER 11

Rayleigh Scattering

Elastic Photon Scattering by Bound Electrons

Lynn Kissel and R. H. Pratt

1. INTRODUCTION

By the term "Rayleigh scattering" we shall understand the elastic scattering of photons from atomic electrons, leaving the target atom unchanged after scattering. Except in the forward direction, this scattering (for any given angle) is increasingly dominated by scattering from the inner atomic shells as one proceeds from X-ray toward γ-ray photon energies. Depending therefore on the choice of energy and scattering angle, the process can be used as a sensitive probe of either inner-shell or outer-shell (and atomic environment) structure.

We illustrate in Figs. 1 and 2, as a function of photon energy, the cross sections for the main elastic and inelastic photon–atom processes. The data are taken from standard tabulations (Hubbell, 1974), and, as we shall see in this chapter, while generally appropriate, they do not entirely correctly represent Rayleigh scattering, particularly in edge regions. The data do show that elastic scattering dominates inelastic scattering until well above the K edge. Except at much lower energies (not shown), Rayleigh scattering never dominates the total (thin-beam) absorption cross section σ_{TOT}. But it does provide a significant (10%) contribution in the hard X-ray regime, both just below the K edge in high-Z elements (75 keV in Pb) and at somewhat lower energies in light-Z elements (20 keV in C).

Rayleigh scattering, as illustrated in Figs. 1 and 2, has commonly been estimated in a simple form-factor approximation that we will describe later. The first efforts toward a more precise, albeit numerical, calculation were

Lynn Kissel • Test Planning and Diagnostics Division, Sandia National Laboratories, Albuquerque, New Mexico 87185. R. H. Pratt • Department of Physics and Astronomy, University of Pittsburgh, Pittsburgh, Pennsylvania 15260.

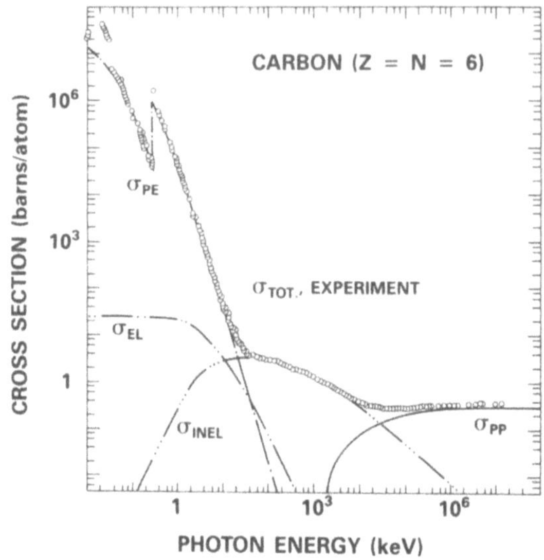

FIGURE 1. Cross sections for the main elastic and inelastic photon–atom scattering processes in carbon. (Reprinted by permission from Hubbell *et al.*, 1980.)

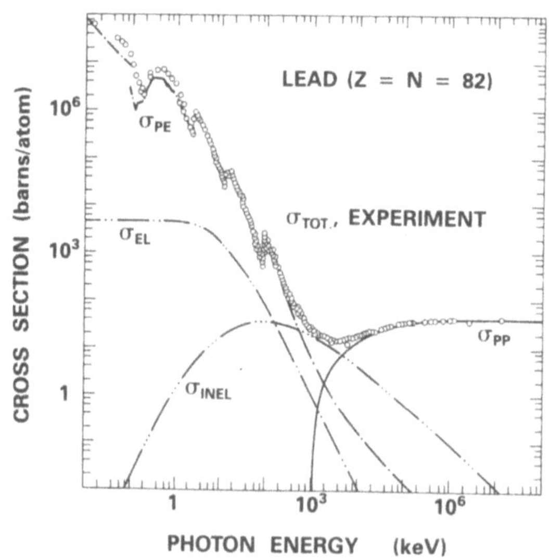

FIGURE 2. Cross sections for the main elastic and inelastic photon–atom scattering processes in lead. (Reprinted by permission from Hubbell *et al.*, 1980.)

initiated by Brown *et al.* (1955, 1957). With improvements in computers, this program of calculation has begun to yield a substantial body of detailed predictions; this information provides the underlying basis for the discussions in this chapter. Simultaneously, improvements in technology have permitted far more precise and detailed experimental observations, as we shall describe. Quite good agreement is now being achieved between theory and experiment, though we shall note a variety of areas in which further work appears of interest.

Recent successes began with resolution of the factor-of-2 discrepancy between large-angle measurements in Pb at 1.33 MeV and the pioneering numerical calculations of Brown *et al.* Subsequently, general agreement has been achieved between theory and experiment in Pb at the 5% level in the hard X-ray and γ-ray regimes. Precision data on Rayleigh scattering and precision experiments on elastic scattering have now made it possible to identify the Delbrück scattering component, due to scattering off virtual electron–positron pairs produced in the atomic field (related to photon–photon scattering), for energies both above and below the real pair-production threshold.

We begin Sec. 2 with a somewhat simplified discussion of the physical features of elastic scattering, including why Rayleigh scattering is an amplitude while elastic scattering is a process, a discussion of features that can be understood from a classical model of the scattering of radiation from a charge bound by a spring, and the assumptions and predictions of the form-factor approximation. In Sec. 3 we discuss theory in more detail, the general formalism, the numerical approach, and various simpler theories, and also the various elastic scattering amplitudes that must be added to obtain a cross section. Section 4 is devoted to a discussion of experimental methods, both for the direct observation of elastic scattering and for indirect determinations of properties of the amplitude. In Sec. 5 we briefly sketch the status of comparisons of theory and experiment, and in Sec. 6 we note some applications of our knowledge of Rayleigh scattering in solid state and nuclear structure studies, in X-ray diagnostics and thin-beam attenuation, and in the study of Delbrück scattering. Finally, in Sec. 7, we identify several areas for further research, both in the basic atomic theory and in its comparison with experiment, its characterization and tabulation, applications to ions and solids, and the generalization to include polarization features.

We mention other reviews of elastic photon–atom scattering by Henke (1981), Kane (1981), Gavrila (1982), and Pratt (1982). We also mention the older reviews by Moon (1950), James (1962), and Nath (1966). For a more general review of elastic photon–atom scattering at higher energies in the X-ray region, see Schumacher *et al.* (1981), and Moreh (1982).

2. PHYSICAL FEATURES OF ELASTIC SCATTERING

In this section we give an elementary overview of elastic scattering of photons by atoms. We mention some simpler theories that are useful for understanding the qualitative behavior of the scattering process for photon energies 100 eV to 10 MeV.

2.1. The Blue Sky of John William Strutt

John William Strutt (1842–1919), the third Baron Rayleigh, was "one of the greatest ornaments of British science..." and "...covered every field of what in the twentieth century is commonly referred to as 'classical' physics..." (Lindsay, 1976). In 1871, Lord Rayleigh published his first works (Strutt, 1871; also see Lindsay, 1976, and Young, 1982) on the scattering of light by small particles (i.e., assuming the wavelength λ is large in comparison to particle size), a subject on which he continued to publish numerous articles throughout his prodigious scientific career. Of particular note in Rayleigh's first work on light scattering was his (1) derivation of the λ^{-4} scattering law (which leads to a blue sky and a red sunset), (2) restatement of Stokes' (see Young, 1982) argument predicting polarization of initially unpolarized light scattered through the scattering angle $\theta = 90°$, and (3) derivation of the $(1 + \cos^2 \theta)$ factor for the angular distribution in scattering of unpolarized light. While these early results were obtained by using the elastic-solid theory of the luminiferous ether, a later derivation used Maxwell's electromagnetic theory (Lord Rayleigh, 1881). In elastic photon scattering by bound electrons (EPSBE), coherence and unchanged frequency are central features. However, it is interesting to note that a hard X-ray or γ-ray sky would be "red" with a "blue" sunset, and important to realize that 100% plane polarization at $\theta = 90°$ as well as the unpolarized scattering distribution $(1 + \cos^2 \theta)$ are only approximations.

In a recent article Young (1982) has detailed the conflicting uses of Lord Rayleigh's name in the field of light scattering by gas molecules. This has prompted us to examine how the nomenclature "Rayleigh scattering" also came to be associated with the contribution to elastic X-ray and γ-ray scattering identified with scattering from bound atomic electrons, a subject that Rayleigh (1906) only briefly discussed, and at a time when neither the wave nature nor the diffraction of X rays had been established. The nomenclature was not used in the *Handbuch der Physik* article of Bothe and Kirschner (1933) on the scattering of X rays, although in one passage (p. 88) the analogy between the scattering of X rays and Rayleigh's theory of light scattering is mentioned. However, the terminology "Rayleigh scattering" to identify EPSBE was "common usage at Sommerfeld's institute."*

* Private communication from F. Smend reporting his correspondence with W. Franz.

The earliest references of which we are aware that use the terminology "Rayleigh scattering" are Franz (1936) and Sommerfeld (1939). While there are many papers on EPSBE during the period 1910–1950, it is not until the work of Moon (1950) and Nelms and Oppenheim (1955) that we have again found the association with Rayleigh's name. (From the footnotes in his review, it is clear that Moon had extended contacts with Franz.) By the 1960s, association of "Rayleigh scattering" with EPSBE seems fairly common, and by the 1970s and early 1980s it appears to be virtually universal. Justification for use of the term "Rayleigh scattering" in connection with EPSBE seems ambiguous. However, since (1) there is no strong conflict with other usage in atomic physics, (2) EPSBE is a cumbersome phrase, and (3) the usage is currently so widespread, we will continue to use the term "Rayleigh scattering" to identify the contribution to elastic X-ray/γ-ray scattering from atoms due to the bound atomic electrons.

2.2. *"Elastic" versus "Coherent" Scattering*

The terms "elastic scattering" and "coherent scattering" have often been used interchangeably in discussing Rayleigh scattering. This leads to some confusion, since with the commonly accepted definitions we shall adopt here, the two concepts are not equivalent. While the Rayleigh scattering amplitude is both elastic and coherent, we shall understand these words to refer to different properties of the amplitude. By "elastic scattering" we shall understand that in the center-of-mass system of projectile and target, the incident and final projectile energy are the same; this implies in particular that there is no transfer of energy between projectile and internal degrees of freedom of the target (Messiah, 1961). If $\hbar\omega$ is the energy of the incident photon and $\hbar\omega'$ is the energy of the scattered photon as measured in the center-of-mass frame, to say that scattering is "elastic" means that $\hbar\omega = \hbar\omega'$. (For elastic scattering of 100-eV to 10-MeV photons by atoms, the distinction between the center-of-mass frame and the laboratory frame may generally be ignored since the atomic nucleus is so massive—even for hydrogen the proton has a rest-mass energy of $\sim$931 MeV.)

On the other hand, by "coherence" we shall understand that it is only possible to define a cross section for a process that corresponds to a transition between *observable* initial and final states. It may be convenient to define such states in terms of their constituents, and in some cases the transition between states can be described in terms of *amplitudes* for transitions among consitutents, but all such amplitudes corresponding to the same initial and final states must be summed *before* squaring to obtain a cross section. Note that coherence here does *not* refer to coherent states of photons; we will assume there is only one photon in the initial and in the final state. Coherence in the Rayleigh scattering amplitude first refers to the fact that in an independent-electron description of an atom one must add the *amplitudes* A^i for photon scattering off each of the atomic electrons. Furthermore, the Rayleigh scattering amplitude itself must be combined with other amplitudes to obtain the total atom elastic scattering amplitude A. In a commonly used approximation

$$A = A^{\mathrm{R}} + A^{\mathrm{NT}} + A^{\mathrm{D}} + A^{\mathrm{NR}}, \tag{1}$$

and the total atom cross section is proportional to $|A|^2$, where $A^{\mathrm{R}} = \sum_i A^i$ is the Rayleigh scattering amplitude, A^{NT} is the nuclear Thomson scattering amplitude (see Sec. 3.3.2), A^{D} is the Delbrück scattering amplitude (see Sec. 3.3.3), and A^{NR} is the nuclear resonance scattering amplitude (see Sec. 3.3.2). While this separation of amplitudes is neither unique nor exact (see, for example, Gavrila, quoted in Kissel *et al.* 1980), it is convenient and believed adequate in the X-ray and γ-ray regimes under consideration here. However, in some circumstances (as in a lattice) the amplitude A for scattering from one atom must be added coherently to scattering amplitudes from other atoms, before squaring, to obtain a total scattering amplitude.

With these definitions, amplitudes can be inelastic but coherent, or elastic but incoherent. But the Rayleigh amplitudes are elastic, and they are coherent with other elastic amplitudes. It is important to distinguish a coherent sum of scattering amplitudes, in which relative phase information is retained, from an incoherent sum in which individual amplitudes (corresponding to observable states) are first squared and then summed to form a total cross section. Incoherence is clearly associated with total cross sections for inelastic ($\hbar\omega' \neq \hbar\omega$) scattering processes, as for example in Compton scattering from bound electrons, where (at the independent-electron level) one can distinguish which electron has been ejected. However, incoherent, practically elastic, processes must sometimes be considered in the discussion of coherent elastic processes. This arises from an experimental view due to the finite energy width of photon sources and finite energy resolution of detectors. But there is also a more fundamental association of elastic and inelastic processes, since once one considers the radiative corrections of quantum electrodynamics and the possibility of emitting very soft photons, one must realize that all processes (even if they do not excite or ionize the target) are in fact inelastic in the sense that $\hbar\omega' \neq \hbar\omega$ (even neglecting target recoil and so assuming that energy has not been transferred to the target).

2.3. *Scattering by Free Electrons: Thomson and Compton Scattering*

In discussing scattering from bound electrons we find it useful to have for reference and comparison the expressions for scattering from free electrons. The theory of scattering of X rays by free electrons was studied by J. J. Thomson, who first derived the classical result for the total (integrated over angles) scattering cross section (Thomson, 1906) that bears his name:

$$\sigma^{\mathrm{T}}=\frac{8\pi}{3}r_e^2, \tag{2}$$

where the classical electron radius

$$r_e=\frac{e^2}{mc^2}\approx 2.82\times 10^{-13}\ \mathrm{cm},$$

with the elementary charge $e\approx 1.60\times 10^{-19}$C, the electron rest mass $m\approx 9.11\times 10^{-28}$g, and the speed of light in vacuum $c\approx 3.00\times 10^{10}$ cm/s. The Thomson formula gives a cross section that is independent of the frequency of radiation (photon energy); in this classical theory the frequency would be unchanged in scattering. Thomson (1906) used σ^{T} to estimate the number of electrons in nitrogen from X-ray scattering measurements. Assuming each bound electron scattered independently (incoherently), so that the cross section was $N\sigma^T$, he erroneously concluded that the number of electrons N in an atom approximately equaled the atomic weight A, not the atomic number Z, an error of a factor of 2 too large. (When the photon energy is large in comparison to binding energy, inelastic scattering from an electron dominates elastic scattering, and binding effects can be neglected. Had Thomson assumed that each electron scattered coherently, as would be appropriate at lower energies—but had still neglected binding effects—he would have made an error of a factor of 2 too small.)

The corresponding Thomson differential cross section (see, for example, Thomson and Thomson, 1933) for scattering of unpolarized radiation by a classical free electron, also not observing the polarization of the scattered radiation, is

$$\frac{d\sigma^{\mathrm{T}}}{d\Omega}=\frac{r_e^2}{2}(1+\cos^2\theta). \tag{3}$$

Here the scattering angle θ is the angle through which the wave vector of the radiation (propagation direction) has been changed. Note that the Bragg angle Θ for reflection from a plane surface is related to the scattering angle θ by $\theta=2\Theta$.

Considering radiation as quantized in photons, Compton (1923b) realized that photon energy would change in scattering from electrons, and calculated that change; the general scattering of photons from free electrons is hence called Compton scattering. The same terminology, however, is also used for inelastic photon scattering from bound electrons (Compton scattering from bound electrons), which generally approaches the behavior of Compton scattering from free electrons for photon energies large compared to binding energies. Within lowest-order relativistic quantum electrodynamics, the full expression for scattering of a photon by a free electron at rest was first derived by Klein and Nishina (1929). In the Klein–Nishina scattering cross section σ^{KN}, the energy of the scattered photon $\hbar\omega'$ is related to the energy of the incident photon $\hbar\omega$ through Compton's relation

$$\hbar\omega' = \frac{\hbar\omega}{1 + (\hbar\omega/mc^2)(1 - \cos\theta)}. \tag{4}$$

This energy is the same as that of the incident photon $\hbar\omega$ (as in Thomson scattering) only for forward scattering or in the limit $\hbar\omega \ll mc^2$. We note the following low-energy approximation of the Klein–Nishina formula (see, for example, Jackson, 1975) for Compton scattering from free electrons:

$$\sigma^{\mathrm{KN}} = \sigma^{\mathrm{T}}\left(1 - \frac{2\hbar\omega}{mc^2} + \cdots\right), \qquad \hbar\omega \ll mc^2, \tag{5}$$

showing that with increasing energy the cross section decreases from the energy-independent Thomson σ^{T} value.* The corresponding differential cross section for Compton scattering from free electrons, averaged over initial and summed over final photon and electron polarizations, is given by the simple expression

$$\frac{d\sigma^{\mathrm{KN}}}{d\Omega} = \frac{r_e^2}{2}\left(\frac{\omega'}{\omega}\right)^2\left(\frac{\omega}{\omega'} + \frac{\omega'}{\omega} - \sin^2\theta\right), \tag{6}$$

which reduces to Eq. (3) when $\omega' \to \omega$.

2.4. *Photon Polarization Effects*

We can define the overall normalization (up to a phase) of the differential scattering amplitude A for elastic scattering of a photon from an atom by requiring that

$$\frac{d\sigma}{d\Omega} = |A|^2. \tag{7}$$

However, we must recognize that there is more than one experimentally distinguishable elastic scattering amplitude A, corresponding to the possibility of specifying the (complex) polarization vectors $\hat{\varepsilon}$ and $\hat{\varepsilon}'$ of the incident and

* For a more detailed discussion of low-energy and intermediate-energy expansions of the Klein–Nishina formula, see Hubbell (1969).

scattered photon. If we assume the atom is not polarized or aligned, it can be seen that there are only *two* independent amplitudes. The general scattering amplitude, which is a scalar, can then only be a function of the vectors, $\hat{\varepsilon}$, $\hat{\varepsilon}'$, $\mathbf{k}$, $\mathbf{k}'$ satisfying

$$\hat{\varepsilon} \cdot \mathbf{k} = \hat{\varepsilon}' \cdot \mathbf{k}' = 0. \tag{8}$$

Since the amplitude must be linear in both $\hat{\varepsilon}$ and $\hat{\varepsilon}'^*$ (from the general form of the interaction of radiation and matter in external field quantum electrodynamics) and must be rotationally invariant and parity conserving, it must have the form

$$A = \hat{\varepsilon} \cdot \hat{\varepsilon}'^* \mathscr{M}(\mathbf{k} \cdot \mathbf{k}') + \hat{\varepsilon} \cdot \hat{k}' \hat{\varepsilon}'^* \cdot \hat{k} \mathscr{N}(\mathbf{k} \cdot \mathbf{k}'), \tag{9}$$

in terms of the invariant amplitudes $\mathscr{M}$ and $\mathscr{N}$, which depend on the vectors of this problem only through the inner product $\mathbf{k} \cdot \mathbf{k}'$. (We define the unit vectors $\hat{k}$, $\hat{k}'$ as $\hat{k} \equiv \mathbf{k}/|\mathbf{k}|$ and $\hat{k}' \equiv \mathbf{k}'/|\mathbf{k}'|$.) Thus there exist only two independent elastic scattering amplitudes, which are, in general, complex. We will now describe the two common choices for the set of independent amplitudes, corresponding to characterization of linear or circular photon polarization.

In the linear basis, we resolve the incident and scattered photon polarizations into components perpendicular and parallel to the plane of scattering (see Fig. 3). The plane of scattering (x–z plane of Fig. 3) is defined by the incident- and scattered-wave vectors $\mathbf{k}$ and $\mathbf{k}'$. The incident-photon wave vector $\mathbf{k}$ and the real polarization basis vectors $\hat{\varepsilon}_{\parallel}$ and $\hat{\varepsilon}_{\perp}$ define the x, y, and z axes, respectively, of the coordinate system in Fig. 3. Similarly,

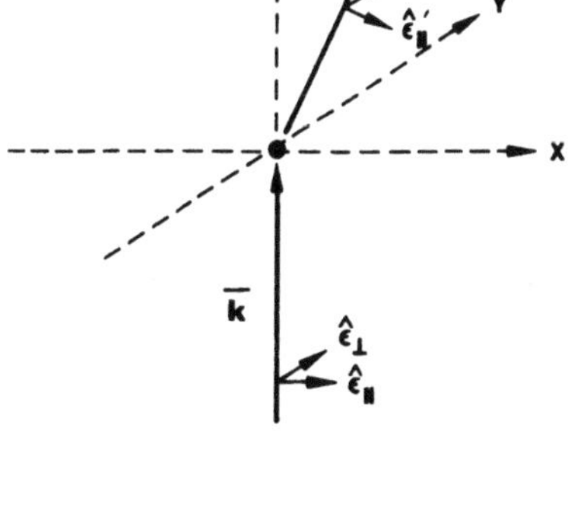

FIGURE 3. Geometry for elastic scattering. The plane of scattering is defined by the incident and scattered photon momenta $\mathbf{k}$ and $\mathbf{k}'$. The linear photon-polarization unit vectors $\hat{\varepsilon}_{\parallel}$, $\hat{\varepsilon}_{\perp}$, and the scattering angle θ are also shown.

the scattered-photon wave vector $\mathbf{k}'$, and polarization basis vectors $\hat{\varepsilon}'_{\parallel}$ and $\hat{\varepsilon}'_{\perp}$ define the x', y', and z' axes, respectively, of a second coordinate system rotated by θ about the y axis. (The y' axis is identical with the y axis.) We note that

$$\begin{aligned} \hat{\varepsilon}_{\perp} \cdot \hat{\varepsilon}'_{\perp} &= 1, \\ \hat{\varepsilon}_{\parallel} \cdot \hat{\varepsilon}'_{\parallel} &= \cos\theta, \\ \hat{\varepsilon}_{\perp} \cdot \hat{\varepsilon}'_{\parallel} &= \hat{\varepsilon}_{\parallel} \cdot \hat{\varepsilon}'_{\perp} = 0. \end{aligned} \tag{10}$$

With these choices we can write

$$A = \varepsilon_{\parallel} \varepsilon'^{*}_{\parallel} (\mathscr{M} \cos\theta - \mathscr{N} \sin^2\theta) + \varepsilon_{\perp} \varepsilon'^{*}_{\perp} \mathscr{M}, \tag{11}$$

or

$$A \equiv \varepsilon_{\parallel} \varepsilon'^{*}_{\parallel} A_{\parallel} + \varepsilon_{\perp} \varepsilon'^{*}_{\perp} A_{\perp}, \tag{12}$$

with

$$\begin{aligned} A_{\parallel} &= \mathscr{M} \cos\theta - \mathscr{N} \sin^2\theta, \\ A_{\perp} &= \mathscr{M}, \end{aligned} \tag{13}$$

and work with the two scattering amplitudes $A_{\perp}$ and $A_{\parallel}$. (We define $\varepsilon_{\parallel}$, $\varepsilon_{\perp}$ as the components of the polarization vector $\hat{\varepsilon} \equiv \varepsilon_{\parallel} \hat{\varepsilon}_{\parallel} + \varepsilon_{\perp} \hat{\varepsilon}_{\perp}$.) Note that this form indicates that polarization initially parallel or perpendicular to the scattering plane continues to have that property after scattering. From symmetry considerations, we conclude that for $\theta = 0°$, $180°$ (forward and backward scattering) the magnitudes of $A_{\perp}$ and $A_{\parallel}$ must be equal, since in these cases the two amplitudes are not physically distinguishable. Explicitly we see from above that

$$\begin{aligned} A_{\perp} &= A_{\parallel} = \mathscr{M}, \qquad \theta = 0°, \\ A_{\perp} &= -A_{\parallel} = \mathscr{M}, \qquad \theta = 180°. \end{aligned} \tag{14}$$

While $A_{\parallel}$ must change sign with respect to $A_{\perp}$ between $0°$ and $180°$, it is not true in general that this must occur at $\theta = 90°$, nor is it true in general that the relative sign change in the real part of the scattering amplitudes must occur at the same angle as the change in the imaginary parts.

The cross section for scattering of photons polarized perpendicular to the plane of scattering is

$$\frac{d\sigma_{\perp}}{d\Omega} = |A_{\perp}|^2, \tag{15}$$

while the cross section for scattering of photons polarized parallel to the plane of scattering is

$$\frac{d\sigma_{\parallel}}{d\Omega} = |A_{\parallel}|^2. \tag{16}$$

The unpolarized scattering cross section, corresponding to an average over incident-photon polarizations and a sum over scattered-photon polarizations, is then given by

$$\frac{d\sigma}{d\Omega} = \frac{1}{2}(|A_{\perp}|^2 + |A_{\parallel}|^2). \tag{17}$$

For Thomson scattering (Sec. 2.3) these amplitudes are

$$A_{\perp}^{\mathrm{T}} = -r_e,$$
$$A_{\parallel}^{\mathrm{T}} = -r_e \cos\theta, \tag{18}$$

with a sign convention that will be discussed in Sec. 3.1.2. The corresponding unpolarized-scattering cross section is

$$\frac{d\sigma^{\mathrm{T}}}{d\Omega} = \frac{1}{2}(|A_{\perp}^{\mathrm{T}}|^2 + |A_{\parallel}^{\mathrm{T}}|^2) = \frac{r_e^2}{2}(1 + \cos^2\theta), \tag{19}$$

as previously stated. For $\theta = 90°$, we have $A_{\parallel}^{\mathrm{T}} = 0$ and the scattered photons from an unpolarized incident beam are 100% plane-polarized perpendicular to the plane of scattering.

The second popular choice for photon-polarization basis vectors characterizes circular polarization. In terms of the linear vectors we write the circular-polarization basis vectors*

$$\hat{\varepsilon}_{\pm} = \frac{1}{\sqrt{2}}(\hat{\varepsilon}_{\parallel} \pm i\hat{\varepsilon}_{\perp}). \tag{20}$$

With these choices for polarization, the general amplitude is

$$A = (\varepsilon_+ \varepsilon_+'^* + \varepsilon_- \varepsilon_-'^*)\left[\frac{\mathscr{M}}{2}(\cos\theta + 1) - \frac{\mathscr{N}}{2}\sin^2\theta\right]$$
$$+ (\varepsilon_+ \varepsilon_-'^* + \varepsilon_- \varepsilon_+'^*)\left[\frac{\mathscr{M}}{2}(\cos\theta - 1) - \frac{\mathscr{N}}{2}\sin^2\theta\right], \tag{21}$$

* While this choice is common in discussions of Rayleigh scattering, in many contexts (e.g., Sakurai, 1967), it is common to introduce a minus sign in the definition of $\hat{\varepsilon}_+$. With such a sign convention, additional minus signs need to be inserted with each use of ε_+ and A_{SF} in the subsequent formulas of this subsection.

or

$$A \equiv (\varepsilon_+ \varepsilon_+'^* + \varepsilon_- \varepsilon_-'^*) A_{\mathrm{NSF}} + (\varepsilon_+ \varepsilon_-'^* + \varepsilon_- \varepsilon_+'^*) A_{\mathrm{SF}}, \tag{22}$$

in terms of the amplitude A_{NSF} for no change in circular polarization (NSF = no spin flip)

$$A_{\mathrm{NSF}} = A_{++} = A_{--} = \frac{\mathscr{M}}{2}(\cos\theta + 1) - \frac{\mathscr{N}}{2}\sin^2\theta, \tag{23}$$

and the amplitude A_{SF} for a change in circular polarization (SF = spin flip)

$$A_{\mathrm{SF}} = A_{+-} = A_{-+} = \frac{\mathscr{M}}{2}(\cos\theta - 1) - \frac{\mathscr{N}}{2}\sin^2\theta. \tag{24}$$

(We define $\varepsilon_+, \varepsilon_-$ as the components of the polarization vector $\hat{\varepsilon} \equiv \varepsilon_+ \hat{\varepsilon}_+ + \varepsilon_- \hat{\varepsilon}_-$.) The circular-polarization amplitudes and linear-polarization amplitudes are related:

$$\begin{aligned} A_{\mathrm{NSF}} &= \tfrac{1}{2}(A_{\parallel} + A_{\perp}), \\ A_{\mathrm{SF}} &= \tfrac{1}{2}(A_{\parallel} - A_{\perp}); \end{aligned} \tag{25}$$

or

$$\begin{aligned} A_{\parallel} &= A_{\mathrm{NSF}} + A_{\mathrm{SF}}, \\ A_{\perp} &= A_{\mathrm{NSF}} - A_{\mathrm{SF}}. \end{aligned} \tag{26}$$

The circular-polarization scattering amplitudes have the following symmetry properties:

$$A_{\mathrm{NSF}}(180^\circ) = A_{\mathrm{SF}}(0^\circ) = 0. \tag{27}$$

The corresponding scattering cross sections are

$$\frac{d\sigma_{++}}{d\Omega} = \frac{d\sigma_{--}}{d\Omega} = \frac{d\sigma_{\mathrm{NSF}}}{d\Omega} = |A_{\mathrm{NSF}}|^2, \tag{28}$$

and

$$\frac{d\sigma_{+-}}{d\Omega} = \frac{d\sigma_{-+}}{d\Omega} = \frac{d\sigma_{\mathrm{SF}}}{d\Omega} = |A_{\mathrm{SF}}|^2. \tag{29}$$

The unpolarized scattering cross section, where we average over incident-photon polarizations and sum over scattered-photon polarizations, is given by

$$\frac{d\sigma}{d\Omega} = |A_{\text{NSF}}|^2 + |A_{\text{SF}}|^2. \tag{30}$$

For Thomson scattering, we have the following circularly polarized scattering amplitudes:

$$A_{\text{NSF}}^{\text{T}} = -\frac{r_e}{2}(\cos\theta + 1),$$

$$A_{\text{SF}}^{\text{T}} = -\frac{r_e}{2}(\cos\theta - 1), \tag{31}$$

and again the unpolarized Thomson-scattering cross section is

$$\frac{d\sigma^{\text{T}}}{d\Omega} = |A_{\text{NSF}}^{\text{T}}|^2 + |A_{\text{SF}}^{\text{T}}|^2 = \frac{r_e^2}{2}(1 + \cos^2\theta). \tag{32}$$

2.5. *Classical Scattering by a Bound Charge*

Several important features of elastic photon scattering by bound electrons can be explicitly and simply exhibited in the calculation of electromagnetic scattering from a classical nonrelativistic electron bound by a spherically symmetric linear restoring force $m\omega_0^2\mathbf{x}$. In this case the unpolarized (or polarized) cross section for scattering of an electromagnetic wave of frequency ω is (Jackson, 1975)

$$\frac{d\sigma^c}{d\Omega} = \frac{d\sigma^{\text{T}}}{d\Omega}\left[\frac{\omega^4}{(\omega_0^2 - \omega^2)^2}\right], \tag{33}$$

where $d\sigma^{\text{T}}/d\Omega$ is the corresponding Thomson scattering result. For low frequencies, $\omega \ll \omega_0$, this gives Rayleigh's law of scattering

$$\frac{d\sigma^c}{d\Omega} = \frac{d\sigma^{\text{T}}}{d\Omega}\left(\frac{\omega}{\omega_0}\right)^4, \qquad \omega \ll \omega_0; \tag{34}$$

the electron moves only slightly as the radiation scatters from it. The scattering varies as λ^{-4}; for $\theta = 90°$, the Thomson formula predicts 100% polarization of the scattered photons and for unpolarized photons it varies as $(1 + \cos^2\theta)$. For high-frequency scattering $\omega \gg \omega_0$, we obtain the Thomson formula for classical scattering by free charges

$$\frac{d\sigma^c}{d\Omega} = \frac{d\sigma^{\text{T}}}{d\Omega} = \frac{r_e^2}{2}(1 + \cos^2\theta); \quad \omega \gg \omega_0, \tag{35}$$

i.e., the scattering occurs as though the electron were free, independent of its binding by the force. The cross section is decreasing toward the Thomson value as one goes to high frequencies. For frequencies in the neighborhood of the natural frequency of the oscillator, $\omega \sim \omega_0$, our simple expression contains a singularity. (In the corresponding quantum case we will identify such frequencies with possible bound–bound transitions corresponding to excitation of the ground-state atomic structure.) If damping terms (corresponding to finite level widths) had been included, the cross section at $\omega \sim \omega_0$ would show a large but finite resonance. This dramatic resonance scattering for $\omega \sim \omega_0$ is called resonance fluorescence and is observed both for scattering from bound electrons and for scattering from the nucleus.

2.6. *An Independent-Electron Atomic Model*

For elastic scattering of 100-eV to 10-MeV photons by atoms, we view the atom as a bound system consisting of N electrons each of negative charge $-e$ and mass m, and a nucleus of positive charge $+Ze$ and mass M. For neutral atoms $N = Z$. The relatively light electrons occupy the bulk of the atomic volume of dimension $1 \text{Å} = 10^{-8}$ cm, while the heavy nucleus of dimension $1 \text{F} = 10^{-13}$ cm appears as a point charge containing the bulk of the atomic mass (for hydrogen $M/m \approx 1840$). Important atomic features are described by independent electrons interacting electrostatically with a central potential resulting from the charge distribution of the nucleus and electrons. Higher-order electron-electron correlation effects and details of outer-electron wave functions will be important only at lower photon energies. Effects of relativity will be important at intermediate and higher photon energies for light atoms, and at all photon energies for heavy atoms.

2.7. *High-Energy Scattering and the Form-Factor Approximation*

As discussed in Sec. 2.2, elastic photon–atom scattering is described by a coherent sum of amplitudes. The total-atom scattering amplitude A is the coherent sum of component scattering amplitudes from the N bound electrons (Rayleigh scattering A^{R}) and the nucleus (nuclear Thomson scattering A^{NT}, Delbrück scattering A^{D}, and nuclear resonance scattering A^{NR}). We might expect (and it is partially true) that high-energy photons $\hbar\omega \gg \varepsilon$, where $\varepsilon > 0$ is an electron binding energy, would scatter off such atomic electrons as though they were free charges. We have in mind σ^{T}, not σ^{KN}, since we do not want to eject the electrons from the atom. Certainly for forward scattering the momentum transfer to the atom is small and the process takes place at large distances (for which the bound electron within the atom does look like a point charge rather than a charge distribution).

This may be contrasted with scattering for which electrons are ejected, for which momentum transfers are larger, and high-energy cross sections approach σ^{KN}. (In fact, binding corrections do persist to high energies, as is discussed in Sec. 3.1.5.)

If for the moment we adopt this viewpoint, treat all the electrons and the nucleus as free point charges, and ignore the generally small contribution of A^{D} and A^{NR}, we have the elastic scattering amplitudes

$$A_{\perp}^{\mathrm{R}} = \sum_{i=1}^{\mathcal{N}} A_{\perp}^{\mathrm{T}} = -r_e \mathcal{N},$$

$$A_{\parallel}^{\mathrm{R}} = \sum_{i=1}^{\mathcal{N}} A_{\parallel}^{\mathrm{T}} = -r_e \mathcal{N} \cos\theta,$$

$$A_{\perp}^{\mathrm{NT}} = -\frac{(Ze)^2}{Mc^2} = -r_e Z^2 \frac{m}{M},$$

$$A_{\parallel}^{\mathrm{NT}} = -\frac{(Ze)^2}{Mc^2}\cos\theta = -r_e Z^2 \frac{m}{M}\cos\theta, \tag{36}$$

where $\mathcal{N} = Z$ in the case of a neutral atom, and the total-atom amplitudes

$$\begin{aligned} A_{\perp}^{\mathrm{R+NT}} &= -r_e\left(\mathcal{N} + Z^2 \frac{m}{M}\right), \\ A_{\parallel}^{\mathrm{R+NT}} &= -r_e\left(\mathcal{N} + Z^2 \frac{m}{M}\right)\cos\theta. \end{aligned} \tag{37}$$

The unpolarized scattering cross section is

$$\begin{aligned} \frac{d\sigma^{\mathrm{R+NT}}}{d\Omega} &= \frac{1}{2}\left(|A_{\perp}^{\mathrm{R+NT}}|^2 + |A_{\parallel}^{\mathrm{R+NT}}|^2\right), \\ &= \frac{r_e^2}{2}\left(\mathcal{N} + Z^2 \frac{m}{M}\right)^2 (1 + \cos^2\theta). \end{aligned} \tag{38}$$

While for the photon-energy range 100 eV to 10 MeV ($\lambda \sim 10^{-6}$ to 10^{-11} cm) the nucleus (of size 10^{-13} cm) appears as a point charge, the charge distributions of the atomic electrons (of size 10^{-8} cm) corresponding to the fact that they cannot be localized on a finer scale of distance at discrete positions on classical orbits, implies that the electrons cannot be treated as point charges, except for forward scattering. Finite momentum transfers correspond to scattering at smaller distances, for which each electron appears as an extended charge distribution, leading to a modification of the point-charge cross sections that may be described in terms of a form factor. Note that *each* of the atomic electrons, of mass m and charge e, is being characterized by a charge distribution; form factors for each electron are summed to obtain a total-atom form factor corresponding to the total-atom charge distribution. (This may be contrasted with the nucleus, which at these energies can be treated as *one* charge distribution of charge Ze and mass M, with its corresponding form factor, so that the nuclear Thomson *amplitude* is proportional to Z^2 rather than Z as the Rayleigh amplitude and is inversely proportional to the nuclear mass rather than the proton mass.)

The atomic form factor can be derived classically to correct the point-charge Thomson formula for scattering from a charge distribution (see, for example, James, 1962). We can interpret the atomic form factor as the effective charge that scatters a given photon. The atomic form factor is defined as

$$f(q) = 4\pi \int_0^\infty \rho(r) \frac{\sin(qr)}{qr} r^2 \, dr, \tag{39}$$

where $\rho(r)$ is the electron number distribution of the atom (assumed spherically symmetric) and

$$\hbar q = 2 \frac{\hbar\omega}{c} \sin\frac{\theta}{2} \tag{40}$$

is the momentum transferred to the atom by the scattered photon. The form factor is the Fourier transform of the atomic charge distribution $\rho(r)$. For near forward-angle scattering ($\sin\theta \sim 0$) or low-energy scattering ($\hbar\omega \sim 0$), the effective number of electrons that scatter is $f(0) = N$, the number of electrons in the atom. At large values of q (finite angles at high energy), the form factor approaches zero, i.e., one is scattering only from the (very small) portion of the electron charge distribution concentrated at very small distance. In form-factor approximation we write the Rayleigh scattering amplitudes

$$\begin{aligned} A_\perp^R &= -r_e f(q), \\ A_\parallel^R &= -r_e f(q)\cos\theta, \end{aligned} \tag{41}$$

with the resulting unpolarized total-atom cross section

$$\frac{d\sigma}{d\Omega} = \frac{r_e^2}{2}\left[f(q) + Z^2 \frac{m}{M}\right]^2 (1 + \cos^2\theta). \tag{42}$$

The atomic form-factor approximation can also be derived within the framework of quantum mechanics (see Sec. 3.2), in which case $\rho(r)$ in Eq. (39) is the electron number density derived from the atomic ground-state wave function. Within the independent-electron model, we can decompose the total-atom electron charge distribution $\rho(r)$ in Eq. (39) as a sum over one-electron, subshell (all electrons of fixed principal quantum number n and angular momentum quantum number j), or shell (all electrons of fixed n) charge distributions. Similarly, the total-atom electron form factor can be decomposed as a sum of individual electron, subshell, or shell form factors. Inner electrons have the most compact spatial distribution, which results in a form-factor contribution (Fourier transform) that is more extended and persists to higher q. The outermost electrons have a much more diffuse spatial distribution and so a form factor with a more compact distribution in q. All electrons contribute equally to the form factor for small q; only the innermost electrons still contribute for large q.

In Fig. 4 we plot the sum of atomic-shell form factors $f(x)$ as a function of the traditional quantity $x = \lambda^{-1} \sin(\theta/2)$, where λ is normally expressed

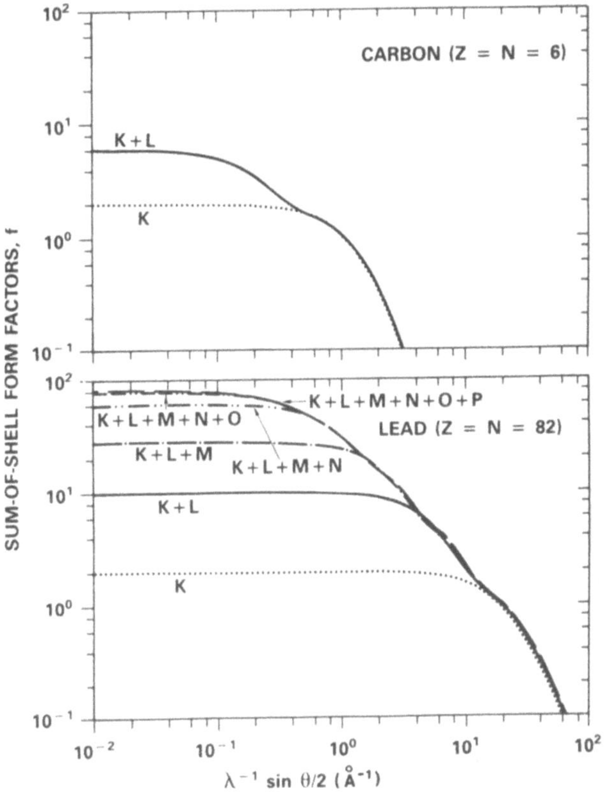

FIGURE 4. Sum of atomic-shell form factors f, as a function of the conventional momentum-transfer variable, $x = \lambda^{-1} \sin(\theta/2)$, for neutral carbon and lead.

in Å. (The momentum transfer $\hbar q$ in units of mc is obtained by dividing x in Å^{-1} units by the factor $20.60744 = 1/2 \times 511.0034/12.39852$.) We note that for small values of x (small q), $f(x)$ approaches the number of electrons N. As the value of x increases, $f(x)$ for a given shell remains relatively constant until momentum transfers corresponding to the typical momentum $\hbar q_{\text{typ}}$ of the electron in that shell are reached. We may estimate this momentum, at least for inner electrons, by using the Bohr radius a_n of an electron of principal quantum number n and the Heisenberg uncertainty principle, as

$$\hbar q_{\text{typ}} \simeq \frac{1}{a_n} = \frac{Z\alpha}{n^2} mc, \tag{43}$$

or

$$x_{\text{typ}} \simeq 20.61 \frac{Z\alpha}{n^2}. \tag{44}$$

For values of x much larger than x_{typ}, the magnitude of $f(x)$ for a given shell decreases. In some cases there exists a region, for $x \approx x_{\text{typ}}$, in which, for the given shell, $f(x)$ oscillates in sign before monotonically decreasing at large values of x.

Nonrelativistically, the radial-charge number density of a given bound electron can be written as

$$\rho(r) = [N_e r^l s(r)]^2, \tag{45}$$

where N_e is the bound-state normalization of the electron wave function and l is the orbital angular-momentum quantum number. For small radial distances, Pratt and Tseng (1972) showed that, independent of the details of the screening of the potential,

$$s(r) = 1 - \frac{Z\alpha r}{l+1} + \cdots. \tag{46}$$

For large q, the corresponding nonrelativistic-electron form factor therefore approaches the form

$$\lim_{q \to \infty} f(q) = (-1)^l \frac{8\pi Z\alpha N_e^2 (2l+2)!}{(l+1)\, q^{2l+4}}. \tag{47}$$

For large q, $f(q)$ decreases least rapidly for $l=0$ (s states) and for inner shells where N_e is largest. The K shell ($n=1, l=0$) contribution eventually dominates, although the L_{I} subshell contribution remains significant. (Considering relativistic effects, contributions from the L_{II} subshell would persist in high-Z elements.)

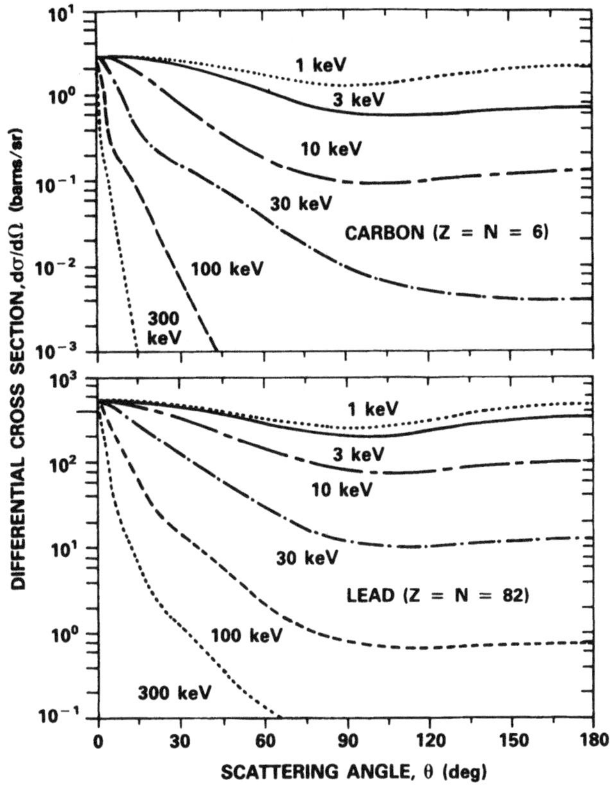

FIGURE 5. Differential elastic photon–atom scattering cross sections at selected photon energies for neutral carbon and lead, obtained from the amplitudes for nuclear Thomson scattering (NT) and Rayleigh scattering (R). The Rayleigh scattering amplitudes are computed in form-factor approximation using the data from Fig. 4.

In Figs. 5 and 6, we show results for the unpolarized cross section Eq. (42) obtained by using the total-atom form factors of Fig. 4. The range of momentum transfers accessible for a given photon energy $\hbar\omega$ is $0 \leqslant \hbar q \leqslant 2(\hbar\omega/c)$. For low photon energies $\hbar\omega$, the form factor used in Eq. (42) is nearly constant and equal to $\mathcal{N}$ for all scattering angles θ. Thus for low photon energies, the form-factor approximation predicts little variation in $d\sigma/d\Omega$ beyond the $(1+\cos^2\theta)$ term resulting from the sum over polarizations. For high photon energies, however, while the form factor is equal to $\mathcal{N}$ in the forward direction $(x = q = 0)$, it becomes very small at larger angles, giving a highly forward-peaked angular distribution. For small x (all θ for small $\hbar\omega$, or small θ for large $\hbar\omega$), the cross section will go as $(\mathcal{N}^2/m^2)$; for large x (large θ for large $\hbar\omega$), the cross section will go as $\mathcal{Z}^4/M^2$,

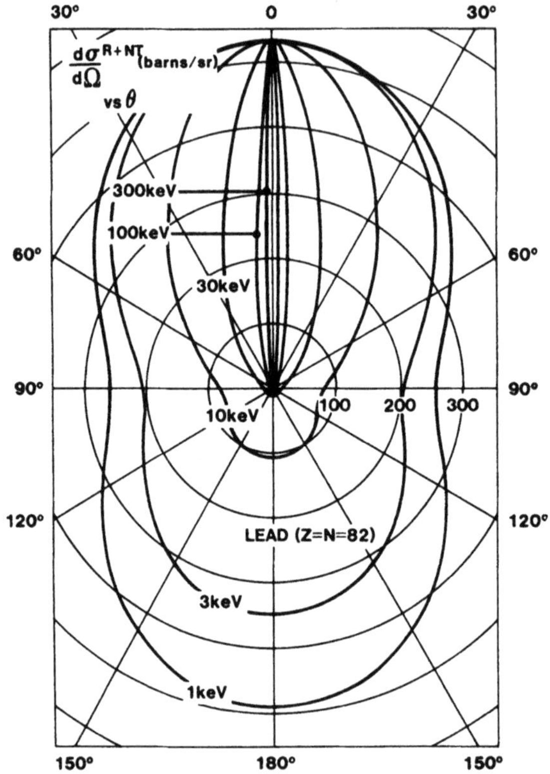

FIGURE 6. Differential elastic photon–atom scattering cross sections for neutral lead of Fig. 5, redrawn as a polar plot.

dominated by nuclear Thomson scattering. However, as q continues to increase, a nuclear form factor must be considered, as well as Delbrück and nuclear resonance amplitudes.

The form-factor approximation can be used to make simple estimates of the importance of electron screening (sensitivity to choice of atomic potential), of relativistic and of exchange effects, in each case based on knowledge of a sensitivity of the charge distribution ρ to these effects. For small momentum transfers (small x) the form factor reduces to the number of electrons, and so is independent of the potential used, relativity, and exchange. For large momentum transfers, small radial distances are important, and screening corrections to Coulomb wave functions appear primarily in their normalizations (Pratt and Tseng, 1972). Thus for large q, the form-factor approximation predicts that screening effects in the scattering amplitude can be expressed as the squared ratio of screened to unscreened bound-state wave-function normalizations. These ratios are always less than unity, and become small for outer electrons, since screening weakens the strength of the potential that brings electrons in to small distances. Relativity and exchange both have the effect of causing the electron charge distribution for inner-shell electrons to contract. So, for large q, relativity and exchange have the effect of increasing the form factor and hence the cross section.

2.8. *Scattering from a Compound System*

Given an understanding of scattering from a particle, one can discuss scattering from a composite systems of interacting particles. (If particles do not interact, the scattering is simply the sum of amplitudes for scattering off the individual particles.) The general principles can be understood by considering a two-particle system. Here the results follow from the separation of the Hamiltonian for the two particles in the presence of (dipole) radiation into center-of-mass and relative coordinates:

$$\frac{1}{2m_1}\left(\mathbf{p}_1-\frac{q_1}{c}\hat{\varepsilon}\right)^2+\frac{1}{2m_2}\left(\mathbf{p}_2-\frac{q_2}{c}\hat{\varepsilon}\right)^2+V(\rho)$$
$$=\frac{1}{2M}\left(\mathbf{P}-\frac{Q}{c}\hat{\varepsilon}\right)^2+\frac{1}{2\mu}\left(\mathbf{p}-\frac{q}{c}\hat{\varepsilon}\right)^2+V(\rho), \tag{48}$$

where total mass and reduced mass are

$$M=m_1+m_2, \qquad \frac{1}{\mu}=\frac{1}{m_1}+\frac{1}{m_2}, \tag{49}$$

total momentum and relative momentum are

$$\mathbf{P}=\mathbf{p}_1+\mathbf{p}_2, \qquad \mathbf{p}=\frac{m_2\mathbf{p}_1-m_1\mathbf{p}_2}{m_1+m_2}, \tag{50}$$

and the total charge and reduced charge are

$$Q=q_1+q_2, \qquad q=\frac{m_2q_1-m_1q_2}{m_1+m_2}. \tag{51}$$

(For the neutral atom we have $q_2=-q_1$, $Q=0$, and $q=q_1$.)

For low energy, as we have seen, scattering off a particle bound in a potential vanishes, and thus in general the scattering off the composite system reduces to scattering off a free particle of mass M and charge Q. However, for the neutral hydrogen atom, the charge is $Q=0$ and this term is identically zero, so, as previously noted, the correct low-energy result (vanishing) is the cross section for an electron bound in a potential, but with *reduced* mass, and with no separate addition of a nuclear Thomson term. For high energy, as we have also seen, scattering off a particle bound in a potential is the same as scattering off a free particle. The result is the *sum* of scattering off two noninteracting particles, which can be taken *either* as (p_1, m_1, q_1), (p_2, m_2, q_2) *or* (P, M, Q), (p, μ, q), yielding the same result. One thus sees the expected transition with increasing energy from scattering off a compound object characterized by compound properties to scattering off individual constituents.

3. DEVELOPMENT OF THEORY

In this section we present a somewhat more detailed description of the quantum-mechanical theory of elastic x-ray and γ-ray scattering by atoms. The ultimate goal is the development of accurate total-atom predictions suitable for use in comparisons with experiment and in the preparation of tabulations.

3.1. *The Rayleigh Scattering Amplitude*

We have defined Rayleigh scattering as the contribution made by the bound electrons to elastic photon scattering by atoms, reflecting a separation of the scattering amplitude into scattering off noninteracting electron and nuclear constituents appropriate at x-ray and γ-ray energies. The Rayleigh scattering amplitude from neutral atoms dominates the total-atom elastic scattering amplitude at all angles for photon energies $\hbar\omega \lesssim 1$ MeV and dominates the forward-angle scattering at higher photon energies. (These statements must be modified for highly ionized targets, in which there are few or no electrons to scatter off.) Here we present a brief review of the theories beyond form-factor approximation appropriate in obtaining an accurate description of this important elastic scattering amplitude.

3.1.1. The Scattering Matrix. The scattering matrix (or simply the S matrix) is an operator that connects the final (scattered) states of a time-dependent system with the initial states. If $\phi(t)$ represents the state of the system at time t, which is subject to a time-dependent perturbation $V_I(t)$, then the initial state corresponds to $\phi(-\infty)$ and the final state corresponds to $\phi(\infty)$, The S matrix is defined as the operator that for all states ϕ satisfies

$$\phi(\infty) = S\phi(-\infty). \tag{52}$$

For Rayleigh scattering in our independent-electron model we can obtain the amplitude for scattering off each atomic electron separately, using an

S-matrix calculation. Thus the initial state $\phi_i = \phi(-\infty)$ may be taken as a stationary state consisting of a single electron in a specific bound orbital and a single photon of wave vector $\mathbf{k}$ and polarization $\hat{\varepsilon}$. The S-matrix element S_{fi}, defined as

$$S_{fi} = \langle \phi_f | \, S \, | \phi_i \rangle \tag{53}$$

represents the amplitude of a specific stationary state ϕ_f in the initial state $\phi(\infty)$ that evolved through the interaction (scattering) from the initial state ϕ_i. If V_I represents the interaction that is responsible for scattering, then we can write a perturbation expansion for the S matrix as

$$\begin{aligned} S &= S^{(0)} + S^{(1)} + S^{(2)} + \cdots \\ &= 1 - i \int_{-\infty}^{\infty} dt_1 \, V_I(t_1) + (-i)^2 \int_{-\infty}^{\infty} dt_1 \int_{-\infty}^{t_1} dt_2 \, V_I(t_1) V_I(t_2) + \cdots. \end{aligned} \tag{54}$$

For further details of the formalism, see Sakurai (1967).

Nonrelativistically, through use of semiclassical radiation theory, the interaction is obtained from the substitution for $\mathbf{p}$ of $\mathbf{p} - (e\mathbf{A}/c)$ in the Schrödinger equation, where $\mathbf{p}$ is the electron momentum operator and $\mathbf{A}$ is the operator describing the radiation field, a linear combination of creation and annihilation operators that operate on photon states. This substitution (minimal electromagnetic coupling) results in an interaction with terms proportional to $(e/2mc)\,\mathbf{A} \cdot \mathbf{p}$, $(e/2mc)\,\mathbf{p} \cdot \mathbf{A}$, and $(e^2/2mc^2)\,\mathbf{A} \cdot \mathbf{A}$. In the matrix element of $S^{(1)}$ corresponding to elastic scattering, terms proportional to $\mathbf{A} \cdot \mathbf{p}$ or $\mathbf{p} \cdot \mathbf{A}$ vanish (since they only involve one incident *or* final photon) and only the $\mathbf{A} \cdot \mathbf{A}$ term survives, corresponding to the "sea-gull" diagram shown in Fig. 7a. If only this contribution to the scattering is considered, the usual form-factor approximation can be obtained.

While the $\mathbf{A} \cdot \mathbf{p}$ terms in V_I do not contribute in $S^{(1)}$ to the elastic scattering matrix element, taken twice, as in $S^{(2)}$, they are still of the same order in e as the $\mathbf{A} \cdot \mathbf{A}$ term in $S^{(1)}$, and so they must be included to obtain the scattering to lowest nonvanishing order in e. The contributions from the $\mathbf{A} \cdot \mathbf{p}$ and $\mathbf{p} \cdot \mathbf{A}$ interaction in $S^{(1)}$ that correspond to the two-step scattering processes are shown in Fig. 7b and 7c. Kramers and Heisenberg (1925) obtained an expression for the differential cross section that included all these lowest-order-in-e terms in nonrelativistic dipole approximation. This Kramers–Heisenberg matrix element has been evaluated analytically in terms of known transcendental function for the case of the Coulomb potential (Gavrila, 1967; Gavrila and Costescu, 1970, 1971; Costescu, 1976). At high energies the $\mathbf{A} \cdot \mathbf{A}$ term dominates, while the two-step process is responsible for the resonance and other anomalous features near thresholds; at low energies the two terms cancel, so that the cross section vanishes.

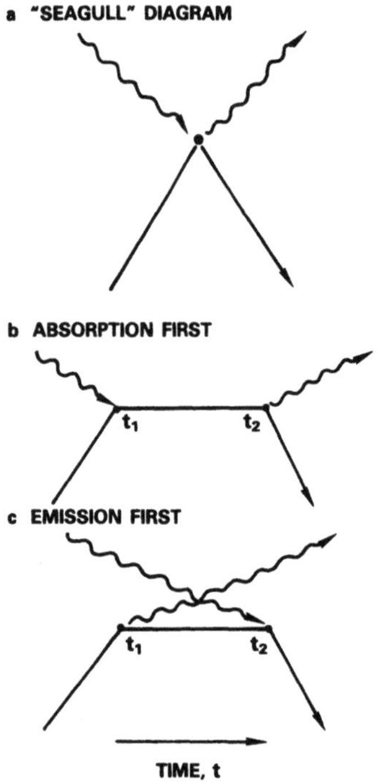

FIGURE 7. Diagrams of lowest-order contributions to the nonrelativistic Rayleigh scattering amplitude. (a) The "sea-gull" diagram corresponding to the $S^{(1)}$ term of the S-matrix expansion. (b) The "absorption-first" contribution to $S^{(2)}$. (c) The "emission-first" contribution to $S^{(2)}$.

Relativistically, in external-field quantum electrodynamics, minimal electromagnetic coupling leads to the interaction term in the Hamiltonian

$$V_I = -ie \int \bar{\phi}\gamma_\mu \phi A_\mu \, d^3x, \tag{55}$$

whose consequences will be calculated in perturbation theory, where ϕ is the Dirac electron-field operator in the external field of the atom. There is no relativistic analog of the "sea-gull" diagram shown in Fig. 7a; $S^{(1)}$ only involves emission *or* absorption diagrams and so has no matrix elements for scattering. To lowest order, relativistic scattering of photons by electrons is a two-step process. In the absorption-first process (Fig. 7b), the incident photon of energy $\hbar\omega$ is first absorbed by the initial-state electron of energy E_i at time t_1. In the intermediate state (time $t_1 \rightarrow t_2$), only an electron of energy $E_i + \hbar\omega$ is present. At time t_2, the final (scattered) photon is emitted. In the emission-first process (Fig. 7c), the final photon of energy $\hbar\omega$ is emitted at time t_1. In the intermediate state an electron of energy $E_i - \hbar\omega$ and the incident and scattered photons are present. At time t_2, the incident photon is absorbed and the electron returns to the stationary bound state. The form-factor approximation can be recovered in a relativistic framework if binding effects in the intermediate state are ignored along with the additional assumption that $\hbar q \ll mc$ (see Levinger, 1952).

Waller (1929) derived the relativistic one-electron matrix element for Rayleigh scattering. Because of its complexity the expression cannot be evaluated analytically even when the external field is taken as a Coulomb potential, unlike the nonrelativistic case. The lowest-order Rayleigh amplitude for scattering of a photon of energy $\hbar\omega$ by a bound electron in the state $n\kappa m_j$ of energy $E_{n\kappa} = 1 - \varepsilon_{n\kappa}$ is obtained relativistically from $S^{(2)}$ and is proportional to the matrix element

$$M_{n\kappa m_j} = \int_p \left[\frac{\langle n\kappa m_j | A^* | p\rangle\langle p| A | n\kappa m_j\rangle}{E_{n\kappa} - E_p + \hbar\omega} + \frac{\langle n\kappa m_j | A | p\rangle\langle p| A^* | n\kappa m_j\rangle}{E_{n\kappa} - E_p - \hbar\omega} \right] \tag{56}$$

The symbol $\int_p$ represents the generalized sum-integral over all states p of energy E_p, both bound and continuum. The operators $A(A^*)$ represent absorption (emission) of a photon of energy $\hbar\omega$. The first term in the expression for $M_{n\kappa m_j}$ describes the absorption-first process; the second term describes the emission-first process.

3.1.2. Forward-Angle Scattering: The Optical Theorem and Dispersion Relation. Within the framework of quantum field-theory, Gell-Mann *et al.* (1954) proved the dispersion relation for the forward-angle scattering of light:

$$\operatorname{Re} A(\omega, 0) = \frac{2\omega^2}{\pi} \int_0^\infty \frac{\operatorname{Im} A(\omega', 0)}{\omega'(\omega'^2 - \omega^2)} d\omega', \tag{57}$$

from which

$$\operatorname{Re} A(\infty, 0) = -\frac{2}{\pi} \int_0^\infty \frac{\operatorname{Im} A(\omega', 0)}{\omega'} d\omega', \tag{58}$$

where $A(\omega, \theta)$ is the complex elastic scattering amplitude; this follows directly from the analyticity of $A(\omega, 0)/\omega$. This relation is often referred to as the Kramers–Kronig relation, as it was written for the complex index of refraction $n(\omega)$ of light [related to $A(\omega, 0)$ by $n(\omega) = 1 + 2\pi(c/\omega)^2 n_{AT} A(\omega, 0)$, where n_{AT} is the number of scatterers per unit volume] by H. A. Kramers and R. Kronig in 1926–1927 (see, for example, Sakurai, 1967). The dispersion relation is exact and is a direct consequence of the causality principle. As Sakurai points out, another dispersion relation may also be written that connects these amplitudes, based on the analyticity of A rather then A/ω. Since the real part of A remains finite at $\omega = \infty$, this alternative relation should be written for $A(\omega, 0) - A(\infty, 0) \equiv f'(\omega) + if''(\omega)$ with f' and f'' real, $f'(\infty)$ finite, and $f''(\infty) = 0$. In terms of the anomalous amplitudes f' and f'', this relation is

$$f'(\omega) = \frac{2}{\pi} \int_0^\infty \frac{\omega' f''(\omega')}{(\omega'^2 - \omega^2)} d\omega', \tag{59}$$

which may also be obtained from Eq. (57).

The optical theorem relates the imaginary part of the forward scattering amplitude and the total cross section σ_{tot} (including all elastic and inelastic processes) as

$$\text{Im}\, A(\omega, 0) = \frac{\omega}{4\pi c} \sigma_{\text{tot}}. \tag{60}$$

The choices of sign in Eq. (57) and Eq. (60) determine our phase conventions for the scattering amplitudes. The optical theorem can be derived from the unitarity of the S matrix and is a consequence of the conservation of probability. For the lowest-order terms in e^2 in scattering by a single bound electron,* we have the following relationship between the second-order Rayleigh forward-scattering amplitude and the first-order one-photon absorptive processes of photoeffect σ_{PE} and pair production into a bound state σ_{BPP}:

$$\text{Im}\, A^{\text{R}}(\omega, 0) = \frac{\omega}{4\pi c} (\sigma_{\text{PE}} - \sigma_{\text{BPP}}) + \frac{\pi r_e}{2} \sum_n \omega_{ni}\, f_{ni} \delta(\omega_{ni} - \omega). \tag{61}$$

Here f_{ni} is the relativistic multipole oscillator strength for the transition $i \to n$ ($\hbar\omega_{ni} \equiv E_n - E_i$) of the given electron from its state i to another single-particle bound state n, written under the assumption that the level widths of all states are being ignored. The Dirac delta function $\delta(\omega_{ni} - \omega)$ is zero except at $\omega = \omega_{ni}$. At each such ω_{ni}, $\text{Re}\, A(\omega, 0)$ will diverge. $\text{Im}\, A^{\text{R}}(\omega, 0)$ vanishes for $\hbar\omega < \varepsilon$ except at bound–bound transition energies, since the photoeffect cross section for a bound electron is zero below the energy threshold $\hbar\omega = \varepsilon$ (ε is the binding energy of the electron i) and the total pair-production cross section σ_{BPP} is zero below the energy threshold

* Note that in next order in e^2 the total cross section on the right-hand side of Eq. (60) would include processes such as Compton scattering from bound electrons, which through Eq. (59) would contribute also to the real scattering amplitude. Such terms are of course not included in our lowest-order calculation of the Rayleigh amplitude. The fact that there are circumstances in which the Compton scattering cross section dominates the photoeffect cross section indicates that a lowest-order calculation may not always be adequate. However, generally in the forward scattering amplitude, where Eq. (60) is applicable, the imaginary amplitude is small in comparison with the real amplitude.

$\hbar\omega = mc^2 - \varepsilon$. The pair-production cross section σ_{BPP} involved here, first identified by Brown, is not the normal pair-production process. Instead, σ_{BPP} describes the case where the electron of the electron–positron pair is created in the bound state i of the atom. (See Kissel *et al.*, 1980, for further discussion and references concerning σ_{BPP} and the reason for the sign difference between contributions of σ_{PE} and σ_{BPP}.) At very high energies, we have $\sigma_{PE} \sim 1/\omega$ and $\sigma_{PE} = \sigma_{BPP}$ so that $\text{Im}\, A(\omega, 0) \sim 1/\omega$. This behavior is needed so that $\text{Re}\, A(\omega, 0)$ approaches a finite limit at high energy.

The sum over bound–bound transitions from i will, in a many-electron atom, include transitions into states j that are occupied. Evidently, when we sum contributions to A from all bound electrons, the $j \to i$ term will cancel the $i \to j$ term ($f_{ij} = f_{ji}$ but $\omega_{ji} = -\omega_{ij}$). Thus the total atom amplitude will only exhibit resonances at energies corresponding to transitions into unoccupied states, as would be expected physically. The canceling single-particle resonance terms are sometimes called "spurious resonances." Sometimes the dispersion relation is used to obtain $\text{Re}\, A$ without including the real bound–bound δ-function contributions in $\text{Im}\, A$. While such terms are generally unimportant, their omission leads to an (incorrect) divergent behavior for $\text{Re}\, A$ as threshold is approached from above, as well as to the omission of the resonance terms just below threshold; in a light-Z isolated atom their contribution remains significant at most energies between the K and L edges.

3.1.3. Binding and the Intermediate State: Numerical Partial-Wave Calculations. We have noted that if binding effects in intermediate states are ignored (representing them by plane waves), and if nonrelativistic momentum transfers are assumed, the form-factor approximation is obtained and Rayleigh scattering can be described solely in terms of the charge distribution of the atomic electrons. The effect of binding in the intermediate state (times between t_1 and t_2 in Fig. 7b and 7c) was explored analitically in the relativistic theory by Brown and Woodward (1952), who calculated the first Born correction in Ze^2, in an expansion of the potential. (This should be distinguished from the basic perturbative expansions in e^2 of quantum electrodynamics). Brown and Woodward found that higher-order Born-approximation terms are important; in the case they considered, the first-order Born-approximation contribution to scattering was larger than the zero-order contribution.

This realization that binding effects in the intermediate state of Rayleigh scattering are important led to major progress when Brown *et al.* (1955), abandoning the slowly convergent Born series approach, developed a numerical partial-wave procedure for calculating Rayleigh scattering from heavy atoms. In their approach, the interaction of the electron with the atomic field in the intermediate state is not expanded in a perturbation series, but this interaction is included exactly to all orders, within a partial-wave expansion. That is, the scattering is calculated in the bound-interaction picture (Furry picture). This implies a calculation of the propagator or alternatively (in most work) solution of inhomogeneous radial Dirac equations. The matrix element reduces to a sum of double integrals over radial variables when a multipole expansion is made for both the incident and scattered photon. One radial integral over a propagator, a photon multipole, and a bound state can be obtained directly as an appropriate solution of an inhomogeneous Dirac equation, a technique first applied in Rayleigh scattering by Brown. The codes are nonetheless lengthy and require many hours of computer time even on fast machines. Brown and co-workers (Brenner *et al.*, 1955; Brown and Mayers, 1956, 1957) applied this numerical partial-wave procedure to the elastic scattering of 0.32, 0.64, 1.28, and 2.56-mc^2 photons ($\sim$164, 327, 654, and 1310 keV) by the K shell of Hg ($Z = 80$), generally assuming a point Coulomb potential. Cornille and Chapdelaine (1959) applied the procedure of Brown *et al.* to the case of 2.62-MeV photons scattered by the K shell of Hg. Later, more systematic partial-wave calculations were performed for K-, L-, and M-shell electrons for $30 \leqslant Z \leqslant 82$ and 145 keV $\leqslant \hbar\omega \leqslant$ 889 keV by Johnson and Cheng (1976), and more recently for K-, L-, M-, and N-shell electrons for $30 \leqslant Z \leqslant 82$ and 22.1 keV $\leqslant \hbar\omega \leqslant$ 2754 keV by Kissel *et al.* (1980). Additional results are now available, including small-momentum-transfer scattering in Pb (Roy *et al.*. 1983), the X-ray regime in Al and Pb (Parker and Pratt, 1984), and 59.5 keV $\leqslant \hbar\omega \leqslant$ 1.33 MeV for Zn, Sn, and U (Kissel *et al.*, 1982); additional calculations in the X-ray regime have been initiated.

While these calculations may be considered to represent the current state of the art in predictions of Rayleigh scattering, their limitations should be understood. First, not all subshell contributions are normally being calculated directly, owing to the great computer-time requirements. Rather, simpler approximations (particularly use of modified form factor) are used to obtain outer-shell contributions well above their thresholds; the validity of this approach is discussed subsequently. Second, these calculations are within an independent-electron model, the single-electron states and the propagator obtained as solutions of the Dirac equation in a Hartree–Slater potential with local Kohn–Sham exchange and a Latter tail (see, for example, Liberman *et al.*, 1965). Full nonlocal exchange and many-electron correlations are not included. Widths of levels, not to mention the environment of outer electrons (as in a solid) are not included—these matters will appreciably modify scattering near threshold.

3.1.4. Correlation Effects. Lin *et al.* (1975) considered certain correlation effects beyond the Hartree–Fock model. They considered elastic scattering by He ($Z = 2$) and included electron–electron correlation effects coming from the fourth-order S matrix. They found that use of Dirac–Hartree–Fock–Slater (DHFS) wave functions in second-order S-matrix calculations introduces contributions to the fourth-order S matrix that tend to cancel other fourth-order contributions. That is, the inclusion of the approximate exchange potential of the DHFS potential does indeed approximately represent the electron-electron interaction. The importance of the residual fourth-order correlation effects, which are expected to be largest for the case of He, decrease as the incident photon energy increases. For total cross sections, these effects of correlations become $<10\%$ by 100 eV, $<5\%$ by 200 eV, and $<1\%$ by 400 eV. Correlation corrections to the angular distributions are insignificant by ~ 1 keV. Correlation effects are expected to be more correctly included within the DHFS model for the inner shells of heavier atoms because of the larger number of electrons present, better justifying the statistical assumptions of the Slater exchange model. In addition, the importance of correlations is expected to decrease as the atomic number increases. Therefore, correlation effects, neglected by considering only contributions of the second-order S matrix with DHFS potentials, are expected to be small in most cases for the heavier atoms and higher photon energies.

3.1.5. The High-Energy Limit. Special attention has been devoted to the scattering of very-high-energy photons by bound electrons. In the forward direction, the form-factor approximation (Sec. 2.7) reduces to elastic scattering off a free charge. At high energies, it has been suggested (Gell-Mann *et al.*, 1954) that the exact forward Rayleigh-scattering amplitude should be that from a free charge, independent of binding. This suggestion was soon questioned by Levinger and Rustgi (1956), who considered forward-angle scattering by the K shell of Pb and showed that binding effects persist to high energies and give results that are reduced from those of a free charge. The modified form-factor approximation suggested by Franz (1936), which we will discuss in the next section, also predicts this reduction. High-energy scattering at finite momentum transfer has been considered by Goldberger and Low (1968) and by Florescu and Gavrila (1976), who have presented numerical results. Florescu and Gavrila were able to rederive the form-factor approximation from their high-energy results in the small-Z limit when $\hbar q \ll mc$.

3.2. Simple Atomic Scattering-Factor Theories

Historically, the terms "atomic scattering factor," "atomic form factor," and "atomic structure factor" have all been used in describing the elastic scattering of photons by atoms. The "atomic scattering factor" F was defined by Hartree (1925) as the "ratio of the amplitude of the wave scattered by this atom to the amplitude scattered by an electron." Here we will define F as

$$A = FA^{\mathrm{T}}, \tag{62}$$

the ratio of the isolated-atom total-atom elastic scattering amplitude A to the Thomson scattering amplitude A^{T}. Since the Thomson amplitude is the low-energy limit of the Klein–Nishina formula (Sec. 2.3), the Thomson amplitude is the correct low-energy quantum-mechanical prediction for elastic scattering by a free electron. Hence, our F represents only a modest redefinition of Hartree's F. In general F depends on the atom or ion that scatters the photon, the photon energy $\hbar\omega$, and the scattering angle θ. In principle, there are two independent atomic scattering factors, say $F_{\perp}$ and $F_{\parallel}$ or F_{NSF} and F_{SF}, corresponding to the two independent photon polarization amplitudes (Sec. 2.4). In general, the atomic scattering factor is complex, with both real (dispersive) and imaginary (absorptive) parts.

The "atomic form factor" $f(q)$ is a particular approximation to the atomic scattering factor F. Here F is approximated by a real function $f(q)$, which depends upon the atom or ion that scatters the photon, and on the momentum transfer to the atom $\hbar\mathbf{q}=\mathbf{k}_i-\mathbf{k}_f$. The magnitude of the momentum transfer is

$$\hbar q = 2\frac{\hbar\omega}{c}\sin\left(\frac{\theta}{2}\right), \tag{63}$$

so that in form-factor approximation the dependence of F on the photon energy $\hbar\omega$ and the scattering angle θ are combined into a single universal variable. The form factor $f(q)$ is defined as the Fourier transform of the electron charge density $\rho(r)$; this approximation to the Rayleigh scattering amplitude is based upon high-energy, small-momentum-transfer, low-Z approximations. We consider $f(q)$ as generally appropriate for the description of scattering of X rays from gases and for γ-ray scattering. Usefulness of this type of approximation, as we shall discuss subsequently, can be extended to larger atomic number Z and to higher momentum transfer $\hbar q$ by introduction of a modified form factor $g(q)$; usefulness of form-factor approximation can be extended to lower energies through inclusion of anomalous scattering-factor corrections f' and if''. Often one defines $f(q)+f'=\mathrm{Re}\,F$ and $f''=\mathrm{Im}\,F$, though we shall see that a different definition of f' is required if f' and f'' are to be related through a dispersion relation.

3.2.1. Atomic Form-Factor Theories. Form-factor theories are important for several reasons:

- They are simple; yet they exhibit many central features of the high-energy Rayleigh scattering amplitudes.
- They are connected with classical concepts, correcting the classical, point-charge, free-electron, Thomson scattering formula for scattering from an extended charge distribution.

- They can also be identified with approximations to a quantum-mechanical description, having been derived through the use of both nonrelativistic and relativistic quantum mechanics.
- As compared with more accurate theories, form-factor predictions are relatively easy to calculate, involving the evaluation of an integral over the atomic-electron charge distribution.
- The basic form-factor expression is a function of the single variable $\hbar q$, the momentum transfer to the atom, instead of $\hbar\omega$ and θ separately. This qualitatively, often quantitatively, correct scaling feature permits a greatly simplified tabulation and a large reduction in computation.
- All present comprehensive tabulations involving Rayleigh scattering use the form-factor approximation.

In all derivations of the atomic form factor $f(q)$ it is assumed both that the electrons are lightly bound to the atom ($mc^2 \gg \varepsilon$) and that the photon energy $\hbar\omega$ is much greater than the electron binding energy ε ($\hbar\omega \gg \varepsilon$). In these circumstances, the scattering can be calculated for a charge distribution according to the Thomson formula (Sec. 2.3). The Thomson formula is valid classically for scattering by a free charge, assuming nonrelativistic motion of the charge. Here the charge, remaining bound in the atom, has nonrelativistic velocity, while the photon at high energy scatters from it as though it were free. The requirement of nonrelativistic electron orbital velocity, estimated for the Bohr atom as

$$\left(\frac{v}{c}\right)^2 \approx \left(\frac{Z\alpha}{n}\right)^2 \ll 1, \tag{64}$$

is equivalent to the requirement of weak binding to the atom. Classical derivations (Hartree, 1925; Compton, 1930; also James, 1962) and nonrelativistic quantum-mechanical derivations (for examples, Sommerfeld, 1939; James, 1962) require these same conditions. Relativistic quantum-mechanical derivations [Franz, 1935 and 1936*; Bethe (see Levinger, 1952); Goldberger and Low, 1968; Florescu and Gavrila, 1976] also require the condition (guaranteed in the nonrelativistic case) of nonrelativistic momentum transfer to the atom,

$$\hbar q \ll mc. \tag{65}$$

The "atomic form factor" $f(q)$ can be defined for an atom with $\mathcal{N}$ elec-

* A factor of 2 is missing in Fig. 2 and in the equations starting with Eq. (12) in the 1936 paper. (Private communication from F. Smend reporting correspondence from W. Franz.)

trons as the matrix element (see, for example, Nelms and Oppenheim, 1955):

$$f(\mathbf{q}) = \int \sum_{n=1}^{N} \exp(i\mathbf{q}\cdot\mathbf{r}_n)\,|\psi|^2\,dr_1\cdots dr_n, \tag{66}$$

where ψ is the initial ($\equiv$final)-state wave function of the atom, $\hbar\mathbf{q} = \mathbf{k}_i - \mathbf{k}_f$ is the momentum transferred to the atom, ($\mathbf{k}_i/\mathbf{k}_f$ are the initial/final photon momenta), and $\mathbf{r}_n$ are the radial coordinates of the nth electron. In form-factor approximation, $F_{\parallel} = F_{\perp} = F_{\rm SF} = F_{\rm NSF} = f(q)$.

In our model of independent electrons interacting electrostatically with a single central potential, the total-atom wave function ψ may be written in terms of a product over single-electron wave functions ψ_n as

$$\psi = \prod_{n=1}^{N} \psi_n. \tag{67}$$

(Since our operator

$$\sum_n e^{i\mathbf{q}\cdot\mathbf{r}_n}$$

is symmetric, we do not need to symmetrize the states). We can now rewrite the form factor as

$$f(q) = \int \rho(\mathbf{r})\,e^{i\mathbf{q}\cdot\mathbf{r}}\,d\mathbf{r}, \tag{68}$$

where the atomic-electron charge density is

$$\rho(r) = \sum_{n=1}^{N} |\psi_n(r)|^2 = \sum_{n=1}^{N} \rho_n(r). \tag{69}$$

Thus the atomic form factor is a Fourier transform of the atomic charge distribution. This leads directly to the description of classical electromagnetic radiation from a charge distribution.

If we assume that $\rho(r)$ depends only on the magnitude of r, that is $\rho(r)$ is spherically symmetric, then we can write

$$f(q) = 4\pi \int_0^{\infty} \rho(r)\,\frac{\sin(qr)}{qr}\,r^2\,dr. \tag{70}$$

For very small values of q, we have $\sin(qr)/qr \approx 1$ and $f(0) = N$, the number of electrons. For very large values of q, $\sin(qr)$ oscillates rapidly, so that $f(\infty) = 0$. See Sec. 2.7 for a discussion of the qualitative behavior of $f(q)$.

We note the simple closed-form expression for the form factor in the case of a single electron bound in the 1*s* state of a nonrelativistic hydrogenlike atom of charge Ze (Pirenne, 1946). The ground-state wave function is

$$\psi=\left(\frac{Z^3}{\pi a_0^3}\right)^{1/2} e^{-Zr/a_0}, \tag{71}$$

with the first Bohr radius

$$a_0=\frac{\hbar^2}{me^2}\approx 5.29\times 10^{-9}\ \text{cm}, \tag{72}$$

yielding the form factor

$$f_H(q)=\left[\frac{1}{1+(a_0 q/2Z)^2}\right]^2. \tag{73}$$

Corresponding expressions can be obtained for excited states. Form-factor calculations for molecular hydrogen H_2 and terminally bonded hydrogen (such as C–H) are reviewed in Hubbell *et al.* (1975).

Levinger (1952) has considered relativistic corrections to the nonrelativistic form factor for K-shell electrons, for small momentum transfers $\hbar q\lesssim mc$. (He did not include corrections due to binding in the intermediate state, known to be of the same order of magnitude, and made further approximations.) Levinger found that these relativistic corrections increase the value of the (already too large) form factor. An expression was also obtained for the relativistic Coulomb form factor for two noninteracting K-shell electrons in a Coulomb potential. This Bethe–Levinger form factor is

$$f_{\text{BL}}(Q)=\frac{\sin(2\gamma\tan^{-1}Q)}{\gamma Q(1+Q^2)^{\gamma}}, \tag{74}$$

where $Q=q/(2amc)$, $\gamma=(1-a^2)^{1/2}$, and $a=Z\alpha$. Hubbell *et al.* (1975) use $f_{\text{BL}}(Q)$ without correction for the change in normalization due to screening to extend their nonrelativistic form-factor tabulation to large momentum transfer. Øverbø (1977, 1978) used $f_{\text{BL}}(Q)$ to parametrize his large-q DHFS form factors (used at large q in the relativistic form-factor tabulation of Hubbell and Øverbø, 1979). Smend and Schumacher (1974) derived analytic expressions for the form factor by using relativistic wave functions in the Coulomb potential. Their results for the K shell reduce to expression (74).

Values for the atomic form factor have been obtained in several atomic models, including results based on configuration-interaction (CI) or Hartree–Fock (HF) wave functions. Hubbell *et al.* (1975) review the wide variety of calculations that have been performed. Comprehensive tabulations of atomic form factors have been published by Hubbell *et al.* (1975) and Hubbell and Øverbø (1979). These tabulations include atomic numbers $Z = 1$–100, and x values 0–10^9 Å^{-1}. None of the available more accurate form factors calculated from CI or HF wave functions directly cover this entire range in x, so that in these tabulations mixed approximations are given for $f(q)$. While Hubbell *et al.* (1975) provide a predominantly nonrelativistic tabulation, the relativistic K-shell Bethe–Levinger formula (Levinger, 1952) was used to extend the tabulation to large x for $Z = 2$–100. For $Z = 7$–100, the nonrelativistic HF form factors stop at $x = 10$ Å^{-1} because of large roundoff oscillations in the calculations for $x = 10$–80 Å^{-1}. Roy *et al.* (1983) have noted smaller $O(6\%)$ oscillations in the values tabulated for Pb ($Z = 82$) near $x = 4$ Å^{-1}. Hubbell and Øverbø (1979) gave a relativistic tabulation of $f(q)$ based on relativistic HF and HFS wave functions; for $Z = 1$, they used Pirenne's nonrelativistic hydrogenic form factors.

The most accurate form factors are assumed to be those based upon the most accurate atomic wave functions available, such as CI wave functions (Brown, 1974) or HF wave functions (Mann, 1967). Unfortunately this does not mean that the best form factors yield the best predictions for scattering, and in fact there is evidence to the contrary. While form factors calculated with less accurate wave functions will differ in places from CI or HF form factors (Cromer, 1965), the interpretation of these differences is far from clear. As an illustration, we compare in Fig. 8 predictions of the differential cross section for Pb according to various form-factor theories with high-energy predictions based on the use of accurate numerical Rayleigh amplitudes (Sec. 3.3.1), which also omit nonlocal exchange and correlation effects. The nonrelativistic Hartree–Fock (labeled HF) form factors (Hubbell *et al.*, 1975) provide a better prediction than the relativistic Dirac–Hartree–Fock–Slater (labeled DHFS) form factors even for a high-Z atom. On the scale of Fig. 8, the differences between form factors based on differing exchange models is not large. The dramatic difference results from use of relativistic versus nonrelativistic wave functions in evaluating the form factors. It should also be noted that the nonrelativistic K-shell form factor due to Pirenne [labeled PIRENNE-K, expression (73)] provides a better prediction at large x values than does the relativistic Bethe–Levinger K-shell form factor (labeled BL-K). It is not totally surprising that the form factors from a relativistic wave function might produce poorer elastic-scattering predictions than those from more approximate wave functions, since other effects are being omitted when the form-factor approximation is used. It is known that there is substantial cancellation among relativistic, retardation, and higher multipole effects, so it is a mistake to include some of these effects

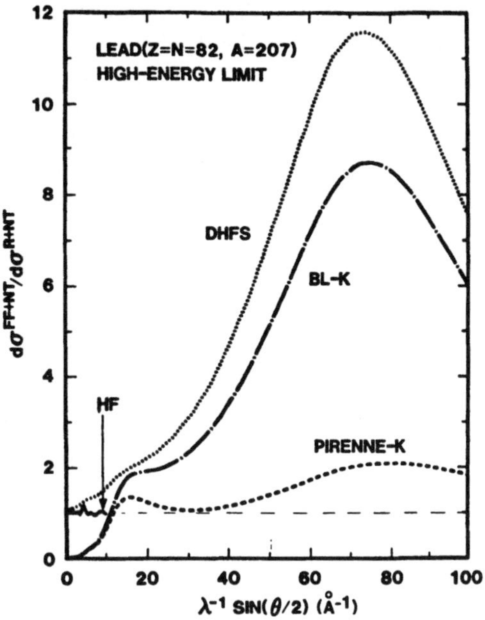

FIGURE 8. Comparison of various form factors for lead with high-energy numerical S-matrix predictions. Numerical data for 5-MeV scattering have been used, that appear to have converged to the high-energy limit for these momentum transfers. The prediction using the nonrelativistic Hartree–Fock form factor is labeled HF, the relativistic Dirac–Hartree–Fock–Slater form factor is labeled DHFS, the nonrelativistic K-shell form factor is labeled PIRENNE-K, and the relativistic K-shell form factor is labeled BL-K. Note the last two cases only include the K shell, which is why their ratio to the full calculation is so small for small momentum transfers.

if not including all. Recently, Roy *et al.* (1983) and Kissel *et al.* (1980) have shown that relativistic form-factor predictions are poorer than nonrelativistic ones in the case of scattering from a heavy atom (Pb, $Z = 82$). Consequently, the nonrelativistic tabulation of Hubbell *et al.* (1975) for $x \leqslant 10$ Å^{-1} is expected to better predict elastic scattering from heavy atoms than the more recent relativistic tabulation of Hubbell and Øverbø (1979). For $x > 10$ Å^{-1}, the validity of the Bethe–Levinger K-shell formula used in Hubbell *et al.* (1975) has not been established. From an experimental viewpoint, Kane *et al.* (1978) had already stressed the greater utility of a nonrelativistic form factor in comparison with a relativistic form factor; they have discussed this point in greater detail subsequently (Kane *et al.*, 1981, 1983).

Detailed comparisons of shell and total-atom differential amplitudes obtained from the second-order QED S-matrix formalism have been made with corresponding form-factor predictions calculated using DHFS wave functions (Kissel *et al.*, 1980). At high energies (where the agreement is best), form-factor values universally deteriorate for momentum transfers $\hbar q \gtrsim 0.2mc$ (a manifestation of the $\hbar q \ll mc$ restriction), failing to decrease as rapidly as the better predictions. For $\hbar q \lesssim 0.2mc$, the error in high-energy form-factor predictions is about $\frac{1}{2}(Z\alpha/n)^2$ [a manifestation of the $(v/c)^2$ restriction]. Thus at small momentum transfers $\hbar q \ll mc$ and for scattering of high-energy photons $\hbar\omega \gg \varepsilon$, errors in form-factor amplitudes for a lighter atom such as aluminum ($Z = 13$) are 0.5% for the K shell and 0.1% for the L shell. For a heavy atom such as Pb ($Z=82$), errors are 20% for the K shell, 5% for the L shell, and 2% for the M shell.

The form factor can be used to make useful predictions of the total-atom differential scattering amplitudes for aluminum at all angles for photon energies from several thousand electron volts ($\varepsilon_K \approx 1.5$ keV) through ~ 50 keV. Above ~ 50 keV, form-factor predictions for Al must be restricted to progressively smaller angles. For Pb, form-factor predictions are useful for similar energies, since the more poorly described inner electrons do not make a major contribution. By 100 keV, requiring $\hbar q \lesssim 0.2mc$ restricts us to scattering angles $\lesssim 60°$. Useful form-factor predictions for Pb are also restricted to increasingly forward angles for photon energies $\hbar\omega \gtrsim 100$ keV. For large momentum transfers $\hbar q \gg 0.2mc$ and high photon energies $\hbar\omega \gg \varepsilon$, the relativistic form-factor approximation leads to predictions that are generally too large, often by factors exceeding 2 to 10.

The form-factor approximation for a given subshell fails in two ways as the photon energy decreases from high energies $\hbar\omega \gg \varepsilon$. First, for lightly bound electrons where the form factor provides reasonable predictions for $\hbar\omega \gg \varepsilon$ and $\hbar\omega \ll mc$, the validity of such predictions for Re A^{R} decreases as the photon energy approaches the electron binding energy. Form-factor predictions for a given shell become completely unreasonable for $\hbar\omega \ll \varepsilon$, as they remain finite while Re A^{R} vanishes. Second, the form-factor approximation makes no prediction for Im A^{R}, which increases in size as the photon energy decreases toward threshold. The anomalous scattering factors discussed in Sec. 3.2.3 provide a framework for correcting the form-factor approximation at lower energies.

3.2.2. The Modified Form Factor. In his derivation of the relativistic form factor, Franz (1936) suggested a corrected or modified form factor based on the charge distribution $\rho_i(r)$ of a given electron,

$$g^i(q) = 4\pi \int_0^\infty \rho(r) \frac{\sin(qr)}{qr} \left[\frac{mc^2}{E - V(r)} \right] r^2\, dr, \tag{75}$$

where E is the total energy of the ith electron ($E = mc^2 - \varepsilon$, ε positive) and $V(r)$ is the atomic potential. The total-atom modified form factor is found by summing $g^i(q)$ over all electrons. The additional denominator factor has its origin in the intermediate-state propagator and represents a binding correction. It has been shown (Kissel *et al.*, 1980) that this modified form factor tends to correct the $O(Z\alpha)^2$ errors observed in the ordinary form-factor approximation; at high energies it nearly coincides with high-energy-limit predictions. However, for large momentum transfers $\hbar q \gtrsim mc$, $g^i(q)$ fails to give good predictions at backward angles $\theta \gtrsim 90°$, being too *small* [while $f(q)$ is too *big*]. From an experimental viewpoint, Kane *et al.* (1981, 1983) also concluded that the relativistic modified-form-factor approach gave a preferred approximation to small-angle Rayleigh scattering. Smend and Schumacher (1974) have derived analytic expressions for the modified form factor (75) by using relativistic wave functions in the Coulomb potential. They find that their analytic results for the Coulomb potential can be extended by use of empirical electron binding energies and by replacing the atomic number Z with appropriate effective charge numbers Z_{eff}.

Brown and Mayers (1957) compared their numerical partial-wave K-shell amplitudes for Hg ($Z=80$) with predictions of the modified form factor $g(q)$ and the ordinary form factor $f(q)$. They found that $g(q)$ better predicted their A_{NSF} amplitudes; $f(q)$ better predicted their A_{SF} amplitudes. They also noted that this version of the modified form factor was suggested by the Born-approximation calculations of Brown and Woodward (1952) on effects of binding.

The modified form-factor approximation proves very useful in our state-of-the-art calculations, where it is used systematically in the calculation of outer shells at high energy. For back angles at high energy where it fails, nuclear Thomson and Delbrück amplitudes dominate the Rayleigh amplitude. Recently, Schaupp *et al.* (1983) have tabulated relativistic modified form factors for atomic numbers $1 \leqslant Z \leqslant 100$ and momentum transfers $0 \leqslant x \leqslant 100$ Å^{-1}. In addition to the total-atom (summed over electrons) modified form factor, they also tabulate the total-atom-minus-K-shell modified form factor in anticipation of the availability of more accurate numerical S-matrix predictions for the K shell.

3.2.3. Anomalous Scattering Factors. All derivations of the form-factor approximation assume that photon energy is large compared with electron binding, $\hbar\omega \gg \varepsilon$, for all electrons of concern in the scattering. But in many cases of interest the photon energy is comparable to electron binding energy, especially for medium- and high-Z atoms where K-shell binding is tens of thousands of electron volts. Near any such threshold there are major deviations from form-factor approximations; well below such a threshold the given subshell ceases to participate in the scattering.

Writing the atomic scattering factor as

$$F = f_1 + if_2 \tag{76}$$

with f_1 and f_2 real or, as is commonly done,

$$F = [f(q) + f'] + if'', \tag{77}$$

we explicitly identify real (dispersive) and imaginary (absorptive) components of the complex atomic scattering factor F. In fact not $f(q)$, but the high-energy limit of f_1 should enter this definition if f' and f'' are to be related through the usual dispersion relation Eq (57). The high-energy limit of f_1, to order $(Z\alpha)^2$, is predicted by $g(0)$, the value of the modified form factor at zero momentum transfer. In the forward direction the difference in definition is of some consequence in high-Z elements. The corrections f' and f'' have been referred to as the dispersion corrections or anomalous-dispersion corrections to the form factor $f(q)$. The term "anomalous" alludes to the observed rapid variation in the scattering for photon energies $\hbar\omega \approx \varepsilon$. James (1962) reviews the older work on this subject.

Most discussion of anomalous scattering factors has concerned the forward-scattering case. Cromer and Liberman (1970, 1981) have published an extensive tabulation of f', f'' values for Cr, Fe, Cu, Mo, and Ag $K\alpha$ radiation ($\sim$5.4, 6.4, 8.0, 17.5, and 22.1 keV) for neutral atoms $Z = 3$–98, based on DHFS wave functions and photoeffect cross sections in nonrelativistic dispersion relations with correction terms. Comparisons of these values with our numerical results have shown good agreement (Kissel *et al.*, 1980). Bremer (1979) has calculated the anomalous scattering of X rays from a nonrelativistic formalism, but has included the effect of the finite lifetime of the intermediate state as well as the influence of strong exciton states. Numerical results are given within electron volts of the absorption edges for Ne, Ar, Kr, and HCl. Henke *et al.* (1982) have tabulated the scattering factors f_1, f_2 for elements $Z = 1$–94 for photon energies 100 eV to 2 keV. Henke's values were derived from the optical theorem and dispersion relation by using the best available experimental and theoretical photoionization cross sections. Omission of bound–bound transition contributions causes some error near threshold, and also in between thresholds, for light-Z elements.

All these anomalous scattering factors have been derived for the case of forward scattering. It has been the practice to transform these zero-angle corrections to nonzero values of θ by multiplying the zero-angle value of f' and f'' for each electron by its corresponding form factor, which is unity at $\theta = q = 0$. That is, the angular distribution for the anomalous corrections to scattering for a given electron has been assumed to be the same as that of the form factor for that electron. Our analysis (Parker and Pratt, 1984) of the anomalous corrections derived from numerical partial-wave amplitudes has shown that their angular distributions are *not* those predicted by the form factor. In fact, a better approximation is to assume that f' and f'' are *independent* of angle—they remain nearly constant with q as the form factor drops, and then at back angles do begin to drop. (However, if f' has been referenced to $f(q)$ rather than to the high-energy limit, it will then have the same q dependence as the form factor). There is evidence that the q dependence in f'' is independent of subshell, but this requires further study. Since f'' drops *much* more slowly than f with q, at intermediate angles and high energies the imaginary part of the Rayleigh amplitude can become more nearly comparable with the real part, even though at high energies in the forward direction it is much smaller.

3.3. Total-Atom Elastic Scattering Amplitudes

In this section, we use our numerical calculations together with our assessment of the validity of form-factor approaches to discuss a practical method for the accurate calculation of total-atom Rayleigh scattering amplitudes for the energy range 100 eV through 10 MeV. We then briefly indicate methods available for estimating the other elastic amplitudes so that theoretical total-atom elastic scattering cross sections appropriate for comparison with experiment can be prepared. We discuss the relative magnitudes of the different elastic amplitudes under various circumstances, both to understand when experiment is sensitive to certain of the amplitudes and to understand the accuracy with which each amplitude must be known if the total amplitude is to be obtained.

3.3.1. Total-Atom Rayleigh Scattering Amplitudes. Accurate Rayleigh scattering amplitudes must include the effects of binding (interaction of the electrons with the atomic potential) in the initial, intermediate, and final states (Fig. 7). The numerical-partial-wave (NPW) calculations discussed in Sec. 3.1.3 are sometimes referred to as "exact" calculations because binding effects in the intermediate state are included exactly to all orders. However, the NPW calculations are only the lowest-order terms of the S-matrix expansion in powers of e, and correlation effects (beyond approximate electron–electron correlation included by means of a Slater-type exchange potential) are ignored. Nonetheless, NPW amplitudes are the most accurate predictions now available that span the entire photon energy range 100 eV through 10 MeV. For scattering of photons with energies within tens of electron volts of an ionization threshold, or for lower-energy photons scattered by low-Z atoms, calculations based upon more detailed atomic models are appropriate (Bremer, 1979; also see Sec. 3.1.4).

While the NPW method is expected to accurately predict the Rayleigh scattering amplitudes for a specific electron, the method is practical by itself for predicting total-atom Rayleigh amplitudes only for light-Z atoms and lower photon energies, except in occasional test cases. At higher photon energies or for high-Z atoms with many electrons, the NPW calculations can generally be applied only to the scattering from inner-shell electrons because of the prohibitive computational cost. Kissel *et al.* (1980) developed a prescription for generating accurate total-atom Rayleigh amplitudes that is based upon NPW predictions for inner electrons and form-factor-based estimates for the outer electrons. For a heavy atom such as Pb ($Z=82$), Kissel *et al.* found that predictions accurate to $O(2\%)$ in $A_{\perp}^{R}$ could be made with only K-shell NPW amplitudes for $\hbar\omega \gtrsim 900$ keV, but that $K+L$-shell

NPW amplitudes were required for $O(1\%)$ amplitudes. As the photon energy decreased from $\hbar\omega \approx 900$ keV, increasingly more of the outer-electron contributions needed to be calculated from the NPW method in order to maintain highly accurate Rayleigh amplitudes.

To estimate the contribution A^i of a given outer electron in the total-atom Rayleigh amplitudes for Pb, Kissel *et al.* (1982) use the following expressions:

$$\begin{aligned}
\operatorname{Re} A^i_\perp &= -g^i(x),\\
\operatorname{Im} A^i_\perp &= \left(\frac{\sigma^i_{\mathrm{PE}}}{\sigma^j_{\mathrm{PE}}}\right) \operatorname{Im} A^j_\perp,\\
\operatorname{Re} A^i_\parallel &= -g^i(x)\cos\theta,\\
\operatorname{Im} A^i_\parallel &= \left(\frac{\sigma^i_{\mathrm{PE}}}{\sigma^j_{\mathrm{PE}}}\right) \operatorname{Im} A^j_\parallel,
\end{aligned} \tag{78}$$

where A^j is the contribution of the inner shells that is calculated through use of the NPW method. The estimate for $\operatorname{Re} A^i$ is simply the modified form-factor approximation discussed in Sec. 3.2.2. The form used for $\operatorname{Im} A^i$ is suggested at forward angles by the optical theorem (Sec. 3.1.2). Remarkably, Kissel *et al.* (1980) find that the imaginary-amplitude shell ratios $\operatorname{Im} A^i/\operatorname{Im} A^j$ are very nearly independent of angle and are accurately predicted by the forward-angle optical theorem. This is in sharp contrast to the historical assumption that the angular dependence of shell ratios is given by ratios of shell form factors, which are not independent of angle (Sec. 3.2.3). Recently, we have confirmed the very-near angle independence of $\operatorname{Im} A^{L_1}/\operatorname{Im} A^K$ at $\hbar\omega = 2.75$ MeV for Pb, well above the pair-production threshold. In addition, at 2.75 MeV the optical theorem predicts at forward angles

$$\frac{\operatorname{Im} A^{L_1}}{\operatorname{Im} A^K} = \frac{\sigma^{L_1}_{\mathrm{PE}} - \sigma^{L_1}_{\mathrm{BPP}}}{\sigma^K_{\mathrm{PE}} - \sigma^K_{\mathrm{BPP}}}, \tag{79}$$

but we have found that simply the ratio $\sigma^{L_1}_{\mathrm{PE}}/\sigma^K_{\mathrm{PE}}$ still provides an accurate estimate of the shell ratio $\operatorname{Im} A^{L_1}/\operatorname{Im} A^K$. This is fortunate since values of σ^i_{BPP} (pair production with the electron of the electron–positron pair created in the bound state i) are not generally available. For the circumstances of interest, some cross sections can be obtained from results for the inverse process (Johnson *et al.*, 1964, 1967; Tseng and Pratt, 1973). At high energies the ratio σ^{L_1}/σ^K equals $(\mathcal{N}^{L_1}_e/\mathcal{N}^K_e)^2$ for both PE and BPP, and indeed one finds that Eq. (79) is given by this ratio of bound-state normalizations, and not just at forward angles.

3.3.2. Nuclear Scattering Amplitudes. Below photon energies of 1 to 4 MeV, the nuclear contribution to elastic photon scattering occurs mainly because of scattering by the nuclear charge. Treating the nucleus as a rigid spin-zero charge Ze of mass M and radius r, we write the nuclear-Thomson scattering amplitudes as

$$A_{\perp}^{NT} = -\frac{f_{nuc}}{Mc^2} = -r_e \frac{m}{M} f_{nuc},$$
$$A_{\parallel}^{NT} = A_{\perp}^{NT} \cos\theta, \tag{80}$$

where at these energies it is adequate to approximate the nuclear form factor as

$$f_{nuc} \sim Z^2 \left(1 - \frac{\omega^2}{3c^2} \langle r^2 \rangle\right) \tag{81}$$

(Ericson and Hüfner, 1973). Some measurements of this form factor are available (Wright and Debevec, 1982). For suitable choices of photon energy, scattering angle, and target nucleus, the nuclear-Thomson contribution can dominate. Experiments (Alvarez *et al.*, 1958; Berant *et al.*, 1977) have confirmed the validity of the nuclear-Thomson contribution given above for low-energy γ-ray scattering with $f_{nuc} = Z^2$.

However, for photon energies higher than ~1 MeV, the complex structure of the nucleus introduces contributions to the total-atom scattering amplitude in addition to the simple nuclear-Thomson amplitude. In the giant-dipole-resonance (GDR) region, the photonuclear absorption cross section can be well fit by one or two Lorentzian lines (e.g. Berman, 1975). Scattering from the low-energy tail of the GDR is referred to as nuclear resonance scattering, and the following form can be derived from the optical theorem and dispersion relations (e.g. Rullhusen *et al.*, 1981):

$$A_{\perp}^{NR} = \frac{E^2}{4\pi\hbar c r_0} \sum_{\nu=1}^{2} \frac{\sigma_\nu \Gamma_\nu (E_\nu^2 - E^2)}{(E_\nu^2 - E^2)^2 + E^2 \Gamma_\nu^2},$$
$$A_{\parallel}^{NR} = A_{\perp}^{NR} \cos\theta, \tag{82}$$

with $E = \hbar\omega$. It must be noted that A^{NR} is purely real for photon energies below the (γ, n) photonuclear threshold (~7.4 MeV for ^{208}Pb). For photon energies above ~4 MeV, the contribution of nuclear structure to elastic scattering is more complex; the interested reader is referred to recent reviews by Schumacher *et al.* (1981) and by Moreh (1982).

3.3.3. Delbrück Scattering Amplitudes. Delbrück scattering is the contribution to the elastic photon scattering amplitude due to scattering of photons off virtual electron–positron pairs created in the field of the nucleus.

The imaginary part of the amplitude is related to the absorptive process of pair production and is zero for $\hbar\omega < 2mc^2$. The real part is related to vacuum polarization and is one of the nonlinear processes predicted by QED that has no classical analog. For high-Z neutral atoms such as U, the contribution of Delbrück scattering is small below ~1 MeV (though still observable) and becomes unimportant below ~500 keV. For 1- to 4-MeV photons scattered by U, strong interference occurs between the contributions of Delbrück, nuclear Thomson, and Rayleigh scattering. Above ~4 MeV, through the giant dipole region, scattering by the nuclear structure dominates the total-atom scattering amplitudes; Delbrück scattering is again observable at higher energies.

The most accurate calculations of the Delbrück amplitudes available below 10 MeV are done in the lowest-order Born approximation, yielding amplitudes that scale as $\alpha(Z\alpha)^2$. In our energy range, numerical results obtained in this way have been reported by Papatzacos and Mork (1975), and Bar-Noy and Kahane (1977). Recent experimental studies (Basavaraju *et al.*, 1979; Rullhusen *et al.*, 1979a, 1981, 1982, 1983b) indicate that these lowest-order Born predictions are accurate to a few percent for $Z < 60$, while it appears that corrections (so-called Coulomb corrections, not yet calculated) have to be included for higher-Z atoms. For scattering of 2.75-MeV photons by Pu, it appears that the total-atom differential cross sections are modified by 30% to 40% due to these corrections, and the Delbrück amplitudes are modified by more than 50% (Rullhusen *et al.*, 1979b).

As discussed in Sec. 2.2, an important aspect of elastic photon scattering is coherence; relative phase information is retained when adding component scattering amplitudes. We point out an error in the phase of the tabulation of Bar-Noy and Kahane. While the real amplitudes of their tabulation were calculated using the circular photon-polarization phase conventions presented in Sec. 2.4, their imaginary amplitudes use the alternate phase convention mentioned in the footnote in Sec. 2.4. To correct the tabulation for use with the phase convention presented here, Bar-Noy and Kahane's Table 1 should be interpreted as $\operatorname{Im} A_{+-}$ values in which all numeric values are positive. To use their tabulation in the alternate phase, Bar-Noy and Kahane's Table 3 should be interpreted as $-\operatorname{Re} A_{+-}$ values (not as $\operatorname{Re} A_{+-}$ values as published), and their Table 1 should be interperted as $-\operatorname{Im} A_{+-}$ values in which all the numerical values are positive. This sign error was first brought to our attention by Schumacher (see Rullhusen *et al.*, 1983a).

3.3.4. Interference between Amplitudes. For photon energies below ~1 MeV, the Rayleigh scattering amplitudes dominate the total-atom scattering amplitudes at all scattering angles for large-Z atoms, while with decreasing Z, nuclear Thomson amplitudes begin to dominate in an

TABLE I

Relative importance of contributions from the various elastic scattering amplitudes in $d\sigma/d\Omega$.

Photon energy range	Rayleigh A^{R}	Nuclear Thomson A^{NT}	Delbrück A^{D}	Nuclear resonance A^{NR}
< 100 keV	Dominates at all θ	Unimportant	Unimportant	Unimportant
100 keV–1 MeV	Dominates at small θ, important at intermediate and large θ	Unimportant at small θ, important at intermediate and large θ, dominates at large θ for smaller Z	Important at intermediate θ for larger Z	Unimportant
> 1 MeV	Importance restricted to smaller θ as energy increases	Important at all but smallest θ	Important at all but smallest θ	Increasingly important at intermediate and large θ as energy increases

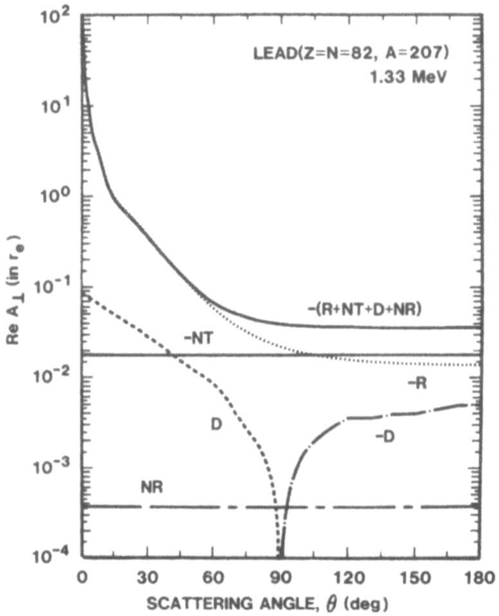

FIGURE 9. Contributions to Re $A_\perp$ for elastic scattering of 1.33-MeV photons by neutral lead, from the Rayleigh (R), nuclear Thomson (NT), Delbrück (D), and nuclear resonance (NR) amplitudes.

expanding back-angle region. For photon energies above several million electron volts, the Rayleigh scattering amplitudes dominate the total-atom scattering only in the forward direction (small scattering angles). In the energy interval of one to several million electron volts, there exists a region of strong interference between the various scattering amplitudes. In Table I we summarize the importance of the various elastic scattering amplitudes.

In Figs. 9, 10, and 11 we display the total-atom and component scattering amplitudes for lead ($Z=82$) at 1.33 MeV. The Rayleigh amplitudes (labeled R) are calculated by the method of Sec. 3.3.1 using S-matrix predictions for the K and L shells; the contribution of the remaining electrons is estimated. We believe that these Rayleigh amplitudes are accurate to $O(1\%)$ in Re $A_\perp$. The nuclear-Thomson amplitudes (labeled NT) and nuclear-resonance amplitudes (labeled NR) are computed from the formulas described in Sec. 3.3.2. The Delbrück amplitudes (labeled D) are the lowest-order Born amplitudes due to Papatzacos and Mork (1975) described in Sec. 3.3.3. At this photon energy, the Rayleigh amplitudes dominate Im A; they are more than a factor of 10 smaller than the corresponding Re A except near the zero in Re $A_\parallel$. For this case we see that Rayleigh amplitudes dominate at forward angles, nuclear Thomson amplitudes dominate at back angles, and Delbrück amplitudes contribute at the 10% level at most angles (except in the small-angle region).

We observe that the Rayleigh and nuclear-Thomson amplitudes are

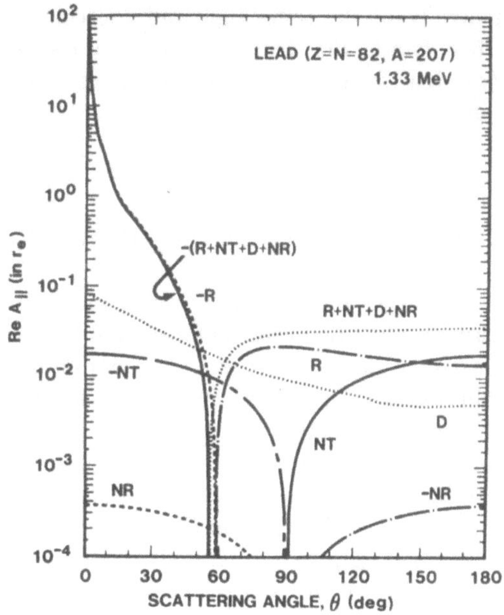

FIGURE 10. Contributions to Re $A_{\parallel}$ for elastic scattering of 1.33-MeV photons by neutral lead, as in Fig. 9.

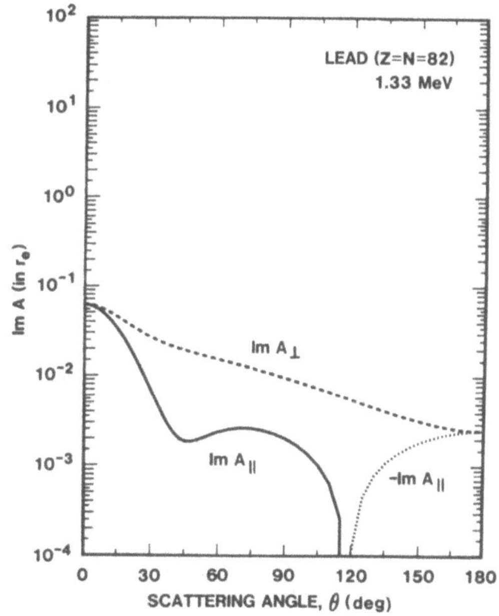

FIGURE 11. Imaginary parts of the Rayleigh scattering amplitudes for elastic scattering of 1.33-MeV photons by neutral lead. The imaginary parts of the Delbrück amplitudes, while nonzero, are assumed to be much smaller than the Rayleigh scattering amplitudes. The imaginary parts of the nuclear Thomson and nuclear resonance amplitudes are zero.

generally of the same sign and therefore interfere constructively. On the other hand, the Delbrück amplitudes destructively interfere with the Rayleigh amplitudes at smaller angles and constructively interfere at larger angles.

4. DEVELOPMENT OF EXPERIMENT

Here we discuss experimental methods for the measurement of elastic scattering in circumstances for which the dominant contribution comes from inner-shell atomic electrons. This emphasis reflects our interest in the contribution of inner-shell electrons to Rayleigh scattering and elastic photon–atom scattering. Hence, we restrict our attention to experimental kinematics and targets for which our isolated-atom assumptions are satisfied. Considering the conditions for which the Rayleigh amplitudes will be dominant, we find primary interest in experiments with photon energies 10 keV through 1 MeV. At higher energies, we must restrict our attention to smaller and smaller angles to ensure that the Rayleigh amplitudes dominate over the other coherent amplitudes. At lower energies, we must remember that solid-state effects in most targets can cause significant distortion of the outer-electron orbitals, violating our isolated-atom assumptions when these orbitals become dominant.

First we review experiments that directly measure the differential elastic-scattering cross section $d\sigma/d\Omega$. Next we review experiments that directly measure the anomalous scattering factors f' and f''. We postpone until Sec. 6 discussion of experiments involving applications of Rayleigh scattering.

4.1. Direct Measurement of $d\sigma/d\Omega$

Here we review experiments that directly measure the elastic photon–atom scattering cross section differential in scattering angle, in circumstances for which the Rayleigh scattering amplitudes are dominant. Experiments of interest utilize sources of medium- and high-energy X rays and low-energy γ rays with energies 10 keV through several million electron volts.

4.1.1. Detectors. In principle, it is necessary to measure the number of monoenergetic photons that are incident on a sample, and the number of elastically scattered photons, to determine the differential cross section $d\sigma/d\Omega$. Originally, the major experimental problem was to separate elastically scattered photons from inelastically scattered photons, and this is still a major problem for small-angle scattering. Inelastic scattering of photons by bound electrons is often called bound Compton scattering, or simply "Compton scattering." Unlike Compton scattering by free electrons (Sec. 2.3), where the scattered-photon energy is determined in terms of the scattering angle by Eq. (4), bound-Compton-scattered photons emitted at a given angle have a range of energies, leading to an energy spectrum often called the "Compton profile." In most cases, the Compton profile peaks at or near the energy of the free-electron Compton-scattered photon.

The primary limitation in the experimental separation of elastically and inelastically scattered photons is determined by the energy resolution of the photon detector, i.e., the ability to distinguish the energy of incident (and elastically scattered) photons from the energy of Compton scattered photons. As detector energy resolution has improved, so has the accuracy with which the separation has been made, reducing the experimental uncertainties. Early gas-ionization detectors using absorption filters (Meitner and Kösters, 1933) had low detection efficiency and poor energy resolution. Starting in the early 1950s (Moon, 1950), scintillation detectors such as NaI(Tl) with improved energy resolution of typically 80 keV at 1.33 MeV and improved detection efficiency were employed. At medium scattering angles, an elaborate and carefully done analysis of NaI pulse-height distributions is necessary to separate the elastically and inelastically scattered photons. For smaller scattering angles ($\theta \lesssim 13°$ for 1.33-MeV photons), the Compton shift is so small that the elastically and inelastically scattered photons cannot be experimentally separated with an NaI detector. In such cases an auxiliary experiment with a low-Z scatterer, for which at the same energy elastic scattering is negligible in comparison to inelastic scattering, could be performed. The theoretical ratio (assumed to be sufficiently well known from the incoherent scattering-factor approximation) of the number of inelastically scattered photons from the target of interest to the number from the low-Z auxiliary scatterer is then used to estimate and subtract the inelastically-scattered photon contribution in the main experiment. This complicated procedure, although probably more satisfactory than a direct subtraction of a theoretical prediction for the inelastic component, makes it difficult to determine precisely the elastic scattering cross section.

In the next generation of experiments (Dixon and Storey, 1968), solid-state detectors such as Ge(Li) have achieved a resolution of 3 to 4 keV at 1.33 MeV and relative accuracies of several percent in measurement of the absolute cross section. The factor-of-20 improvement in the resolution of Ge detectors over NaI detectors allows a direct separation of the elastic and inelastic components of the scattered photons down to quite small scattering angles θ. Although agreement with previous work is not good in all cases, many of the scintillation-detector measurements (for example, Standing and Jovanovich, 1962; Basavaraju and Kane, 1970) agree quite well with the more accurate solid-state detector measurements (e.g., Dixon and Storey,

1968; Hardie *et al.*, 1970; Basavaraju *et al.*, 1979). In other cases, scintillation-detector measurements claiming 20% accuracy (Hara *et al.*, 1958; Bernstein and Mann, 1958) differ from each other by a factor of 2.

4.1.2. Sources. A variety of photon sources have been developed to support experiments in this energy range. Sources based upon characteristic line radiation (e.g., $K\alpha$ radiation) have long been used. Such photon sources are relatively inexpensive but are limited to selected photon energies and in the purity with which the radiation of a given energy can be produced. Furthermore, the width of the lines is nonnegligible. Characteristic line-radiation sources span the energy range from ~ 100 eV through ~ 100 keV.

Radioactive isotopes are γ-ray sources with very high monochromaticity (typical energy spreads of a fraction of an electron volt). Radioactive isotope sources can be naturally occurring, but they are more commonly produced by neutron activation in a fission reactor, by bombarding samples with accelerator-produced particles, or by selecting isotopes from fission products. While the energy width of these lines is narrow, the available energies and intensities of these sources are limited. In addition, these sources are always "on" and therefore require special handling. Radioactive isotopes emit γ rays with energies in the range of tens of thousands through millions of electron volts.

A small amount of almost any naturally occurring isotope when placed in the high neutron flux of a fission reactor can be used to produce γ rays. With an appropriate choice of isotope, these in-beam neutron-capture sources can produce photons close to any desired X-ray energy (on a scale of several thousand electron volts). The advantage of this type of source is high intensity. The disadvantage of this type of source is the relatively small number of reactors that are available to support this type of experiment.

More recently, synchrotron radiation has become more readily available (cf. Chap. 7). Synchrotron radiation sources that are equipped with a monochromator are a very bright, continuously tunable source of photons. Another important characteristic is that synchrotron radiation is also highly polarized. This should allow for a new series of experiments that can provide a strong test for theoretical predictions of polarized scattering cross sections. Experiments at tens of thousands of electron volts have been performed (see, for example, Fuoss *et al.*, 1981), while experiments up to 8 MeV have been proposed (Schumacher, 1980).

4.1.3. Experimental Arrangements. Figure 12 shows a typical medium-angle experimental scattering geometry.* In this figure, the point source of

* Ring geometries can also be used, as is discussed subsequently in the case of small-angle scattering. As an alternative to a ring geometry, Ghose (1977) has suggested the use of a surface of revolution geometry. An experiment using this arrangement has been implemented by Bradley and Ghose (1983).

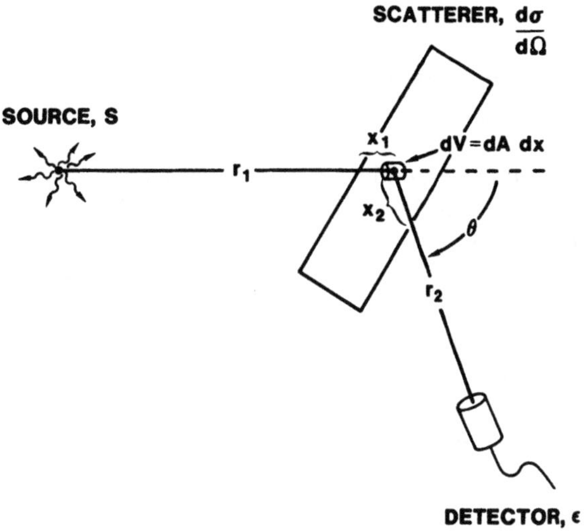

FIGURE 12. A typical medium-angle geometry for measurement of the differential elastic photon–atom scattering cross section.

monoenergetic photons is characterized by S, the number of photons of energy $E = \hbar\omega$ which the source emits per unit time and per solid angle. The number of photons $d\mathcal{N}_S$ incident per second on the scattering volume $dV = dA\,dx$ is

$$d\mathcal{N}_S = S\frac{dA}{r_1^2}\,e^{-\mu_1^{\text{air}}(r_1 - x_1)}e^{-\mu_1 x_1}. \tag{83}$$

(Attenuation, characterized by the linear attenuation coefficients μ_1^{air} and μ_1, in passing through air of thickness $r_1 - x_1$ and scattering medium of thickness x_1 has been included. If the experiment is performed in vacuum, $\mu_1^{\text{air}} = 0$.) The fraction df_{sc} of photons incident on the scattering volume dV that are scattered into the solid angle $d\Omega$ is

$$df_{\text{sc}} = \frac{(d\sigma/d\Omega)\,d\Omega(\mathcal{N}_{\text{AT}}/V)\,dA\,dx}{dA}. \tag{84}$$

Here $\mathcal{N}_{\text{AT}}$ is the number of scattering atoms in the target ($\mathcal{N}_{\text{AT}} = \mathcal{N}_0 M/\mathscr{A}$, with Avogadro's number $\mathcal{N}_0 \approx 6.02 \times 10^{23}$ particles per mole, M the mass of the target, and $\mathscr{A}$ the mass of the scattering atoms of the target in mass per mole). The magnitude of the signal generated per photon in the detector by a beam of photons that has been scattered into the solid angle $d\Omega$ by the elemental volume dV is

$$\mathscr{E} = e^{-\mu_2 x_2}e^{-\mu_2^{\text{air}}(r_2 - x_2)}\Delta\varepsilon_D, \tag{85}$$

where

$$\Delta = \begin{cases} 1 \text{ if the scattered ray passes through the detector,} \\ 0 \text{ otherwise,} \end{cases}$$

and where ε_D is the corresponding detector efficiency in signal strength per photon. (Attenuation has been characterized by linear attenuation coefficients μ_2 and μ_2^{air} as photons pass through a scattering medium of thickness x_2 and air of thickness $r_2 - x_2$. For elastic scattering, $\mu_2 = \mu_1$ and $\mu_2^{\text{air}} = \mu_1^{\text{air}}$. However, since attenuation is frequency dependent, for inelastic scattering, $\mu_2 \neq \mu_1$ and $\mu_2^{\text{air}} \neq \mu_1^{\text{air}}$.) Thus the total signal strength generated in the detector due to photons scattered by dV into the solid angle $d\Omega$ is $dI_{\text{sc}} = dN_S \, df_{\text{sc}} \, \mathscr{E}$. The total scattered-photon signal is found by integrating over the scatterer volume V and the scattering solid angle Ω:

$$I_{\text{sc}} = N_{\text{AT}} S \varepsilon_D \int d\Omega \frac{1}{V} \int dV \frac{e^{-\mu_1^{\text{air}}(r_1 - x_1)} e^{-\mu_1 x_1}}{r_1^2} \frac{d\sigma}{d\Omega} e^{-\mu_2 x_2} e^{-\mu_2^{\text{air}}(r_2 - x_2)} \Delta. \quad (86)$$

Such a scattering experiment does not measure the cross section at a precise scattering angle Ω, but measures a weighted average of the cross section in the neighborhood of Ω.

If polarization properties of the radiation and target are not being considered, then only θ dependence remains in the differential scattering cross section. In this case, the signal generated in the detector can be written as

$$I_{\text{sc}} = \int_{-\pi/2}^{\pi/2} d\theta \, W(\theta) \frac{d\sigma}{d\Omega}, \quad (87)$$

where $d\Omega = \sin\theta \, d\theta \, d\phi$ and

$$W(\theta) = N_{\text{AT}} S \varepsilon_D \sin\theta \int_0^{2\pi} d\phi \frac{1}{V} \int dV \frac{e^{-\mu_1^{\text{air}}(r_1 - x_1)} e^{-\mu_1 x_1}}{r_1^2} e^{-\mu_2 x_2} e^{-\mu_2^{\text{air}}(r_2 - x_2)} \Delta.$$

This expression, given the signal generated in the detector and wishing to solve for $d\sigma/d\Omega$, is in the form of a Fredholm integral equation of the first kind. Such an equation has an infinite number of solutions. Additional information concerning the properties of the cross section (such as nonnegativity, smoothness, shape, etc.,) must be specified to narrow the space of solutions. Since $W(\theta)$ will only be nonnegligible over a small range in θ, a possible approach is to use theory to predict the *shape* of the differential cross section as a function of scattering angle and use the measurement to determine the normalization (i.e., a multiplicative constant).

Over the range of solid angles from which rays will pass through the detector, one generally proceeds by ignoring, for the moment, the variation in differential cross section and rewrites Eq. (86) as

$$I_{sc} = \mathcal{N}_{AT} S \varepsilon_D \frac{\overline{d\sigma}}{d\Omega} \frac{1}{V} \int dV \frac{e^{-\mu_1^{air}(r_1 - x_1)} e^{(\mu_1 - x_1)}}{r_1^2} e^{-\mu_2 x_2} e^{-\mu_2^{air}(r_2 - x_2)} \int d\Omega \, \Delta. \quad (88)$$

We have written $\overline{d\sigma/d\Omega}$ to remind us that this cross section is a weighted-average cross section. To obtain $d\sigma/d\Omega$, it will be necessary to correct this measured cross section (e.g., Rullhusen *et al.*, 1979a); if the cross section is slowly varying or the angular width of $W(\theta)$ is small, then the correction will be small. The integral over solid angles reduces to A_D/r_2^2, the effective solid angle subtended by the detector with respect to the elemental scattering volume dV; A_D is the effective detector cross-sectional area, and r_2 is measured from an appropriate effective center inside the detector (e.g., Rullhusen *et al.*, 1979a). This equation can then be rewritten as

$$I_{sc} = \mathcal{N}_{AT} S A_D \varepsilon_D \frac{\overline{d\sigma}}{d\Omega} G, \quad (89)$$

where the geometrical factor is

$$G = \frac{1}{V} \int dV \frac{e^{-\mu_1^{air}(r_1 - x_1)} e^{-\mu_1 x_1}}{r_1^2} \frac{e^{-\mu_2 x_2} e^{-\mu_2^{air}(r_2 - x_2)}}{r_2^2}.$$

Here G is the Göttingen-group G factor (Rullhusen *et al.*, 1979a) if $\mu_1^{air} = \mu_2^{air} = 0$.

If the source-scatterer distance, r_1, and scatterer-detector distance, r_2, are large with respect to the scatterer dimension, then Eq. (89) can be rewritten as

$$I_{sc} = \mathcal{N}_{AT} S A_D \varepsilon_D \frac{\overline{d\sigma}}{d\Omega} \frac{e^{-\mu_1^{air} r_1}}{r_1^2} \frac{e^{-\mu_2^{air} r_2}}{r_2^2} T, \quad (90)$$

where the transmission factor is

$$T = \frac{1}{V} \int dV \, e^{-\mu_1 x_1} e^{-\mu_2 x_2}.$$

This is the T factor used by the Bombay group (e.g., Basavaraju and Kane, 1970).

To eliminate the product $SA_D\varepsilon_D$ of the source strength, the detector active area, and the detector efficiency, we might consider an auxiliary experiment. The signal generated in the detector when it directly views the source from a distance r is

$$I_S = S\frac{A_D}{r^2}e^{-\mu_1^{\text{air}}r}\varepsilon_D. \tag{91}$$

Using this result to eliminate the product $SA_D\varepsilon_D$ in Eq. (89), we write

$$\overline{\frac{d\sigma}{d\Omega}} = \frac{1}{\mathcal{N}_{\text{AT}}G}\frac{e^{-\mu_1^{\text{air}}r}}{r^2}\frac{I_{\text{sc}}}{I_S}. \tag{92}$$

Another technique aimed at eliminating dependence of the measured cross section on knowledge of the detector efficiency ε_D and the detector solid angle with respect to the scatterer, A_D/r_2^2, is the weak-auxiliary-source method (e.g., Standing and Jovanovich, 1962). The detector signal I_s, when a weak auxiliary photon source of strength s and of the same geometrical shape as that of the scatterer is placed at the position of the scatterer, is

$$I_s = s\frac{A_D}{r_2^2}e^{-\mu_2^{\text{air}}r_2}\varepsilon_D. \tag{93}$$

Using this measurement, we can rewrite Eq. (90) as

$$\overline{\frac{d\sigma}{d\Omega}} = \frac{1}{\mathcal{N}_{\text{AT}}T}\frac{r_1^2}{e^{-\mu_1^{\text{air}}r_1}}\frac{s}{S}\frac{I_{\text{sc}}}{I_s}. \tag{94}$$

Another approach, not requiring a knowledge of s/S, is the low-Z comparison-scatterer method (Kane *et al.*, 1978), which is useful for medium- and high-Z scatterers, especially at small scattering angles θ where the Compton shift is small. A typical experimental geometry for the measurement of small-angle scattering is shown in Fig. 13. (An alternate mechanical arrangement for small-angle experiments has been used by Nath and Ghose, 1964.) The low-Z comparison-scatterer method is also used to avoid the need for a direct measurement of the detector solid angle A_D/r_2^2 and the source strength S. This method relies on the validity of the nonrelativistic incoherent scattering function (Hubbell *et al.*, 1975) for inelastic scattering (Compton scattering) by bound electrons of low-Z atoms for $x = \lambda^{-1}\sin(\theta/2) > 3\text{Å}^{-1}$, as has been demonstrated in experiment (Ramanathan *et al.*, 1979). An auxiliary inelastic scattering experiment with a low-Z target (typically carbon or aluminum) for which the elastic component is negligible is performed; self-scattering in the source and multiple scattering in the low-Z target must be considered. (Unlike the low-Z auxiliary experiments used in NaI detector measurements, the aim now is not to *separate* the inelastic component, which has already been done, but to

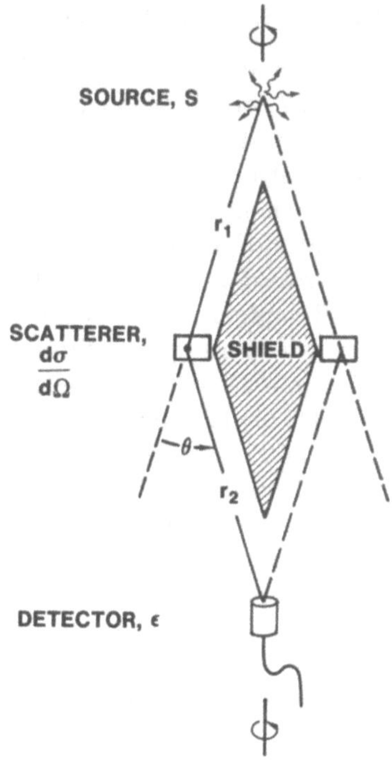

FIGURE 13. A typical small-angle geometry for measurement of the differential elastic photon–atom scattering cross section.

calibrate the measurement of the elastic component.) Using Eq. (90), we can write the source strength and detector solid angle in terms of the comparison-scatterer count rate n_c and the theoretical inelastic scattering cross section $d\sigma_c/d\Omega$ as

$$S\frac{A_D}{r_2^2}\frac{e^{-\mu_1^{\text{air}}r_1}}{r_1^2}=\frac{I_c}{\mathcal{N}_{\text{AT}}^c\,T^c\varepsilon_D^c(\overline{d\sigma^c}/d\Omega)\,e^{-\mu_2^{\text{air}-c}r_2}}. \tag{95}$$

Remember that the scattered photon in the above experiment is at the reduced, Compton-shifted energy. This affects the choice of μ_2 in T^c, and ε_D^c. Thus the elastic scattering cross section for the medium- to high-$\mathcal{Z}$ scatterer is given by

$$\frac{\overline{d\sigma}}{d\Omega}=\frac{\overline{d\sigma}^c}{d\Omega}\frac{\mathcal{N}_{\text{AT}}^c}{\mathcal{N}_{\text{AT}}}\frac{T^c}{T}\frac{\varepsilon_D^c}{\varepsilon_D}\frac{e^{-\mu_2^{\text{air}}r_2}}{e^{-\mu_2^{\text{air}-c}r_2}}\frac{I_{\text{sc}}}{I_c}. \tag{96}$$

In this type of experiment knowledge of the absolute detector efficiency is not necessary. The ratio $\varepsilon_D^c/\varepsilon_D$ can be determined with a small error when the Compton shift is small.

4.2. The Anomalous Scattering Factors

Several techniques have been developed to measure the scattering factors f_1, f_2 or, alternately, the anomalous scattering factors f', f'' defined in Sec. 3.2.3. These include (1) fitting Bragg-peak intensities from crystals of known structure, (2) calculation from total-reflection measurements, (3) measurement of the photoabsorption cross section, and (4) measurement of the index of refraction. Most of these efforts have been devoted to studies of the forward-angle factors only.

Templeton and Templeton (1978) suggested that f_1, f_2 could be determined by least-squares fitting to a set of measured Bragg reflection intensities from a crystal of known structure. This procedure was used by Phillips *et al.* (1978) near the L_3 edge of Cs, by Templeton *et al.* (1980) near the L_3 edges of Pr and Sm, and by Yakel (1980) near the K edge in Fe. This method has the added benefit that it is possible in principle to investigate the angular distribution of f_1, f_2 beyond form-factor approximation.

Compton (1923a) realized that total external reflection should occur at the air/sample interface if the index of refraction of the sample material is less than unity, as is generally the case with X rays. In terms of the anomalous scattering factors f_1, f_2, the complex index of refraction is given for a material composed of a pure element with n_{AT} atoms per unit volume as

$$n(\omega) = 1 - \delta - i\beta, \tag{97}$$

where

$$\delta = 2\pi \left(\frac{c}{\omega}\right)^2 n_{\mathrm{AT}} r_e f_1,$$

$$\beta = 2\pi \left(\frac{c}{\omega}\right)^2 n_{\mathrm{AT}} r_e f_2. \tag{98}$$

If for the moment we assume that f_2 is zero, then X rays are totally reflected for all angles less than

$$\cos \Theta_c = 1 - \delta. \tag{99}$$

The angle Θ is measured between the propagation direction $\mathbf{k}$ of the incident photons and the surface. The fraction of X rays reflected drops rapidly for angles larger than Θ_c. The rolloff in reflection is not so abrupt if absorption is nonzero (i.e., $f_2 \neq 0$). Therefore, measurement of Θ_c gives the value of f_1; the shape of the angular dependence of the reflection gives f_2 (James, 1962). Fukamachi *et al.* (1978, 1979) used this technique for measurements of the anomalous scattering factors near the Cu and Ni K

edges and the Yb L_3 edge; Fuoss (1980) obtained data near the Ge and Se K edges. This method requires high geometrical precision as very small angles are involved. Furthermore, surface details can affect the angular dependence of reflection, which degrades the measurement of f_2.

The third technique involves the determination of the imaginary anomalous scattering factor f_2 by measurement of the photoabsorption cross section. From the optical theorem (61), we have

$$f_2 = -\frac{\omega}{4\pi c r_e}(\sigma_{\text{PE}} - \sigma_{\text{BPP}}) - \frac{\pi}{2}\sum_n \omega_{ni} f_{ni}\,\delta(\omega_{ni} - \omega). \tag{100}$$

For energies lower than the pair-production threshold $2mc^2$, and for energies different from the bound–bound transitions energies $\hbar\omega_{ni}$, the value of f_2 is given as

$$f_2 = -\frac{\omega}{4\pi c r_e}\sigma_{\text{PE}}. \tag{101}$$

The signs in the preceding expressions are a consequence of our definition of the scattering factor in terms of the Thomson amplitude, which has been defined as negative. Many techniques have been developed for the measurement of photoabsorption (e.g., Creagh, 1978). Fuoss (1980) and Fuoss *et al.* (1980) measured f_2 near the K edges of Ge and Se, Lye *et al.* (1980) near the L edges of Eu.

In general, interferometers measure the index of refraction of a material. X-ray interferometers have been designed that allow direct measurement of Re $n(\omega)$ and thus f_1 with accuracies at the fraction of a percent level. Creagh (1980) has reviewed the relative accuracy of measurements by using the LLL-type interferometer and Ångstrom-ruler interferometer. Both these types of devices are constructed from a monolithic boule of perfect single-crystal silicon.

The dispersion relation (57) can be used to compare the independent measurements of f_1 by means of interferometry and f_2 by attenuation. This can provide a sensitive test of the agreement between theory and experiment. But care must be taken in evaluating the integral over f_2 as the contribution to f_1 from bound–bound transitions can be important, especially near absorption edges. Another common error is to assume that the high-energy limit of f_1 is given by $f(0)$, the value of the form factor at zero momentum transfer. As discussed in Secs. 3.1.5, 3.2.2, and 3.2.3, the high-energy limit of f_1 to $O((Z\alpha)^2)$ is given by $g(0)$, the value of the modified form factor at zero momentum transfer.

5. A COMPARISON OF THEORY AND EXPERIMENT

Here we consider experiments in which the inner-shell Rayleigh scattering amplitudes make a dominant contribution to the total-atom elastic scattering cross section. Experiments in the photon energy range of 10 keV through 1 MeV are of special interest. The comparison between theory and experiment is far from complete. Yet the incomplete comparisons to date of medium- and high-Z targets have been encouraging.

5.1. Differential Scattering of Low-Energy Gamma Rays

As discussed in the previous section, historically the principal experimental difficulty in the direct measurement of the differential elastic-scattering cross section was the separation of inelastically scattered photons. Consequently, the higher energy resolution of solid-state detectors has generally allowed for more precise measurements. This is especially true in smaller-angle experiments on medium- and high-Z targets where the validity of theoretical predictions of the inelastic-scattering cross section is not as certain. Experiments based on the theoretical knowledge of the inelastic cross section at small angles are open to some question.

In Fig. 14, we compare our predictions of the total-atom elastic cross section with three experiments for Pb at 145 keV. The medium-angle experiment of Schumacher (1969) agrees well in its overlap with the newer, smaller-angle experiment of deBarros *et al.* (1981b). Both of these experiments use Ge(Li) solid-state detectors. We compare with these experiments the recent NaI detector measurements of SenGupta *et al.* (1982). At these small scattering angles, the NaI detector does not have enough energy resolution to resolve the inelastically scattered photons from the elastically scattered photons. SenGupta *et al.* use the nonrelativistic Hartree–Fock incoherent scattering function (Hubbell *et al.*, 1975) to predict the number of inelastic photons in their detected count rate. For the 1°-to-25° region measured by SenGupta *et al.*, the incoherent scattering function is significantly different from its high-energy, large-angle value. While the incoherent scattering function is expected to give accurate predictions for low-Z scatterers when it is not too different from its high-energy, large-angle value, its accuracy for medium- and high-Z scatterers is open to question. SenGupta *et al.* assume a 5% uncertainty because of this source of error; yet their measured cross section differs significantly from theory and other experiments.

We are continuing with a systematic comparison of theory and experiment for medium- and high-Z atoms in the photon energy range 59.5 keV through 1.33 MeV. At present (Roy *et al.*, 1983; Kissel *et al.*, 1982), our predictions agree at the 5% level with the bulk of the Ge(Li) solid-state

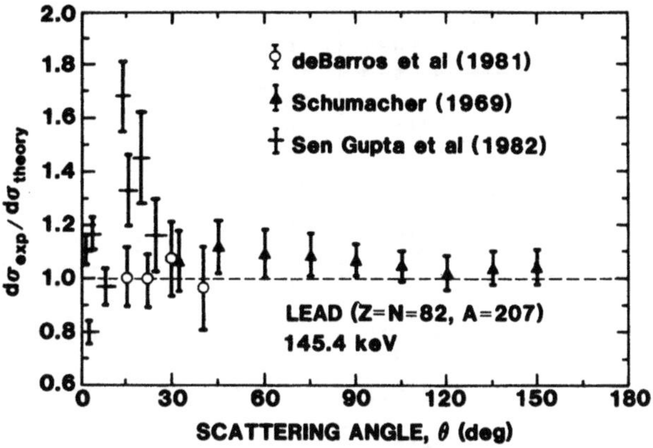

FIGURE 14. Comparison of theory and experiment for 145-keV photons scattered by neutral lead at medium and small scattering angles. The Ge-detector measurements (deBarros *et al.*, 1981b, and Schumacher, 1969), which can experimentally separate elastically scattered photons from inelastically scattered photons, show good agreement with theory and each other. The small-angle NaI-detector measurements (SenGupta *et al.*, 1982), which cannot experimentally separate the elastic and inelastic contributions, disagree with theory and with the Ge-detector experiments.

detector measurements for Zn, Pb, and U. In most remaining cases, experiments disagree among themselves and no clear discrepancies are evident. Gone are the factor-of-2 differences (Dixon and Storey, 1968) found at 1.33 MeV at large scattering angles, which were caused by numerical difficulties in the original pioneering theoretical calculations. Furthermore, the smaller differences found at 30° to 60° scattering angles at 662 keV (Kissel *et al.*, 1980; Mückenheim and Schumacher, 1980) have been resolved. These smaller differences were corrected by including the Delbrück amplitudes for Pb, and also including outer electrons by means of modified form-factor estimates for U.

5.2. Small-Angle Scattering

In recent years a large number of experiments has been performed that measure elastic scattering at small angles. Roy *et al.* (1983) have compared experiment with theory for the small momentum-transfer range $x = 0$ to $10\ \text{Å}^{-1}$. Comparisons with available experiments for Pb at energies from 145 keV to 1408 keV are shown in Fig. 15. Only experiments using Ge(Li) detectors were considered, since for a small angle of scattering the inelastically scattered photons lie very close in energy to the elastically scattered photons. This comparison indicates no systematic discrepancy between theory and the significant scatter among different experimental results. Until any such discrepancy between theory and experiment is more definitely

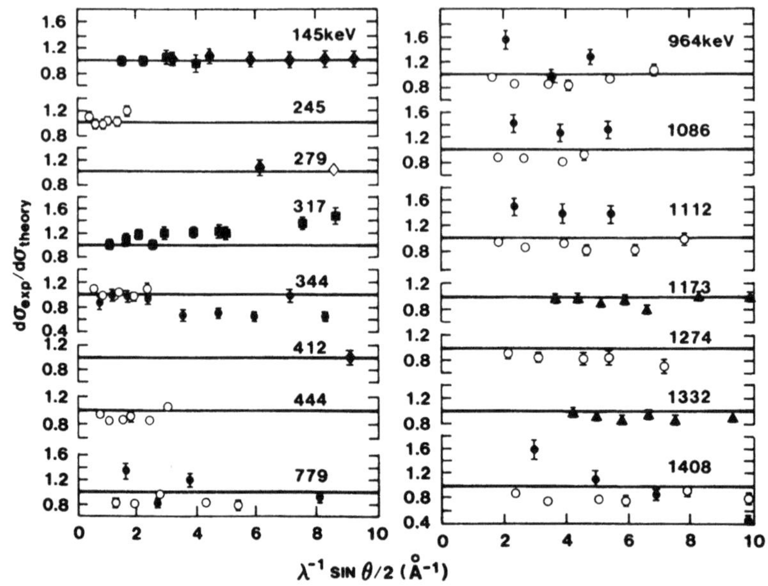

FIGURE 15. Comparison of small-momentum-transfer Ge-detector measurements and theory for elastic scattering of low-energy γ rays by neutral lead. The symbols used in this figure indicate the following experiments: ■, deBarros *et al.* (1981b); ●, Chitwattanagorn *et al.* (1980); ○, Ramanathan *et al.* (1979); ▲, Kane *et al.* (1978, 1981), ◇, Smend *et al.* (1973); and ◆, Schumacher (1969). This comparison indicates no systematic discrepancy between theory and experiment, but does indicate significant scatter among the different experimental results. The theoretical cross sections used in this comparison include the contribution from the Rayleigh scattering and nuclear Thomson scattering amplitudes, only, that dominate the total-atom scattering for λ^{-1} sin $(\theta/2) < 10$ Å^{-1} for photon energies < 1 MeV. For energies > 1 MeV, the inclusion of the Delbrück scattering amplitude would tend to reduce the theoretical cross section slightly at these momentum transfers. For example, at 1.33 MeV the theoretical cross sections used here are too large by 2% at 4 Å^{-1} and 7% at 10 Å^{-1}.

established, it seems premature to consider further refinements of the numerical calculations.

Kahane *et al.* (1977) have measured the elastic scattering cross sections at 1.5° for photons of energies 6.84 to 11.39 MeV incident on Pb and U. Kissel *et al.* (1980) have obtained predictions for these cases and find agreement between theory and experiment within the experimental errors.

6. RAYLEIGH-SCATTERING APPLICATIONS

Here we mention some applications of predictions for inner-shell Rayleigh-scattering amplitudes. We consider situations in which the accurate evaluation of the inner-shell scattering amplitudes is important.

6.1. *Solid-State Structure Studies*

Elastic photon scattering has held a central position in studies of the structure of materials in the solid state. Traditionally, in studies of crystalline materials, predictions of X-ray diffraction from a proposed model of the structure are compared with the observed diffraction pattern. The different possible arrangements of atoms in most simple crystal structures yield diffraction patterns that are measurably different, and thus the crystal structure can be determined (James, 1962). This application assumes that the scattering from a given structure can be accurately predicted. Theoretical efforts to support this application by predicting the scattering from atoms in solids date back to the 1910s and continue today.

For photon energies above several thousand electron volts, inner-shell Rayleigh scattering makes an important contribution to the total scattering from the atom. However, our isolated-atom prescription for the total-atom scattering presented in Sec. 3.3 is no longer appropriate, since the outer-electron wave functions are significantly distorted in the solid. Nevertheless, our inner-shell predictions should still be useful when combined with predictions for the scattering from the distorted outer electrons. This is especially important for high-Z atoms, where the departure from the form-factor approximation is large for inner electrons. However, care must be taken when combining estimates of outer-electron scattering with S-matrix amplitudes for the inner electrons. This is because of the existence of nonphysical bound–bound transitions that are included in the inner-shell amplitudes. In principle, it should be possible to subtract these bound–bound transitions from the inner-electron amplitudes. In addition, the neglect of the finite lifetime of excited states reached in transitions from inner shells will lead to an incorrect description of scattering above and below ionization thresholds (Bremer, 1979).

The study of the short-range and long-range structure of amorphous materials is also of interest. The technique of EXAFS (extended x-ray fine structure) analysis has been used to study the nearest-neighbor distances. However, EXAFS is of limited utility when the near-neighbor atoms are not too different in atomic number. Recently, Fuoss *et al.* (1981) have applied anomalous scattering to the investigation of the structure of amorphous Ge–Se systems with tunable synchrotron radiation. By selecting photons near the K absorption edge of Ge or Se, Fuoss *et al.* were able to selectively investigate the nearest-neighbor distances from Ge atoms or Se atoms since the anomalous scattering factors are large and vary rapidly with energy near the edges. In Fig. 16 we illustrate their results for the derivative of the intensity of scattered X rays from samples of amorphous GeSe and $GeSe_2$. The solid curve represents the difference between scattering intensities at 11.000 and 11.090 keV (the Ge K absorption edge is at 11.103 keV); the

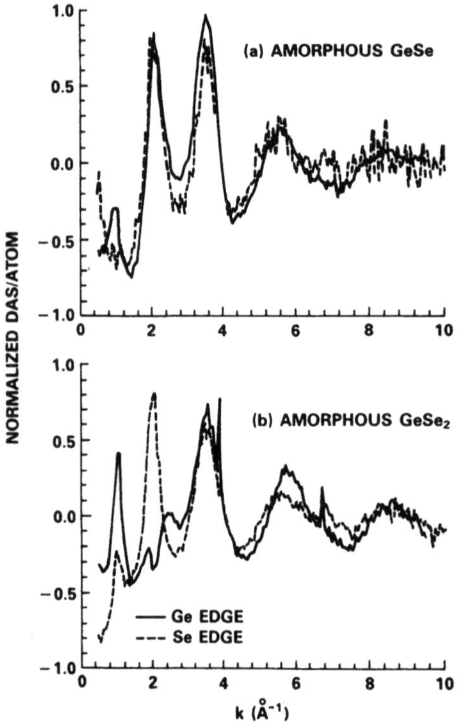

FIGURE 16. The differential anomalous scattering from amorphous GeSe and $GeSe_2$ is shown for photon energies near the K absorption edges of Ge and Se. These results were used to investigate the short-range coordinations of Ge and Se in amorphous GeSe. (Reprinted by permission from Fuoss *et al.*, 1981.)

dashed curve is a similar difference between scattering intensities at 12.555 keV and 12.645 keV (the Se K absorption edge is at 12.656 keV). This technique resolved a ten-year debate over the short-range coordinations of Ge and Se in amorphous GeSe. Two different models for the structure of GeSe had been proposed that were both consistent with the normal radial-distribution-function analysis. These two models did not predict the same differential anomalous scattering, and the results of Fig. 16 were consistent with the known structure of $GeSe_2$ and with only one of the proposed models for GeSe. However, the technique used a differential approach that is very sensitive to error. In addition, the accuracy of the method is limited by inadequate knowledge of the anomalous scattering factor f'.

6.2. *Nuclear Structure Studies*

Measurement of nuclear resonance fluorescence (NRF) photons is useful in the study of the structure of the nucleus. Elastic photon scattering is a background in these measurements, and it is desirable to have predictions for the strength of this background. For photon energies less than several million electron volts, Rayleigh scattering is an important component. Predictions for the Rayleigh contribution based upon form-factor approximation are wrong by a factor of 2. Yet for many cases in the energy range of several million electron volts, the NRF photon intensity is only about twice the coherently scattered photon background (Schumacher *et al.*, 1981). For photon energies of several hundred thousand electron volts, NRF photons can be observed above the coherently scattered photon background only by use of polarized photon sources (Schumacher, 1980), using situations in which the coherent background can thereby be suppressed.

6.3. *Observation of Delbrück Scattering*

Delbrück scattering is the contribution to the elastic photon–atom scattering amplitude due to scattering of photons from virtual electron–positron pairs created in the Coulomb field of the nucleus (Sec. 3.3.3). Delbrück scattering is one of the nonlinear processes in quantum electrodynamics (QED) that are a direct consequence of vacuum polarization. An important test of the perturbative approach to QED in high orders is possible in a careful study of Delbrück scattering, as the lowest-order nonvanishing amplitudes are of the order $Z^2e^6 = \alpha(Z\alpha)^2$. The Delbrück amplitudes are dominated by their imaginary parts for photon energies well above the pair-production threshold of $2mc^2$; they are purely real for photon energies below this threshold.

Recently there have been two major efforts to investigate Delbrück scattering, one at photon energies 1 to 7 GeV (Jarlskog *et al.*, 1973) primarily sensitive to the imaginary part of the amplitude. The other experiments, at photon energies 1 to 11 MeV (Basavaraju *et al.*, 1979; Rullhusen *et al.*, 1981, and references therein), have been designed to indicate the existence of the real, or dispersive, part of the Delbrück scattering amplitude. Now, Rayleigh scattering makes an important contribution at all angles for photon energies near 1 MeV, and remains important for forward angles at higher energies. In the past, Delbrück scattering investigations have been significantly hampered by the lack of accurate Rayleigh amplitudes and the identification of the real Delbrück amplitude was only achieved once these Rayleigh amplitudes became available. More recently, for example in studies at 2.75 MeV (Rullhusen *et al.*, 1981), precise investigations have become possible, thanks to the availability of improved Rayleigh predictions, and this has allowed for the experimental investigation of contributions from higher-order corrections (so-called Coulomb corrections) to the Delbrück amplitudes. The contribution of Delbrück amplitudes to elastic scattering has now been observed (at the 7% level) for energies as low as 889 keV in scattering from U (experimental results in Mückenheim and Schumacher, 1980, theoretical predictions and comparisons in Kissel *et al.*, 1982).

6.4. *X-Ray Diagnostics*

Availability of accurate Rayleigh amplitudes is important in X-ray diagnostics experiments. In these experiments it is desired to determine various properties such as temperature, density, spectrum, etc., by measuring the intensity of X rays emitted by a source such as a fusion plasma. Detector systems with X-ray filters, mirrors, scatterers, crystals, etc., are used. It is very important to be able to predict the performance of these detector components in the design and calibration stages of such experiments. Henke (1981) has summarized the role of Rayleigh scattering in the low-energy X-ray interactions of scattering, reflection, and refraction. Henke finds differences between sample comparisons of theory and experiment at the 10% to 30% level in the performance of complex diffraction structures.

6.5. *Narrow-Beam Attenuation*

Elastic photon–atom scattering removes photons from a narrow beam. Thus elastic scattering has been included in many tabulations of X-ray and γ-ray attenuation coefficients (Hubbell *et al.*, 1974). In most cases, elastic scattering makes a small contribution to the total attenuation. However, elastic scattering contributes as much as 10% to narrow-beam attenuation for photon energies just below the K-shell ionization threshold of high-Z atoms. Changes in the attenuation coefficients at $\lesssim 1\%$ level can be expected from the use of improved Rayleigh scattering predictions. In most cases this level of change is probably at the limit of experimental verification.

7. FURTHER RESEARCH TOPICS

In concluding this chapter, we wish to indicate several new areas in which work is in progress or proposed, and that may be expected to contribute to our further understanding of the properties and uses of the Rayleigh scattering amplitudes. This section is therefore to be understood as outlining an agenda for further work.

Work is in progress (Kissel *et al.*, 1982) to compare theory and experiment over a broader range in elements and energy. Data have been obtained for Zn, Sn, Pb, and U in the range 59.5 keV through 1.33 MeV, by using numerical calculations for inner shells and modified form-factor approximation for outer shells, and by adding in the other elastic amplitudes. A separate effort has begun to see if theory can reproduce the high-precision (17.479-keV) X-ray scattering results of Chipman and Jennings (1963) on the noble gases Ne, Ar, Kr, and Xe, and other similar work in the X-ray regime. An accuracy of $\sim 0.5\%$ was claimed; results are not in agreement with the calculations of Cromer and Liberman (1970). Since the anomalous terms (deviation from form factor) are small, this provides a stringent test of the accuracy of the numerical calculations, as well as the single-electron models on which it is based. This work should also provide an opportunity to assess the accuracy of various approaches based on nonrelativistic dispersion relations and correction factors, the contribution of bound-bound transitions, and the importance of the Z dependence of the high-energy limit beyond form factor. Nearer to edges, the anomalous terms become much larger, and at the same time a single-electron calculation becomes less reliable. A testing ground is thus provided for many-electron effects; it also becomes necessary to take account of the finite lifetimes (widths) of excited states. It should also be interesting to study the importance of the anomalous terms for attenuation, where generally only form-factor values have been used. As noted earlier, it would appear that at certain energies anomalous terms should contribute at the level of accuracy that is claimed for the measured attenuation coefficients. However, in obtaining their results, Chipman and Jennings subtracted their estimate of the inelastic contribution to scattering. Before their data can be used in precision discussion of elastic scattering, it may be necessary to revise and improve this subtraction.

Efforts are in progress (Wang, 1981, 1982) to characterize and parametrize the Rayleigh amplitudes. Such approaches, based on simple theoretical ideas, make it possible to identify the underlying mechanisms responsible for the main features of the amplitude; they form the basis for parametrization schemes that can be used as small, rapid, component subprograms in large computer codes that repeatedly (many millions of times) must call for information about the amplitudes; and, finally, they provide an interpolating form that can be used together with an accurate basis data set in the preparation of systematic tabulations of the amplitudes. Medical applications of such parametrizations have been discussed by Jackson and Hawkes (1981). Wang and Goldberg (1982) began with a parametrization of photoeffect based on ideas of an analytical perturbation theory. Through the use of dispersion relations, a corresponding parametrization was developed for f'. Finally, treating f'' and f' as angle-independent terms to be added to the form factor, expressions of moderate accuracy were obtained for the low-Z Rayleigh amplitudes. An extension to higher-Z elements is in process. When combined with the accurate set of basis data described in the previous paragraph, this should permit the development of high-accuracy tables of the Rayleigh amplitudes.

The Rayleigh contribution to elastic scattering can be modified when the atomic process takes place in an environment, rather than for an isolated neutral atom. A simple example is to consider Rayleigh scattering from ions, as would occur in thin hot plasmas. Scattering factors have also often been calculated for ions as well as for neutral atoms, intended for application to ionic crystals, as for example in the Ibers (1962), the work of Cromer *et al.* (1964, 1965), and of Doyle and Turner (1968). Parker has been using Kissel's code to study the modification due to ionization. As we would expect, the largest effects are simply the decreases in amplitude resulting from the removal of electrons from which to scatter. Thresholds move, without the subshell amplitudes showing much change. The resonance regions enlarge, as more subshells are not full and bound–bound transitions are allowed for energies much farther below thresholds. An important environment is an atom in a solid, with covalent bonding, and outer atomic levels replaced by bands. Elastic scattering from such a solid yields information on properties of the solid. Efforts are now under way to improve the interpretation of such experiments [for example, in the scattering of 17.479- and 22.163-keV X rays from Si by Aldred and Hart (1973a, b) and by Bonse and Teworte (1980), with a claimed accuracy at the 0.1% level] by including a more accurate evaluation of the contribution from the inner atomic shells. Other environments that could be considered include those of high magnetic or elastic fields, or of high temperature and density.

In this chapter we have not devoted much attention to the polarization properties of Rayleigh scattering (Sec. 2.4). Kissel's code contains such information, but it has not yet been studied. A survey of experimental studies, none of them very recent, has been given by Kane (1981); these are of two types. One group of experiments has determined the linear polarization of photons scattered through 90° from an unpolarized beam, through the use of a polarimeter based on Compton scattering; the experiments have been performed at 412, 662, and 1250 keV in Hg and Pb. A second class of experiments, performed for Pb at 1250 keV, used polarized photons and detected photons scattered through 90° without determining their polarization. Clearly, the improvements of the last two decades in both theory and experiment could be brought to bear on these circumstances. A rather different situation is represented in the experiment of Simon and Daniel (1977); these authors make reference to relevant earlier literature. In their experiment Simon and Daniel studied *spin* dependence, scattering 129-keV circularly polarized γ rays from a magnetized iron–cobalt alloy target, and observing γ rays scattered through angles of 21° to 55°; they found an 0.2% effect. Theoretical interpretation of such an experiment requires a more complete treatment of the basic formulation than that given here or employed in the numerical calculations, proceeding without summing (averaging) over the magnetic substates of the bound electrons.

ACKNOWLEDGMENTS

This work has been supported in part by the US Department of Energy under contract No. DE-AC04-76DP00789 and contract No. W-7405-Eng-48, subcontract No. 9907401, and by the National Science Foundation under grant No. PHY-8120785. Both authors would like to acknowledge many helpful discussions with M. Gavrila, A. M. Ghose, J. H. Hubbell, P. P. Kane, S. C. Roy, M. Schumacher, and F. Smend, who read and commented on earlier drafts of this manuscript.

REFERENCES

Aldred, P. J. E., and M. Hart, 1973a, *Proc. R. Soc. London, Sec. A* **332**, 223.
Aldred, P. J. E. and M. Hart, 1973b, *Proc. R. Soc. London, Sec. A* **332**, 239.
Alvarez, L. W., F. S. Crawford, Jr., and M. L. Stevenson, 1958, *Phys. Rev.* **112**, 1267.
Anand, S., and B. S. Sood, 1965, *Nucl. Phys.* **73**, 368.
Bar-Noy, T., and S. Kahane, 1977, *Nucl. Phys. A* **288**, 132.
Basavaraju, G. and P. P. Kane, 1970, *Nucl. Phys. A* **149**, 49.
Basavaraju, G., P. P. Kane, and K. M. Varier, 1979, *Pramana* **12**, 665.
Berant, Z., R. Moreh, and S. Kahane, 1977, *Phys. Lett.* **69**B, 281.
Berman, B. L., 1975, *At. Data Nucl. Data Tables* **15**, 319.
Bernstein, A. M., and A. K. Mann, 1958, *Phys. Rev.* **110**, 805.
Bonse, U., and R. Teworte, 1980, *J. Appl. Crystallogr.* **13**, 410.
Bothe, W., and F. Kirschner, 1933, in *Handbuch der Physik*, Vol. 23 Part 2, edited by H. Geiger and K. Scheel, (Springer, Berlin).
Bradley, D. and A. M. Ghose, 1983, private communication.
Bremer, J., 1979, *J. Phys. B* **12**, 2797.
Brenner, S., G. E. Brown, and J. B. Woodward, 1955, *Proc. R. Soc. London, Sec. A* **227**, 59.
Brown, G. E., and D. F. Mayers, 1957, *Proc. R. Soc. London, Sec. A* **242**, 89.
Brown, G. E., and D. F. Mayers, 1956, *Proc. R. Soc. London, Sec A* **234**, 387.
Brown, G. E., R. E. Peierls, and J. B. Woodward, 1955, *Proc. R. Soc. London, Sec. A* **227**, 51.
Brown, G. E., and J. B. Woodward, 1952, *Proc. Phys. Soc. London, Sec. A* **65**, 977.
Brown, R. T., 1974, *Phys. Rev. A* **10**, 438.
Chipman, D. R., and L. D. Jennings, 1963, *Phys. Rev.* **132**, 728.
Chitwattanagorn, W., R. B. Taylor, P. Teansomprasong, and I. B. Whittingham, 1980, *J. Phys. G* **6**, 1147.
Cindro, N., and K. Ilakovac, 1958, *Nucl. Phys.* **5**, 647.
Compton, A. H., 1930, *Phys. Rev.* **35**, 925.
Compton, A. H., 1923a, *Phil. Mag.* **45**, 1121.
Compton, A. H., 1923b, *Phys. Rev.* **22**, 409.
Cornille, H., and M. Chapdelaine, 1959, *Nuovo Cimento* **14**, 1386.
Costescu, A., 1976, *Rev. Roum. Phys.* **21**, 3.
Creagh, D., 1980, *Phys. Lett. A* **77**, 129.
Creagh, D., 1978, in *Advances in X-Ray Analysis*, edited by C. S. Barret, and D. E. Leyden (Plenum, New York).

CROMER, D. T., 1965, *Acta Crystallogr.* **19**, 224.
CROMER, D. T., A. C. LARSON, AND J. T. WABER, 1964, *Acta Crystallogr.* **17**, 1044.
CROMER, D. T., AND D. A. LIBERMAN, 1981, *Acta Crystallogr. Sec. A* **37**, 267.
CROMER, D. T., AND D. A. LIBERMAN, 1970, *J. Chem. Phys.* **53**, 1891.
CROMER, D. T., AND J. T. WABER, 1965, *Acta Crystallogr.* **18**, 104.
DEBARROS, S., J. EICHLER, M. GASPAR, AND O. GONÇALVES, 1981a, *Phys. Rev. C* **24**, 1765.
DEBARROS, S., J. EICHLER, O. GONÇALVES, AND M. GASPAR, 1981b, *Z. Naturforsch. Teil A* **36**, 595.
DIXON, W. R., AND R. S. STOREY, 1968, *Can. J. Phys.* **46**, 1153.
DOYLE, P. A., AND P. S. TURNER, 1968, *Acta Crystallogr. Sec. A* **24**, 390.
ERICSON, T. E. O., AND J. HÜFNER, 1973, *Nucl. Phys. B* **57**, 604.
FLORESCU, V., AND M. GAVRILA, 1976, *Phys. Rev. A* **14**, 211.
FRANZ, W., 1936, *Z. Phys.* **98**, 314.
FRANZ, W., 1935, *Z. Phys.* **95**, 652.
FUKAMACHI, T., S. HOSOYA, T. KAWAMURA, S. HUNTER, AND Y. NAKANO, 1978, *Jpn. J. Appl. Phys.* **17** [Suppl 17-2], 326.
FUKAMACHI, T., O. SHIMOMURA, T. KAWAMURA, M. TOKONAMI, P. H. FUOSS, AND S. HOSOYA, 1979, Stanford Synchrotron Radiation Laboratory Activity report 79/03:VII-43.
FUOSS, P. H., 1980, Structural Studies of Amorphous Materials Using X-Ray Anomalous Scattering, PhD Thesis, Stanford University.
FUOSS, P. H., P. EISENBERGER, W. K. WARBURTON, AND A. BIENENSTOCK, 1981, *Phys. Rev. Lett.* **46**, 1537.
FUOSS, P. H., W. K. WARBURTON, AND A. BIENENSTOCK, 1980, *J. Non-Cryst. Solids* **35&36**, 1233.
GAVRILA, M., 1982, *in X-Ray and Atomic Inner-Shell Physics—1982*, edited by B. Crasemann, AIP Conference Proceedings No. 94 (American Institute of Physics, New York).
GAVRILA, M., 1967, *Phys. Rev.* **163**, 147.
GAVRILA, M., AND A. COSTESCU, 1971, *Phys. Rev. A* **4**, 1688.
GAVRILA, M., AND A. COSTESCU, 1970, *Phys. Rev. A* **2**, 1752.
GELL-MANN, M., M. L. GOLDBERGER, AND W. THIRRING, 1954, *Phys. Rev.* **95**, 1612.
GHOSE, A. M., 1977, *Radiation Physics*, NBS Special Publication 461, edited by A. M. Ghose, D. V. Gopinath, J. H. Hubbell, and S. C. Roy (National Bureau of Standards, Washington) p. 47.
GOLDBERGER, M. L., AND F. E. LOW, 1968, *Phys. Rev.* **176**, 1778.
HARA, E., J. BANAIGS, P. EBERHARD, L. GOLDZAHL, AND J. MEY, 1958, *J. Phys. Rad.* **19**, 668.
HARDIE, G., W, J. MERROW, AND D. R. SCHWANDT, 1970, *Phys. Rev. C* **1**, 714.
HARTREE, D. R., 1925, *Phil. Mag.* **50**, 289.
HENKE, B. L., 1981, *Low-Energy X-Ray Diagnostics—1981 (Monterey)*, edited by D. T. Attwood and B. L. Henke, AIP Conference Proceedings No. 75 (American Institute of Physics, New York).
HENKE, B. L., P. LEE, T. J. TANAKA, R. SHIMABUKURO, AND B. K. FUJIKAWA, 1982, *At. Data Nucl. Data Tables* **27**, 1.
HUBBELL, J. H., 1969, *Nat. Stand. Ref. Data Ser.* NSRDS-NBS **29**.
HUBBELL, J. H., H. A. GIMM, AND I. ØVERBØ, 1980, *J. Phys. Chem. Ref. Data* **9**, 1023.
HUBBELL, J. H., W. H. MCMASTER, N. KERR-DEL GRANDE, AND J. H. MALLETT, 1974, in *International Tables for X-Ray Crystallography* (Kynoch Press, Birmingham, England). Vol. IV.
HUBBELL, J. H., AND I. ØVERBØ, 1979, *J. Phys. Chem. Ref. Data* **8**, 69.
HUBBELL, J. H., W. J. VEIGELE, E. A. BRIGGS, R. T. BROWN, D. T. CROMER, AND R. J. HOWERTON, 1975, *J. Phys. Chem. Ref. Data* **4**, 471; erratum in 1977, *J. Phys. Chem. Ref. Data* **6**, 615.
IBERS, J. A., 1962, in *International Tables for X-Ray Crystallography* (Kynoch Press, Birmingham, England) Vol. III.
JACKSON, D. F., AND D. J. HAWKES, 1981, *Phys. Rep.* **70**, 169.
JACKSON, J. D., 1975, *Classical Electrodynamics* (Wiley, New York), 2nd edition.

JAMES, R. W., 1962, *The Optical Principles of the Diffraction of X-Rays* (Bell, London).

JARLSKOG, G., L. JÖNSSON, S. PRÜNSTER, H. D. SCHULZ, H. J. WILLUTZKI, AND G. G. WINTER, 1973, *Phys. Rev. D* **8**, 3813.

JOHNSON, W. R., D. J. BUSS, AND C. O. CARROLL, 1967, *Phys. Rev.* **159**, 61.

JOHNSON, W. R., D. J. BUSS, AND C. O. CARROLL, 1964, *Phys. Rev.* **135**, A1232.

JOHNSON, W. R., AND K. T. CHENG, 1976, *Phys. Rev. A* **13**, 692.

KAHANE, S., O. SHAHAL, AND R. MOREH, 1977, *Phys. Lett. B* **66**, 229.

KANE, P. P., J. MAHAJANI, G. BASAVARAJU, AND A. K. PRIYADARSINI, 1983, *Phys. Rev. A* **28**, 1509.

KANE, P. P., 1981, Elastic Scattering of Gamma Rays, in *Proc. 4th National Symposium on Radiation Physics, Patiala, India, 5–7 October* (unpublished).

KANE, P. P., J. MAHAJANI, G. BASAVARAJU, AND A. K. PRIYADARSINI, 1981, in *Abstracts of Contributed Papers to the XII-th International Conference on the Physics of Electron and Atomic Collisions*, Gatlinburg, Tennessee, July 15–21, edited by S. Datz, Vol. 2, p. 1091 (unpublished), and private communications.

KANE, P. P., G. BASAVARAJU, J. MAHAJANI, AND A. K. PRIYADARSINI, 1978, *Nucl. Instrum. Methods* **155**, 467.

KISSEL, L., R. H. PRATT, AND S. C. ROY, 1980, *Phys. Rev. A* **22**, 1970.

KISSEL, L., S. C. ROY, AND R. H. PRATT, 1982, in Abstracts of Contributed Papers, International Conference on X-Ray and Atomic Inner-shell Physics, Eugene, Oregon.

KLEIN, O., AND Y. NISHINA, 1929, *Z. Phys.* **52**, 853.

KRAMERS, H. A., AND W. HEISENBERG, 1925, *Z. Phys.* **31**, 681.

LEVINGER, J. S., 1952, *Phys. Rev.* **87**, 656.

LEVINGER, J. S., AND M. L. RUSTGI, 1956, *Phys. Rev.* **103**, 439.

LIBERMAN, D. A., J. T. WABER, AND D. T. CROMER, 1965, *Phys. Rev.* **137**, A27.

LIN, C., K. T. CHENG, AND W. R. JOHNSON, 1975, *Phys. Rev. A* **11**, 1946.

LINDSAY, R. B., 1976, Strutt, John William, Third Baron Rayleigh in *Dictionary of Scientific Biography* (Scribner's, New York), Vol. XIII.

LORD RAYLEIGH (STRUTT, J. W.), 1881, *Phil. Mag.* **12**, 81.

LORD RAYLEIGH (STRUTT, J. W.), 1906, *Phil. Mag.* **11**, 123.

LYE, R. C., J. C. PHILLIPS, D. KAPLAN, S. DONIACH, AND K. O. HODGSON, 1980, *Proc. Nat. Acad. Sci. USA* **77**, 5884.

MANN, J. B., 1967, Atomic Structure Calculations: I. Hartree-Fock Energy Results for the Elements Hydrogen to Lawrencium, Los Alamos Scientific Laboratory report LA-3690.

MEITNER, L., AND H. KÖSTERS, 1933, *Z. Phys.* **84**, 137.

MESSIAH, A., 1961, *Quantum Mechanics* (North-Holland, Amsterdam).

MOON, P. B., 1950, *Proc. Phys. Soc. A* **63**, 80.

MOREH, R., 1982, Low Energy Photon Scattering, in *Proceedings of the International School of Intermediate Energy Nuclear Physics*, edited by R. Bergere, S. Costa, and C. Schaerf (World Scientific Publishing, Verona, Italy).

MÜCKENHEIM, W. AND M. SCHUMACHER, 1980, *J. Phys. G* **6**, 1237.

NATH, A., 1966, *Trans. Bose Res. Inst.* **29**[2], 51.

NATH, A., AND A. M. GHOSE, 1964, *Nucl. Phys.* **57**, 547.

NELMS, A. T., AND I. J. OPPENHEIM, 1955, *J. Res. Nat. Bur. Std. (US)* **55**, 53.

ØVERBØ, I., 1977, *Nuovo Cimento* **40B**, 330.

ØVERBØ, I., 1978, *Phys. Scr.* **17**, 547.

PAPATZACOS, P., AND K. MORK, 1975, *Phys. Rev. D* **12**, 206.

PARKER, J. C., AND R. H. PRATT, 1984, *Phys. Rev. A* **29**, 152.

PHILLIPS, J. C., D. H. TEMPLETON, L. K. TEMPLETON, AND K. O. HODGSON, 1978, *Science* **201**, 257.

PIRENNE, M. H., 1946, *The Diffraction of X-Rays and Electrons by Free Molecules* (Cambridge University Press).

PRATT, R. H., 1982, in Proceedings of the 2nd International Symposium on Radiation Physics, Universiti Sains Malaysia, Penang, May 25–29 (to be published).

PRATT, R. H., AND H. K. TSENG, 1972, *Phys. Rev. A* **5**, 1063.

RAMANATHAN, N., T. J. KENNETT, AND W. V. PRESTOVITCH, 1979, *Can. J. Phys.* **57**, 343.

ROY, S. C., L. KISSEL, AND R. H. PRATT, 1983, *Phys. Rev. A* **27**, 285.

RULLHUSEN, P., F. SMEND, AND M. SCHUMACHER, 1979a, *Nucl. Phys. A* **313**, 307.

RULLHUSEN, P., F. SMEND, AND M. SCHUMACHER, 1979b, *Z. Phys. A* **293**, 287.

RULLHUSEN, P., W. MÜCKENHEIM, F. SMEND, M. SCHUMACHER, G. P. A. BERG, K. MORK, AND L. KISSEL, 1981, *Phys. Rev. C* **23**,1375.

RULLHUSEN, P., U. ZURMÜHL, F. SMEND, M. SCHUMACHER, H. G. BÖRNER, AND S. A. KERR, 1983a, *Phys. Rev. D* **27**, 1962.

RULLHUSEN, P., U. ZURMÜHL, F. SMEND, M. SCHUMACHER, H. G. BÖRNER, AND S. A. KERR, 1983b, *Phys. Rev. C* **27**, 559.

RULLHUSEN, P., U. ZURMÜHL, W. MÜCKENHEIM, F. SMEND, M. SCHUMACHER, AND H. G. BÖRNER, 1982, *Nucl. Phys. A* **382**, 79.

SAKURAI, J. J., 1967, *Advanced Quantum Mechanics* (Addison-Wesley, Reading, Massachusetts).

SCHAUPP, D., M. SCHUMACHER, F. SMEND, P. RULLHUSEN, AND J. H. HUBBELL, 1983, *J. Phys. Chem. Ref. Data* **12**, 467.

SCHUMACHER, M., 1969, *Phys. Rev.* **182**, 7.

SCHUMACHER, M., 1980, in Proceedings of the Workshop on Physics with a Synchrotron Radiation Photon Beam at LEP, CERN, Geneva, 8-9 December.

SCHUMACHER, M., F. SMEND, P. RULLHUSEN, W. MÜCKENHEIM, AND H. G. BÖRNER, 1981, in Proceedings of the Fourth International Symposium on Neutron-Capture Gamma-Ray Spectroscopy and Related Topics, Grenoble, September 7–11 (unpublished.)

SENGUPTA, S. K., N. C. PAUL, J. BOSE, G. C. GOSWAMI, S. C. DAS, AND N. CHAUDHURI, 1982, *J. Phys. B* **15**, 595.

SENGUPTA, S. K., N. C. PAUL, S. C. ROY, AND N. CHAUDHURI, 1979, *J. Phys. B* **12**, 1211.

SIMON, T., AND H. DANIEL, 1977, *Phys. Rev. A* **15**, 1015.

SMEND, F., AND M. SCHUMACHER, 1974, *Nucl. Phys. A* **223**, 423.

SMEND, F., M. SCHUMACHER, AND I. BORCHERT, 1973, *Nucl. Phys. A* **213**, 309.

SOMMERFELD, A., 1939, *Atombau und Spektrallinien* (Vieweg, Braunschweig), Vol. 2.

STANDING, K. G., AND J. V. JOVANOVICH, 1962, *Can. J. Phys.* **40**, 622.

STRUTT, J. W., 1871, *Phil. Mag.* **41**, 107, 274, 447.

TEMPLETON, L. K., AND D. H. TEMPLETON, 1978, *Acta Crystallogr. Sec. A* **34**, 368.

TEMPLETON, D. H., L. K. TEMPLETON, J. C. PHILLIPS, AND K. O. HODGSON, 1980, *Acta Crystallogr. Sec. A* **36**, 436.

THOMSON, J. J., 1906, *Conduction of Electricity Through Gases*, (Cambridge U. P., Cambridge), 2nd ed.

THOMSON, J. J., AND G. P. THOMSON, 1933, *Conduction of Electricity Through Gases* (Cambridge University Press, Cambridge) 3rd Ed., Vol. II.

TSENG, H. K., AND R. H. PRATT, 1973, *Phys. Rev. A* **7**, 1423.

WALLER, I., 1929, *Z. Phys.* **58**, 75.

WANG, M. S., 1981, Analytic parameterization of the imaginary part of the Rayleigh amplitude at forward angles, Internal Report PITT-271, University of Pittsburgh.

WANG, M. S., 1982, Analytic parameterization of the real part of the forward-angle total Rayleigh scattering amplitude, Internal Report PITT-281, University of Pittsburgh.

WANG, M. S., AND I. B. GOLDBERG, 1982, Program for calculating Rayleigh scattering amplitudes from a simple parameterization, Internal Report PITT-295, University of Pittsburgh.

WRIGHT, D. H., AND P. T. DEBEVEC, 1982, *Bull. Am. Phys. Soc.* **27**, 529.

YAKEL, H. L., 1980, *J. Phys. Chem. Solids* **41**, 1097.

YOUNG, A. T., 1982, *Phys. Today* **35**(1), 42.

CHAPTER 12

Electron–Atom Bremsstrahlung

R. H. Pratt and I. J. Feng

1. INTRODUCTION

In this chapter we discuss electron bremsstrahlung, the radiation of a photon in the scattering of an electron from an atom or ion. We shall focus our attention on incident electrons with kinetic energies T_i in the 100-eV to 10-MeV range (and especially for 1 keV to 1 MeV) and radiated photons in the vacuum ultraviolet, X-ray, and soft γ-ray regimes. Electron bremsstrahlung is only one particular consequence of the general coupling of the electromagnetic field and matter fields. The process is closely related, as we shall see, both to elastic electron scattering and to direct radiative recombination, in which the incident electron does not scatter but is captured by the atom. It is also closely related to pair production. Other charged particles also radiate in scattering, as follows from Maxwell's equations, and we shall discuss the most closely related case of positron bremsstrahlung. More generally, the electromagnetic coupling to matter fields is responsible for radiative versions of all nonradiative processes, including "inverse bremsstrahlung" processes in which electromagnetic radiation is absorbed rather than emitted.

We shall see in this chapter that electron bremsstrahlung can teach us about many fundamental aspects of the coupling of radiation and matter fields, and that it serves also as a probe of the (atomic) structures of the matter fields. But we should note here that electron bremsstrahlung is also of practical interest and importance for a variety of applications. These of course include the needs of astrophysics, where knowledge of bremsstrahlung is needed both for diagnostics (interpretation of astrophysical spectra) and

R. H. Pratt • Department of Physics and Astronomy, University of Pittsburgh, Pittsburgh, Pennsylvania 15260. I. J. Feng • Varian Associates, Palo Alto, California 94303.

for modeling hot plasmas (Tucker, 1975). Similar needs arise from the fusion plasma programs, both in magnetic confinement and inertial confinement efforts, as in determination of plasma temperature, prediction of plasma properties, and estimation of plasma heating (More, 1984; Weisheit, 1984). Other applications concern transport and energy loss of charged-particle beams (Seltzer and Berger, 1982), shielding and radiation damage (both to materials and biological tissues), and the study of solids through isochromat spectroscopy.

Our discussion here is based in part on several earlier short review articles (Pratt, 1981a, b, 1982; Pratt and Feng, 1984), to which the reader is referred. The experimental situation has recently been reviewed by Nakel (1983); see also the earlier discussion of the triply differential cross section (Nakel, 1980). The older review of Koch and Motz (1959) still contains much useful information, as do the books of Heitler (1954) and of Bethe and Salpeter (1957). Classical bremsstrahlung is discussed in some detail in the texts of Jackson (1975) and of Landau and Lifshitz (1971). The bremsstrahlung spectrum in the energy range of concern here has been tabulated by Pratt *et al.* (1977, 1981); tables of the shape of the angular distribution have been published by Kissel *et al.* (1983) and Bernhardi *et al.* (1983) for screened and point Coulomb cases, respectively. We can summarize the current status of work in electron bremsstrahlung by saying that neither theory nor experiment is as advanced as for elastic scattering or for photoionization. Most theoretical work is based on the model of bremsstrahlung as a single electron transition in a screened central potential, generally summed over a number of the experimental variables. Experiments are limited in number. A general problem is that the process involves more independent variables, in comparison to elastic scattering or photoeffect, and so it is difficult to efficiently characterize either theoretical or experimental data.

Our objective in this chapter is to give an overview of our current understanding of the electron bremsstrahlung process and its major features, with some indication of the major steps in achieving this understanding and with some identification of issues of current concern and for future work. We will not give a detailed historical review and we will not give a detailed derivation of the various theoretical predictions. We will begin in the next section with a description of the observables of the process and the assumptions which are basic to its discussion. This will give us enough understanding of the process that we can then give, in Sec. 3, a brief sketch of the main historical steps in the development of that understanding, which also serves to introduce some of the results and concepts to be discussed subsequently. In Sec. 4 we complete these preliminaries with a discussion of the characteristic distances relevant to the electron bremsstrahlung process and the consequent formulation of appropriate theoretical descriptions, classical, quantum, and relativistic formulations, and a discussion of which formulations are the best now available in various circumstances. In Sec. 5 we discuss predictions for the spectrum in scattering from a point Coulomb potential. In Sec. 6 we examine the modifications which result when that potential is screened, including the applicability of form factors. In Sec. 7 we discuss the special features of the *end points* of the spectrum (when almost none or almost all of the incident electron kinetic energy is radiated), and their relation to the related processes of elastic electron scattering and direct radiative recombination (inverse to photoionization). In Sec. 8 we describe the angular distributions and polarization correlations of the bremsstrahlung radiation. Given this theoretical understanding of the process, in Sec. 9 we review the available experimental information and the status of the comparison and extent of agreement between theory and experiment. Finally, in Sec. 10, we note some newer developments, including studies of positron bremsstrahlung, bremsstrahlung from ions, and bremsstrahlung in a hot dense plasma environment.

2. OBSERVABLES AND ASSUMPTIONS

We shall mean by electron bremsstrahlung the process in which, within the limits of observation, *one* photon is radiated in the scattering of *one* electron off an isolated atom or ion whose internal state remains unchanged in the process

$$e + \text{atom} \rightarrow e + \gamma + \text{atom (unchanged)}. \tag{1}$$

As we know, such a description is always an idealization, in view of the general problem of measurement in quantum mechanics and the nonexistence of totally isolated subsystems. Description of electromagnetic radiation in photon number states introduces a further problem, since if the photon mass is indeed zero an indefinite number of indefinitely soft photons must be considered to be included in any physically produced state. And while in principle excitation or ionization of the atom in radiative scattering can be determined, in practice it usually will not be. Nonetheless there are a variety of circumstances in which these assumptions appear appropriate, as when the density of electrons in a beam is low, when atoms in a solid or gas or plasma are sufficiently separated that the environment of an atom does not affect the process, when multiphoton effects and atomic excitation and ionization may be ignored. We may call all these presumptions *physical* assumptions, since in principle each is subject to experimental verification. (For lattice effects see Barbinellini *et al.*, 1962; Komar *et al.*, 1971; for molecular effects see Borie *et al.*, 1970.)

There is a second set of *model* assumptions which we make for convenience (or to achieve feasibility) in our theoretical calculations of the bremsstrahlung process; in principle each of these assumptions can be tested against improved calculations. Thus in most work bremsstrahlung in scattering from an atom is separated into radiation in scattering from a screened central potential (due to the nuclear charge as screened by the atomic electrons), and into radiation in inelastic scattering from bound electrons. The electron screened-nucleus bremsstrahlung will be our main subject of discussion in this chapter. Omission of electron–electron bremsstrahlung is a physical assumption, in the sense that emission of electrons can be observed (Nakel, 1982); its calculation generally involves model assumptions (Selzer and Berger, 1982; Haug, 1975) describing the process in terms of electron bremsstrahlung from *free* electrons, combined with incoherent scattering factors. Except in the lightest elements electron–electron bremsstrahlung is negligible in the keV range and approaches Z^{-1} of electron–nucleus bremsstrahlung in the MeV range. However, the separation into electron–nucleus and electron–electron bremsstrahlung itself involves model assumptions and omits the many-electron correlation effects which have been discussed by Amusia (1982) under the name "atomic" bremsstrahlung. He has indicated that these effects can become dominant for photon energies corresponding to an ionization threshold at very small momentum transfers (see also the work of Golovinskii, 1982). Other model assumptions, appropriate for the situations of concern here, include the neglect of finite nuclear size and mass (Berg and Laudner, 1958) and magnetic moments (Ginsberg and Pratt, 1964, 1965; Goldenberg and Pratt, 1966), the neglect of the radiative corrections of quantum electrodynamics (Mork and Olsen, 1965, 1968; Gavrila and McEnnan, 1977; McEnnan and Gavrila, 1977; Botto and Gavrila, 1982). A common model assumption, which we shall discuss but not rely on, is the description of electromagnetic coupling in nonrelativistic dipole approximation.

Under these physical and model assumptions electron bremsstrahlung is characterized (Fig. 1) by the nuclear charge Z and ionic charge Z_i of the target (normally taken to be in its ground state), the momentum $\mathbf{p}_i$ and spin $\mathbf{s}_i$ of the incident electron, the momentum $\mathbf{p}_f$ and spin $\mathbf{s}_f$ of the scattered electron, the momentum $\mathbf{k}$ and polarization $\boldsymbol{\varepsilon}$ of the radiated photon. The spin $\mathbf{s}$ of a continuum electron state, described in the rest system of that electron, is written in the form $\mathbf{s} = \sum_n \zeta_n \hat{\mathbf{e}}_n$, with $\hat{\mathbf{e}}_n$ an orthogonal set of unit vectors and the real ζ_n satisfying $\sum_n \zeta_n^2 = 1$. Photon polarization can also be described in terms of the real Stokes parameter ξ_n satisfying $\sum_n \xi_n^2 = 1$ instead of the complex polarization vector $\boldsymbol{\varepsilon}$ satisfying $\boldsymbol{\varepsilon} \cdot \boldsymbol{\varepsilon}^* = 1$ and $\mathbf{k} \cdot \boldsymbol{\varepsilon} = 0$; the two descriptions are connected through the relation $\xi_n = \varepsilon^+ \sigma_n \varepsilon$ with the row vector $\varepsilon^+ = (\varepsilon_1^*, \varepsilon_2^*)$, with $\boldsymbol{\varepsilon} = \varepsilon_1 \mathbf{e}_1 + \varepsilon_2 \mathbf{e}_2$, σ_n the Pauli spin matrices, and $\mathbf{e}_1$, $\mathbf{e}_2$, $\mathbf{k}$ an orthonormal set.

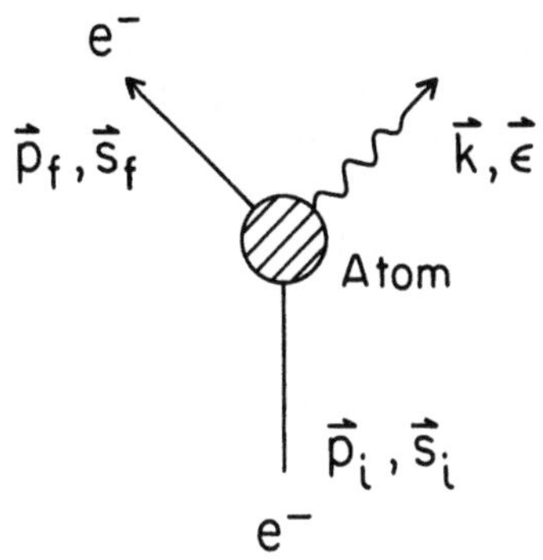

FIGURE 1. Schematic diagram of electron bremsstrahlung in an atomic field.

Relativistic energy conservation requires

$$(1+p_i^2)^{1/2}=k+(1+p_f^2)^{1/2} \tag{2}$$

(with units $\hbar=m_e=c=1$), where we are assuming the nuclear mass M is large enough and the electron and photon energies small enough that the energy transfer to the atom ($q^2/2M$ when the momentum transfer to the atom $q \ll M$) is negligible in comparison to electron and photon energies. In the nonrelativistic case (2) reduces to $k=\frac{1}{2}(p_i^2-p_f^2)$, so that the photon momentum is generally small in comparison with electron momentum, while in the extreme relativistic case it becomes $k=p_i-p_f$. An important quantity for our subsequent discussions is k/T_i, the fraction of kinetic energy radiated, where the incident electron kinetic energy $T_i=(1+p_i^2)^{1/2}-1$. The fraction varies from 0 at the soft photon end of the spectrum to 1 at the hard photon ("tip") end of the spectrum. Momentum conservation requires

$$\mathbf{p}_i=\mathbf{k}+\mathbf{p}_f+\mathbf{q}, \tag{3}$$

with $\mathbf{q}$ the momentum transfer to the atom, generally comparable (unlike the energy transfer) to the other momenta of the problem. The momentum transfer $\mathbf{q}$ (unlike that in photoeffect) can never vanish as long as k is finite: kinematics do not allow a free electron–photon vertex. For a given energy split between radiated photon and final electron the minimum momentum transfer occurs in a colinear alignment, with final electron and photon along the incident electron direction; in the extreme relativistic limit this transfer $q_{\min}$ is $\frac{1}{2}(p_f^{-1}-p_i^{-1})$ and in the nonrelativistic limit it is (p_i-p_f). For a given point on the photon spectrum, the momentum transfer is maximum when the scattered electron and photon are emitted backward to the incident electron direction. Thus $q_{\max}$ varies from $2p_i$ to p_i+p_f as the incident energy decreases from extreme relativistic to nonrelativistic.

In the bremsstrahlung process as we have described it the atomic target remains unchanged. The most detailed independently observable quantity is specified by giving, in the initial state, (Z, Z_i) for the target and kinetic energy T_i and spin ζ_n^i for the incident electron, and by giving, in the final state, the energy split k/T_i between photon and electron, the angles of photon and final electron emission with respect to the incident electron direction, the angle between the scattering plane of the two electrons and the plane of $(\mathbf{p}_i\mathbf{k})$, the photon polarization ξ_n, and the spin ζ_n^f of the final electron. (Energy conservation and rotational invariance eliminate other independent variables.) This observable, which implies a coincidence measurement between radiated photons and scattered electrons emerging when a collimated electron beam of definite energy is incident on a specific target, is conventionally called the triply differential bremsstrahlung cross section $d^3\sigma \equiv d^3\sigma/dk\, d\Omega_k\, d\Omega_f$, corresponding to energy split, photon, and electron angles (although, as we have seen, there are actually *three* angular variables). When all polarization states are specified we write

$$d^3\sigma^{\rm pol} = d^3\sigma\left[\frac{1}{4}\sum_{lmn=0}^{3} C_{lmn}\zeta_l^i\xi_m\zeta_n^f\right], \tag{4}$$

where $\zeta_0 = \xi_0 = 1$, the C_{lmn} (functions of the independent energies and angles) are the general polarization correlations, with $C_{000} \equiv 1$. Here $d^3\sigma$ is the unpolarized cross section corresponding to an average over incident electron spin directions (unpolarized beam) and a sum over final photon polarizations and electron spins. While $d^3\sigma^{\rm pol}$, the most general observable, has not been measured and is not soon likely to be, $d^3\sigma$ has been measured and also C_{030}, as discussed in Sec. 9.

A second, and more common, type of experiment involves the detection of only one of the two final particles, thereby implying a sum over the direction of emission and spin or polarization of the unobserved particle. The energy of the unobserved particle will still be known through energy conservation. For our purposes in this chapter we shall only consider the case in which it is the radiated photon which is observed. (See Jauch and Rohrlich, 1976, for a discussion of the measurement of electrons which have lost energy in bremsstrahlung production; note also Weber *et al.*, 1963.) We then obtain the doubly differential cross section $d^2\sigma^{\rm pol}/dk\, d\Omega_k \equiv d^2\sigma^{\rm pol}$ defined as

$$\begin{aligned} d^2\sigma^{\rm pol} &= \int d\Omega_f\, d^3\sigma^{\rm unpol}\left(\frac{1}{2}\sum_{lm=0}^{3} C_{lm0}\zeta_l^i\xi_m\right) \\ &\equiv d^2\sigma^{\rm unpol}\left(\frac{1}{2}\sum_{lm=0}^{3} C_{lm}\zeta_l^i\xi_m\right), \end{aligned} \tag{5}$$

where $C_{00} \equiv 1$ and the polarization correlations C_{lm} depend on photon energy and angle. (Note that $C_{lm0} \neq C_{lm}$, still depending on final electron angles.) While $d^2\sigma$ is written in terms of $d\Omega_k$, from rotational invariance it has no dependence on the azimuthal angle. Features of $d^2\sigma$ are discussed in Sec. 8. A fair number of observations of $d^2\sigma$ are available, as well as measurements of C_{03}, C_{20}, and C_{32}; these are discussed in Sec. 9.

Often one wishes to know (and use in applications) the spectrum of photon energies radiated, independent of angle of emission or polarization, independent of incident electron spin. We obtain this from $d^2\sigma$ as $d\sigma/dk \equiv d\sigma$, with

$$d\sigma = \int d\Omega_k \, d^2\sigma^{\text{unpol}}. \tag{6}$$

In our initial discussion of the various theoretical formulations, in Sec. 5–7, we shall concentrate our attention on $d\sigma$, even though it is not normally accessible to direct experimental measurement. Making use of $d\sigma$, a *shape function* $S \equiv d^2\sigma/d\sigma$ for the angular distribution of unpolarized electrons can be defined, as is discussed in Sec. 8. Instead of $d\sigma$, which gives the *number* of photons radiated in the range dk, it is often of interest to consider $k\,d\sigma$, which gives the *energy* radiated, often tabulated as $\sigma \equiv \beta_i^2 Z^{-2}\, k d\sigma/dk$ (in mb), where $\beta_i = v_i/c$ for the incident electron, or as the Gaunt factor $g = \sigma/5.61$ mb. Here, 5.61 mb is $16\pi\alpha^3 3^{-3/2} \lambda_e^2$, with α the fine-structure constant and λ_e the electron Compton wavelength, a classical Coulomb result we shall examine subsequently.

Finally, one sometimes wishes to define the total energy loss due to radiation, obtained by integrating σ over all photon energies, conventionally defined as (Bethe and Salpeter, 1957)

$$E^{-1} \int_0^{T_i} k \frac{d\sigma}{dk} \, dk, \tag{7}$$

where E is the total energy (including rest mass) of the incident electrons. Note that the total cross section, corresponding to integrating $d\sigma/dk$ rather than $k\,d\sigma/dk$, does not exist: due to the zero rest mass of the photon the *number* of photons of very low energy radiated grows at least as k^{-1}. It is for this reason that there are fundamental difficulties in speaking of states of specified total numbers of photons.

3. HISTORY

We will not attempt a systematic survey of the history of the study of electron bremsstrahlung, but it is instructive to note a few of the highlights. Useful information is found in the 1911 *Encyclopedia Britannica* article on

"Röntgen Rays" written by J. J. Thomson (1911), the *Handbuch der Physik* reviews of Kulenkampff (1926) on "Röntgenstrahlung" and of Stephenson (1955) on "The Continuous X-Ray Spectrum," the standard texts of Compton and Allison (1935) and Richtmeyer and Kennard (1947), and the book of Hermann (1971) on "The Genesis of Quantum Theory."

One may begin with the identification by Maxwell (1862) of light as an electromagnetic wave:

> ...we can scarcely avoid the inference that light consists in the transverse undulations of the same medium, which is the cause of electric and magnetic phenomena....

which he systematized (1865) and presented in his treatise (1873). Hertz (1887) discovered waves of electromagnetic origin, and later showed that they propagate at the speed of light. Lorentz in the 1890s extended Maxwell's theory, developing his electron theory, taking into consideration the atomic structure of matter, and (working on black-body radiation) calculating the radiation of linearly oscillating electrons.

X rays were discovered by Roentgen (1895); Buguet (1897) found indications that X-ray beams were not homogeneous but consisted of components of different absorptivities. Thomson (1897) discovered the electron, and Stokes (1898) and Thomson (1898) developed a theory of X rays as classical electromagnetic pulses due to the sudden stopping of charged particles. Barkla (1905) demonstrated polarization properties of X rays, and Sommerfeld (1909) used special relativity to explain features of their angular distributions. However, it was not yet settled whether X rays were an electromagnetic wave, and Thomson (1911) notes the opposing corpuscular theory of Bragg (1907) which described X rays as neutral electric doublets of opposite charge held together by attraction. The quantum nature of X-rays was also unsettled. Stark (1907) had argued that explanation of bremsstrahlung required a quantum theory and gave the (believed) short wavelength limit of the bremsstrahlung spectrum as one of three phenomena supporting the validity of Planck's quantum law; Wien (1907) also discussed these issues. However, Sommerfeld (1909) believed he had explained the bremsstrahlung phenomena in classical terms and that there was no room for the quantum of action, changing his mind in 1910 and withdrawing the remarks in 1911.

Friedrich, Knipping, and Laue (1913) observed the diffraction of X rays by a crystal, indicating they were electromagnetic waves of definite wave lengths, and almost immediately W. H. and W. L. Bragg (1913) demonstrated the existence of monochromatic characteristic X-ray radiation superimposed on a continuous spectrum, as shown in Fig. 2. The separate study of these two components could then begin. Duane and Hunt (1915) studied the 37-keV spectrum of W (entirely continuous) and discovered the

1913.] *The Reflection of X-rays by Crystals.* 431

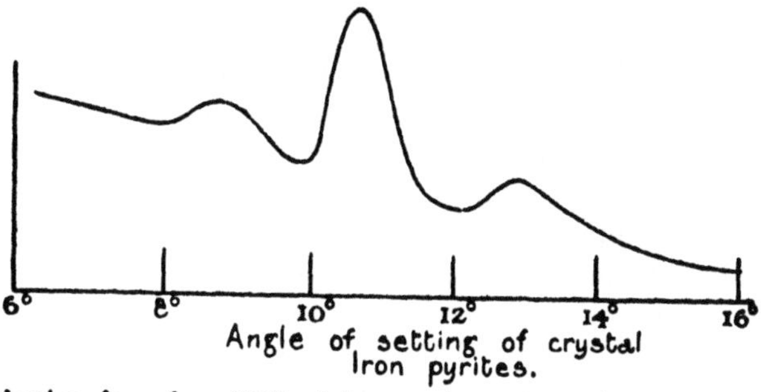

FIG. 2.—Reflection from face (100) of iron pyrites, at varying angles of incidence. Abscissa—Angle of incidence of rays on crystal face; Ordinate—Strength of reflected beam, arbitrary scale.

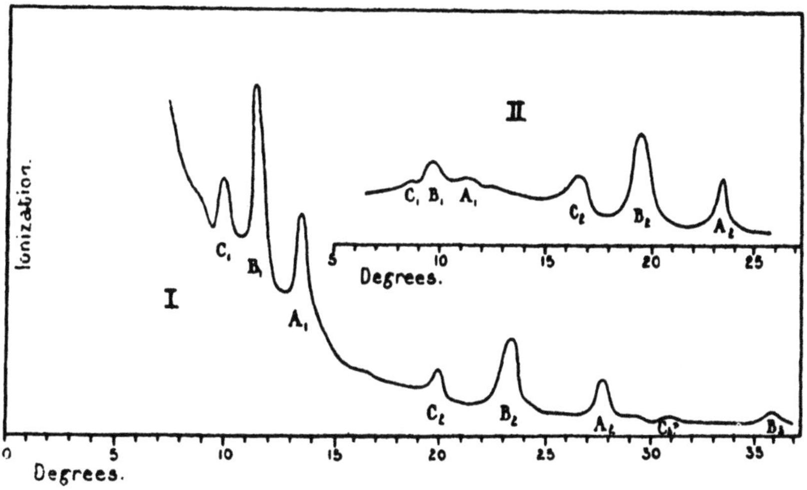

FIG. 3.—Reflection (I) from face (100) and (II) from face (111) of rock-salt. The curves show the variation of strength of reflected beam with angle of incidence.

FIGURE 2. The original curves (from Bragg, 1913) showing the variation of strength of reflected beam with angle of incidence.

short wavelength cutoff; systematic (still thick-target) studies were reported by Ulrey (1918). Thin-target work began a few years later with Duane (1927, 1928) and Kulenkampff (1928). Bohm (1937, 1938) reported extensive studies of angular distributions; Hinz (1941) extended measurements to the 150–300 keV range; Harworth and Kirkpatrick (1942) improved techniques with the use of very thin films. After World WarII, studies were extended to higher energies (circa 20 MeV) by Bosley *et al.* (1948), Koch and Carter (1950), and Lanzl and Hanson (1951), and to still higher energies, using synchrotrons, by Powell *et al.* (1951), McDiarmid (1952), and Muirhead *et al.* (1952).

A successful semiclassical theory of bremsstrahlung was developed by Kramers (1923), calculating classical radiation on Coulomb orbits but invoking energy conservation, considering the radiation as implying the emission of photons (particles) of definite energy. Quantum mechanics, which we may now understand as at the level of the semiclassical theory of radiation in dipole approximation, was applied by several authors in 1929, and the results for radiation in the Coulomb potential are now usually associated with the work of Sommerfeld (1929, 1931), further developed by Sauter (1933, 1934), Sommerfeld and Maue (1935), and others. A relativistic Born approximation treatment in the quantum theory of radiation was developed by Bethe and Heitler (1934). The high-energy limit, without the restriction of Born approximation, was obtained by Bethe and Maximon (1954).

It is also interesting to trace the history of the use of the term "bremsstrahlung," which goes back (in Germany) to the first decade of the century but did not become universal (even in German writing) until the 1950s. It will be noted that neither the 1926 (in German) or 1955 (in English) *Handbuch* articles use this title; Stephenson in 1955 uses "bremsstrahlung" only in connection with the relativistic Bethe–Heitler regime, not for the continuous X-ray spectrum, and German authors more often used "kontinuierliches Röntgenspektrum," later also "Röntgen-bremsstrahlung." In 1909 Sommerfeld does use "Bremsstrahlung" several times within one paragraph, but not otherwise in his paper, which talks of "Röntgenstrahlung in der Bremsung des Kathodenstrahlelektrons," of "Bremsanteil der Strahlung," etc. and mentions the "Bremstheorie" of Wien (1905). The earliest usage we have found in English is in the textbook of Ruark and Urey (1930); the first English usage in research papers we have found is in a paper of Anderson *et al.* (1934) on cosmic-ray showers, but it remains uncommon until much later. It appears in the *title* of *Physical Review* papers in 1948 and after, and is first listed as a subject index category in Science Abstracts in 1951. Sommerfeld (1939) states that "the author [Sommerfeld] proposed the name Bremsstrahlung with the approval of Röntgen himself."

4. CHARACTERISTIC DISTANCES AND APPROPRIATE FORMULATIONS

The bremsstrahlung process illustrates many of the regimes of physics and so also the transitions between those regimes. We can trace the transition from classical mechanics to quantum mechanics, the transition from nonrelativistic quantum mechanics to the full relativistic theory. It is not always obvious when a given formulation is, or ceases to be, appropriate. Nonrelativistic classical features can persist to 10 keV in the photon spectrum, an energy for which relativistic and retardation effects are already visible in the angular distribution. The soft photon end-point predictions for the spectrum in a point Coulomb potential are the same in all formulations. However, quantum mechanical features are also visible at low energies: shape resonances can become visible in the spectrum for incident electron kinetic energies below 1 keV, not to mention the "atomic structure" beyond the potential approximation discussed by Amusia (1982). Nonrelativistic dipole quantum mechanics can suffice to explain much of the spectrum up to 50 keV, but the persistence of tip region cross sections is a relativistic effect. At higher energies both relativity and quantum mechanics are needed to understand the spectrum, as well as the angular distribution, away from the soft photon end point.

One can therefore begin the discussion of bremsstrahlung with the classical coupling of electromagnetic dipole radiation and charged particles which follows from Maxwell's equations. The intensity $dI/d\Omega$ of radiation emitted into a given solid angle $d\Omega$ due to the acceleration of a charge e, and the total radiation intensity I, are given in terms of the dipole moment $\mathbf{d} \equiv e\mathbf{r}$ associated with that charge as

$$\frac{dI}{d\Omega} = \frac{\ddot{\mathbf{d}}^2}{4\pi c^3} \sin^2 \theta, \qquad I = \frac{2}{3c^3} \ddot{\mathbf{d}}^2. \tag{8}$$

The energy spectrum $d\varepsilon/d\omega$ of the emitted radiation is obtained by Fourier transforming as

$$\frac{d\varepsilon}{d\omega} = \frac{8\pi\omega^4}{c^3} |\mathbf{d}_\omega|^2, \tag{9}$$

with a corresponding expression for the angular distribution. Here $\mathbf{d}_\omega$ is to be obtained from the equation of motion $\mathbf{r}(t)$ of the charged particle, and it is assumed that the amount of radiation emitted is small enough not to affect the trajectory. In a beam of such particles scattering from a potential each trajectory is characterized by an impact parameter ρ; the energy spectrum of the radiation produced as the beam scatters is obtained by integrating over impact parameters. It is striking that this approach can yield good results when, considering the quantization of radiation one is in fact looking at a region of the spectrum in which an incident electron loses much (or most) of its energy in radiation. However, it is still true that the total energy radiated from the incident beam (as distinguished from a particular electron) is small. The nonrelativistic Hamiltonian for a charged particle interacting with the electromagnetic field **A** is obtained from the Hamiltonian in the absence of the field through the substitution $\mathbf{p} \rightarrow \mathbf{p} - (e/c)\mathbf{A}$ for the canonical momentum or momentum operator, generating $p \cdot A$ and A^2 interaction terms. Applying first-order time-dependent perturbation theory in nonrelativistic quantum mechanics in the case of a single electron in a potential leads to the transition matrix element (assuming Coulomb gauge)

$$M \sim \int \psi_f^* \mathbf{p} \cdot \boldsymbol{\varepsilon} e^{-i\mathbf{k}\cdot\mathbf{r}} \psi_i \, d^3r, \tag{10}$$

for radiation of a photon of momentum **k** and polarization $\boldsymbol{\varepsilon}$ in a transition from state ψ_i to state ψ_f. In dipole approximation the exponential should be omitted. The triply differential cross section $d^3\sigma$ is proportional to $|M|^2$. The corresponding relativistic matrix element resulting from the same substitution in the Dirac Hamiltonian is of the same form, now understanding the ψ's as solutions of the Dirac equation in a potential and replacing $\mathbf{p} \cdot \boldsymbol{\varepsilon}$ by $\boldsymbol{\alpha} \cdot \boldsymbol{\varepsilon}$, with $\boldsymbol{\alpha}$ the Dirac matrices.

Given these formulations, we may ask which regions of the atomic system are probed in the bremsstrahlung process. This is important in assessing the accuracy of an approximation scheme for calculation in one of these formalisms; it is also important in understanding what one may hope to learn in a particular experimental arrangement. Classically, the question is which impact parameter gives the dominant contribution to a given region of the spectrum or angular distribution. In the Coulomb case one finds that the soft photon region of the spectrum is obtained from large impact parameters (small angle scatterings, reflecting the long range of the potential) while the hard-photon region of the spectrum is obtained from small impact parameters (large-angle scattering). However, with a short-range potential there is no contribution from impact parameters larger than the cutoff of the potential (as such particles will not scatter, they will not be accelerated), and so the soft-photon end point of the spectrum is much reduced, no longer diverging logarithmically but remaining finite. Subject to this upper bound, the most common (quantum) estimate for the distance probed in a particular bremsstrahlung event is q^{-1}, the inverse of the momentum transfer to the atom: the integrand of the matrix element M oscillates as $e^{iq\cdot r}$ and so cuts off the integral at $\sim q^{-1}$. Now for a given point on the photon spectrum q ranges from a small lower limit $q_{\min}$ (which we saw varied from $p_i - p_f$ to $\frac{1}{2}p_f^{-1} - \frac{1}{2}p_i^{-1}$ as one goes from nonrelativitistic to extreme relativistic energies) to an upper limit $q_{\max}$ corresponding to backward emission (varying from $p_i - p_f$ to $2p_i$). Thus for low incident electron kinetic energies one expects the process to be sensitive to a large distance region (but within the range of the potential), while bremsstrahlung from high-energy electrons can sample either a small or large distance region, depending on the choice of kinematics in a coincidence measurement. (The quantum analog, in a full calculation, of the earlier classical cutoff outside the range of the potential, can be understood by realizing that the matrix element would vanish for free-particle states, and that outside the range of the potential the states behave as free.)

We may expect that the triply differential cross section will generally be largest in the situations of small q, since then a larger volume (larger distances) contribute to the matrix element. Thus, at high energy, cross sections, considered for fixed electron and photon energies, drop for increasing scattering angles. Further, in integrated cross sections the regions probed by the smallest momentum transfers (up to the edge of the atom) included in the integration can be expected to dominate. Values of $r_{max} \equiv q_{min}^{-1}$ are shown in Fig. 3. Thus for the energy spectrum σ large distances are always important near the soft photon end point. At intermediate energies most of the spectrum will be determined at smaller distances, while at high energy only the tip of the spectrum will be determined at small distances. However, this last statement must be qualified, in the sense that the value of M depends not

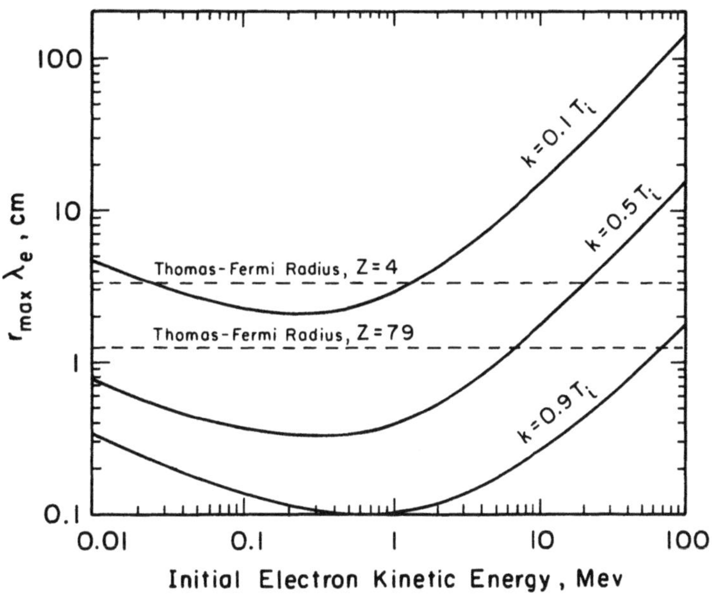

FIGURE 3. Dependence of r_{max} on the initial electron kinetic energy for k equal to $0.1T_i$, $0.5T_i$, and $0.9T_i$; r_{max} is the reciprocal of the minimum momentum transferred to the nucleus $(=p_i-p_f-k)$, and is given in units of Compton wavelength, λ_c. The dashed lines give the values of the Thomas–Fermi radius $(=137Z^{-1/3})$ for beryllium and gold (from Koch and Motz, 1959).

only on the range of integration but also on the normalization (at the origin) of the wave functions (and the phase shifts). For a low-energy continuum wave function the normalization and phase are determined at larger distances, and in this way the larger distance regions also affect the tip region cross sections, even at high energy, and at lower energies are responsible for shape-resonance features (Lee and Pratt, 1975a; Olsen and Maximon, 1978a, b; D'yachkov, 1981).

The formulations of relativistic quantum mechanics include as limiting cases the other simpler formulations and are appropriate for the various circumstances discussed in this chapter. We will now describe the two best available calculational schemes within this formulation, which provide our predictions for comparison with experiment and for the assessment of the validity of simpler calculational methods in the various formulations. The best available results for σ and $d^2\sigma$, for incident energies 500 eV to 10 MeV and all elements, have been obtained from a relativistic numerical calculation of M in partial waves and multipoles, treating the process in a Dirac–Slater self-consistent central potential, with local Kohn–Sham exchange and no Latter tail in the neutral-atom case. The code was developed by Tseng (1970), who also obtained most of the base data which have subsequently made tabulation of predictions possible (Pratt *et al.*, 1977, 1981; Kissel *et al.*, 1983). Results are obtained in the form of *many* (up to 10^6) radial integrals of the form $\int g j_l(kr) f\, dr$, where g and f are large and small component radial wave functions of initial and final states. The wave functions and integrals are evaluated numerically (convergence, due only to oscillatory functions, is delicate) and the radial matrix elements summed over angular momentum states (convergence of the sums is slow at high energy). In such sums partial waves or multipoles do not interfere for a particle whose momentum *direction* is not observed, improving convergence, and this is why $d^3\sigma$ has not been obtained. (For technical reasons the code has not been run at lower energies, but other, nonrelativistic codes are available (Scofield, 1981; Green, 1979; Mertz, 1976; Geltman, 1973). A similar code was also developed by Brysk *et al.* (1969), but only limited results have been reported.

The best available results for $d^3\sigma$ and for the high-energy limit of σ have been obtained with Sommerfeld–Maue (SM) wave functions (Sommerfeld and Maue, 1935), a closed-form approximation to the full three-dimensional Dirac continuum Coulomb wave functions (which are not known in closed form). The SM wave function, which has the form of the nonrelativistic Coulomb wave function (with altered parameters) multiplying a (near) free spinor function, is obtained by dropping terms of order $(Z\alpha)^2/k$ in each term of a partial-wave expansion, where $\kappa = \mp(j \pm 1/2)$ as $j = l \pm 1/2$. It is thus valid for small Z or when large κ dominate (as at high energies). Bethe and Maximon (1954), Olsen (1955), and Olsen *et al.* (1957)

used SM wave functions in M to obtain the high-energy limit of bremsstrahlung (and pair production) cross sections; they were also able to show that in the high-energy limit screening could be included in the form of a form factor, as we shall discuss subsequently. Elwert and Haug (1967) later repeated the calculation without making high-energy and small-angle assumptions, obtaining a result which should be correct for low Z when screening is not important, but which has been used more generally, particularly for $d^3\sigma$, in the absence of anything better. The Bethe–Maximon calculation fails in the tip region of the spectrum, where the final electron is not of high energy. A separate calculation of the high-energy limit of the tip was made by Jabbur and Pratt (Pratt, 1960c; Jabbur and Pratt, 1963, 1964), using a modified plane wave (eikonal) which is a simplification of the SM function adequate for the purpose (for the initial state), a low-energy expansion of the final state (only the first partial waves contribute, but *not* just s waves), together with techniques originally developed for the high-energy limit of atomic photoeffect (Pratt, 1960a, b). This work provides the best available predictions for the bremsstrahlung tip at high energies. One can interpolate fairly well between high-energy limit predictions and numerical data at lower energies. Seltzer and Berger (1982) have recently interpolated between the predictions of Bethe and Maximon (1954) and of Jabbur and Pratt (1963, 1964), and the extensive numerical data at 2 MeV and below; their method produced good agreement with the fragmentary numerical data available at 5 and 10 MeV (Tseng and Pratt, 1979).

5. THE COULOMB SPECTRUM

The general features of the bremsstrahlung spectrum in a point Coulomb potential are illustrated in Figs. 4 and 5. In the soft photon region there is a logarithmic divergence, while for harder photons the spectrum flattens. At low energies most of the spectrum is flat; for higher energies the logarithmic region dominates more and more of the spectrum, and for still higher energies the hard-photon region of the spectrum drops, then recovers (though the tip value remains small in low-Z elements). The behavior is thus fairly complex, and not all the responsible factors have been identified, particularly in the relativistic region.

In the Coulomb case several analytic results are available, giving some additional insights into the process as well as serving as a basis for interpolation and parametrization. These analytic expressions are as follows:

1. Classical predictions, given (in a footnote) by Kramers (1923); the derivation is given in some detail by Landau and Lifshitz (1971). The soft-photon limit is the universal logarithm, the hard-photon limit the (constant) Kramers' formula.

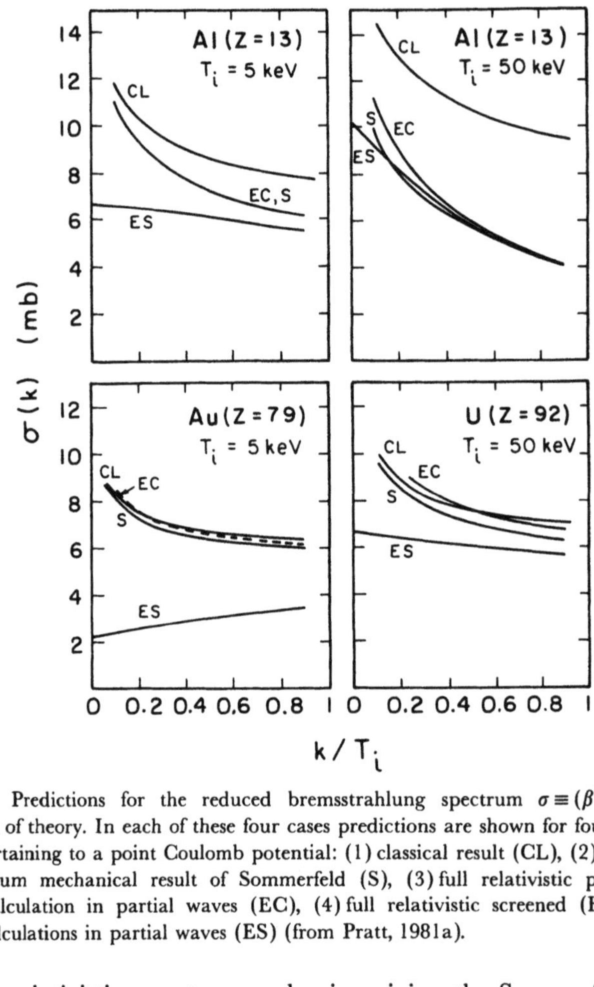

FIGURE 4. Predictions for the reduced bremsstrahlung spectrum $\sigma \equiv (\beta^2 k/Z^2) d\sigma/dk$ at various levels of theory. In each of these four cases predictions are shown for four theories, the first three pertaining to a point Coulomb potential: (1) classical result (CL), (2) nonrelativistic dipole quantum mechanical result of Sommerfeld (S), (3) full relativistic point Coulomb numerical calculation in partial waves (EC), (4) full relativistic screened (HFS potential) numerical calculations in partial waves (ES) (from Pratt, 1981a).

2. Nonrelativistic quantum mechanics, giving the Sommerfeld formula in dipole approximation. For small $Z\alpha/\beta$ this reduces to nonrelativistic Born approximation, and from the two the Elwert factor (1939) is derived as a simple correction to Born approximation.
3. Relativistic Born approximation, including the Bethe–Heitler formula (Bethe and Heitler, 1934; Sauter, 1934; Racah, 1934a, b). The Elwert factor again leads to improvement, even for situations not corresponding to its original derivation.
4. High-energy limit, obtained by Bethe and Maximon (1954), (Davies *et al.*, 1954; Olsen, 1955) with SM wave functions as we have described, and the subsequent efforts of Elwert and Haug (1967) to provide an interpolating form at intermediate energies.

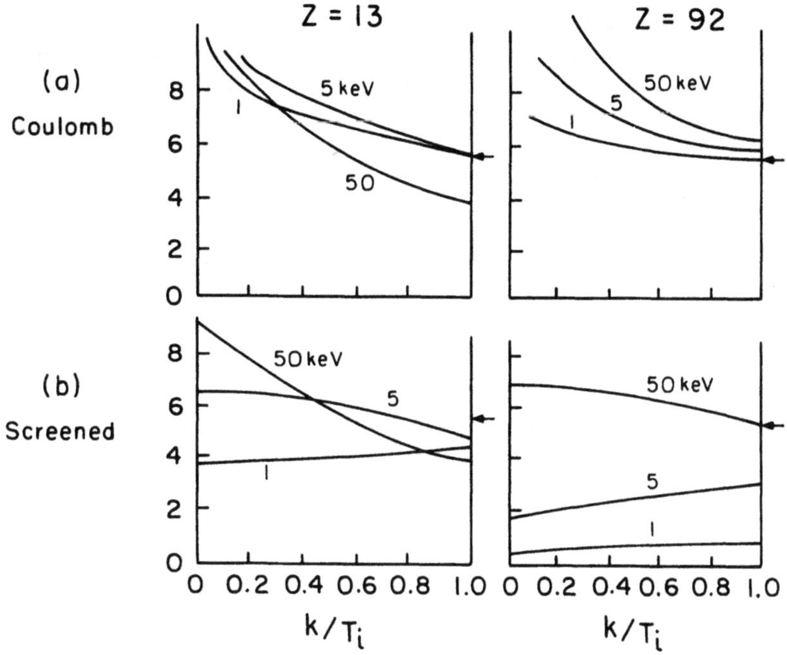

FIGURE 5. Bremsstrahlung spectrum σ as a function of k/T_i from a relativistic numerical calculation in partial waves, for Coulomb and screened potentials, for $Z = 13$ and 92, for incident electron kinetic energies $T_i = 1$, 5, and 50 keV (from Pratt, 1983).

Recently Feng and Pratt (1981), in developing a parametrization scheme, have shown that simple versions of these expansions suffice to describe the Coulomb spectrum into the soft gamma-ray regime. The first observation was that throughout these energies nonrelativistic dipole approximation is valid for high-Z elements, while for low-Z elements once nonrelativistic dipole approximation fails relativistic Born approximation combined with the Elwert factor is good. The second observation was that the application of nonrelativistic dipole approximation can be divided into a classical regime and a quantum regime, and when the classical expression fails a nonrelativistic Born–Elwert result is adequate. The third observation was that the classical bremsstrahlung spectrum can be well characterized by the first terms of its two limiting soft-photon and hard-photon expansions. We shall return to the quantitative illustration of these observations after we give the various analytic expressions for the Coulomb spectrum.

The full classical result, given by Kramers (1923), and derived in some detail by Landau and Lifshitz (1971), is

$$\frac{4\pi^2\alpha^3}{3}\, i\mu H^{(1)}_{i\mu}(i\mu)\, H^{(1)\prime}_{i\mu}(i\mu), \tag{11}$$

written in terms of Hankel functions of complex argument and index, a function of the one variable $\mu = \nu_i k/2T_i$ ($\nu_i = Z\alpha/\beta_i$). Florescu and Costescu (1978) have obtained the large-μ and small-μ expansions

$$\sigma = \frac{16\pi\alpha^3}{3\sqrt{3}}[1 + d_1\mu^{-2/3} + d_2\nu^{-4/3} + 2d_1 d_2 \nu^{-2} + O(\mu^{-8/3})] \qquad \text{(large } \mu\text{)}$$

$$d_1 \simeq 0.217747, \qquad d_2 \simeq 0.0131214, \tag{12}$$

$$\sigma = \frac{16\alpha^3}{3}(1 + \pi\mu)\ln\frac{2}{\gamma\mu} + O(\mu^2) \qquad \text{(small } \mu\text{)},$$

$$\gamma = e^c = 1.78, \qquad c = 0.577 \qquad \text{(Euler's constant)}. \tag{13}$$

The first term in the large-μ expansion is the Kramers formula, the constant value ~5.61 mb; the first term in the small μ expansion is the universal logarithm resulting from the long-range character of the Coulomb potential. However, the next terms in the two expansions are large and are important in achieving a quantitative characterization.

The prediction of nonrelativistic quantum mechanics in dipole approximation, obtained by Sommerfeld (1931, 1939), is

$$\sigma = \frac{16\pi^2\alpha^3}{3}\frac{1}{(e^{2\pi\nu_i} - 1)(1 - e^{-2\pi\nu_f})} x_0 \frac{d}{dx_0}\left| {}_2F_1(i\nu_i, i\nu_f, 1; x_0) \right|^2, \tag{14}$$

where

$$-x_0 = \frac{4\nu_i\nu_f}{(\nu_f - \nu_i)^2}, \tag{15}$$

and is thus a function of the two variables ν_i and ν_f. Florescu and Costescu have examined the circumstances in which this begins to deviate from the classical prediction. For $\nu_i \geqslant 5$ the difference is less than 5% for almost the entire spectrum, while for higher energies (smaller ν_i) the validity of classical results is increasingly restricted to the soft photon region of the spectrum, and the quantum results lie below classical. For high energies (ν_i and ν_f small) the Sommerfeld formula reduces to the nonrelativistic Born approximation result

$$\sigma = \frac{16}{3}\alpha^3 \ln\left(\frac{p_i + p_f}{p_i - p_f}\right) \tag{16}$$

(depending only on the one variable p_f/p_i), while at the tip (ν_i small but ν_f large) it approaches

$$\sigma = \frac{64\pi}{3}\alpha^3\gamma_i, \tag{17}$$

showing, in comparison to the Kramers value at low energy, a decrease with decreasing Z and with increasing incident energy. The Elwert factor

$$f_E = \frac{\nu_f}{\nu_i} \frac{1 - e^{-2\pi\nu_i}}{1 - e^{-2\pi\nu_f}} \tag{18}$$

(which approaches unity in Born approximation) well characterizes the ratio of the Sommerfeld formula to Born approximation for small ν_i *and* at the hard-photon end of the spectrum (though this does not correspond to its original derivation). Note that for small ν_i and large ν_f, $f_E = 2\pi\nu_f$, which indeed converts Eq. (16) into Eq. (17). Except for very light elements, however, the $2\pi\nu$ will always be large in the nonrelativistic domain and $f_E = \nu_f/\nu_i$. Note that $f_E \equiv (N_f/N_i)^2$, the square of the ratio of final and initial electron wave-function normalizations. The consequence of f_E is to replace the vanishing Born approximation tip end point by a finite value.

The relativistic Born approximation also assumes that ν_i and ν_f are small, but since β_i and β_f now cannot exceed unity this is only possible for low-Z elements. The relativistic Born approximation of Bethe and Heitler (1934) for the spectrum is

$$\begin{aligned}\sigma_{\mathrm{BH}} = \alpha^3 \Bigg\{ &\frac{4}{3} - 2E_iE_f \frac{p_i^2 + p_f^2}{p_i^2 p_f^2} + \frac{\varepsilon_i E_f}{p_i^3} + \frac{\varepsilon_f E_i}{p_f^3} - \frac{\varepsilon_i \varepsilon_f}{p_i p_f} \\ &+ L\left[\frac{2E_iE_f}{3p_ip_f} + k^2\frac{E_i^2E_f^2 + p_i^2p_f^2}{p_i^3p_f^3} + \frac{k}{2p_ip_f}\left(\frac{E_iE_f + p_i^2}{p_i^3}\varepsilon_i \right.\right. \\ &\left.\left. - \frac{E_iE_f + p_f^2}{p_i^3}\varepsilon_f + \frac{2kE_iE_f}{p_f^2p_i^2}\right)\right]\Bigg\} \frac{p_f}{p_i}\beta_i^2,\end{aligned} \tag{19}$$

where

$$L = 2\ln\left(\frac{E_iE_f + p_ip_f - 1}{k}\right), \qquad \varepsilon_i = \ln\frac{E_i + p_i}{E_i - p_i}, \qquad \varepsilon_f = \ln\frac{E_f + p_f}{E_f - p_f}, \tag{20}$$

depending on the two variables p_i, p_f with relativistic total energies $E = (1 + p^2)^{1/2}$ and photon energy $k = E_i - E_f$. This expression can be multiplied by the Elwert factor f_E of Eq. (18) to improve results for the hard-photon end of the spectrum, for the same reason (in terms of correcting Born approximation slow electron normalization) that f_E is useful in the non-relativistic case. The finite value at the tip of the spectrum now does not decrease with T_i, since ν_i remains finite. The relativistic Born–Elwert approximation is a function of three variables. The nonrelativistic limit of σ_{BH} is Eq. (16). The high-energy limit of the Bethe–Heitler formula is

$$\sigma_{\mathrm{BH}}^{\mathrm{h.e.}} = 4\alpha^3 \left(1 + \frac{p_f^2}{p_i^2} - \frac{2}{3}\frac{p_f}{p_i}\right)\left(\ln \frac{2p_i p_f}{k} - \frac{1}{2}\right). \tag{21}$$

Bethe and Maximon (1954) used SM wave functions to obtain the relativistic high-energy limit of the spectrum without assuming Born approximation. The result was an additional Z-dependent term:

$$\sigma_{\mathrm{BM}} = \sigma_{\mathrm{BH}}^{\mathrm{h.e.}} - \frac{4\alpha^3}{p_i^2}\left(E_i^2 + E_f^2 - \tfrac{2}{3}E_i E_f\right) f(Z),$$

$$f(Z) = a^2 \sum_{\gamma=1}^{\infty} \frac{1}{\gamma(\gamma^2 + a^2)} \sim 1.2021 a^2 \qquad \text{for small } a, \tag{22}$$

$$a \equiv Z\alpha.$$

There has been uncertainty as to how high an energy is required before this result is valid. Our recent numerical data (Tseng and Pratt, 1979) indicate that for most of the spectrum ($k/T_i \leqslant 0.6$) by $T_i = 10$ MeV accuracy to 2% is achieved for low Z and to 10% for high Z.

We may now discuss the accuracy of these various expressions in representing the Coulomb spectrum and the ability to characterize the spec-

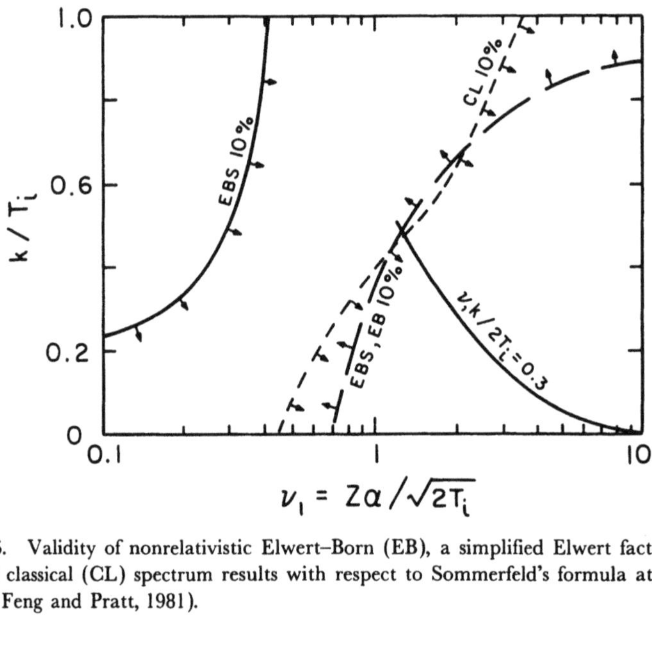

FIGURE 6. Validity of nonrelativistic Elwert–Born (EB), a simplified Elwert factor version (EBS), and classical (CL) spectrum results with respect to Sommerfeld's formula at the 10% level (from Feng and Pratt, 1981).

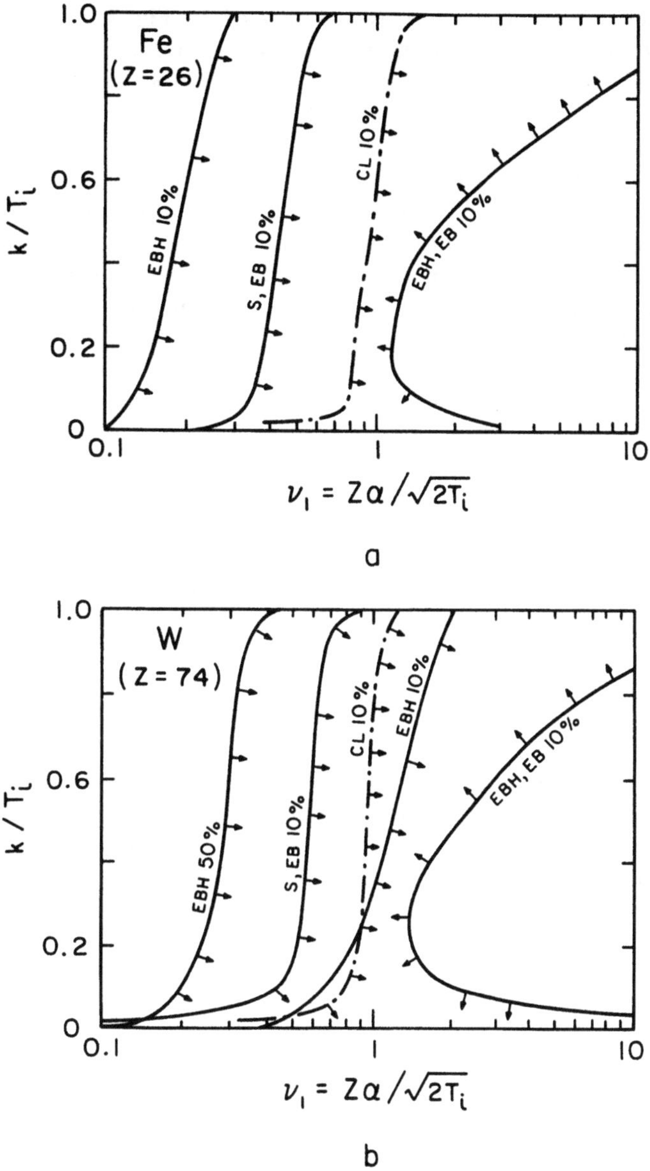

FIGURE 7. Validity of nonrelativistic Elwert–Born (EB), Sommerfeld (S), classical (CL), and relativistic Elwert–Bethe–Heitler (EBH) spectrum predictions with respect to exact numerical Coulomb results (EC), for $Z=26$ and 74 (from Feng and Pratt, 1981).

trum in terms of a small set of simple expressions and simple switching criteria. The essential conclusions are charted in Figs. 6 and 7, from the studies of Feng and Pratt (1981) which led to a simple parametrization scheme for the spectrum. Here Fig. 6 shows the validity of simpler approximations relative to the Sommerfeld formula and Fig. 7a (for $Z = 26$) and Fig. 7b (for $Z = 74$) show validity relative to exact numerical point Coulomb calculations. Switching between small μ and large μ expansions of the classical result Eq. (11) at $\mu \equiv 0.3$ leads to an approximation of the full classical result within 3% in the worst case. The region of validity of these classical results and of the nonrelativistic Born–Elwert formula overlap at the 10% level, so that at this level of accuracy the full Sommerfeld expression never needs to be used. (The failure of the simplified Elwert factor without exponentials generally occurs only when the nonrelativistic approximation itself is failing.) The validity of the nonrelativistic dipole calculation is illustrated in Fig. 7. For high Z it could be used throughout the range considered (showing a cancellation of relativistic and retardation effects), while for Fe it failed at the 10% level for $v_i < 0.5$. It was recommended that one switch to relativistic Born–Elwert approximation for $T_i \equiv (Z\alpha)^4$; for light elements this will lead to a quantitatively correct result at all energies, while in high-Z elements further modifications would be needed at higher energies.

6. SCREENING

Figures 4 and 5 also illustrate how screening of the nucleus by the charge distribution of atomic electrons affects the bremsstrahlung spectrum. The general consequence of screening is a reduction in cross section, corresponding to a reduction in the effective charge seen by the electron being scattered by the atom. The reduction is largest near the soft-photon end point of the spectrum, characterized by large distances for which the nuclear charge is nearly completely shielded; the reduction is least in the hard-photon region of the spectrum characterized (particularly for hard x rays and soft γ rays) by interior atomic distances. The spectra are flatter and lower than in the Coulomb case, with a finite soft-photon endpoint. With increasing energy they become more Coulomb-like, except near the soft-photon end point, but the Coulomb rise at still higher energies is suppressed since (as in Fig. 8) a larger distance region again dominates except at the tip. There are special screening features (most visible for low incident energies) very close to the tip of the spectrum, as illustrated in Fig. 9 and discussed in the next section, connected with the very different normalizations at the origin of low-energy continuum screened and Coulomb wave functions.

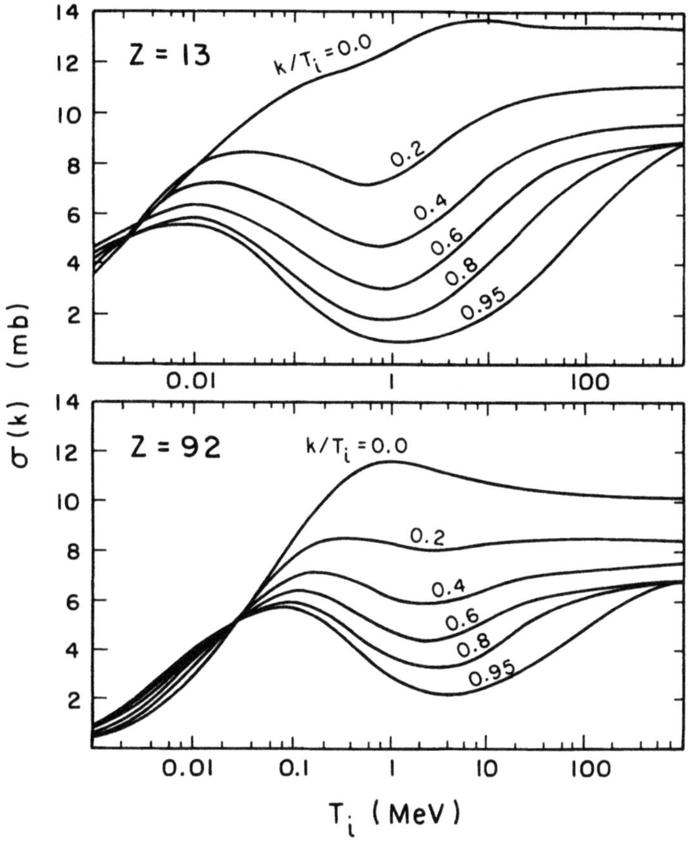

FIGURE 8. Bremsstrahlung cross section $\sigma(k)$ for $Z = 13$ and 92, $T_i = 1$ keV to 1000 MeV, obtained from numerical calculations in partial waves (from Tseng and Pratt, 1979).

In the Born approximation (both relativistic and nonrelativistic) the effect of screening is to multiply $d^3\sigma$ by $|F(q)|^2$, where the form factor is

$$F(q) = 1 - \frac{1}{Z}\int \rho(\bar{r})\, e^{i\mathbf{q}\cdot\mathbf{r}}\, d^3r$$

with (23)

$$\int \rho(\mathbf{r})\, d^3r = Z,$$

where $\mathbf{q}$ is the momentum transfer to the nucleus and ρ is the (here assumed spherically symmetric) charge density of the atomic electrons. For large q

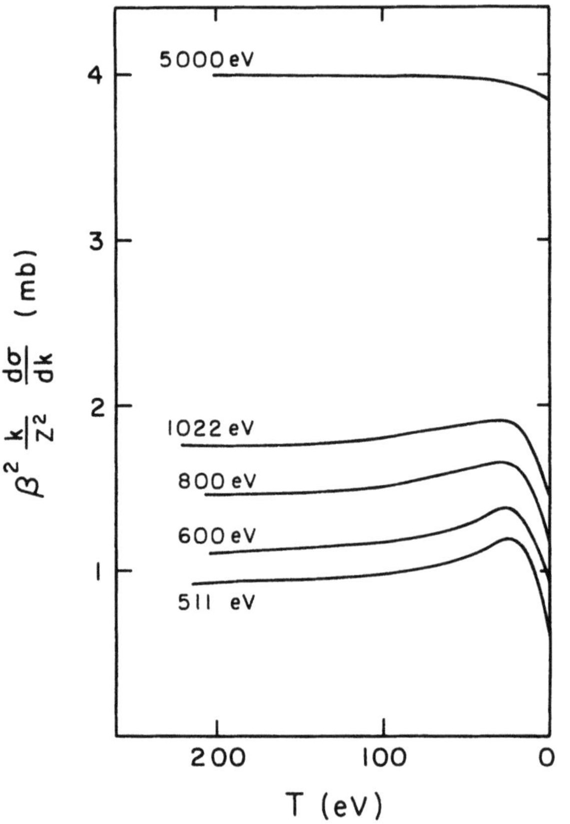

FIGURE 9. Tip-region bremsstrahlung spectrum for lanthanum for several initial-electron energies T_i. $\beta^2 k Z^{-2} d\sigma/dk$ in mb (10^{-27} cm^2) is plotted vs. final scattered-electron kinetic energy T_f in eV, i.e., the photon energy is $k = T_i - T_f$ (from Lee and Pratt, 1975).

(small distance dominance) $F \simeq 1$ and there is no screening; for small q (large distance dominance) $F \simeq 0$ and shielding is complete. Form-factor approximation requires the validity of Born approximation and therefore, in view of our previous discussion, is quantitatively useful in the nonrelativistic quantum regime and for light Z elements in the relativistic regime. The form factor-approximation does not give a correct account of screening in the classical regime or for heavy elements at relativistic energies. However Bethe and Maximon (1954) showed that, for all Z, in the high-energy limit the effect of screening on the spectrum is to modify $\sigma_{\text{BH}}^{\text{h.e.}}$ by $|F(q)|^2$ while leaving the second term of Eq. (22) unchanged—i.e., in the high-energy limit the higher-order Coulomb corrections to the Born approximation are determined at small distances. To obtain $d^2\sigma$, σ, etc. in the Born approximation one must now integrate the Coulomb $d^3\sigma \times |F(q)|^2$.

Still less can be said analytically about screening effects in the classical regime. The soft-photon end point of the spectrum is finite and, as discussed in the next section, simply related to the elastic electron scattering cross section. For not too low an energy this value has been estimated (Jackson, 1975) from the Thomas–Fermi atomic model as

$$\sigma = \frac{16}{3}\alpha^3 \ln\left[\lambda\left(\frac{1.4}{\alpha}\right)\left(\frac{M\beta_i}{Z^{1/3}}\right)\right], \tag{24}$$

where λ depends weakly on Z and T_i and is of order unity and M is the mass of the atom. This result *decreases* as $\ln T_i$ as T_i decreases and *increases* as $-\ln Z$ as Z decreases. For very low energies one may generally write the soft end-point value in terms of the s-wave scattering length R as $(64/3)\,\pi\alpha\beta_i^2R^2$, where $(d\sigma/d\Omega)_{\mathrm{el}} = 4\pi R^2$. In the large-$\mu$ region of the classical spectrum (and particularly for small k/T_i), it is possible to argue (Tucker, 1975; Lamoureux *et al.*, 1983) that the Gaunt factor should be modified by $|g(b_m)|^2$, where b_m is determined from $[g(b_m)/b_m] = [mv^2]/[2Ze^2]$ and g is defined in terms of the potential $Vr^2[(dV)/(dr)] = -Ze^2g(r)$, based on a classical estimate of the maximum impact parameter corresponding to a small angle scattering. The suppression from the Kramers value predicted in this way is in reasonable accord with the numerical calculations.

7. END POINTS OF THE SPECTRUM

The soft-photon and hard-photon (tip) regions of the bremsstrahlung spectrum are related to other atomic processes. In some situations these regions provide information sufficient to characterize the entire spectrum, while in other circumstances the behavior in these end-point regions can be quite different from that for less unequal energy splits. Qualitatively the soft-photon end point, related to elastic electron scattering, diverges as $-\ln k$ for low photon energy k in the point Coulomb potential, but remains finite in a screened potential. By contrast the hard-photon end point of the spectrum, related to direct radiative recombination and atomic photoeffect, remains finite in the point Coulomb potential (while the Born approximation would predict that it goes to zero) but does go to zero in a screened potential.

The bremsstrahlung cross section in the soft photon region is related to the cross section $(d\sigma/d\Omega_f)_{\mathrm{el}}$ for elastic electron scattering through the low-energy theorem (Low, 1958; Jauch and Rohrlich, 1954, 1976; Rohrlich, 1955; Burnet and Kroll, 1968)

$$\lim_{k \to 0} k^{-1} d^3\sigma = \frac{\alpha}{4\pi^2} \left(\frac{\varepsilon \cdot p_i}{k \cdot p_i} - \frac{\varepsilon \cdot p_f}{k \cdot p_f} \right)^2 \left(\frac{d\sigma}{d\Omega_f} \right)_{\text{el}}, \tag{25}$$

where in the parentheses of the right-hand side ε, k, p_i, and p_f are 4-vectors, not 3-vectors. The relation can also be written between the respective matrix elements, and it can be extended to include next-order terms in small k (Burnett and Kroll, 1968), although there are differences between screened and Coulomb potential cases (Rosenberg, 1983). These results enter the discussion of the radiative corrections to elastic scattering. Integration of Eq. (25) over $d\Omega_f$ gives the angular distribution $d^2\sigma$ in this limit (Tseng *et al.*, 1979), as discussed in the next section. Also integrating over $d\Omega_k$, the result for $k\, d\sigma$ may be written

$$\lim_{k \to 0} k \frac{d\sigma}{dk} = 4\alpha \int_0^\pi d\theta \sin\theta \left(\frac{d\sigma}{d\Omega} \right)_{\text{el}} \left[A(A^2 - B^2)^{-1/2} \cosh^{-1}\left(\frac{A}{B} \right) - 1 \right], \tag{26}$$

where $A = 1 - \beta_i^2 \cos\theta$, $B = 1 - \beta_i^2$, $\beta_i = p_i/E_i$, and θ is the electron scattering angle. The low-energy theorem in these forms has been used (Pratt and Lee, 1977) in obtaining the soft-photon end point of $d^2\sigma$ and σ by a numerical integration over the differential cross section for elastic electron scattering (obtained with partial-wave convergence-enhancing techniques). This was convenient since such convergence-enhancement techniques are not in the present numerical bremsstrahlung code and in this region of the spectrum the expansions converge very slowly. In the Coulomb case $(d\sigma/d\Omega)_{\text{el}}$ diverges as θ^{-4} in the forward direction, so that the integral of Eq. (26) diverges, corresponding to the fact that for the Coulomb case the spectrum σ diverges as $-\ln k$ in this limit. For the Coulomb case one may indeed show (Pratt and Lee, 1977) from Eq. (25) that σ diverges as $-(16\alpha^3/3)\ln k$, as in Eqs. (13) and (16). The low-energy theorem tells us that the soft-photon end point reflects only the information of the elastic scattering cross section, and in particular the Rutherford forward angle singularity in the Coulomb case; alternatively, this region of the spectrum provides an alternate way to obtain elastic scattering information. However, Jung and Taylor (1981) have emphasized that the limit in Eq. (25) should not be taken for any (resonance) feature in $(d\sigma/d\Omega_f)_{\text{el}}$ varying rapidly on the scale of observed k. They proposed exploiting this with laser fields to suppress background terms in $d\sigma/d\Omega_f$ and study only the rapidly varying features accessible due to absorption of the laser-beam photons.

The bremsstrahlung cross sections in the hard-photon (tip) region of the spectrum can be related to the cross sections for direct radiative recombination (DRR) into negative energy states (so that $k/T_i > 1$) in the same potential and so also to the inverse process of photoionization. The bremsstrahlung matrix resulting in a final low-energy electron may be analytically continued to give the DRR matrix element for capture into a negative energy state of the same angular momentum in the same potential, since (due to Poincaré's theorem) low-energy negative and positive energy wave functions are analytic continuations, apart from normalization factors. It is a separate question whether the bound state in that potential is available (unoccupied), and in some applications (particularly to inner shells) one may ignore differences between potentials differing by one electron. The characteristic region of the matrix element determines over how broad an energy range the bound and continuum matrix elements are nearly the same (apart from normalization). This is illustrated in Fig. 10, which on the one hand shows characteristic regions and distances over which wave functions are the same in shape and on the other hand shows the behavior of these matrix elements. The key fact is that the shapes of wave functions of energy T are the *same* (independent of their energy) over the range of distances for which T is small compared to the potential energy V. Thus wave functions of small T have the same shape to large distances, while wave functions of large T (positive or negative) approach this common shape only at small distances. For high incident kinetic energies T_i the characteristic distance for the matrix element is small and so, except for normalization, the matrix element is the same over a wide range of final energies T_f; in this case tip region bremsstrahlung and even inner-shell DRR (or photoeffect) are basically the same. For low T_i the characteristic region is large and the matrix elements are equivalent only over a narrow range of T_f; thus in the Coulomb case the Kramers formula for bremsstrahlung is related to outer-shell DRR (Kramers formula). The connection between tip-region bremsstrahlung and inner-shell photoeffect was first noticed by Fano (Fano, 1959; McVoy and Fano, 1959; Fano *et al.*, 1958), then exploited in obtaining predictions for the tip (Lee and Pratt, 1975b, 1976), and more recently used (Kim, 1981; Kim and Pratt, 1983) in obtaining sums over all the DRR cross sections.

It can already be seen in Kramers' classical calculation (Kramers, 1923), and has been established even in relativistic quantum mechanics (Pratt and Tseng, 1975), that in the tip region of the spectrum from a *high-energy incident* electron only the lowest partial waves in the *final* electron contribute (s waves but also p waves in the high-Z case), while for *lower-energy incident* electrons more and more partial waves contribute in the *final* electron. In the Coulomb case the tip value of σ remains finite, as in the Kramers formula. But in a screened potential the spectrum goes to zero at the tip (albeit only in the last 10's of eV), even in the high-incident-energy situations for which the characteristic regions of the matrix element are

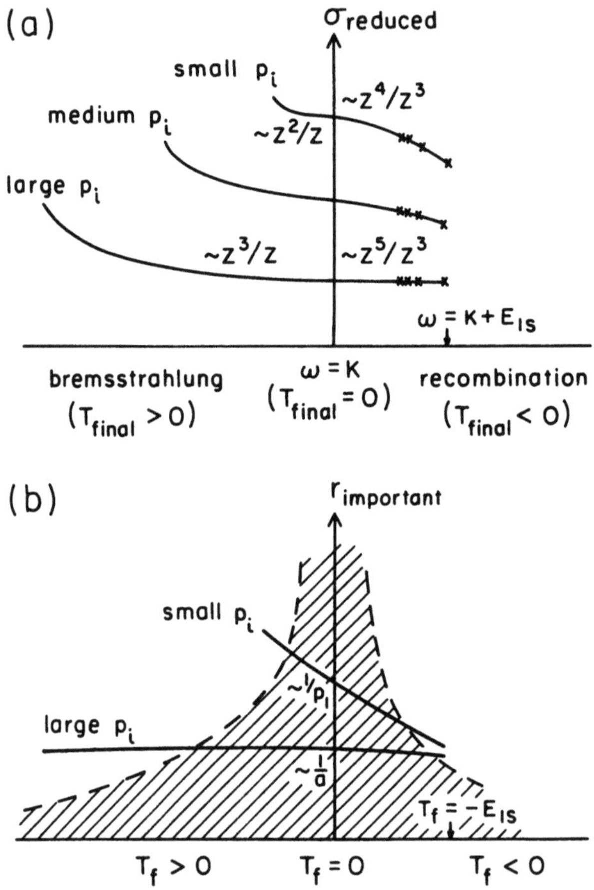

FIGURE 10. Schematic diagrams. (a) Illustration of the smooth connections between the reduced cross sections of positive and negative final electron kinetic-energy regions. While these reduced cross sections are independent of final-state angular momentum, the cross sections and normalization Z dependences shown are for a final s state. (b) Illustration of important regions of matrix elements for high and low incident electron energies as a function of final electron energy. Downward arrows indicate that the important distances are below each solid curve. Shaded area indicates the ranges of r where the low-energy wave-function shapes are similar, and the outer pair of the dashed curves represents the typical distance which determines a bound state of energy E. Horizontal barred lines indicate, for the two choices of p_1, the ranges of E_f in which the reduced cross sections will be similar.

Coulombic. The reason for this, as has been mentioned earlier, is in the normalization at the origin of the final low-energy continuum wave function. Down to some 10's of eV, at least for s waves, these Coulomb and screened normalizations are the same. (And of course the two wave-function shapes are the same in the interior of the atom.) But ultimately, an l-wave continuum Coulomb normalization continues to behave as $p^{-1/2}$, a screened normalization as p^l; this difference leads to the vanishing of the screened end point. The effect becomes more pronounced for smaller T_i, where more final l contribute. However, a countervailing feature for small T_i is the appearance of shape resonances, as illustrated in Fig. 9. Shape resonances, corresponding to trapping of continuum electrons in inner wells, occur for higher l, and lead to enhanced inner normalizations (as well as rapid phase shift changes) and so cross sections. Such features are well known in photoabsorption, but they have been observed in bremsstrahlung as well. Evidently they will first become visible in the tip region—for an appropriate small T_f—but they will also become visible for lower T_i, where the higher *final* angular momentum states (in which a shape resonance can occur) play a more important role in the cross section. More generally, and especially for lower T_i, the tip region of the spectrum becomes sensitive to the details of the spectrum of accessible states, as in a solid. This leads to the general subject of isochromat spectroscopy (Stephenson, 1957; Silberbach *et al.*, 1981; Sobczak *et al.*, 1981).

8. ANGULAR DISTRIBUTIONS AND POLARIZATION CORRELATIONS

While the bremsstrahlung spectrum, depending on the three variables Z, T_i, k/T_i, is relatively simple to discuss, important in its own right, and sufficient to illustrate much of the underlying dynamics of the process, it is the bremsstrahlung angular distributions $d^2\sigma$ and $d^3\sigma$, as well as the polarization correlations, which are generally accessible to direct experimental observation. However, introduction of photon scattering angle θ as a fourth variable (and perhaps further variables as well) complicates the presentation of data and makes it more difficult to identify and discuss systematically the further features in these observations. There has recently been some progress in coping with this situation, at least in the case of $d^2\sigma$. It has long been known that at low energies the angular distribution can be characterized by one "particle parameter" a_2, a function of the *three* variables Z, T_i, k/T_i; methods have now been developed to extend this approach to appreciably higher energies in terms of a few such parameters. Here we will begin with a discussion of $d^2\sigma$ as a function of θ following the organization of the previous sections: the point Coulomb case, the effects of screening, and the end points of the spectrum. Illustrative numerical data are shown in Fig. 11. We will then more briefly discuss the corresponding polarization correlations $C_{\mu\nu}$. We will have rather little to say about $d^3\sigma$ and its associated correlations. The main reference data and discussion of $d^2\sigma$ have been given by Tseng, Pratt, and Lee (1979), with further data and

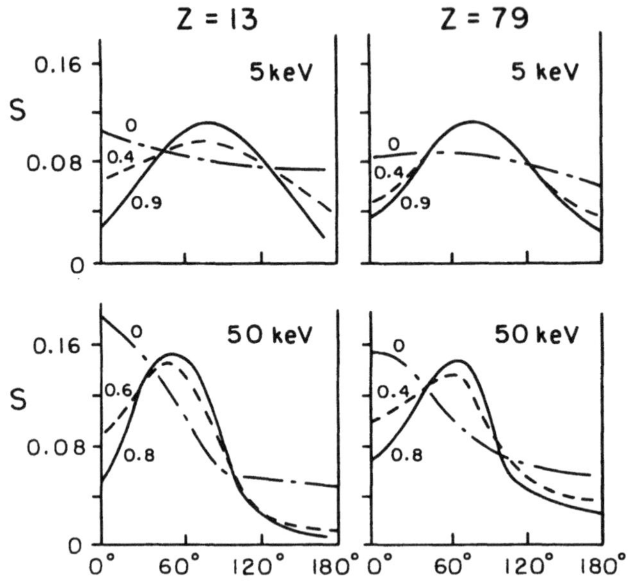

FIGURE 11. Shape function $S(\theta)$ for $Z = 13$ and 79, $T_i = 5$ and 50 keV for various values of k/T_i (from Pratt, 1983).

examination of the behavior of the parameter characterization by Kissel (1982); a tabulation of the shape function S has been given by Kissel, Quarles, and Pratt (1983). Data on the polarization correlations were presented by Tseng and Pratt (1973). The behavior of $d^2\sigma$ and C_{03} at the tip of the spectrum has recently been discussed by Feng, Goldberg, Kim, and Pratt (1983).

In dipole approximation (classical or quantum) the shape function $S \equiv d^2\sigma/d\sigma$ has the form

$$S \propto 1 + \tfrac{1}{2} a_2 P_2(\cos\theta), \tag{27}$$

where a_2 is the particle parameter mentioned above. In quantum mechanics this form is easily understood from the fact that the matrix element is linear in ε^*. Once one has summed and averaged over electron spin directions, the cross section must have the form $a + b(\mathbf{p} \cdot \boldsymbol{\varepsilon})^2$, since $\boldsymbol{\varepsilon}^2 = 1$ and $\mathbf{p}$ is the only other vector (since in dipole approximation one has set $k = 0$ in the matrix element, and so also $\mathbf{p}_i = \mathbf{p}_f$). Summing over photon polarization directions then leads to the form above.

Florescu and Steiner (1982) have obtained and tabulated expressions for the classical Coulomb a_2 parameter; the corresponding quantum Coulomb results have been tabulated and discussed by Thaler *et al.* (1956),

Banuelus and Rodriguez-Trelles (1978, 1979), and Florescu and Steiner (1982). Unlike for the spectrum, simple analytic expressions for a_2 are generally not available. Also unlike for the spectrum, quantum results differ from classical results much sooner, failing as soon as one leaves the classical regime. In the Coulomb case the exact soft-photon end-point value (as we shall discuss) is $a_2 = +1$ corresponding to the forward and backward peaked distribution $1+\cos^2\theta$, while at the tip of the spectrum the Born result approaches $a_2 = -2$ corresponding to a $\sin^2\theta$ distribution symmetric about $90°$ and vanishing at forward and backward angles.

However, a particle-parameter tabulation is not especially useful at X-ray energies and above because, again unlike for the spectrum, the non-relativistic dipole approximation for the angular distribution fails at fairly low energies—the Coulomb Born corrections are of order $4v/c \equiv 4\beta_i$. Due to relativistic, retardation, and higher multipole effects the shape function is no longer symmetric about $90°$ and becomes increasingly forward peaked for increasing incident energy—the peaking behaves as $\Delta = E_i(1-\beta_i\cos\theta)^{-4}$ in the Coulomb Born case. Bethe and Heitler (1934) obtained the full analytic expression for S in this case; the result is complicated but elementary, involving the photon angle θ in Δ and in $q^2 \propto p_i^2 + k^2 - 2p_i k\cos\theta$. The work of Elwert and Haug (1967), tabulated by Bernhardi *et al.* (1983) in a publication adjoining that of Kissel, Quarles, and Pratt (1983), provides an improvement over the Bethe–Heitler results for low-Z Coulomb cases at low energies. At higher energies and for higher Z elements numerical calculation indicates that either approach is fairly good at angles where the shape function is large, poorer at small and large angles (Tseng, Pratt, and Lee, 1979); the predictions are better for S than for the spectrum.

The screening effect of atomic electrons becomes important when the matrix element involves large distances, i.e., when momentum transfers q are small. In the Born approximation the result is to multiply $d^3\sigma$ by $[1-F(q)]^2$, where F is the atomic form factor; if for the range of allowed q in the integration to obtain $d^2\sigma$,

$$(d^2\sigma)_{\text{Born}} = \int d\Omega_f (d^3\sigma)_{\substack{\text{Coulomb}\\ \text{Born}}} [1-F(q)]^2, \tag{28}$$

$F(q)$ may be neglected in comparison to 1, one obtains the point-Coulomb Bethe–Heitler result for S. Screening effects will always be large at low energy, and large at all energies in $d^2\sigma$ at the soft-photon end point of the spectrum where $q=0$ is included in the range of integration. Although in an *ad hoc* fashion one could also multiply the Elwert–Haug $d^3\sigma$ by such a form factor, one finds that the superiority of EH over BH is reduced by the (inconsistent) use of the form factor approach for screening. At low energies, where the Born approximation fails for the spectrum, it would also be expected to fail for S, but this does not appear to have been studied.

Since the bremsstrahlung matrix element, at the soft- and hard-photon end points of the spectrum, has simple properties, the discussion of the end points of the spectrum can be extended to discussions of the shape functions at those end points. For the point Coulomb case, or at high energy in the screened case

$$S(k/T_i=0)=\frac{1-\beta_i^4}{4\pi(1-\beta_i\cos\theta)^4}\times\left[1-\frac{3\beta_i}{1+\beta_i^2}P_1(\cos\theta)+\frac{1}{2}P_2(\cos\theta)\right]; \quad (29)$$

an explicit result may also be given for the low β_i screened case in terms of integrals over the elastic electron scattering cross section (Tseng, Pratt, and Lee, 1979). At the hard-photon limit of the spectrum S for bremsstrahlung can be related to corresponding shapes of DRR (or photoeffect) cross sections, as has been demonstrated for an example by Feng, Goldberg, Kim, and Pratt (1983a). One must add appropriately final angular momentum states in DRR corresponding to the angular momentum of the final low-energy continuum state in bremsstrahlung.

A general characterization of S would be in the coefficients B_n of a Legendre expansion

$$S=\frac{1}{4\pi}\sum_{n=0}^{\infty}B_nP_n(\cos\theta), \quad (30)$$

with $B_n\equiv 1$. In nonrelativistic dipole approximation only B_2 ($\equiv\frac{1}{2}a_2$) is considered, and indeed the other B_n become small at low energy. In the softer X-ray regime only B_1 and B_3 need to be added for an adequate characterization and so, as for photoeffect, two or three such parameters suffice to characterize S. With increasing energy all B_n are large, but not independent, since the poor convergence of the series simply reflects the existence of forward peaking factors like the $(1-\beta\cos\theta)^{-4}$ term found in the Born approximation. This suggests a representation of S in the form

$$S=\frac{A}{4\pi}(1-\beta_i\cos\theta)^{-m}\sum_{n=0}B_nP_n(\cos\theta), \quad (31)$$

with $B_0\equiv 1$ and A defined by

$$\int S\,d\Omega_k=1. \quad (32)$$

It is not clear whether there is a fundamental choice of m (the first Coulomb Born approximation would suggest $m=4$, but the second Born approximation introduces half-integer powers), but in any event it is possible to achieve a rapidly converging series in this way at least up to MeV energies. Data of this type were obtained by Kissel (1982) and formed the basis for the systematic tabulation of Kissel, Quarles, and Pratt (1983).

The theory of $d^3\sigma$ is not yet especially satisfactory. Relativistic Born approximation results are available following the approach of Bethe and Heitler (1934), to which form-factor screening may be applied. Predictions have also been obtained with the approach of Elwert and Haug, which can be expected to be useful for light-Z elements in Coulomb-like or form-factor screening situations, and for heavier elements at very high energies. Results have been used in other situations, in the absence of anything better, with mixed success.

In principle an entirely similar discussion could be given of the various polarization correlations associated with these cross sections. Various relativistic Born approximation results are available. Some numerical results have been reported by Tseng and Pratt (1973) for the seven correlations associated with $d^2\sigma$, and these are the basis for generally satisfactory comparisons with experiment. Only C_{03}, corresponding in most situations to emission of radiation polarized parallel to the emission plane, exists in the nonrelativistic dipole approximation; it (like C_{12} and C_{32}) remains large for high energy and small Z. However, at higher energies there is a cross-over, and then at back angles polarization perpendicular to the plane is favored. C_{20}, which serves as an analyzer for transversely polarized electrons, has also been observed; it (like C_{23}, C_{31}, C_{11}) vanishes for small and large energy and for small Z. C_{32}, which has also been observed, is a transmitter of helicity, producing circularly polarized photons from longitudinally polarized electrons. The correlations C, like S, can be characterized by sets of B's; in nonrelativistic dipole approximations the particle parameter a_2 suffices to determine C_{03}. It was also demonstrated by Feng, Goldberg, Kim, and Pratt (1983a) that C_{03} in the tip region of bremsstrahlung can be determined from the polarization information of DRR.

9. COMPARISON OF THEORY AND EXPERIMENT

The most recent survey of electron bremsstrahlung experiments has been given by Nakel (1982; also 1980); a summary of the situations studied has been prepared by Tseng and Pratt (1983). Fairly extensive data are available for $d^2\sigma$, while coverage of all other circumstances is highly fragmentary. Earlier theory and experiment were compared in the review article of Koch and Motz. Comparisons between experiment and our numerical calculations were reported for $d^2\sigma$ by Tseng and Pratt (1971), for $d^2\sigma$ in the tip region by Pratt and Tseng (1975), for the polarization correlations c_{ij} by Tseng and Pratt (1973), and in the subsequent experimental papers. Nakel (1980) has also discussed the study of the various polarization correlations. Here we will briefly note some of the general experimental considerations and then discuss the status of work on $d^2\sigma$. We will note what is known about the correlations c_{ij}, $d^3\sigma$, and the correlations c_{ijk}.

Following Nakel, the discussion of experimental arrangements may be separated into discussion of electron beam sources, targets, photon detectors, photon polarimeters, and electron detectors. In addition, the measurements of $d^3\sigma$ and its correlations involve coincidence counting for the simultaneous detection of the radiated photon and recoil electron. Electrons have been produced with electrostatic accelerators, polarized beams of electrons from radioactive emitters, Mott scattering, or the Fano effect. Thin solid foils or gaseous targets are used, so as to avoid multiple thick-target bremsstrahlung (which is of practical interest in its own right). Photons are detected by counting, especially with the high-resolution semiconductor detectors now available: lithium-drifted germanium, intrinsic germanium, lithium-drifted silicon. Polarimeters for linear polarization depend on the polarization sensitivity of Compton scattering. Magnetic selectors are used to isolate the inelastically scattered electrons from the much larger elastically scattered background; they are then counted with scintillation or surface barrier detectors.

A survey of the more recent experimental studies of $d^2\sigma$ is shown in Fig. 12. The coverage in Z and T_i is good, the coverage in θ and in k/T_i (only partically indicated) is fairly good, with the important exception that only $\theta = 90°$ data are available at X-ray energies. (While some of these experiments are presented as a measure of σ, some version of theory is being used in going from a measurement at 90° to a value for the spectrum.) Some representative comparisons of experiment and best available theory are shown in Figs. 13 and 14. In general theory and experiment agree within their stated errors; there are a number of discrepancies, but none which are confirmed by more than one experiment. Two groups have been active recently. Quarles and Heroy (1981) reported data from 50-, 100-, and 140-keV electrons incident on Al, Cu, Ag, and Au; some of their data are compared with older results in Fig. 13. Semaan and Quarles (1981, 1982) have reported data for 6–10-keV electrons incident on neon, argon, krypton, and radon. Aydinol *et al.* (1980) and Hippler *et al.* (1981, 1982) have reported data for 6–12-keV electrons incident on a variety of elements; some of their results are shown in Fig. 14 and compared with theory. Agreement is found except in their only high-Z case ($Z = 92$), whereas at a comparable energy the high-Z case of Semaan and Quarles ($Z = 80$) does agree with

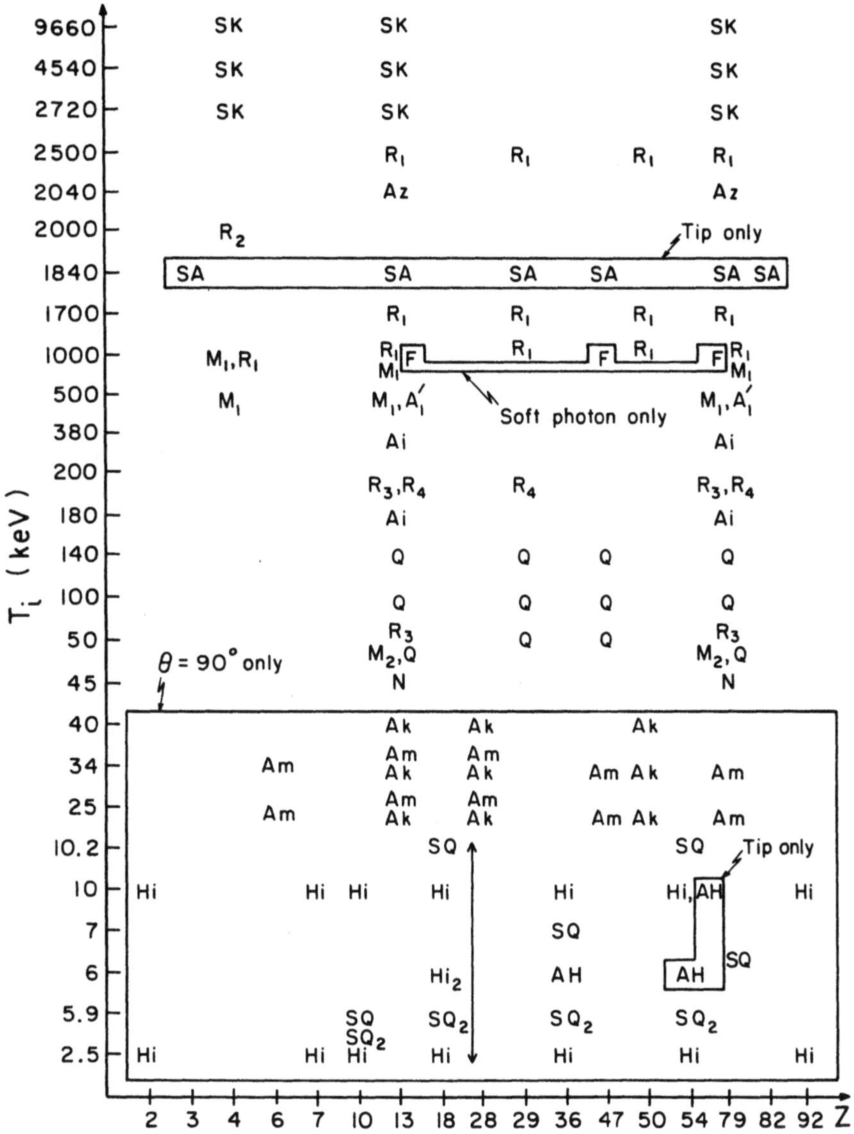

FIGURE 12. Summary of the available experimental data for bremsstrahlung cross sections differential both in photon energy and angle $d^2\sigma$. Symbols denote the following references: AH (Aydinol *et al.*, 1980), Ai (Aiginger, 1960), AK (Amrehn and Kulenkampff, 1955), Am (Amrehn, 1956), AZ (Aiginger and Zinke, 1966), F (Feber and Knop, 1963), Hi (Hippler *et al.*, 1981), Hi2 (Hippler *et al.*, 1982), M1 (Motz, 1955), M2 (Motz and Placious, 1958), N (Nicholas, 1929), Q (Quarles and Heroy, 1981), R1 (Rester and Dance, 1967), R2 (Rester, 1968), R3 (Rester *et al.*, 1970), R4 (Rester and Edmonson, 1972), SA (Starek *et al.*, 1972), SK (Starfelt and Koch, 1956), SQ (Semaan and Quarles, 1981), SQ2 (Semaan and Quarles, 1982).

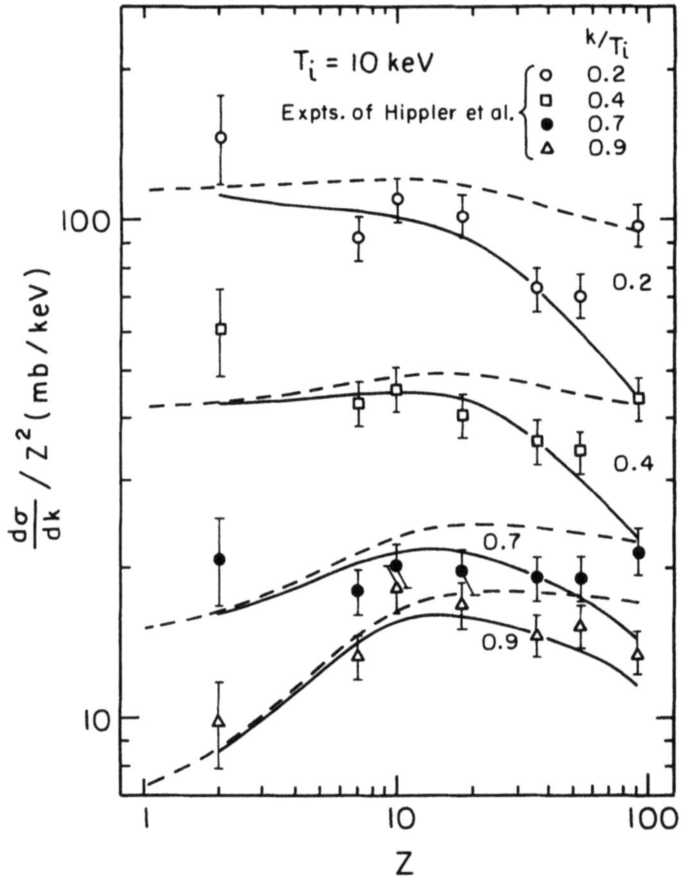

FIGURE 13. Comparisons of the bremsstrahlung cross section differential in photon energy $d\sigma$ for $Z = 2$, 7, 10, 18, 36, 54, 92, $T_i = 10$ keV, $k/T_i = 0.2$, 0.4, 0.7, 0.9 between experimental data of Hippler *et al.*, the Sommerfeld results (broken lines), and our partial-wave results (solid lines) (from Hippler *et al.*, 1981).

theory. Nakel has noted that further experiments would be desirable to resolve this question.

Three of the polarization correlations associated with $d^2\sigma$ (detecting only the radiated photon) have been observed: C_{03}, C_{20}, C_{32}. Most work has focused on C_{03}, the linear polarization of bremsstrahlung photons radiated from an unpolarized electron beam. This polarization is large and in the scattering plane for intermediate angles, though with increasing energy the large polarization moves to forward angles and reverses to polarization perpendicular to the plane at back angles. Early work was summarized by Motz and Placious (1960), subsequent work by Tseng and Pratt (1973) in

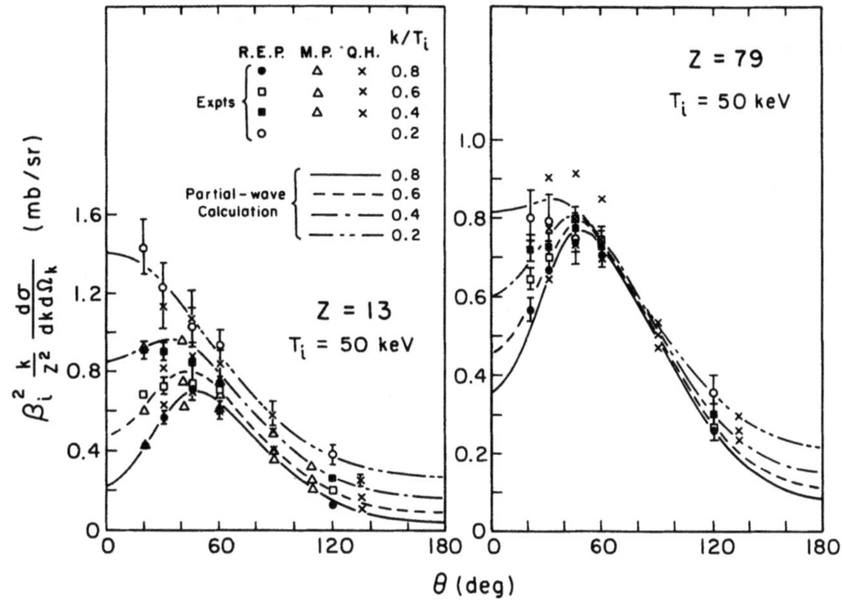

FIGURE 14. Comparisons of the bremsstrahlung angular cross sections $\beta_i^2[k/Z^2][(d\sigma)/dk\,d\Omega]$ for $Z = 13$, 79, $T_i = 50$ keV, $k/T_i = 0.2$, 0.4, 0.6, 0.8 between experimental data of Rester *et al.* (1970) (REP), of Quarles and Heroy (1981) (QH), and numerical calculation in partial waves (from Tseng and Pratt, 1983).

comparison with their numerical calculations. Good agreement was generally achieved with the fairly systematic studies of Scheer *et al.* (1968) for $Z = 6$, 13, 79 at 50 and 180 keV and of Kuckuck and Ebert (1973) for $Z = 13$, 29, 47, 79 at 50, 75, 100 keV. Agreement with theory was afterwards reported by Lichtenberg *et al.* (1975) for $Z = 4$, 13, 47, 79 at 533, 1000, 1500 keV. For low Z at these energies screened Born calculations were adequate, while for high Z the numerical approach is necessary.

Studies of C_{20}, causing an azimuthal asymmetry in bremsstrahlung due to an incident beam of electrons transversely polarized perpendicular to the production plane, have been reported in recent years by Guthner (1965), Pencynski and Wehner (1970), Aehlig (1980), and Schaefer *et al.* (1982). The two new experiments, both for Au at 128 keV, generally agree with the predictions of Tseng and Pratt (1973), though in the second experiment there is a discrepancy for $\theta = 145°$. The correlation is generally fairly small. Studies of C_{32} (transmission of helicity), production of circularly polarized photons from longitudinally polarized electrons, were reported by Galstor (1964) and others; they have not been compared with theoretical predictions beyond Born approximation. The correlation becomes large (with a sign reversal) at high energies and the calculations of Tseng and Pratt (1973) indicate that it is then fairly well predicted in Born approximation.

Coincidence experiments for $d^3\sigma$, observing both the radiated photon and the recoil electron, are far more limited, and no theoretical predictions are available beyond Born approximation or the work of Elwert and Haug (1969). The first such experiment was reported by Nakel (1966); subsequent work from the groups of Nakel (Nakel, 1967, 1968; Hub and Nakel, 1967; Nakel and Sailer, 1970; Kreuzer and Nakel, 1971) (for certain geometries and energy splits from 300-keV electrons on $Z = 6$, 13, 79), of Aehlig and Scheer (1972) and Aehlig *et al.* (1977), (180 keV on $Z = 47$, 300 keV on $Z = 79$), and Faulk and Quarles (1974) (140 keV on $Z = 13$, 79) has been reviewed by Nakel (1980). Some examples of these data are shown in Fig. 15. A subsequent experiment for $Z = 6$, 29, 47, 79 at 300 keV has been

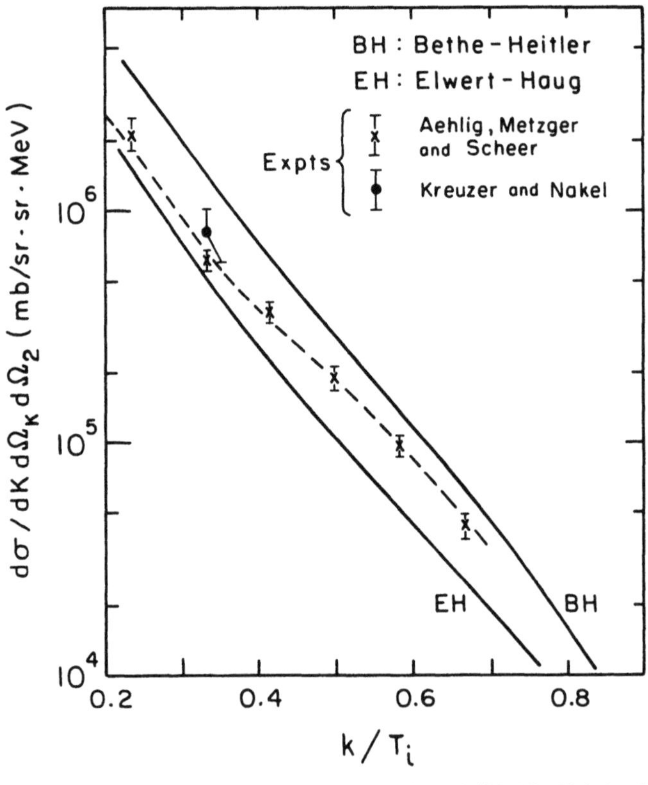

FIGURE 15. Comparisons of bremsstrahlung cross sections $d\sigma/(dk\, d\Omega_k\, d\Omega_f)$ for $Z = 79$, $T_i =$ 300 keV, electron scattering angle 0° and photon emission angle 10° between experimental data of Aehlig, Metzger, and Scheer (1977), experimental data of Kreuzer and Nakel (1971), the Bethe–Heitler calculation (BH), and the Elwert–Haug calculation (EH) (from Tseng and Pratt, 1983).

reported by Komma and Nakel (1982). In the lower-Z cases at these energies Bethe–Heitler or Elwert–Haug predictions should be adequate while in the higher-Z cases no necessarily valid approaches are available. For the cases of Komma and Nakel the EH approach was adequate, while for the situations of Fig. 15 it is not.

One experiment has been reported on a polarization correlation associated with $d^3\sigma$, by Behncke and Nakel (1978); the work is being continued by Bleier and Nakel (1982). The correlation is C_{030}, the linear polarization of bremsstrahlung photons observed in coincidence with the recoil electron. Behncke and Nakel scattered 300-keV electrons from carbon, holding the recoil electron direction fixed and varying the photon direction. The polarization was nearly complete except when $d^3\sigma$ went through a minimum, reflecting the importance of spin–flip radiation. The behavior of the polarization was in good agreement with EH preditions, as would be expected for a light-Z high-energy situation. Bleier and Nakel have now performed the same experiments on Cu and Au targets; for Au the EH predictions are poor in the region of rapidly varying polarization. Nakel (1982) comments that this once again illustrates the need for full numerical predictions of $d^3\sigma$ and its associated correlations.

10. SOME NEW DEVELOPMENTS

Here, in conclusion, we briefly mention and illustrate some newer aspects of the bremsstrahlung process which are now being studied, but for which experimental results are not yet available. These include (1) the changes which result if our target is an ion rather than a neutral atom, (2) the changes which result if our target is in a hot dense plasma and thereby compressed as well as ionized, and (3) the changes which result if our projectile is a positron rather than an electron. We could also mention radiative electron capture in ion–atom collisions, which viewed in the rest system of the projectile (and ignoring the binding of the electrons to the target) is another example of DRR and has been calculated with the same numerical codes (Lee, 1978). We could also mention the free–free transitions (inverse bremsstrahlung, including multiphoton) which are of current interest in studies of laser-plasma interactions (Gavrila and Van der Wiel, 1978).

We show in Fig. 16 examples of results obtained for the bremsstrahlung spectrum from atomic ions, of interest for example in magnetic fusion plasma where at sufficiently high energies (not yet reached) bremsstrahlung can become a dominant energy-loss mechanism. We see that even rather highly ionized species, except the few-electron systems, stay quite close to the neutral spectrum, except near the soft-photon end point or for low incident

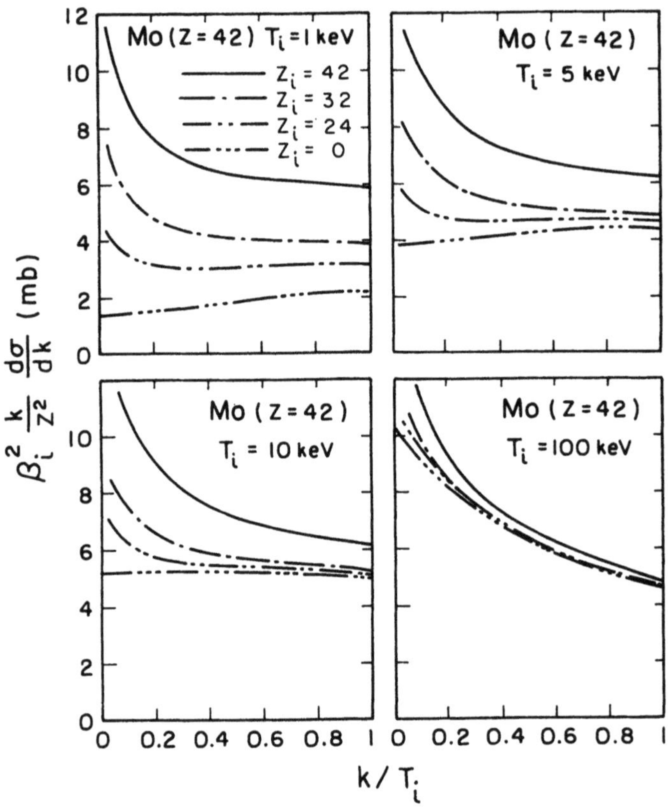

FIGURE 16. Bremsstrahlung energy spectrum from completely stripped, Ne-like, Ar-like Mo ions and the neutral Mo atom at $T_i = 1$, 5, 10, and 100 keV (from Lee and Pratt, 1977).

electron kinetic energies. This reflects the fact that a process taking place at some interior distance is not sensitive to the presence or absence of exterior screening charge. Thus the effects of ionization on these spectra will grow for still lower kinetic energies, disappear at higher energies where the process takes place interior to most atomic electrons, and increase again at still higher energies. In characterizing the effects of ionization it was useful (Lee *et al.*, 1977) to introduce an ionization factor

$$I = \frac{\sigma(k/T_i, T_i, Z, Z_i) - \sigma(k/T_i, T_i, Z, 0)}{\sigma(k/T_i, T_i, Z, Z) - \sigma(k/T_i, T_i, Z, 0)}, \tag{33}$$

describing how for fixed k/T_i, T_i, and Z the spectrum values are distributed as a function of Z_i between the neutral and screened limits. For $k/T_i = 0$ one can show that $I = (Z_i/Z)^2$; in general I was found to be a function only of

Z_i/Z rather than Z and Z_i separately, and only weakly dependent on k/T_i and T_i. In parametrization (Feng and Pratt, 1981) one made the simple choice $I = (Z_i/Z)^3$.

The importance of bremsstrahlung emission by superthermal electrons in laser targets has been discussed by More (1984). Free–free transitions in dense plasmas have been reviewed by Weisheit (1984). In a hot dense (high $T\rho$) plasma, as in inertial confinement fusion (ICF), atoms are ionized, but if the plasma is dense enough the continuum electrons are confined to the ionic volume and may screen the atomic nucleus even more than in the case of an isolated neutral atom. Such a study has been reported (Lamoureux *et al.*, 1982; Feng *et al.*, 1983). The main features of the bremsstrahlung spectrum in such a plasma will follow if one has a suitable model for the atomic potential and its modification due to the plasma environment. It is generally believed that in ICF conditions ion correlations restrict ion interpenetration, so that, for example, a confined-atom Thomas–Fermi (TF) model ion is more appropriate than an extended-atom Debye–Hückel–TF (DHTF) model. One may be concerned because the TF

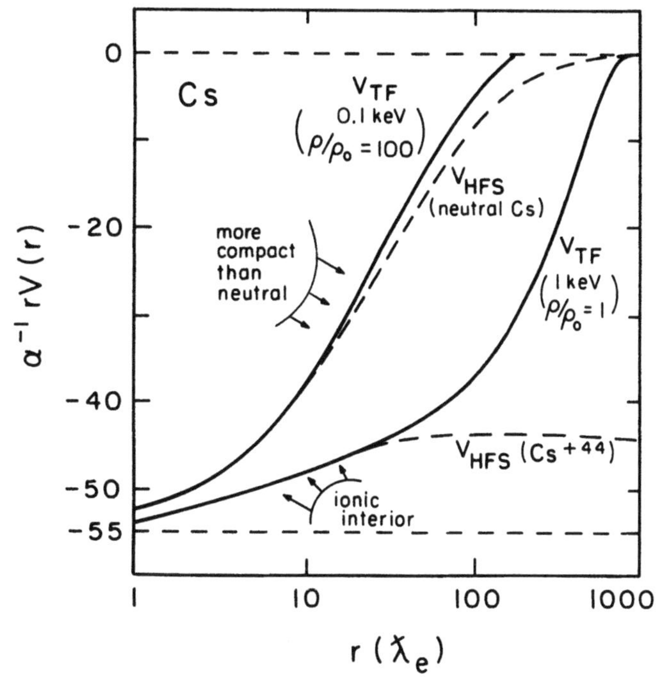

FIGURE 17. Comparison of TF potentials with isolated-atom or ion potentials for Cs ($Z = 55$), for $kT = 0.1$ and 1 keV, $\rho/\rho_0 = 1$ and 100. We plot rV, which goes to $Z = 55$ for $r = 0$ and to 0 (neutral case) or the ionic charge for large distances (from Pratt, 1983).

model is an average-atom (AA) model (averaging the atomic potential over ionic species and their excitation stages in the plasma), and at least for inner shells (not subject to rapid fluctuation) a detailed configuration approach would seem more appropriate. This AA approach may be more acceptable for free–free transitions, which do not depend on detailed configuration properties, than for bound–free and bound–bound transitions. We show such potentials for Cs in Fig. 17, including TF potentials for solid density $\rho = \rho_0$ and temperature of 1 keV and for a much denser case $\rho = 100\rho_0$ at a temperature 0.1 keV. For comparison we also show the isolated neutral-atom potential and the isolated ion potential corresponding to the same degree of ionization as in the solid density case. We see that the interior of the potential for $\rho = \rho_0$ is the same as for the isolated ion, so that the

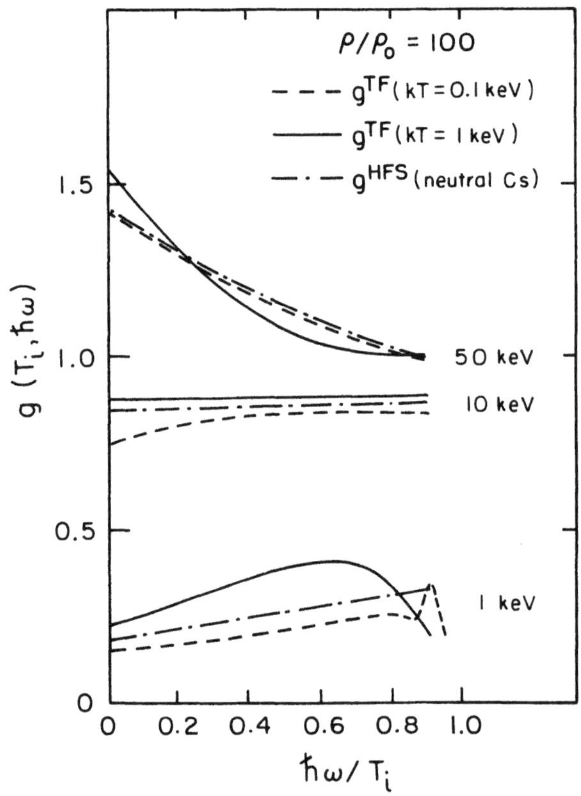

FIGURE 18. Free–free Gaunt factors obtained by full relativistic calculation in TF potentials (g^{TF}), as a function of $\hbar\omega/T_i$ (ratio of photon energy radiated to incident electron kinetic energy) for Cs plasmas at $kT = 0.1$ keV, $\rho/\rho_0 = 100$, for incident electron energies $T_i = 1$, 10, 50 keV, compared with neutral results (from Feng *et al.*, 1983).

bremsstrahlung spectrum should be the same when determined in the interior region. However, we see that for $\rho = 100\rho_0$ the potential is more compact than for the neutral atom: even though the atom is ionized the screening of its nuclear charge will be greater. We show in Fig. 18 some of the bremsstrahlung spectra resulting from these potentials. In the $\rho = \rho_0$ case the spectra are above neutral, especially for low kinetic energy, and agree with the isolated-ion spectra except at the soft-photon end (dominated by large-distance behaviors which are different in the two cases). For the $\rho = 100\rho_0$ case which we show we see superscreening—suppression below neutral-atom values—a shape resonance far back from the tip (supported by the more compact potential) and substantial suppression in the end-point region (not observed so far back from the tip in isolated neutral atoms). Other calculations of bremsstrahlung in hot dense plasmas have been reported by Green (1979, 1981), Collins and Merts (1981), and Rozsnyai (1979).

We show in Fig. 19 comparative results for the spectra of positron and

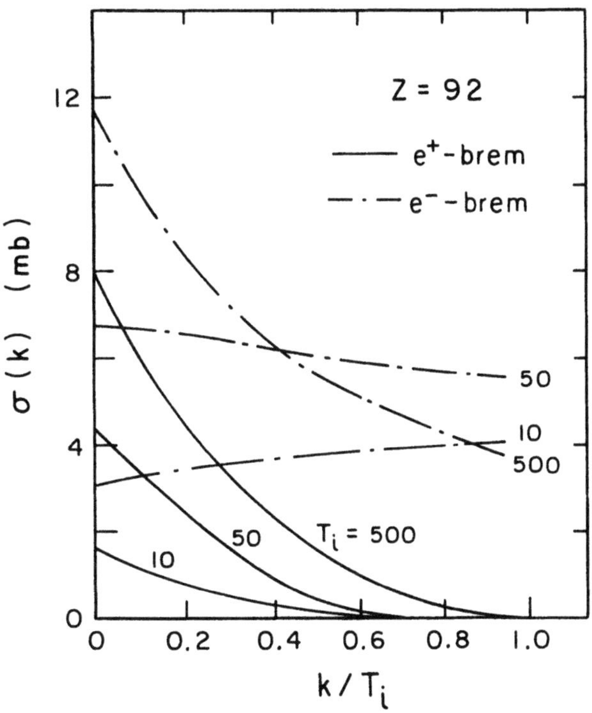

FIGURE 19. Comparison of positron and electron bremsstrahlung energy spectra $\sigma(k)$ for $Z = 92$; $T_i = 10$, 50, and 500 keV (from Feng *et al.*, 1981).

electron bremsstrahlung (Feng *et al.*, 1981). With increasing energy or lower Z the two spectra become more similar, while at low energies the positron spectrum shows dramatic suppression. We note that the electron– and positron–atom interactions are very similar at high energies and very different at low energies, reflecting attraction by (repulsion from) the nuclear charge. It is also interesting to note that the positron approaches the nucleus more closely in the presence of screening, the electron less so. In the Coulomb case the ratio of the spectra we have obtained numerically is well described by $e^{2\pi(\nu_i - \nu_f)}$, as would follow from the Sommerfeld formula or the Elwert factor and its ratio of normalization; in the screened case the use of the ratio of *s*-wave screened normalizations instead gives a qualitatively correct prediction. Use of these data leads to significant differences from earlier estimates of positron stopping powers, which were based on the assumption that electron and positron energy-loss cross sections were the same (Seltzer, 1983).

ACKNOWLEDGMENTS

This work has been supported in part by the National Science Foundation under grant No. PHY-8120785. Both authors are indebted to Dr. H. K. Tseng for many discussions and for use of his bremsstrahlung code.

REFERENCES

AEHLIG, A., 1980, *Z. Phys.* **A294**, 291.
AEHLIG, A., AND M. SCHEER, 1972, *Z. Phys.* **250**, 235.
AEHLIG, A., L. METZGER, AND M. SCHEER, 1977, *Z. Phys.* **A281**, 205.
AIGINGER, H., 1960, *Z. Phys.* **917**, 8.
AIGINGER, H., AND H. ZINKE, 1966, *Acta Phys. Austriaca* **23**, 76.
AMREHN, H., AND H. KULENKAMPFF, 1955, *Z. Phys.* **140**, 452.
AMREHN, H., 1956, *Z. Phys.* **144**, 529.
AMUSIA, M. Y., 1982, *Comments At. Mol. Phys.* **11**, 123.
ANDERSON, C. D., R. A. MILLIKAN, S. NEDDERMEYER, AND W. PICKERING, 1931, *Phys. Rev.* **45**, 352.
AYDINOL, M., R. HIPPLER, I. MCGREGOR, AND H. KLEINPOPPEN, 1980, *J. Phys. B* **13**, 989.
BANUELOS, A., AND F. RODRIGUES-TRELLES, 1978, *J. Phys. B* **11**, 1663.
BANUELOS, A., AND F. RODRIGUES-TRELLES, 1979, *J. Phys. B* **12**, 2249.
BARBIELLINI, G., G. BOLOGNA, G. DAIMBRINI, AND G. P. MURTAS, 1962, *Phys. Rev. Lett.* **8**, 454.
BARKLA, G. C., 1905, *Phil. Trans. R. Soc. London* **204**, 467.
BEHNCKE, H. H., AND W. NAKEL, 1978, *Phys. Rev. A* **17**, 1679.
BERG, R. A., AND C. N. LINDER, 1958, *Phys. Rev.* **112**, 2072.
BERNHARDI, K., E. HAUG, AND K. WIESEMANN, 1983, *At. Data Nucl. Data Tables* **28**, 461.
BETHE, H. A., AND W. HEITLER, 1934, *Proc. R. Soc. London* **A146**, 83.
BETHE, H. A., AND L. C. MAXIMON, 1954, *Phys. Rev.* **93**, 768.

BETHE, H. A., AND E. S. SALPETER, 1957, *Quantum Mechanics of One- and Two-Electron Atoms* (Academic, New York).
BLEIER, W., AND W. NAKEL, 1982, Program Abstracts, Int. Conf. on X-Ray and Atomic Inner Shell Physics, University of Oregon, p. 29.
BOHM, K., 1937, *Z. Phys.* **38**, 334.
BOHM, K., 1938, *Ann. Phys.* **33**, 315.
BORIE, E., L. C. MAXIMON, AND H. OLSEN, 1970, *Phys. Rev. A* **2**, 1443.
BOSLEY, W., J. D. CRAGGS, W. F. NASH, AND R. M. PAYNE, 1948, *Nature* **161**, 1022.
BOTTO, D. J., AND M. GAVRILA, 1982, *Phys. Rev. A* **26**, 237.
BRAGG, W. H., 1907, *Phil. Mag.* **14**, 429.
BRAGG, W. H., AND W. L. BRAGG, 1913, *Proc. R. Soc. London* **88**, 428.
BRYSK, H., C. D. ZERBY, AND S. K. PENNY, 1969, *Phys. Rev.* **180**, 104.
BUGUET, A., 1897, *C. R. Acad. Sci.* **125**, 398.
BURNET, T. H., AND N. M. KROLL, 1968, *Phys. Rev. Lett.* **20**, 86.
COLLINS, L. A., AND A. L. MERTS, 1981, *J. Quant. Spectrosc. Radiat. Transfer* **26**, 443.
COMPTON, A. H., AND S. K. ALLISON, 1935, *X-Rays in Theory and Experiment* (Van Nostrand, Princeton).
D'YACHKOV, L. G., 1981, *J. Phys. B* **14**, L695.
DAVIES, H., H. A. BETHE, AND L. C. MAXIMON, 1954, *Phys. Rev.* **93**, 788.
DUANE, W., 1927, *Proc. Natl. Acad. Sci. USA* **13**, 662.
DUANE, W., 1928, *Proc. Natl. Acad. Sci. USA* **14**, 450.
DUANE, W., AND F. L. HUNT, 1915, *Phys. Rev.* **6**, 166.
ELWERT, G., 1939, *Ann. Phys. (NY)* **34**, 178.
ELWERT, G., AND E. HAUG, 1967, *Phys. Rev.* **183**, 90.
FANO, U., 1959, *Phys. Rev.* **116**, 1156.
FANO, U., H. W. KOCH, AND J. W. MOTZ, 1958, *Phys. Rev.* **112**, 1679.
FAULK, J. D., AND C. A. QUARLES, 1974, *Phys. Rev. A* **9**, 732.
FEBEL, A., AND G. KONP, 1963, *Z. Phys.* **174**, 257.
FENG, I. J., AND R. H. PRATT, 1981, University of Pittsburgh, report No. PITT-266.
FENG, I. J., I. B. GOLDBERG, Y. S. KIM, AND R. H. PRATT, 1983a, *Phys. Rev. A* **28**, 609.
FENG, I. J., M. LAMOUREUX, R. H. PRATT, AND H. K. TSENG, 1983b, *Phys. Rev. A* **27**, 3209.
FENG, I. J., R. H. PRATT, AND H. K. TSENG, 1981, *Phys. Rev. A* **24**, 1358.
FLORESCU, V., AND A. COSTESCU, 1978, *Rev. Roum. Phys.* **23**, 131.
FLORESCU, V., AND V. STEINER, 1982, private communication.
FRIEDRICH, W., P. KNIPPING, AND M. LAUE, 1913, *Radium (Paris)* **10**, 47.
GALSTER, S., 1964, *Nucl. Phys.* **58**, 72.
GAVRILA, M., AND J. MCENNAN, 1977, *Phys. Lett.* **59A**, 441.
GAVRILA, M., AND M. VAN DER WIEL, 1978, *Comments At. Mol. Phys.* **8**, 1.
GELTMAN, S., 1973, *J. Quant. Spectrosc. Radiat. Transfer* **13**, 601.
GINSBERG, E. S., AND R. H. PRATT, 1964, *Phys. Rev. A* **134**, B773.
GINSBERG, E. S., AND R. H. PRATT, 1965, *Phys. Rev.* **137**, B1500.
GOLDEMBERG, J., AND R. H. PRATT, 1966, *Rev. Mod. Phys.* **38**, 311.
GOLOVINSKII, P. A., 1982, *Opt. Spectrosc. (USSR)* **53**, 365.
GREEN, J. M., 1979, report No. RDA-TR-108600-003, R & D Assoc., Marina del Rey, California.
GREEN, J., 1981, report No. RDA-TR-108600-003 R. 1, R & D Assoc., Marina del Rey, California.
GUTHNER, K., 1965, *Z. Phys.* **182**, 278.
HARWORTH, P., AND P. KIRKPATRICK, 1942, *Phys. Rev.* **62**, 334.
HAUG, E., 1975, *Z. Naturforsch.* **30A**, 1099.
HEITLER, W., 1954, *Quantum Theory of Radiation* (Oxford University Press), 3rd edition.

HERMANN, A., 1971, *The Genesis of Quantum Theory 1899–1913*, translated by Claude W. Nash (MIT Press, Cambridge).
HERTZ, H., 1887, *Ann. Phys. (Paris)* **31**, 421; **34**, 373, 551.
HINZ, H., 1941, *Ann. Phys. (Leipzig)* **39**, 573.
HIPPLER, R., K. SAEED, I. MCGREGOR, AND H. KLEINPOPPEN, 1981, *Phys. Rev. Lett.* **46**, 1622.
HIPPLER, R., K. SAEED, I. MCGREGOR, AND H. KLEINPOPPEN, 1982, *Z. Phys.* **A307**, 83.
HUB, R., AND W. NAKEL, 1967, *Phys. Lett.* **31A**, 601.
JABBUR, R. J., AND R. H. PRATT, 1963, *Phys. Rev.* **129**, 184.
JABBUR, R. J., AND R. H. PRATT, 1964, *Phys. Rev.* **133**, B1090.
JACKSON, J. D., 1975, *Classical Electrodynamics* (Wiley, New York), 2nd edition.
JAUCH, J. M., AND F. ROHRLICH, 1954, *Helv. Phys. Acta* **27**, 613.
JAUCH, J. M., AND R. ROHRLICH, 1976, *The Theory of Photons and Electrons* (Springer, Berlin, 2nd edition).
JUNG, C., AND H. S. TAYLOR, 1981, *Phys. Rev. A* **23**, 1115.
KIM, Y. S., 1981, Ph.D. dissertation, University of Pittsburgh.
KIM, Y. S., AND R. H. PRATT, 1983, *Phys. Rev. A* **27**, 2913.
KISSEL, L., 1982, Sandia Report SAND 81-2154.
KISSEL, L., C. A. QUARLES, AND R. H. PRATT, 1983, *At. Data Nucl. Data Tables* **28**, 381.
KOCH, H. W., AND R. E. CARTER, 1950, *Phys. Rev.* **77**, 165.
KOCH, H. W., AND J. W. MOTZ, 1959, *Rev. Mod. Phys.* **31**, 920.
KOMAR, A. P. *et al.*, 1971, *Sov. Phys.-Tech. Phys.* **16**, 631, and references therein.
KOMMA, M., AND W. NAKEL, 1982, *J. Phys. B* **15**, 1433.
KRAMERS, H. A., 1923, *Phil. Mag.* **46**, 836.
KREUZER, K., AND W. NAKEL, 1971, *Phys. Lett.* **34A**, 407.
KUCKUCK, R. W., AND P. J. EBERT, 1973, *Phys. Rev. A* **7**, 456.
KULENKAMPFF, H., 1926, *Handbuch der Physik* **23**, edited by Geiger-Scheel (Springer, Berlin), p. 433.
KULENKAMPFF, H., 1928, *Ann. Phys. (Leipzig)* **87**, 597.
LAMOUREUX, M., I. J. FENG, R. H. PRATT, AND H. K. TSENG, 1982, *J. Quant. Spectrosc. Radiat. Transfer* **27**, 227.
LAMOUREUX, M., I. J. FENG, H. K. TSENG, AND R. H. PRATT, 1983 (unpublished).
LANDAU, L. D., AND L. M. LIFSHITZ, 1971, *The Classical Theory of Fields*, (Pergamon, New York), 3rd edition.
LANZL, L. H., AND A. O. HANSON, 1951, *Phys. Rev.* **83**, 959.
LEE, C. M., 1978, *Phys. Rev. A* **17**, 566.
LEE, C. M., AND R. H. PRATT, 1975a, *Phys. Rev. A* **12**, 707.
LEE, C. M., AND R. H. PRATT, 1975b, *Phys. Rev. A* **13**, 1325.
LEE, C. M., AND R. H. PRATT, 1976, *Phys. Rev. A* **14**, 990.
LEE, C. M., R. H. PRATT, AND H. K. TSENG, 1977, *Phys. Rev. A* **16**, 2169.
LICHTENBERG, W., A. PRZYBYLSKI, AND M. SCHEER, 1975, *Phys. Rev. A* **11**, 480.
LOW, F. E., 1958, *Phys. Rev.* **110**, 974.
MAXWELL, J. C., 1862, *Phil. Mag.* **23**, 12, 85.
MAXWELL, J. C., 1865, *Phil. Trans.* **155**, 459.
MAXWELL, J. C., 1873, *A Treatise on Electricity and Magnetism*
MCDIARMID, I. B., 1952, *Phil. Mag.* **43**, 1003.
MCENNAN, J., AND M. GAVRILA, 1977, *Phys. Rev. A* **15**, 1537.
MCVOY, K. W., AND U. FANO, 1959, *Phys. Rev.* **116**, 1168.
MERTS, A., 1976, unpublished Los Alamos National Laboratory Report.
MORE, R. M., 1984, Atomic Physics in Inertial Confinement Fusion, (unpublished).
MORK, K., AND H. OLSEN, 1965, *Phys. Rev.* **140**, B1661.
MORK, K., AND H. OLSEN, 1968, *Phys. Rev.* **166**, 1862.

Motz, J. W., 1955, *Phys. Rev.* **100**, 1560.
Motz, J. W., and R. C. Placious, 1958, *Phys. Rev.* **109**, 235.
Motz, J. W., and R. C. Placious, 1960, *Nuovo Cimento* **15**, 571.
Muirhead, E. G., B. M. Spicer, and H. Lichtblau, 1952, *Proc. Phys. Soc. London* **A65**, 59.
Nakel, W., 1966a, *Phys. Lett.* **22**, 614.
Nakel, W., 1967, *Phys. Lett.* **25A**, 569.
Nakel, W., 1968, *Z. Phys.* **214**, 168.
Nakel, W., 1980, Electron–Photon Correlations in Bremsstrahlung Processes, in *Coherence and Correlations in Atomic Collisions*, edited by H. Kleinpoppen and J. F. Williams (Plenum, New York).
Nakel, W., 1982, Electron Bremsstrahlung Experiments, in *X-Ray and Atomic Inner-Shell Physics—1982*, edited by B. Crasemann, (American Institute of Physics, New York), p. 389.
Nakel, W., and U. Sailer, 1970, *Phys. Lett.* **31A**, 181.
Nicholas, W. W., 1929, *J. Res. NBS* **2**, 837.
Olsen, H., 1955, *Phys. Rev.* **99**, 1335.
Olsen, H. A., and L. C. Maximon, 1978a, *Phys. Lett.* **69A**, 90.
Olsen, H. A., and L. C. Maximon, 1978b, *Phys. Rev. A* **18**, 2517.
Olsen, H., L. C. Maximon, and H. Wergeland, 1957, *Phys. Rev.* **106**, 27.
Pencynski, P. E., and H. L. Wehner, 1970, *Z. Phys.* **237**, 75.
Powell, W. M., W. Hartsargh, and M. Hill, 1951, *Phys. Rev.* **81**, 213.
Pratt, R. H., 1960a, *Phys. Rev.* **117**, 1017.
Pratt, R. H., 1960b, *Phys. Rev.* **119**, 1619.
Pratt, R. H., 1960c, *Phys. Rev.* **120**, 1717.
Pratt, R. H., 1981a, *Comments At. Mol. Phys.* **10**, 121.
Pratt, R. H., 1981b, Electron Bremsstrahlung X-80: Theory and Recent Developments, in *Inner-Shell and X-Ray Physics of Atoms and Solids*, edited by D. Fabian, H. Kleinpoppen, and L. Watson (Plenum Press, New York),. p. 367.
Pratt, R. H., 1982, Electron Bremsstrahlung, in *Advances in X-Ray Spectroscopy* (contributions in Honor of Professor Y. Cauchois), edited by C. Bonnelle and C. Mande, (Pergamon, New York), p. 411.
Pratt, R. H., and I. J. Feng, 1984, Bremsstrahlung, in *Applied Collision Physics*, edited by C. J. Barnett (Academic, New York), Vol. II, p. 307.
Pratt, R. H., and C. M. Lee, 1977, *Phys. Rev. A* **16**, 1733.
Pratt, R. H., and H. K. Tseng, 1975, *Phys. Rev. A* **11**, 1797.
Pratt, R. H., H. K. Tseng, C. M. Lee, L. Kissel, C. MacCallum, and M. Riley, 1977, *At. Data Nucl. Data Tables* **20**, 175.
Pratt, R. H., H. K. Tseng, C. M. Lee, L. Kissel, C. MacCallum, and M. Riley, 1981, *At. Data Nucl. Data Tables* **26**, 477, erratum.
Quarles, C. A., and D. B. Heroy, 1981, *Phys. Rev. A* **24**, 48.
Racah, G., 1934a, *Nuovo Cimento* **11**, 461.
Racah, G., 1934b, *Nuovo Cimento* **11**, 467.
Rester, D. H., 1968, *Nucl. Phys.* **A118**, 129.
Rester, D. H., and W. E. Dance, 1967, *Phys. Rev.* **161**, 85.
Rester, D. H., and N. Edmonson, 1972, *Phys. Rev. A* **6**, 1405.
Rester, D. H., N. Edmonson, and Q. Peasley, 1970, *Phys. Rev. A* **2**, 2190.
Richtmyer, F. K., and E. H. Kennard, 1947, *Introduction to Modern Physics* (McGraw-Hill, New York).
Roentgen, W. C., 1945, Sitzungsber. der Würzburger Physik-Medic. Gesellsch. Jahrg. reprinted 1898, *Ann. Phys. (Paris)* **64**, 1.
Rohrlich, F., 1955, *Phys. Rev.* **98**, 181.
Rosenberg, L., 1983, *Phys. Rev. A* **27**, 1879.

ROZSNYAI, B. F., 1979, *J. Quant. Spectrosc. Radiat. Transfer* **22**, 337.
RUARK, A., AND H. UREY, 1930, *Atoms, Molecules, and Quanta* (McGraw-Hill, New York).
SAUTER, F., 1933, *Ann. Phys. (Leipzig)* **18**, 486.
SAUTER, F., 1934, *Ann. Phys. (Leipzig)* **20**, 404.
SCHAEFER, H. R., W. VON DRACHENFELS, AND W. PAUL, 1982, *Z. Phys.* **A305**, 213.
SCHEER, M., E. TROTT, AND G. ZAHS, 1968, *Z. Phys.* **209**, 68.
SCOFIELD, J., 1981, private communication.
SELTZER, S. M., 1983, private communication.
SELTZER, S. M., AND M. J. BERGER, 1982, National Bureau of Standard, Report No. NBSIR 82-2572.
SEMAAN, M., AND C. A. QUARLES, 1981, *Phys. Rev. A* **24**, 2280.
SEMAAN, M., AND C. A. QUARLES, 1982, *Phys. Rev. A* **26**, 3152.
SHAKESHAFT, R., AND L. SPRUCH, 1979, *Rev. Mod. Phys.* **51**, 369.
SILBERBACH, H., D. FUNNEMAN, AND H. MERZ, 1981, in *Inner Shell and X-Ray Physics of Atoms and Solids*, edited by D. Fabian, H. Kleinpoppen, and L. Watson (Plenum, New York), p. 525.
SOBCZAK, E., R. GOLDBERG, J. PELKA, AND J. AULEYTNER, 1981, in *Inner Shell and X-Ray Physics of Atoms and Solids*, edited by D. Fabian, H. Kleinpoppen, and L. Watson (Plenum, New York), p. 529.
SOMMERFELD, A., 1909, *Z. Phys.* **10**, 969.
SOMMERFELD, A., 1929, *Proc. Natl. Acad. Sci. USA* **15**, 393.
SOMMERFELD, A., 1931, *Ann. Phys. (Leipzig)* **11**, 257.
SOMMERFELD, A., 1939, *Atombau und Spectrallinien* (Vieweg und Sohn, Braunschweig).
SOMMERFELD, A., AND A. W. MAUE, 1935, *Ann. Phys. (Leipzig)* **22**, 629.
STAREK, B., H. AIGINGER, AND E. UNFRIED, 1972, *Phys. Lett.* **30A**, 151.
STARFELT, N., AND H. W. KOCH, 1956, *Phys. Rev.* **102**, 1598.
STARK, J., 1907, *Z. Phys.* **8**, 882.
STEPHENSON, S. T., 1957, *Handbuch der Physik* **30** (Springer-Verlag, Berlin), p. 337.
STOKES, G., 1898, *Proc. Manch. Lit. Phil. Soc.*,
THOMSON, J. J., 1897, *Phil. Mag.* **44**, 293.
THOMSON, J. J., 1898, *Phil. Mag.* **45**, 172.
THOMSON, J. J., 1910–11, Röntgen Rays, in *The Encyclopedia Britannica*, 11th edition.
TSENG, H. K., 1970, Ph.D. disseration, University of Pittsburgh.
TSENG, H. K., AND R. H. PRATT, 1971, *Phys. Rev. A* **3**, 100.
TSENG, H. K., AND R. H. PRATT, 1973, *Phys. Rev. A* **7**, 1502.
TSENG, H. K., AND R. H. PRATT, 1979, *Phys. Rev. A* **19**, 1525.
TSENG, H. K., AND R. H. PRATT, 1983 (unpublished).
TSENG, H. K., R. H. PRATT, AND C. M. LEE, 1979, *Phys. Rev. A* **19**, 187.
TUCKER, W. H., 1975, *Radiation Processes in Astrophysics* (MIT Press, Cambridge).
ULREY, C. T., 1918, *Phys. Rev.* **11**, 401.
WEBER, T. A., R. T. DECK, AND C. J. MULLIN, 1963, *Phys. Rev.* **130**, 660.
WEISHEIT, J. C., 1984, Atomic Phenomena in Dense Plasmas, in *Applied Atomic Collision Physics* (Academic, New York), Vol. II, p. 441.
WIEN, W., 1905, *Ann. Phys. (Leipzig)* **18**, 999.
WIEN, W., 1907, *Nachr. Göttinger Akad.*, 598.

CHAPTER 13

X-Ray and Bremsstrahlung Production in Nuclear Reactions

R. Anholt

1. INTRODUCTION

While the role of the atomic electrons in nuclear physics has been explored thoroughly, nuclear effects on atomic interactions have remained relatively unstudied. Atomic electrons can play an important role in γ-decay (internal conversion) or beta decay (electron capture) especially when the normal nuclear decay rates are small, e.g., when they are suppressed by unfavorable selection rules or energies. Aside from providing the all-important Coulomb field, nuclei have little effect on the atom, however. Through hyperfine interactions, nuclei perturb atomic energy levels slightly (and vice versa). These interactions have been the subject of many studies. In the case of atomic collisions, nuclear interactions play no role unless the two atomic nuclei come together, a relatively unlikely process. Atomic excitation normally occurs at impact parameters of the order of the atomic or inner-shell radii, which are enormous compared with the size of the nuclei. Also, in spite of this, atomic physicists generally avoid making measurements of electron excitation cross sections or probabilities in collisions where nuclear reactions are possible. Most measurements are made with low-energy projectiles that cannot overcome the nuclear Coulomb barrier.

Nevertheless, as early as 1960 experiments were proposed to detect nuclear-reaction effects in atomic collisions (Gugelot 1962; Ciochetti and Molinari, 1963, 1965; Eisberg *et al.*, 1960). Basically two different kinds of experiments were proposed. Gugelot (1962) suggested that it should be possible to detect united-atom (UA) X rays in collisions where relatively long-lived compound nuclei (CN) are formed. United-atom X-ray produc-

R. Anholt • Department of Physics, Stanford University, Stanford, California 94305.

tion requires that an inner-shell electron be excited while the projectile impinges on the target nucleus, and that the lifetime of the resulting CN should be approximately equal to the lifetime of the inner-shell vacancy. Gugelot's main interest in this experiment lay in the possibility of measuring extremely short (10^{-18} to 10^{-15} s) nuclear lifetimes. Atomic physicists are interested in determining the zero-impact-parameter inner-shell ionization probability for one-half of a collision. Normally only the full collision probability is obtainable, but the UA X-ray production probability is proportional to the probability of ionizing K electrons while the projectile is on the way into the collision. In addition, if X rays from short-lived superheavy CN could be observed, a new spectroscopic tool would be available to study atomic systems in which extreme relativistic effects ($Z \gtrsim 137$) are expected (Reinhardt and Greiner, 1977).

The other kind of experiment proposed is an interference experiment. Ciochetti and Molinari (1963, 1965) suggested looking for nuclear time-delay effects on inner-shell ionization due to the formation of CN. Here one determines the target K-shell ionization probability, by measuring target K X rays in coincidence with scattered particles (Burch *et al.*, 1974). Due to the formation of a long-lived CN, the amplitude for ionizing electrons while the projectile is on the way in can interfere with the amplitude for the way out, giving an altered ionization probability. Like the UA X-ray experiments, CN lifetimes and information about atomic inner-shell ionization amplitudes can be obtained in these measurements. CN lifetimes of the order 10^{-20} to 10^{-16} s can be determined. Information on the relative magnitudes of CN and direct scattering can also be obtained.

A similar kind of interference experiment was proposed in which nucleus–nucleus bremsstrahlung in nuclear reactions is measured instead of inner-shell ionization (Eisberg *et al.*, 1960). Interference effects on bremsstrahlung and ionization can be calculated using the same techniques; only the perturbation Hamiltonian is different, namely, the electromagnetic interaction for bremsstrahlung (Reinhardt *et al.*, 1976) and the Coulomb (plus recoil) interaction for ionization. Bremsstrahlung measurements have the advantage of being slightly more tunable than ionization measurements; one can measure different photon energies (different energy transfers) to investigate different nuclear time delays. In a K-shell ionization measurement, the energy transfer is approximately fixed for a given nucleus, so only a narrow range of time delays can be measured. Unfortunately, bremsstrahlung production is a less likely process than inner-shell ionization, so measurements are difficult.

Nevertheless all of these experiments have been done successfully in the past few years, and several similar new experiments have also been proposed. In addition, new theoretical developments have contributed greatly to our understanding of nuclear effects on atomic inner-shell ionization and bremsstrahlung production. Hence it is highly appropriate to review this material at the present time.

This review was originally to have been written by John S. Blair, a theoretical physicist instrumental in developing our understanding of the K-shell ionization interference experiments and in encouraging pertinent experiments (Blair *et al.*, 1978; Blair and Anholt, 1982). John's death in August, 1982 occurred before he could write this chapter. The present review is dedicated to his memory. The present author will sorely miss his guidance and inspiration in the further development of the theory of nuclear effects on inner-shell ionization in ion–atom collisions.

2. THEORY OF INTERFERENCE EXPERIMENTS

2.1. Basic Idea

The idea behind the time-delay interference experiments can be understood by using first-order time-dependent perturbation theory (Ciochetti and Molinari, 1963, 1965; Eisberg *et al.*, 1960). The probability for excitation from a state i to f in a collision is given by the square of the amplitude a_{if} (Merzbacher, 1967; Bang and Hansteen, 1959):

$$a_{if} = \int_{-\infty}^{\infty} dt \, \langle f | H' | i \rangle \, e^{i(\omega_i - \omega_f)t}, \qquad (1)$$

where H' is the perturbation Hamiltonian, $\hbar\omega_i$ and $\hbar\omega_f$ are the energies of states i and f, and the time integral is done along a Coulomb trajectory with relative ion velocity v and impact parameter b.

Now suppose that during the collision a CN is formed which lives for a time T. The integral over t can be separated into an interval $-\infty < t < 0$ during which time the projectile is coming into the collision (the projectile and target nuclei merge at $t = 0$), an interval $0 < t < T$ in which the two nuclei stick together, and an interval $t > T$ when the projectile is on the way out of the collision. While the projectile and target nuclei are merged into a UA, no excitation can take place. Hence the amplitude can be written as

$$a_{if} = a_{\text{in}} + e^{i\Delta\omega T} a_{\text{out}}, \qquad (2)$$

where

$$a_{\text{in}} = \int_{-\infty}^{0} dt \, \langle f | H' | i \rangle \, e^{i(\omega_i - \omega_f)t},$$

$$a_{\text{out}} = \int_{0}^{\infty} dt \, \langle f | H' | i \rangle \, e^{i(\omega_i - \omega_f)t}, \qquad (3)$$

and $\Delta\omega(0)=\omega_i-\omega_f$ is the energy difference. Upon squaring a_{ij}, the resulting probability

$$P_{if}(T)=|a_{\rm in}+e^{i\Delta\omega T}a_{\rm out}|^2 \tag{4}$$

is an oscillating function of the time delay T. A CN will usually decay with a distribution of sticking times, the most common being an exponential distribution, so that P_{if} must be averaged according to

$$P_{if}(\tau)=\tau^{-1}\int_0^\infty dT\exp(-T/\tau)\,P_{if}(T), \tag{5}$$

where τ is the CN lifetime.

It is most useful to consider two limits. If $\Delta\omega T\approx\Delta\omega\tau$ is much less than unity, P_{ij} is given by

$$P_{if}=|a_{\rm in}+a_{\rm out}|^2=\left|\int_{-\infty}^{\infty}dt\,\langle f|H'|i\rangle\,e^{i(\omega_i-\omega_f)t}\right|^2, \tag{6}$$

which is the normal excitation probability measured in the absence of nuclear interactions. If $\Delta\omega\tau\gg1$, however, terms in $\cos(\Delta\omega T)$ and $\sin(\Delta\omega T)$ go to zero upon averaging over T, hence we have

$$P_{if}=|a_{\rm in}|^2+|a_{\rm out}|^2. \tag{7}$$

The latter expression is an incoherent sum of incoming and outgoing excitation probabilties which, in general, is not equal to the square of the coherent sum of amplitudes for the case $\Delta\omega T\ll1$.

To realize this difference in a practical experiment it is best to compare the measured excitation probabilities in reactions with $\Delta\omega T\ll1$ and $\Delta\omega T\gg1$. This can be done by measuring the excitation probability as function of projectile energy E in the vicinity of a nuclear scattering resonance. For a resonance whose width Γ is much smaller than $\Delta\omega$ ($\Delta\omega/\Gamma=\Delta\omega\tau\gg1$) the excitation probability at the resonance energy E_R will differ from that away from the resonance energy where $\Delta\omega\tau\approx0$.

The main thing we learn from semiclassical perturbation theory is that the excitation probability will vary across nuclear scattering resonances if $\Delta\omega/\Gamma\gg1$. For K-shell ionization, $\Delta\omega$ is approximately equal to the K-shell binding energy E_K. For bremsstrahlung production $\Delta\omega$ is equal to the photon frequency. Unfortunately this is all that semiclassical theory predicts. A quantum-mechanical theory is needed to describe quantitatively how the excitation probability varies.

2.2. Survey of Atomic Collision Processes

Before turning to the quantum-mechanical theories, we briefly survey atomic collision processes to specify the initial and final states and the interaction Hamiltonian. We take the molecular-model view of ion–atom collisions (Lichten, 1967; Anholt and Meyerhof, 1977). Normally, the collisions of interest are sufficiently slow for the inner-shell electrons to adjust their motion to the presence of both the projectile and target nuclei, forming diatomic molecular orbitals (MOs). Excitation from MOs is analyzed using correlation diagrams that show how the molecular energy levels vary with internuclear distance R or time t (Fig. 1). In studies of K-shell ionization during proton- or alpha-particle-resonance nuclear scattering, the correlation diagram in Fig. 1b is generally qualitatively valid, since light nuclei perturb high-Z target atoms only slightly. It is important to recognize, however, that the UA electron binding energies (at $t=0$) are larger than the separated-atom (SA) energies (at $t=\pm\infty$). This not only shifts the energy of the UA K X ray, making UA X rays observable, but it also makes it difficult to excite the K electron to the continuum (Anholt and Meyerhof, 1977; Andersen *et al.*, 1976b).

Although convenient in principle, MOs do not allow calculations to be made easily, hence approximations are made in practice. The dependence of the K-shell ionization probability on impact parameter b implies that ionization occurs at internuclear distances R much smaller than the K-shell radius a_K, but larger than the nuclear radius R_N (Bang and Hansteen, 1959). One can therefore use UA wave functions and binding energies E_K at all R. There are several versions of this approximation (Andersen *et al.*, 1976b; Briggs, 1975; Anholt, 1980a). The simplest uses UA wave functions centered on the target nucleus (Andersen *et al.*, 1976b). The perturbation Hamiltonian is then given by the sum of the Coulomb interaction between the projectile nucleus (Z_1) and electron, and a recoil contribution:

$$H' = \frac{Z_1 e^2}{r} - \frac{Z_1 e^2}{|\mathbf{R}-\mathbf{r}|} + H_R. \tag{8}$$

In this equation, $Z_1 e^2/r$ is subtracted because the wave functions are eigenfunctions of $(Z_1+Z_2)e^2/r$ and the projectile-electron Coulomb interaction must not be overcounted. The recoil Hamiltonian H_R is just the potential due to the force of the acceleration of the target nucleus when struck by the projectile nucleus on the electron coordinate (Amundsen, 1978; Blair and Anholt, 1982):

$$H_R = \frac{m}{M_2} \nabla_R V_N(R) \cdot \mathbf{r}, \tag{9}$$

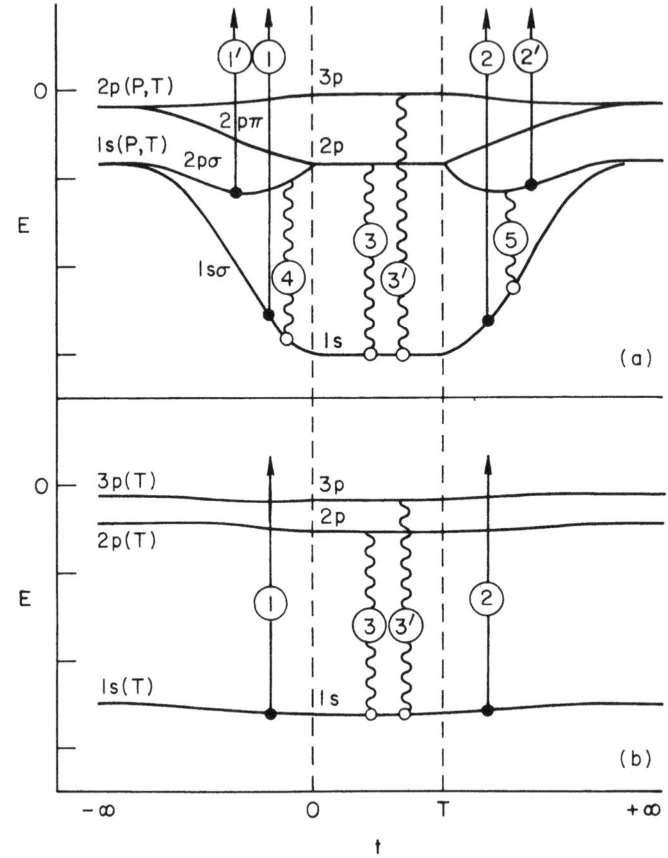

FIGURE 1. Correlation diagrams showing the electronic energy versus time as the orbitals evolve from the projectile and target $1s$ and $2p$ atomic orbitals, to the molecular $1s\sigma$, $2p\sigma$, and $2p\pi$ orbitals, then into UA $1s$, $2p$, and $3p$ orbitals. K-shell ionization occurs if electrons are excited out of the $1s\sigma$ or $2p\sigma$ MO's, either on the way in (processes 1 or 1′), or on the way out (2 or 2′) of the collision. UA $K\alpha$ (3) or $K\beta$ (3′) X-ray emission can occur if a vacancy is present during the time $0 < t < T$. At intermediate internuclear distances, MO X-ray emission can occur on the way in (4) or on the way out (5), (see Sec. 4.3). Figure 1a is for a symmetric collision $Z_1 = Z_2$. Figure 1b is for a very asymmetric collision, where $Z_1 \ll Z_2$.

where m is the electron mass, M_2 is the mass of the target nucleus, and the acceleration has been written as the derivative of the projectile–target nuclear interaction $V_{\mathcal{N}}(R)$.

Unfortunately, MOs are not appropriate in some nuclear reactions in which the projectile velocity exceeds the velocity of the K electron. No fundamental difficulty arises since one can also use separated-atom wave functions and binding energies E_K with the same perturbation Hamiltonian.

Inclusion of the Z_1e^2/r term in Eq. (8) is optional, as discussed below. Ionization probabilities calculated with UA and SA wave functions differ by 5% to 30% for most proton energies we consider.

For nucleus–nucleus bremsstrahlung the initial and final wave functions are projectile–target reltive-motion wave functions (Coulomb waves in the absence of nuclear interactions). The electromagnetic Hamiltonian in the dipole approximation (Eisberg *et al.*, 1960; Reinhardt *et al.*, 1976) is

$$H' \sim M\left(\frac{Z_1}{M_1} - \frac{Z_2}{M_2}\right)\sum_\alpha \hat{\varepsilon}_\alpha \cdot \mathbf{P}$$
$$\sim M\left(\frac{Z_1}{M_1} - \frac{Z_2}{M_2}\right)\hat{k} \times \mathbf{P}, \tag{10}$$

where $\mathbf{k}$ is the photon momentum, ω is the photon energy, $\mathbf{P}$ is the relative momentum, Z_2 is the target atomic number, M_1 and M_2 are the projectile and target masses, M is the reduced mass, and $\hat{\varepsilon}_\alpha$ is the photon polarization. For proton resonance scattering ($MZ_1/M_1 \approx 1$), dipole bremsstrahlung emission dominates, but for heavy ions with $Z_1/M_1 \approx Z_2/M_2$, quadrupole bremsstrahlung is also important (Reinhardt *et al.*, 1976). By Ehrenfest's theorem, $\mathbf{P}$ can be written as the derivative of the internuclear potential $\nabla V_N(R)$, consequently the treatment of bremsstrahlung emission is similar to that of the recoil contribution to K-shell ionization.

2.3. *Quantum-Mechanical Theory*

In the distorted-wave Born approximation (DWBA), the amplitude for exciting a K electron into a continuum state with energy ε and angular-momentum quantum numbers λ and μ is (Blair and Anholt, 1982)

$$A_{\varepsilon\lambda\mu}(E, \theta) = \frac{(KK')^{1/2}}{2\pi\hbar v}\langle \chi^-(\mathbf{K}', \mathbf{R})\,\phi_{\varepsilon\lambda\mu}(\mathbf{r})|\,H'\,|\chi^+(\mathbf{K}, \mathbf{R})\,\phi_{1s}(\mathbf{r})\rangle, \tag{11}$$

where $\mathbf{K}$ and $\mathbf{K}'$ are initial and final relative particle momenta, v is the initial ion velocity, $\phi_{\varepsilon\lambda\mu}$ and ϕ_{1s} are electronic wave functions, and $\chi^\pm(\mathbf{K}, \mathbf{R})$ are relative nuclear distorted waves. The differential cross section for K-shell ionization by projectiles with energy E scattering into a center-of-mass solid angle $d\Omega$ at θ is obtained by summing over continuum states:

$$\frac{d\sigma_K}{d\Omega} = \sum_{\lambda\mu}\int_0^\infty d\varepsilon\,|A_{\varepsilon\lambda\mu}(E, \theta)|^2. \tag{12}$$

The K-shell ionization probability is given by

$$P_K = \frac{d\sigma_K}{d\Omega} \bigg/ |f(E, \theta)|^2, \tag{13}$$

where $f(E, \theta)$ is the nuclear reaction scattering amplitude.

Several quantum-mechanical calculations of K-shell ionization have been made, which differ mainly in the nuclear wave functions used. McVoy and Weidenmüller (1982) treat the resonance as a bound state embedded in the projectile (usually proton) continuum. The scattering wave functions are then written as sums of Coulomb continuum wave and bound-state wave functions of the CN. Although this view is not strictly valid when the resonance is a nuclear shape resonance, McVoy and Weidenmuller (1982) show that the wave function can be cast into a similar form so that no separate treatment is necessary.

Feagin and Kocbach (1981) use asymptotic wave functions:

$$\chi^{\pm} = e^{i\mathbf{K}\cdot\mathbf{R}} + f^{\pm}(E, \Omega)\, R^{-1}\, e^{\pm iKR}. \tag{14}$$

This approximation greatly simplifies calculations, and is not too inaccurate because the K-shell ionization occurs at large internuclear distances (compared to the nuclear radii). The wave functions differ significantly from plane waves only at small internuclear distances where the scattering occurs. Blair and Anholt (1982) use a partial-wave expansion

$$\chi^{\pm} = 4\pi \sum_{lm} i^l Y^*_{lm}(\hat{K})\, Y_{lm}(\hat{R})(KR)^{-1}\, e^{i\delta_l} \mathscr{F}_l(KR), \tag{15}$$

where δ_l is the scattering phase shift (Coulomb plus a resonance contribution), K is the momentum, and $\mathscr{F}_l$ is the exact eigenfunction of the nuclear Hamiltonian containing $V_N(R)$. Here V_N is the Coulomb potential modified at small R by the nuclear interaction.

Although the forms of nuclear wave functions differ, all three theories use asymptotic waves to calculate K-shell ionization. With UA electronic wave functions, most perturbation matrix elements approach zero at small internuclear distance. One can hypothesize a matching radius R_m inside of which the electronic matrix elements vanish, giving no contribution to the ionization amplitude, but outside of which asymptotic wave functions can be used. For example, one can use Eq. (14) or (Blair and Anholt, 1982)

$$\mathscr{F}_l(KR) = \sin\left[KR - \eta \ln 2KR - l(\pi/2) + \delta_l(E)\right], \tag{16}$$

where η is equal to $Z_1 Z_2 e^2/v$. With this approximation the ionization amplitudes can be calculated simply. Furthermore the partial-wave series can be summed exactly.

The resulting K-shell ionization probability is

$$P_K=\sum_{\lambda\mu}\int_0^\infty \frac{d\varepsilon\,|b_\lambda D^\lambda_{\mu 0}(\theta)\,f(E,\theta)+(-1)^\lambda\,\delta_{\mu 0}b^*_\lambda f(E-\Delta E,\theta)|^2}{|f(E,\theta)|^2}, \tag{17}$$

where $D^\lambda_{\mu 0}(\theta)$ is the rotation matrix element, $\Delta E=E_K+\varepsilon_f$ is the energy transferred to the electron, b_λ is the ionization amplitude given by

$$b_\lambda=\frac{(2\lambda+1)^{-1/2}}{\hbar v}\int_{R_m}^\infty dR\,G_\lambda(R)\exp(i\bar{q}R), \tag{18}$$

$$\bar{q}R=(K-K')R-\eta\ln 2KR+\eta'\ln 2K'R, \tag{19}$$

G_λ is the radial electronic matrix element (Kocbach, 1976)

$$G_\lambda(R)=\int_0^\infty r^2\,dr\,R_{\varepsilon\lambda}R_{1s}\left[-\frac{r^\lambda_<}{r^{\lambda+1}_>}+\delta_{\lambda 0}\frac{1}{r}+\delta_{\lambda 1}mr\frac{M}{M_2}(\Delta E)^2R\right], \tag{20}$$

$R_{\varepsilon\lambda}$ and R_{1s} are radial electronic UA wave functions, and $r_<=\min(r,R)$. The $\delta_{\lambda 0}$ contribution to G_λ comes from the Z_1e^2/r contribution to the interaction Hamiltonian in Eq. (8). For $\lambda=0$, $G_\lambda(R)$ appropriately goes to zero as R vanishes, enabling one to use asymptotic waves. With SA electronic wave functions this contribution is not present, and $G_\lambda(R)$ does not vanish at small R. Nevertheless, Blair and Anholt (1982) show that nearly identical results are obtained if the required interior contribution to the monopole excitation amplitude is then evaluated.

The last term in Eq. (19) comes from the recoil Hamiltonian, where Ehrenfest's theorem or commutator relationships have been used to write

$$\left\langle\left|\frac{\partial V_N}{\partial R}\right|\right\rangle=M(\Delta E)^2\langle|R|\rangle. \tag{21}$$

Special consideration must be given to the recoil contribution because, although the matrix element proportional to R goes to zero for $R<R_m$, the original matrix element of $\partial V/\partial R$ proportional to R^{-2} is largest for $R<R_m$. This leads one to question the use of asymptotic wave functions here. Fortunately, a theorem due to Gerjuoy (1965) stating

$$\lim_{R_m\to\infty}\int_0^{R_m} dR\,\mathscr{F}_{l'}(KR)\,\mathscr{F}_l(KR)\frac{\partial V_N}{\partial R}=-E\cos\left[\delta_l-\delta_{l'}-(l-l')\frac{\pi}{2}\right] \tag{22}$$

rescues us from the dilemma. Equation (22) allows the partial-wave series to be evaluated exactly, giving the same results as one obtains using asymptotic waves and the matrix element proportional to R. The resulting recoil contribution to the ionization amplitude can be written as

$$b_\lambda^R \approx \frac{1}{\sqrt{3}} \frac{Mv}{M_2} m\langle r\rangle \delta_{\lambda 1}, \tag{23}$$

where $\langle r\rangle$ is the dipole matrix element given by

$$\langle r\rangle = \int_0^\infty r^3\, dr\, R_{\varepsilon 1} R_{1s}. \tag{24}$$

Since the electromagnetic interaction [Eq. (10)] involves a matrix element of $\partial V_N/\partial R$, the Gerjuoy theorem can also be employed to calculate bremsstrahlung emission, and the partial-wave series can be summed in the same way in which the recoil contribution was evaluated by Blair and Anholt (1982). With this method, one obtains (Anholt, to be published)

$$\frac{d^3P(\omega, \theta_\gamma, \theta)}{d\omega\, d\Omega\, d\Omega_\gamma} = \frac{\alpha}{(2\pi)^2} \left(\frac{Z_1}{A_1} - \frac{Z_2}{A_2}\right)^2 \frac{\beta^2}{\omega} \times \frac{|\hat{k} \times \hat{K}' f(E, \theta) - \hat{k} \times \hat{K} f(E - \hbar\omega, \theta)|^2}{|f(E, \theta)|^2}, \tag{25}$$

where $\hat{k}$ is the photon direction, α is the fine-structure constant, $\beta = v/c$, and ω is the photon frequency. Equation (25) was orginaly obtained by Feshbach and Yennie (1962) using a T-matrix approach.

The application of the Gerjuoy theorem is valid only if $K \approx K'$. For K-shell ionization $q = K - K'$ is always much less than K so the use of this theorem is valid. For bremsstrahlung emission where $q/K \sim \omega/2E$ may be large sometimes, e.g., $\omega = 300$ keV in 700- to 2000-keV $p + {}^{12}C$ collisions (Lesser *et al.*, 1982; Trial *et al.*, 1980), the Gerjuoy theorem cannot be used. To account for projectile energy loss is such cases Liou *et al.* (1978, 1980) go beyond the Feshbach–Yennie approximation, expanding the T matrix to obtain a correction term which depends on derivatives of the nuclear scattering amplitude. An approach not yet attempted might be to evaluate the matrix elements of $\partial V/\partial R$ numerically using exact Coulomb waves, as is done in calculations of Coulomb-nuclear interference (Griffy and Biedenharn, 1962). Evaluation of $E1$ nuclear Coulomb excitation also requires matrix elements of $\partial V/\partial R$.

2.4. *Discussion*

Equations (17) and (25) have a simple interpretation. The numerator of each gives the differential excitation cross section for a collision in which the projectile scatters to an angle θ while the denominator, $|f(E, \theta)|^2$, gives the scattering cross section into the angle θ. The ratio therefore appropriately expresses the excitation probability. If ionization or bremsstrahlung emission occurs while the projectile is coming into the collision (with an amplitude proportional to b_λ^* or $\hat{k} \times \hat{K}$), the energy ΔE is lost before the projectile scatters from the target nucleus, so the energy at which f is evaluated is appropriately reduced. If ionization or emission occurs on the way out of the collision (with an amplitude b_λ or $\hat{k} \times \hat{K}'$), the projectile has already scattered with its initial energy.

Either Eq. (17) or (25) can be written as

$$P_{if} \sim |a_{\text{out}} + a_{\text{in}} f(E - \Delta E, \theta)/f(E, \theta)|^2, \tag{26}$$

where a_{out} and a_{in} are outgoing and incoming ionization or emission amplitudes. Equation (26) can be compared to the semiclassical approximation (Blair and Anholt, 1982; Liu *et al.*, 1982)

$$P_{ij} = |a_{\text{out}} + a_{\text{in}} \exp(-i\Delta\omega T)|^2. \tag{4'}$$

The ratio of scattering amplitudes and the exponential can be expanded:

$$\begin{aligned} \frac{f(E-\Delta E)}{f(E)} &\approx 1 - \Delta E \frac{\partial \ln f(E)}{\partial E} = 1 - i\Delta\omega\tau_{\text{QM}}(E, \theta), \\ e^{-i\Delta\omega T} &\approx 1 - i\Delta\omega T. \end{aligned} \tag{27}$$

These expansions indicate the relation between the semiclassical time-dependent and quantum-mechanical energy-dependent approaches. We note, however, that, whereas the semiclassical time delay T is a real number (that must be averaged over an exponential or some other distribution of decay times), τ_{QM} is a complex energy- and angle-dependent quantity. The quantum-mechanical theory allows the excitation probability to be calculated quantitatively at any energy and angle.

To analyze measured excitation probabilities, the nuclear-scattering cross section $d\sigma/d\Omega \sim |f(E, \theta)|^2$ must also be determined. For elastic scattering $f(E, \theta)$ can be written as the sum of a Coulomb scattering amplitude and a resonant Breit–Wigner contribution:

$$f = f_{\text{Coul}} + f_{\text{res}}, \tag{28}$$

where, for $s_{\frac{1}{2}}$ resonances

$$f_{\text{res}} \approx \Gamma_p (E - E_R + \tfrac{1}{2} i\Gamma)^{-1}. \tag{29}$$

The nuclear scattering amplitude is fitted to the measured scattering cross section by adjusting the total width Γ, partial width Γ_p, resonance energy E_R, and perhaps background or optical phase shifts (Chemin *et al.*, 1981). The fitted amplitudes are then inserted into Eqs. (17) or (25) to obtain the ionization or emission probabilities. One-electron Dirac UA wave functions are often used to calculate the ionization amplitudes.

Equations (17) and (25) should be used only to calculate excitation probabilities in collisions with large scattering angles (small impact parameters). The use of asymptotic wave functions at large l values, important for small scattering angles, is not expected to be valid.

3. INTERFERENCE EXPERIMENTS

3.1. *K-Shell Ionization in Elastic Proton Scattering Reactions*

The K-shell ionization probability is determined by measuring coincidences between target K X rays or Auger electrons and elastically scattered protons. These measurements are difficult because generally the probabilities are small ($\sim 10^{-3}$ for $p+$Ni, Sr collisions) but the K-shell ionization cross sections are enormous, leading to a relatively large number of random coincidences. To maximize the coincidence counting rate with a minimum random rate, large X-ray and proton-detector solid angles are used. Fast–slow timing techniques are used, so that all of the true coincidences fall into a narrow time window. With all of these tricks one nevertheless obtains a true-to-random-coincidence rate of only approximately 1:1 with a 20-pA proton beam in 6-MeV ^{88}Sr(p, p_0) collisions. Each ionization probability requires 48 h to measure. If one could measure coincidences with only those particles which lose energy ΔE to electronic excitation or bremsstrahlung emission, the random coincidence rate would be lower, allowing a much larger beam current to be used, hence the measurement could be completed more rapidly. Unfortunately, for K-shell ionization the energy loss is much smaller than the proton energy. Solid-state proton detectors cannot resolve this energy loss, so that magnetic spectrometers are required (Laegsgaard *et al.*, 1972). For bremsstrahlung measurements, on the other hand, where ΔE is much larger, this technique works well.

K-shell ionization has been measured in $p+{}^{58}$Ni reactions at 3.15 MeV (Blair *et al.*, 1978), $p+{}^{88}$Sr reactions at 5.01 MeV (Chemin *et al.*, 1981), $p+{}^{12}$C reactions at 0.461 MeV (Duinker *et al.*, 1980; Meyerhof *et al.*, 1983) and $p+{}^{88}$Sr reactions at 6 MeV (Chemin *et al.*, 1982). The $p+{}^{58}$Ni results (Fig. 2) were the first to demonstrate the nuclear time delay effect on K-shell ionization. Here $\Delta\omega\tau \sim E_K/\Gamma$ is approximately equal to 1.5, and accordingly the ionization probability varies strongly across the resonance. The ^{58}Ni

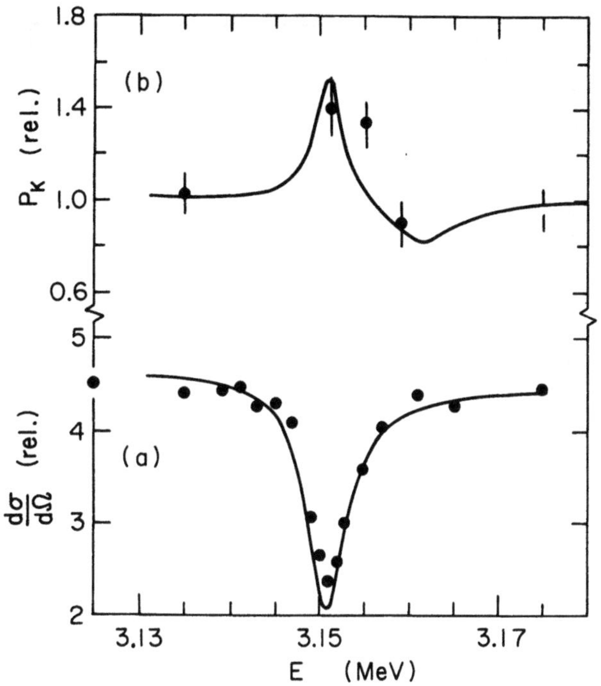

FIGURE 2. (a) Observed and calculated elastic scattering cross section for $p+{}^{58}Ni$ collisions. (b) Relative K-shell ionization probability. The solid curve is calculated with Eq. (17) using atomic ionization amplitudes computed with Dirac one-electron wave functions. (From Blair *et al.*, 1978.)

resonance is an $s_{\frac{1}{2}}$ nuclear resonance. Measurements were also made for ${}^{88}Sr(p, p_0)$ at 5.06 MeV, a $d_{\frac{5}{2}}$ resonance with a similar value of E_K/Γ ($\sim$0.84), to investigate whether theory adequately accounts for the spin of the resonance (Chemin *et al.*, 1981). Both sets of measurements agree well with theory.

The results of Duinker *et al.* (1980) for $p+{}^{12}C$ collision near 461 keV (Fig. 3) caused much consternation among theorists, because in this case in which E_K/Γ is very small (0.008), no measurable time delay effect on K-shell ionization is expected. The reaction has recently been remeasured by Meyerhof *et al.* (1983) using a gaseous instead of solid target; their new results are consistent with theory. Meyerhof *et al.* (1983) suggested that Duinker's results might be due to proton scattering from something other than ${}^{12}C$ in coincidence with background electrons.

In addition two other measurements were made for which $E_K/\Gamma \ll 1$. K-shell ionization was measured in 6.06-MeV $p+{}^{88}Sr$ collisions with $E_K/\Gamma \simeq 0.22$ (Chemin *et al.*, 1983). Their results agree well with theory.

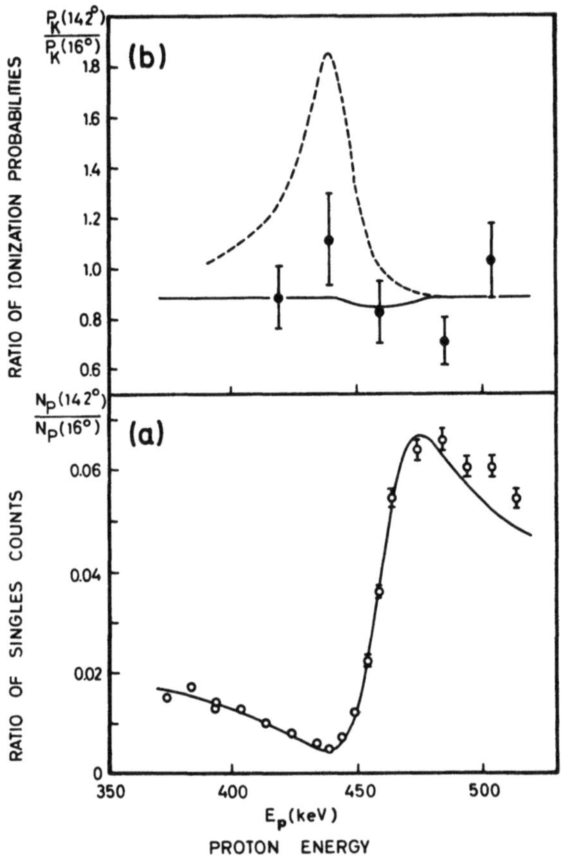

FIGURE 3. (a) Ratio of observed and calculated scattering cross sections at 142° and 16° for $p + {}^{12}C$ collisions. (b) Ratio of ionization probabilities at 142° and 16°. In each curve, the resonance scattering contribution is negligible at 16°, so that a 16° measurement serves to normalize the curve. The points show the data of Meyerhof *et al.* (1983). The solid curves were calculated using Eq. (17). The measurements of Duinker *et al.* (1980) were made at a different angle, 125°. The dashed curve was calculated with a proposed modification of Eq. (17) that fit the 125° data of Duinker *et al.* (1980).

Horsdal-Pedersen *et al.* (1982) measured target K-electron capture by protons in $p + {}^{12}C$ collisions near the resonances at 1.7 MeV. For electron capture, the relevant energy transfer is $\Delta E \approx E_K + \frac{1}{2}mv^2$, where $\frac{1}{2}mv^2$ is the relative kinetic energy of the electron when captured by the proton traveling at a velocity v. Although the quantum-mechanical theory of electron capture has not been worked out, one expects that the capture probability should be nearly constant across this resonance where $\Delta E/\Gamma(\simeq 0.008) \ll 1$. Indeed, this was found to be so.

3.2. *Bremsstrahlung Emission near Elastic Proton Resonances*

Although bremsstrahlung emission probabilities are much smaller than K-shell ionization probabilities ($\sim 10^{-9}\,\mathrm{keV}^{-1}$ versus $\sim 10^{-3}$), measurements with larger beam currents (0.4 μA) can be done thanks to a smaller random coincidence rate. Fast–slow timing techniques are still required, however. Bremsstrahlung has been measured near resonances at 1.7 and 0.46 MeV in $p + {}^{12}\mathrm{C}$ collisions (Lesser *et al.*, 1982; Trail *et al.*, 1980; Maroni *et al.*, 1976; Taketani *et al.*, 1982). At 1.7 MeV there are two closely spaced resonances corresponding to the $5/2^{+}$ and $3/2^{-}$ levels in the CN ^{13}N. At photon energies of 100 keV, ω/Γ is approximately 1.5 in these collisions. The results for the 1.7-MeV resonance are in good agreement with the extended theory of Feshbach and Yennie (1962; Liou *et al.*, 1980) which includes corrections for the projectile-energy loss. These corrections are small for the 1.7-MeV resonances.

The results for the 0.46-MeV resonance in ^{12}C do not agree well with theory. Figure 4 illustrates that the leading term alone in the Feshbach–Yennie approximation [Eq. (25)] agrees better with experiment than the full result including the energy-loss corrections (Lesser *et al.*, 1982). The corrections involve derivatives of the reaction scattering amplitudes with respect to ion velocity, weighted by the ratio of the energy loss ΔE to the total energy E. As these derivatives vary strongly near the resonance, and the weighting factor $\Delta E/E$ is large, one might question whether the expansion of the T-matrix element is accurate here. Also, projectile Coulomb-deflection effects become important when energy-loss effects are significant (Anholt *et al.*, 1979). It is not clear how accurately these are taken into account in the full Feshbach–Yennie theory.

3.3. *Ionized Electron Measurements*

Bremsstrahlung measurements are advantageous because one can measure the emission probability as a function of ω at a single projectile energy and traverse the resonance, instead of making separate measurements at several projectile energies as one does for K-shell ionization. An analogous technique can be applied for K-shell ionization by measuring the ionized electron distribution, $dP_K/d\varepsilon$. Triple coincidences between target K X-rays, scattering protons, and electrons must be measured to establish that the secondary electrons come from K-shell ionization. Figure 5 shows calculations of $dP_K/d\varepsilon$ for the $s_{\frac{1}{2}}$ resonance in ^{58}Ni. There are two problems with such measurements. Unlike the bremsstrahlung spectrum, which falls off slowly (as $1/\omega$) with photon energy, the ionized-electron spectrum falls off very quickly, so that only a region from 0 to 10 keV is measurable in ^{58}Ni collisions (where $E_K \sim 8$ keV). The resonance in ^{58}Ni is 5.6 keV wide;

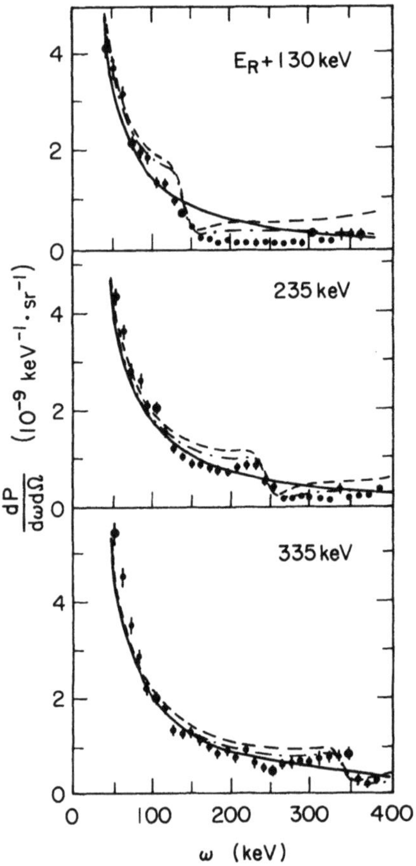

FIGURE 4. The bremsstrahlung emission probability in $p+{}^{12}C$ collisions near the resonance at $E_R = 461$ keV. Three sets of data are plotted on the same γ-ray energy scale in order to show the shift of the resonance structure with the incident-proton energy. The dashed curves represent the cross section calculated from the full Feshbach–Yennie approximation (Liou *et al.*, 1980), which includes both the principal term and the correction term. The chain curves represent a calculation using only the principal term of the Feshbach–Yennie approximation, Eq. (25). The solid curve was calculated neglecting resonance effects. (From Lesser *et al.*, 1982.)

therefore only a very broad variation on the steeply falling distribution is seen. The resonance effect could be observed more readily if the resonance were narrower, for example if the CN had a width $\Gamma = 1$ keV, as shown in Fig. 5b. The second difficulty is that the differential probability, $\sim 10^{-5}$ keV^{-1}, is much smaller than the total K-shell ionization probability (10^{-3}) so that the triple-coincidence measurement will be much more difficult to do.

3.4. *Isobaric Analog Resonances*

Isobaric analog resonances (IARs) such as those seen in $p+{}^{88}Sr$ scattering are made up of many narrow, unresolved resonances (Bilpuch *et al.*, 1974). In theories of CN reactions involving IARs, it is customary to write

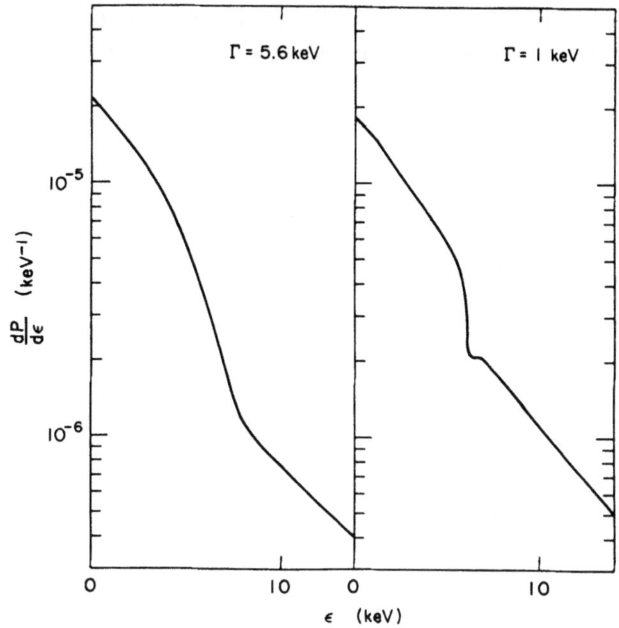

FIGURE 5. Ionized electron-production probability, $dP_K/d\varepsilon$, versus electron kinetic energy for $E = 3165$ keV $p + {}^{58}Ni$ collisions, calculated with Eq. (17) using the actual resonance width width of $\Gamma = 5.6$ keV and assuming a ficticious width $\Gamma = 1$ keV. The resonance energy is $E_R = 3151$ keV, so the elastic resonance scattering dip occurs at $\varepsilon = 6$ keV where $E = E_R + E_K + \varepsilon$.

the nuclear reaction amplitude as (Mahaux and Weidenmüller, 1979; Anholt *et al.*, 1982)

$$f(E, \theta) = \bar{f} + f_{fl}, \tag{30}$$

where $\bar{f}$ is the average of the total reaction amplitude over the beam energy resolution I,

$$\bar{f} = \langle f \rangle = I^{-1} \int_{E - I/2}^{E + I/2} dE\, f(E, \theta), \tag{31}$$

and f_{fl} is a rapidly fluctuating amplitude which averages to zero in the same energy interval. The nuclear scattering cross section averaged over the energy resolution is

$$\left\langle \frac{d\sigma}{d\Omega} \right\rangle = |\bar{f}(E, \theta)|^2 + \langle |f_{fl}(E, \theta)|^2 \rangle = \sigma_D + \sigma_{CN}, \tag{32}$$

where σ_D is the differential direct elastic scattering cross section, and σ_{CN} is the compound elastic cross section.

Inserting these amplitudes into Eq. (17) and averaging the differential K-shell ionization cross section, the ionization probability is given by

$$P_K = \frac{\sigma_K^D + \sigma_K^{\mathrm{CN}}}{\sigma_D + \sigma_{\mathrm{CN}}}, \tag{33}$$

where

$$\sigma_K^D = \sum_{\lambda\mu} \int_0^\infty d\varepsilon \, |\bar{f}(E,\theta)\, b_\lambda D^\lambda_{\mu 0}(\theta) + (-1)^\lambda \delta_{\mu 0} b_\lambda^* \bar{f}(E-\Delta E,\theta)|^2, \tag{34}$$

and

$$\sigma_K^{\mathrm{CN}} = \sum_\lambda \int_0^\infty d\varepsilon \, |b_\lambda|^2 [\sigma_{\mathrm{CN}}(E,\theta) + \sigma_{\mathrm{CN}}(E-\Delta E,\theta)]. \tag{35}$$

For the direct part of P_K, σ_K^D must be evaluated using a coherent sum of amplitudes since $\bar{f}$ varies over a width $\Gamma_{\mathrm{IAR}} \approx \Delta E$. Since the widths of the narrow resonances contributing to f_{fl} are always much smaller than ΔE, however, implying that the time delay is always so long that the information about the phase of the ionization amplitude is lost, we form the incoherent sum of incoming and outgoing ionization probabilities to obtain σ_K^{CN}. In deriving Eqs. (33) to (35) the expression

$$\langle f_{\mathrm{fl}}^*(E) f_{\mathrm{fl}}(E-\Delta E)))\rangle = \frac{\Gamma_c}{\Gamma_c + i\Delta E}, \tag{36}$$

is used (Ericson and Mayer-Kuckuk, 1966). Here Γ_c is the Ericson correlation width, which is related to the average width of the many narrow resonances contributing to f_{fl} (Mahaux and Weidenmüller, 1979). Near $Z = 38$, Γ_c is only a few eV and ΔE is larger than 16 keV, so these terms can be neglected. For smaller target atomic numbers or larger CN excitation energies, the correlation width increases, so that this approximation may break down.

Figure 6 shows calculations for $p + {}^{86}\mathrm{Sr}$ elastic scattering. For elastic scattering from IARs, $\bar{f}$ is written as the sum of a Coulomb-scattering contribution and a single Breit–Wigner resonance contribution with a width Γ_{IAR} which here is of the order of ΔE. The compound elastic cross section also has a resonance shape (Hofmann *et al.*, 1975; King and Slater, 1977)

$$\sigma_{\mathrm{CE}}(\theta) = \sigma_0 \left(\frac{\Gamma_{\mathrm{IAR}}}{2}\right)^2 \left[(E-E_R)^2 + \left(\frac{\Gamma_{\mathrm{IAR}}}{2}\right)^2\right]^{-1} \tag{37}$$

with a similar width Γ_{IAR}. At $\theta = 90°$, the compound elastic contribution

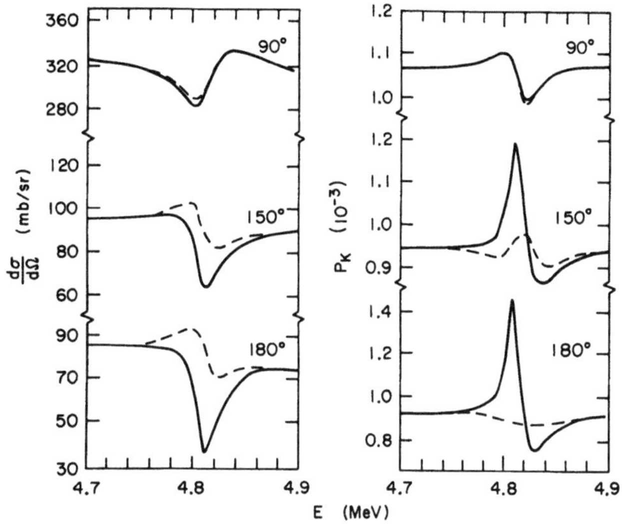

FIGURE 6. Elastic scattering differential cross sections and K-shell ionization probabilities for $p + {}^{86}Sr$ reactions for three scattering angles. Solid curves: direct scattering only; dashed curves: compound elastic + direct scattering. (From Anholt *et al.*, 1982.)

affects the total elastic scattering cross section only slightly. P_K is affected similarily, hence one can neglect CN scattering in analyzing the $\theta = 90°$ ${}^{88}Sr$ measurements (Chemin *et al.*, 1981, 1982). At large angles, however, σ_{CE} contributes significantly to the total scattering cross section. The ionization probability calculated omitting the CN contribution has a large peak, but when σ_{CE} is included the resulting total ionization probability is nearly constant across the resonance.

Chemin *et al.* (1983) have pointed out that measurements of K-shell ionization near IARs are sensitive to the magnitude of the compound-elastic cross section. This is illustrated in Fig. 7, where the plotted ionization probability for 5.06-MeV $p + {}^{88}Sr$ was calculated assuming different values of σ_{CE}. The P_K measurements indicates that the peak value of σ_{CE} [denoted by σ_0 in Eq. (37)] is approximately equal to 40 mb/sr. For IARs with $j\pi \neq \frac{1}{2}^+$, σ_{CE} can also be obtained from a measurement of the analyzing power scattering cross section with polarized proton beams (King and Slater, 1975). For the $5/2^+$ ${}^{88}Sr$ resonance at 5.06 MeV, the polarized-beam results agree well with the ionization measurements. K-shell ionization measurements can give compound-elastic cross section for $s_{\frac{1}{2}}$ IARs, also, whereas the analyzing power method cannot be applied in those cases.

Bremsstrahlung emission probabilities also are sensitive to CN scattering. Consider a measurement where protons are backscattered ($\theta = 180°$, $\hat{k} = -\hat{K}$), and the photon is emitted at 90° to the collision plane

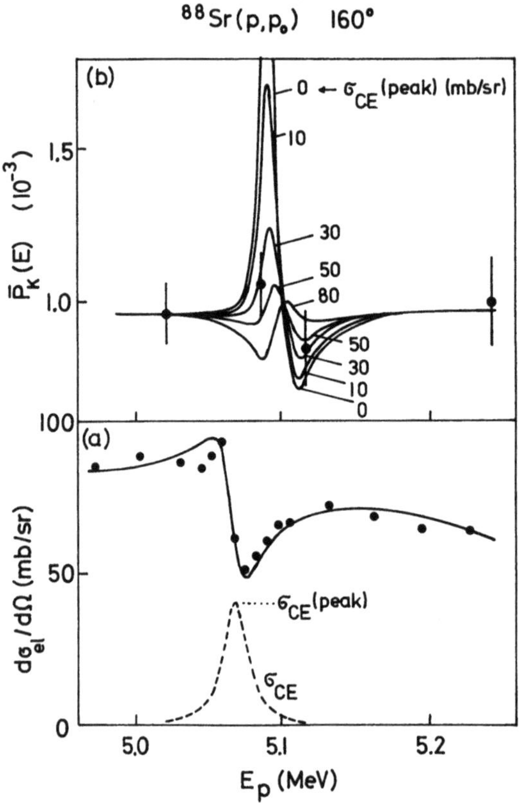

FIGURE 7. (a) K-shell-ionization probability for 160° elastically scattered protons in the $p + {}^{88}Sr$ reaction near 5.06 MeV. The curves are calculated with Eq. (33) using various assumed peak values for the compound-elastic cross section. (b) Elastic scattering cross section at 160° as a function of proton energy. The cross section has been fitted assuming a compound-elastic cross section with a peak value of 40 mb/sr in accord with King and Slater (1975). (From Chemin *et al.*, 1983.)

($|\hat{k} \times \hat{K}| = 1$). From Eq. (25) the relative bremsstrahlung emission probability is (Anholt, unpublished):

$$\frac{\omega\, d^2P}{d\Omega\, d\Omega_\gamma}(\theta = 180°, \theta_\gamma = 90°)$$

$$\sim \frac{|\bar{f}(E,\theta) + \bar{f}(E-\omega,\theta)|^2 + \sigma_{\mathrm{CE}}(E) + \sigma_{\mathrm{CE}}(E-\omega)}{|\bar{f}(E,\theta)|^2 + \sigma_{\mathrm{CE}}(E)}. \quad (38)$$

Figure 8 shows calculations for the ${}^{86}Sr$ resonance at 4.81 MeV. (For the scattering cross section see Fig. 6.) The projectile energy here is 4.9 MeV; therefore, the resonance pattern appears in the bremsstrahlung spectrum at ~90 keV. Here the photon energy is much smaller than the projectile energy, hence the first-order Feshbach–Yennie approximation [1962; Eq. (25)] should be valid. Like K-shell ionization, the bremsstrahlung-emission probability varies when different magnitudes, σ_0, of the compound elastic cross section are assumed.

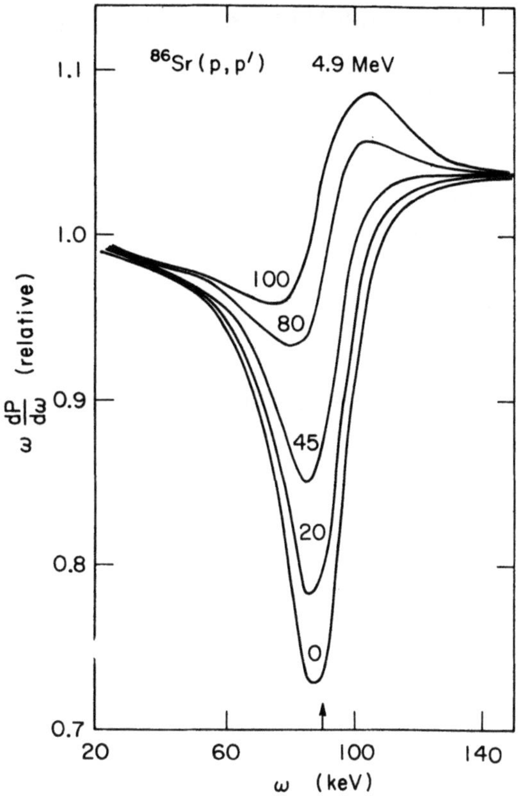

FIGURE 8. Bremsstrahlung-emission probability in $E = 4900$ keV $p + {}^{86}$Sr collisions calculated assuming various relative magnitudes of the compound elastic scattering cross section (0 to 100 mb/sr at the peak). The dip occurs where the photon energy is equal to $E - E_R$.

3.5. K-Shell Ionization by Neutrons

Although with neutral projectiles there is no Coulomb interaction with target electrons, K electrons can nevertheless be excited through the recoil interaction. We first consider nonresonant scattering where $f(E) \approx f(E - \Delta E)$. The neutron-induced ionization probability, from Eqs. (17) and (22), is

$$P_K = \frac{4}{3} \int_0^\infty d\varepsilon \left(\frac{Mv}{M_2}\right)^2 |m\langle r\rangle|^2 \sin^2 \frac{\theta}{2}. \tag{39}$$

This probability can be calculated accurately using tabulated K-shell photoelectric cross sections (McMaster *et al.*, 1969) σ_{KPE} which are related to $\langle r \rangle$ [Eq. (23)]:

$$\sigma_{\mathrm{KPE}}(\omega = E_K + \varepsilon) = (2\pi)^2\, \alpha(E_K + \varepsilon)\,|\langle r \rangle|^2. \tag{40}$$

Furthermore since $\sigma_{\mathrm{KPE}}(\omega)$ varies as ω^{-s}, where s is a constant ($\sim$2.5), we can integrate over continuum energies ε, obtaining for $\theta = 180°$

$$P_K(180°) = \frac{4}{3}\left(\frac{Mv}{M_2}\right)^2 \frac{m^2}{(2\pi)^2 \alpha s}\, \sigma_{K\mathrm{PE}}(\varepsilon_K). \tag{41}$$

Figure 9 shows neutron-induced K-shell ionization probabilities for 1-MeV neutrons plotted against the target atomic number Z_2. One can also

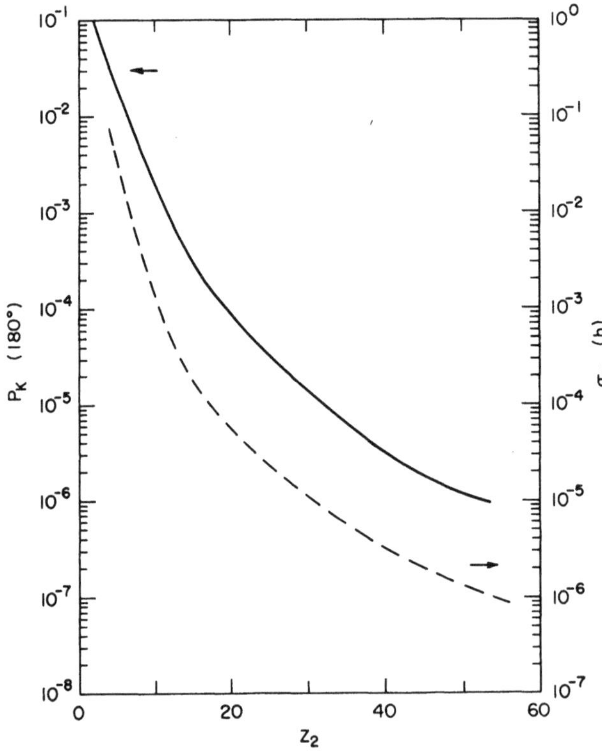

FIGURE 9. Probability at $\theta = 180°$ (solid curve) and total cross section (dashed curve) for K-shell ionization by 1-MeV neutrons versus target atomic number. For other neutron energies, these values need only be multiplied by the neutron energy in MeV.

integrate over the scattering angle to obtain the total neutron-induced K-shell ionization cross section:

$$\sigma_{Kn} = \tfrac{1}{2}\sigma_n P_K(180°), \tag{42}$$

where σ_n is the elastic neutron scattering cross section. Figure 9 also shows the neutron-induced K-shell ionization cross section, calculated assuming that $\sigma_n = \pi R_N^2$, with $R_N = 1.4(1 + A_2^{1/3})$ fm, where A_2 is the mass number of the target nucleus.

For $Z_2 < 30$, the K-shell ionization cross sections are greater than 5 μb. Proton-induced K X-ray cross sections of this magnitude are not too difficult to measure. Unfortunately, intense well-focused neutron beams are not obtained as easly, hence the K X-ray count rate will undoubtedly be extremely small, and the magnitude of the room- and neutron-induced low-energy γ-ray background will determine whether such a measurement is feasible. One unsuccessful measurement has been attempted (McMurray and Lutz, 1981).

For resonance scattering many neutron resonances tend to have $j\pi = \frac{1}{2}^+$ so that the nuclear reation amplitude $f(E, \theta)$ is (Blatt and Weisskopf, 1962)

$$f(E, \theta) = K^{-1}\left[\sin(KR_N)e^{iKR_N} + \frac{\Gamma/2}{E - E_R + i\Gamma/2}\right] \tag{43}$$

The most important feature here is that $f(E, \theta)$ is independent of the scattering angle θ. Hence we can integrated over θ to obtain the total neutron-induced K-vacancy production cross section. Defining R_K as the K X-ray cross section divided by the neutron scattering cross section, we find

$$R_K = \int_0^\infty d\varepsilon \left(\frac{Mv}{M_2}\right)^2 \frac{|m\langle r\rangle|^2}{3}\{1 + |f(E - \Delta E)/f(E)|^2\}. \tag{44}$$

Although the neutron-induced K X-ray production cross section is difficult to measure, it is no more difficult to measure the time-delay effect, since one has just to determine the total neutron-induced K-shell ionization cross section instead of a probability.

Figure 10 illustrates the expected probability for a typical neutron resonance reaction, for various values of $y = E_K/\Gamma$. There are two peaks in P_K. One peak is always at the point where σ_n goes to zero due to interference between the hard-sphere and resonance contributions to the scattering amplitude. Although $f(E)$ vanishes here, $f(E - \Delta E)$ does not, hence the ratio $|f(E - \Delta E)/f(E)|^2$ peaks. The second peak in R_K occurs where $|f(E - \Delta E)|^2$ peaks; it is shifted by an amount ΔE from $E = E_R$. The dip in R_K occurs at $E = E_R$ where $f(E)$ peaks but $f(E - \Delta E)$ does not, so that $|f(E - \Delta E)/f(E)|^2 \approx 0$.

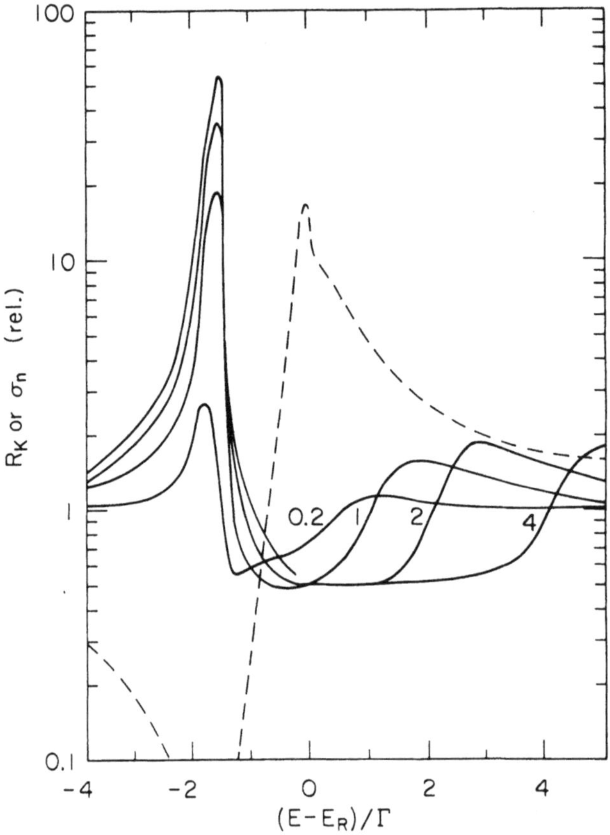

FIGURE 10. Total neutron scattering cross section (dashed curve) and ratios R_K of the K-shell ionization cross section to the neutron scattering cross section for various values of $y=E_K/\Gamma$ (solid curves). For this resonance we took $KR_N=0.3$ in Eq. (3). On these relative scales the results are approximately independent of atomic number.

3.6. Cross-Section Measurements

In most of the collisions discussed so far, coincidences between K X-rays, Auger electrons, or bremsstrahlung and the scattered particle were measured to select the events in which excitation and nuclear scattering occur simultaneously. However, electronic effects in nuclear reactions have been observed also in noncoincidence experiments. The earliest example of this is in $^4He^+ + {}^4He$ scattering near the ^8Be resonance at ~ 180 keV (Benn *et al.*, 1968; Maglich, 1973). Here, the position of the resonance is shifted due to the rearrangement of the electrons during the collision. To illustrate this effect, we suppose there is one electron ($He^+ + He^{2+}$) and that the collision is sufficiently slow so that the MO picture is valid (Feagin and Kocbach, 1982). The relevant correlation diagram would then be that shown in Fig. 1a. The single electron can go into the $1s\sigma$ MO, correlating to the $1s$ orbital of the UA, or into the $2p\sigma$ MO, correlating to the UA $2p$ level. If it goes into the $1s\sigma$ MO, the electron becomes more tightly bound at $R=0$ by an amount $\Delta E = E_K(\text{UA}) - E_K(\text{SA}) = 163$ eV, which causes the projectile center-of-mass energy to increase. If the electron goes into the $2p\sigma$ MO, the binding energy change is $E_L(\text{UA}) - E_K(\text{SA})$, which is zero. Thus the 6.3-eV-wide resonance will split into two resonances, 163 eV apart. Where $He^+ + He$ scattering was measured, there are three electrons with other possible energy differences, and the observed resonance pattern splits into the peaks as shown in Fig. 11. [In calculating the shape of this resonance, the sudden approximation was used instead of the MO model, because the velocity of the projectile is large relative to that of the He $1s$ electrons. The possibility that the recoil contribution to the perturbation Hamiltonian excites dipole electronic transitions was also included (Meyer *et al.*, 1964).]

In 14.2-MeV $p + {}^{12}C$ collisions, the K-shell ionization probability, due mostly to the recoil contribution, is 0.22 for $\theta = 180°$ (Briggs and Lane, 1982; see also Fig. 9). The measured resonance-scattering pattern is broadened due to the superposition of two patterns: one of relative magnitude 0.78 centered at the resonance energy and another of relative magnitude 0.22 shifted by an amount $\Delta E \approx E_K \approx 450$ eV that can be calculated from Eq. (17). The width of the resonance at 14.2 MeV has been measured in (p, p) scattering reactions and in (p, γ) reactions using a thick-target-yield branching-ratios method. Thompson *et al.* (1981) have suggested

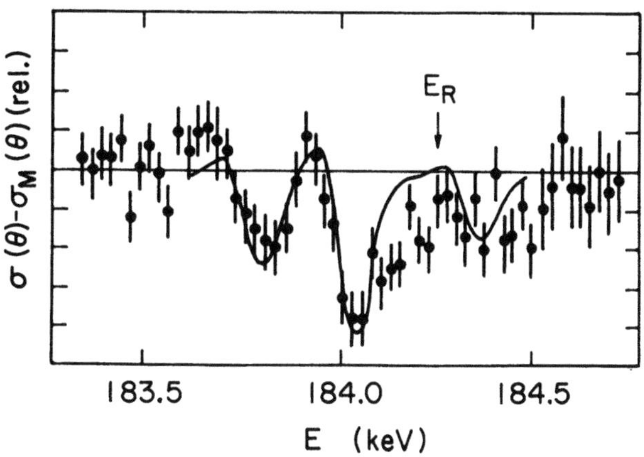

FIGURE 11. $He^+ + He$ elastic scattering cross section (from which a smooth Coulomb scattering cross section has been subtracted) versus projectile energy. The 6.3-eV-wide ^8Be resonance splits into three resonances due to electron rearrangement. (From Benn *et al.*, 1968.)

that the larger (p, p) width [1.6 keV for (p, γ)], after correcting for the beam-energy resolution and the Doppler broadening caused by the vibration of the ^{12}C nucleus in the carbon solid, is due to the electronic ionization contribution.

Briggs and Lane (1982) have suggested that the additional width is caused by the recoil of the ^{12}C nucleus correlated to the unperturbed motion of the inner-shell electrons. For a K electron of velocity $\mathbf{v}_K$, the ^{12}C nucleus recoils with a velocity $\mathbf{v}_T = m/M_2 \mathbf{v}_K$, which adds to the projectile velocity $\mathbf{v}$, so that the projectile energy is

$$E = \tfrac{1}{2} M_1 (\mathbf{v} + \mathbf{v}_T)^2. \tag{45}$$

Upon averaging over the electron-velocity distribution, a negligible energy shift is obtained, but the resonance is broadened to a full width at half-maximum Γ given by

$$\Gamma = \hbar \frac{M_1}{M_2} [\langle \phi_{1s} | (\mathbf{v} \cdot \nabla)^2 | \phi_{1s} \rangle]^{1/2}, \tag{46}$$

for the K-shell electron. For two electrons, the width becomes larger by a factor of $\sqrt{2}$, if the electronic motion of the two $1s$ electrons is assumed to be uncorrelated. For 14.2-MeV $p + {}^{12}$C, this width is 460 eV, which may explain part of the $(p, \gamma) - (p, p)$ width difference. It should be noted that this effect is a kind of Doppler broadening. Because a correction has already been made for the Doppler broadening due to the lattice vibration of the ^{12}C nuclei (Thompson *et al.*, 1981), it is unclear how this inner-shell electron Doppler broadening should be added. Furthermore, since the K electrons in the atom will attempt to avoid each other as much as possible, their motion might be sufficiently correlated so that the recoil of the target nucleus from one cancels that of the other.

3.7. K-Shell Ionization in Inelastic Nuclear Reactions

3.7.1. Theory. Studies of K-shell ionization in elastic resonance scattering have emphasized the time-delay effect on inner-shell vacancy production. For inelastic nuclear reactions, there are three additional effects to consider. First, the projectile loses energy in an inelastic nuclear reaction, so the outgoing relative velocity is different from the incoming velocity. This energy loss is in addition to the energy given up to electronic excitation, which we have assumed to be negligible. Since K-shell ionization amplitudes vary steeply with ion velocity, the inelastic nuclear energy loss can cause interference between the incoming and outgoing amplitudes even in the absence of time delay. Secondly, in nucleon transfer or change exchange reactions, such as (α, p) or (p, n) reactions, the incoming and outgoing projectile charge varies. Since electronic excitation is caused by the Coulomb interaction between the projectile nucleus and the electron, the different incoming and outgoing strengths of the interaction can also lead to interference. Finally, an additional perturbation is caused by the recoil of the target nucleus when struck by the projectile. The recoil perturbation is proportional to the product of the acceleration of the target nucleus and the electron coordinate r. In an inelastic nuclear reaction, the incoming projectile strikes the target nucleus to form a CN, causing an initial "incoming acceleration." Then the CN breaks up into a different projectile–target combination, causing a second "outgoing acceleration."

Analogous to Eq. (17), the most general expression for the K-shell ionization probability is (Anholt, to be published):

$$P_K(\theta) = \sum_{ss'vv'\lambda\mu} \int_0^\infty d\varepsilon \frac{v'}{v} |b_\lambda(q', Z_1')\, D^\lambda_{\mu 0}(\theta)\, f_{ss'vv'}(E, \theta)$$
$$+ (-1)^\lambda\, \delta_{\mu 0}\, b_\lambda(-q, Z_1)\, f_{ss'vv}(E - \Delta E, \theta)|^2 \Big/ \sum_{ss'vv'} |f_{ss'vv'}(E, \theta)|^2. \qquad (47)$$

Here $f_{ss'vv'}(E, \theta)$ is the nuclear reaction amplitude evaluated at the incoming projectile energy E, $\Delta E = \varepsilon + E_K$ is the energy lost by the projectile exciting the K electron with binding E_K into a continuum state with energy ε, $\lambda\mu$ are the angular-momentum quantum numbers for the ionized electron, sv and $s'v'$ are channel spin quantum numbers for the nuclear reaction (Lane and Thomas, 1958), θ is the center-of-mass scattering angle, $D^\lambda_{\mu 0}$ is the rotation matrix element, and v and v' are the incoming and outgoing relative velocities. The ionization amplitudes b_λ are given by

$$b_\lambda(q', Z_1') = \int_0^\infty \frac{dR}{v'} Z_1' G_\lambda(R)\, e^{i\bar{q}'R} + \frac{m\langle r\rangle M_1' v'}{\sqrt{3}(M_1' + M_2')} \delta_{\lambda 1},$$
$$b_\lambda(-q, Z_1) = \int_0^\infty \frac{dR}{v} Z_1 G_\lambda(R)\, e^{-i\bar{q}R} + \frac{m\langle r\rangle M_1 v}{\sqrt{3}(M_1 + M_2)} \delta_{\lambda 1}. \qquad (48)$$

Here Z_1, Z_1', M_1, M_1', and M_2' are initial and final projectile charges, projectile masses, and target masses; $q = \Delta E/v$ and $q' = \Delta E/v'$ are the momenta transferred to the electron, R is the internuclear distance, and G_λ and $\langle r\rangle$ are the radial matrix elements of the Coulomb interaction that can be gleaned from Eqs. (20) and (24).

Although formidable in appearance, Eq. (47) has a simple interpretation. The amplitude $b_\lambda(q', Z_1')\, D^\lambda_{\mu 0} \equiv a_{\text{out}}$ is just that for ionizing an electron on the outgoing trajectory, hence is evaluated with the outgoing nuclear masses, and velocity. The amplitude $b_\lambda(-q, Z_1)\,\delta_{\mu 0}(-1)^\lambda \equiv a_{\rm in}$ is that for ionizing on the incoming trajectory, appropriately evaluated with the incoming nuclear parameters. If no time delay is present $f(E)$ is approximately equal to $f(E-\Delta E)$ and the have

$$P_K = \sum_{\lambda\mu} \int_0^\infty d\varepsilon \frac{v'}{v} |a_{\rm out} + a_{\rm in}|^2, \tag{49}$$

which is to be expected from semiclassical considerations, except for the factor v'/v.

In deriving Eq. (47) we have assumed that the energy ΔE lost to electronic excitation is much less than the projectile center-of-mass energy, therefore v differs from v' only in that energy lost to nuclear excitation. In general, v' is given by

$$v' = v\left(\frac{M}{M'}\right)^{1/2}\left[\frac{1+(Q-E_{\rm exc})M_1}{EM}\right]^{1/2}, \tag{50}$$

where M and M' are the initial and final reduced nuclear masses, Q is the Q value for the nuclear reaction, $E_{\rm exc}$ is any excitation energy lost to the projectile or target nuclei, and E is the initial laboratory projectile energy.

Nuclear reactions can be classified as direct reactions and those which involve a CN mechanism (Blatt and Weiskopf, 1962; Austern, 1970). CN reactions may exhibit a time-delay effect on K-shell ionization (if $E_K/\Gamma > 1$), but direct reactions will have smoothly varying amplitudes so that Eq. (49) can be used to calculate K-shell ionization. The elastic reactions $p + {}^{12}C$ at 0.46 and 1.7 MeV and the $p + {}^{88}Sr$ reactions at 5 and 6 MeV involve CN formation. The CN scattering cross section sometimes exhibits a strongly varying E dependence, but one may observe only a smoothly varying E dependence if there are many fine unresolved resonances and no isobaric analog state is nearby. Nuclear Coulomb excitation (Alder *et al.*, 1956) and most heavy-ion transfer reactions are direct reactions.

Often both direct and CN reactions are present simultaneously, so that the reaction amplitude can be written as a sum of a direct part $f_{\rm DI}$ and a CN part $f_{\rm CN}$. This is analogous to elastic scattering, for which the amplitude is written as the sum of a Coulomb scattering contribution $f_{\rm Coul}$ (direct) and a Breit–Wigner resonance scattering contribution $f_{\rm res}$(CN). For inelastic scattering near IARs one has two additional amplitudes: an enhanced rapidly fluctuating amplitude $f_{\rm fl}^E(E, \theta)$ that is largest at $E = E_R$, and an unenhanced fluctuating amplitude $f_{\rm fl}^{\rm U}(E, \theta)$ (King and Slater, 1977). The enhanced amplitude contains those fine-structure resonances that have the appropriate $j\pi$ to couple to the analog state. The unenhanced amplitude contains all other $j\pi$ levels; and the unenhanced CN cross section $\langle |f_{fi}^{U}|^2 \rangle$, varies smoothly with the projectile energy.

3.7.2. Experiments. Although few measurements of K-shell ionization in inelastic nuclear reactions have been made until now, we present a few calculations to illustrate some of the possible effects. Figure 12 shows calculated K-shell ionization probabilities in 6-MeV $p+\text{Sr}$ and 4-MeV $p+\text{Ni}$ collisions, where we have assumed the loss of various energies E_{exc} due to direct nuclear Coulomb excitation (Alder *et al.*, 1956). Any possible nearby resonances are neglected. In these calculations we have set $f(E)$ equal to $f(E-\Delta E)$, hence have used Eq. (49). The calculated ionization probability P_K increases with scattering angles for $\theta > 5°$. The velocity loss reduces P_K at all scattering angles, though the relative anisotropy, $B = [P_K(0°) - P_K(180°)]/2P_K(90°)$, is also reduced in absolute magnitude.

Experiments to determine P_K in inelastic reactions might measure triple coincidences between X rays, scattered particles, and γ rays from the decay of the excited nuclei. The energy of the particles and γ rays established that an inelastic reaction took place and that no internal conversion of the γ ray (giving background K X rays) has occurred. One might also consider a

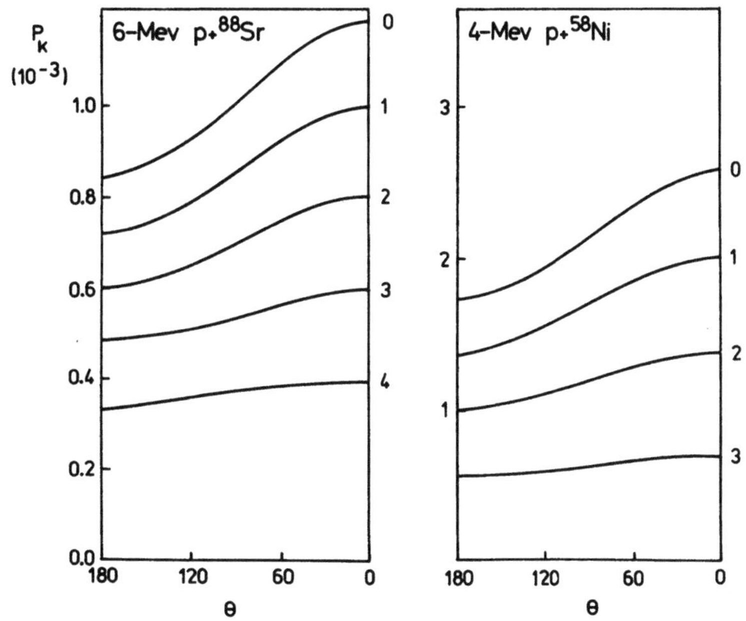

FIGURE 12. Ionization probabilities in elastic scattering reactions 6-MeV ^{88}Sr(p, p') for various center-of-mass scattering angles θ and various values of the energy loss (in MeV).

measurement involving just X-ray–γ-ray coincidences. If no multiple nuclear excitation is present one would measure

$$\bar{P}_K \sim \text{Number of x–}\gamma\text{ coincidences/number of }\gamma\text{ rays}$$
$$= \int_0^\infty d\varepsilon \Big\{ |b_0(v') + b_0^*(v)|^2 + |b_1(v')|^2 + |b_1(v)|^2$$
$$- [b_1^*(v')\, b_1^*(v) + b_1(v')\, b_1(v)] \int d\Omega\, |f(E, \theta)|^2 \cos\theta$$
$$\div \int d\Omega\, |f(E, \theta)|^2 \Big\}. \tag{51}$$

In a heavy-ion collision in which $v' \approx v$ and monopole excitation is dominant, hence the anisotropy term proportional to $\cos\theta$ is small (Andersen *et al.*, 1976a), such a measurement could yield a value of the excitation probability at impact parameter $b = 0$:

$$\bar{P}_K(0) = \int_0^\infty d\varepsilon\, \{ |b_0^*(v) + b_0(v)|^2 + 2|b_1(v)|^2 \}. \tag{52}$$

For an inelastic nuclear sattering reaction involving CN formation, if the direct reaction contribution to the total amplitude is negligible we might have

$$f_{ss'vv'}(E, \theta) = \frac{a_{ss'vv'}(\theta)}{E - E_R + i\Gamma/2}, \tag{53}$$

so that the ionization probability is

$$P_K = \int_0^\infty d\varepsilon \frac{v'}{v} \left| \frac{b_\lambda(v')\, D^\lambda_{\mu 0}(\theta)}{E - E_R + i\Gamma/2} + \frac{(-1)^\lambda \delta_{\mu 0} b_\lambda^*(v)}{E - E_R - \Delta E + i\Gamma/2} \right|^2 \left(E - E_R + i\frac{\Gamma}{2} \right)^2 \tag{54}$$

The constant $a_{ss'vv'}(\theta)$ containing many angular-momentum and geometrical factors cancels when P_K is calculated, hence we can omit the sum over channel spins. One inelastic reaction where the direct contribution appears negligible and the inelastic cross section is unusually large is the $^{28}Si(p, p'(1.78))$ reaction at 3.1 MeV (Belote *et al.*, 1967). The width of the resonance (12.9 keV) is larger than the Si K-shell binding energy ($E_K \sim 2$ keV) but the time-delay effect on P_K is nevertheless quite large (Fig. 13).

Figure 14 shows calculated ionization probabilities for several transfer reactions. In these reactions the incoming and outgoing relative projectile velocity, Coulomb interaction, and recoil interaction all change. Transfer reactions often tend to be direct reactions, so we have assumed that f_{CN} is

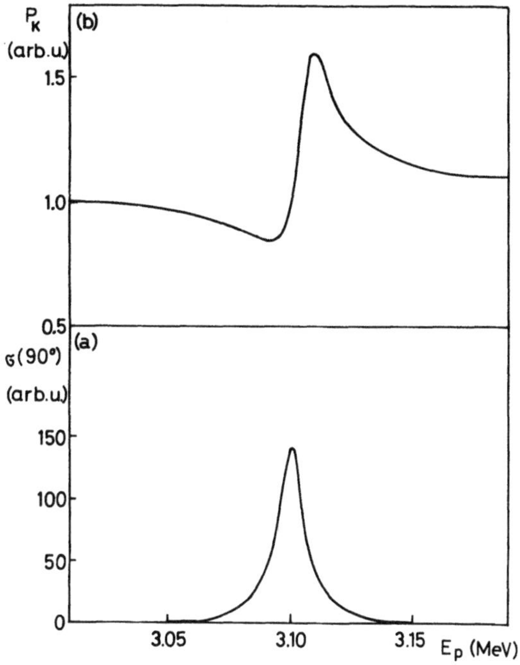

FIGURE 13. (a) Differential scattering cross sections and (b) ionization probability P_K for the reaction $^{28}Si[p, p'(1.78\ MeV)]^{28}Si^*$.

negligible, and used Eq. (49). Generally, one can measure the probability for K-shell ionization in the elastic reaction (solid line in Fig. 14), simultaneously with that for the transfer reaction (dashed line). For the $^{88}Sr(p, \alpha)$ reaction, the probability for the transfer reaction is much less than that for the elastic one, due to the smaller outgoing velocity. Here, Q/E is much less than unity, and E_{exc} in Eq. (50) is assumed to be zero in all cases, so that v' is approximately equal to $\frac{1}{2}v$ because of the larger reduced mass of the alpha particle. For the $^{88}Sr(\alpha, p)$ reaction, v' is larger than v and the corresponding transfer-reaction probability is larger than the elastic (α, α_0) one. For the endothermic $^{88}Sr(p, d)$ reaction, v' is much smaller than v, hence the transfer-reaction probability is much smaller than the elastic one, similar to the (p, α) case. In the (p, n) reaction, the incoming and outgoing velocities are similar, but due to the lack of the Coulomb interaction between the neutron and electron in the outgoing channel, the transfer-reaction probability is smaller.

Also shown in Fig. 14 is the K-shell ionization probability for a (p, γ) reaction, which is approximately one-half of the elastic scattering probability (p, p_0) at $\theta = 180°$. Since the recoil of the γ ray gives a negligible

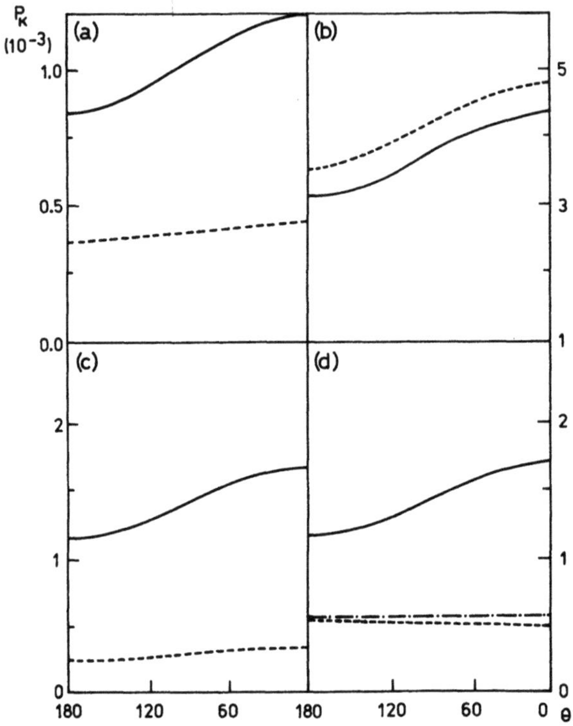

FIGURE 14. *K*-shell ionization probabilities during elastic scattering (solid lines) and transfer reactions (dashed lines) for various reactions: (a) ^{88}Sr(p, α) at 6 MeV, $Q = -0.87$ MeV, $v'/v = 0.47$; (b) ^{88}Sr(α, p) at 24 MeV, $Q = -6.4$ MeV, $v'/v = 1.73$; (c) ^{88}Sr(p, d) at 12 MeV, $Q = -8.9$ MeV, $v'/v = 0.35$; (d) ^{88}Sr(p, n) at 12 MeV, $Q = -4.4$ MeV, $v'/v = 0.77$ and ^{88}Sr(p, γ) at 12 MeV (chain curve).

contribution to *K*-shell ionization (Anholt, to be published), the (p, γ) ionization probability is just the incoming ionization probability

$$P_{\text{in}} = \int_0^\infty d\varepsilon \sum_\lambda |b_\lambda|^2 \tag{55}$$

and is independent of the angle. Duinker and Boersma (1981) measured *K*-shell ionization in ^{27}Al(p, γ)^{28}Si and ^{26}Mg(p, γ)^{27}Al resonance-capture reactions. They measured the γ-ray yield as a function of proton energy. The yield peaks near the resonance energy (992 or 1288 keV), and the width of the resonance is much smaller than the *K*-shell binding energy. A smaller peak is also observed that is shifted upward in energy by an amount $\sim E_K + \varepsilon$, due to those particles that ionize *K* electrons before capture. The area of the smaller peak gives the (p, γ) *K*-shell ionization probability P_{in}, which Duinker and Boerma (1981) find is approximately one-half of the elastic scattering probability, calculated with the semiclassical approximation (Hansteen *et al.*, 1975).

3.7.3. Deep Inelastic Nuclear Reactions. Deep inelastic nuclear reactions between very heavy nuclei are direct reactions that do not involve CN formation, but exhibit a time-delay effect on K-shell ionization (Anholt, 1979b; Müller *et al.*, 1980). In these collisions, e.g., 1160-MeV Xe + Pb or 2000-MeV U + U, the projectile nucleus strikes the target nucleus and the radial center-of-mass kinetic energy dissipates quickly as the nuclei come into contact (Schröder *et al.*, 1978; Riedel *et al.*, 1979). Due to the incoming orbital angular momentum, the composite system rotates through an angle $\Delta\phi$ (Fig. 15), during which time the internuclear distance stays nearly constant, and an exchange of nucleons between the projectile and target nuclei takes place. The composite system then breaks up into highly excited projectile- and targetlike fragments, which may in turn fission or release a large number of γ rays, neutrons, and light charged particles.

The time required for the rotation of the transient composite system is approximately 10^{-21} s in Xe + Pb collisions (Schröder *et al.*, 1979; Riedel *et al.*, 1980). This time can be interpreted as the delay time T in the semiclassical theory [Eq. (2)]. In these collisions, the near symmetric correlation diagram shown in Fig. 1a is most appropriate. In Xe + Pb collisions, Pb K vacancies are made by excitation of electrons from the $1s\sigma$ MO (processes 1 and 2); in U + U collisions, U K vacancies are made by excitation of electrons mainly from the $2p\sigma$ MO (processes 1′ and 2′). The relevant energy transfer is at least the UA $1s$ or $2p$ binding energy which leads to $\Delta\omega T$ values of >0.5 (Xe + Pb) and >1 (U + U).

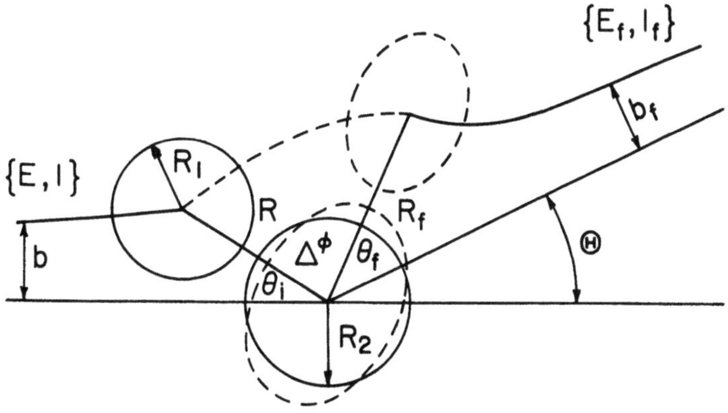

FIGURE 15. Illustration of the classical model of deep-inelastic nuclear reaction. (From Riedel *et al.*, 1980.)

A detailed model of deep-inelastic collisions is needed to relate the interaction time T to a variable measurable in the laboratory. Not surprisingly, T is a decreasing function of the impact parameter. In grazing collisions the rotation angle $\Delta\phi$ is small, so T is small. (In collisions where the nuclei do not come together, T is zero.) In central collisions, $\Delta\phi$ is large, so it takes longer to rotate to the final angle. The kinetic energy loss $(-Q)$ is also a function of the impact parameter. Most of the lost energy is radial kinetic energy, which is largest in central collisions and smallest in grazing collisions. While the impact parameter cannot be measured, the kinetic energy loss is easily determined. To calculate K-shell ionization therefore we use the semiclassical approximation [Eq. (4)] to find the excitation probability as a function of impact parameter and T. Nuclear models are used to relate T to the energy loss $(-Q)$.

The only differences with Eqs. (2) and (4) and the interaction Hamiltonian discussed so far are (i) we use initial and final MO wave functions; hence, the binding energies $\omega_{1s\sigma}$ or $\omega_{2p\sigma}$ are functions of internuclear distance $R(t)$ and the interaction Hamiltonian is the $\partial/\partial t$ operator written as $\dot{R}\partial/\partial R$ (Briggs, 1976; Soff *et al.*, 1981; Müller *et al.*, 1979). (ii) Due to centrifugal stretching, $R(t)$ changes slightly during the interaction time, by a factor of 1.3 in Xe + Pb collisions (Riedel *et al.*, 1980). This allows for excitation while to two nuclei are in contact, giving the a_{UA} contribution to the amplitude below, not found in Eq. (2). (iii) The outgoing radial velocity v' is reduced due to radial kinetic-energy loss. The amplitude for exciting a $1s\sigma$ electron into a continuum state with energy ε and zero angular momentum $(\lambda\mu = 00)$ is given by

$$a_{\varepsilon 00} = a_{\mathrm{in}} + e^{i\Delta(T)} a_{\mathrm{out}} + a_{\mathrm{UA}}, \tag{56}$$

where

$$a_{\mathrm{in}} = -\int_{R_0}^{\infty} dR\, V(R) \exp\left\{ i \int_{R_0}^{R} dR'/v_R[\omega_{1s\sigma}(R') + \varepsilon] \right\},$$

$$a_{\mathrm{out}} = \int_{R_f}^{\infty} dR\, V(R) \exp\left\{ i \int_{R_f}^{R} dR'/v'_R[\omega_{1s\sigma}(R') + \varepsilon] \right\},$$

$$a_{\mathrm{UA}} = \int_0^T dt\, \dot{\mathscr{R}}\, V(R) \exp\,[i\Delta(t)], \tag{57}$$

$$\Delta(t) = \int_0^t dt'\, \{\omega_{1s\sigma}[\mathscr{R}(t')] + \varepsilon\},$$

$$\mathscr{R}(t) = R_0\{1 + \alpha[1 - \exp(-t/\tau_\alpha)]\},$$

$\alpha = 0.3$ for Xe + Pb), τ_α is a deformation relaxation time (Riedel *et al.*, 1980), $V(R)$ is the transition matrix element $\langle 1s\sigma | \partial/\partial R | \varepsilon \rangle$ (Müller *et al.*,

1979), and $R_f = (1 + \alpha)R_0$. Through the radial velocity v_R and the time T the ionization amplitude is a function of the impact parameter b, which is then related to the kinetic energy loss $(-Q)$. The relation between impact parameter and Q is not one-to-one, however. The calculated ionization probability $P_K(-Q)$ is folded with a Gaussian distribution of energy losses $-Q$ with variance equal to $(2\ \text{MeV}) \times (-Q)$ (Riedel *et al.*, 1980).

Figure 16 compares calculations made for 1160-MeV Xe + Pb and

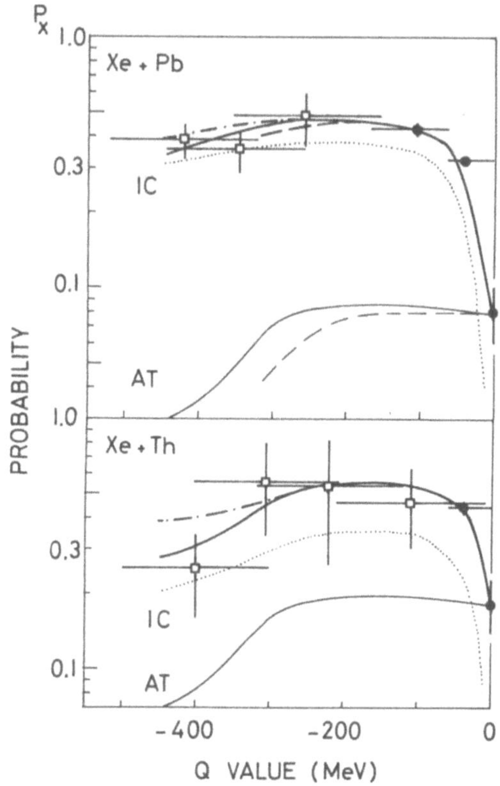

FIGURE 16. Atomic (AT) and internal-conversion (IC) contributions to target K-shell ionization in 1160-MeV ^{136}Xe + Pb and 1160-MeV ^{136}Xe + ^{232}Th reactions. The solid circles represent coincidences with projectilelike fragments; the open squares with targetlike fragments. The widths of the Q distribution result from the widths of the E bins and the particle detector opening. The atomic ionization probabilities were calculated from time-delay models of Schröder *et al.* (1978) (thin solid line) and Riedel *et al.* (1980) (dashed line). Dotted curves labeled IC represent fitted internal-conversion contributions, obtained from measurements of γ-ray production. The thick solid curves and dashed curves indicate the summed probability calculated with the models of Schröder *et al.* and Riedel *et al.*, respectively. The chain curve indicates the total probability calculated in the absence of a time-delay effect. (From Stoller *et al.*, 1983.)

Xe + Th collisions with experiment (Stoller *et al.*, 1983). Two calculations of the time-delay effect have been made: one based on the nuclear model of Schröder *et al.* (1978) and the other based on the more detailed model of Riedel *et al.* (1979). The interaction times do not differ much, but the relationship between energy loss and impact parameters differs. Using the model of Riedel *et al.* (1979) $P_{1s\sigma}$ falls off faster with $-Q$.

Unfortunately, the measured ionization probabilities are dominated by internal conversion of γ rays emitted by the targetlike fragments (Stoller *et al.*, 1983). More than 30 γ rays may be emitted following an energy loss of $-Q \gtrsim 100$ MeV, so it is hardly surprising that the observed probability for at least one of these γ rays to convert to the K shell is greater than 0.3. With this large γ-ray internal conversion background, the time-delay effect cannot be observed in Xe + Pb collisions. The effect might be seen in U + U collisions, however, where the atomic ionization probability is much larger ($P_{2p\sigma} > 1$) (Müller *et al.*, 1980).

Reinhardt *et al.* (1979) suggested that one measure the ionized-electron distribution $dP_K/d\varepsilon$, which should oscillate like $\exp(i\varepsilon T)$ with ε for fixed impact parameter. The internal-conversion electron probability distribution $dP_{IC}/d\varepsilon$ would undoubtedly be much greater than the ionized-electron probability, however. Using measured γ-ray distributions and $E2$ internal-conversion coefficients, $dP_{IC}/d\varepsilon$ was estimated to be two orders of magnitude larger than $dP_K/d\varepsilon$ in 1160-MeV Xe + Pb collisions (Anholt, 1980b).

4. UNITED-ATOM X RAYS

4.1. *Theory*

If a long-lived CN is formed during a collision, and if a K vacancy is made on the incoming part of the collision with a probability P_{in}, the vacancy may decay while the projectile and target nuclei are together, emitting a shifted K X ray, characteristic of the UA. The semiclassical probability of emitting UA $K\alpha$ X rays was originally given by Gugelot (1962):

$$P_{UAK\alpha} = \frac{P_{in}\Gamma_{K\alpha}\tau}{1+\Gamma_K\tau} r_{CN}(\theta), \tag{58}$$

where Γ_K is the decay rate of the UA K vacancy, $\Gamma_{K\alpha}$ is the partial $K\alpha$ X-ray emission rate, and τ is the CN lifetime. In Eq. (58) it is assumed the $P_{UAK\alpha}$ is determined by measuring coincidences between CN K X rays and reaction products detected at a scattering angle θ. Not all of the detected reaction products may have undergone a CN reaction, however. A significant number might have undergone direct scattering, in which case the CN lifetime is zero. Therefore, r_{CN} is introduced in Eq. (58) to represent the fraction of reaction products which have undergone a CN reaction.

Gugelot's main interest in UA K X-ray production was in the determination of CN lifetimes τ. This requires a precise calculation of the K-shell ionization probability P_{in} and an understanding of what is meant by r_{CN}.

Quantum-mechanical calculations of UA K X-ray production have been made by McVoy *et al.* (1981), von Brentano and Kleber (1980), and Anholt (1982). To account both for the formation of the K vacancy and its radiative decay, second-order T matrix (McVoy *et al.*, 1981) or distorted-wave Born approximation (Anholt, 1982) calculations have been made. The cross section for UA $K\alpha$ X-ray emission with frequency ω in a collision in which the projectile scatters into a center-of-mass solid angle $d\Omega$ is

$$\frac{d^2\sigma_{UAK\alpha}}{d\omega\, d\Omega}=\frac{1}{2\pi}\frac{\Gamma_{K\alpha}}{|\omega-\omega_{K\alpha}|^2}\int_0^\infty d\varepsilon \sum_\lambda |b_\lambda(v, Z_1)|^2 \times |f(\theta, E-E_K-\varepsilon)-f(\theta, E-E_K-\varepsilon-\omega+\omega_{K\alpha})|^2, \quad (59)$$

where E_K is the united-atom K-shell binding energy, and b_λ is the K-shell ionization amplitude given by Eq. (18). To account for the decay of the UA K vacancy, we take $\omega_{K\alpha}$ to be complex and equal to $E_{K\alpha}-i\Gamma_K/2$, where $E_{K\alpha}$ is the UA $K\alpha$ transition energy.

UA X-ray emission is most likely in CN reactions and reactions involving IARs where f is given by Eq. (30). The difference in amplitudes in Eq. (59) is just the difference in the fluctuating amplitudes. Therefore, the UA X-ray production cross section must be averaged over the beam energy resolution. Making use of Ericson correlation functions [Eq. (36); Ericson and Mayer-Kuckuk, 1966], we obtain

$$\left\langle\frac{d^2\sigma_{UAK\alpha}}{d\omega\, d\Omega}\right\rangle\approx\frac{1}{2\pi}\frac{\Gamma_{K\alpha}P_{in}\sigma_{CN}(E-E_K)(2\Gamma_c+\Gamma_K)}{(\Gamma_c+\Gamma_K)[(\Gamma_c+\frac{1}{2}\Gamma_K)^2+(\omega-E_{K\alpha})^2]}, \quad (60)$$

where P_{in} is the incoming ionization probability [Eq. (55)], $\sigma_{CN}=\langle|f_{fl}(E,\theta)|^2\rangle$ is the CN cross section, and Γ_c is the correlation width, which is related to the average width of the many narrow resonances. In deriving Eq. (60), we assume that σ_{CN} varies slowly with E, hence $\sigma_{CN}(E-E_K-\varepsilon)$ can be taken outside of the integral over ε. Upon integrating over the photon frequency one obtains for the K X-ray emission probability

$$P_{UAK\alpha}=P_{in}\frac{\Gamma_{K\alpha}}{\Gamma_c+\Gamma_K}\frac{\sigma_{CN}(E-E_K)}{\sigma_{CN}(E)+|\bar{f}(E,\theta)|^2}. \quad (61)$$

Comparing with Gugelot's equation [Eq. (58)] we see that the CN lifetime is just the inverse of the correlation width Γ_c. The quantum-mechanical formula gives r_{CN} in terms of the ratio of the CN cross section to the total scattering cross section, including any direct contribution $|\bar{f}|^2$.

United-atom X rays are not easily seen in elastic-scattering reactions where the Coulomb scattering cross section $|\bar{f}|^2$ is much greater than the compound-elastic cross section. The exception is found in IAR reactions where at large scattering angles σ_{CE} can be of the same order of magnitude as $|\bar{f}|^2$, as shown in Fig. 6. An interesting feature of IAR reactions is that Γ_c is expected to vary across the resonance (Hofmann *et al.*, 1975; Anholt, 1982). Due to the mixing between the CN states and the analog state, the CN levels of the same $j\pi$ as the analog level pick up additional width in the vicinity of the analog level, giving a larger Γ_c value. The simplest interpretation of this widening is that the analog state, in addition to providing a doorway into the CN (giving large cross sections), also allows the particle to escape out of the doorway quickly, leading to shorter CN lifetimes of larger widths. Blocking experiments have verified smaller CN lifetimes on IARs than off IARs (Gibson *et al.*, 1979). A measurement of UA K X rays should find a similar behavior.

4.2. *Experiments*

Successful measurements of CN decay widths by observing UA K X rays must satisfy several requirements: (1) The CN scattering cross section should be large compared to the direct scattering cross section, so that r_{CN} is large. (2) The K-shell ionization probability should be large. (3) The CN width should be of a similar order of magnitude as the K vacancy decay width. If $\Gamma_c \gg \Gamma_K$, the UA K X-ray production probability will be small. If $\Gamma_c \ll \Gamma_K$, the measurement will not be sensitive to the magnitude of Γ_c. (4) γ-ray background should be small. (5) The X-ray detector should have good resolution. Measurements of target K X rays can be made with large-area NaI detectors; only target K X rays are observed, so resolution is not required. However, for proton CN reactions, the UA K X rays to be measured are shifted by only $\sim$0.5 to 2 keV from the target K X rays. Good detector resolution improves the peak- to γ-ray background intensity ratio, also.

Many measurements of UA K X rays in (heavy ion, xn) reactions have been made. These do not yield CN widths, however, since Γ_c is essentially zero for the β-unstable nuclei. Most of the UA K X rays are formed by γ-ray internal conversion, so these measurements give discrete (Anholt *et al.*, 1973) or average (Karwowski *et al.*, 1979; Meyerhof *et al.*, 1979) internal-conversion probabilities, from which γ-ray multipolarities can be determined.

Measurements of UA K X rays in collisions where superheavy CN are formed (e.g., Ar + Th, Xe + Pb, U + U) would be extremely interesting if they yielded spectroscopic information in the region where electronic relativistic effects play an important role (Reinhardt and Greiner, 1977). Unfortunately, deep-inelastic nuclear reactions are dominant in this region, and the lifetimes of the composite systems ($<10^{-20}$ s) are much smaller than the K-vacancy lifetimes ($>10^{-19}$ s). Although the ionization probability can be large, γ-ray background following these reactions is very intense, so UA K X rays are probably not observable (Meyerhof *et al.*, 1980).

Successful measurements of CN lifetimes have been made in $p + {}^{106}$Cd (Chemin *et al.*, 1979) and in $p + {}^{112}$Sn (Röhl *et al.*, 1979, 1981) reactions. Coincidences between UA K X rays and inelastically scattered protons were measured. For inelastic scattering, the CN cross section is larger than the direct scattering cross section, so r_{CN} is nearly unity. Si(Li) X-ray detectors were used to measure the UA K X rays. Figure 17 shows the X-ray and scattered-proton spectra in the ^{106}Cd$(p, p')^{106}$Cd measurement. A peak due to UA In K X rays is clearly visible in this spectrum; the area of this peak was used to determine the CN lifetime.

The CN lifetime is found by measuring the number of UA K X-ray–inelastic-proton coincidences relative to the number of SA K X-ray–elastic-proton coincidences. For $r_{\text{CN}} \approx 1$, the ratio gives

$$\frac{P_{\text{UA}K\alpha}}{P_{\text{SA}K\alpha}} = \frac{\omega^u_{K\alpha}}{\omega^s_{K\alpha}} \frac{P_{\text{in}}}{P(\theta)} \frac{\Gamma_K}{\Gamma_K + \Gamma_c}, \tag{62}$$

where $\omega^u_{K\alpha}$ and $\omega^s_{K\alpha}$ are UA and SA $K\alpha$ fluorescence yields, P_{in} is the probability of ionizing K electrons on the way in, and $P(\theta)$ is that for ionizing K electrons during the entire collision. For lack of detailed calculations, the ratio P_{in}/P is usually assumed to have its classical value 1/2; half of the electrons are ionized while the projectile is coming in; half while the projectile is going out. Quantum-mechanical calculations of P_{in}/P yield values between 0.2 and 0.6 in this energy range (Rohl *et al.*, 1981). Depending on the CN excitation energy, CN decay widths between 4 and 20 eV were found in these reactions, in agreement with statistical model calculations (Chemin *et al.*, 1979) and fluctuation-averaging and crystal-blocking measurements (Rohl *et al.*, 1981).

Measuring UA X rays in coincidence with inelastic protons makes r_{CN} large, but also worsens the γ-ray background. This limits the largest measureable CN width. For 12-MeV $p + {}^{106}$Cd, if Γ_c were ten times larger, the In $K\alpha$ X-ray peak in Fig. 17 would be approximately 0.1 as large, hence indistinguishable from the γ-ray background. Most of the background comes from high-energy γ rays undergoing small-angle Compton scattering in the Si(Li) X-ray detector. Compton suppressors that veto X-ray-detector pulses when γ rays are detected behind the X-ray detector are being constructed to

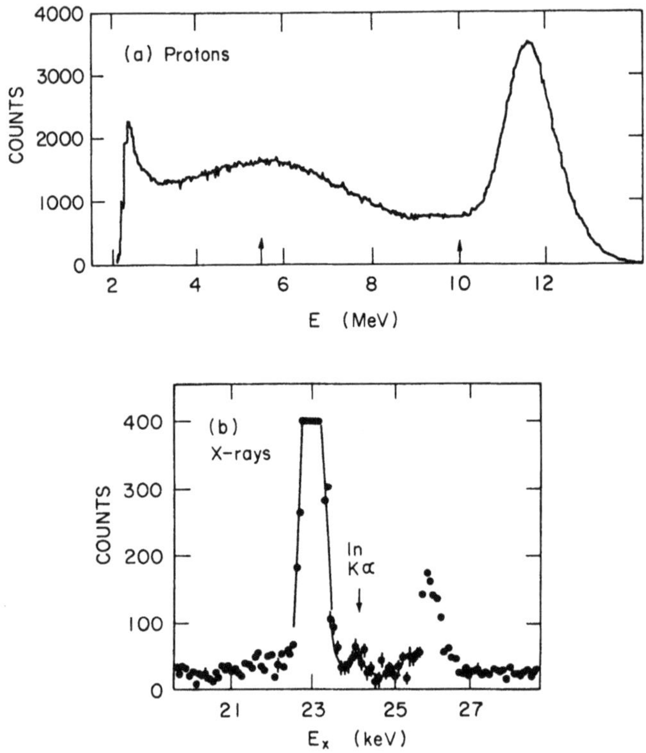

FIGURE 17. (a) Singles proton spectrum measured with a plastic scintillator detector in 12-MeV $p + {}^{106}$Cd collisions. The spectrum shows an elastic scattering peak near 12 MeV and a wide distribution of inelastically scattered protons. UA K X rays were measured in coincidence with the inelastic protons in the window denoted by the arrows. (b) Spectrum of X rays in coincidence with inelastically scattered particles. Most of the X rays are target $K\alpha$ ($\sim$23 keV) and $K\beta$ ($\sim$26 keV) X rays. The UA K X rays are at $\sim$24 keV. (From Chemin *et al.*, 1979.)

decrease the γ-ray background (Chemin, 1982; Dost *et al.*, 1982). Measurements of UA K X rays in coincidence with elastically scattered protons in reactions involving IARs might have less γ-ray background, and therefore might obviate the need for special spectrometers.

4.3. Widths of United-Atom K X-Ray Lines

Although the magnitude of the γ-ray background often decides the smallest measurable CN lifetime, it is interesting to examine the fundamental limit. This can be determined by the Heisenberg uncertainty principle

$$\Delta\omega\tau > \hbar. \tag{63}$$

A UA K X-ray peak is likely to be observable only if the FWHM $\Delta\omega$ is less than one-twentieth of the transition frequency $\omega_{K\alpha}$. The product $\omega_{K\alpha}\tau$ must therefore be greater than 20.

The same requirement is also reflected in the quantum-mechanical UA K X-ray production probability. Equation (60) describes a Lorentzian line shape centered on the UA $K\alpha$ energy $E_{K\alpha}$ with a full width at half-maximum, $\Delta\omega$, given by $2\Gamma_c + \Gamma_K$. We have

$$\frac{\hbar\Delta\omega}{E_{K\alpha}} < 0.05 \Rightarrow \frac{E_K}{2\Gamma_c + \Gamma_K} \approx \frac{E_{K\alpha}}{2\Gamma_c} > 20. \tag{64}$$

One finds the equivalent requirement as long as one interprets the CN τ lifetime as the inverse of twice the Ericson correlation width Γ_c.

Semiclassical calculations of UA K X-ray production in near symmetric collisions have also been made (Anholt, 1979a). These calculations emphasize the interference between UA K X-ray and MO X-ray production. A vacancy in the $1s\sigma$ MO can decay while the projectile and target

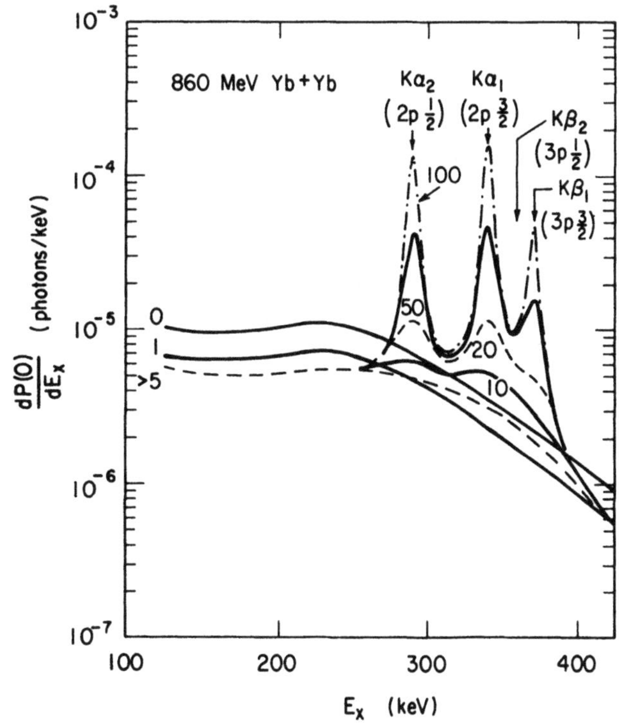

FIGURE 18. Calculated molecular-orbital–united-atom K X-ray emission probabilities in 860-MeV Yb + Yb $b = 0$ collisions for various values of $\omega_{K\alpha}\tau$ (Anholt, 1979a).

nuclei are coming together (processes 4 or 5 in Fig. 1a) giving a continuous X-ray distribution between the SA K X-ray energy and roughly the UA K X-ray energy. If $\omega_{K\alpha}\tau$ is much less than 20, the UA K X-ray peak is broad and merges into the MO X-ray continuum. If $\omega_{K\alpha}\tau$ is greater than 20, sharp UA K X-ray lines are seen.

5. CONCLUSIONS

Time-delay effects on inner-shell ionization and bremsstrahlung emission in atomic collisions can be divided into those in which $\Delta\omega\tau$ is approximately unity and those in which $\Gamma_K\tau$ is approximately unity. Collisions in which the product of the atomic excitation energy $\Delta\omega$ and the CN lifetime τ is approximately unity should exhibit interference between incoming and outgoing excitation amplitudes. Measurements of the altered excitation probability yield information about atomic ionization amplitudes for the atomic physicist, and information about the relative magnitude of the CN scattering cross sections for the nuclear physicist. In collisions in which the product of the UA K-vacancy decay rate Γ_K and the CN lifetime is approximately unity, X rays characteristic of the united atom may be observed if an inner shell vacancy is created prior to CN formation. Such measurements lead to information about average decay widths in CN reactions involving many unresolved fine-structure resonances.

Measurements of time-delay effects on atomic collisions are presently being made at only a few laboratories. Given the difficulty of these experiments, it is hardly surprising that few measurements have been done. The theory of time-delay effects in elastic and inelastic nuclear reactions is presently much further advanced than experiment. There are, however, a large number of models of nuclear reactions, appropriate for different physical situations, and the validity of the application of Eqs. (17), (25), or (47) to the different cases is not known exactly. In this chapter we have predicted results for a few simple inelastic reactions where measurements have not yet been made. No doubt if measurements are done in inelastic collisions, more subtle nuclear-physics points will have to be taken into account.

ACKNOWLEDGMENTS

The author would like to thank W. E. Meyerhof and B. Crasemann for careful reading of the manuscript. In addition to collaboration with John Blair, this author has greatly profited over the past three years from interactions with Professors W. E. Meyerhof, G. Temmer, H. A. Weidenmüller, K.

McVoy, G. Soff, P. A. Amundsen, J. R. Huizenga, and L. G. Moretto. Special thanks go to J. F. Chemin, who raised many of the questions discussed in this work. This work was supported in part by the National Science Foundation under grant No. PHY-80-15348.

REFERENCES

ALDER, K., A. BOHR, T. HUUS, B. MOTTELSON, AND A. WINTHER, 1956, *Rev. Mod. Phys.* **28**, 432.

AMUNDSEN, P. A., 1978, *J. Phys. B* **11**, 3197.

ANDERSEN, J. U., L. KOCBACH, E. LAEGSGAARD, M. LUND, AND C. D. MOAK, 1976a, *J. Phys. B* **9**, 3247.

ANDERSEN, J. U., E. LAEGSGAARD, M. LUND, AND C. D. MOAK, 1976b, *Nucl. Instrum. Methods* **132**, 507.

ANHOLT, R., 1979a, *Z. Phys.* **A292**, 123.

ANHOLT, R., 1979b, *Phys. Lett.* **88B**, 262.

ANHOLT, R., 1980a, *Z. Phys.* **A295**, 201.

ANHOLT, R., 1980b, unpublished report.

ANHOLT, R., 1982, *Z. Phys.* **A306**, 285.

ANHOLT, R., to be published.

ANHOLT, R., AND W. E. MEYERHOF, 1977, *Phys. Rev. A* **16**, 190.

ANHOLT, R., J. O. RASMUSSEN, AND I. REZENKA, 1973, *Nucl. Phys.* **A209**, 72.

ANHOLT, R., D. P. WANG, AND Y. C. LIU, 1979, *J. Phys. Soc. Jpn* **47**, 1260.

ANHOLT, R., J. F. CHEMIN, AND P. A. AMUNDSEN, 1982, *Phys. Lett.* **118B**, 245.

AUSTERN, N., 1970, *Direct Nuclear Reaction Theories* (Wiley, New York).

BANG, J., AND J. M. HANSTEEN, 1959, *K. Dan. Vidensk. Selsk. Mat. Fys. Medd.* **31**, No. 13.

BELOTE, J. A., E. KASHY, AND J. R. RISSEN, 1967, *Phys. Rev.* **122**, 920.

BENN, J., E. B. DALLY, H. H. MÜLLER, R. E. PIXLEY, H. H. STAUB, AND H. WINKLER, 1968, *Nucl. Phys.* **A106**, 296.

BILPUCH, E. G., J. D. MOSES, F. O. PURSER, H. W. NEWSEN, G. E. MITCHELL, R. O. NELSON, AND D. A. OUTLAW, 1974, *Phys. Rev. C* **9**, 1589.

BLAIR, J. S., AND R. ANHOLT, 1982, *Phys. Rev. A* **25**, 907.

BLAIR, J. S., P. DYER, K. A. SNOVER, AND T. A. TRAINER, 1978, *Phys. Rev. Lett.* **41**, 1712.

BLATT, J. M., AND V. F. WEISSKOPF, 1962, *Theoretical Nuclear Physics* (Wiley, New York).

BRIGGS, J. S., 1975, *J. Phys. B* **8**, L485.

BRIGGS, J. S., 1976, *Rep. Prog. Phys.* **39**, 217.

BRIGGS, J. S., AND A. M. LANE, 1981, *Phys. Lett.* **106B**, 436.

BURCH, D., W. B. INGALLS, H. WIEMAN, AND R. VANDENBOSCH, 1974, *Phys. Rev. A* **10**, 1254.

CHEMIN, J. F., 1982, private communication.

CHEMIN, J. F., S. ANDRIAMONJE, J. ROTURIER, B. SABOYA, J. P. THIBAUD, S. JOLY, S. PLATTARD, H. UZUREAU, H. LAURENT, J.-M. MAISON, AND J. P. SHAPIRA, 1979, *Nucl. Phys.* **A331**, 407.

CHEMIN, J. F., R. ANHOLT, CH. STOLLER, W. E. MEYERHOF, AND P. A. AMUNDSEN, 1981, *Phys. Rev. A* **24**, 1218.

CHEMIN, J. F., W. E. MEYERHOF, R. ANHOLT, J. D. MOLITORIS, AND CH. STOLLER, 1982, *Phys. Rev. A* **26**, 1239.

CHEMIN, J. F., S. ANDRIAMONJE, E. MORENZONI, CH. STOLLER, W. E. MEYERHOF, J. D. MOLITORIS, AND O. K. BAKER, 1983, to be published.

CIOCCHETTI, G., AND A. MOLINARI, 1963, *Nuovo Cimento* **24**, 1263.

CIOCCHETTI, G., AND A. MOLINARI, 1965, *Nuovo Cimento* **40B**, 69.

DOST, M., S. ROHL, AND S. SEIDEL, 1982, *Report Arbeitsgruppe Energiereiche Atomare Stösse*, edited by B. Fricke, Vol. 3, p. 102.

DUINKER, W., AND C. R. BOERSMA, 1981, *Phys. Lett.* **100B**, 13.

DUINKER, W., J. VAN ECK, AND A. NIEHAUS, 1980, *Phys. Rev. Lett.* **45**, 2102.

EISBERG, R. M., D. R. YENNIE, AND D. H. WILKENSON, 1960, *Nucl. Phys.* **18**, 330.

ERICSON, T., AND T. MAYER-KUCKUK, 1966, *Ann. Rev. Nucl. Sci.* **16**, 183.

FEAGIN, J. M., AND L. KOCBACH, 1981, *J. Phys. B* **14**, 4349.

FEAGIN, J. M., AND L. KOCBACH, 1983, *I.E.E.E. Trans. Nucl. Sci.* **NS30**, 1086.

FESHBACH, H., AND D. R. YENNIE, 1962, *Nucl. Phys.* **37**, 150.

GERJUOY, E., 1965, *J. Math. Phys.* **6**, 993.

GIBSON, W. M., M. MARUYAMA, Y. HASHIMOTO, E. P. KANTER, G. M. TEMMER, R. J. KEDDY, R. W. MINGAY, AND J. P. F. SELLSCHOP, 1979, *Nucl. Phys.* **A317**, 313.

GRIFFY, T. A., AND L. C. BIEDENHARM, 1962, *Nucl. Phys.* **32**, 273.

GUGELOT, P. C., 1962, in *Proc. Int. Symp. on Direct Interactions and Nuclear Reaction Mechanisms* (Padua), edited by E. Clemental.

HANSTEEN, J. M., O. M. JOHNSON, AND L. KOCBACH, 1975, *At. Data Nucl. Data Tables* **15**, 305.

HOFMANN, H. M., J. RICHERT, J. W. TEPEL, AND H. A. WEIDENMULLER, 1975, *Ann. Phys. (NY)* **90**, 403.

HORSDAL-PEDERSEN, E., P. LOFTAGER, AND J. L. RASMUSSEN, 1982, *J. Phys. B* **15**, 4423.

KARWOWSKI, H. J., S. E. VIGDOR, W. W. JACOBS, S. KAILAS, P. P. SINGH, F. SOGA, AND W. D. PLOUGHE, 1979, *Phys. Rev. Lett.* **42**, 1723.

KING, H. T., AND D. C. SLATER, 1977, *Nucl. Phys.* **A283**, 365.

KOCBACH, L., 1976, *Z. Phys.* **A279**, 233.

LAEGSGAARD, E., J. U. ANDERSEN, AND L. C. FELDMAN, 1972, *Phys. Rev. Lett.* **29**, 1206.

LANE, A. M., AND R. G. THOMAS, 1958, *Rev. Mod. Phys.* **30**, 257.

LESSER, P. M. S., C. C. TRAIL, C. C. PERNG, AND M. K. LIOU, 1982, *Phys. Rev. Lett.* **48**, 308.

LICHTEN, W., 1967, *Phys. Rev.* **164**, 131.

LIOU, M. K., 1978, *Phys. Rev. D* **18**, 3350.

LIOU, M. K., C. K. LIU, P. M. S. LESSER, AND C. C. TRAIL, 1980, *Phys. Rev. C* **21**, 518.

LIU, C. K., M. K. LIOU, C. C. TRAIL, AND P. M. S. LESSER, 1982, *Phys. Rev. C* **26**, 723.

MAGLICH, B., 1973, ed. *Adventures in Experimental Physics* (γ Volume), p. 1.

MAHAUX, C., AND H. A. WEIDENMULLER, 1979, *Ann. Rev. Nucl. Sci.* **29**, 1.

MARONI, C., I. MASSA, AND G. VANNINI, 1976, *Nucl. Phys.* **A273**, 429.

MCMASTER, W. H., N. KERR DEL GRANDE, J. M. MALLETT, AND J. H. HUBBEL, 1969, *Compilation of X-Ray Cross Sections*, unpublished report UCRL-50174, Sec. II, rev. 1.

MCMURRAY, W. R., AND H. O. LUTZ, 1981, in Annual Report, Southern University Nuclear Institute, Faur, South Africa.

MCVOY, K. W., AND H. A. WEIDENMÜLLER, 1982, *Phys. Rev. A* **25**, 1462.

MCVOY, K. W., X. T. TANG, AND H. A. WEIDENMÜLLER, 1981, *Z. Phys.* **A299**, 195.

MEYER, V., H. MÜLLER, AND H. H. STAUB, 1964, *Helv. Phys. Acta* **37**, 611.

MEYERHOF, W. E., R. ANHOLT, AND W. WOELFLI, 1979, *Phys. Lett.* **84B**, 59.

MEYERHOF, W. E., J. F. CHEMIN, CH. STOLLER, AND R. ANHOLT, 1980, in Proc. Int. Conf. on Nucl. Phys. LBL-11118, Berkeley, p. 647.

MEYERHOF, W. E., G. ASTNER, D. HOFMANN, K. O. GROENVELD, AND J. F. CHEMIN, 1982, *Z. Phys.* **A309**, 93.

MERZBACHER, E., 1961, *Quantum Mechanics* (Wiley, New York), Chap. 18.

MÜLLER, B., G. SOFF, W. GREINER, AND V. CEAUSESCU, 1979, *Z. Phys.* **A285**, 27.

MÜLLER, U., J. REINHARDT, G. SOFF, B. MÜLLER, AND W. GREINER, 1980, *Z. Phys.* **A297**, 357.

REINHARDT, J., AND W. GREINER, 1977, *Rep. Prog. Phys.* **49**, 219.

REINHARDT, J., G. SOFF, AND W. GREINER, 1976, *Z. Phys.* **A276**, 285.

REINHARDT, J., B. MÜLLER, W. GREINER, AND G. SOFF, 1979, *Z. Phys.* **A292**, 211.

Riedel, C., G. Wolschin, and W. Norenberg, 1979, *Z. Phys.* **A290**, 47.
Röhl, S., S. Hoppenau, and M. Dost, 1979, *Phys. Rev. Lett.* **43**, 1300.
Röhl, S., S. Hoppenau, and M. Dost, 1981, *Nucl. Phys.* **A369**, 301.
Schröder, W. U., J. R. Birkelund, J. R. Huizenga, K. L. Wolf, and V. E. Viola, 1978, *Phys. Rep.* **45**, 301.
Soff, G., B. Müller, and W. Greiner, 1981, *Z. Phys.* **A299**, 189.
Stoller, Ch., J. F. Chemin, R. Anholt, W. E. Meyerhof, and W. Woelfli, 1983, *Z. Phys.* **A310**, 9.
Taketani, H., M. Adachi, N. Endo, and T. Suzuki, 1982, *Phys. Lett.* **113B**, 11.
Thompson, W. J., J.-F. Wilkerson, T. B. Clegg, J. M. Feagin, E. J. Ludwig, and E. Merzbacher, 1980, *Phys. Rev. Lett.* **45**, 703.
Trail, C. C., P. M. S. Lesser, A. H. Bond, M. K. Liou, and C. K. Liu, 1980, *Phys. Rev. C* **21**, 2131.
von Brentano, P., and M. Kleber, 1980, *Phys. Lett.* **92B**, 5.

CHAPTER 14

Positron Production in Heavy-Ion Collisions

H. Backe and B. Müller

1. INTRODUCTION

Atomic systems with a nuclear charge Z much greater than 100 exhibit a number of unique features not otherwise found in nature. Two characteristic properties are illustrated in Figs. 1 and 2. In Fig. 1 we have plotted the binding energy of a K-shell electron around hypothetical nuclei up to $Z \approx 200$. For $Z > 150$ the binding energy exceeds the rest energy $m_e c^2$ of the electron; i.e., adding the electron to the nucleus actually diminishes the total mass of the system. At the *critical charge* $Z_C \approx 170$–175 the binding energy reaches twice the electron rest mass, the threshold for spontaneous creation of an electron–positron pair. As has been discussed extensively in the literature (Pieper and Greiner, 1969; Zeldovich and Popov, 1972; Müller, Rafelski, and Greiner, 1972; Greiner, Müller, and Rafelski, 1985), this signals the transition to a new ground state of quantum electrodynamics: for $Z > Z_C$ the static electric field surrounding a bare nucleus spontaneously creates two electrons to fill the vacant K shell, while two positrons are emitted to balance the overall charge. Thus, the ground state is not any more a bare nucleus but one that is surrounded by two electrons. Since the new ground state carries charge, one speaks of a *charged vacuum* (Rafelski, Müller, and Greiner, 1974). This situation offers the unique opportunity to study by laboratory experiments a phase transition, i.e., a discontinuous change, in the ground state according to relativistic quantum field theory.

A related, curious property of atomic systems with $Z \gg 100$ is illustrated

H. Backe • Institut für Physik, Johannes Gutenberg-Universität, D-6500 Mainz, Federal Republic of Germany. B. Müller • Institut für Theoretische Physik, Johann Wolfgang Goethe-Universität, D-6000 Frankfurt am Main, Federal Republic of Germany.

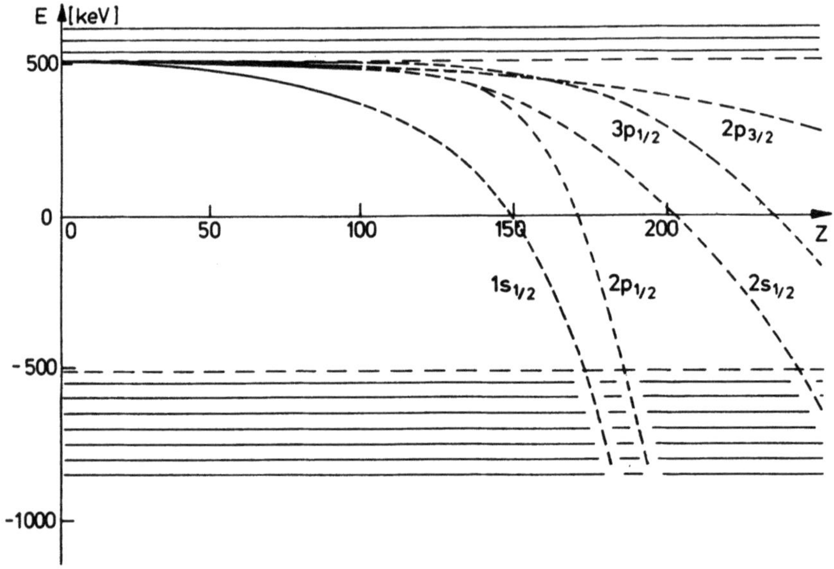

FIGURE 1. Extrapolation of energy eigenvalues of deeply bound states for atomic numbers up to $Z = 250$. The energies were calculated in the Dirac–Hartree–Fock–Slater approximation with finite-size nuclei. The $1s$ state acquires negative total energy at $Z = 150$ and becomes supercritically bound at $Z_C \approx 173$.

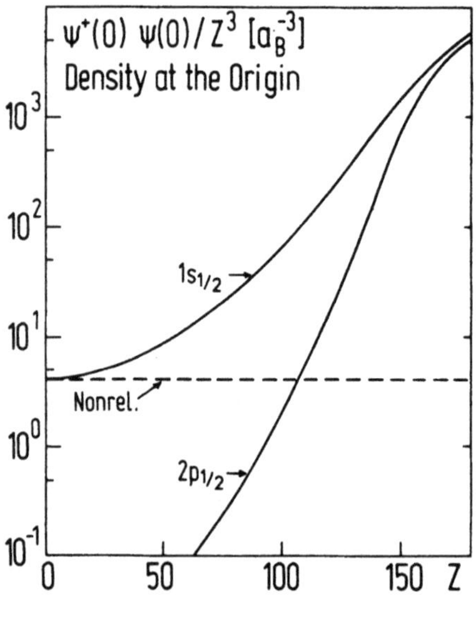

FIGURE 2. Density of $1s$ and $2p_{1/2}$ electrons at the nuclear center, as function of nuclear charge Z. The density has been scaled by the nonrelativistic dependence Z^3. (From Soff *et al.*, 1982b.)

in Fig. 2. Due to relativistic effects, the electronic wave functions become localized at the nucleus in a rather dramatic way. For instance, at the critical charge $Z_C \approx 170$, the density of a K electron inside the nucleus is increased about 5000 times above the extrapolation of the nonrelativistic Schrödinger theory. This high density is caused by a phenomenon that has been called "collapse to the center" (Popov, 1971). By deriving an effective potential for the relativistic Dirac equation, one can show that the Coulomb potential always overwhelms the centrifugal barrier if $Z\alpha$ is larger than $(j+\frac{1}{2})$, where j is the angular momentum of the electron and α is the fine-structure constant. In fact, the electronic wave function would collapse to a point, were it not for the finite size of the nucleus (Gärtner *et al.*, 1980, 1981). The high degree of localization of the wave function can be expressed in a complementary way by saying that the high-momentum components of the electronic wave function are greatly enhanced (Backe, 1983). This effect is of particular importance for an understanding of ionization processes in atomic systems with $Z \gg 100$.

For a long time it was not clear whether modifications of the single-particle picture introduced by the interaction with virtual electron–positron pairs (vacuum polarization, Fig. 3a) and virtual photons (self-energy, Fig. 3b) might change the value of the critical charge and, possibly, prevent the binding energy from exceeding $2mc^2$ altogether (Rafelski, Fulcher, and Greiner, 1971; Soff, Rafelski, and Greiner, 1973). Recent numerical calculations of these corrections to first order in α but to all orders in the Coulomb field characterized by the value of $Z\alpha$ have shown the following results. At $Z = Z_C$ vacuum polarization causes an increase in the binding energy by 10.7 keV (Pieper and Greiner, 1969; Gyulassy, 1974, 1975), whereas self-energy corrections reduce the binding by 11.0 keV (Soff *et al.*, 1982). The net effect is a tiny change of 0.3 keV in the total binding energy of 1022 keV at the critical charge, which may be neglected for all practical purposes.

The study of such superheavy atomic systems is complicated by the fact that, in spite of considerable efforts, no stable nucleus with $Z \gg 100$ has been found to this day. The essential step towards an experimental realization of the phenomena discussed above was taken independently by groups at

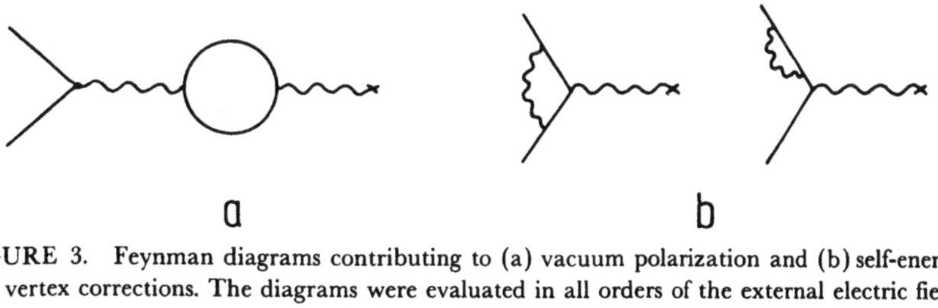

FIGURE 3. Feynman diagrams contributing to (a) vacuum polarization and (b) self-energy and vertex corrections. The diagrams were evaluated in all orders of the external electric field.

Frankfurt and Moscow some ten years ago, when they suggested that a supercritical nuclear charge could be assembled for an appreciable interval of time during the sub-barrier collisions of two very heavy atoms, such as U + U yielding a total $Z = 92 + 92 = 184$. The argument for this idea is based on the observation that the relative velocity of the two nuclei is of the order of $c/10$, where c is the speed of light, while the K electrons move with a velocity essentially equal to the velocity of light, due to the strong binding. Thus it was expected that the K-shell electrons can follow the nuclear motion and adjust to the instantaneous relative position of the nuclei. Such a collision system is called a *quasimolecule* or, in the limit when the internuclear distance is much smaller than the K-shell radius, a *quasiatom*. We shall see that this expectation has been confirmed to a large degree.

We make a few order-of-magnitude estimates of the characteristic time scales involved in such a heavy-ion–atom collision (Table I). Here τ_{coll} is the typical collision time required for the nuclei to approach from a distance equal to the electron Compton wavelength (386 fm) and to recede again; τ_{crit} is the time the nuclei spend together while the K-shell is supercritical (for U + U); τ_{sp} is the characteristic decay time with respect to spontaneous positron emission, and τ_{rad} is the time required to fill a K vacancy through an x-ray transition. It is clear from Table I that, while positron emission acts as the fastest probe of the physics of quasi atoms, the critical collision time τ_{crit} is still shorter by almost two orders of magnitude. This has two main effects: it gives a K vacancy only a 1% chance of being filled through positron emission and—worse—it introduces an energy uncertainty of about 500 keV into the positron spectrum. The latter effect has motivated investigations of possible delays in the collision process due to nuclear interactions, which have already yielded interesting results (Sec. 3).

TABLE I

Characteristic time scales for various processes occurring in a collision between very heavy ions: collision time (τ_{coll}), period of supercritical binding in the absence of nuclear contact (τ_{crit}), spontaneous positron emission time (τ_{sp}), and radiative lifetime of a K vacancy (τ_{rad}).

Process	Time (s)
τ_{coll}	2×10^{-20}
τ_{crit}	2×10^{-21}
τ_{sp}	10^{-19}
τ_{rad}	$10^{-18} \cdots 10^{-17}$

In Sec. 2, we present a short outline of the theoretical treatment of excitation processes in quasimolecular collisions, with special emphasis on the inclusion of supercritically bound states. We then discuss the possible effects of nuclear time delay on positron spectra (Sec. 3). A discussion follows of measurable quantities and of the experimental equipment that has been used so far (Sec. 4). After a description of the methods of data analysis (Sec. 5), we present experimental results and compare these with recent calculations (Sec. 6). This chapter concludes with a short summary and outlook towards future experiments, and some theoretical speculations (Sec. 7).

A detailed discussion of the historical development of the field and its bearing on related problems goes beyond the scope of this chapter; we refer the reader to various earlier review articles, by Müller (1976), Reinhardt and Greiner (1977, 1982), Rafelski *et al.* (1978), Backe (1983), and Greiner *et al.* (1985), which contain many of the details that have been omitted here.

2. ELECTRON–POSITRON EXCITATIONS IN SUPERHEAVY QUASIMOLECULES

2.1. Time-Dependent Perturbation Theory

As pointed out in Sec. 1, a crude comparison of the velocities of inner-shell electrons and of the colliding nuclei indicates that electronic excitations in the scattering process can be successfully treated by starting from an adiabatic picture. Assuming that the electrons (or positrons) adjust themselves instantaneously to the nuclear motion described in terms of the internuclear distance R, one obtains a complete basis of states φ_i, the so-called quasimolecular states, which are eigenstates of the instantaneous relativistic two-center Hamiltonian (Müller, Rafelski, and Greiner, 1973):

$$H_{\mathrm{TCD}}(r, R)\, \varphi_i(r, R) = \varepsilon_i(R)\, \varphi_i(r, R). \tag{1}$$

In Fig. 4 we show the molecular correlation diagram for the heaviest system so far investigated, U + Cm ($Z_1 + Z_2 = 92 + 96 = 188$). The continuous gain in binding energy of the lowest-angular-momentum states is conspicuous, the $1s\sigma$ state diving into the negative-energy continuum at $R_{\mathrm{cr}} \sim 40$ fm.

The true wave function of an electron is next described as an arbitrary superposition of these quasimolecular states. As long as there is no restriction on the number of basis states, no approximation is involved in this scheme.

It is important to note that the quantum-mechanical wavelength associated with nuclear motion in the collision of very heavy ions is much smaller (~ 0.1 fm) than the typical distance of closest approach (~ 10 fm). It

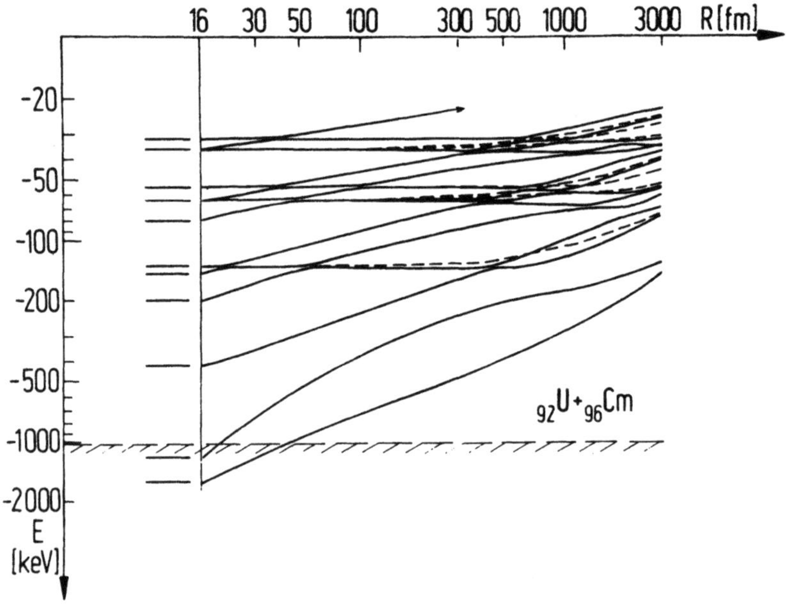

FIGURE 4. Molecular-correlation diagram for the supercritical system U + Cm. The critical radius is ~40 fm.

is therefore possible, at least in the absence of direct nuclear contact, to treat the collision semiclassically, thereby reducing the relative separation R between the two nuclei from a dynamical variable to a parameter $R(t)$ which depends upon time according to the classical equations of motion (Smith, Müller, and Greiner, 1975; Heinz, Müller, and Greiner, 1981). The electron motion is then described by a solution of the time-dependent two-center Dirac equation,

$$i\frac{\partial}{\partial t}\phi_k(r,t)=H_{\mathrm{TCD}}(r,R(t))\,\phi_k(r,t). \tag{2}$$

If the bombarding energy is not too high, the wave functions $\phi_k(t)$ can be expanded in terms of adiabatic basis states

$$\phi_k(r,t)=\sum_i a_{ki}(t)\,\varphi_i(r,R(t))\exp[-i\chi_i(t)], \tag{3}$$

where

$$\chi_i(t)=\int_{t_0}^{t} dt'\,\langle\varphi_i|\,H_{\mathrm{TCD}}\,|\varphi_i\rangle \tag{4}$$

is a conveniently chosen phase. The initial condition

$$a_{ki}(t_0) = \delta_{ki} \tag{5}$$

states that the electron starts out in the stationary state φ_k. There is one more simplification that can be safely made. A collision between a uranium atom and an incoming uranium ion includes far more than 100 electrons. The two-center Hamiltonian, H_{TCD}, therefore involves the two-body electron–electron interaction. Accordingly, the states ϕ_k are, in principle, correlated many-electron states. The amount of corrrelation can, however, be shown to be small for inner-shell electrons (Rihan *et al.*, 1978). The many-electron wave function can, therefore, be represented by a single Slater determinant. Then H_{TCD} is a one-body operator that contains effective screening due to the presence of the other electrons.

Under these conditions, Fock-space methods allow one to work with single-particle wave functions φ_i, ϕ_k instead of Slater determinants. It can be shown that all relevant observables in the final state can be expressed in terms of the single-particle occupation amplitudes a_{ki} at $t = \infty$, the only criterion being whether a state φ_i was initially occupied or vacant (Reading, 1973; Reinhardt *et al.*, 1980a; Reading and Ford, 1980). For example, the number of particles in an initially vacant state p is

$$\mathcal{N}_p = \sum_{k,\text{occ.}} |a_{kp}|^2, \tag{6}$$

while the number of holes in an initially occupied state h is

$$\mathcal{N}_h = \sum_{k,\text{vac.}} |a_{kh}|^2. \tag{7}$$

Similar expressions hold for particle–hole coincidences, where proper care must be taken in counting the spin states (Soff *et al.*, 1980).

By inserting the ansatz (3) into Eq. (2), a system of coupled equations for the amplitudes a_{ki} is obtained:

$$\dot{a}_{ki} = -\sum_{j \neq i} a_{kj} \langle \varphi_i \mid \dot{\varphi}_j \rangle \exp[i(\chi_i - \chi_j)]. \tag{8}$$

The temporal variation of the adiabatic basis states can be rewritten in terms of the variation due to a change in the internuclear distance and in the orientation of the quasimolecule. For K-shell ionization and positron emission in very heavy collision systems, the rotational coupling is only of minor importance and will be neglected in the following (Betz *et al.*, 1976, but see Heiligenthal *et al.*, 1978, for exceptions). The matrix elements $\langle \varphi_i \mid \dot{\varphi}_j \rangle$ in Eq. (8) must be properly corrected for the asymptotic translational motion of the separated atoms (Bates and McCarroll, 1958; Thorson and Delos, 1978). This need arises from the fact that in the asymptotic region the relative nuclear coordinate R must be replaced by the relative distance of the centers of mass of the full atoms or ions. The creation of positrons, which takes place at closest approach of the nuclei, is, however, not greatly influenced by this effect (Heinz, Müller, and Greiner, 1981).

Early attempts to solve Eqs. (8), based on first-order perturbation theory, led to considerable quantitative disagreement with experimental results. These discrepancies were resolved by Reinhardt *et al.* (1979), who observed that multistep processes make large contributions to the various excitations. All calculations discussed in this chapter are based on solving the coupled equations (8) numerically, after splitting the positive- and negative-energy continua into a sufficiently large number of discrete states ($\sim$30–50). As discussed briefly in Sec. 2.2, these calculations agree well with presently available experimental data.

2.2. *Ionization and Electron Emission*

Before discussing the problem of positron creation in more detail, we must make sure that the use of the adiabatic basis for the treatment of electronic excitations is justified. This precaution is required for two reasons:

1. Pair creation itself can be understood as a special kind of electronic excitation in which an electron is knocked out of a negative-energy continuum state, leaving a hole that is observed as positron (see Fig. 5a).
2. The creation of a vacancy in the K shell of the united quasiatom in the initial phase of the collision is an essential prerequisite for spontaneous positron emission (see Fig. 5b).

We must therefore see how well we understand K-shell vacancy creation and ejection of electrons into high momentum states, so-called delta electrons. The test of theory has been carried out by checking whether the results are consistent with the underlying assumptions, and by comparison with experimental data.

We first look at theoretical self-consistency. The basic assumption underlying the quasimolecular excitation model outlined above is that nuclear motion can be treated as adiabatic. This means that an electron in the K shell of one of the colliding atoms indeed ends up, with overwhelming probability, in the K shell of the united quasiatom. If we denote the $1s$ state by $k=0$ and the time of closest approach by $t=0$, we need to test whether or not

$$|a_{00}(0)|^2 \sim 1, \qquad |a_{0i}(0)|^2 \ll 1 \qquad \text{for } i \neq 0. \tag{9}$$

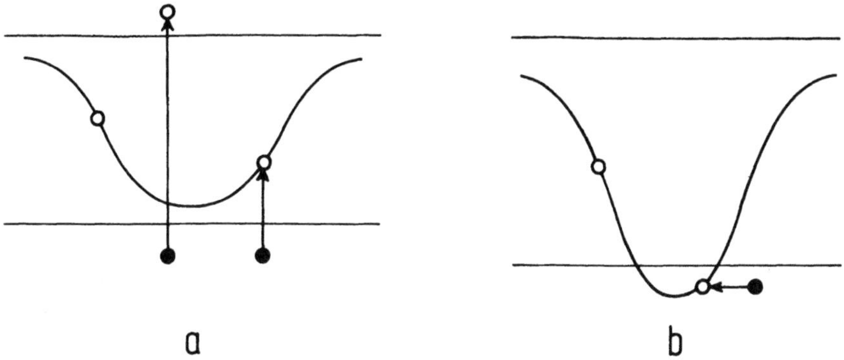

FIGURE 5. Schematic representation (a) of dynamically induced pair production and (b) of spontaneous positron creation.

The crucial question concerning supercritical binding, however, is whether the electron really becomes bound by more than $2mc^2$, i.e., whether its energy expectation value

$$\bar{E}_0(t) = \langle \phi_0(t) | \, H_{\text{TCD}} \, | \phi_0(t) \rangle = \sum_i \varepsilon_i \, |a_{0i}|^2 \tag{10}$$

actually reaches the critical value $-mc^2$.

Figure 6 contains a plot of this mean energy $E_0(t)$ of a K-shell electron in a U + U collision. It can be seen that, although the electron is somewhat less strongly bound than would be predicted in a strictly adiabatic picture, binding becomes supercritical at small nuclear separation. We conclude that the adiabatic picture is, indeed, a very good starting point for the description of dynamical processes, and that these processes introduce modifications on the 10% level. Results are similar if one tries to find a better basis for treating the excitations than is provided by the dynamical one. Again, it turns out that only minor modifications of the basis states are permitted (Theis, Reinhardt, and Müller, 1979; Theis and Müller, 1981).

The crucial test, however, is provided by a comparison with experimental data. To do this systematically would require an extensive review. We refer the reader to the original literature: Soff *et al.* (1980, 1981a); Liesen *et al.* (1980b); deReus *et al.* (1983). Here, a few highlights will suffice. The Pb + Cm system ($Z = Z_1 + Z_2 = 178$) is one of the best-studied collision systems with regard to K-shell ionization (Liesen *et al.*, 1980a). The system is sufficiently asymmetric to separate the contributions from the $1s\sigma$ and the $2p_{1/2}\sigma$ quasimolecular states by means of Meyerhof's (1973) sharing correction. Special experiments (Ito *et al.*, 1982; Bosch, 1982) have been performed

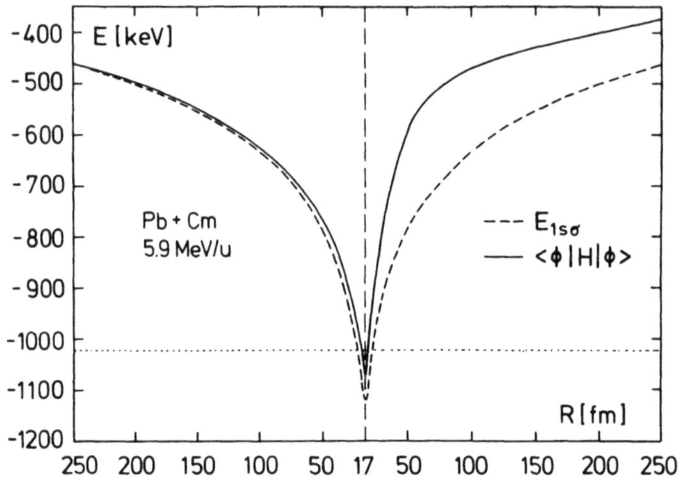

FIGURE 6. Binding energy of the quasimolecular 1s state in a Pb + Cm collision, as function of internuclear distance R. The adiabatic energy eigenvalue (dashed curve) is compared with the expectation value of the Hamiltonian with the fully time-dependent wave function (solid curve). The difference corresponds to excitation of the 1s electron into higher orbitals. The adiabatic approximation can be seen to be very good until well after the point of closest approach. (From Soff *et al.*, 1980.)

for very small impact parameters ($b \leqslant 30$ fm), making use of a clever Doppler-shift attentuation technique to separate vacancies created during the collision from those formed much later due to conversion of excited states in the Cm nucleus. The agreement found at 5.4 MeV/u bombarding energy is impressive (Fig. 7). At 3.6 MeV/u there is some disagreement in absolute magnitude between theory and experiment. A general, yet unexplained, feature also found in other collision systems, such as Xe + Pb (Meyerhof, 1983), is that theory predicts too steep a rise of the ionization probability with increasing bombarding energy. However, it has been pointed out (Soff, Müller, and Greiner, 1978) that it is the impact-parameter dependence which carries most information on the structure of the united quasiatomic system, and this dependence is well reproduced in all cases.

In a different type of experiment, the spectra of delta electrons are measured. These spectra are of particular interest when measured in coincidence with the K-shell vacancy left behind by the emitted electron (Kozhuharov, 1982; Güttner *et al.*, 1982). Again, asymmetric systems are required to resolve the K shell of the united quasiatom. Delta-electron spectra are known to provide information about the localization of the quasiatomic wave function, or more precisely, about the wave function in

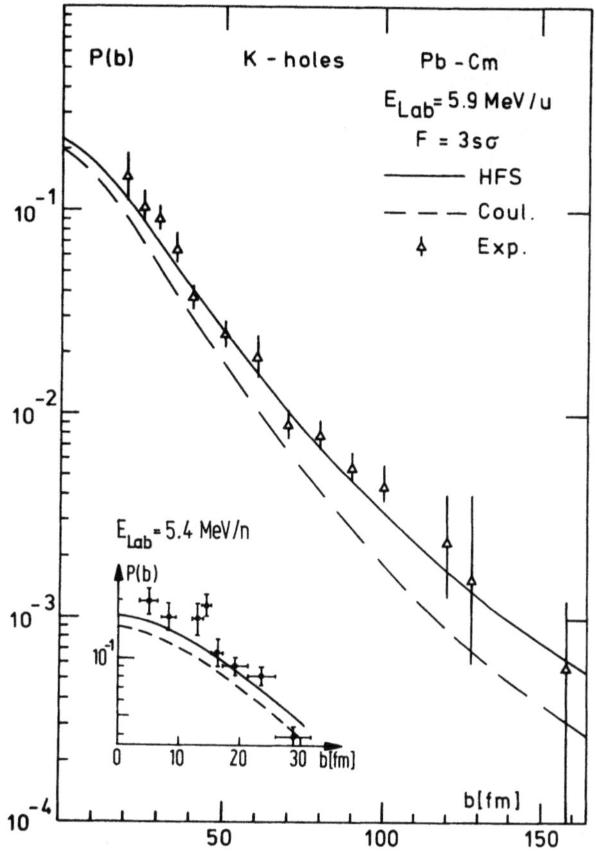

FIGURE 7. 1sσ vacancy production in Pb + Cm collisions versus impact parameter σ. When electron screening is included, there is good agreement between theory and experimental data at 5.9 MeV/u or or 5.4 MeV/u bombarding energy. (Data from Ito *et al.*, 1982.)

momentum space. Experiments by the Heidelberg group (Koenig, 1982) have shown that there is a distinct transition from nonrelativistic behavior to a highly relativistic regime near a total Z of ~ 100. As an example we show in Fig. 8 the coincidence data for the system I + Pb ($Z_1 + Z_2 = 135$), which is in excellent qualitative and quantitative agreement with calculations (Soff *et al.*, 1983). This result constitutes clear evidence of the high degree of localization of the inner-shell electrons in atomic systems with charge $Z \gg 100$. With these successes in mind we now turn to the subject of creation of positrons.

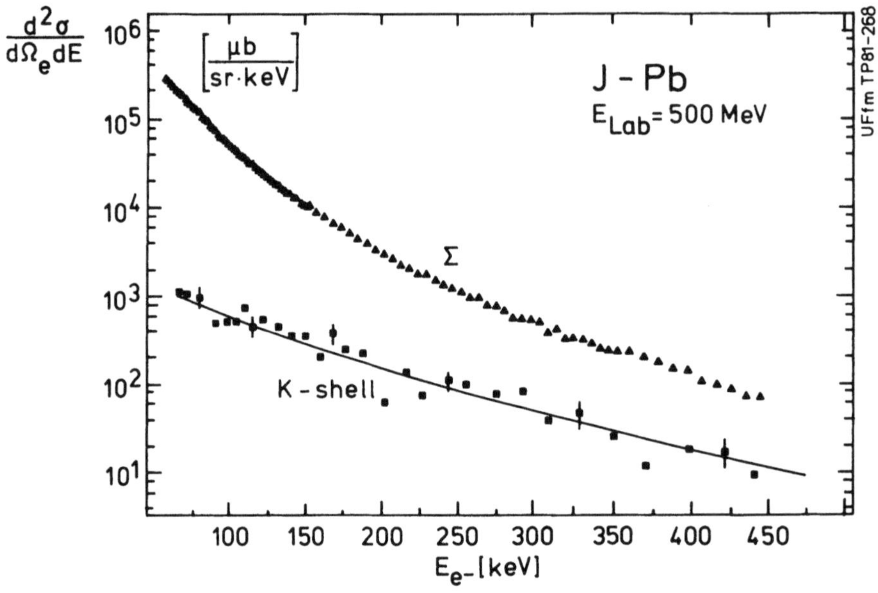

FIGURE 8. Delta-electron spectra of I–Pb collisions at 500 MeV. The spectrum in coincidence with a Pb K vacancy (lower set of points) agrees well with a theoretical calculation.

3. POSITRON CREATION

3.1. Inclusion of Supercritically Bound States

When a quasimolecular state becomes supercritically bound, i.e., when its binding energy exceeds $2mc^2$, it disappears from the discrete spectrum of the Hamiltonian H_{TCD} and reappears as a resonance in the negative-energy continuum of the Dirac equation (Müller *et al.*, 1972a, b). The resonance, in general, is narrow ($\Gamma \sim$ a few keV) and changes position as a function of internuclear distance, hence it is impractical to treat it by splitting the continuum into energy bands. Moreover, such a treatment would obscure much of the physics involved, since the charge density associated with the resonance remains well localized. Reinhardt, Müller, and Greiner (1980a, 1981a) proposed giving up the strict requirement that the basis states be eigenstates of the full instantaneous Hamiltonian. Instead, they chose a localized wave function φ_R to represent the bound-state part of the supercritical resonance. This is a reasonable concept because the resonance is narrow and the amplitude of the bound state part is two orders of magnitude larger than that of the continuum tail at the resonance. As a consequence, the characteristic time scale for positron emission—as listed in

Table I—is much larger than the typical collision time (except in collisions in which a compound nucleus is formed). The resonance decay is, therefore, not appropriately treated as a quasistatic process.

To obtain a suitable basis for expansion of the time-dependent wave functions ϕ_k, a new set of continuum states $\tilde{\varphi}_E$ must be constructed that is orthogonal to the "resonance" state φ_R. This is achieved by finding eigenstates of the projected Hamiltonian

$$\tilde{H}_{\mathrm{TCD}} = H_{\mathrm{TCD}} - |\varphi_R\rangle\langle\varphi_R|\, H_{\mathrm{TCD}}. \tag{11}$$

As φ_R and $\tilde{\varphi}_E$ are not eigenstates of the full Hamiltonian H_{TCD}, additional couplings of the form

$$V_E = \langle\tilde{\varphi}_E|\, H_{\mathrm{TCD}}\, |\varphi_R\rangle \tag{12}$$

arise in the coupled equations for the amplitudes. These couplings, being independent of dR/dt, describe spontaneous positron emission. The couplings would still be present if the collision were artificially slowed down to occur infinitely slowly, i.e., their action does not originate in the collision dynamics but is caused by the supercritical binding.

The choice of the localized resonance state φ_R is not unique. Different choices result in a different distribution of the coupling strength between dynamical and spontaneous couplings away from resonance, but not in the immediate vicinity of the resonance. When the coupled equations for the amplitudes are solved, however, the final values $|a_{ki}(\infty)|^2$ are independent of the choice of φ_R for all practical purposes. This fact is corroborated by the recent calculations of Tomoda and Weidenmüller (1982), who have used a completely different method for constructing φ_R but found results agreeing with those of Reinhardt *et al.* (1981) within the numerical accuracy.

We now look at some results for collisions in which the nuclei do not come into contact (Fig. 9). We see that with nuclear charge increasing from $Z = 164$ (Pb + Pb) to $Z = 188$ (U + Cm), the positron creation probability increases enormously, but there is no sign of a qualitative change of the spectrum in going to the supercritical systems U + U and U + Cm. The reason for this disappointing result is that the collision time, as estimated in Table I, is simply too short to allow for a significant number of positrons to be created spontaneously. In other words, there is too much pair creation from dynamical processes. The situation is made even worse by the large energy uncertainty introduced by the shortness of the period of supercritical binding, causing the spontaneous positrons to be spread out over the entire spectral range. We are thus forced to conclude that there does not seem to

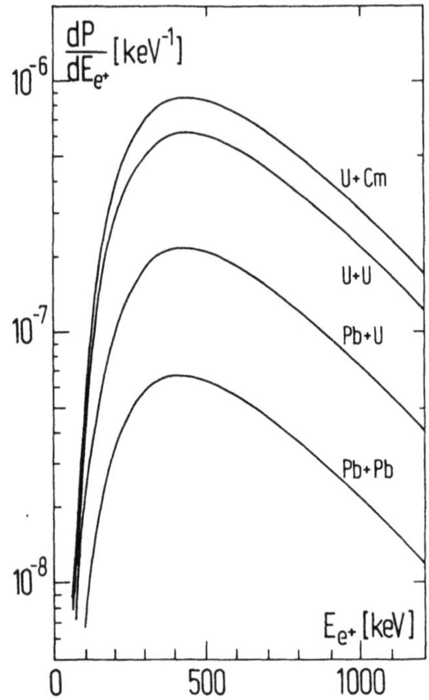

FIGURE 9. Theoretical positron spectra in head-on ($b=0$) collisions for four different systems. No qualitative change is seen to occur in the transition from subcritical systems (Pb + Pb, Pb + U) to supercritical systems (U + U, U + Cm) because the period of supercritical binding is too short. (From U. Müller, private communication.)

exist a clear-cut way of identifying spontaneous positron creation in Rutherford scattering.

3.2. *Collisions with Nuclear Delay*

The difficulty of identifying spontaneous positron creation in low-energy nuclear collisions led Rafelski, Müller, and Greiner (1978) to suggest that it might be useful to look at collisions with somewhat higher energy where nuclear reactions occur. Two scenarios could be envisaged in the energy region from the Coulomb barrier up to ~10 MeV/u. In the first scenario, one would make use of the long interaction time between two nuclei in a deep inelastic collision. The large mass transfer observed in such collisions (up to 20 nucleons and more) has been understood to be associated with an extended period of nuclear contact that may reach $10\text{–}15 \times 10^{-21}$ s for medium-heavy nuclei. Estimates of this reaction time are based on microscopic models for nucleon exchange (Schmidt *et al.*, 1978). The system U + U has been the object of a fair amount of work, but the situation here is not so clear because the excited fragments undergo subsequent fission. Still, several models exist that explain the gross features of mass transfer and energy dissipation in the U + U reaction, yielding contact times of the order of 10^{-21} s (Schmidt *et al.*, 1978). The shortness of the period of contact is mainly due to the strong Coulomb repulsion between the nuclei.

A second, more spectacular scenario is based on the assumption that there is a certain chance of forming a quasistable nuclear composite system with about 184 protons and 290 neutrons, if a collision is close to the Coulomb barrier (otherwise, too much excitation energy would be present, causing immediate fission). In the extreme, a compound nucleus or even a quasistable superheavy nucleus might be formed. In this scenario, the time available for spontaneous positron emission might be very long, of the order of 10^{-19} s or even more. It should be pointed out, however, that until recently this second scenario was considered as extremely unlikely by specialists in the field.

It is clear that in the case of a very long nuclear contact time, $T \gtrsim 10^{-19}$ s, the spontaneous emission of positrons will proceed essentially in the same way as in a static situation. In other words, the spectrum of spontaneously emitted positrons will have a Lorentzian shape, of a width determined by the width of the bound-state resonance in the negative-energy continuum. Since the collision process is still present, with the nuclei approaching and finally separating, a flat dynamically generated background will be superimposed on the tails of the Lorentzian. If, however, the contact time T is comparable to, or shorter than, the spontaneous emission time τ_{sp}, contributions to the spectrum from both mechanisms become thoroughly mixed. In fact, their amplitudes must be added coherently because there is no way of distinguishing, in a given collision, whether a positron has been emitted spontaneously or due to dynamical processes.

Within the framework of the semiclassical model, the effect of a given delay time T on the electron–positron excitation amplitudes is found in the following way: Before and after nuclear contact, the amplitudes $a_{ki}(t)$ are determined by the coupled equations (8), which contain no information about the nuclear interaction. We call the excitation amplitudes during the approach of the nuclei a_{ki}^-, and those of the separation phase a_{ki}^+. The amplitudes a_{ki}^- and a_{ki}^+ are solutions of Eq. (8) with initial conditions $a_{ki}(-\infty)=\delta_{ki}$ and $a_{ki}(T)=\delta_{ki}$, respectively. During nuclear contact the internuclear distance R remains constant; therefore, the electronic wave functions propagate undisturbed between $t=0$ and $t=T$. For eigenstates of the Hamiltonian, this stage corresponds to a mere phase change $\exp(-iET)$, whereas in the presence of the spontaneous coupling (12), transitions between the resonance state φ_R and the positron states $\tilde{\varphi}_E$ occur. One finds (Müller and Oberacker, 1980; Reinhardt *et al.*, 1981b) that the excitation amplitude after the collision is given by

$$a_{ki}(\infty)=\sum_{lm} a_{kl}^+ A_{lm} a_{mi}^-, \tag{13}$$

where the matrix

$$A_{lm}(T) = \delta_{lm} \exp(-iE_l T - \tfrac{1}{2}\Gamma T \delta_{lR})$$
$$+ \delta_{lR} \frac{V_{E_m}}{E_m - E_R + (i/2)\Gamma} [\exp(-iE_m T) - \exp(-iE_R T - \tfrac{1}{2}\Gamma T)]$$
$$+ \delta_{mR} \frac{V^*_{E_l}}{E_R - E_l - (i/2)\Gamma} [\exp(-iE_R T - \tfrac{1}{2}\Gamma T) - \exp(-iE_l T)] \tag{14}$$

describes the time evolution during nuclear contact. The quantity $\Gamma = 2\pi\, |V_{E_R}|^2$ is the spontaneous decay width of the supercritical resonant state due to positron production.

Calculations which take the foregoing considerations into account have been performed by Reinhardt and U. Müller (Reinhardt *et al.*, 1981a,b). To realize the importance of the spontaneous creation process, it is useful to compare a subcritical system, such as Pb + Pb, with a supercritical system, e.g., U + U (Fig. 10). In the subcritical system the introduction of a nuclear contact time T only results in a change in the pattern of interference

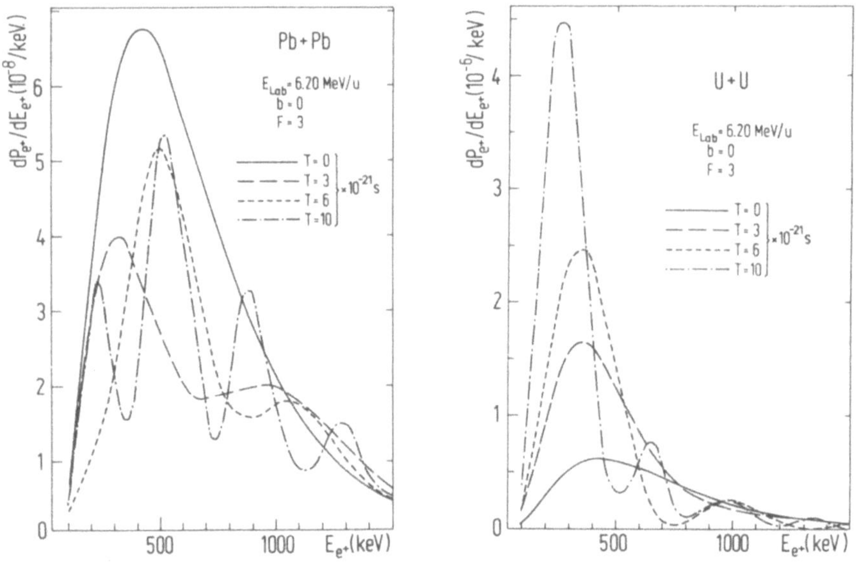

FIGURE 10. A comparison between Pb + Pb (left) and U + U (right) shows the dramatically different behavior of positron spectra in coincidence with nuclear time delay for a subcritical and a supercritical collision system. In the supercritical (U + U) system, a pronounced spontaneous emission line grows out of the spectrum for large reaction times T. The position of the line in the limit of very long T is given by the $1s$ binding energy minus twice the electron rest mass.

between the dynamic processes that occur during approach and separation of the nuclei, respectively. No enhancement of positron production is found. In the supercritical collision system, on the other hand, the curve representing the spectrum of spontaneously emitted positrons grows gradually in amplitude as the time of contact becomes longer and longer (note the difference between the ordinate scales in the two parts of Fig. 10). It seems clear that with a contact time of $\sim 5 \times 10^{-21}$ s or more the spontaneous process could be experimentally detected.

One must keep in mind, however, that nuclear reactions are not classical deterministic processes which can be assigned a definite duration T. In principle, these processes must be treated on a quantal level, in which the relevant quantity is the scattering amplitude. For "dissipative" collisions between heavy, complex nuclei, such a description is unpracticable and is usually replaced by a statistical treatment on a classical or semiclassical level (Agassi *et al.*, 1977, 1979; Ko *et al.*, 1979). The contact time then has a statistical distribution $f(T)$, usually taken to be a Gaussian of the form

$$f(T) \sim \exp\left[-\frac{(T-T_0)^2}{2\Delta T^2}\right]. \tag{15}$$

The relative spread $\Delta T/T_0$ has been estimated, within a Monte Carlo trajectory simulation model, to be of the order of 25% (Mathews *et al.*, 1980). The situation is quite different for quasielastic collisions near the Coulomb barrier. Here, nuclear resonance scattering can account for the main part of the time delay. Then the nuclear scattering must be formulated as a fully quantum-mechanical process (Heinz *et al.*, 1983; see also Chap. 13). In many cases it is possible to separate the space determined by the relative nuclear coordinate R into an inner region (small R), in which nuclear forces act but electronic excitations can be neglected, and an outer region (large R), where the nuclear motion can be treated semiclassically. Then one can show (Tomoda, 1984) that Eq. (13) remains valid if the matrix $A_{lm}(T)$ is replaced by the nuclear scattering matrix $S(E-E_l)\,\delta_{lm}$ in the absence of spontaneous decay, where E is the total scattering energy. Spontaneous decay can be included in more complicated expressions similar to Eq. (14). The physical meaning is obvious: if the electron occupies an intermediate excited state l, the energy available for *nuclear* scattering is reduced by the excitation energy E_l.

To relate these results to the semiclassical expression, the final excitation probabilities must be compared. Without spontaneous decay, we have

$$|a_{ki}(\infty)|^2 = \sum_{lm} a_{kl}^{-}\, a_{km}^{-*}\, a_{li}^{+}\, a_{mi}^{+*} \cdot \begin{cases} \int_0^\infty dT\, f(T)\, e^{i(E_m-E_l)T} \\ \langle S(E-E_l)\, S^*(E-E_m)\rangle, \end{cases} \tag{16}$$

where the quantum-mechanical result has been averaged over the energy spread of the beam. We thus arrive at the result that the classical delay-time distribution function $f(T)$ must be identified with the Fourier transform of the autocorrelation function of nuclear scattering (Reinhardt *et al.*, 1983). Both formulations are completely equivalent. The limiting cases of an isolated resonance and of many overlapping resonances both correspond to an exponential distribution of delay times, but almost anything else is possible, depending on the details of the nuclear reaction.

How a spread of reaction times affects the positron spectrum in a U + U collision is shown in Fig. 11. The line shape is drawn for three different time distributions $f(T)$: a sharply defined reaction time (*a*), a Gaussian distribution with 25% smearing (*b*), and an exponential time distribution (*c*). The main effect of the time smearing is to eliminate interferences (oscillations) from the spectrum, but there is also some change in the energy of the spontaneous-emission line. However, high-precision data would be needed to distinguish between these different distributions.

4. EXPERIMENTAL CONFIGURATIONS FOR IN-BEAM POSITRON SPECTROSCOPY

4.1. *Requirements for the Detection Systems*

The objective of positron spectroscopy in heavy-ion collisions is, in principle, to investigate a triple differential cross section

$$\frac{d^3\sigma_{e^+}}{dE_{e^+}\, d\Omega_{e^+}\, d\Omega_p}(E_{e^+}, \theta_{e^+}, \phi_{e^+}, \theta_p, E_1, M_1, Z_1, M_2, Z_2) \tag{17}$$

which is differential with respect to the kinetic positron energy E_{e^+}, the solid-angle element $d\Omega_{e^+}$ of the emitted positron, and the solid-angle element $d\Omega_p$ of scattered projectiles or recoils. The quantities θ_{e^+} and ϕ_{e^+} describe the emission angles of the positron relative to the beam axis and the scattering plane, respectively. The scattering angle θ_p is measured with respect to the beam direction of scattered projectiles or recoils. All quantities are measured in the laboratory system. The cross section is determined as a function of the charge numbers Z_1, Z_2 and masses M_1, M_2 of the projectile and target nuclei, respectively..

In addition, the cross section in close collisions with nuclear contact can also depend on the charge and mass numbers of the colliding nuclei in the exit channel and on the dissipated energy. Since these quantities have not

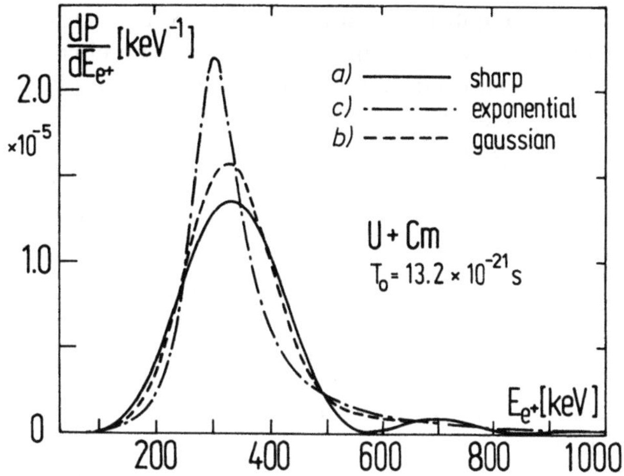

FIGURE 11. Shape of the spontaneous positron emission line for three different distributions of nuclear delay times. In all cases the average time was fixed at $T_0 = 13.2 \times 10^{-21}$ s.

been measured in most of the experiments described below we have not explicitly included them in Eq. (17).

The positron-production cross sections in heavy-ion collisions are fairly small, therefore positron spectrometers for these experiments must accept large solid angles to obtain reasonable counting rates. Therefore what is measured is a cross section

$$\frac{d^2\sigma'_{e^+}}{dE_{e^+}\, d\Omega_p} = \frac{\int_{\Omega_{e^+}} d\Omega_{e^+}\, \tilde{\varepsilon}(E_{e^+}, \theta_{e^+}, \phi_{e^+})\, d^3\sigma_{e^+}/(dE_{e^+}\, d\Omega_{e^+}\, d\Omega_p)}{\int_{\Omega_{e^+}} d\Omega_{e^+}\, \tilde{\varepsilon}(E_{e^+}, \theta_{e^+}, \phi_{e^+})}, \tag{18}$$

whereby $\tilde{\varepsilon}(E_{e^+}, \theta_{e^+}, \phi_{e^+})$ is the positron detection efficiency and Ω_{e^+} the acceptance solid angle of the spectrometer (1.9–9.3 sr for the various instruments in use).

From an experimental standpoint it is more convenient, however, to measure the energy differential probability for positron production,

$$\frac{dP^{\text{exp}}_{e^+}}{dE_{e^+}} = \frac{d^2\sigma'_{e^+}}{dE_{e^+}\, d\Omega_p}\left(\frac{d\sigma_R}{d\Omega_p}\right)^{-1} \approx \frac{\Delta N^{\text{exp}}_{e^+}}{\varepsilon(E_{e^+})\, \Delta E_{e^+}\, N_p}, \tag{19}$$

which is obtained by normalizing the cross section in Eq. (18) with respect to the Rutherford cross section for projectiles and target recoils:

$$\frac{d\sigma_R}{d\Omega_p} = \left(\frac{d\sigma_R}{d\Omega_p}\right)^{\text{proj}} + \left(\frac{d\sigma_R}{d\Omega_p}\right)^{\text{rec}} \tag{20}$$

Experimentally, the number of positrons $\Delta\mathcal{N}_{e^+}^{\exp}$ in an energy interval ΔE_{e^+} (at energy E_{e^+}) is measured in coincidence with the number $\mathcal{N}_p$ of scattered particles. Efficiency and dead-time corrections must be applied.

A few remarks should be added concerning the transformation of the positron-production probabilities or cross sections to the center-of-mass (CM) system. This complication arises because the positrons are emitted from fast-moving systems. Unfortunately, there is a further complication because atomic and nuclear positrons are emitted from different systems. Atomic positrons originate from a quasi-"compound" system that moves with velocity $v = \beta c$ along the beam direction, while nuclear positrons originate from scattered particles and recoils which have velocity components perpendicular to the beam direction. Even if positrons with discrete energies are emitted in the CM system, large Doppler broadening effects are expected in the laboratory system depending on the angle θ_{e^+} observed by the positron spectrometer. To first order in β, the energy shift is given by

$$E_{e^+} = E'_{e^+}[1 + \beta(1 + 2m_e c^2/E'_{e^+})^{1/2} \cos\theta_{e^+}], \tag{21}$$

where E'_{e^+} is the kinetic energy of a positron line in the CM system.

Spectrometers used for in-beam positron spectroscopy must have a large positron collection efficiency and a broad energy acceptance band, while at the same time effectively suppressing other background radiation such as electrons, γ rays, and neutrons. These requirements are fairly well met by the two different types of spectrometers described in this section, viz., the magnetic "orange"-type β spectrometer and the positron transport systems with Si(Li) detectors for energy determination. We will not describe the first-generation spectrometers here (Backe *et al.*, 1978; Kozhuharov *et al.*, 1979). Since the time when these spectrometers were in use, positron detection techniques have been refined considerably; we concentrate on recent progress with second-generation spectrometers.

4.2. *The Orange-Type β Spectrometer*

In Fig. 12, the improved setup for measurements of positrons (Bердermann *et al.*, 1980; Kienle, 1983a) with an "orange"-type β spectrometer (Moll and Kankeleit, 1965) is shown. Positrons emitted from a target between 30° and 70° relative to the beam direction can be focused by the toroidal magnetic field produced by 60 coils, onto a special detection system which accepts a momentum bite of 14%. The positron detection system shown in Fig. 12 consists of a 12-cm-long conical (diameter = 12–16 mm) plastic scintillator, placed along the axis of the spectrometer in the region of the momentum focus. The positrons are stopped in the scintillator and produce an energy-proportional signal with an energy resolution of 17% at 500 keV.

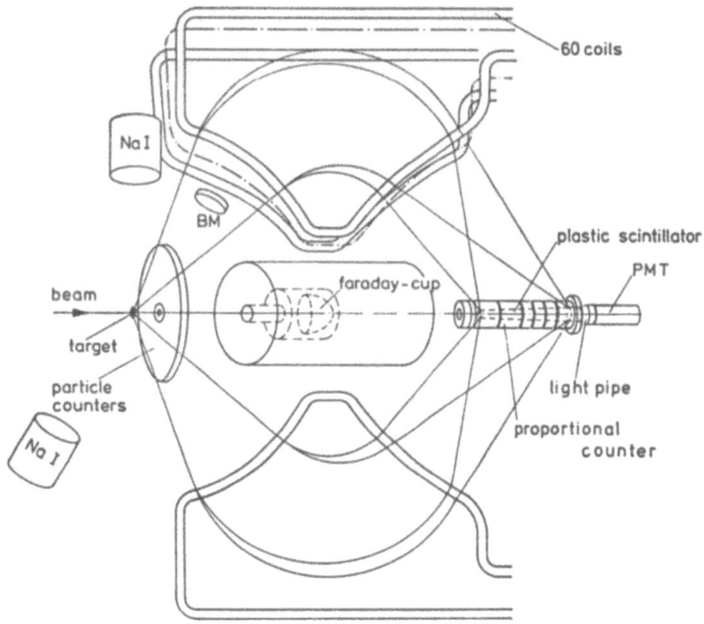

FIGURE 12. Schematic view of the "orange"-type β spectrometer setup to measure positron production in heavy-ion collisions. Positrons are focused by the toroidal field on a conically shaped plastic scintillator (55 cm away) surrounded by a position-sensitive proportional counter. Scattered particles are detected by an annular parallel-plate avalanche counter with 16 concentric anode rings. BM indicates a Si beam monitor. Gamma rays are detected by NaI scintillation counters (Kienle, 1983a; Berdermann *et al.*, 1980).

To suppress the low-energy background which originates mainly from Compton-scattered γ rays, and also to improve the momentum resolution of the detection system, an annular position-sensitive proportional counter surrounds the plastic scintillator. The total positron detection efficiency is up to 15% of 4π in the selected momentum interval, and is energy independent. A positron spectrum is measured by varying the magnetic field up and down in a selected number of steps to span a certain momentum range.

Scattered ions and recoils from the target are detected by an annular parallel-plate avalanche counter, made angle sensitive by subdivision of the anode into 8 to 16 concentric rings. Each ring is read out independently in order to handle high counting rates (up to $10^7\ s^{-1}$). Of great importance are Si detectors, mounted in a cone 45° relative to the beam, which measure the energy spectra of the scattered ions. In this way, the extent of target deterioriation can be continuously monitored. The γ spectra in coincidence with scattered particles are recorded by two 3×3-in. NaI scintillation counters placed at 90° and 150° relative to the beam, respectively.

4.3. Solenoidal Positron Transport Systems

(a) Another type of spectrometer used for positron spectroscopy at GSI in Darmstadt is the solenoidal transport system with normal conducting coils and Si(Li) detectors for energy determination. An improved version of the solenoid used for the pioneering positron experiments (Backe *et al.*, 1978) is shown schematically in Fig. 13. The solenoid axis is perpendicular to the beam. Positrons created in the target will spiral through the magnetic field to a Si(Li)-detector assembly 88-cm away. The magnetic-mirror field configuration focuses onto the detector all positrons within an emission angle

$$\theta < \theta_{\max} = \pi - \arcsin(B_T/B_{\max})^{1/2} \tag{22}$$

with respect to the ζ axis. Here, B_T is the magnetic field at the target position, and $B_{\max}$ its maximum value at $\zeta = 6.5$ cm. With the magnetic field configuration shown in Fig. 13, the acceptance angle is $\theta_{\max} = 148.7°$, corresponding to a solid angle of $\Omega/4\pi = 0.74$.

Unfortunately, all electrons resulting from Coulombic projectile–target interactions or internal-conversion transitions of excited nuclear levels are also focused onto the detector. To suppress this very high electron background without appreciably decreasing the positron detection efficiency, the detector system shown in the insert of Fig. 13 has been used

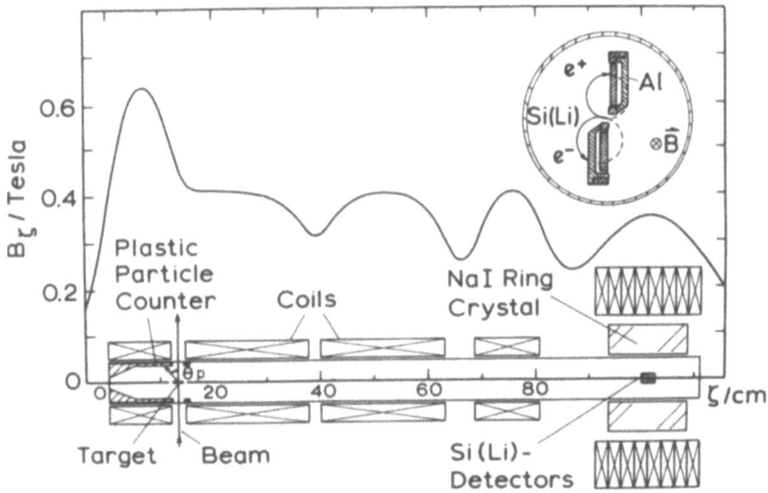

FIGURE 13. Schematic drawing of the solenoid spectrometer. Also shown is the magnetic field on the ζ axis. The heavy-ion beam enters the vacuum chamber perpendicular to the solenoid axis. The insert shows the circular Si(Li) detectors in a cut at $\zeta = 102$ cm. The fourfold segmented NaI ring crystal has an inner diameter $= 90$ mm, an outer diameter $= 204$ mm, and a length of 105 mm. (From Backe *et al.*, 1980.)

(Backe *et al.*, 1980). The latter system consists of two Si(Li) diodes with a diameter of 19.5 mm and a thickness of 3 mm, mounted with their surfaces parallel to the solenoid axis. One of these diodes is placed above the axis, its sensitive face looking to the left, while the other diode is below, looking to the right. With this configuration, positrons whose trajectory is a right-handed spiral have a good probability of entering the sensitive area of one of the Si(Li) diodes. Electrons, spiraling in the opposite sense, are absorbed or scattered in the 3-mm-thick aluminum back cover of the detector. Only electrons with an energy greater than ~ 1.4 MeV can penetrate the aluminum absorber. A suppression factor of ~ 150 was measured for 365-keV electrons.

In spite of this strong suppression of electrons, the counting rate observed in the Si(Li) detectors during in-beam positron experiments is still caused nearly entirely by electrons. In order to select the very few positrons, the Si(Li) diodes were surrounded by a four-fold segmented NaI(Tl) ring crystal detecting 511-keV annihilation radiation from the positron. To characterize a positron, a coincidence was required between an event registered in the 511-keV region of one of the NaI(Tl) segments and in the total energy region of the opposite one. The positron detection efficiency amounts to 6% in the energy region of interest between 0.2 and 1.0 MeV.

Positrons are measured in coincidence with scattered particles or target recoils using a plastic scintillation counter with a thickness of 50 μm, also shown in Fig. 13. Recoils of light target contaminants, such as carbon or oxygen, are excluded from detection electronically by pulse-height discrimination. Projectile fission products resulting from fusion reactions with light target contaminants were kinematically excluded from entering the sensitive region of the plastic counter at angles between 35° and 55°. For U-beam energies below 5.9 MeV/u, a small fraction of fission fragments from target–projectile interactions were also detected. However, positrons associated with these events were estimated to be negligible. Target monitoring and simultaneous recording of the γ-ray spectra was achieved as previously described for the orange-type spectrometer.

(b) The new solenoid positron detection system EPOS at GSI shown in Fig. 14 is based on the same principle of operation using the magnetic mirror effect, but has been improved over the older spectrometer in several ways (Balanda *et al.*, 1980). In particular, the tube diameter of 260 mm is large enough to allow for insertion of two position-sensitive parallel-plate avalanche detectors with delay-line readout for particle detection. The angular acceptance of each of the particle counters is $20° \leqslant \theta_p \leqslant 70°$ at constant $\Delta\phi = 60°$, with angular resolution better than 1°. This results in an excellent definition of the scattering kinematics and a good suppression of unwanted reaction channels. In addition to the particle-scattering-angle information, the time-of-flight difference between the two particles is obtained, as is rough Z information via the energy loss of the particles in the

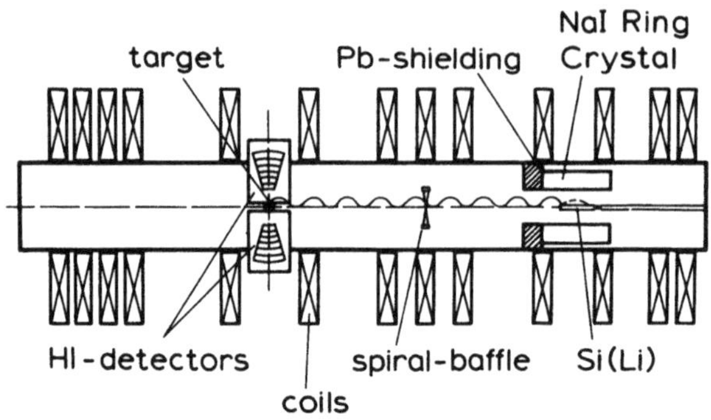

FIGURE 14. Schematic view of the solenoidal spectrometer EPOS. The heavy-ion beam axis is perpendicular to the plane of the drawing. The four coils to the left produce a magnetic-mirror field configuration similar to that shown in Fig. 13. Target electrons are excluded from detection in the two pencil-like positron Si(Li) detectors (outer diameter 10 mm, sensitive thickness 2.5 mm, length 50 mm), by means of an Al spiral baffle. The eightfold NaI detector has the following dimensions: inner diameter 100 mm, outer diameter 200 mm, length 200 mm. (From Balanda *et al.*, 1980.)

detector gas. The coincident detection of both particles is especially advantageous. For asymmetric systems such as U + Pb, the distance of closest approach R_0 can be determined uniquely, under the assumption of Rutherford-scattering conditions, from the kinematic θ_1, θ_2 correlation. For symmetric systems, such as U + U, the distinction of close from distant collisions is impossible in principle.

Electrons are suppressed in the EPOS system by a spiral-shaped Venetian-blind baffle between target and positron detector, which makes use of the different helicities of positron and electron trajectories. Background counts in the e^+ detector due to leakage of scattered electrons through the baffle and due to other possible sources outside of the target are reduced by using a pencil-like e^+ detector located along the solenoid axis. The coaxial geometry of the e^+ detector combines maximum efficiency for target-produced positrons with minimum sensitivity for positrons and electrons which have been scattered on their way from the target or which do not originate in the target. Positrons are identified by demanding coincidences between the e^+ detector and at least one of the two 511-keV annihilation quanta of the positron, as registered in an eightfold cylindrical NaI-detector array surrounding the e^+ counter.

The pencil-like positron detector consists of two coaxially drifted Si(Li) detectors, mounted closely one behind the other on a cooling rod. The energy resolution of the e^+ detector is better than 10 keV for the 662-keV conversion-electron line of ^{137}Cs. This resolution is much better than the expected width of an e^+ line measured in-beam, a width determined mainly by the maximum Doppler broadening [Eq. (21)]. As in the other experiments described above, the γ-ray spectrum is measured with a 3×3 in. NaI detector in coincidence with the heavy-ion detectors.

The problem of δ-ray suppression in transport systems can be solved very efficiently when positrons and electrons are transported in *S*-shaped magnetic fields. The main advantage of such a field configuration is that electrons and positrons can be separated and measured in different counters. The *S*-shaped magnetic field can be realized by pancake coils surrounding two quarters of a toroidal tube ("Tori" spectrometer, Kankeleit *et al.*, 1985). In the first quarter torus, electrons and positrons leaving the target experience a drift in the center of their orbits, perpendicular to the spectrometer plane. This drift is opposite for electrons and positrons and can give complete separation of the two. The electrons are absorbed in a detector, while the positrons are left free to pass into the second quarter torus in which the drift is reversed. This spectrometer has image characteristics for positrons which are very similar to those of the solenoid systems. However, the rejection of electrons is achieved in a way that does not affect the detection efficiency of positrons.

5. DATA EVALUATION AND BACKGROUND SUBTRACTION

The major difficulty in the evaluation of the positron spectra is that nuclear levels which decay by transitions with energies greater than $2m_ec^2$ may be populated, e.g., by Coulomb excitation. Positrons due to internal pair conversion of these transitions cannot be electronically distinguished from the atomic contribution because the time delay is only of the order of 10^{-16} to 10^{-13} s. The nuclear positron background is estimated by converting the observed γ-ray spectra into positron spectra. This procedure was first proposed by W. Meyerhof (1977) and is described here only for experiments performed with the solenoid spectrometer shown in Fig. 13. The evaluation procedure for other experiments is described in the papers of Bokemeyer *et al.* (1983) and Kienle (1983a).

To calculate the atomic positron spectrum $dP_{e^+}^{\text{atom}}/dE_{e^+}$ from Eq. (19), the number of measured positrons $\mathcal{N}_{e^+}^{\text{exp}}$ must be corrected for contributions from positrons produced in the internal pair decay of excited levels in the projectile and target nuclei [$\Delta\mathcal{N}_{e^+}^{\text{int}}(\mathcal{Z}_1, M_1)$ and $\Delta\mathcal{N}_{e^+}^{\text{int}}(\mathcal{Z}_2, M_2)$] as well as for positrons from external pair conversion of γ rays in the target, $\Delta\mathcal{N}_{e^+}^{\text{ext}}(T)$, and in the solenoid, $\Delta\mathcal{N}_{e^+}^{\text{ext}}(S)$, and positrons from the conversion of bremsstrahlung quanta, $\Delta\mathcal{N}_{e^+}^{\text{BS}}$:

$$\Delta\mathcal{N}_{e^+}^{\text{atom}} = \Delta\mathcal{N}_{e^+}^{\text{exp}} - \Delta\mathcal{N}_{e^+}^{\text{int}}(\mathcal{Z}_1, M_1) - \Delta\mathcal{N}_{e^+}^{\text{int}}(\mathcal{Z}_2, M_2) - \Delta\mathcal{N}_{e^+}^{\text{ext}}(T) - \Delta\mathcal{N}_{e^+}^{\text{ext}}(S) - \Delta\mathcal{N}_{e^+}^{\text{BS}}. \tag{23}$$

From test experiments and calculations it appears that $\Delta\mathcal{N}_{e^+}^{\text{ext}}(S)$ and $\Delta\mathcal{N}_{e^+}^{\text{BS}}$ are negligible. On the other hand, the external pair creation in a 1-mg/cm^2 Pb or U target requires corrections of the order of 5% in the positron yield.

As already mentioned, the internal pair conversion contribution of positrons is derived from γ-ray spectra. For the $^{208}Pb + ^{208}Pb$ scattering system, the γ-ray spectrum is straightforward because only the 3^- level at 2.614 MeV and the 2^+ level at 4.09 MeV are populated by Coulomb excitation. However, to obtain the highest possible united charge $Z_u = Z_1 + Z_2$, ^{238}U beams have been used, resulting in the excitation of a large number of nuclear levels with unknown decay characteristics. In this case, the fractions of atomic and nuclear positrons are determined by comparing the positron und γ-ray production from various targets. To calculate these fractions it is assumed that in systems with $Z_u < 150$ only nuclear positrons are emitted, and that the multipolarity of the observed γ rays is independent of the atomic charge of the target. Thus extrapolation to systems with $Z_u > 150$ is possible. Corrections for the nuclear positrons are carried out as

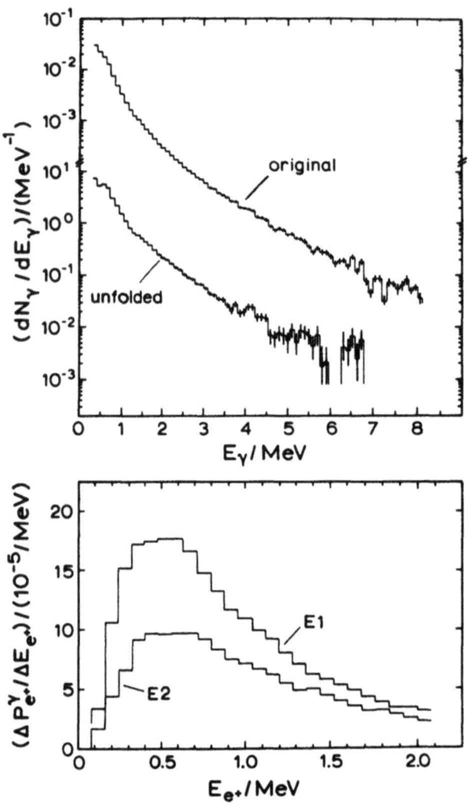

FIGURE 15. Target γ-ray spectrum taken with a 7.5 × 7.5 cm NaI detector. A ^{108}Pd target was bombarded with ^{238}U projectiles at an energy of 5.9 MeV/u. The γ-ray spectra were recorded in coincidence with the ^{108}Pd recoils, detected in the particle counter at $\theta_p = 45° \pm 10°$. The spectrum labeled "unfolded" was obtained from the original spectrum by unfolding the detector response and applying the efficiency correction. The lower part of this figure exhibits positron spectra calculated from the unfolded γ-ray spectrum, assuming $E1$ and $E2$ multipolarity.

follows. A typical γ spectrum is shown in Fig. 15. After unfolding the response of the NaI detector from this spectrum, the resultant γ distribution $(d\mathcal{N}_\gamma/dE_\gamma)(E_\gamma)$ can be transformed into a positron spectrum through theoretical pair conversion coefficients $d\beta_{M\lambda}/dE_{e^+}(E_\gamma, E_{e^+}, \mathcal{Z})$ for multipolarity $M\lambda$ (Schlüter *et al.*, 1981):

$$\frac{d\mathcal{N}_{e^+}}{dE_{e^+}}(E_{e^+}, M\lambda) = \int_{2m_ec^2}^{\infty} dE_\gamma \frac{d\mathcal{N}_\gamma}{dE_\gamma}(E_\gamma) \frac{d\beta_{M\lambda}}{dE_{e^+}}. \tag{24}$$

In Fig. 15, we show two positron spectra in the form

$$\frac{\Delta P^\gamma_{e^+}}{\Delta E_{e^+}} = \frac{\Delta\mathcal{N}_{e^+}/\Delta E_{e^+}(E_{e^+}, M\lambda)}{\mathcal{N}_P}, \tag{25}$$

calculated from the γ distribution on the assumption that $E1$ and $E2$ multipolarities occur, respectively. The remarkable difference between these two spectra emphasizes the difficulty in correcting for nuclear background positrons when the multipolarity of the γ-ray transition is unknown. However, the experimentally measured positron spectrum shown below in Fig. 17 strongly suggests an $E1$ multipolarity if we make the plausible assumption that there are negligible amounts of atomic positrons created in the U + Pd scattering system ($\mathcal{Z}_u = 138$).

Further information on the γ-ray multipolarity can be gained from energy-integrated positron yields, which can be determined with much less beam time. We calculate

$$\mathcal{N}^{\text{calc}}_{e^+,\gamma}(M\lambda) = \int_0^\infty dE_{e^+}\, \varepsilon^{\text{tot}}_{e^+}(E_{e^+}) \frac{d\mathcal{N}_{e^+}}{dE_{e^+}}(E_{e^+}, M\lambda), \tag{26}$$

and compare this number with the experimental value $\mathcal{N}^{\text{exp}}_{e^+}$ by forming the ratio $\mathcal{N}^{\text{exp}}_{e^+}/\mathcal{N}^{\text{calc}}_{e^+,\gamma}(M\lambda)$. The result is shown in Fig. 16. The data have been normalized to unity in the region $\mathcal{Z}_u < 160$ by introducing a factor f close to unity. The ratio clearly and increasingly exceeds unity for the U + Pb, U + U, and U + Cm scattering systems. This fraction of excess positrons is believed to originate from atomic processes for which we can write

$$\mathcal{N}^{\text{atom}}_{e^+} = \mathcal{N}^{\text{exp}}_{e^+} - \mathcal{N}^{\text{ext}}_{e^+}(T) - \mathcal{N}^{\text{calc}}_{e^+,\gamma}(E1)f. \tag{27}$$

From Eq. (23) we can identify the last term of Eq. (27) as the nuclear contribution of positrons:

$$\mathcal{N}^{\text{calc}}_{e^+}(E1)f = \mathcal{N}^{\text{int}}_{e^+}(\mathcal{Z}_1, M_1) + \mathcal{N}^{\text{int}}_{e^+}(\mathcal{Z}_2, M_2). \tag{28}$$

It should be emphasized that this procedure is only correct if the stated assumptions are valid. The approximations introduced by these assumptions can lead to an error of up to 30%.

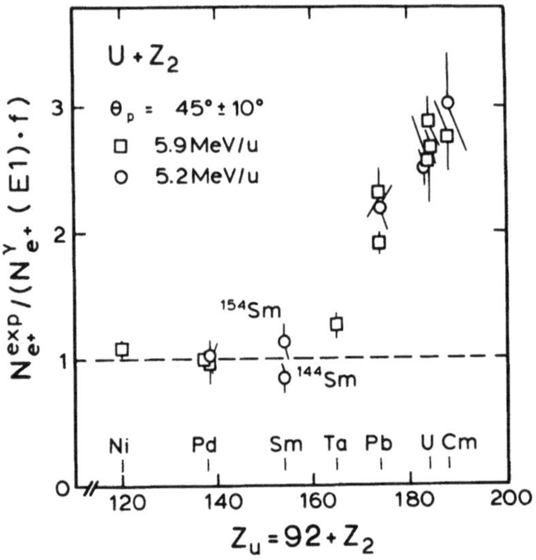

FIGURE 16. Ratio of measured-to-calculated positron yields $\mathcal{N}^{exp}_{e^+}/\mathcal{N}^{\gamma}_{e^+}(E1)f$ as a function of the united nuclear charge Z_u, assuming $E1$ multipolarity for the nuclear target γ rays. For $Z_u \geqslant 174$ there is a clear indication of a large number of excess positrons which are attributed to atomic processes.

6. EXPERIMENTAL RESULTS AND DISCUSSION

6.1. First Results and Gross Features

In the first positron experiments it had to be demonstrated that the separation of atomic and nuclear positrons is possible. This could be done strikingly in a ^{208}Pb + ^{208}Pb scattering experiment at beam energies well below the Coulomb barrier (Backe *et al.*, 1978). A forerunner of the solenoid spectrometer shown in Fig. 13 was used with a positron detection system only able to measure the energy-integrated positron yield. The principal source of nuclear positrons in this scattering system originates from the pair-conversion decay of the Coulomb-excited 3^- (2.614-MeV) state of ^{208}Pb, which is the first excited state of this doubly magic nucleus. This state is also populated in the decay of ^{212}Pb. Determination of the background was therefore directly possible from the primary experiment.

Parallel to these investigations it was demonstrated that atomic and nuclear positrons can also be separated for the ^{238}U + ^{208}Pb and the ^{238}U + ^{238}U scattering systems (Kozhuharov *et al.*, 1979). The positron-production probability was investigated systematicly in a fixed positron energy window ($E_{e^+} = 478 \pm 54$ keV) for various scattering systems including very light ones, for which only the emission of nuclear positrons is expected. A forerunner of the orange spectrometer set-up (Fig. 12) was used for these experiments. A clear indication of excess positrons for high-Z_u systems was found, similar to that shown in Fig. 16.

More information about the positron production process is expected to be contained in the complete positron spectrum. Careful investigations are currently under way; we discuss a few interesting aspects of these experiments.

The first measurements of positron spectra were made with the solenoid spectrometer shown in Fig. 13. The emphasis of this experiment was to compare the supercritical U + U scattering system with the subcritical U + Pb system at a U-beam energy of 5.9 MeV/u and a scattering angle of $\theta_p^{\text{lab}} = 45° \pm 10°$. In Figs. 17a–17c we show the measured spectra, corrected for detector response and efficiency. The corresponding atomic positron spectra are shown in Fig. 17d on a semilogarithmic plot. Comparing the supercritical U + U spectrum with the subcritical U + Pb spectrum, no distinct effects of a diving level in the U + U system can be observed. Especially at low positron energies, the two spectra agree within statistics.

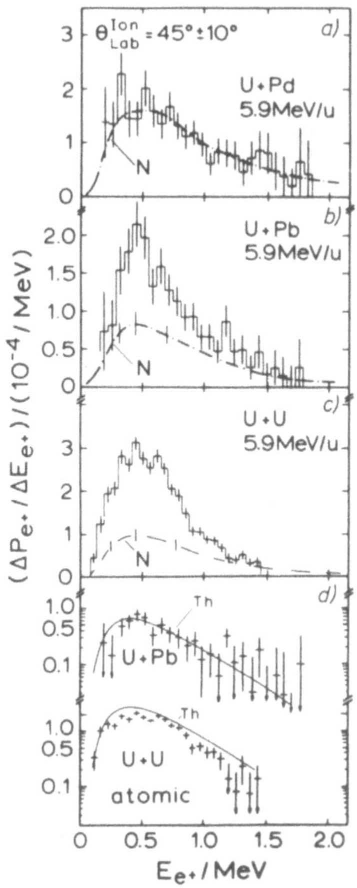

FIGURE 17. Preliminary positron spectra $\Delta P_{e^+}/\Delta E_{e^+}$ for (a) the nuclear U + Pd, (b) the subcritical U + Pb, (c) the supercritical U + U systems at a U-beam energy of 5.9 MeV/u and $\theta_{\text{lab}}^{\text{Ion}} = 45° \pm 10°$, taken with the solenoid spectrometer shown in Fig. 13. The dash-dotted curves "$\mathcal{N}$" indicate nuclear positrons as derived from γ-ray spectra, assuming $E1$ multipolarity; (d) is a semilogarithmic presentation of the spectra of atomic positrons compared with theoretical calculations indicated by "Th". (From Backe *et al.*, 1983a, b.)

The nearly exponential drop of the intensity at high positron energies reflects the Fourier-frequency spectrum of the transition matrix element for pair creation. At small internuclear distances this matrix element is proportional to $\dot{R}/R$, where $\dot{R}$ is the radial velocity and R the distance between the colliding nuclei. It can be shown that the high-energy tail is exponential (Kankeleit, 1980). The drop-off at low energy is caused by the Coulomb repulsion of the positrons from the nuclear charges. Finally, we compare the positron probabilities with theoretical calculations as described in Sec. 3. The calculated shape of the spectra shown in Fig. 17d agrees quite well with the experimental data. As discussed in Sec. 3.1, the calculations do not predict a significant effect due to the diving $1s\sigma$ state. The diving time is too short when the colliding nuclei follow pure Rutherford trajectories. There is some indication that the calculations overestimate the positron production probability, but it should be stressed that the absolute experimental errors are at present about 30%.

6.2. *Positron Spectra in Deep Inelastic Collisions*

If the colliding nuclei are brought into contact, so that nuclear forces become important, the time in which the electrons are exposed to the combined Coulomb potential can be considerably prolonged. From a detailed analysis of deflection functions and cross sections for the total kinetic-energy loss in deep inelastic collisions, nuclear interaction times of up to a few times 10^{-20} s have been deduced (Wollersheim *et al.*, 1982; Gobbi and Nörenberg, 1980). For such long sticking times [time for which $\dot{R}(t)=0$], the spontaneous positron peak should be clearly visible, as shown in Fig. 10.

In the experiment about to be described (Backe *et al.*, 1983a), nuclear contact was signaled by fission of at least one of the collision partners. ^{238}U targets were bombarded with ^{238}U beams at energies between 5.9 and 8.4 MeV/u, and the resulting positron spectra recorded with the solenoid detection system are shown in Fig. 18. In a plastic-scintillator ring counter accepting scattering angles $35° \leqslant \theta_p \leqslant 55°$, the full-energy peak from quasielastic events allowed a separation from the broad distribution of fission reactions. At projectile energies of 5.9, 7.5, and 8.4 MeV/u, the quasielastic contributions varied from $\sim$100% to $\sim$30% and $\sim$0%, respectively.

In the presence of fission the separation of quasiatomic and nuclear positrons is a major problem. The technique applied was essentially the same as that described in Sec. 5, namely, a conversion of measured γ-ray spectra into nuclear positron spectra. This technique was complemented by a measurement of electron spectra coincident with the positrons. It turned out that the high-energy tails of the electron spectra are of nuclear origin, which supports the γ-ray analysis procedure. Nuclear background and

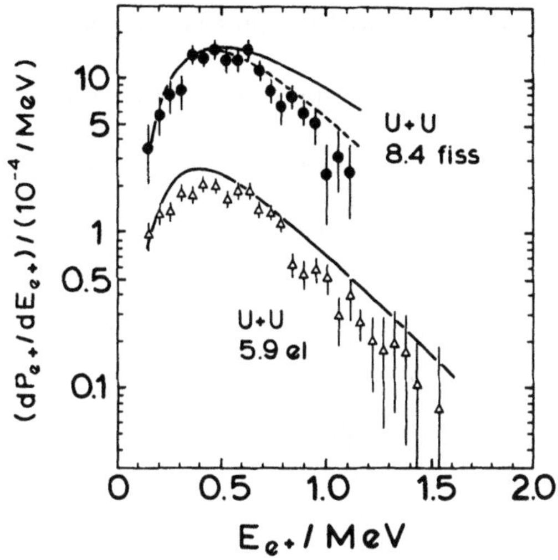

FIGURE 18. Number of positrons per detected plastic fission event and positron energy interval dP_{e^+}/dE_{e^+} for the U + U scattering system at 8.4 MeV/u. For comparison, the spectrum at 5.9 MeV in coincidence with elastically scattered particles is also shown. Nuclear background positrons have been subtracted. (From Backe *et al.*, 1983a.) The heavy curves represent calculations based on the assumption of Rutherford trajectories. The dashed curve includes effects of a nuclear contact time of $\lesssim 10^{-21}$ s. (From Müller *et al.*, 1982.)

efficiency-corrected positron spectra are shown in Fig. 18. The error bars include, besides statistics, a 10% uncertainty due to background subtraction. The spectra exhibit the typical exponential drop-off at high energies indicating the importance of collision dynamics in the positron production process. For the U + U system at 5.9 MeV/u taken in coincidence with elastically scattered particles, calculations on the basis of pure Rutherford trajectories reproduce the shape of the spectrum. This, however, is not the case for the spectra taken in coincidence with fission products (i.e., after nuclear contact), which drop off much more steeply. Such an effect can be caused by a time delay during the nuclear contact (Kankeleit, 1980; Reinhardt *et al.*, 1980b).

A detailed coupled-channel calculation of the positron spectra, in the presence of a nuclear time delay on the basis of the friction model of Schmidt *et al.* (1978), has recently been performed by U. Müller *et al.* (1982). These calculations indicate the presence of a nuclear time delay $\Delta T = 10^{-21}$ s. Such small contact times in deep inelastic collisions again rule out an observable contribution of spontaneous positrons. This prediction agrees with the experimental finding that the positron spectrum for the U + Cm system differs from that for the U + U system only by a constant factor but not in shape (Backe *et al.*, 1983a).

6.3. *Peak Structures in the Positron Spectra from "Elastic" U + U and U + Cm Scattering*

It was a big surprise when detailed studies of positron emission in "elastic" U + U and U + Cm scattering systems at beam energies around

5.9 MeV/u showed the presence of peaklike structures in the positron spectrum.

In Fig. 19 we present various positron spectra resulting from bombardement of ^{238}U foils with ^{238}U ions (Berdermann *et al.*, 1982; Clemente *et al.*, 1984), as recorded by the orange spectrometer (Fig. 12). Nuclear background, as determined from the simultaneously measured γ spectra, has not been subtracted. The positron spectrum (B) taken at 5.7 MeV/u in coin-

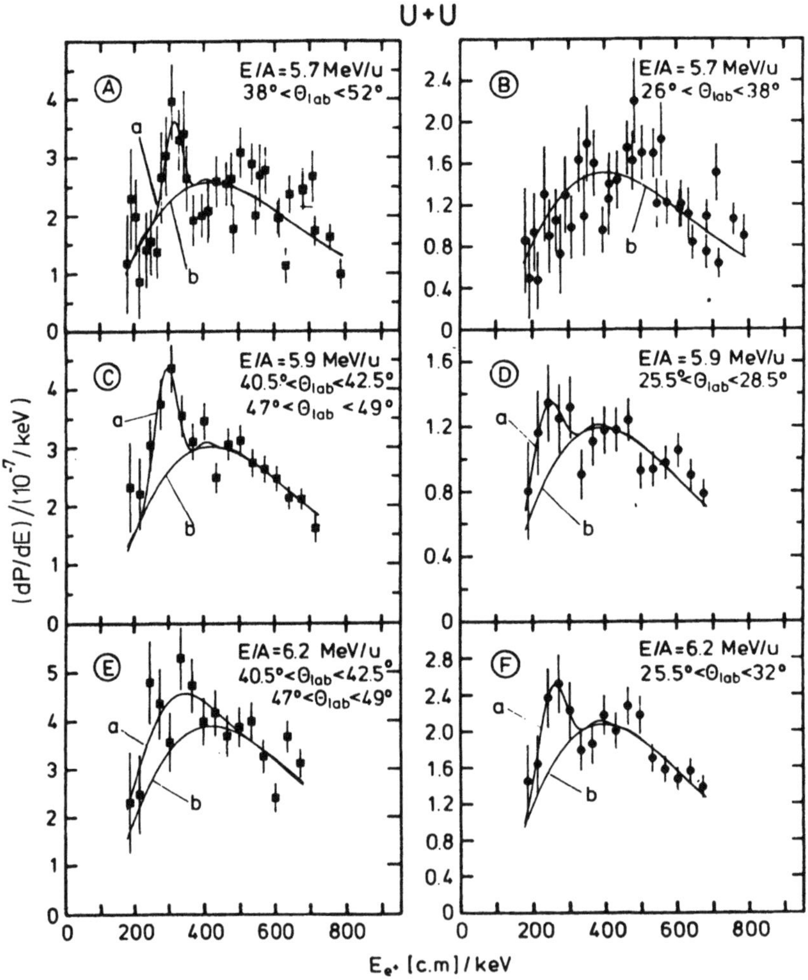

FIGURE 19. Positron spectra from U + U collisions taken with the "orange" spectrometer (Fig. 12) at three different beam energies. At each bombarding energy, spectra measured at two different scattering angles are shown. The solid curve (b) represents spectra expected from Rutherford trajectories; curve (a) includes in addition positrons emitted from a giant nuclear complex. (From Clemente *et al.*, 1984.)

cidence with ions scattered to angles between 26° and 38° in the laboratory system is broad and mostly structureless. However, the positron spectrum (A) in coincidence with ions scattered between the angles of 38° and 52° shows a peak at 311 keV with a width of ~60 keV. From the other spectra in Fig. 19 it can be seen that position and width of the peak may depend slightly on scattering angle and beam energy.

With the new solenoid system EPOS (Fig. 14), the U + U scattering system was investigated at a beam energy of 5.9 MeV/u (Bokemeyer *et al.*, 1983). Two peaklike structures were observed, one centered around 320 keV and another at 590 keV. The peak at 320 keV obviously corresponds to the structure observed with the orange spectrometer. It is not completely clear why the peak structures were not seen in the third experiment shown in Fig. 17c.

Before discussing the possible reasons for such peak structures we mention another very interesting result obtained with the new solenoid spectrometer EPOS. In Fig. 20, positron spectra emitted from the U + Cm collision system at 6.05 MeV/u are shown (Schweppe *et al.*, 1983). It is particularly striking to observe a well-defined peak in Fig. 20a centered at an energy of 316 keV. This peak is correlated with essentially backward scattering. Figure 20b shows a spectrum for mostly forward scattering in which this peak is not present. However, a clear separation between close and distant collisions could not be achieved in this nearly symmetric system. The width of the line in Fig. 20 is 80 keV.

What could be the origin of the peak structures observed in the different experiments? First of all we would like to mention that the strong dependence of the peak intensities on the ion scattering angle largely excludes the possibility that the lines originate from instrumental effects. An obvious explanation could be that the structures originate from nuclear transitions in the separate ions which were populated by Coulomb excitation or by nuclear transfer in combination with Coulomb excitation. Such a mechanism could explain the strong dependence of the positron-line intensity on the scattering angle. That nucleon transfer cannot be excluded *a priori* follows from the fact that, e.g., in a U + Cm collision at 6.05 MeV/u and $\theta_P^{\mathrm{CM}} = 120°$ the distance of closest approach is ~19.5 fm. This distance is well below the "safe" distance

$$R_{\mathrm{safe}} = 1.5(A_1^{1/3} + A_2^{1/3})\ \mathrm{fm} + 3\ \mathrm{fm}, \tag{29}$$

which for the U + Cm system is $R_{\mathrm{safe}} = 21.7$ fm. As pointed out, decay with an energy E_t greater than $2m_ec^2$ can proceed via pair decay. The resulting positron spectrum in high-Z atoms like uranium or curium exhibits a nearly triangular shape with a sharp cut-off at the threshold energy $E_t - 2m_ec^2$. In the experiment under discussion, this sharp cutoff will not be observed

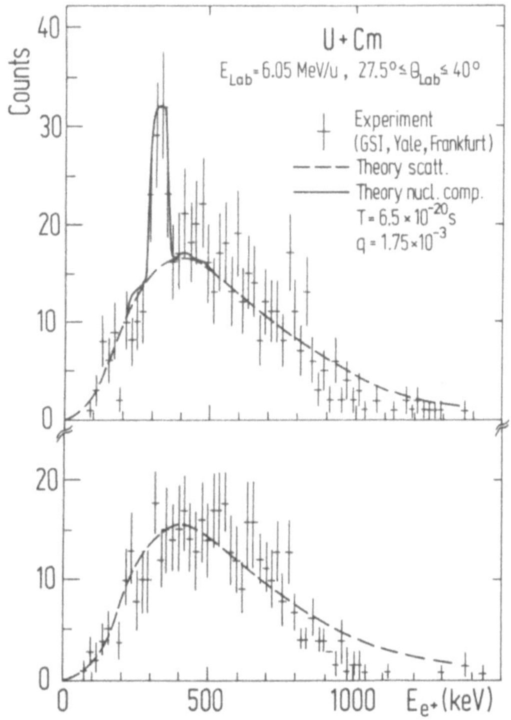

FIGURE 20. Preliminary positron spectra from 6.05-MeV/u U + Cm collisions, recorded with the solenoid spectrometer EPOS. The spectra are not corrected for spectrometer transport efficiency. In (a) and (b), kinematic conditions were selected to emphasize backward and forward scattering events, respectively, in the center-of-mass frame for the collision. (From Schweppe *et al.*, 1983.)

because of the Doppler broadening produced by the in-flight emission of the positrons and the detector resolution. If the statistics in the positron spectrum are not sufficient, such a triangular shape can easily be misinterpreted as a line. However, the shapes of the observed positron lines are very dissimilar to the expected distributions (Fig. 21a) and therefore render this explanation questionable (Schweppe *et al.*, 1983). In addition, a γ ray with an energy $E_\gamma = E_t$ will be emitted with much larger probability, yet a corresponding line was not observed in the NaI spectrum. This fact requires the assumption that the transition is of E0-type, in which case strong conversion-electron lines will be emitted. Electron spectra recorded up to now do not show any indication of the presence of such lines.

We next discuss another process which has long been known (Sliv, 1953) to yield *monochromatic* positrons. This is a pair-decay process in which the electron occupies the K or L vacancy instead of going into the

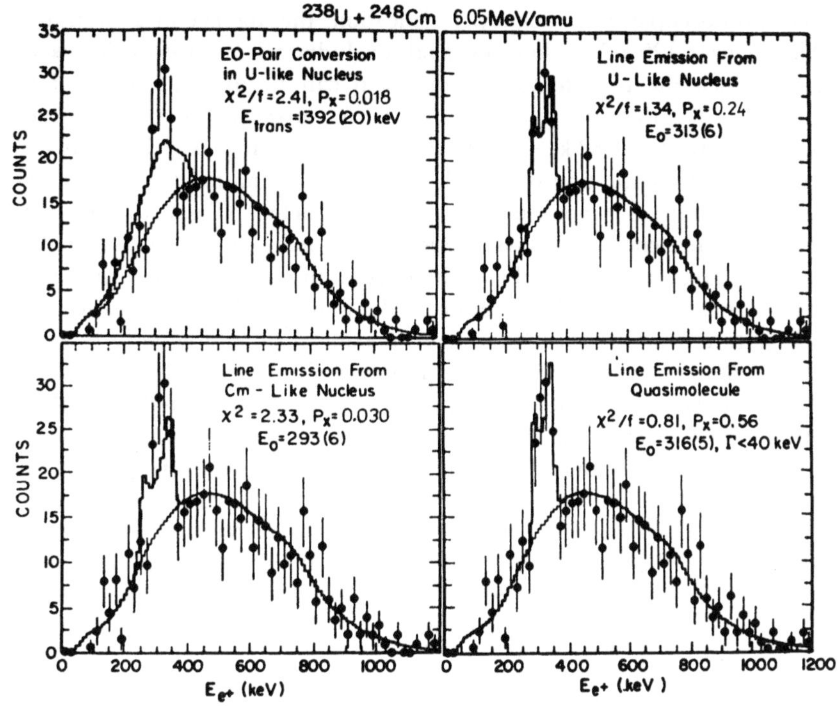

FIGURE 21. Four different attempts to fit the positron line structure shown in Fig. 20: (a) E0 pair conversion; (b) and (c) monoenergetic pair conversion of an excited nuclear level in the U- or Cm-like fragments, respectively; and (d) positron emission from the combined nuclear complex. P_χ is the confidence level for each specific mechanism. (From Schweppe *et al.*, 1983.)

continuum. Indeed, K or L vacancies are produced in a close collision of very heavy nearly symmetric partners with probability near unity. To observe these monochromatic positrons of energy $E_{e^+} = E_t - 2m_e c^2 + |E_B|$ with sufficient intensity, the lifetimes τ of the K or L holes must be of the order of the lifetime τ_N of the nuclear level. If we assume that a K vacancy has a lifetime of 10^{-17} s, typical for heavy neutral atoms, the emission of monochromatic positrons is strongly suppressed (Soff *et al.*, 1981b), unless we make the extreme assumption that the corresponding nuclear level also has a lifetime of $\sim 10^{-17}$ s. The possibility should also be considered, however, that the K- or L-vacancy lifetime in a highly ionized atom could be much longer than in the neutral atom. If we assume that the lifetime of a K vacancy is 10^{-15} s or even longer, monochromatic positrons could result from fast $E1$ or $E0$ transitions. This mechanism, however, also seems to be a very unlikely explanation of the origin of the observed positron lines, since the triangular continuum below the positron lines should be present but is not observed. This fact largely excludes the possibility of pair decay in which the electron occupies a L vacancy. In addition, the absence of a continuum in Fig. 20 near 120 keV below the peak excludes all multipolarities except $E0$, for which the peak-to-continuum ratio can be as large as 10 if the lifetime τ_N is less than the lifetime τ_K of the K vacancy. The experimentally observed linewidth of 80 keV in the U + Cm scattering experiment is consistent with the assumption that such positrons are emitted from the U-like nucleus (Fig. 21b) but not from the Cm-like nucleus (Fig. 21c). However, the average presence of 1.85 isomeric K vacancies must be assumed to explain the large intensity of the positron peak in combination with the fact that a corresponding transition could not be found in the electron spectrum.

In future experiments it should be possible to unambigously establish the origin of the positron line by measurement of the linewidth as a function of θ_{e^+} or of the line profile as a function of the emission angle ϕ_{e^+} of the positron with respect to the scattering plane. If ϕ dependence is observed, a line is proven to be of nuclear origin. Such experiments are in preparation (Kienle, 1983b; Cowan *et al.*, 1985).

We now discuss the possibility that the observed positron structures originate from the combined system. Different explanations have been proposed. It was discussed (Kienle, 1983a) that the structure could be due to interference phenomena, as suggested by the "oscillation" shown in some positron spectra presented in Fig. 19.

On the other hand, Fig. 20 exhibits a clear line. The observed line shape is consistent with the assumption that this line originates from in-flight emission from the combined nuclear complex (Fig. 21d). Such lines cannot be explained in the framework of dynamically induced positron creation based on Rutherford scattering alone. As discussed in Sec. 3.2, Rafelski, Müller, and Greiner (1978) suggested, long before such narrow lines were found experimentally, that a peak structure due to spontaneous positrons is expected if the time for which the collision systems remain critical is prolonged due to nuclear interactions. Unexpectedly long nuclear sticking times of $\sim 10^{-20}$ s are required to explain the observed intrinsic linewidth. In addition, the intermediate superheavy nuclear complex must be very cold, since otherwise the separated nuclei would undergo fission. The observed energy of the positron line in the U + Cm system agrees with theoretical calculations (Reinhardt *et al.*, 1981; U. Müller *et al.*, 1983), if one assumes that the nuclei in the molecule have a separation of ~ 16 fm. On the other hand, this explanation clearly implies that the line observed in the U + U system at 5.9 MeV/u must have a different origin or must be emitted from an intermediate nuclear complex of far less elongated shape. According to the same theoretical calculations, the spontaneous-positron line should have an energy of 210 keV, assuming a separation of 16 fm of the uranium nuclei in the molecule.

7. SUMMARY AND OUTLOOK

Since 1976 when the UNILAC accelerator went into operation, experimental results have confirmed the theoretical ideas of the nature of superheavy atomic systems in a spectacular way. Experiments have firmly established the (transient) existence of binding energies in excess of the electron rest mass and the localization of strongly bound electrons in a region much smaller than the electron Compton wavelength. Both effects were predicted from relativistic quantum mechanics but had never before been tested. The measurements of positron production in very heavy but subcritical collision systems (like Pb + Pb and Pb + U) have exhibited an extremely strong dependence on nuclear charge—roughly like Z^{20}—in full agreement with calculations. This result, by itself, constitutes an important test of nonperturbative quantum field theory, in this case quantum electrodynamics.

The most fundamental aspect of superheavy quasiatoms, however, is the spontaneous pair-creation process in super-critical fields which is associated with the transition from a neutral to a charged vacuum state. This process does not occur at any finite order of perturbation theory, i.e., it is an essentially nonperturbative phenomenon. The recently found line structures in positron spectra from U + U and U + Cm collisions seem to be evidence for the existence of spontaneous pair creation which was discussed already more than 50 years ago (Klein, 1929; Sauter, 1931; Heisenberg and Euler, 1936; Hund, 1940), but never before seen in an experiment. It is the task of the ongoing experiments to definitely establish the origin of the observed positron lines.

Important steps have recently been taken in this direction. A new measurement of the electron spectrum in U–Cm collisions indicates that the process of bound-electron–free-positron conversion discussed in Sec. 6.3, which would yield monoenergetic positrons, is quite unlikely. It would require the assumption of 1.85 K vacancies to be present in one of the nuclei at least 10^{-15} s after the collision to retain this explanation. No reasonable mechanism is known which can lead to these numbers.

On the other hand, height, location, and shape of the line structures have been found to depend very sensitively on the beam energy. In U + Cm collisions, the positron line seems to be present only in a window $\sim$40 MeV wide (at 1450 MeV beam energy). Some dependence on bombarding energy is also seen in the U + U system (Fig. 19). A natural explanation for such a narrow enhancement in the excitation function can be given if a pocket is assumed to be present in the internuclear potential $V(R)$. At too-low energies, the pocket is missed, and the nuclei scatter off the Coulomb potential. When the beam energy is just right to hit the pocket, resonance scattering takes place and the two nuclei may spend a long time together forming a long-lived nuclear molecule. At energies much above the barrier, the resonances become very wide so that the delay is too short to give rise to a peak in the positron spectrum.

Recent calculations (Rhoades-Brown *et al.*, 1983; Seiwert, 1983) of the nuclear potential for the systems U + U and U + Cm have in fact revealed the possible presence of a shallow pocket at ~1450 MeV laboratory energy, corresponding to molecular configurations of the two deformed nuclei. The system can vibrate around the local minimum in the potential and it can rotate. Taking into account the rotational bands of resonances built on several vibrational states of the nuclear molecule, Heinz, Reinhardt *et al.* (1983) have recently shown that contact times of 10^{-19} s and more are possible. For scattering just below the top of the barrier of the pocket, positron lines with a width of 50–70 keV have been predicted. The total calculated yield is still much below the resonance cross section of several millibarns required to explain the height of the observed line structures, but it can be hoped that inclusion of inelastic scattering processes will bring the calculated absolute magnitude in line with experiment. Furthermore, the strong dependence on beam energy has been quantitatively confirmed in this model (Heinz *et al.*, 1984a, b).

A definite proof that the observed positron lines are signals for spontaneous pair creation would be the keystone for atomic physics of very high-Z systems. The interest of nuclear physics, however, starts right at this point.

Note added in proof. This review is based on the status of experiment and theory in this field at the end of 1983. More recent experimental results have been partly published by Cowan *et al.* (1985) and by Kienle (1984). A more comprehensive review of the status of the theory of Strong Fields can be found in Greiner *et al.* (1985).

REFERENCES

Agassi, D., C. M. Ko, and H. A. Weidenmüller, 1977, *Ann. Phys. (NY)* **107**, 140.

Agassi, D., C. M. Ko, and H. A. Weidenmüller, 1979, *Ann. Phys. (NY)* **117**, 407.

Backe, H., L. Handschug, F. Hessberger, E. Kankeleit, L. Richter, F. Weik, R. Willwater, H. Bokemeyer, P. Vincent, Y. Nakayama, and J. S. Greenberg, 1978, *Phys. Rev. Lett.* **40**, 1443.

Backe, H., W. Bonin, E. Kankeleit, W. Patzner, P. Senger, and F. Weik, 1980, GSI Scientific report No. 80-3, p. 168.

Backe, H., 1983, *Atomic Physics of High-Z Systems*, preprint, Mainz.

Backe, H., P. Senger, W. Bonin, E. Kankeleit, M. Krämer, R. Krieg, V. Metag, N. Trautmann, and J. B. Wilhelmy, 1983a, *Phys. Rev. Lett.* **50**, 1838.

Backe, H., W. Bonin, E. Kankeleit, M. Krämer, R. Krieg, V. Metag, P. Senger, N. Trautmann, F. Weik, and J. B. Wilhelmy, 1983b, in *Quantum Electrodynamics of Strong Fields*, edited by W. Greiner (Plenum, New York), p. 107.

Balanda, A., H. J. Beeskow, K. Bethge, H. Bokemeyer, H. Folger, J. S. Greenberg, H. Grein, A. Gruppe, S. Ito, S. Matsuki, R. Schulé, R. Schultz, D. Schwalm, J. Schweppe, R. Steiner, P. Vincent, and M. Waldtschmidt, 1980, GSI Scientific report No. 80-3, p. 161.

Bates, D. R., and R. McCarroll, 1958, *Proc. R. Soc. London* **A245**, 175.

Berdermann, E., F. Bosch, M. Clemente, F. Güttner, P. Kienle, W. Koenig, C. Kozhuharov, B. Martin, W. Potzel, B. Povh, C. Tsertos, W. Wagner, and Th. Walcher, 1980, GSI Scientific report No. 80-3, p. 103.

Berdermann, E., F. Bosch, M. Clemente, P. Kienle, W. Koenig, C. Kozhuharov, H. Tsertos, and W. Wagner, 1982, GSI Scientific report No. 82-1, p. 138.

Betz, W., G. Soff, B. Müller, and W. Creiner, 1976, *Phys. Rev. Lett.* **37**, 1046.

Błocki, J., J. Randrup, W. J. Swiatecki, and C. F. Tsang, 1977, *Ann. Phys (N.Y.)* **105**, 427.

Bokemeyer, H., K. Bethge, H. Folger, J. S. Greenberg, H. Grein, A. Gruppe, S. Ito, R. Schulé, D. Schwalm, J. Schweppe, N. Trautmann, P. Vincent, and M. Waldschmidt, 1983, in *Quantum Electrodynamics of Strong Fields*, edited by W. Greiner (Plenum, New York), p. 273.

Bosch, F., 1982, *Phys. Bl.* **38**, 205.

Clemente, M., E. Berdermann, P. Kienle, H. Tsertos, W. Wagner, C. Kozhuharov, F. Bosch, and W. Koenig, 1984, *Phys. Lett.* **137B**, 41.

Cowan, T., H. Backe, M. Begemann, K. Bethge, H. Bokemeyer, H. Folger, J. S. Greenberg, H. Grein, A. Gruppe, Y. Kido, M. Klüver, D. Schwalm, J. Schweppe, K. E. Stiebing, N. Trautmann, and P. Vincent, 1985, *Phys. Rev. Lett.* **54**, 1761.

Gärtner, P., B. Müller, J. Reinhardt, and W. Greiner, 1980, *Phys. Lett.* **95B**, 181.

Gärtner, P., U. Heinz, B. Müller, and W. Greiner, 1981, *Z. Phys.* **A300**, 143.

Gobbi, A., W. Nörenberg, 1980, in: *Heavy Ion Collisions 2*, edited by R. Bock (North-Holland, Amsterdam), p. 127.

Greiner, W., B. Müller, and J. Rafelski, 1985, *Quantum Electrodynamics of Strong Fields* (Springer, Heidelberg).

Gross, D. H. E., and H. Kalinowski, 1978, *Phys. Rev.* **45**, 175.

Güttner, F., W. Koenig, B. Martin, B. Povh, H. Skapa, J. Soltani, Th. Walcher, F. Bosch, C. Kozhuharov, 1982, *Z. Phys.* **A304**, 207.

Gyulassy, M., 1974, *Phys. Rev. Lett.* **33**, 921.

Gyulassy, M., 1975, *Nucl. Phys.* **A244**, 497.

Heiligenthal, G., W. Betz, G. Soff, B. Müller, and W. Greiner, 1978, *Z. Phys.* **A285**, 105.

Heinz, U., W. Greiner, and B. Müller, 1981, *Phys. Rev. A* **23**, 562.

Heinz, U., B. Müller, and W. Greiner, 1983, *Ann. Phys. (NY)*, **151**, 227.

Heinz, U., J. Reinhardt, B. Müller, W. Greiner, and U. Müller, 1983, *Z. Phys.* **A314**, 125.

Heinz, U., J. Reinhardt, B. Müller, W. Greiner, and W. T. Pinkston, 1984, *Z. Phys.* **A316**, 341.

Heinz, U., U. Müller, J. Reinhardt, B. Müller, and W. Greiner, 1984b, *Ann. Phys. (NY)* **158**, 476.

Heisenberg, W., and H. Euler, 1936, *Z. Phys.* **38**, 714.

Hund, F., 1940, *Z. Phys.* **117**, 1.

Ito, S., P. Armbruster, H. Bokemeyer, F. Bosch, H. Emling, H. Folger, E. Grosse, R. Kulessa, D. Liesen, D. Maor, D. Schwalm, J. S. Greenberg, R. Schulé, and N. Trautmann, 1982, in GSI Scientific report No. 82-1, Darmstadt, p. 141.

Kankeleit, E., 1980, *Nukleonika* **25**, 253.

KANKELEIT, E., U. GOLLERTHAN, G. KLOTZ, M. KOLLATZ, M. KRÄMER, R. KRIEG, U. MEYER, H. OESCHLER, AND P. SENGER, 1985, *Nucl. Instr. Meth.* **A234**, 81.
KIENLE, P., 1983a, in *Quantum Electrodynamics of Strong Fields*, edited by W. Greiner (Plenum, New York), p. 293.
KIENLE, P., 1983b, private communication.
KIENLE, P., 1984, *Proceedings of the INS-RIKEN International Symposium on Heavy-Ion Physics, Part II*, p. 549 (Physics Society of Japan), GSI preprint 84/60 (Darmstadt).
KLEIN, O., 1929, *Z. Phys.* **53**, 157.
KO, C. H., D. AGASSI, AND H. A. WEIDENMÜLLER, 1979, *Ann. Phys. (NY)* **117**, 273.
KOZHUHAROV, CH., P. KIENLE, E. BERDERMANN, H. BOKEMEYER, J. S. GREENBERG, Y. NAKAYAMA, P. VINCENT, H. BACKE, L. HANDSCHUG, AND E. KANKELEIT, 1979, *Phys. Rev. Lett.* **42**, 376.
KOZHUHAROV, CH., 1982, in *Physics of Electronic and Atomic Collisions* (Proc. XII ICPEAC), edited by S. Datz (North-Holland, Amsterdam), p. 179.
LIESEN, D., P. ARMBRUSTER, F. BOSCH, S. HAGMANN, P. H. MOKLER, H. SCHMIDT-BÖCKING, R. SCHUCH, J. B. WILHELMY, AND W. J. WOLLERSHEIM, 1980a, *Phys. Rev. Lett.* **44**, 983.
LIESEN, D., P. ARMBRUSTER, H. H. BEHNCKE, F. BOSCH, S. HAGMANN, P. H. MOKLER, H. SCHMIDT-BÖCKING, AND R. SCHUCH, 1980b, in *Electronic and Atomic Collisions* (Proc. XIth ICPEAC), edited by N. Oda, K. Takayanagi (North-Holland, Amsterdam), p. 337.
MATHEWS, G. J., B. MÜLLER, J. RANDRUP, AND L. G. MORETTO, 1980, Fluctuations of the Reaction Time in Deep-Inelastic Heavy-Ion Collisions, Caltech preprint.
MEYERHOF, W. E., 1973, *Phys. Rev. Lett.* **31**, 1341.
MEYERHOF, W. E., R. ANHOLT, Y. EL MASRI, D. CLINE, F. S. STEPHENS, AND R. DIAMOND, 1977, *Phys. Lett.* **B69**, 41.
MEYERHOF, W. E., 1983, Collisions of Fast Heavy Ions—an Overview, Stanford University preprint.
MOLL, E. AND E. KANKELEIT, 1965, *Nukleonik* **7**, 180.
MÜLLER, B., W. PEITZ, J. RAFELSKI, AND W. GREINER, 1972a, *Phys. Rev. Lett.* **28**, 1245.
MÜLLER, B., J. RAFELSKI, AND W. GREINER, 1972b, *Z. Phys.* **257**, 62, 183.
MÜLLER, B., J. RAFELSKI, AND W. GREINER, 1973, *Phys. Lett.* **47B**, 5.
MÜLLER, B., 1976, *Ann. Rev. Nucl. Sci.* **26**, 351.
MÜLLER, B., AND V. OBERACKER, 1980, *Phys. Rev. C* **22**, 1909.
MÜLLER, U., J. REINHARDT, T. DE REUS, P. SCHLÜTER, G. SOFF, W. GREINER, AND B. MÜLLER, 1982, in *X-Ray and Atomic Inner-Shell Physics—1982*, edited by B. Crasemann (American Institute of Physics, New York), p. 206.
MÜLLER, U., G. SOFF, T. DE REUS, J. REINHARDT, B. MÜLLER, AND W. GREINER, 1983, *Z. Phys.* **A313**, 263.
PIEPER, W., AND W. GREINER, 1969, *Z. Phys.* **218**, 327.
POPOV, V. S., 1971, *Sov. J. Nucl. Phys.* **12**, 235.
RAFELSKI, J., L. P. FULCHER, AND W. GREINER, 1971, *Phys. Rev. Lett.* **27**, 958.
RAFELSKI, J., B. MÜLLER, AND W. GREINER, 1974, *Nucl. Phys.* **B68**, 585.
RAFELSKI, J., L. P. FULCHER, AND A. KLEIN, 1978, *Phys. Rep.* **38**, 227.
RAFELSKI, J., B. MÜLLER, AND W. GREINER, 1978, *Z. Phys.* **A285**, 49.
READING, J. F., 1973, *Phys. Rev. A* **8**, 3262.
READING, J. F., AND A. L. FORD, 1980, *Phys. Rev. A* **21**, 124.
REINHARDT, J., AND W. GREINER, 1977, *Rep. Prog. Phys.* **40**, 219.
REINHARDT, J., B. MÜLLER, W. GREINER, AND G. SOFF, 1979, *Phys. Rev. Lett.* **43**, 1307.
REINHARDT, J., B. MÜLLER, AND W. GREINER, 1980, in *Coherence and Correlation in Atomic Collisions*, edited by H. Kleinpoppen and J. F. Williams (Plenum, New York), p. 331.
REINHARDT, J., B. MÜLLER, AND W. GREINER, 1981a, *Phys. Rev. A* **24**, 103.
REINHARDT, J., U. MÜLLER, B. MÜLLER, AND W. GREINER, 1981b, *Z. Phys.* **A303**, 173.

REINHARDT, J., AND W. GREINER, 1985, Heavy-Ion Atomic Physics, in *Treatise on Heavy-Ion Science*, Vol. III, edited by D. A. Bromley (Plenum, New York).

REINHARDT, J., B. MÜLLER, W. GREINER, AND U. MÜLLER, 1983, *Phys. Rev. A* **28**, 2558.

DE REUS, T., J. REINHARDT, B. MÜLLER, W. GREINER, G. SOFF, AND U. MÜLLER, 1983, *J. Phys. B* **16**, in print.

RHOADES-BROWN, M. J., V. OBERACKER, M. SEIWERT, W. GREINER, 1983, *Z. Phys.* **A310**, 287.

RIHAN, T. H., N. S. ALY, E. MERZBACHER, B. MÜLLER, AND W. GREINER, 1978, *Z. Phys.* **A285**, 397.

SAUTER, F., 1931, *Z. Phys.* **69**, 742.

SCHLÜTER, P., G. SOFF, AND W. GREINER, 1981, *Phys. Rep.* **75**, 329.

SCHMIDT, R., V. D. TONEEV, AND G. WOLSCHIN, 1978, *Nucl. Phys.* **A311**, 247.

SCHWEPPE, J., A. GRUPPE, K. BETHGE, H. BOKEMEYER, T. COWAN, H. FOLGER, J. S. GREENBERG, H. GREIN, S. ITO, R. SCHULE, D. SCHWALM, K. E. STIEBING, N. TRAUTMANN, P. VINCENT, AND M. WALDSCHMIDT, 1983, *Phys. Rev. Lett.* **51**, 2261.

SEIWERT, M., 1983, private communication (to be published).

SLIV, L. A., 1955, *Sov. Phys. JETP* **25** (1953) 7 and *J. Phys. Rad.* **16**, 589.

SMITH, K., B. MÜLLER, AND W. GREINER, 1975, *J. Phys. B* **8**, 75.

SOFF, G., J. RAFELSKI, AND W. GREINER, 1973, *Phys. Rev.* **A7**, 903.

SOFF, G., B. MÜLLER, AND W. GREINER, 1978, *Phys. Rev. Lett.* **40**, 540.

SOFF, G., J. REINHARDT, B. MÜLLER, AND W. GREINER, 1980, *Z. Phys.* **A294**, 137.

SOFF, G., B. MÜLLER, AND W. GREINER, 1981a, *Z. Phys.* **A299**, 189.

SOFF, G., P. SCHLÜTER, AND W. GREINER, 1981b, *Z. Phys.* **A303**, 189.

SOFF, G., U. MÜLLER, T. DE REUS, P. SCHLÜTER, A. SCHÄFER, J. REINHARDT, B. MÜLLER, AND W. GREINER, 1983b, in *Atomic Physics of Highly Ionized Atoms*, edited by R. Marrus (Plenum, New York), p. 177.

SOFF, G., T. DE REUS, B. MÜLLER, AND W. GREINER, 1982a, *Phys. Lett.* **88A**, 398.

SOFF, G., P. SCHLÜTER, B. MÜLLER, AND W. GREINER, 1982, *Phys. Rev. Lett.* **48**, 1465.

THEIS, J., J. REINHARDT, AND B. MÜLLER, 1979, *J. Phys. B* **12**, L479.

THEIS, J., AND B. MÜLLER, 1981, *Phys. Rev. A* **24**, 89.

THORSON, W. R., AND J. B. DELOS, 1978, *Phys. Rev. A* **18**, 117, 135.

TOMODA, T., 1982, *Phys. Rev. A* **26**, 174.

TOMODA, T., AND H. A. WEIDENMÜLLER, 1982, *Phys. Rev. A* **26**, 162.

TOMODA, T., 1984, *Phys. Rev.* **A29**, 536.

TOMODA, T., AND H. A. WEIDENMÜLLER, 1983, *Phys. Rev. C* **28**, 739.

WOLLERSHEIM, H. J., W. W. WILCKE, J. R. BIRKELUND, AND J. R. HUIZENGA, 1982, *Phys. Rev. C* **25**, 338.

ZEL'DOVICH, YA. B., AND V. S. POPOV, 1972, *Sov. Phys. Usp.* **14**, 673.

CHAPTER 15

X-Ray Processes in Heavy-Ion Collisions

P. Vincent

1. INTRODUCTION

During the past decade, heavy-ion accelerators have become increasingly available to the atomic physics community. Within the past five years, extensive measurements on high-energy atomic collisions using the heaviest possible collision partners have been carried out. Although some of the features exhibited by the lighter encounters simply scale uniformly with increasing atomic number Z (Meyerhof *et al.*, 1977a; Mokler *et al.*, 1978, 1983), the large Coulomb fields associated with the heaviest collision systems induce extensive relativistic effects which result in qualitatively new phenomena. Information gathered through studies of lower-Z encounters has provided the framework within which these high-Z relativistic effects can be understood.

A common aspect of energetic heavy-ion encounters across the Periodic Table is the fact that the velocities of the projectiles are generally much less than typical orbital velocities of inner-shell electrons. From the point of view of these electrons, such collisions are adiabatic and lead to the formation of transient quasimolecular structures. The time-dependent electron wave functions can be theoretically described as expansions in sets of molecular-orbital (MO) wave functions which are solutions of the two-center Dirac or Schrödinger equations at the internuclear separations R that occur during the course of collision (Kessel *et al.*, 1973; Lichten, 1976; Greenberg *et al.*, 1985).

Figure 1 displays the solutions to the two-center, one-electron Dirac equation versus internuclear separation R in the Ni + Ni system (W. Betz *et*

P. Vincent • Department of Physics and Astronomy, Rutgers University, New Brunswick, New Jersey 08903. *Present address*: Department of Physics, Brookhaven National Laboratory, Upton, New York 11973.

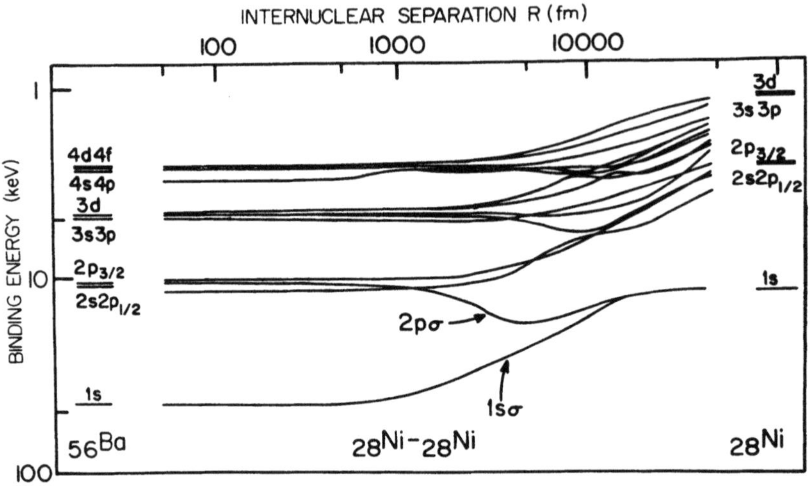

FIGURE 1. Molecular orbital correlation diagram for the Ni + Ni quasimolecule showing the dependence of molecular binding on internuclear separation R for nonrelativistic (lighter) systems. (W. Betz *et al.*, 1975; Vincent, 1983.)

al., 1975). The electrons evolve from separated-atom (SA) states to the united atomic (UA) configuration along molecular orbitals. In this system, relativistic effects are small and the states are labeled with the principal quantum number and orbital angular momentum of the UA orbital to which they correlate. The projection of this orbital angular momentum along the internuclear axis is indicated by the notation $\sigma = 0$, $\pi = 1$, $\delta = 2$, etc., in units of angular momentum.

These results can be compared and contrasted to similar calculations for the heaviest collision systems. Figure 2 shows the molecular orbital diagram for the Pb + Pb system calculated with the self-consistent relativistic Dirac–Fock method (Sepp *et al.*, 1981). The molecular orbitals are now labeled with the principal quantum number, orbital angular momentum, and total angular momentum $(\mathbf{j} = \mathbf{l} + \mathbf{s})$ of the united atomic orbital to which they connect as well as the projections m_j of $\mathbf{j}$ along the internuclear axis ($\sigma = 1/2$, $\pi = 3/2$, $\delta = 5/2$, etc.).

The most important feature seen in the correlation diagram of Fig. 2 is simply the large magnitude of binding which can be achieved. The $2p_{1/2}\sigma$ level has energies approaching the rest-mass energy of the electron $(m_e c^2)$, and the $1s_{1/2}\sigma$ level in still heavier collision systems, such as U + U, attains binding in excess of $2m_e c^2$. Heavy-ion collisions therefore offer a unique opportunity to explore the influence of extreme Coulomb fields on bound states. In particular, the opportunity is provided to search for the spontaneous production of electron–positron pairs from vacant bound states

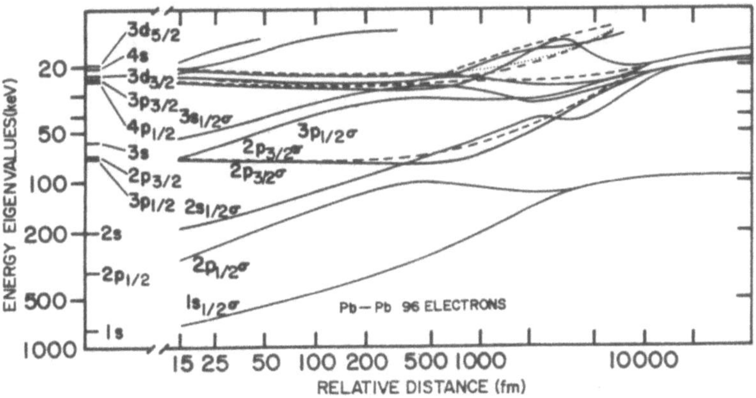

FIGURE 2. Relativistic Dirac–Fock calculations of molecular orbitals in the Pb + Pb quasimolecule calculated for 96 electrons. The left-hand labels indicate the united-atom state to which the orbitals correlate. (Sepp *et al.*, 1981.)

which are degenerate with the negative-energy Dirac continuum (Greenberg *et al.*, 1985; Reinhardt *et al.*, 1985).

A typical example of X-ray spectra generated in these heavy-ion collisions is shown in Fig. 3 (Vincent, 1977, 1981, 1983). The data were taken with an intrinsic germanium X-ray spectrometer using a 0.85-mm Al absorber in Ni + Ni collisions at 64.8 and 12.6 MeV beam energies. Visible

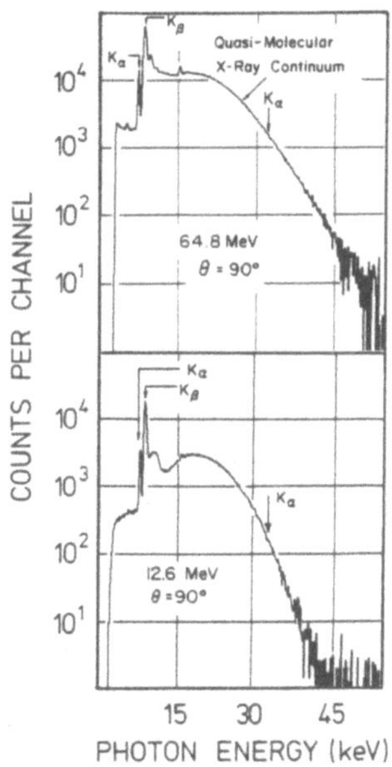

FIGURE 3. X-ray spectra from Ni + Ni collisions measured with an intrinsic germanium X-ray spectrometer. A thick Al absorber was interposed. (Vincent, 1977, 1981, 1983.)

above the strongly absorbed characteristic Ni K_α and K_β X-ray lines near 8 keV is a continuum band extending well above the UA K_α energy of Ba. This band is due to quasimolecular (MO) X-ray transitions which take place during the course of the collision between the transiently formed molecular orbitals.

The extension of the MO X-ray spectra of Fig. 3 to energies above the UA limit and the dependence of this extension on incident bombarding energy indicate that, although increased molecular binding is clearly present, dynamic effects play a significant role in the collisions. The amount of information on strong binding phenomena which can be extracted from such measurements will therefore be determined by the extent to which electrons are "really" bound in quasimolecular orbitals. The overwhelming success of the quasimolecular model lies in its ability to describe wide ranges of inner-shell excitation phenomena using truncated sets of molecular orbital basis states with as few as one, two, or three members.

The focus of this review is on the heaviest possible collision systems where strong binding effects may be observed. Therefore, treatment of lighter collision systems has been restricted to cases which provide important information on the physics of the quasimolecule, and of heavy-ion *K*-shell X-ray emission. Section 2 describes results of characteristic *K* X-ray studies in lighter and heavier systems, and Sec. 3, results of MO X-ray measurements. Section 4 contains a brief treatment of selected X-ray phenomena characteristic of heavy-ion collisions.

2. *K* X RAYS

2.1. *Lighter Systems*

Large amounts of information on reaction mechanisms responsible for inner-shell excitation cross sections in symmetric and asymmetric collision systems have been obtained in recent years. A convenient way to categorize the wide range of available data is by Z dependence. In Fig. 4 (Meyerhof *et al.*, 1976), total production cross sections (ordinate) for *K*-vacancy generation between 202-MeV Kr collisions with various targets (abscissa) are shown. Triangles represent the projectile and circles the target X rays. The dashed vertical line is at $Z_{\text{target}} = Z_{\text{projectile}}$. Four regions of excitation are distinguished:

Region i. Region i refers to excitation of the heavy-Z collision partner in very asymmetric collisions. For light targets, this corresponds to projectile excitation, and for heavy targets, target ionization. For very asymmetric collisions between low-Z targets and projectiles, a perturbation-theory approach has been successful in describing excitation of the heavy-Z

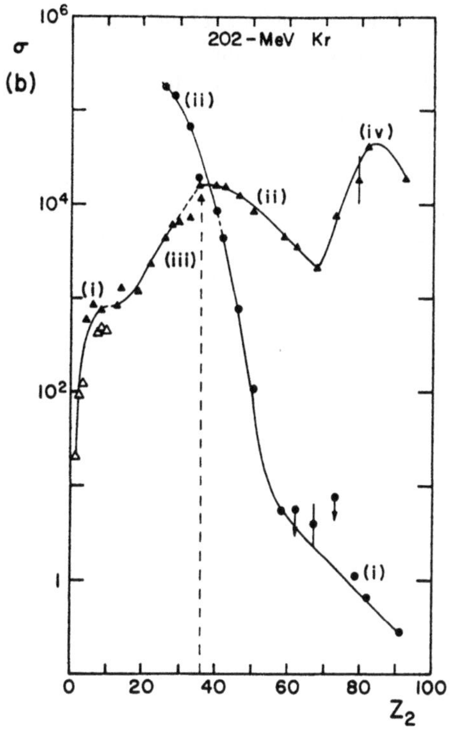

FIGURE 4. Projectile and target K-vacancy production cross sections in 202-MeV ^{84}Kr collisions. Symbols: △, ▲, thick-target extracted projectile cross sections; ●, thick-target extracted target cross sections. Curves are drawn only to guide the eye. (Meyerhof *et al.*, 1976.)

collision partner (Bang *et al.*, 1959; Merzbacher *et al.*, 1958). In these calculations, the lighter-Z collision partner induces ionization by perturbing the *atomic* wave function of the heavy-Z K-shell electron. Therefore, the concept of *quasimolecular* formation is not utilized. As the Z of the lighter collision partner increases, this perturbation becomes larger and larger, eventually causing the approximations to break down. For Region i at high Z, incorporation of relativistic effects becomes important (Jamnik *et al.*, 1957; Amundsen, 1976a).

Alternatively, "Region i" collisions can be described within the context of the molecular-orbital picture. For symmetric collisions, the K shells of the projectile and target atoms each connect to both $2p\sigma$ and $1s\sigma$ orbitals (Figs. 1 and 2). For asymmetric collisions, however, the $1s\sigma$ orbital connects to the K shell of the heavy collision partner and the $2p\sigma$ level to that of the lighter collision partner (Barat *et al.*, 1972). The binding-energy separation between these two molecular orbitals inhibits their mixing, and heavy collision-partner K X rays can therefore be said to originate from the $1s\sigma$ orbital. This molecular-orbital approach is especially useful for less asymmetric cases and for very heavy target–projectile combinations.

Regions ii and iii. These regions are dominated by excitations of the $2p\sigma$ molecular orbital followed by vacancy sharing on the outgoing part of the collision between the light ($2p\sigma$) and heavy ($1s\sigma$) collision partner (Meyerhof *et al.*, 1977b). The $2p\sigma$ vacancies themselves can be generated by direct excitation of the $2p\sigma$ electron to an empty bound state or to the continuum. Alternatively, rotational coupling of the $2p\sigma$ to a vacant $2p\pi$ orbital at small internuclear separations will also create $2p\sigma$ holes. In cases where a $2p\pi$ vacancy is present on the ingoing part of the collision, this rotational coupling process dominates. For nearly symmetric collisions, $2p\sigma$ vacancy sharing is the dominant means of generating heavy-Z K vacancies (Meyerhof, 1973a).

Region iv. This is the range of collisions where the K-shell binding energy of the lighter collision partner matches that of the L shell ($2p\sigma$) of the higher-Z partner. Large increases in the K-vacancy cross sections of the lower-Z partner then occur. A number of explanations for this phenomena have been made, all within the molecular-orbital picture (Barat *et al.*, 1972; Meyerhof *et al.*, 1978). The essential physical picture is that, for such cases, the K shell of the lower-Z partner is either connected with or strongly coupled to orbitals whose binding energies decrease sharply with decreasing internuclear separation ($3d\sigma$ orbital). Electrons in such states have large ionization probabilities, thereby accounting for the observed increase in K-vacancy production.

Of primary concern in this review are the cases of K-shell excitation in nearly symmetric systems, and excitation of the heavy collision partner in asymmetric systems. Processes involving the $2p\sigma$ and $1s\sigma$ molecular orbitals are therefore of primary interest. For these reasons, direct and rotational $2p\sigma$ excitation, $2p\sigma$ vacancy sharing (Demkov, 1964; Meyerhof, 1973a), and direct $1s\sigma$ excitation are the topics of central concern (Regions i, ii, and iii).

We must furthermore note that the quantity of theoretical interest is the cross section for *vacancy* production in the various collision systems. Experiment determines the cross section for K X-ray production or, for very light systems, K Auger-electron emission. Conversion of measured K X-ray data into K-vacancy data requires knowledge of the K-shell fluorescence yields. Calculations and measurements of the neutral-atom yields have been carried out (Bambynek *et al.*, 1972). However, for symmetric or nearly symmetric collisions, excitation of the K shell is usually accompanied by multiple excitations of the outer shells. The atoms are rarely, if ever, "neutral" at the time of X-ray emission. Determination of these fluorescence yields can be problematic. Since fluorescence yields increase with increasing atomic number the problem is particularly serious in the lighter-Z collision systems. Statistical arguments can be used to estimate the yields (Larkins, 1971a, b; McGuire, 1969) or, in some cases, theoretical calculations can be utilized (Bhalla, 1973; Bhalla, 1975; Chen *et al.*, 1975).

We now turn to a brief discussion of direct excitation phenomena (Region i in Fig. 4) (Basbas *et al.*, 1973, 1978; Anholt *et al.*, 1977a). As a first example, Fig. 5 summarizes results of target K-shell excitation using protons as projectiles (Basbas *et al.*, 1973). The data are plotted in the form of scaled cross sections (ordinate) versus scaled projectile energies (abscissa) and are compared to plane-wave Born approximation (PWBA) calculations for direct excitation of the electrons from their initial atomic $1s$ orbital. In Fig. 5, we have $\theta_K = \hbar\omega_{2K}/Z_{2K}^2 R$, where $\hbar\omega_{2K}$ is the binding energy of the target K shell, $Z_{2K} = (Z_2 - 0.3)$ the effective Z of the screened target K shell, and R is the rydberg $= 1/2$ hartree $= 13.6$ eV. Therefore θ_K is a dimensionless quantity which measures the nonhydrogenic aspects of the K-shell ionization energy; θ_K varies from 0.6 to 0.9 throughout the range of collision systems plotted in Fig. 5. The quantity ε is the binding-energy correction factor which modifies θ_K to correct for the increased binding caused by the projectile's presence within the atomic target K shell during the collision. The quantity $\eta_K = v_1^2/v_{2K}^2$, where v_1 is the projectile velocity, and v_{2K} is the Bohr velocity of the screened projectile K-shell electron, is proportional to the incident projectile energy; a_0 is the Bohr radius. The function $E_{10}(x)$ is the exponential integral of order 10; it arises from corrections to straight-line

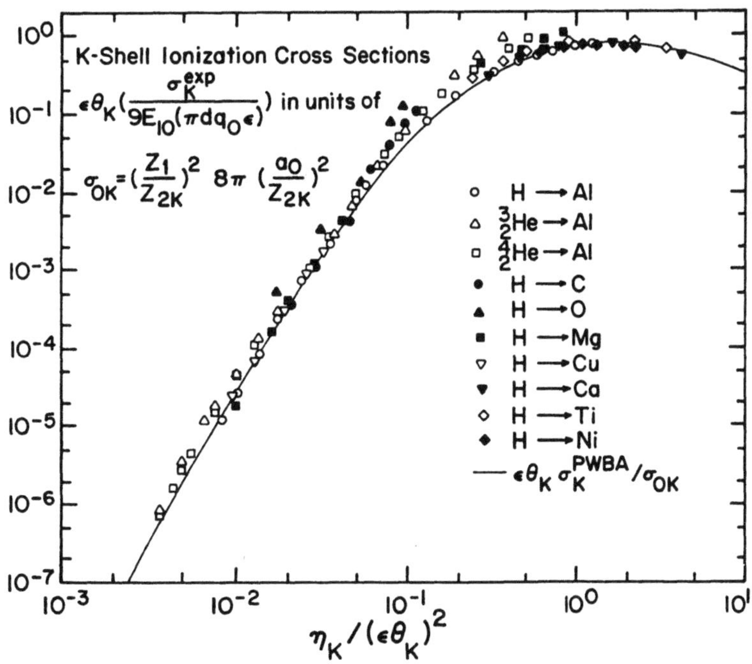

FIGURE 5. Universal graph of K-shell ionization cross sections. The solid curve represents the PWBA prediction. (Basbas *et al.*, 1973.)

trajectories to take account of nuclear Coulomb deflection of the projectile. The argument of the exponential integral function, $x = \pi d q_0$, contains the variables d, which is one-half the distance of closest approach in a head-on collision, and $q_0 = \omega_{2K}/v_1$.

Corrections to the PWBA calculations of Fig. 5 for nuclear Coulomb scattering and target K-electron binding turn out to be important because excitation usually takes place during collisions which deeply penetrate the target K shell. This occurs because the minimum momentum transfer necessary to ionize the K shell is

$$\hbar q_0 = \frac{\hbar \omega_{2K}}{v_1}. \tag{1}$$

For slow projectiles we have $q_0^{-1} \ll a_{2K}$, where a_{2K} denotes the target K-shell atomic radius. The projectile is therefore deep inside the target K shell during excitation, causing an effective increase in the target binding energy. In Fig. 5 the poorer agreement between measurement and theory at higher values of the abscissa is due to polarization of the target K shell in the field of the projectile, and to contributions from projectile-electron capture.

The binding-energy correction necessary to produce good agreement between the measured and calculated cross sections has its origins in the quasimolecular aspects of the collision. From a molecular standpoint, excitation of the high-Z collision partner in highly asymmetric systems proceeds from the $1s\sigma$ orbital. The $1s\sigma$ binding energy increases with decreasing internuclear separation. Since excitation occurs inside the K-shell radius, the actual binding energy of the $1s\sigma$ orbital at the time of excitation is therefore intermediate between the separated and united-atom K-shell limits.

Similar effects are observed in studies of the impact-parameter dependence of K-shell excitation. For example, Fig. 6 shows results of Cu K X-ray impact-parameter dependence in a variety of asymmetric collision systems (Anderson *et al.*, 1976). Within the context of a perturbed atomic picture of target ionization, the K-shell excitation probabilities should increase with the square of the projectile charge (Z_1) for projectiles of the same velocity. The ordinate of Fig. 6 is the ratio of the ionization probability to Z_1^2. Contrary to the expectations of simple perturbation theory, the scaled ionization probabilities fail to form a universal curve, but exhibit decreasing scaled intensities with increasing Z_1. This tendency is simply due to increases in the K-shell ($1s\sigma$) binding energy during the collision ("binding energy correction") (Brandt *et al.*, 1966; Brandt, 1973).

The solid lines in Fig. 6 indicate calculations of the ionization probabilities based on a scaling procedure which utilizes experimental results for protons on Cu (Lund, 1974) at various bombarding energies to

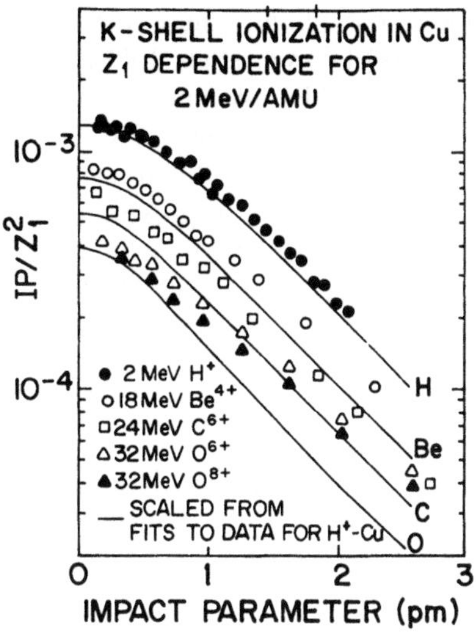

FIGURE 6. Impact-parameter dependence of Cu K-shell ionization probabilities (IP) reduced by a factor $Z_{\text{projectile}}^{-2}$, for H^+, Be^{4+}, C^{8+}, O^{6+}, and O^{8+} beams at 2 MeV/amu. Solid curves are scaled from parameterized proton-beam results. (Anderson *et al.*, 1976.)

derive semiempirical binding-energy corrections. The scaling is successful at small impact parameters, but significant discrepancies occur for more distant collisions. Nevertheless, the results provide strong, relatively direct evidence for increases in K-shell binding during the course of these encounters, and as such illustrate the quasimolecular aspects of these collisions (Cocke *et al.*, 1973; Randall *et al.*, 1976).

Extensions of the measurements shown in Fig. 5 to collisions involving heavier projectiles have been performed. In such studies (Presser *et al.*, 1980), comparisons have been made with Coulomb perturbed-stationary-state (CPSS) (Basbas *et al.*, 1978) calculations with relativistic corrections (CPSSR) (Amundsen *et al.*, 1976b) and with results from the statistical model (Mittleman *et al.*, 1967; Brandt *et al.*, 1976). The statistical-model calculations are essentially made within the framework of the molecular-orbital picture. Inner-shell excitation is viewed as a random-walk process through which the electrons travel upwards in binding energy through a complicated series of molecular-orbital crossings and couplings. It is the very complexity of these many couplings which allows the ionization process to be described as a diffusion of the electron "upward" in energy space to unoccupied molecular orbitals or the continuum.

The CPSSR calculations have been moderately successful in reproducing the beam-energy dependence of measured data. Overall agreement is, in general, better for increasingly asymmetric collisions and higher bombarding energies. At low beam energies, the statistical model does quite well, but falls below the measured cross sections at higher beam energies. Taken together, these results demonstrate the importance of molecular excitation processes under conditions where the collision can be considered adiabatic (low energies) (Presser *et al.*, 1980).

In highly asymmetric collision systems, the total cross sections for heavy-collision-partner K-shell excitation have also been successfully described using the binary encounter (Garcia *et al.*, 1973; Hansen, 1973) and semiclassical approximations (Bang *et al.*, 1959; Choi *et al.*, 1969; Hansteen *et al.*, 1975). In the usual BEA calculation, the cross sections can be written (Garcia *et al.*, 1973; Hansen, 1973)

$$\sigma = \frac{Z_1^2}{U_K^2} f\left(\frac{E}{\lambda U_K}\right), \tag{2}$$

where Z_1 is the projectile atomic number, U_K is the binding energy of the target K shell, E the projectile energy, λ the ratio of the projectile mass to that of the electron, and f is a universal function. A plot of $U_K^2\sigma/Z_1^2$ versus $E/\lambda U_K$ should therefore yield a universal curve.

In adiabatic collisions, molecular orbitals will be formed. The inner-shell electrons are therefore not bound by an amount U_K at the time of their excitation, but rather by the binding energy of either the $1s\sigma$ or $2p\sigma$ molecular orbital at the internuclear distance where excitation occurs. Since excitation usually takes place deep within the atomic K shells, the binding energy at the time of excitation can be approximated by the binding energy of the united atom $1s\sigma$ or $2p\sigma$ states for heavy and light collision-partner excitation, respectively.

Figure 7 (Foster *et al.*, 1976) shows results of scaled cross-section measurements versus the scaled BEA bombarding energy for a variety of collision systems. In Fig. 7, the usual binding energy U_K has been replaced with the value U_{ua} indicating the united-atom binding energy of the appropriate ($1s\sigma$ or $2p\sigma$) molecular orbital. The solid line is the BEA prediction. Agreement between experiment and modified BEA theory is quite good when the united-atom molecular binding energies are used to scale the cross-section measurements.

The results presented in Fig. 7 provide an intuitively pleasing unification of the atomic and molecular aspects of these collisions. Nevertheless, from a theoretical point of view, the scaling procedure is without justification, since it implies the equivalence of both the lighter and heavier collision partners, while the BEA formulation uses the target nucleus to define the electron-velocity distribution and binding energy and the projectile nucleus to transfer momentum and energy to that target electron. A theoretical justification for such a description has, however, been suggested by the results of Briggs (Briggs *et al.*, 1975), who showed that excitation in slow collisions can be described as Coulomb ionization from united-atom molecular orbitals by the coherent field of both nuclei.

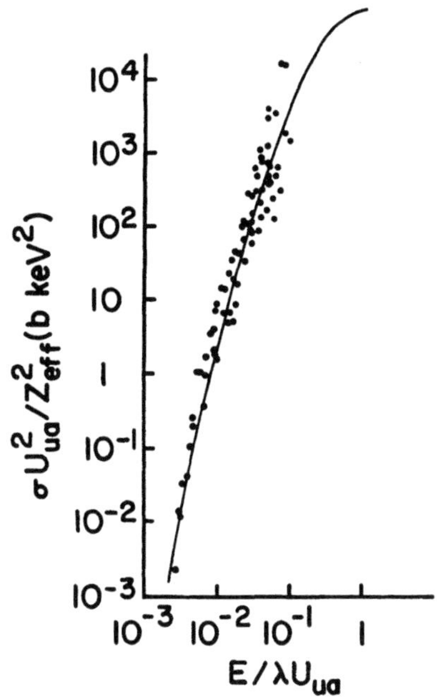

FIGURE 7. Comparison of reduced K-vacancy production cross sections (data points) with BEA calculations based on the united-atom binding energy (solid curve). For $1s\sigma$ excitation we have $Z_{eff} = Z_{projectile}$; for $2p\sigma$ excitation, $Z_{eff} = [1/2(Z_{targ.}^2 + Z_{proj.}^2)]^{1/2}$. (Foster *et al.*, 1976.)

We now turn to a discussion of Regions ii and iii in Fig. 4. In these regions, the molecular-orbital description is most likely to be the best. For symmetric or nearly symmetric collisions, total-yield (singles) measurements have indicated that excitation should proceed by rotational coupling of the $2p\sigma$ and $2p\pi$ orbitals. This hypothesis can be tested through measurements of the impact-parameter dependence of the excitation probabilities. Figure 8 (Sackmann *et al.*, 1974) shows results of one such study using Ne^+–Ne collisions at 235 and 363 keV. In such symmetric collisions one cannot distinguish between the projectile and target atoms. At forward scattering angles, contributions from both forward scattered projectiles and recoiling target atoms will therefore be present. Usually, the rapid decrease in the Rutherford scattering cross section with increasing projectile scattering angles strongly attenuates target-recoil contributions. For cases in which the excitation probability of a process decreases rapidly with increasing impact parameter, recoil contributions can, however, be important.

The solid lines in Fig. 8 indicate results of calculations (Briggs *et al.*, 1972) of the symmetrized excitation probabilities due to a simple two-state coupling of the initial $2p\pi$ vacancy to the $2p\sigma$ orbital. The striking change in the measured excitation-probability shapes in going from 235 to 363 keV is very well reproduced by the calculations. The results provide strong support for the dominance of two-state $2p\pi$–$2p\sigma$ rotational coupling in the production of K-shell vacancies during these collisions.

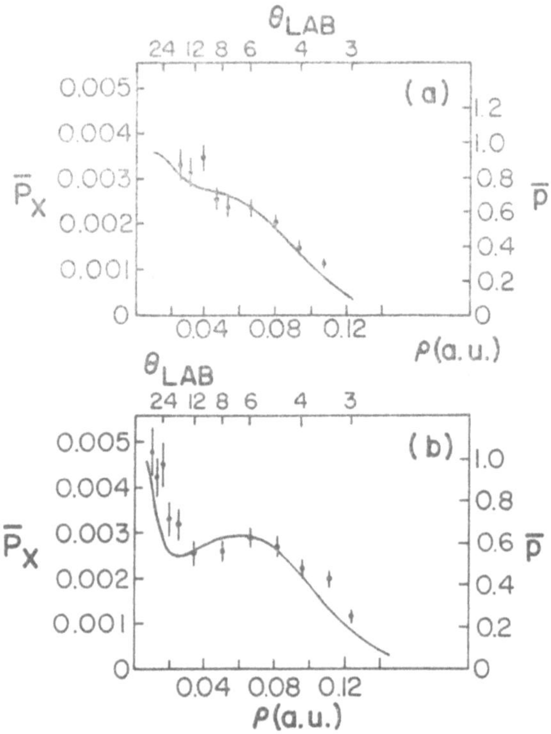

FIGURE 8. Experimental X-ray production (left-hand scale) and theoretical vacancy-production probabilities (solid curve, right-hand scale) in Ne^+–Ne collisions, versus impact parameter ρ. (a) Ion energy 235 keV; (b) ion energy 363 keV. (Sackmann *et al.*, 1974.)

The success of the molecular-orbital description of K-shell excitation in terms of only three states ($2p\pi$, $2p\sigma$, $1s\sigma$) has been one of the most convincing demonstrations of its validity. The range of validity of the three-state description has been explored in a variety of systematic studies of K X-ray impact-parameter dependence. In symmetric or nearly symmetric collisions, the two-state description ($2p\pi$, $2p\sigma$) has been successful in reproducing experimental results from lighter collision partners. For heavier systems, however, systematic deviations from the two-state calculations begin to occur.

Under certain simplifying approximations, the impact-parameter dependences of $2p\pi$–$2p\sigma$ rotational coupling in different collision systems can be scaled to form a universal curve, which exhibits a characteristic (adiabatic) peak (Taulberg *et al.*, 1976). The results of such scaling are shown in Fig. 9 (Schuch *et al.*, 1980a). The data are plotted against a reduced impact parameter $b' = [\alpha(Z_1, Z_2)/\hbar v]^{1/3}\, b$; Z_1 and Z_2 are the projectile and target nuclear charges, v is the projectile velocity, and b is the impact parameter. The function $\alpha(Z_1, Z_2)$ scales the energy splitting between the $2p\sigma$–$2p\pi$ molecular orbitals for various systems, and is taken from calculations. The solid line in Fig. 9 is the prediction of the simplified $2p\pi$–$2p\sigma$ rotational coupling calculation. The absolute heights of the data and calculation have been normalized to each other. The scaling of the data from this wide variety of collision systems (Luz, 1979; Cocke, 1976; Nolte *et al.*, 1980) according to the predictions of the simplified two-state description and the good agreement in shape between calculation and scaled experimental results serve to justify the $2p\pi$–$2p\sigma$ molecular-orbital description of these collisions. It should be noted that collision systems up to a combined Z of 35 are included in Fig. 9.

A similar scaled plot for heavier collision systems is indicated in Fig. 10 (Schuch *et al.*, 1980a), where united-atom Z values between 56 and 100 are

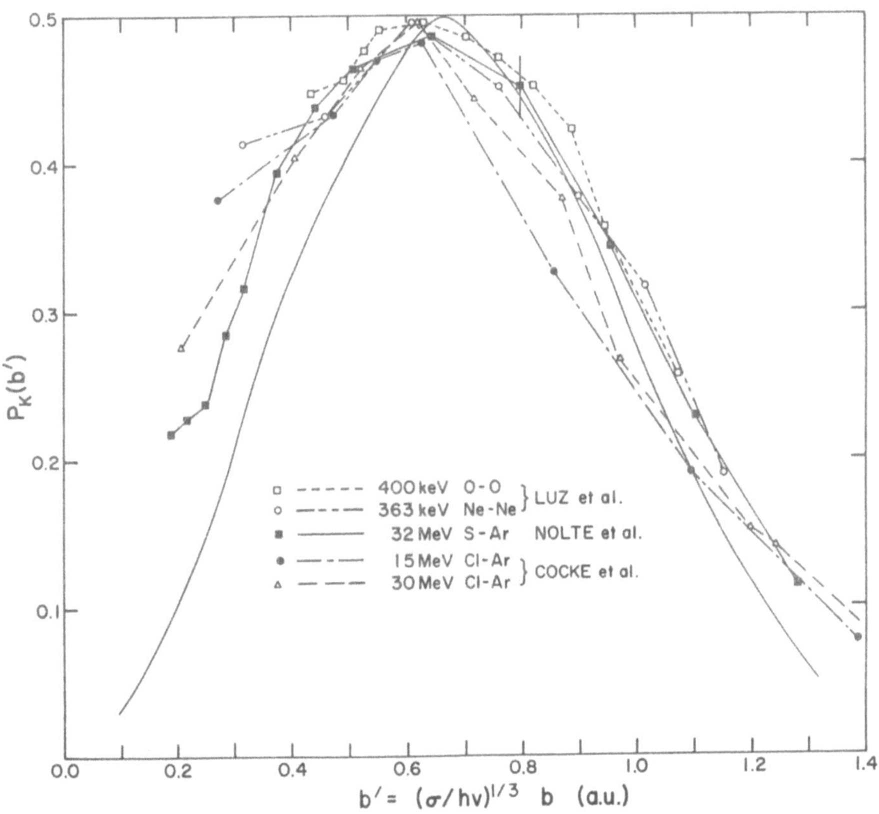

FIGURE 9: Normalized K-shell vacancy-production probabilities for $Z_1 \cong Z_2 \leqslant 20$ versus reduced impact parameter b'. Solid curve represents universal $2p\pi$–$2p\sigma$ rotational-coupling calculations. (Schuch *et al.*, 1980a.)

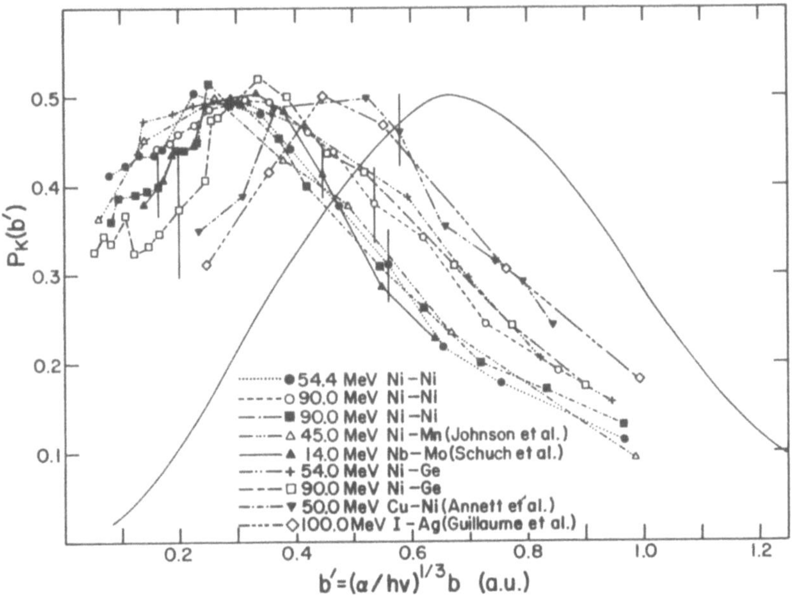

FIGURE 10. Same as Fig. 9, except for $Z_1 \cong Z_2 \geqslant 25$. (Schuch *et al.*, 1980a.)

shown. The data (Annett *et al.*, 1979; Schuch *et al.*, 1979; Johnson *et al.*, 1979; Guillaume *et al.*, 1978) display a common scaling in quantitative agreement with the two-state description, but the centroid of the reduced-impact-parameter distributions is shifted noticeably downward from the position of the adiabatic peak. Several authors (Anholt *et al.*, 1977b; Jakubassa, 1980) have examined relativistic corrections to the assumed $2p\pi$–$2p\sigma$ energy gap, and have concluded that such corrections cannot account for the observed shift.

In addition to the adiabatic peak mentioned above, $2p\pi$–$2p\sigma$ rotational coupling exhibits a second characteristic feature at smaller impact parameters corresponding to the region of 90° scattering in the center of mass. At this $\theta_{\text{c.m.}}$ angle, the molecular axis rotates through 90° during the course of the collision, thereby causing strong mixing of the $2p\pi$ and $2p\sigma$ states. Two-state rotational coupling calculations predict a second peak (referred to as the "kinematic" peak) in the impact-parameter dependence centered at 90° scattering with a $\sin^2 \theta_{\text{c.m.}}$ shape. Subsequent measurements of K-shell excitation in 30-MeV Ca + Ti (Pflanz, 1980; Schuch, 1981a) and 60-MeV I + Ag (Anholt, 1980) collisions in the "kinematic" peak region agree well with the rotational-coupling prediction. Furthermore, investigations of the shape and size of K X-ray impact-parameter dependence in nearly symmetric systems for various target thicknesses τ (Annett,

1979; Stähler, 1981) and incident projectile charge states q (Hagmann, 1979; Schmidt-Böcking *et al.*, 1982b) have been carried out. The shape of the impact-parameter dependences was found to be independent of target thickness and incident charge state although the absolute height increases with increasing τ and q.

The picture emerging from all these and other studies is that, although $2p\pi$–$2p\sigma$ rotational coupling dominates K-shell excitation in nearly symmetric collisions, other couplings of the $2p\sigma$ orbital to higher-lying bound and continuum states become increasingly important at higher Z. This latter point has been illustrated nicely with measurements of K-shell excitation in the 12 MeV Ar + Ar system (Schmidt-Böcking *et al.*, 1982b). The ratio of incident projectile $q = 9^+$ to $q = 4^+$ charge-state yields was studied versus impact parameter. The 9^+ yields were observed to increase over the 4^+ yields in the kinematic ($b = 100$ fm) and adiabatic ($b = 1000$ fm) peak regions. The valley region showed no increase in yield at higher q. The kinematic and adiabatic regions therefore increased with the number of $2p\pi$ vacancies brought into the collision whereas the "valley" region did not. The authors attributed this "valley" region behavior to contributions from direct excitation of the $2p\sigma$ state into vacant higher-lying molecular orbitals and the continuum (Foster *et al.*, 1976; Briggs, 1975). These additional excitations add to $2p\pi$–$2p\sigma$ rotational coupling (adiabatic and kinematic peak regions), to produce the measured results.

We conclude this section with a brief treatment of target-thickness and charge-state effects on K X-ray yields. This subject is in general quite complicated; the complications stem principally from two important facts: the large size of the numerous atomic collision cross sections, and the dependence of these cross sections on the atomic configuration of the projectile. If the projectile atomic configuration is to be maintained throughout the collision, extremely thin targets are needed. This requires gas targets at pressures of 10^{-3} Torr or less. Even with the use of such targets, however, it is often not possible to maintain single-collision conditions for measuring the K X-ray cross sections of interest. This is true since the vacancy configuration of the outer shells during the collision influences inner-shell vacancy production and decay, and collision cross sections for these loosely bound outer electrons are very large. Futhermore, gas targets are only available for certain Z, can be difficult to prepare experimentally, or simply are not thick enough to produce sufficiently high counting rates.

For these reasons, solid targets are often employed. Thin-foil experiments (several $\mu g/cm^2$), however, can strictly speaking only produce determinations of yields and not cross sections, since multiple collisions always occur (Meyerhof *et al.*, 1977b). Even under conditions where the change in beam energy of the incident projectile in traversing the foil is negligible, target-thickness effects can result in cross sections for inner-shell phenomena which differ by factors of 2 or more. The question of target-thickness effects can therefore be quite subtle, and each measurement must be considered on an individual basis (Gray, 1980).

In Fig. 11 we plot the results of one study (Gardner *et al.*, 1977) which involves a variety of collision systems. Depending on the incident charge state, the observed target *K*-shell excitation cross sections either increase or decrease with target thickness. These results can be interpreted within a

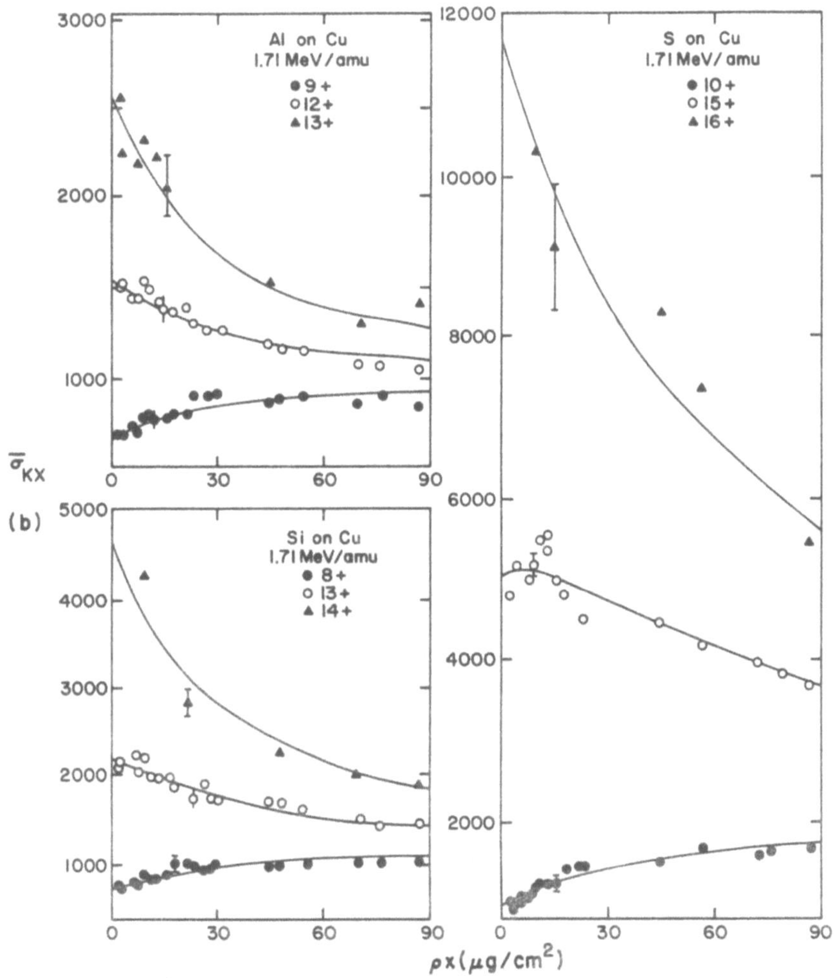

FIGURE 11. Cu *K* X-ray cross section $\bar{\sigma}_{Kx}$ versus target thickness, for Al, Si, and S ions on thin solid Cu targets. Solid curves represent calculations based on the three-component model. (Gardner *et al.*, 1977.)

three-component model of K-vacancy production (Allison, 1958; Hopkins, 1975; Gray *et al.*, 1976). The model allows for an enhancement of target-vacancy production by projectiles with a K vacancy, over that from projectiles without a K vacancy, by a factor of α. The enhancement for projectiles that carry two K vacancies is given by a factor β. Changes in the projectile K-vacancy fractions during traversal of the target are used to explain the observed target-thickness dependence. In the Al + Cu data of Fig. 11, for example, the 13^+ Al projectile cross sections decrease with increasing target thickness due to quenching of the projectile K vacancies as the target is traversed. The increase in the Al 9^+ cross sections with target thickness reflects the excitation of Al K-shell vacancies in the target.

The solid curves in Fig. 11 are results of least-squares fits to the data, of the several parameters comprising the three-component model. These parameters include the various projectile K-shell charge-changing cross sections as a function of target thickness and the α and β enhancement factors. The results can be compared with theoretical calculations. Results suggest that projectile quenching is primarily due to charge exchange from the target to the projectile K shell (Nikolaev, 1967). The enhancement factors α and β can be successfully described by a vacancy-sharing calculation of vacancy transfer from the $1s\sigma$ state of the projectile directly to the target K shell (Meyerhof, 1973a). We should note here that in cases where only one projectile vacancy is involved a simpler two-component model can be used instead of the three-component description in the example of Fig. 11 (Betz *et al.*, 1973; Hopkins, 1975).

We therefore see that the spectrum of K-shell excitation phenomena in heavy-ion collisions is very wide indeed. In adiabatic encounters between symmetric or nearly symmetric target and projectile atoms, the molecular-orbital picture best describes the evolution of inner-shell electrons. Charge-state and multiple-collision effects, as well as other excitation mechanisms of a nonmolecular nature, can, however, also play important roles in these collisions (Greenberg *et al.*, 1985). Each target–projectile combination must therefore be examined on an individual basis in order to determine the degree to which these various models are applicable. Nevertheless, the concept of increased inner-shell electron binding in heavy-ion encounters through the establishment of transient quasimolecular orbitals is clearly verified by extensive K X-ray data from lighter collision systems.

2.2. *Heavy Systems*

In extending our discussion from lighter systems to the very heavy collision partners, it is important to recall the similarities and differences between encounters exhibiting relativistic and nonrelativistic characteristics.

Clearly, in both cases, adiabatic molecular orbital levels are involved, and for the heavy systems the adiabatic condition is extremely well satisfied by the $1s_{1/2}\sigma$ and $2p_{1/2}\sigma$ orbitals. However, the large spin–orbit splitting of the molecular levels in the very heavy systems (compare and contrast Figs. 1 and 2) leads to a different topography of the levels in energy space, resulting in the dominance of different couplings among different orbitals in the two cases. In particular, the rapid increase in inner-shell binding energies with decreasing internuclear separation and the strong localization of the $1s_{1/2}\sigma$ wave function near the nuclei for small internuclear distances (Soff *et al.*, 1979) are aspects of the relativistic quasimolecule which lead to qualitatively different behavior of measured excitation and decay probabilities.

Important aspects of the very heavy quasimolecule are illustrated nicely by measurements of the Z dependence of total K X-ray cross sections. Figure 12 (Behncke *et al.*, 1979) contains a plot of the results of such measurements using 3.6-, 4.7-, and 5.9-MeV/amu Pb projectiles on various targets. This graph can be compared directly with the results from lighter systems (Fig. 4). The two plots are qualitatively similar, both illustrating the regions of direct $1s\sigma$ excitation (see 3.6-MeV/amu data of Fig. 12 and labeled data points—Region i in Fig. 4), of K–K vacancy sharing (Region iii in Fig. 4), of $2p\sigma$ excitation (Region ii in Fig. 4), and of K–L level matching (Region iv in Fig. 4). The major qualitative difference between Figs. 12 and 4 is simply the displacement of the entire characteristic Z-dependent excitation profile from lower to higher target atomic numbers.

One particularly significant aspect of the total cross-section data of Fig. 12 is their large absolute magnitude. The role of increased $1s\sigma$ orbital binding in decreasing the magnitude of vacancy-production probabilities from lighter collision systems has been discussed already (Figs. 5, 6). This "binding-energy" effect led to early predictions that K X-ray production cross sections in collisions where $Z_{\text{projectile}} + Z_{\text{target}} > 100$ would be several orders of magnitude smaller than the actual measured results of Fig. 12 (Foster *et al.*, 1976; Meyerhof, 1974). Although the increased binding of the $1s\sigma$ orbital in very heavy collision systems does, in fact, make $1s\sigma$ ionization more difficult, the strong relativistic contraction of the $1s\sigma$ wave function associated with this strong binding and the rapid increase in this localization with decreasing internuclear separation act to increase the ionization probability (Betz *et al.*, 1976; Anholt, 1979c). The net effects result in cross sections between 1 and 10,000 barns as shown in Fig. 12.

These results are illustrated by the success of a semiempirical calculation (Anholt, 1978c) which explains the observed magnitudes of the extracted $1s\sigma$ contribution to the K X-ray cross sections. Within the context of such a semiempirical model, we have

$$\sigma_{1s\sigma} = Z_L^2 \sigma_{\text{PWBA}} CBR, \tag{3}$$

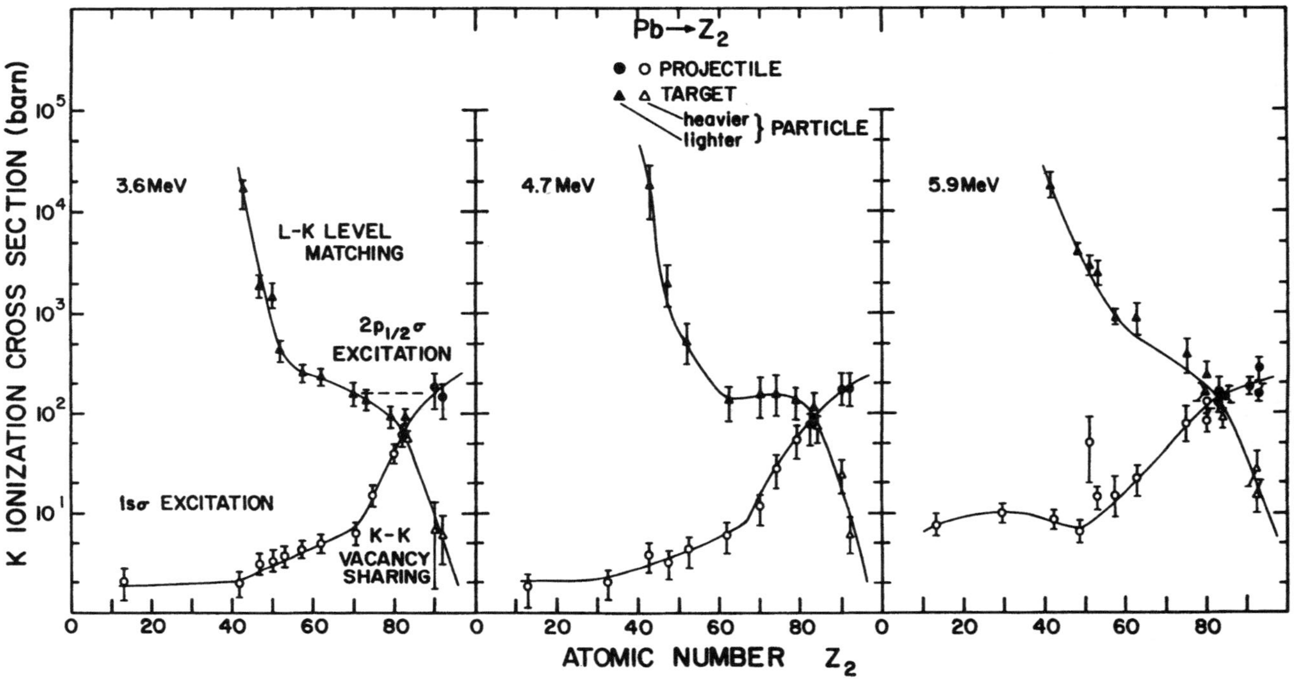

FIGURE 12. Projectile and target K-shell ionization cross sections for 3.6-, 4.7-, and 5.9-MeV/amu Pb ions. Solid curves guide the eye. Regions where $1s\sigma$ and $2p_{1/2}\sigma$ vacancy production and K–K, L–K vacancy sharing dominate are indicated. (Behncke *et al.*, 1979.)

where σ_{PWBA} is the cross section calculated in the plane-wave Born approximation, Z_L is the atomic number of the lighter collision partner, and C, B, and R are large correction factors for Coulomb deflection, binding, and relativistic effects, respectively. For 3.6-MeV/amu Xe + Pb, we have $C = 0.64$, $B = 6.3 \times 10^{-3}$, and $R = 3.4 \times 10^3$. Although the magnitude of these corrections sheds considerable doubt on the validity of the physical picture of excitation associated with the PWBA when applied to these collisions, it is useful to point out the fact that B is very small and R is very large. Hence we see that the large decrease in production cross sections caused by the increased binding of the $1s\sigma$ orbital is canceled effectively by the relativistic collapse of the $1s\sigma$ wave function, resulting in large production cross sections. The very fact that these cross sections are large therefore directly implies the dominant role played by relativistic effects on the molecular wave functions.

Before determinations of $1s\sigma$ and $2p\sigma$ contributions to the K X-ray production cross sections shown in Fig. 12 can be made, analysis of vacancy sharing between these orbitals must be carried out (Demkov, 1964; Meyerhof, 1973a). Within the context of the vacancy-sharing picture, we write the K vacancy-production cross sections for the lighter (σ_L) and heavier (σ_H) collision partners

$$\begin{aligned} \sigma_L &= (1-W)\,\sigma_{2p_{1/2}\sigma} + W\sigma_{1s\sigma}, \\ \sigma_H &= W\sigma_{2p_{1/2}\sigma} + (1-W)\,\sigma_{1s\sigma}, \end{aligned} \tag{4}$$

(Behncke *et al.*, 1979), where $\sigma_{2p_{1/2}\sigma}$ and $\sigma_{1s\sigma}$ are the cross sections for the production of vacancies in the $2p_{1/2}\sigma$ and $1s\sigma$ quasimolecular orbitals and W is the vacancy transfer probability. If $\sigma_{1s\sigma} \ll \sigma_{2p_{1/2}\sigma}$ and $W \ll 1$, we find

$$\begin{aligned} R = \frac{\sigma_H}{\sigma_L} &= \frac{W}{1-W} + \left[1 - \left(\frac{W}{1-W}\right)\right]^2 \frac{\sigma_{1s\sigma}}{\sigma_{2p_{1/2}\sigma}} \\ &\cong \frac{W}{1-W} + \frac{\sigma_{1s\sigma}}{\sigma_{2p_{1/2}\sigma}} \end{aligned} \tag{5}$$

The theory of vacancy sharing leads to

$$W = (1 + e^{2X})^{-1} \tag{6}$$

(Demkov, 1964; Meyerhof, 1973a), where

$$\begin{aligned} 2X &= \frac{\pi\,|I_1^{1/2} - I_2^{1/2}|}{\left(\dfrac{mv_i^2}{2}\right)^{1/2}} \\ &\cong \frac{\pi\,|Z_1 - Z_2|}{(v_i/v_0)}, \end{aligned} \tag{7}$$

I_1 and I_2 are the K-shell binding energies of the heavier and lighter partners, m is the electron mass, and v_i is the incident-projectile velocity. Combining Eqs. (5) and (6) we find

$$R \cong e^{-2X} + \frac{\sigma_{1s\sigma}}{\sigma_{2p_{1/2}\sigma}}. \tag{8}$$

The usual expression for the vacancy-sharing ratio ($R = e^{-2X}$) therefore, will contain an additional term $\sigma_{1s\sigma}/\sigma_{2p_{1/2}\sigma}$ (Meyerhof, 1973a). When we have $\sigma_{1s\sigma} \ll \sigma_{2p_{1/2}\sigma}$ as is the case in very light collision systems, the exponential term dominates. However, in very heavy systems, $\sigma_{1s\sigma}$ becomes large and the second term in Eq. (8) becomes important. Figure 13 illustrates these remarks, containing a plot of R (ordinate) versus $(Z_1 - Z_2)/v_i$ (abscissa). The vacancy-sharing prescription including only the exponential term is plotted as a solid curve in Fig. 13. The dashed curve represents a calculation of R including estimates of the second term of Eq. (8). These results show that, in contrast to lighter systems where R follows an exponential form over a range from 1 to 10^{-4} (Meyerhof, 1973a), for very heavy systems vacancy sharing dominates only for values of $R \geqslant 10^{-2}$, i.e., $|Z_1 - Z_2|/v_i \leqslant 1$. These deviations, in turn, give a direct indication of the relative strength of the $1s\sigma$ and $2p\sigma$ cross sections as functions of the asymmetry and total Z of the collision systems.

Relatively direct information on strong-binding phenomena in quasimolecular collisions is provided by studies of the impact-parameter dependence of K-shell excitation. We refer to the earlier discussion of the minimum momentum transfer $\hbar q_0$ necessary to ionize the K shell in lighter

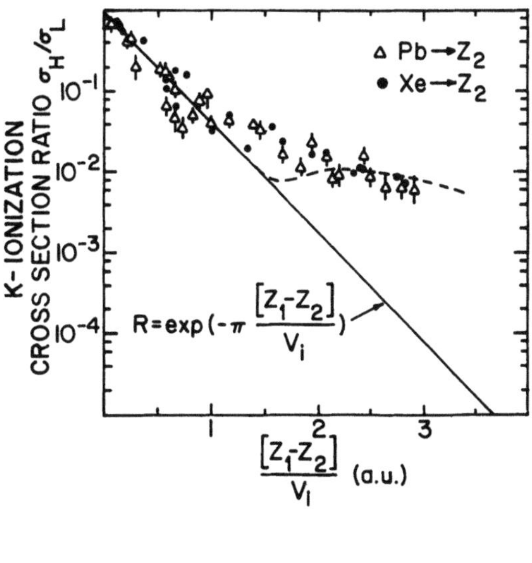

FIGURE 13. K-shell vacancy-sharing ratio versus $|\Delta Z|/v_i$. (Behncke *et al.*, 1979.)

collision systems (Eq. (1)). This minimum momentum transfer is directly proportional to the binding energy and inversely proportional to the projectile velocity. For slow projectiles, the characteristic distance associated with this minimum momentum transfer is $q_0^{-1} \ll a_K$ where a_K is the K-shell radius. If excitation takes place out of the combined system ($1s\sigma$ or $2p_{1/2}\sigma$), the appropriate binding energy to use in calculating this minimum momentum transfer is the binding energy of the *molecular* orbital at the time of excitation. The higher this binding is, the more localized the excitation must be at small internuclear separations. Excitation from highly bound orbitals is therefore signified by impact-parameter (b) dependent excitation probabilities which decrease rapidly with increasing b; excitation from loosely bound orbitals has an impact-parameter dependence which decreases slowly with increasing b. The slope of the impact-parameter dependence of the excitation process is therefore directly related to the degree of binding of the electron at the time of excitation. Clearly, the legitimacy of applying these arguments, which are based on the properties of nonrelativistic atomic wave functions and perturbation-theory calculations, to the heaviest collision systems must be examined (Bang *et al.*, 1979).

The first evidence that K-shell excitation in very heavy collision systems does, in fact, take place at extremely small impact parameters was provided by analysis of the Doppler-shifted line shape of the X-ray lines in Xe + Pb and U + Pb collisions (Greenberg *et al.*, 1977a). If an X-ray detector is placed at forward angles, Pb K X rays excited from, for example, Xe + Pb encounters will be Doppler-shifted to higher X-ray energies due to the recoiling velocity of the target atom in the laboratory frame at the time of X-ray emission. If appreciable amounts of Pb target excitation take place at small impact parameters, the line shape of the Pb K X rays emitted at zero degrees to the beam direction will exhibit large modifications due to Doppler-shifted X rays. An analysis of the observed line shape can therefore be utilized to extract the experimental impact-parameter dependence of K-shell excitation.

When the detector is placed exactly at zero degrees, and the X rays are emitted isotropically in the laboratory, the relationship between the intensity of X rays as a function of Doppler-shifted energy and the impact parameter can be shown to be

$$\frac{dN(\Delta E_x)}{d(\Delta E_x)} = \frac{2\pi c}{E_{x0}\, v_{\text{c.m.}} K_i} P[\rho(\Delta E_x)] \frac{a^2}{4} \left\{1 + \left[\frac{\rho(\Delta E_x)}{a}\right]^2\right\}^2 \left(\frac{E_x}{E_{x0}}\right)^2 \xi(E_x) \tag{9}$$

for Rutherford scattering (Greenberg *et al.*, 1977a; Schwalm *et al.*, 1972), where ΔE_x is the laboratory Doppler shift of an X-ray of center-of-mass energy E_{x0}. Here, $v_{\text{c.m.}}$ is the center-of-mass velocity, and the indices $i = 1, 2$ refer to the projectile and target, respectively, with $K_1 = A_1/A_2$ and $K_2 = 1$.

The function $\xi(E_x)$ is the photon-frequency response of the detector at $E_{x0} + \Delta E_x$; a is half the distance of closest approach in a head-on collison, $\rho(\Delta E_x)$ is the impact parameter, and $P(\rho)$ is the impact-parameter dependence of the excitation probability. Equation (9) is correct to first order in v/c.

Figure 14 shows results of such a line-shape study of $1s\sigma$ excitation in Xe + Pb collisions (Greenberg *et al.*, 1977a). The various dotted curves show (from top to bottom, respectively) the background-corrected $K\alpha$ line shapes at 90° and 0°, the uncorrected shape at 0°, and the background contribution from the thick Ni backings employed in the measurement. The solid curves drawn in the raw 0° data are the $K\alpha$ line shapes from Pb given by a ^{207}Bi source. The deviations of the measured line shapes from that of radiation from the radioactive source are large, thereby indicating significant amounts of ionization at small impact parameters.

Utilizing Eq. (9), Greenberg *et al.* (1977a) extracted the impact-parameter dependence shown as an inset in Fig. 14. The total intensity was normalized to independent measurements of the total cross section for Pb

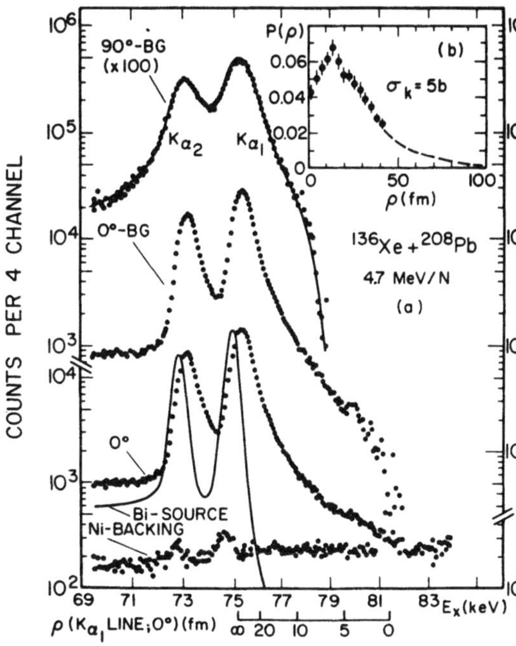

FIGURE 14. K_{α_1} and K_{α_2} line profiles of Pb from $^{136}Xe \rightarrow {}^{208}Pb$ collisions, observed at 0°, before and after background (BG) subtraction, and at 90° after BG subtraction. The solid curve through the 90° spectrum represents the 90° line shape calculated with $P(\rho)$ (inset) extracted from the 0° line shape. (Greenberg *et al.*, 1977a.)

excitation ($\sigma_K = 5$ barns) (Behncke *et al.*, 1979). These results show that the $1s\sigma$ excitation probabilities climb rapidly for impact parameters below 100 fm, reaching values of 7% for central collisions. The decrease in $P(\rho)$ below 15 fm is not well understood. Subsequently, particle–X-ray coincidence studies of $P(\rho)$ in the Xe + Pb system have yielded nearly constant excitation probabilities for $\rho < 20$ fm which are in better agreement with theory (Anholt *et al.*, 1979c; Reinhardt *et al.*, 1985; Amundsen, 1978).

These line-shape measurements have been extended to direct particle X-ray coincidence studies. The impact-parameter dependence is particularly interesting, largely due to the discovery of a universal, systematic scaling law for $1s\sigma$ excitation. The scaling behavior might be utilized to extract spectroscopic information on the large transient binding energies generated during heavy-ion–atom encounters. The simplicity of the empirically observed scaling has been an especially attractive feature of these experiments, and has generated considerable study of its theoretical origins.

In Fig. 15 (Liesen *et al.*, 1980; Bosch *et al.*, 1980) we plot values of the

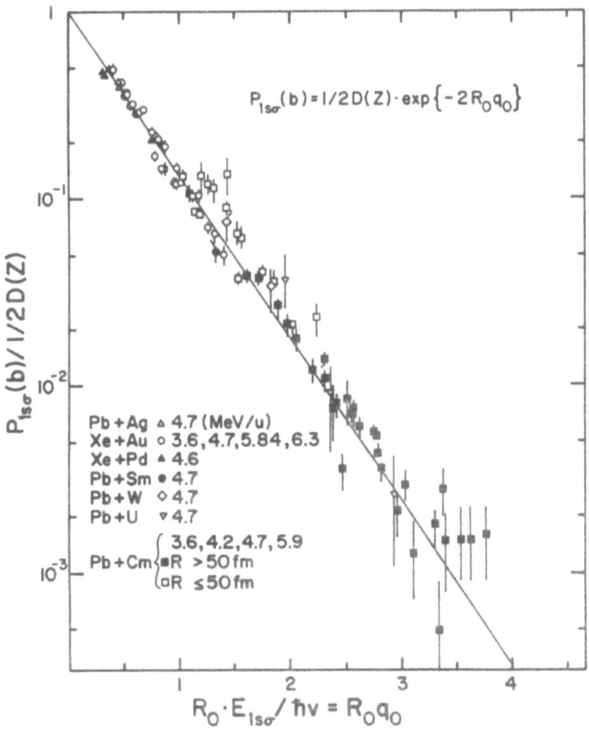

FIGURE 15. Reduced $1s\sigma$ excitation probabilities versus R_0q_0, for systems with $Z_{\text{united}} > 137$. The solid curve is the result calculated from Eq. (12). (Liesen *et al.*, 1980; Bosch *et al.*, 1980.)

reduced experimental $1s\sigma$ excitation probabilities $P_{1s\sigma}(b)$ (ordinate) versus the reduced impact parameter R_0 (abscissa), from a variety of heavy asymmetric collision systems. The impact parameter expressed in terms of the distance of closest approach R_0 at the pertinent value of b is

$$R_0 \equiv a + (a^2 + b^2)^{1/2}, \tag{10}$$

where we have

$$a = \frac{Z_1 Z_2 e^2}{2E_{\text{c.m.}}}. \tag{11}$$

Here, Z_1 is the target atomic number, Z_2 is the atomic number of the projectile, e is the electron charge, and $E_{\text{c.m.}}$ is the bombarding energy in the center-of-mass system; a is therefore half the distance of closest approach at zero impact parameter. In Fig. 15, $E_{1s\sigma}(R_0)$ is the binding energy of the $1s\sigma$ molecular orbital at internuclear separation R_0, and v is the projectile velocity. In analogy with the previous discussion of direct excitation processes [Eq. (1)] $\hbar q_0 = E_{1s\sigma}/v$ is the minimum momentum transfer necessary to ionize the $1s\sigma$ level. The solid line in Fig. 15 indicates the relationship

$$P_{1s\sigma}(b) = \tfrac{1}{2} D(Z) \exp(-2R_0 q_0), \tag{12}$$

where $D(Z)$ is a universal, smoothly increasing function of the united atom Z only, taken from theoretical calculations (Müller *et al.*, 1978).

The striking result illustrated in Fig. 15 is that all these $1s\sigma$ excitation data, which represent a wide variety of collision systems, lie on a universal curve which is extremely well described by a simple exponential function. The theoretical justification for such a scaling law is the subject of the following paragraphs.

Within the context of first-order time-dependent perturbation theory, the transition amplitude $a(E_f)$ of an electron from an initial $1s\sigma$ state of energy $E_i(t)$ to a final state with energy E_f is simply

$$a(E_f) = -\int_{-\infty}^{\infty} dt \, \langle \psi_{E_f} | \, \partial/\partial t \, | \psi_{E_i}(t) \rangle \exp\left\{ i/\hbar \int_{-\infty}^{t} dt' \, [E_f - E_i(t')] \right\} \tag{13}$$

(Müller *et al.*, 1978, 1983; Bosch *et al.*, 1980). Neglecting rotational coupling (Reinhardt *et al.*, 1985), we have

$$a(E_f) = -\int_{-\infty}^{\infty} dt \, \dot{R}(t) \, \langle \psi_{E_f} | \, \partial/\partial R \, | \psi_{E_i}(t) \rangle \exp\left\{ i/\hbar \int_{-\infty}^{t} dt' [E_f - E_i(t')] \right\}. \tag{14}$$

Within the framework of the monopole approximation (Soff *et al.*, 1978a), the coupling matrix element in Eq. (14) can be closely parametrized by

$$\left\langle \psi_{E_f} \left| \frac{\partial}{\partial R} \right| \psi_{E_i}(R) \right\rangle = \left[\frac{D(Z)}{4\pi^2 mc^2} \right]^{1/2} \left(\frac{mc^2}{E_f} \right)^{\eta(Z)/2} \frac{1}{R} \tag{15}$$

(Müller *et al.*, 1978, 1983; Bosch *et al.*, 1980), where $D(Z)$ and $\eta(Z)$ are slowly varying functions of the united-atom Z only. The important point to note here is the proportionality of the matrix element to R^{-1}. If one further assumes that electrons from the $1s\sigma$ orbital are ionized into continuum states just above the ionization threshold, one can approximate $E_f = mc^2$ (Bosch *et al.*, 1980).

The dependence of the monopole matrix element on R^{-1} implies that excitation probabilities are peaked at small internuclear separations. Utilizing this feature, one can replace the initial binding energy $E_i(R)$ with its value at R_0, viz., $E_i(R_0)$ in Eqs. (14) and (15). Taken together, all these ingredients combine to yield the simple scaling relation of Eq. (12).

The agreement between this approximate result from first-order perturbation theory and experiment provides justification for such a semiempirical description of these excitations. It must be emphasized, however, that this description is an approximate calculation based on first-order perturbation theory. Exact calculations within the perturbation-theory picture yield $P(b)$ values which are factors of two to three lower. (Müller *et al.*, 1983; Soff *et al.*, 1981). Agreement in magnitude between measurement and theory is only obtained when full coupled-channel calculations are performed.

The coupled-channel results have two important aspects, both of which result in increased excitation probabilities. First, inclusion of multistep and coherent excitation processes yields enhanced probability of $1s\sigma$ excitation to the *continuum*, by as much as by a factor of 3. The second aspect is that excitation to empty *bound* states can further increase the excitation probabilities by another factor of 2. If one assumes a Fermi level for the electrons above which the quasimolecular levels are considered empty, resulting excitation probabilities can be calculated as a function of this Fermi level (Reinhardt *et al.*, 1979).

Some questions must consequently be raised as to why the approximate perturbation theory is, in fact, so successful in describing the experimental results. This question has recently been explored (Müller *et al.*, 1983); the results of the study are shown in Fig. 16, where the calculated probability of creating $1s\sigma$ vacancies in 5.9 MeV/amu Pb + Cm collisions is indicated. In graph (a) we compare various approximate model calculations of first-order perturbation theory (lower curves) with the predictions of Eq. (12) (upper curve). Graph (b) contains a comparison of an exact perturbation-theory

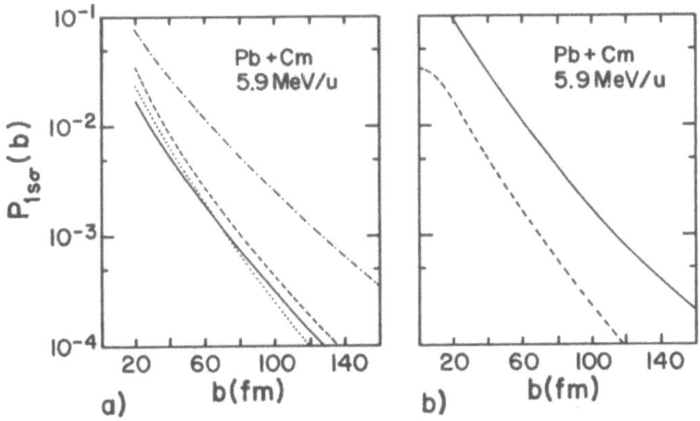

FIGURE 16. Theoretical calculations of $1s\sigma$ vacancy production versus impact parameter in 5.9-MeV/amu Pb + Cm collisions, based on various approximations. (Müller *et al.*, 1983.)

calculation (dashed line) with a coupled-channel calculation of the kind discussed above. All the approximate perturbation-theory results agree well with each other, and with the exact perturbation-theory result *except* for the prediction of Eq. (12). All pertubation-theory results, lie, furthermore, substantially below the coupled-channel result, except for the result of Eq. (12), which displays much better agreement in absolute magnitude with the coupled-channel calculations than the other approximate and exact perturbation-theory curves.

The principal cause of these discrepancies is the approximation made in Eq. (12) that excitation leads only to very low-lying states in the continuum, i.e. that $E_f = \text{const} = mc^2$. Proper integration over the continuum leads to the lower excitation probabilities illustrated in Fig. 16. The choice of $E_f = mc^2$ in Eq. (12) has the result of increasing production probabilities to values in agreement with experiment. Equation (12) must therefore be taken as a semiempirical description of the excitation probabilities which agrees well with experiment (Bosch *et al.*, 1980; Müller *et al.*, 1983).

One use of Eq. (12) is simply to utilize the success of the relationship over a wide range of collisions to attempt to extract information on the binding energy of the $1s\sigma$ orbital at the time of excitation, $E_{1s\sigma}(R_0)$. From measured $P(b)$ one can utilize relation (12) to extract $E_{1s\sigma}(R_0)$. The results are indicated in Fig. 17 for the Pb + Cm system (Liesen *et al.*, 1980). The main graph shows measured $P_{1s\sigma}(b)$ values in the reduced representation of Eq. (12) (solid curve). Agreement between measurement and Eq. (12) is very good for values of $R_0 q_0$ above 2. Below $R_0 q_0 = 2$, the measured excitation probabilities considerably exceed the predictions from Eq. (12). The inset in Fig. 17 contains a plot of the ratio $Q_{1s\sigma}$ of the measure $P_{1s\sigma}(b)$

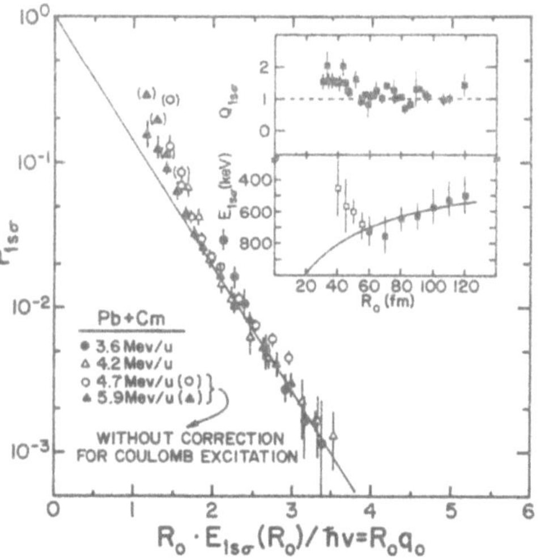

FIGURE 17. $1s\sigma$ excitation probabilities in the Pb + Cm system at different projectile energies versus $R_0 q_0$. Data in brackets and without brackets at the same value of $R_0 q_0$ are obtained without and with subtraction of K X rays from internally converted nuclear states, respectively. The solid curve represents the prediction of Eq. (12). (Liesen *et al.*, 1980.)

to predictions from Eq. (12) versus R_0 (upper part of inset). The deviation below $R_0 = 50$ fm is quite noticeable. The lower part of the inset contains a plot of the extracted binding energies at the time of excitation versus R_0. The solid curve in the inset is the prediction of the solution to the two-center Dirac equation for the binding energies. The agreement between measurement and theory is good for $R_0 > 60$ fm. However, the increased $P_{1s\sigma}(b)$ values for $R_0 < 60$ fm lead to decreasing $E_{1s\sigma}(R_0)$ results, in disagreement with theory.

These results appear to imply that the $1s\sigma$ electrons do not actually become bound by the energies predicted by the solutions of the two-center Dirac equation at small R_0 (Liesen *et al.*, 1980). However, before drawing such conclusions, the approximations leading up to Eq. (12) must be carefully examined. Deviations at low R_0 could just as easily be due to the breakdown of Eq. (12) as to anomalous binding for the $1s\sigma$ orbital. Furthermore, proper subtraction of background X rays from the internal conversion of Coulomb-excited nuclear states must be critically examined. Independent investigations of this background indicate that improper background subtraction is the most likely cause of the increased $P(b)$ values at small R_0, rather than the suggested anomalous binding energies (Ito *et al.*, 1982). Considerable care should therefore be exercised in extracting $1s\sigma$ binding energies from such measurements (Liesen *et al.*, 1980).

The picture of $1s\sigma$ excitation presented in this section is rather detailed. In fact, the size and shape of $P_{1s\sigma}(b)$ in very heavy collision systems is fairly well understood. The same cannot be said of $2p_{1/2}\sigma$ excitation. In Fig. 18, measured values of $P_{2p_{1/2}\sigma}(b)$ from Pb + Au collisions are compared with calculations (Soff *et al.*, 1978b). The agreement at low bombarding energies

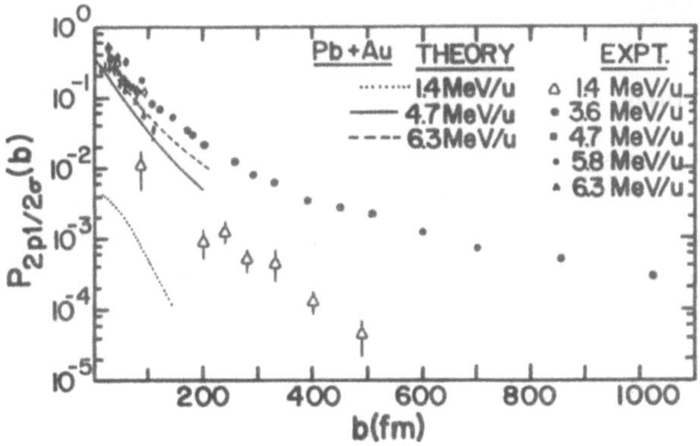

FIGURE 18. Summed Pb, Au K-vacancy production probabilities versus impact parameter b, at different projectile energies. Theoretical curves indicate $2p_{1/2}\sigma$ excitation probabilities for direct excitation to the continuum. (Mokler *et al.*, 1983; Soff *et al.*, 1978b.)

is poor, and at higher energies it is only fair. Unlike the $1s\sigma$ orbital, which is energetically well isolated from other molecular levels and which satisfies the adiabatic condition for quasimolecular orbitals very well, the $2p_{1/2}\sigma$ level is strongly rotational and radially coupled to neighboring $2p$ and $3p$ molecular orbitals. Considerably more theoretical work must be done in order to completely understand the more complicated case of $2p_{1/2}\sigma$ excitation.

In closing this section, we emphasize the simple and direct statement made by measurements of total and differential X-ray production in these collisions about the behavior of the relativistically bound $1s\sigma$ orbital. The large magnitude of the cross sections and their concentration at small impact parameters, taken together, provide direct evidence for strong binding and associated relativistic contraction of the $1s\sigma$ electron wave functions. Details concerning the precise applicability of scaling laws or the feasibility of quantitatively extracting a $1s\sigma$ binding energy from $P_{1s\sigma}(b)$ measurements should not detract our attention from the principal results of these studies. For the case of positron production from the spontaneous decay of the vacuum, the presence of $1s\sigma$ vacancies with large probabilities at small internuclear separations is an essential initial condition for the observation of the spontaneous positron process. The establishment of these large, concentrated $1s\sigma$ excitation probabilities through the K X-ray measurements discussed in this Section is important in the quasimolecular physics of the heaviest possible collision systems.

3. MO X RAYS

3.1. Lighter Systems

We have seen significant amounts of evidence for the molecular orbital nature of atomic collisions based on studies of characteristic K X-ray emission. Although emission of the X rays takes place after the collision, vacancy production during the collision reflects the properties of the molecular orbitals. The evidence for molecule formation is therefore somewhat indirect. More direct evidence for the formation of these orbitals can be gained through studies of MO X-ray transitions which occur actually during the course of the collision itself.

Vacancies generated in the $1s\sigma$ and $2p\sigma$ molecular orbitals during a collision can be filled with electrons from high-lying levels, leading to the emission of X rays whose energies exceed those of the characteristic K lines. The essential features of these MO X rays are their continuum character and extension to energies in excess of the united atom K X-ray transition frequency, both of which reflect the dynamic nature of transient molecule formation (Fig. 3).

Figure 19 (Vincent, 1977, 1983) displays MO transitions from a series of symmetric collision systems after correction of the continuum spectra for external absorption and detector efficiency. The data reveal a two-component continuum structure which scales nicely with the combined projectile–target Z. The low-frequency component has been identified with transitions into the minimum in the $2p\sigma$ orbital (Heinig *et al.*, 1976; Kaun *et al.*, 1977), and the high-frequency part with transitions into the $1s\sigma$ state (Meyerhof *et al.*, 1973b; Davis *et al.*, 1974). These assignments have been greatly facilitated by studies of the angular distributions of the X rays, which display peaks at energies that correspond to the frequencies where the $2p\sigma$ and $1s\sigma$ radiations are of comparable strength, and near the $1s\sigma$ united-atom transition energy (Greenberg *et al.*, 1974). These angular-dependence studies are discussed in detail below. Clearly, transitions to vacancies in higher-lying orbitals will also lead to the emission of lower-energy MO X rays (Kraft *et al.*, 1974; Mokler *et al.*, 1983; Morenzoni 1981; Wölfli *et al.*, 1978a). Because our primary interest is in high-binding phenomena, we restrict our discussion to $2p\sigma$ and $1s\sigma$ MO X rays.

A prerequisite to continuum molecular-orbital X-ray emission is the presence of a vacancy in the $1s\sigma$ or $2p\sigma$ orbitals. The $2p\sigma$ vacancies are generated by rotational coupling to the $2p\pi$ level or by direct excitation to higher-lying molecular states and to the continuum. Production of $1s\sigma$ MO proceeds by either a one- or a two-step process. In the one-collision process, direct excitation of the $1s\sigma$ electron to vacant bound states and to the continuum is followed by MO X-ray emission in the same collision. In the two-

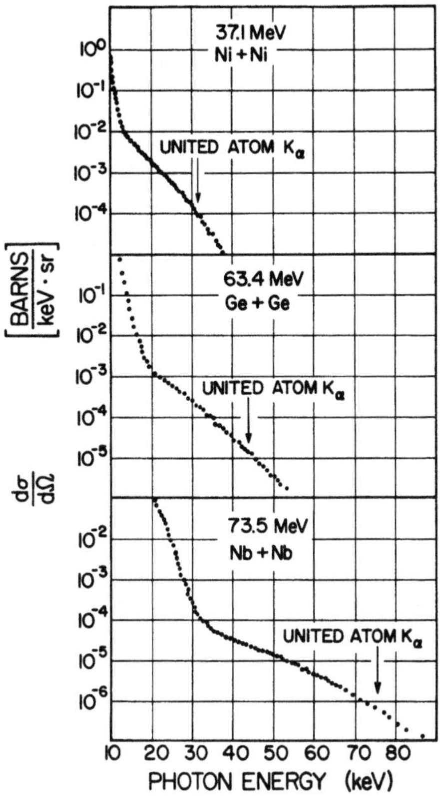

FIGURE 19. Cross sections for continuum X-ray production in three selected collision systems, illustrating the two-component features of the MO X rays. (Vincent, 1977, 1981, 1983.)

collision process, $2p\sigma$ vacancies generated in a first collision are carried into the $1s\sigma$ level in the second collision. Although two collisions are required for the latter production mechanism, the lifetimes of K vacancies and the very small cross section for direct excitation of the $1s\sigma$ orbital actually make the two-collision mechanism more likely than the one-collision process, in lower-Z systems and solid targets. At higher bombarding energies, higher Z, and small impact parameters, the one-collision process grows in importance. The dominance of two-collision excitation is important for understanding the physics of MO X-ray production in lighter-Z systems, because calculations of direct one-collision $1s\sigma$ MO X rays are extremely difficult for light Z and have only been performed for the H^+–H system (Thorson *et al.*, 1977).

The results of systematic studies of $1s\sigma$ MO X-ray production in solid targets are shown in Fig. 20 (Anholt, 1978a) compared with predictions of two-collision theory. Within the two-collision picture, MO $1s\sigma$ X-ray cross sections are expected to scale with the asymmetry of the collision system

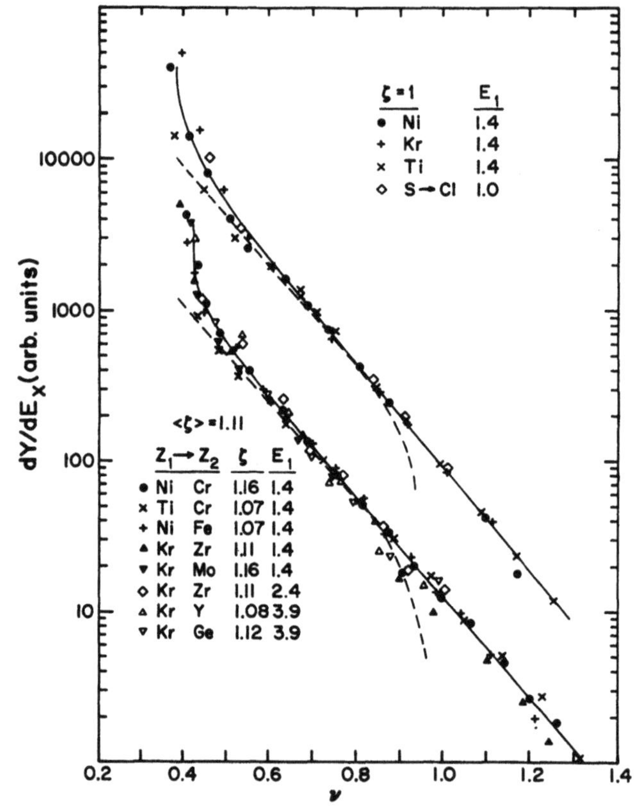

FIGURE 20. Thick-target MO X-ray yields versus the reduced parameter ν. For the upper set of data we have $\xi = 1$. For the lower set of data, $\langle \xi \rangle = 1.11$. The solid curve is fitted to data. The dashed curve represents quasistatic two-step calculations. The projectile energy E_1 is in MeV/amu. (Anholt, 1978a.)

$\xi = Z_{\text{heavy}}/Z_{\text{light}}$, and the photon energy $\nu = \omega/\omega_0$, $\hbar\omega$ being the energy of the X-ray and $\hbar\omega_0$ a characteristic frequency for the quasimolecule which can be taken to be the K_{α_1} transition energy of the united atom. In general, both the shapes and sizes of the $1s\sigma$ MO X-ray yields will also depend on the beam velocity v through the parameter $q = \omega_0 a_{KL}/v$, where $a_{KL} = Z_L^{-1}$ a.u. is the K-shell radius of the lighter collision partner. However, if one restricts study to intermediate X-ray energies $(0.4 \leqslant \nu \leqslant 1)$, it can be shown that the shapes of the MO X-ray yield depend only on ξ and ν (Anholt, 1978a).

In Fig. 20 we compare experimental results for continuum X-ray yields with predictions of the two-step theory for a variety of collision systems. Thick targets have been used in these studies primarily for reasons of experimental convenience. The data have been arbitrarily normalized to each other at $\nu = 0.7$. It should be noted that the scaled experimental results at equal ξ form universal curves, in agreement with the predictions of two-step calculations. The calculations (dashed lines in Fig. 20) agree well with experiment for $0.5 \leqslant \nu \leqslant 0.9$. These results confirm the molecular origins of the radiation and the dominance of the two-step production mechanism for lower-Z encounters with solid targets.

In gas targets, K vacancies generated in a first collision will decay before entering a second collision, thereby resulting in the dominance of one-collision MO X rays. Studies of the bombarding-energy dependence of one- and two-collision MO X-ray production can be carried out using molecular targets (Saris *et al.*, 1977). Figure 21 (Lennard *et al.*, 1981) contains spectra obtained from collisions of 300-keV O with O_2 and H_2O (gas) targets. For the diatomic O_2 target, the close proximity of the two bound O atoms allows transfer of $2p\sigma$ vacancies to the $1s\sigma$ in a secondary collision. For the H_2O target, the large separation between molecules suppresses two-step MO X-ray production. The solid curve in Fig. 21 is a superposition of the O_2 data after normalization to the O K_α peak. The large excess of O_2 over H_2O intensity between 1.3 and 2 keV beautifully illustrates the important role of the two-collision process in these light systems. The large peak at 1.1 keV represents two-electron one-photon X-ray transitions (Sec. 4) in the O atom. Extension of these studies to higher beam energies (1 MeV) showed only slight differences between O_2 and H_2O targets. These latter results demonstrate the increasing dominance of the one-step process at higher projectile velocities, due essentially to the rapid increase in direct $1s\sigma$ excitation with incident bombarding energy.

The extension of continuum X-ray emission to X-ray energies beyond the united-atom static X-ray limit is direct illustration of collision broaden-

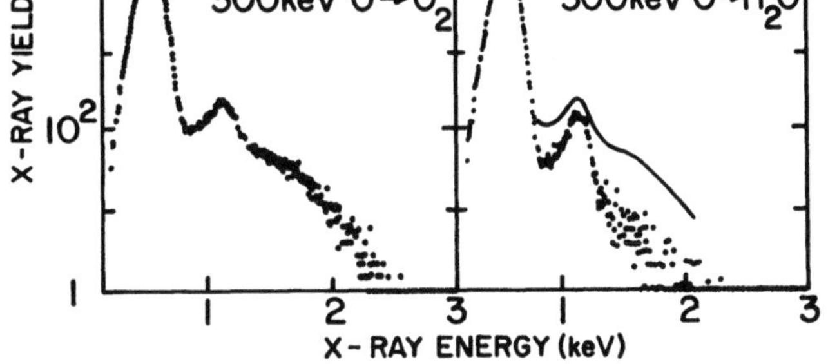

FIGURE 21. X-ray spectra from $O^+ \rightarrow O_2$ and H_2O (gas) collisions. The characteristic K_α lines are at 0.5 keV. Spectra are normalized to equal number of scattered particles at 45° to the beam. The solid curve on the right-hand side is the superposition of the $O^+ \rightarrow O_2$ data. (Lennard *et al.*, 1981.)

ing effects on the quasimolecular orbitals. Since the central role played by molecular orbital formation in very heavy systems is that of a vehicle for the generation of extremely strong binding energies during the course of these collisions, the extent to which these energies are well defined is an extremely important question. In the case of molecular orbital X rays, the continuum $1s\sigma$ radiation extends past the united-atom limit with a spectral energy dependence that can be described by the relation

$$\frac{d\sigma}{d(\hbar\omega)} \propto \hbar\omega \exp\left(\frac{-\hbar\omega}{2\Gamma}\right) \tag{16}$$

(Macek *et al.*, 1974; Müller, 1975), where $\hbar\omega$ is the radiated photon energy and Γ a characteristic fall-off parameter. The experimental projectile-velocity dependence of the half-width H of this experimental tail $[H=(2 \ln 2)\Gamma]$ has been studied for the case of Ni + Ni collisions (Vincent *et al.*, 1979). The half-width is found to increase as $v^{0.59 \pm 0.03}$, where v is the projectile velocity.

A number of theoretical studies have been directed towards the understanding of this velocity dependence (Macek *et al.*, 1974; Müller, 1975; H. D. Betz *et al.*, 1975; Fritsch *et al.*, 1979). In addition to its importance in the treatment of dynamic molecular effects, this tailing of MO emissions beyond the united-atom limit has important consequences for the peaked angular dependence of the molecular orbital X rays alluded to earlier. Within the context of the two-step picture of MO X-ray production, both the angular distribution and intensity of the $1s\sigma$ X rays can be theoretically obtained by Fourier analyzing the dipole velocity matrix element between the initial and final molecular states (Briggs *et al.*, 1977). Results of one dynamical study (Vincent, 1977; Vincent *et al.*, 1979) performed for Ni + Ni collisions show good agreement between measurement and calculation at 12.6 MeV. Theory fails to predict, however, the measured increase in yields beyond the united-atom K_α X-ray energy at higher projectile velocities. These results may be interpreted as indicating additional contributions to the experiment from one-collision MO X-ray production at high X-ray and beam energies (Anholt, 1978b).

Except for the general features of scaling and the two-component structure, measurements of MO X-ray spectral yields display no additional signature or feature which can be exploited as a definitive indication of radiation from the combined projectile–target system. In this connection, the discovery (Greenberg *et al.*, 1974) of the distinctive directional anisotropy of the $1s\sigma$ radiation justified significant optimism for a spectroscopy of the combined atom using MO X Rays. In Fig. 22 we display the anisotropy $[I(90°)/I(0°)]-1$ for selected collision systems (Vincent, 1977, 1981, 1983). The peaking of these spectra at photon energies near the high-

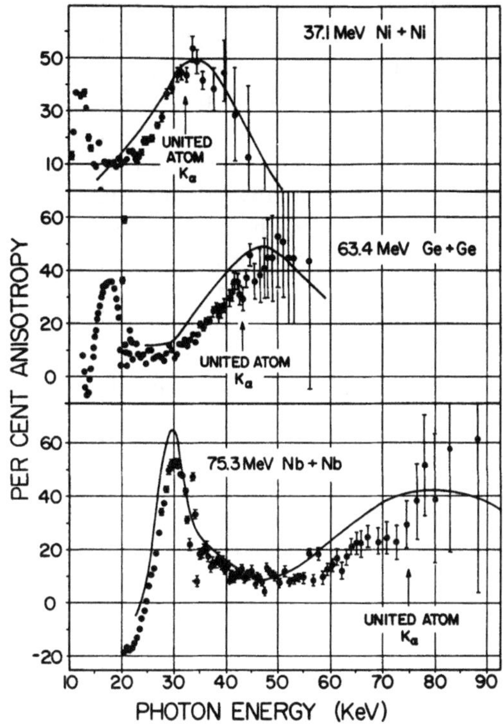

FIGURE 22. The directional anisotropy of MO X rays for three collision systems, compared with theory (solid curves) (Vincent, 1977, 1981, 1983; Anholt, 1978b, 1979a, b).

frequency ends of the $2p\sigma$ and $1s\sigma$ X-ray components (cf. Fig. 19) and the obvious scaling of these peaks with the combined projectile–target Z generated enthusiasm for systematic exploration of this effect in both lighter and heavier systems. The solid curves in Fig. 22 represent calculations of the directional anisotropies (Anholt, 1978b, 1979a, b) which will be discussed below.

Subsequent systematic studies of the $1s\sigma$ MO X-ray anisotropy peak led to a search for semiempirical correlations between the peak positions and the K X-ray energies of the united atomic system. The results of such a systematic investigation have shown (Stoller *et al.*, 1977a) that the first derivatives of the empirical $1s\sigma$ MO anisotropies attain a maximum at the K_α X-ray energies of the united atom. Although as yet lacking theoretical justification, this striking correlation between the X-ray anisotropy points of inflection and the united-atom K_α energies illustrates the degree to which scaling with the combined projectile–target Z is obeyed. It turns out that this peaking of the directional anisotropy is a general feature of MO X-ray transitions which occurs not only in transitions to $1s\sigma$ and $2p\sigma$ final states, but also to higher-lying levels. Extensive studies of MO X-ray transitions to these higher-lying orbitals have been carried out, and have led to semiempirical scaling laws for the peaks (Morenzoni, 1981; Wölfli *et al.*, 1978a).

A number of theoretical explanations for the peaking of the directional anisotropy have been proposed (Wölfli *et al.*, 1978b; Hartung *et al.*, 1978; Anholt, 1978b). The most successful description (Anholt, 1978b) entails a full dynamical treatment of the collision problem which includes a Fourier analysis of the radiative dipole matrix elements within a quasimolecular basis, and coupled-channel calculations of the time development of the strongly mixed $2p\pi$, σ orbitals. The dynamic nature of these collisions leads to the inability of the $2p\pi$, σ molecular orbitals to follow the rapid rotation of the internuclear axis in close collisions. As a result, the electron clouds tend to remain fixed in space, while the molecular axis rotates within them. Since the π and σ states are defined by their angular-momentum projections along this internuclear axis complete "slippage" (Gros *et al.*, 1976) of the electron clouds for 90° scattering in the center of mass causes coupling between the $2p\pi$ and σ states, with unit probability. This phenomenon is simply the kinematic peaking effect in $2p\pi$, $2p\sigma$ rotational coupling which was discussed in Sec. 2.1.

It is furthermore necessary to include transitions from the $3p\pi$ orbital in order to explain the measured $1s\sigma$ directional anisotropies. Although for $1s\sigma$ MO radiation these K_β-like transitions in the quasimolecule are generally weaker than the K_α-like X rays ($2p\pi$–$1s\sigma$), their higher energy leads to their eventual dominance at frequencies above the united-atom K_β X-ray energy. The strong rotational or Coriolis mixing of the $3p\pi$ and $3p\sigma$ levels must be included in the calculations, although the rather weak coupling between the $3p$ and $2p$ MO orbitals can be neglected. The results of such calculations are indicated by the solid curves in Fig. 22. Measurement and calculation agree quite well for both the $1s\sigma$ (Anholt, 1978b; Vincent, 1977, 1983) and $2p\sigma$ (Anholt, 1979a) cases.

The continuum X rays discussed above are emitted from a quasimolecular system which moves with the velocity of the center of mass along the beam axis. The observed laboratory directional anisotropies must therefore be corrected for the Doppler shift of the X rays. This correction enhances the magnitude of the observed peaks in the directional anisotropy by factors of 2 and more and is therefore extremely important (Greenberg *et al.*, 1974). Furthermore, the Doppler effect can be utilized to derive information on the actual velocity of the emitting source of the continuum X rays, and therefore be used to identify them as quasimolecular in origin (Vincent *et al.*, 1978; Meyerhof *et al.*, 1975; Frank *et al.*, 1976).

The results of such an analysis are indicated in Fig. 23 (Vincent *et al.*, 1978). The top panels of Fig. 23 display the directional anisotropies, and the bottom panels, the extracted emitter velocities of the X rays. For photon energies above 12 keV in the 64.8-MeV data, good agreement with a cen-

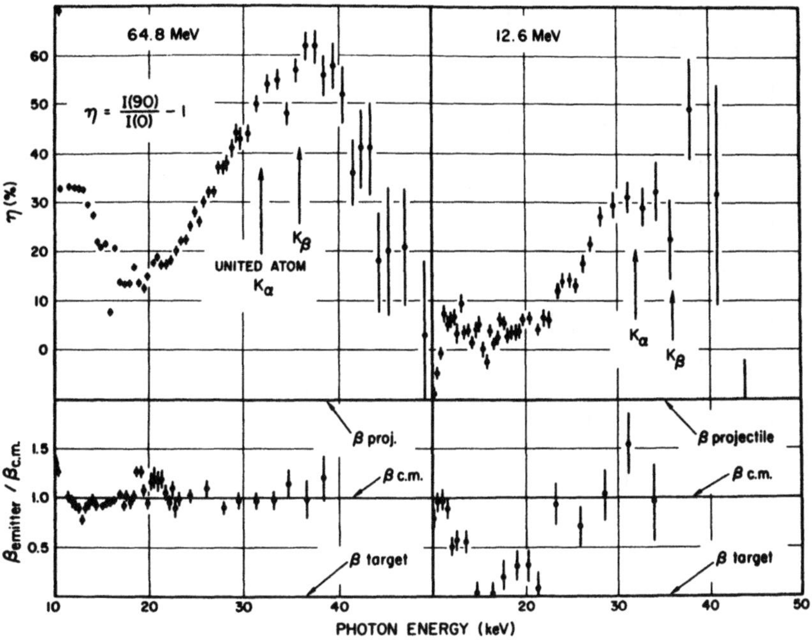

FIGURE 23. Anisotropies η (upper part of figure) and extracted emitter velocities, in units of the center-of-mass velocity (bottom figure), for continuum Ni + Ni X rays, based on the Doppler-shift technique. (Vincent *et al.*, 1978.)

ter-of-mass emitter velocity is obtained. The rise in both the directional anisotropy and its associated emitter velocity near 10 keV X-ray energy is due to contributions from radiative electron capture (Schnopper *et al.*, 1972; Spindler *et al.*, 1979). Above 22 keV, the 12.6-MeV data agree well with the center-of-mass velocity. The origin of the very low emitter velocities in the 12- to 20-keV region remains unexplained. Possible contributions from recoiling target atoms or secondary electron bremsstrahlung to the 12.6-MeV data do not appear to be the cause of the discrepancy. Further discussion of other sources of continuum X-ray emission in these collisions follows in Sec. 4.

Figure 24 shows results of recent experimental comparisons between directional anisotropies under one- and two-collision conditions through measurements in gas and solid targets at four bombarding energies (Stoller *et al.*, 1981). The solid curves guide the eye through the solid-target data, and dashed curves, through the gas-target results. Qualitatively similar behavior in the gas- versus solid-target anisotropies is found, except at low bombarding energies where the gas-target result is small and shifted to the left. Adjustment of vacancy occupation numbers within the context of a

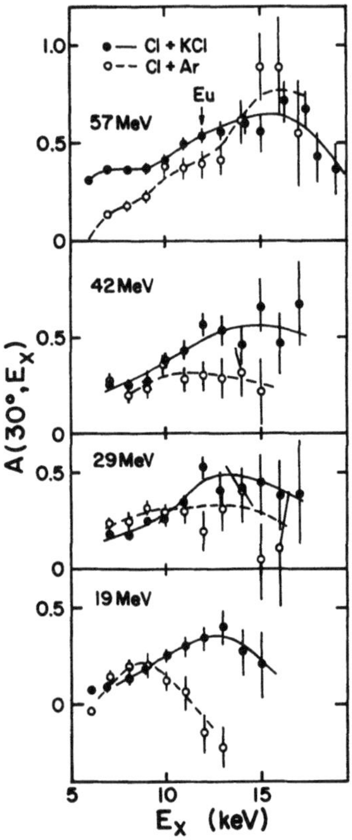

FIGURE 24. Gas (Cl + Ar) and solid (Cl + KCl) target MO anisotropies $[I(90°)/I(30°) - 1]$ for 19- to 57-MeV Cl projectiles. Curves are drawn to guide the eye. (Stoller *et al.*, 1981.)

two-step production mechanism described above cannot account for these low-energy gas-target results. Apparently, a full one-collision calculation is necessary to explain these data.

In addition to the singles measurements discussed thus far, a small number of coincidence studies of MO X-ray production in lighter collision systems have been carried out. The additional detail provided by these studies has allowed simpler and more direct comparisons between experiment and calculation. We present selected examples of these experiments.

The first example involves measurements of the impact-parameter dependence of MO X rays in 90-MeV Ni + Ni collisions (Schmidt-Böcking *et al.*, 1982a; Tserruya *et al.*, 1976; Schmidt-Böcking *et al.*, 1978; Schuch *et al.*, 1980b). The results of these experiments are compared with two-step calculations (Anholt, 1982) of $1s\sigma$ MO X-ray production probabilities at 20 and 32 keV (Fig. 25) (Schmidt-Böcking *et al.*, 1982a). The top part of Fig. 25 [labeled $P_K(b)$] indicates the measured impact-parameter depen-

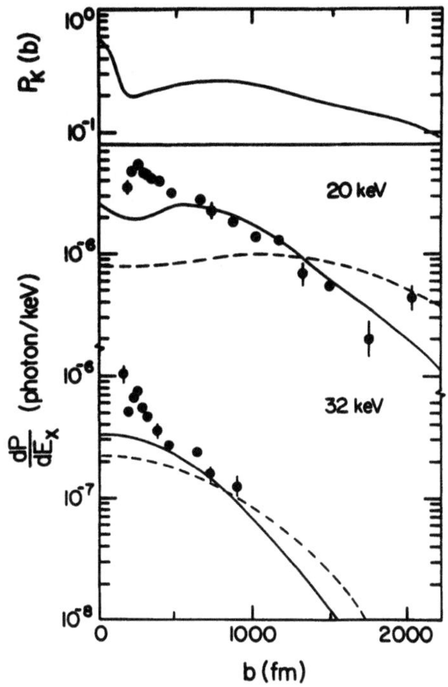

FIGURE 25. Top: Measured impact-parameter dependence of characteristic K X rays. Bottom: Comparison of MO X-ray impact-parameter dependence measurement (dots) with two-step collision calculations for Ni + Ni collisions at 90 MeV. (Schmidt-Böcking *et al.*, 1978, 1982a; Schuch *et al.*, 1980b; Anholt, 1982.)

dence of the characteristic K X rays. The dashed curves in the bottom panel of Fig. 25 are the two-step dependences calculated on the assumption that the K vacancy generated in the first collision is carried into the second collision along the incident-beam axis. This is equivalent to assuming that K vacancies are produced in the first collision only at very large impact parameters. The solid curves represent the effect of folding the actual empirical projectile scattering-angle dependence of K-vacancy production into the two-step MO X-ray calculations. This folding causes a mixing of laboratory azimuthal angles ϕ of the projectile in the first collision with various final observed projectile scattering angles θ with respect to the beam axis; the calculations therefore become somewhat complicated (Anholt, 1982). It is clear, furthermore that the definition of an impact parameter is no longer meaningful under such conditions. Instead, the abscissa in Fig. 25 represents the pseudo-impact-parameter b defined as

$$b = \rho \cot \theta, \tag{17}$$

where ρ is one-half the distance of closest approach in a head-on collision, and θ is the *experimental* scattering angle in the laboratory. The parameter b represents the actual impact parameter given by Rutherford scattering only when K-vacancy production in the *first* collision occurs at very large impact parameters and negligibly small θ. The results of Fig. 25 show that this folding of the empirical impact-parameter dependence of the first collision is necessary to describe the measured results correctly. The remaining excess of experiment over theory at small b can be interpreted as additional contributions from single-collision MO X-ray production. The importance of one-step contributions at high $1s\sigma$ MO X-ray energies and small impact parameters must be noted.

A second type of coincidence study has been performed between characteristic and continuum X rays from the Nb + Nb collision system (O'Brien *et al.*, 1980). Such an experiment exploits the existence of a cascade relationship between certain MO X-ray transitions and separated-atom K X rays. In particular, $2p\sigma$-to-$1s\sigma$ MO X-ray transitions result in the generation of $2p\sigma$ vacancies which will, with high probability, survive the collision and decay under emission of K X rays. Alternatively, contributions to K X-ray, MO X-ray coincidences from $2p\pi$-to$1s\sigma$ MO X-ray transitions with a subsequent rotational coupling of the resulting $2p\pi$ vacancy to the $2p\sigma$ orbital may occur. Coincidences between characteristic and continuum X rays should therefore select $2p\pi$, σ-to-$1s\sigma$ MO transitions and suppress contributions from higher-lying MO orbitals to the $1s\sigma$ final state.

Figure 26 (O'Brien *et al.*, 1980) shows the results of such coincidence measurements in the Nb + Nb system at three bombarding energies. The solid curves and their continuation into data points show spectra of continuum X rays in coincidence with characteristic K X rays. The dot–dashed curve in part (a) of Fig. 26 represents the singles spectrum normalized to the coincidence data in the region 40–70 keV. This latter comparison implies that the intensity of continuum X rays below 40 keV is reduced relative to that above 40 keV, when the coincidence condition is imposed. This result is consistent with the assignment of X rays below 40 keV primarily to $2p\sigma$ final-state transitions.

The appearance of characteristic K X rays in the coincidence data of Fig. 26 reflects the relatively large probability of double-K-vacancy production in these collisions. Such double-vacancy production in the quasimolecule implies the presence of a second source of K X-ray, MO X-ray coincidences due to multiple excitations which complete with the cascade events of interest. Solid curves and their continuation into dashed curves in Fig. 26 represent estimates of these multiple-vacancy contributions to the coincidence yields. The authors conclude that multiple-vacancy processes cannot account for the measured coincidence intensities above 40 keV, and that significant contributions from $2p\sigma$, π-to-$1s\sigma$ cascade events have, in fact, been observed.

A third kind of coincidence study has been performed to test various theoretical descriptions of the angular distribution of the molecular-orbital

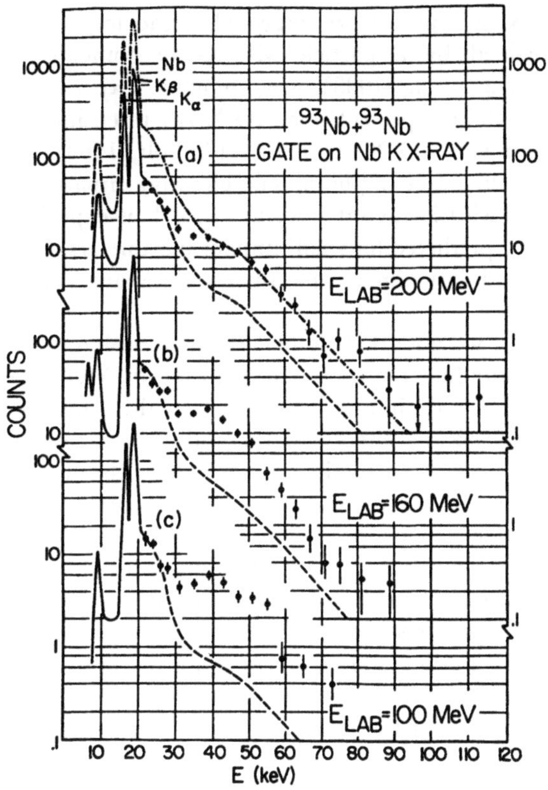

FIGURE 26. *K* X-ray, MO X-ray coincidence spectra in Nb + Nb collisions at 100, 160, and 200 MeV. (O'Brien *et al.*, 1980.)

X rays. Different model calculations of MO X-ray emission predict conflicting dependences for the X rays on the azimuthal angle of emission ϕ, i.e., the angle between the scattering plane formed by the beam axis and scattered particle, and the X-ray detector. The experiment consists of detecting the continuum X-ray yields parallel and perpendicular to the scattering plane through particle, X-ray coincidences (Schuch *et al.*, 1981b; Bürgy *et al.*, 1981). Results of measurements of the Ni + Ni system at 90 MeV (Schuch *et al.*, 1981b) agree better with full dynamical calculations of the azimuthal anisotropies (Anholt, 1981, 1978b) than with the other available model calculations (Wölfli *et al.*, 1978b; Hartung *et al.*, 1978).

Recently, several authors have explored interference effects between the emission of $1s\sigma$ MO X rays during the incoming versus outgoing parts of the collision trajectory. By allowing hydrogenlike 10-MeV Cl ions to impinge on Ar gas targets, essentially a 50% probability of bringing a $1s\sigma$ vacancy into the collision is established. Under such conditions, oscillations in the resulting MO X-ray spectra are expected if the impact parameter of the collision is defined by a particle–MO-X-ray coincidence condition. In Fig. 27 (Tserruya *et al.*, 1983), we compare experimental MO X-ray production probabilities per keV for MO X-ray intensities at various impact parameters with $1s\sigma$ MO X-ray theory (Anholt, 1983). The agreement is quite good over the range of photon energies and impact parameters shown. A striking dip in MO X-ray intensity at ~6 keV is experimentally observed and theoretically predicted for small impact parameters. Corresponding dips in the singles and large-impact-parameter data are neither observed nor predicted.

In summary, we note that significant progress has been made in quantitatively understanding the origins of continuum MO X rays from heavy-ion collisions. The continuum character of the emissions obscures the relationship between the energy of the radiated X rays and the binding energies of the molecular orbitals between which the transitions occur. The structure exhibited by the directional anisotropy of the radiation has made it possible, however, to perform semiempirical spectroscopy of the united atom. As we shall see in the following Section, large relativistic effects in the very heavy collision systems lead to diminished anisotropies at high Z. Nevertheless, comparisons between MO X-ray measurements and calculations across the Periodic Table confirm the essentially molecular nature of heavy-ion collisions and act as a check on our quantitative understanding of the time development of the molecular wave functions during the encounters.

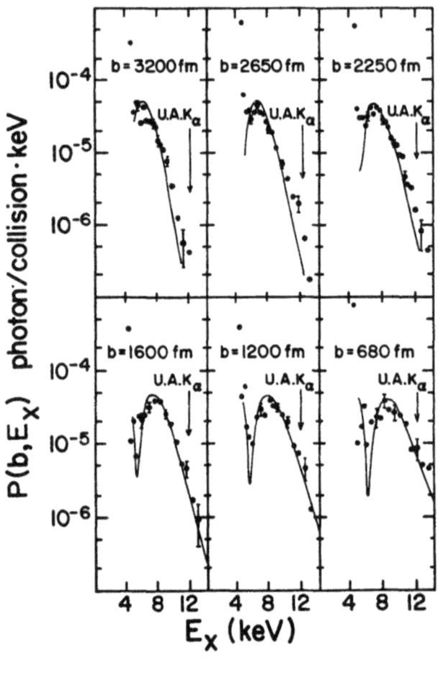

FIGURE 27. X-ray production probability versus photon energy for various impact parameters. Arrows indicate united-atom (U.A.) K_α transition energies. Solid curves are results of dynamical calculations. (Tserruya *et al.*, 1983; Anholt, 1983.)

3.2. Heavy Systems

The principal experimental problem in going from light to heavy systems is the increase of disturbing background from the Coulomb excitation of nuclear levels. Lighter systems are associated with continuum X-ray emission below 100-keV photon energies; in heavy systems, strongly bound inner molecular orbitals generate MO X rays with energies in excess of 1 MeV. Heavy nuclei have many levels in the 100-keV to 1-MeV region of excitation, which are readily Coulomb-excited in energetic heavy-ion encounters. Deexcitation of these states by γ-ray emission is a troublesome source of background in MO X-ray measurements. For close collisions, the strong relativistic contraction of the $1s\sigma$ wave function localizes this electron to the spatial vicinity of the nucleus. As a result, there is a large probability for excitation of the nucleus in collisions which excite the quasiatom, and vice versa, resulting in background which is difficult to suppress. For this reason, up to now only investigation of the spherial nuclear system $^{208}Pb + ^{208}Pb$ has been possible.

A second major difference between MO X-ray production in lighter versus heavier systems is the growing dominance of the one-step production mechanism with increasing united-atom Z. As mentioned previously, both the decreased lifetime of K vacancies and the rapid rise of direct $1s\sigma$ excitation probabilities with increasing Z are responsible for this result. This dominance of the one-collision production process makes it simpler to interpret the MO X-ray impact-parameter dependence, because there is no distortion of the relationship between laboratory scattering angle and impact parameter caused by the two-collision mechanism.

In Fig. 28 we show the measured pulse-height distribution from con-

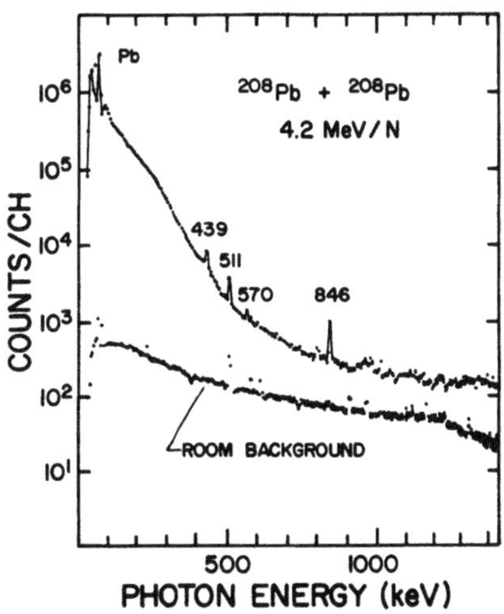

FIGURE 28. Pulse-height distribution of X rays from 4.2-MeV/amu $^{208}Pb + ^{208}Pb$ collisions with highly enriched ^{208}Pb targets. Room background is shown. (Morenzoni, 1981; Wölfli *et al.*, 1978a; Stoller *et al.*, 1980; Meyerhof *et al.*, 1979.)

tinuum X-ray emission in $^{208}Pb + {}^{208}Pb$ collisions at 4.2 MeV/amu (Morenzoni, 1981; Wölfli *et al.*, 1978a; Stoller *et al.*, 1980; Meyerhof *et al.*, 1979). A 3-mm Sn plus 1-mm Cu absorber was used in these measurements to attenuate the K X-ray lines, thereby preventing dead-time and pulse pile-up problems. The continuum band extending from the characteristic Pb K_α and K_β lines to 1.5 MeV photon energy is dominated by MO X rays. Also shown in Fig. 28 is the background originating predominantly from trace radioactivity in the concrete walls and floors of the experimental target room. Several-inches-thick Pb shielding surrounded the counters to minimize this background. The data were taken with a thick, enriched ^{208}Pb target and a large-volume ($\sim$100-cm^3) coaxial Ge (Li) γ-ray detector.

Various contaminant lines from ^{23}Na (439 keV), positrons (511 keV), ^{207}Pb (570 keV), and ^{27}Al (846 keV) are evident in Fig. 28. The Al contamination was caused by stray beam striking the target frame. Despite the use of highly enriched ^{208}Pb targest (0.7% ^{207}Pb, 0.2% ^{206}Pb, and <0.05% ^{204}Pb) and beams (<1% ^{207}Pb), small amounts of the 570-keV ^{207}Pb excitation are noticeable. Clearly, a target with many levels in the 100-keV to 1-MeV range (^{238}U, for example) would yield a spectrum in which the MO X-ray component would be dwarfed by large contributions from nuclear Coulomb excitation. Gamma-decay of the Coulomb-excited ^{208}Pb state produces additional Compton-scattering background at high X-ray energies in Fig. 28.

In the upper panel of Fig. 29 (Stoller *et al.*, 1980), the continuum X-ray data are plotted after correction for absorption and detector efficiency. A

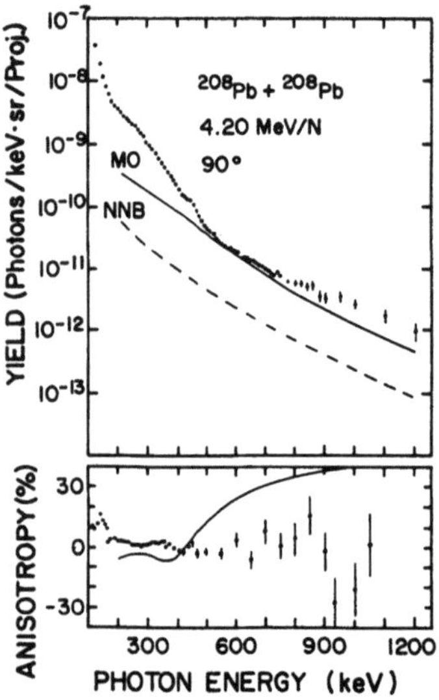

FIGURE 29. Thick-target photon yield for 4.2-MeV/amu $^{208}Pb \rightarrow {}^{208}Pb$ collisions, compared with theoretical predictions for nucleus–nucleus bremsstrahlung (dashed curves) and $1s\sigma$ MO X rays (solid curves). The lower curve shows corresponding anisotropies, compared with theory (solid curve). (Stoller *et al.*, 1980; Kirsch *et al.*, 1978; Kirsch, 1980.)

small (~10%) correction for the tails of the photon peaks, due to Compton scattering, was also unfolded and subtracted. The dashed curve represents the calculated contribution from nucleus–nucleus bremsstrahlung (NNB), and the solid curve, that from MO X rays (Kirsch *et al.*, 1978; Kirsch, 1980). The calculations of the MO X-ray yield include only contributions to $1s\sigma$ final states. In the lower panel, the measured directional anisotropy of the continuum radiation $[I(90°)/(I(0°)-1]$ is compared with the calculated anisotropy (solid curve) (Kirsch, 1980).

Calculation of the MO X-ray intensities from these collisions is very difficult. In certain respects, the one-step process complicates matters considerably since coupled solutions for the vacancy-production and radiative-decay amplitudes as a function of time (internuclear separation) must be found. The effect of including rotational as well as radial coupling in the collision problem has been examined theoretically: rotational coupling was found to strongly enhance the MO intensities above 600 keV. The theoretical MO yields displayed in Fig. 29 include an approximate treatment of both rotational and radial coupling (Kirsch, 1980).

These calculations are subject to a number of very difficult theoretical problems. The long-wavelength approximation was used to calculate the radiative matrix elements, and analysis of the validity of this approximation suggests that its accuracy is questionable. In addition, the matrix elements were attenuated smoothly at large internuclear separations (~2000 fm) to simulate the unknown influence of translation factors. Since two-center Dirac continuum wave functions are unknown, only excitation to continuum s states was included (Kirsch, 1980).

The calculations suffer from the inaccuracies of the long-wavelength approximation at high photon energies. At low photon energies, transitions at large internuclear separations occur, and the precise behavior of the artifically simulated translation factors becomes important. A comparison between calculation and measurement in Fig. 29 shows good agreement at high X-ray energies and poor agreement below 400 keV. At low photon frequencies, transitions into higher-lying molecular orbitals (i.e., $2p_{1/2}\sigma$) will also occur. The results of coincidence studies (see below) indicate that $2p_{1/2}\sigma$ MO X rays dominate the spectrum below approximately 200 keV. The discrepancy between theory and measurement in the 200- to 400-keV region does not, therefore, appear to be attributable to additional contributions from higher-lying states. Nevertheless, rough overall agreement between experiment and theory is found, implying that the observed intensities are principally due to MO X-ray transitions into a $1s\sigma$ final state. The extension of the X-ray intensities to nearly 1.5-MeV photon energy directly illustrates that high binding energies are achieved in these collisions.

The bottom panel in Fig. 29 contains a plot of the directional anisotropy. The small bump at 150 keV most likely reflects contributions from transitions to a $2p_{1/2}\sigma$ final state (Morenzoni, 1981; Wölfli *et al.*, 1978a). The $1s\sigma$ MO X rays exhibit no anisotropy, contrary to theoretical expectations. This disappointing result precludes a straightforward extension of the semiempirical united-atom *K*-shell spectroscopy of lighter systems to these interesting very heavy encounters.

As discussed in connection with Fig. 23, measurements of forward–backward continuum intensities can be utilized to extract the velocity of the emitter of the continuum radiation. In Fig. 30 (Stoller *et al.*, 1980) we show the results of such a Doppler-shift measurement for the 4.8-MeV/amu $^{209}Bi + {}^{209}Bi$ system. In this Figure, the anisotropy ratios $I(137.5°)/I(42.5°) - 1$ are plotted as functions of photon energy for various assumed emitter velocities. The upper data (labeled v_p) are ratios corrected for Doppler shifts and relativistic solid-angle effects corresponding to the projectile velocity; the middle points correspond to the velocity of the center of mass system, and the lower points are not corrected for any Doppler shift. Only when the measured intensities are corrected for Doppler effects and relativistic solid-angle effects corresponding to 90% of the center-of-mass velocity is forward–backward symmetry achieved. The use of 90% $v_{c.m.}$ is based on the use of thick targets in these experiments and simply reflects energy loss of the projectile ions in the thick target. These results provide further evidence that the observed emission is indeed quasimolecular in origin.

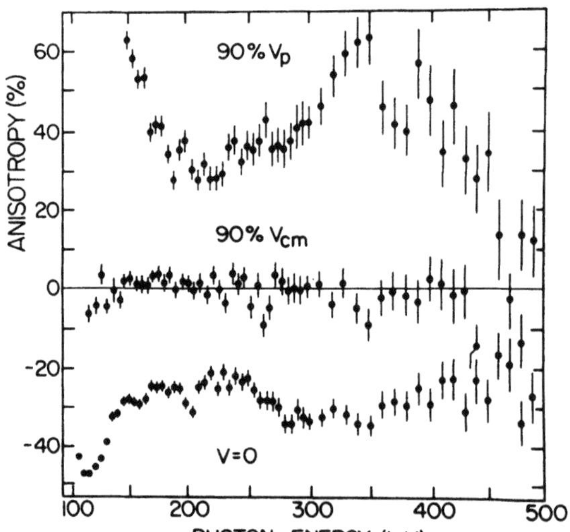

FIGURE 30. Forward–backward continuum X-ray anisotropies calculated for three different emitter velocities of the continuum X rays. Symmetry is obtained only for the case in which the radiation is assumed to be emitted from the projectile–target center of mass. (Stoller *et al.*, 1980.)

Two kinds of coincidence experiments with MO X rays from the ^{208}Pb + ^{208}Pb collision system have been carried out. The first experiment yielded the impact-parameter dependence for the MO X rays shown in Fig. 31 (Vincent, 1981, 1983; Stiebing *et al.*, 1982). The top portion of Fig. 31 displays the function $P_K(b)$ for K X rays, and the bottom portion, $P(b)$ for MO X rays at various photon energies. Close examination of the $P_K(b)$ data in Fig. 31 indicates a slight rise in the probability for b below ~50 fm. At MO X-ray photon energies this rise becomes increasingly apparent, and the two-component structure of the X-ray function $P(b)$ is quite clear.

In the earlier discussion of characteristic K X-ray impact-parameter-dependence studies in heavy-ion collisions we stated that these dependences provide information on the *vacancy excitation* process *during* the collision. Those studies dealt with cases in which X-ray emission takes place *after* the collision, and is therefore well decoupled from the excitation process itself. For emission of MO X rays, on the otherhand, the decay of the vacancies takes place *during* the collision, and therefore cannot in principle be treated independently from vacancy excitation.

Within the context of first-order perturbation theory, the K X-ray $P(b)$ curves scale [Eq. (12)], and a relationship exists between the impact-parameter dependence of the excitation process and the binding energy of

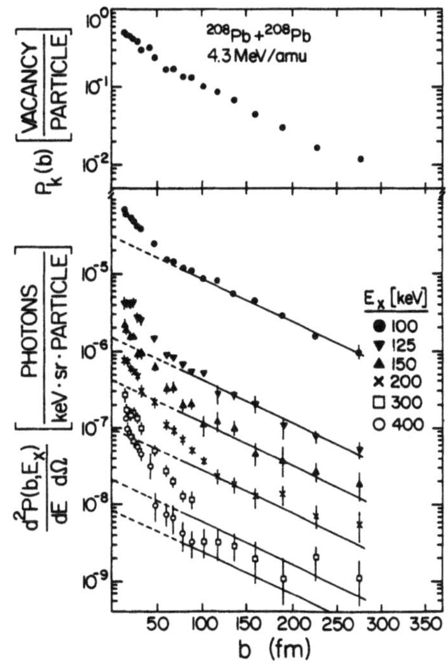

FIGURE 31. Impact-parameter dependence of characteristic (upper figure) and MO (lower figure) X rays in 4.3-MeV/amu ^{208}Pb → ^{208}Pb collisions, showing two-component structure of the $P(b)$ curves. Curves guide the eye. (Vincent, 1981, 1983; Stiebing *et al.*, 1982.)

the molecular orbital from which the vacancy originates. In fact, the relation

$$\langle b \rangle \langle q_0 \rangle \cong 1 \tag{18}$$

can be demonstrated (Bang *et al.*, 1979). Here, $\langle b \rangle$ is the average impact parameter associated with the observed $P(b)$:

$$\langle b \rangle \equiv \frac{\int_0^\infty b^2 P_K(b)\, db}{\int_0^\infty b P_K(b)\, db}, \tag{19}$$

and $\hbar\langle q_0 \rangle = \langle E \rangle / v$ is the mean momentum transfer to the electron bound with mean energy $\langle E \rangle$ at the time of excitation.

In analogy with the K X-ray studies, one can assume that the slopes of the MO X-ray $P(b)$ curves reflect *excitation* of vacancies, rather than their decay; then one can fit exponential functions to the data of Fig. 31, extract an average impact parameter for each component, and interpret the results in terms of Eq. 18 (Stiebing *et al.*, 1982). For the steeper component one finds $\langle b \rangle_1 \cong 50$ fm, and for the flatter component (solid lines in Fig. 31), $\langle b \rangle_2 \cong 160$ fm. From Eq. (18) it then follows that $\langle E \rangle_1 \cong 380$ keV for the steeper component and $\langle E \rangle_2 \cong 120$ keV for the flatter component. Examination of the correlation diagram for the Pb + Pb quasimolecule (Fig. 2) indicates that these energies agree qualitatively with expectations for $1s\sigma$ and $2p_{1/2}\sigma$ excitation, respectively. The two-component structure in Fig. 31 may therefore be interpreted as due to mixed, photon-energy-dependent contributions from $1s\sigma$ (steeper component) and $2p_{1/2}\sigma$ (flatter component) final-state MO X-ray transitions, respectively. Such an analysis is greatly facilitated by the one-step nature of the MO X-ray production mechanism (Fig. 25). A semiempirical separation of mixed contributions from $1s\sigma$ and $2p_{1/2}\sigma$ MO X rays based on studies of their impact-parameter dependence therefore appears quite feasible.

Within the context of such an interpretation, the $2p_{1/2}\sigma$ portion of the $P(b)$ curves can be subtracted, leaving the $1s\sigma$ component only (Fig. 32) (Stiebing *et al.*, 1982). The solid lines in Fig. 32 are fits to the $1s\sigma$ data for $\langle b \rangle_1 \cong 50$ fm, and the broken lines indicate results of theoretical calculations of the $1s\sigma$ MO $P(b)$ at 200, 400, and 600 keV (Kirsch, 1980). Agreement in *slope* between theory and experiment is good. However, if one normalizes theory to experiment at 400 keV, the magnitude of the theoretical results falls short at 200 keV. That the theory underestimates $1s\sigma$ MO yields at low photon energies was already mentioned in connection with Fig. 29.

A second type of coincidence experiment has been performed between K X rays and MO X rays. The idea behind such a study is illustrated in Fig. 33 (Vincent, 1981, 1983) which shows the correlation diagram for the Pb + Pb quasimolecule, indicating various possible excitation and decay

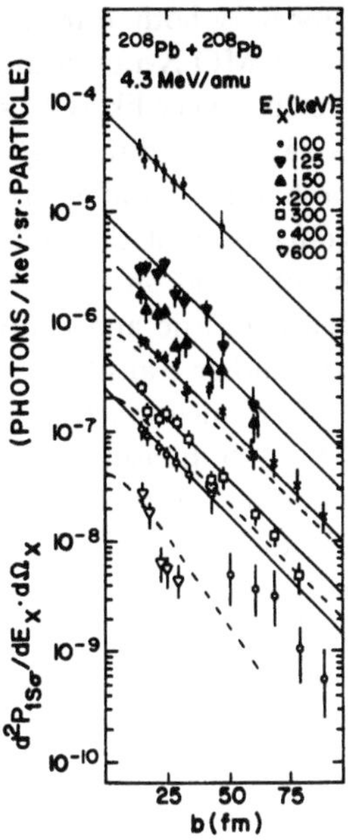

FIGURE 32. Doubly differential probability for the emission of $1s\sigma$ MO radiation in 4.3-MeV/amu $^{208}Pb \rightarrow {}^{208}Pb$ collisions for various X-ray-energy regions. (Stiebing *et al.*, 1982.)

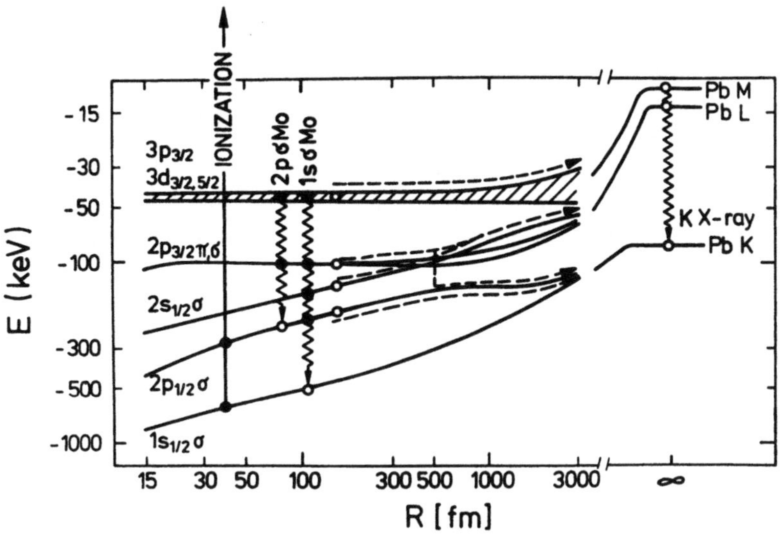

FIGURE 33. The Pb + Pb quasimolecular correlation diagram showing the excitation and decay of inner quasimolecular orbitals during and after the collision. (Vincent, 1981, 1983.)

processes both during and after the collision. In particular, a $2p_{1/2}\sigma$ to $1s_{1/2}\sigma$ MO X-ray transition will produce a vacancy in the $2p_{1/2}\sigma$ orbital which is very likely to survive the collision and decay under emission of a K X-ray. In addition, $2p_{3/2}\sigma$ to $1s_{1/2}\sigma$ transitions with the subsequent (highly probable) coupling of the $2p_{3/2}\sigma$ vacancy to the $2p_{1/2}\sigma$ orbital on the outgoing part of the collision will also result in the production of one continuum and one characteristic K X ray. A coincidence experiment between characteristic and continuum K X rays will therefore isolate these cascades from other transitions in the quasimolecule.

As mentioned in Sec. 3.1 in connection with analogous studies in the Nb + Nb quasimolecule (O'Brien *et al.*, 1980), the actual experimental situation in such coincidence measurements is considerably more complicated than outlined above. In Fig. 33 we schematically indicate that independent excitation of two electrons in the same collision can also occur, with one vacancy decaying into an MO X ray during the collision, and the other vacancy into a K X ray afterwards. The latter process can also produce one continuum and one characteristic X-ray. The coincidence requirement between MO and K X rays will therefore result in a continuum X-ray spectrum with quantitative aspects which reflect the competing strengths of various MO transitions into the $1s_{1/2}\sigma$ and $2p_{1/2}\sigma$ molecular orbitals, as well as the probability of single and double vacancy production in these orbitals.

In Fig. 34 we show the results of such MO X-ray K X-ray coincidence measurements in 4.8 MeV/amu ^{208}Pb + ^{208}Pb collisions (Vincent, 1981, 1983). The *top* part of the figure indicates extracted coincidence and singles MO X-ray cross sections compared with theory (Kirsch, 1980). Only singles radiative transitions to the $1s\sigma$ molecular orbital are calculated theoretically. The bottom part of Fig. 34 shows the ratio of measured coincidence-to-singles cross sections as a function of photon energy. The two parallel lines are coincidence-to-singles ratios calculated from independent excitation of *double vacancies*, with one vacancy decaying into a $1s\sigma$ (top line) or $2p\sigma$ (bottom line) MO X ray (Vincent, 1981, 1983). The calculated ratio for $1s\sigma$ MO radiation is larger than that for $2p_{1/2}\sigma$ MO X rays, reflecting the steeper impact-parameter dependence of $1s\sigma$ excitation.

The bottom panel of Fig. 34 illustrates that the measured cross-section ratios can, in principle, be explained by assuming that coincidences come from independently excited double vacancies only, and simply choosing a suitable mixture of $1s_{1/2}\sigma$ to $2p_{1/2}\sigma$ MO X-ray intensities as a function of photon energy. Moreover, the mixture necessary to explain the measured ratios in Fig. 34 agrees with the mixture which one obtains from MO X-ray impact-parameter-dependence studies through analyses of their two-component structure (Stiebing *et al.*, 1982). In contrast to lighter systems, where dominant contributions from cascades were obtained (O'Brien *et al.*, 1980), a

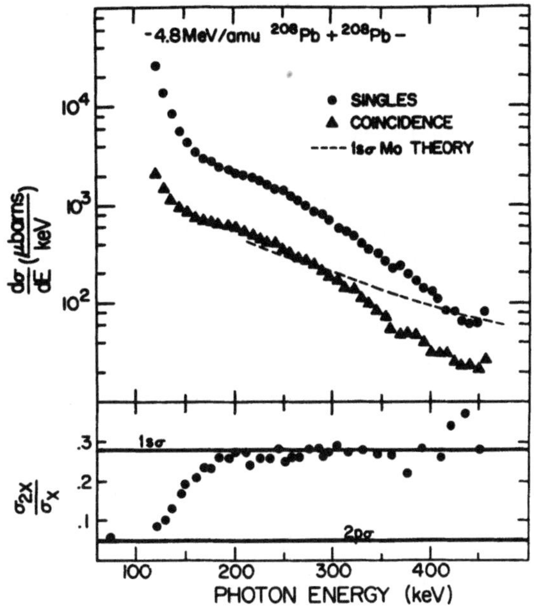

FIGURE 34. Top: Singles and K X-ray, MO X-ray coincidence cross sections from 4.8-MeV/amu $^{208}Pb \rightarrow {}^{208}Pb$ collisions. Dashed line, theoretical $1s\sigma$ MO X-ray singles normalized to the data above 400 keV. Bottom: coincidence-to-singles cross-section ratios compared with double-vacancy calculation (solid curves). (Vincent, 1981, 1983.)

consistent description of the results in Fig. 34 can therefore be made in terms only of double-vacancy production. This description, in turn, indicates a significantly higher cross section for $1s_{1/2}\sigma$ MO X rays than that calculated theoretically at photon energies below 400 keV (Kirsch, 1980).

In conclusion, we again emphasize the unique role played by quasimolecular X-ray emission in providing direct evidence for the formation of strongly bound molecular orbitals in these collisions. The straightforward spectroscopy of the united quasiatom based on directional anisotropy measurements in lighter systems cannot be extended to the superheavy quasimolecule. However, impact-parameter dependence studies provide a means by which $1s_{1/2}\sigma$ and $2p_{1/2}\sigma$ MO X-ray yields can be separated and distinguished from each other, and K-MO X-ray coincidence measurements provide further complementary information on relative excitation and radiative decay probabilities in the innermost molecular orbitals. It is not clear whether or not such studies can be extended to still heavier collision systems where disturbing background from nuclear Coulomb excitation dominates.

4. OTHER X RAYS

Energetic collisions between heavy atoms, although often adiabatic with regard to velocities, can be quite violent for the atomic electrons. Even for quasimolecular orbitals, electron binding energies take up only a small fraction of the total available energy in the center-of-mass system. Much atomic ionization therefore occurs, with associated production of both discrete and continuum X rays of various, sometimes exotic, origins. In addition to characteristic K X rays and quasimolecular X rays, X rays are observed from two-electron–one-photon atomic transitions (TET), radiative electron capture (REC), secondary electron bremsstrahlung (SEB), and nucleus–nucleus bremsstrahlung (NNB).

TET occurs when an atom loses two electrons from the K shell in a heavy-ion collision (Wölfli *et al.*, 1975; Greenberg *et al.*, 1977b). Usually this excited atomic state decays through sequential emission of two K X rays, the first due to the hypersatellite transition (K_α^h or K_β^h), and the second X ray, to a "normal" satellite transition (Chap. 2, Sec. 4.5). The second transition is designated as a satellite because in heavy-ion collisions double-vacancy production in the K shell is nearly always accompanied by simultaneous ionization of the outer shells. Alternatively, double-vacancy decay can proceed by a simultaneous transition of two electrons from higher atomic shells into the empty K shell, under emission of a single photon. These latter events are designated as two-electron–one-photon transitions (TET).

The first observation of TET events in the X-ray regime was made during efforts to measure quasimolecular X-ray production in heavy-ion collisions (Wölfli *et al.*, 1975). An example of typical experimental spectra is shown in Fig. 35 for 17.9- to 91.5-MeV Ni + Ni collisions (Greenberg *et al.*, 1977b). The linear plots show the prominent, intense contributions from (strongly absorbed) characteristic Ni K_α and K_β X rays below 10-keV X-ray energy, and the quasimolecular X-ray continuum extending to higher photon frequencies. The small bumps in the 15- to 16-keV region labeled $K_{\alpha\alpha}$ and $K_{\alpha\beta}$ are due to TET events. The $K_{\alpha\alpha}$ peak originates from transitions of two L-shell electrons, and the $K_{\alpha\beta}$ peak, from transitions of one L-shell and one M-shell electron. The 91.5-MeV data include the spectrum at zero degrees to the beam direction, where the prominent Doppler-shifted projectile $K_{\alpha\alpha}$ transition is superimposed upon the target $K_{\alpha\beta}$ peak. The use of an absorber is essential in these studies, not only to prevent distortion of the TET lines by the high-energy detector-line-shape tails of the K X rays, but also to reduce pulse pile-up.

Theoretical calculations of the branching ratios for double-vacancy decay by TET show that the transitions are many-electron processes in which correlations play an important role in both the initial and final states (Luken *et al.*, 1977). A number of authors have carried out theoretical

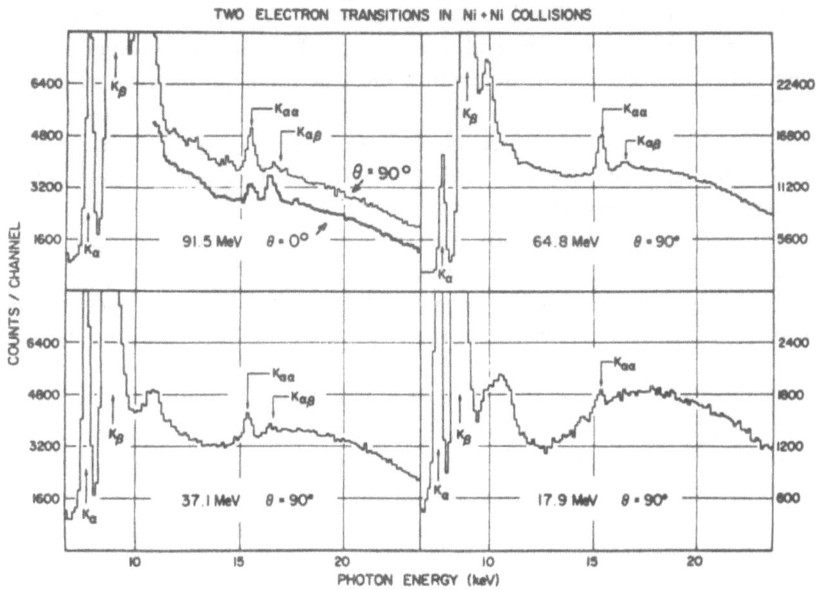

FIGURE 35. X-ray spectra from Ni + Ni collisions. Data pertain to 90° to the beam direction, except for 91.5 MeV energy where both 90° and 0° distributions are shown. Peaks at ~15 keV are due to two-electron–one-photon transitions. (Greenberg *et al.*, 1977b.)

studies of the TET decay rates and transition energies (Vinti, 1932; Åberg *et al.*, 1976; Kelly, 1976; Gavrila *et al.*, 1976). Results show that the observed transition energies can be adequately explained by calculations which take account of the state of *L*-shell ionization at the time of TET emission.

Studies of TET X rays in asymmetric collision systems provide an opportunity to extend understanding of the vacancy-sharing phenomena to the case of double-vacancy transfer. We showed in Sec. 2.2 that the ratio of heavy to light collision-partner vacancies in asymmetric collisions is given by Eqs. (7) and (8). For lighter collision systems, we can neglect contributions from the second term in Eq. (8), because we have $\sigma_{1s\sigma} \ll \sigma_{2p\sigma}$.

If one assumes that in double-vacancy sharing the two vacancies are shared independently, then the probability of double sharing is simply the square of the single-vacancy sharing probability. In such a description the ratio of double vacancies in the heavy collision partner to those in the light partner is

$$R_{kk} \simeq e^{-4X} \tag{20}$$

(Lennard *et al.*, 1978).

TET data can be utilized to test the predictions of Eq. (20). By measuring the ratio of heavy- to light-collision-partner TET intensities over a range of collision systems, one can determine the fractional population of double K-shell vacancies in the two atoms. Clearly, corrections for fluorescence yields must be considered. However, use of a ratio rather than an absolute comparison minimizes errors from fluorescence-yield variations, particularly in nearly symmetric collision systems.

The results of such a study are indicated in Fig. 36 (Lennard *et al.*, 1978). The solid line is the prediction from Eq. (20). Experiment and Eq. (20) agree extremely well over nearly two orders of magnitude in R_{kk}, providing strong evidence for the uncorrelated nature of the double-vacancy transfer process.

The continuum character of MO X rays has induced studies of other possible competing sources of X-ray emission in heavy-ion collisions. Secondary-electron bremsstrahlung (SEB) and radiative electron capture (REC) are two candidates for continuum X-ray production processes which in some collision systems could generate intensities comparable to or greater than those expected from MO X rays. Both as background processes accompanying MO emission as well as for reasons of their intrinsic interest, these competing sources of continuum X rays in heavy-ion–atom encounters have been explored.

REC is a process by which a weakly bound target electron is captured by a projectile atom with an inner-shell vacancy. The released energy is

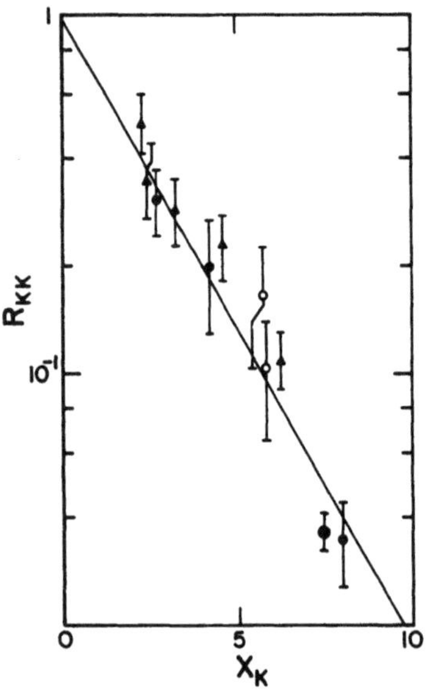

FIGURE 36. Comparison between theoretical (solid curve) and experimental double-K-vacancy sharing ratios, versus the reduced sharing parameter x as determined by TET measurements. (Lennard *et al.*, 1978.) Symbols: ▲, thick targets; •, thin solid targets; ●, gas target; ○, results. (Stoller *et al.*, 1979b.)

expended by emission of an X ray. Conservation of energy leads to the condition

$$\hbar\omega = E_f - E_i + p_0^2/2m + \mathbf{v}\cdot\mathbf{p}, \tag{21}$$

where $\hbar\omega$ is the energy of the radiated photon, E_f the binding energy of the electron in its final state, E_i the initial electron binding energy, m the mass of the electron, $\mathbf{v}$ the projectile velocity, $p_0 = mv$ the momentum of the target electron viewed in the frame of the projectile, due to the motion of the projectile, and $\mathbf{p}$ is the momentum of the electron due to its presence in a target atomic orbital. Thus, the photon-energy distribution for REC exhibits a peak at $\hbar\omega = E_f - E_i + p_0^2/2m$ with a width reflecting the projection of the target-electron momentum distribution (Compton profile) along the projectile direction ($\mathbf{v}\cdot\mathbf{p}$). The cross section for REC is related to the capture cross section and the probability that projectile vacancies exist.

Figure 37 contains REC spectra from studies of F^{n+} + He (gas)

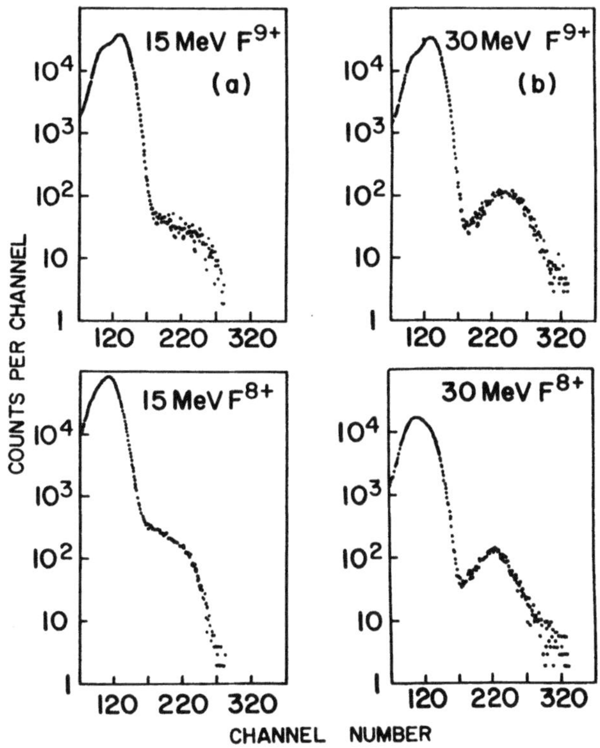

FIGURE 37. Typical X-ray spectra from F^{8+} and F^{9+} ions incident on a He gas target at 15 and 30 MeV under single-collision conditions. The peak at the left is due to F *K* X rays, and the peak at the right is due to REC X rays. (Tawara *et al.*, 1982.)

collisions at 15 and 30 MeV (Tawara *et al.*, 1982). The peaks at low X-ray energies are fluorine K X rays, and those at higher photon energies, REC X rays that arise from capture into the projectile K shell. It should be noted that at 15 MeV REC produces a shoulder which is not resolved from the K X rays. At 30 MeV, the peak has moved out to higher energies and is well resolved from the K X-ray distribution. The centroids of the REC peaks agree well with the energies predicted by Eq. (21). Studies show, however, that the REC peaks are somewhat wider for projectiles that contain two K vacancies rather than one, in contrast to simple theoretical expectations.

The magnitude of REC cross sections for K-shell capture reflects the K-vacancy probability of the projectile. In experiments in which thin solid targets are used, this vacancy probability will change with target thickness, owing to competition between excitation and quenching processes in the foil (Sec. 2.1). The relevant cross section for REC is the cross section per projectile K vacancy, which can be determined experimentally as a function of target thickness when solid targets are used (Tanis *et al.*, 1978, 1981).

In Fig. 38 from Tanis *et al.* (1981) we plot the results of experimental REC cross sections per projectile vacancy for 20–80-MeV Cl beams incident on thin solid C and Cu targets. The dashed curve is the prediction of the free-electron theory of REC (Bethe *et al.*, 1977); the two solid curves represent this prediction multiplied by factors of 4 and 11, respectively. The authors interpret the excellent agreement between experiment and the solid curves as evidence that only the most weakly bound target electrons with velocities

$$\frac{v_e}{v}=\left(\frac{E_B/m_e}{E/M}\right)^{1/2}\leqslant 0.1 \tag{22}$$

contribute to the REC cross section (Tanis *et al.*, 1981). Here, v_e is the target-electron velocity, E_B the target-electron binding energy, m_e the mass of the electron, E the projectile energy, M the projectile mass, and v the projectile velocity. The criterion (22) is satisfied by the four carbon L-shell electrons and the 11 outermost Cu electrons. The plausibility of this interpretation is confirmed by subsequent, more precise gas-target studies (data of Fig. 37), which also found good agreement between cross-section measurements and REC calculations (Tawara *et al.*, 1982).

Another source of continuum X-ray emission in heavy-ion collisions is secondary electron bremsstrahlung (SEB). This is a two-step production mechanism: the electron is ejected in the first collision and bremsstrahlung is produced in a subsequent collision between the electron and a target atom. Problems in estimating the size of SEB in heavy-ion collisions stem principally from theoretical uncertainties in the "δ"-electron intensities and energy distributions. In addition, electron bremsstrahlung can arise from

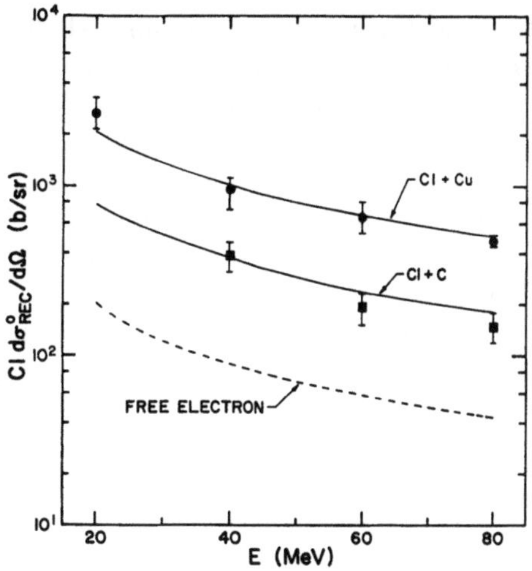

FIGURE 38. Chlorine REC differential cross sections at 90°, per K vacancy versus incident Cl-ion energy. Solid squares, Cl + C. Solid circles, Cl + Cu. The dashed curve represents the free-electron Bethe–Salpeter theory assuming $\sin^2\theta$ REC angular distribution. The solid curves represent the Bethe–Salpeter results multiplied by 4 and 11, respectively, representing the effective number of free electrons in these target atoms. (Tanis *et al.*, 1981.)

collisions with materials surrounding the targets (frames, chamber walls, etc.), thereby leading to SEB yields which depend on experimental geometry. It has generally been assumed that SEB yields are small in studies of MO X-ray production (Vincent *et al.*, 1978; Stoller *et al.*, 1980). However, only recently has experimental evidence in support of this assumption been obtained.

In Fig. 39 we plot measured cross sections for continuum X-ray emission in Ni + Ni collisions at 45° and 135° with respect to the beam direction (Vincent, 1977; Vincent *et al.*, 1978, 1979). The solid curves in Fig. 39 indicate calculated SEB X-ray intensities, based on measured δ-elec-

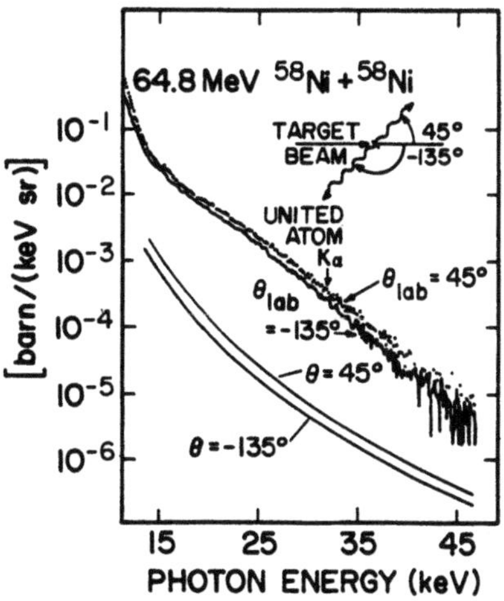

FIGURE 39. Comparison between experimental continuum X rays (upper curves) and extracted SEB yields (lower curves). Both cases exhibit forward–backward asymmetry in the laboratory frame of reference. (Vincent *et al.*, 1978; Bell *et al.*, 1982.)

tron spectra from these collisions (Bell *et al.*, 1982). Clearly, SEB intensities are too small to explain the observed data, indicating that background from electron bremsstrahlung can safely be neglected in these studies of MO X-ray emission from Ni + Ni collisions.

Nucleus–nucleus bremsstrahlung (NNB) can be studied experimentally in very asymmetric systems for which MO X-ray yields are suppressed by the small probability of producing $1s\sigma$ vacancies (Reinhardt *et al.*, 1976; Trautvetter *et al.*, 1976). Examination of the angular distribution of NNB in selected collision systems has illustrated the importance of dipole–quadrupole interference effects on the measured intensities, and has demonstrated the validity of a semiclassical theoretical approach to the calculation of NNB yields. NNB can constitute a large source of background in MO X-ray measurements, particularly at high photon frequencies in asymmetric collision systems.

The cross section for NNB can be expressed as a multipole expansion of the prominent dipole term plus the weaker quadrupole contribution (Trautvetter *et al.*, 1976):

$$\frac{d\sigma}{dE_X d\Omega} \propto \frac{Z_1^2 Z_2^2}{E_x \beta^2}[D^2 W_{E1}(E, E_x, \theta) + \mu^2\beta^2 Q^2 W_{E2}(E, E_x, \theta) + 2\mu\beta D Q W_{E1,E2}(E, E_x, \theta)]. \tag{23}$$

In this equation $D = Z_1/A_1 - Z_2/A_2$ arises from the dipole moment $\mathbf{d} = \sum_\nu \varepsilon_\nu \mathbf{r}_\nu$ transformed to the center-of-mass system, and $Q = Z_1/A_1^2 + Z_2/A_2^2$ is associated with the quadrupole moment. The functions W_{E1}, W_{E2}, and $W_{E1,E2}$ represent the dipole, quadrupole, and interference parts of the angular distribution at bombarding energy E, photon energy E_x, and emission angle θ. The reduced mass, is denoted by μ, and we have $\beta = v_{\text{projectile}}/c$.

Examination of Eq. (23) shows that for equal atomic numbers Z_1 (projectile) = Z_2 (target) the prominent dipole contribution vanishes, leaving only weaker quadrupole contributions. Clean MO X-ray experiments should therefore be performed with isotopically pure, symmetric collision systems. Furthermore, although the quadrupole and dipole terms each are separately symmetric about 90°, the interference term results in a forward–backward asymmetry for the NNB angular distributions.

By changing the sign of the D term in Eq. (23), the sign of the interference term can be changed. This effect is dramatically evident in Fig. 40 (Trautvetter *et al.*, 1976), where comparison between $^{18}O + {}^{58}Ni$ ($D < 0$) and $^{12}C + {}^{58}Ni$ ($D > 0$) show a striking reversal of the forward–backward angular distributions. Alternatively, simple interchange of the projectile with

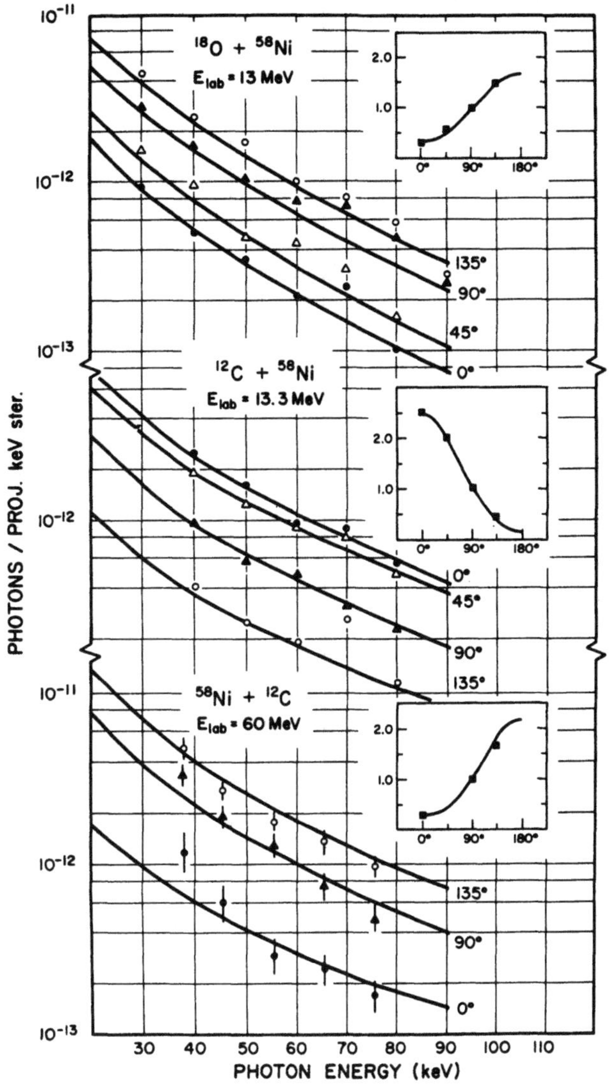

FIGURE 40. Comparisons of calculated NNB absolute thick-target yields with experimental X-ray intensities versus photon energy, at four angles of observation. Each datum point represents an integration over a 10-keV photon-energy interval. The insets (vertical scale relative) show a comparison of calculated angular distribution (solid curves) with the data (squares) for the photon-energy interval 40–60 keV. (Trautvetter *et al.*, 1976.)

the target ($^{58}Ni + {}^{12}C$ versus $^{12}C + {}^{58}Ni$) changes the sign of D and reverses the angular distribution. Experiment and theory agree extremely well in absolute magnitude.

5. CONCLUSIONS

The existence of quasimolecules in heavy-ion collision systems is well established. In addition, significant quantitative understanding of the dynamic aspects of quasimolecular phenomena now exists. Studies of characteristic and MO X rays in the heaviest possible collisions conclusively demonstrate the transient occurrence of binding energies that approach twice the rest-mass energy of the electron. The precision to which spectroscopic information on these binding energies can be extracted is, however, limited by the model dependence of any spectroscopy, and by dynamic effects. It is questionable whether or not spectroscopic precision sufficient to test deviations from the predictions of quantum electrodynamics in strong Coulomb fields can be achieved with the X-ray methods so far employed. Nevertheless, all X-ray information to date indicates that the spontaneous production of positrons should occur in such collisions. In fact, the recent discovery of peaked structures in positron spectra from U + U and U + Cm collisions (Schweppe *et al.*, 1983) may provide a definitive signal for this spontaneous emission. Such an observation would furnish direct information on the structural aspects of the highly bound molecular states, rather than reflecting dynamic collision effects, and could therefore provide the much-needed signal with which the quantum electrodynamics of strong fields can be quantitatively studied (Greenberg *et al.*, 1984; Reinhardt *et al.*, 1984). This very interesting subject is reviewed in detail in Chap. 14 of this volume.

ACKNOWLEDGMENTS

The author is indebted to Mike Jakubowycz for preparing the figures, Donna Kay for typing, and J. S. Greenberg for a critical reading of the manuscript. This work was supported by the National Science Foundation under Contract No. PHY 79-23340.

REFERENCES

ÅBERG, T., K. A. JAMISON, AND P. RICHARD, 1976, *Phys. Rev. Lett.* **37**, 63.
ALLISON, S. K., 1958, *Rev. Mod. Phys.* **30**, 1137.
AMUNDSEN, P. A., 1976a, *J. Phys. B* **9**, 971.
AMUNDSEN, P. A., L. KOCBACH, AND J. M. HANSTEEN, 1976b, *J. Phys. B* **9**, L203.

AMUNDSEN P., 1978, *J. Phys. B* **11**, L737.

ANDERSON, J. U., E. LAEGSGAARD, M. LUND, AND C. D. MOAK, 1976, *Nucl. Instrum. Methods* **132**, 507.

ANHOLT, R., AND W. E. MEYERHOF, 1977a, *Phys. Rev. A* **16**, 190.

ANHOLT, R., W. E. MEYERHOF, AND A. SALIN, 1977b, *Phys. Rev. A* **16**, 951.

ANHOLT, R., 1978a, *Z. Phys. A* **289**, 41.

ANHOLT, R., 1978b, *Z. Phys. A* **288**, 257.

ANHOLT, R., 1978c, *Phys. Rev. A* **17**, 983.

ANHOLT, R., 1979a, *Z. Phys. A* **291**, 289.

ANHOLT, R., 1979b, private communication.

ANHOLT, R., W. E. MEYERHOF, AND CH. STOLLER, 1979c, *Z. Phys. A* **291**, 287.

ANHOLT, R., CH. STOLLER, AND W. E. MEYERHOF, 1980, *J. Phys. B* **13**, 3807.

ANHOLT, R., 1981, *Phys. Lett.* **84A**, 62.

ANHOLT, R., 1982, *Nucl. Instrum. Methods* **198**, 567.

ANHOLT, R., 1983, Molecular Orbital X-ray Production by Low-Velocity Highly Stripped Ions, in *Progress Report No. 1981-82*, Stanford University (to be published).

ANNETT, C. H., B. CURNUTTE, AND C. L. COCKE, 1979, *Phys. Rev. A* **19**, 1038.

BAMBYNEK, W., B. CRASEMANN, R. W. FINK, H. U. FREUND, H. MARK, C. D. SWIFT, R. E. PRICE, AND P. V. RAO, 1972, *Rev. Mod. Phys.* **44**, 716.

BANG, J., AND J. M. HANSTEEN, 1959, *K. Dan. Vidensk. Selsk. Mat.-Fys. Medd.* **31**, No. 13.

BANG, J., AND J. M. HANSTEEN, 1979, *Phys. Lett.* **72A**, 218.

BARAT, M., AND W. LICHTEN, 1972, *Phys. Rev. A* **6**, 211.

BASBAS, G., W. BRANDT, AND R. LAUBERT, 1973, *Phys. Rev. A* **7**, 983.

BASBAS, G., W. BRANDT, AND R. LAUBERT, 1978, *Phys. Rev. A* **17**, 1655.

BEHNCKE, H. H., P. ARMBRUSTER, F. FOLKMAN, S. HAGMAN, J. R. MACDONALD, AND P. H. MOKLER, 1979, *Z. Phys. A* **289**, 333.

BELL, F., G. TROLLMAN, H. D. BETZ, AND E. SPINDLER, 1982, *Phys. Rev. A* **26**, 72.

BETHE, H. A., AND E. E. SALPETER, 1977, *Quantum Mechanics of One-and Two-Electron Atoms* (Plenum Press, New York).

BETZ, H. D., F. BELL, H. PANKE, G. KALKOFFEN, M. SOLZ, AND D. EVERS, 1973, *Phys. Rev. Lett.* **30**, 358.

BETZ, H. D., F. BELL, H. PANKE, W. STEHLING, E. SPINDLER, AND M. KLEBER, 1975, *Phys. Rev. Lett.* **34**, 1256.

BETZ, W., *et al.*, 1975, Important Problems in Future Heavy Ion Physics, in *The Physics of Electronic and Atomic Collisions*, edited by J. S. Risley and R. Geballe (Univ. of Washington Press, Seattle), p. 531.

BETZ, W., G. SOFF, B. MÜLLER, AND W. GREINER, 1976, *Phys. Rev. Lett.* **37**, 1046.

BHALLA, C. P., 1973, *Phys. Rev. A* **8**, 2877.

BHALLA, C. P., 1975, *Phys. Rev. A* **12**, 122.

BOSCH, F., D. LEISEN, P. ARMBRUSTER, D. MAOR, P. H. MOKLER, H. SCHMIDT-BÖCKING, AND R. SCHUCH, 1980, *Z. Phys. A* **296**, 11.

BRANDT, W., R. LAUBERT, AND I. SELLIN, 1966, *Phys. Rev.* **151**, 56.

BRANDT W., 1972, Proc. Int. Conf. Inner-shell Ionization Phenomena and Future Applications, Atlanta, Georgia; (U.S. Atomic Energy Commission Oak Ridge, Tennessee, 1973). p. 948.

BRANDT, W., AND K. W. JONES, 1976, *Phys. Lett.* **57A**, 35.

BRIGGS, J. S., AND J. MACEK, 1972, *J. Phys. B* **5**,579.

BRIGGS, J. S., 1975, *J. Phys. B* **8**, L485.

BRIGGS, J. S., 1976, *Rep. Prog. Phys.* **39**, 217.

BRIGGS, J. S., AND K. DETTMANN, 1977, *J. Phys. B* **10**, 1113.

BÜRGY, P., G. BONANI, E. MORENZONI, M. STÖCKLI, CH. STOLLER, M. SUTER, AND W. WÖLFLI, 1981, *Z. Phys. A* **300**, 105.

CHEN, M. H., B. CRASEMANN, AND D. L. MATHEWS, 1975, *Phys. Rev. Lett.* **34**, 1309.
CHOI, B. H., AND E. MERZBACHER, 1969, *Phys. Rev.* **177**, 233.
COCKE, C. L., AND R. RANDALL, 1973, *Phys. Rev. Lett.* **30**, 1016.
COCKE, C. L., R. R. RANDALL, S. L. VARGHESE, AND B. CURNETTE, 1976, *Phys. Rev. A* **14**, 2026.
DAVIS, C. K., AND J. S. GREENBERG, 1974, *Phys. Rev. Lett.* **32**, 1215.
DEMKOV, Y. N., 1964, *Sov. Phys. JETP* **18**, 138.
FOLKMAN, F., P. ARMBRUSTER, S. HAGMANN, G. KRAFT, P. H. MOKLER, AND J. H. STEIN, 1976, *Z. Phys. A* **276**, 15.
FOSTER, C., T. P. HOOGKAMER, P. WOERLEE, AND F. W. SARIS, 1976, *J. Phys. B* **9**, 1943.
FRANK, W., K. H. KAUN, P. MANFRASS, N. V. PRONIN, AND YU. P. TETRYAKOV, 1976, *Z. Phys. A* **279**, 213.
FRITSCH, W., AND U. WILLE, 1979, *J. Phys. B* **12**, L335.
GARCIA, J. D., R. J. FORTNER, AND T. M. KAVANAGH, 1973, *Rev. Mod. Phys.* **45**, 111.
GARDNER, R. K., T. J. GRAY, P. RICHARD, C. SCHMIEDEKAMP, K. A. JAMISON, AND J. M. HALL, 1977, *Phys. Rev. A* **15**, 2202.
GAVRILA, M., AND J. E. HANSEN, 1976, *Phys. Lett.* **58A**, 158.
GRAY, T. J., P. RICHARD, K. A. JAMISON, J. M. HALL, AND R. K. GARDNER, 1976, *Phys. Rev. A* **14**, 1333.
GRAY, T. J., 1980, Target Ionization and X-ray Production for Ions Incident Upon Solid Targets, in *Methods of Experimental Physics 17*, edited by P. Richard (Academic, New York).
GREENBERG, J. S., C. K. DAVIS, AND P. VINCENT, 1974, *Phys. Rev. Lett.* **33**, 473.
GREENBERG, J. S., H. BOKEMEYER, H. EMLING, E. GROSSE, D. SCHWALM, AND F. BOSCH, 1977a, *Phys. Rev. Lett.* **39**, 1404.
GREENBERG, J. S., P. VINCENT, AND W. LICHTEN, 1977b, *Phys. Rev. A* **16**, 964.
GREENBERG, J. S., AND P. VINCENT, 1985, High Energy Atomic Physics Experiment, in *Treatise on Heavy-Ion Science, III*, edited by D. A. Bromley (Plenum, New York).
GROENEVELD, K. O., B. KOLB, J. SCHADER, AND K. D. SEVIER, 1976, *Z. Phys. A* **277**, 13.
GROS, M., B. MÜLLER, AND W. GREINER, 1976, *J. Phys. B* **9**, 1849.
GUILLAUME, G., P. FINTZ, F. C. JUNDT, K. W. JONES, AND B. M. JOHNSON, 1978, *Phys. Lett.* **68A**, 39.
HAGMANN, S., C. L. COCKE, E. JUSTINIANO, J. R. MACDONALD, AND H. SCHMIDT-BÖCKING, 1979, *J. Phys. B* **12**, L395.
HANSEN, J. S., 1973, *Phys. Rev. A* **8**, 822.
HANSTEEN, J. M., O. D. JOHNSEN, AND L. KOCBACH, 1975, *Atom. Data Nucl. Data Tables* **15**, 305.
HARTUNG, H., AND B. FRICKE, 1978, *Z. Phys. A* **288**, 345.
HEINIG, K. H., H. U. JAEGER, H. RICHTER, AND H. WOITTENNIK, 1976, *Phys. Lett.* **60B**, 240.
HOPKINS, F., 1975, *Phys. Rev. Lett.* **35**, 270.
ITO, S., P. ARMBRUSTER, H. BOKEMEYER, F. BOSCH, H. EMLING, H. FOGLER, E. GROSSE, R. KULESSA, D. LIESEN, D. MAOR, D. SCHWALM, J. S. GREENBERG, R. SCHULÉ, AND N. TRAUTMANN, 1982, GSI Scientific report 1981, p. 141.
JAKUBASSA, D. H., AND K. TAULBJERG, 1980, *J. Phys. B* **13**, 757.
JAMNIK, D., AND C. ZUPANIC, 1957, *K. Dan. Vidensk. Selsk. Mat. Fys. Medd.* **31**, No. 2.
JOHNSON, B. M., K. W. JONES, W. BRANDT, F. C. JUNDT, G. GUILLAUME, AND T. H. KRUSE, 1979, *Phys. Rev. A* **19**, 81.
KAUN, K. H., P. MANFRASS, AND W. FRANK, 1977, *Part. Nucl. (Moskow)* **8**, 1246.
KELLY, H. P., 1976, *Phys. Rev. Lett.* **37**, 386.
KESSEL, Q. C., AND B. FASTRUP, 1973, The Production of Inner-Shell Vacancies in Heavy Ion–Atom Collisions, *Case Studies in Atom Physics 3*, (North-Holland, Amsterdam), p. 137.
KIRSCH, J., W. BETZ. J. REINHARDT, G. SOFF, B. MÜLLER, AND W. GREINER, 1978, *Phys. Lett.* **72B**, 298.
KIRSCH, J., W. BETZ, J. REINHARDT, B. MÜLLER, W. GREINER, AND G. SOFF, 1979, *Z. Phys. A* **292**, 227.

Kirsch, J., 1980, Ph.D. thesis, U. Frankfurt.
Kleber, M., and D. H. Jakubassa, 1975, *Nucl. Phys. A* **252**, 152.
Kraft, G., P. H. Mokler, and J. H. Stein, 1974, *Phys. Rev. Lett.* **33**, 476.
Larkins, F. P., 1971a, *J. Phys. B* **4**, L29.
Larkins, F. P., 1971b, *J. Phys. B* **4**, 14.
Lennard, W. N., I. V. Mitchell, and D. Phillips, 1978, *J. Phys. B* **11**, 1283.
Lennard, W. N., I. V. Mitchell, and D. Phillips, 1981, *J. Phys. B* **14**, 2215.
Lichten, W., 1976, Proc. 2nd International Conf. on Inner Shell Ioniz. Phen., Invited Papers, Freiburg, p. 249.
Liesen, D., P. Armbruster, F. Bosch, S. Hagmann, P. H. Mokler, H. Schmidt-Böcking, R. Schuch, J. B. Wilhemy, and H. J. Wollersheim, 1980, *Phys. Rev. Lett.* **44**, 983.
Luken, W. L., J. S. Greenberg, and P. Vincent, 1977, *Phys. Rev. A* **15**, 2305.
Lund, M., 1974, thesis, University of Aarhus, Denmark.
Luz, N., S. Sackman, and H. O. Lutz, 1979, *J. Phys. B* **12**, 1973.
McGuire, E. J., 1969, *Phys. Rev.* **185**, 1.
Macek, J., and J. S. Briggs, 1974, *J. Phys. B* **7**, 1312.
Merzbacher, E., and H. W. Lewis, 1958, *Handbuch der Physik 34* (Springer, Berlin), p. 166.
Meyerhof, W. E., 1973a, *Phys. Rev. Lett.* **31**, 1341.
Meyerhof, W. E., T. K. Saylor, S. M. Lazarus, W. A. Little, B. B. Triplett, and L. F. Chase, Jr., 1973b, *Phys. Rev. Lett.* **30**, 1279.
Meyerhof, W. E., 1974, *Phys. Rev. A* **10**, 1005.
Meyerhof, W. E., T. K. Saylor, and R. Anholt, 1975, *Phys. Rev. A* **12**, 2641.
Meyerhof, W. E., R. Anholt, T. K. Saylor, S. M. Lazarus, A. Little, and L. F. Chase, 1976, *Phys. Rev. A* **14**, 1653.
Meyerhof, W. E., and K. Taulbjerg, 1977a, *K* Shell Ionization in Heavy-Ion Collisions, *Ann. Rev. Nucl. Sci.* **27**, 279.
Meyerhof, W. E., R. Anholt, and T. K. Saylor, 1977b, *Phys. Rev. A* **16**, 169.
Meyerhof, W. E., R. Anholt, J. Eichler, and A. Salop, 1978, *Phys. Rev. A* **17**, 108.
Meyerhof, W. E., D. L. Clark, Ch. Stoller, E. Morenzoni, W. Wölfli, F. Folkmann, P. Vincent, P. Mokler, and P. Armbruster, 1979, *Phys. Lett.* **70A**, 303.
Mittleman, M. H., and L. Wilets, 1967, *Phys. Rev.* **154**, 12.
Mokler, P. H., and F. Folkmann, 1978, X-Ray Production in Heavy Ion–Atom Collisions, in *Structure and Collisions of Ions and Atoms*, edited by I.A. Sellin (Springer, Berlin), p. 201.
Mokler, P. H., and D. Liesen, 1983, X-Rays from Superheavy Collision Systems, in *Progress in Atomic Spectroscopy, Part C*, edited by Beyer and H. Kleinpoppen (Plenum Press, New York), p. 321.
Morenzoni, E. A. A., 1981, Ph.D. thesis, ETH Zürich.
Müller, B., 1975, Proc 9th Int. Conf. on Physics of Electronic and Atomic Collisions, Seattle; University Washington Press (1975) Abstracts, p. 481.
Müller, B., G. Soff, W. Greiner, and V. Ceausescu, 1978, *Z. Phys. A* **285**, 27.
Müller, B., J. Reinhardt, W. Greiner, and G. Soff, 1983, *Z. Phys. A* **311**, 151.
Nikolaev, V. S., 1967, *Sov. Phys. JETP* **24**, 847.
Nolte, G., J. Volpp, R. Schuch, J. H. Specht, W. Lichtenberg, and H. Schmidt-Böcking, 1980, *J. Phys. B* **13**, 4599.
O'Brien, J. J., E. Liarokapis, and J. S. Greenberg, 1980, *Phys. Rev. Lett.* **44**, 386.
Olsen, R. E., 1972, *Phys. Rev. A* **6**, 1822.
Pflanz, E., R. Schuch, J. Neumann, G. Nolte, and H. Schmidt-Böcking, 1980, Contributed paper International Conference on X Rays and Inner-shell Ionization, Stirling.
Presser, G., E. Scherer, and J. Stahler, 1980, *Z. Phys. A* **295**, 27.
Randall, R. R., J. A. Bednar, B. Curnutte, and C. L. Cocke, 1976, *Phys. Rev. A* **13**, 204.
Reinhardt, J., G. Soff, and W. Greiner, 1976, *Z. Phys. A* **276**, 285.
Reinhardt, J., B. Müller, W. Greiner, and G. Soff, 1979, *Phys. Rev. Lett.* **43**, 1307.

REINHARDT, J., AND W. GREINER, 1985, High-Energy Atomic Physics—Theory, in *Treatise on Heavy-Ion Science III*, edited by D.A. Bromley (Plenum, New York).

SACKMANN, S., H. O. LUTZ, AND J. BRIGGS, 1974, *Phys. Rev. Lett.* **32**, 805.

SARIS, F. W., AND T. P. HOOGKAMER, 1977, *Atomic Physics 5*, edited by R. Marrus, M. Prior, and H.A. Shugart (Plenum, New York), p. 507.

SCHMIDT-BÖCKING, H., R. SCHUCH, I. TSERRUYA, R. SCHULÉ, H. J. SPECHT, AND K. BETHGE, 1978, *Z. Phys. A* **284**, 39.

SCHMIDT-BÖCKING, H., R. ANHOLT, R. SCHUCH, P. VINCENT, K. STIEBING, AND H. U. JÄGER, 1982a, *J. Phys. B* **15**, 3057.

SCHMIDT-BÖCKING, H., H. BAUMANN, K. BETHGE, A. GRUPPE, W. SCHADT, R. SCHUCH, AND K. E. STIEBING, 1982b, *Z. Phys. A* **304**, 177.

SCHNOPPER, H. W., H. D. BETZ, J. P. DELVAILLE, K. KALATA, A. R. SOHVAL, K. W. JONES, AND H. E. WEGNER, 1972, *Phys. Rev. Lett.* **29**, 898.

SCHUCH, R., G. NOLTE, B. M. JOHNSON, AND K. M. JONES, 1979, *Z. Phys. A* **293**, 91.

SCHUCH, R., G. NOLTE, AND H. SCHMIDT-BÖCKING, 1980a, *Phys. Rev. A* **22**, 1447.

SCHUCH, R., G. GAUKLER, G. NOLTE, K. W. JONES, AND B. M. JOHNSON, 1980b, *Phys. Rev. A* **22**, 2513.

SCHUCH, R., 1981a, Invited Talks, XII Inter. Conf. on Phys. Electron and Atom. Collisions, Gatlinburg.

SCHUCH, R., H. SCHMIDT-BÖCKING, AND I. TSERRUYA, 1981b, *Phys. Lett.* **82A**, 453.

SCHWALM, D., A. BAMBERGER, P. G. BIZZETI, B. POVH, G. A. P. ENGELBERTNIK, J. W. OLNESS, AND E. K. WARBURTON, 1972, *Nucl. Phys. A* **192**, 449.

SCHWEPPE, J., A. GRUPPE, K. BETHGE, H. BOKEMEYER, T. COWAN, H. FOLGER, J. S. GREENBERG, H. GREIN, S. ITO, R. SCHULE, D. SCHWALM, K. E. STIEBING, N. TRAUTMANN, P. VINCENT, AND M. WALDSCHMIDT, 1983, *Phys. Rev. Lett.* **51**, 2261.

SEPP, W. D., B. FRICKE, AND T. MOROVIC, 1981, *Phys. Lett.* **81A**, 258.

SOFF, G., J. REINHARDT, W. BETZ, AND J. RAFELSKI, 1978a, *Phys. Scr.* **17**, 417.

SOFF, G., W. BETZ, B. MÜLLER, W. GREINER, AND E. MERZBACHER, 1978b, *Phys. Lett.* **65A**, 19.

SOFF, G., W. GREINER, W. BETZ, AND B. MÜLLER, 1979, *Phys. Rev. A* **20**, 169.

SOFF, G., B. MÜLLER, AND W. GREINER, 1981, *Z. Phys. A* **299**, 189.

SPINDLER, E., H. D. BETZ, AND F. BELL, 1979, *Phys. Lett.* **42**, 832.

STÄHLER, G., AND G. PRESSER, 1981, *J. Phys. B* **14**, 4093.

STIEBING, K., H. SCHMIDT-BÖCKING, W. SCHADT, K. BETHGE, P. H. MOKLER, F. BOSCH, D. LIESEN, P. VINCENT, R. SCHUCH, AND S. HAGMANN, 1984, *Z. Phys.* **A319**, 239.

STOLLER, G., W. WÖLFLI, G. BONANI, M. STÖCKLI, AND M. SUTER, 1977a, *J. Phys. B* **10**, L347.

STOLLER, C., W. WÖLFLI, G. BONANI, M. STÖCKLI, AND M. SUTER, 1977b. *Phys. Rev. A* **15**, 990.

STOLLER, C., W. WÖLFLI, G. BONANI, E. MORENZONI, AND M. STÖCKLI, 1978. *Z. Phys. A* **287**, 33.

STOLLER, C., E. MORENZONI, W. WÖLFLI, W. E. MEYERHOF, F. FOLKMAN, P. VINCENT, P. H. MOKLER, AND P. ARMBRUSTER, 1980, *Z. Phys. A* **297**, 93.

STOLLER, C., R. ANHOLT, AND W. E. MEYERHOF, 1981, *Z. Phys. A* **302**, 95.

TANIS, J. A., AND S. M. SHAFROTH, 1978, *Phys. Rev. Lett.* **40**, 1174.

TANIS, J. A., S. M. SHAFROTH, J. E. WILLIS, AND J. R. MOWAT, 1981, *Phys. Rev. A* **23**, 366.

TAULBJERG, K., J. S. BRIGGS, AND J. VAABEN, 1976, *J. Phys. B* **9**, 1351.

TAWARA, H., P. RICHARD, AND K. KAWATSURA, 1982, *Phys.'Rev. A* **26**, 154.

THORSON, W. R., AND J. H. CHOI, 1977, *Phys. Rev. A* **15**, 550.

TRAUTVETTER, H. P., J. S. GREENBERG, AND P. VINCENT, 1976, *Phys. Rev. Lett.* **37**, 202.

TSERRUYA, I., H. SCHMIDT-BÖCKING, R. SCHULÉ, K. BETHGE, R. SCHUCH, AND J. SPECHT, 1976, *Phys. Rev. Lett.* **36**, 1451.

TSERRUYA, I., R. SCHUCH, H. SCHMIDT-BÖCKING, J. BARRETTE, WANG DA-HAI, B. M. JOHNSON, M. MERON, AND K. W. JONES, 1983, *Phys. Rev. Lett.* **50**, 30.

Vincent, P., 1977, Ph.D. thesis, Yale University.

Vincent, P., C. K. Davis, and J. S. Greenberg, 1978, *Phys. Rev. A* **18**, 1878.

Vincent, P., and J. S. Greenberg, 1979, *J. Phys. B* **12**, L641.

Vincent, P., 1981, Molecular Orbital X-ray Emission in Heavy Ion Collisions, in *Inner Shell and X-Ray Physics of Atoms and Solids*, edited by D.J. Fabian, H. Kleinpoppen, L.M. Watson (Plenum, New York), p. 117.

Vincent, P., 1983, Continuum X-ray Emission in Adiabatic Heavy Ion-Atom Collisions in *Quantum Electrodynamics of Strong Fields*, edited by W. Greiner (Plenum, New York), p. 359.

Vinti, J. P., 1932, *Phys. Rev.* **42**, 632.

Wölfli, W., C. Stoller, G. Bonani, M. Suter and, M. Stöckli, 1975, *Phys. Rev. Lett.* **35**, 656.

Wölfli W., E. Morenzoni, C. Stoller, G. Bonani, and M. Stöckli, 1978a, *Phys. Lett.* **68A**, 217.

Wölfli, W., C. Stoller, G. Bonani, M. Stöckli, M. Suter, W. Däppen, 1978b, *Z. Phys. A* **286**, 249.

Subject Index

Author Index

MIX
Papier aus verantwortungsvollen Quellen
Paper from responsible sources
FSC® C105338

If you have any concerns about our products,
you can contact us on
ProductSafety@springernature.com

In case Publisher is established outside the EU,
the EU authorized representative is:
Springer Nature Customer Service Center GmbH
Europaplatz 3, 69115 Heidelberg, Germany

Printed by Libri Plureos GmbH
in Hamburg, Germany